History Of Hereford Cattle

Timothy Lathrop Miller

REPRESSED
PUBLISHING

ISBN: 978-1-4622-7996-8

Printed in the United States of America
Repressed Publishing LLC
Provo, Utah

www.repressedpublishing.com

ISBN 978-1-4622-7996-8
9 781462 279968

T. L. MILLER

HISTORY

OF

HEREFORD CATTLE

PROVEN CONCLUSIVELY THE OLDEST
OF IMPROVED BREEDS

By T. L. MILLER

WITH WHICH IS INCORPORATED A HISTORY OF THE HEREFORDS IN AMERICA

By WM. H. SOTHAM

1902

T. F. B. SOTHAM, Publisher
CHILLICOTHE, MISSOURI

INTRODUCTION.

"I have just returned from a trip to the Pacific Coast, and everywhere on the western plains found the Whitefaces conspicuous. And they look well wherever they are to be seen. I know they have a magnificent future. I am not a boomer, but this is a great country and I feel assured that our present prosperity will continue. It has always been demonstrated that Herefords are great cattle wherever grass grows, and the settlement of this irrigation problem now agitating the West will convert the desert into pastures and open up vast areas to cattle raising. The more grass, the more Hereford cattle will be needed, and with all due respect to other breeds, I venture the assertion that Herefords will make more beef on grass than any other breed of which I have any knowledge."

When we consider where and by whom the sentence above quoted was uttered, and the unequalled sources of information and experience that led to this conclusion, we feel, that at last, the Hereford breed of cattle is coming into its own.

The speaker was Mr. Alvin H. Sanders, Managing Editor of the greatest live stock publication in the world, namely, the "Breeders' Gazette," Chicago. The occasion was the reduction sale of the Weavergrace Herd, at which, in Chicago, Tuesday, March 25th, 1902, was gathered one of the largest representative companies of Hereford breeders the country has ever seen. With all due respect for the work of individuals in popularizing pure breeds of live stock, their work without the aid of the agricultural press must of necessity be limited in its influence. The leading advocates of Hereford cattle in America ever fully realized the influence of the agricultural press, and as far as they could have endeavored to utilize it in spreading a correct knowledge of the Hereford breed.

The influence that utilized the power of the agricultural press to propagate injurious theories, and their success in that direction, are thoroughly set forth in the body of this work and need not be discussed further here. We are glad to say, however, that conditions and influences, existent at the time the "Breeders' Gazette" was founded, have largely disappeared. Nothing shows plainer the present unprejudiced position of the "Breeders' Gazette" than the fact that it has been accused by prominent speakers in important meetings of Shorthorn breeders of being partial to the Herefords; that on several occasions adherents of other breeds have asserted that the "Gazette" leaned toward the Aberdeen-Angus "Doddies," while complaints that the "Gazette" was a Shorthorn paper have been frequent. Each breed in turn, prompted by some activity of its rivals (fully reported in the "Gazette"), exhibiting its inborn jealousy.

The statement that "Herefords will make more beef on grass than any other breed," made by Mr. Sanders, has been made repeatedly by Col. F. M. Woods, the celebrated live stock

WELLINGTON COURT, NEAR HEREFORD.

auctioneer, whenever he has conducted a sale of Herefords. Similar statements have been made on hundreds of occasions by that other favorite auctioneer, Col. R. E. Edmonson, who for years was the idol of the Kentucky Shorthorn breeders. This "Kentuckian of Kentuckians" not only makes this assertion in words, but backs it up in action, having become an extensive breeder of Hereford cattle on his Texas ranch. These are not to be counted straws to show which way the wind blows, but rather may be considered goodly-sized weather-vanes, known and seen of all men, for without commenting on the influence of these two great auctioneers we can say of the "Breeders' Gazette" under the management

INTRODUCTION

KING'S-PYON CHURCH, NEAR HEREFORD.

Wigmore Grange, born 1713, and the superiority of their cattle long before B. Tomkins, Jr.'s, marriage. These recollections and many facts that will be mentioned clearly prove that the systematic improvement of the Herefords was begun by the elder B. Tomkins in 1742.

It will be seen on referring to Sinclair's History of Hereford Cattle that when Richard Tomkins of the New House Farm, King's-Pyon, died in 1723, he left seven children—six sons and one daughter. The eldest, Miles, was nineteen and the youngest, Thomas, was three years old.

He left to Richard, his second son, a yoke of oxen, named "Spark" and "Merchant," with ten acres of land. To his fourth son, Benjamin, a cottage and land and a cow called "Silver" with her calf. The others were left small sums of money, and his widow, Catherine, the little New House Farm for life, making her his sole executrix. His will was proved in the Consistory Court at Hereford.

He doubtless had good reason for leaving these cattle to Richard and Benjamin. Richard being nineteen was old enough to be teamster and was left his team. Benjamin being nine years old would be old enough to take an interest in the cattle, possibly wishing to possess the Silver cow and calf, which his father afterwards left him by will. This would be calculated to give him additional interest in the cattle and dairy, probably filling the office of cattleman for a time.

Having two brothers younger than himself, he necessarily would have to leave the small farm when one of these were old enough to take his place and seek a situation in the particular line he had followed at home. This he apparently did, as all the sons but one were brought up to be farmers, and Professor Low was informed that Mr. Tomkins married his employer's daughter. Low does not give the source of this information, but it certainly would not come from the Misses Tomkins.

We find by the parish records that B. Tomkins, Sr., married a Miss Ann Preece, of Alton Court Farm, in the parish of Dilwyn, in 1742, and began business at the Court House Farm, Canon-Pyon.

In Low's words:

"Mr. Tomkins when a young man was in the employment of an individual, afterwards his father-in-law, and had the especial charge of the dairy. Two cows had been brought to this dairy, supposed to have been purchased at the fair of Kington, on the confines of Wales. Mr. Tomkins remarked the extraordinary tendency of these animals to become fat. On his marriage he acquired these two cows and commenced breeding them on his own account. The one with more white he called 'Pigeon' and the other of a rich red color with a spotted face he called 'Mottle.'"

Further he says: "He then began a system of breeding that ultimately completely altered the character of the Herefords." Low applied this information to B. Tomkins, Jr., who married his first cousin Sarah, daughter of Richard Tomkins, of Wormsley Grange, in 1772. Family tradition and the King's-Pyon register attest that his father died in 1748, when Mr. B. Tomkins, Jr., was but three years old, consequently he could not have been in the "employment of an individual, afterwards his father-in-law;" nor is it likely that Low's informant would call his first cousin "daughter of an individual." This expression may well have been applied to the elder man's wife, as in the lapse of years her father's name would most likely have been forgotten. It may be retorted that it was the widow who employed him, but Richard Tomkins left three sons, Richard, George and Thomas, aged twelve, eight and two years respectively. All were brought up as farmers, so that the widow could not have at any time required the services of her nephew when old enough even to manage for her. As to his ever being employed as a dairyman, the idea is absurd, for his father, as will be shown, had only two sons, and was a well-to-do man at the time.

Many other circumstances prove this portion of Low's information to refer to B. Tomkins, Sr., on his marriage in 1742. Low not knowing that there had been two of the same name applied it to the son, on his marriage in 1772.

Again, Professor Low in his "Practical Agriculture" (1843) writing of Hereford cattle, says: "The breed owes all its celebrity to changes began about the year 1760. The great improver, or rather it may be said, the founder of the modern breed, was the late Mr. Tomkins of King's-Pyon, near Hereford, who, from a very humble stock of cows, but by means of a long course of skilled selection, communicated to the breed its most valuable distinctive characters."

There is unmistakable evidence that the Misses Tomkins let Low believe that it was all the work of their father, consequently it is not surprising that he did not harmonize facts and dates when applying them all to the younger man. Low here says the changes began about 1760. This was nine years before the younger man commenced business (as will be shown later on) and twelve years before his marriage in 1772, yet, as previously quoted, Low says he commenced the improvement on his marriage, evidently referring to the elder man.

It is greatly to be regretted that Low never seriously attempted to harmonize his facts and dates. If he had done so he must have discovered at once the existence of B. Tomkins, Sr., and given a chronologically correct history of the origin and development of the Tomkins cattle. The correction of his remarkable mistake quite allies the hitherto accepted ideas as to when the systematic improvement of the Herefords began, which was in 1742, not 1772.

COURT HOUSE, CANON-PYON, HEREFORDSHIRE.

Tomkins was at work improving the Herefords nearly a quarter of a century before Bakewell began to improve the Longhorns, and a much longer time before the Brothers Colling, the Shorthorns; consequently Bakewell followed Tomkins, not Tomkins Bakewell, as has been generally asserted and believed.

Thus the Herefords are the oldest improved breed of cattle in the kingdom.

B. Tomkins, Sr., had evidently formed the idea of developing a superior breed of cattle to any then extant some years before his marriage in 1742.

On his marriage he was able to set about it in a systematic way, having already selected the materials. He began with the three distinguishing color types of the Tomkins cattle, namely, the Silvers, reds with white faces, the mottles and the greys. These three type names were applied to the Herefords for more than a century, unfortunately giving rise to endless controversies amongst breeders about breed.

Low says Tomkins acquired the cows

"Pigeon" and "Mottle" on account of their extraordinary tendency to become fat, qualities most likely not manifest in his Silvers, which we infer would be more of the large bony type, suitable for draught purposes, like the majority of the cattle were at that time. The rising importance of our manufacturing and commercial interests would be creating an increased demand for butcher's meat. Young Tomkins saw in this a new sphere of usefulness and profit in cattle other than the yoke and pail. He saw manifest in these two cows some of the characteristics he desired to conserve and intensify for this purpose. Fortunately, we know under what conditions these cows fed so rapidly. Artificial foods were not used in those days, and the Alton Court pastures are by no means feeding lands, yet their cows quickly became fat on them, having most likely come off the still poorer Welsh Hills; at any rate they apparently fed much more rapidly than any of the others under the same conditions.

Tomkins was impressed with this, and conceived the idea of raising up a race of cattle that would readily fatten on inferior food. As the grass at the Court House partook of the same character, he for upwards of twenty years carried out his works on inferior pastures.

Although Richard Tomkins was but a small yeoman farmer, the minute books of King's-Pyon show that the family received a good education at a time when many of the middle classes could not read nor write, and the peasants were wholly uneducated.

All the sons appear to have been enterprising, as they soon became established on farms in the neighborhood: Miles at the Hill in 1727, Richard at Wormsley Grange in 1734, Benjamin at the Court House in 1742, George at Wooton in 1746; John, known as "Butcher Jack," as a butcher in Canon Pyon, afterwards joining the army as a life guardsman. The daughter married Oakley, a farmer in Canon Pyon, and the youngest son, Thomas, remained with his mother, subsequently going with her to Calverhill Farm, in the parish of Norton Canon. This shows that there were a whole colony of the family farming in the neighborhood when B. Tomkins, Sr., was at his work of improvement, and we know that all these men and their descendants acquired the Tomkins cattle.

B. Tomkins, Sr., had only two sons who lived to grow up, Thomas and Benjamin, born 1743 and 1745. He had one daughter, who married Williams of Brinsop Court, brother of Williams of Thinghill Court. Both these men won prizes for cattle at the early Hereford shows. When B. Tomkins, Sr., increased his holding by taking Wellington Court, in addition to the Court House, in 1758, his two sons were fifteen and thirteen years old, by which time he had become comparatively well off, chiefly by the sale of his improved cattle that were then in possession of all the farming members of the family and many others throughout the country. His most intimate friend, William Galliers of Wigmore Grange, had a fine herd founded on the Tomkins breed, as it was already called. This was in the recollection of members of the family still living when Sinclair wrote his history. From the foregoing it will be seen that B. Tomkins, Sr., was not only the first who systematically improved the Herefords as beef producers, but was actually the founder or originator of the improved breed, and pioneer improver of cattle in these islands on systematic lines. He worked for nearly fifty years, 1742 to 1789, and his son B. Tomkins, Jr., worked with him for more than thirty of them, continuing for twenty-five years after his father's death.

Unlike Bakewell, he was a quiet, steady worker, accumulating means rapidly, without pushing himself for public notice. This characteristic was even more pronounced in his son, whose name seldom appeared, but when challenging the boastings of contemporary breeders, whom he never failed to silence. This, in some measure, accounts for the great value of their work not being publicly recognized until years after, when the glamour of the famous sale at the Brook House in 1819 drew universal attention to the younger man's work, quite obliterating the elder, by merging his life into that of the more conspicuous figure of his son. In the meantime their improved cattle had been quietly absorbed and helped to build up without exception all the famous old herds that it has been possible to trace back to their foundation.

Soon after going to live at Wellington Court in 1758 B. Tomkins, Sr., relinquished the Court House to his eldest son, Thomas. The younger son, Benjamin, appears to have remained with his father until on his contemplated marriage with his first cousin he took the Blackhall farm, King's Pyon, 1769.

Writers have been uncertain as to the exact date when B. Tomkins, Jr., began business at the Blackhall. The parish books of King's-Pyon show this date, as the last signature of his predecessor appears in 1768 and B. Tomkins, Jr.'s, first signature in 1769. B. Tomkins, Jr., undoubtedly had his select breeding cattle from his father's herd at Wellington

Court. Being the favorite son he would have his pick of the best of them, and the two herds were afterwards bred conjointly for twenty years, when on his father's death, in 1789, he acquired the stock and farm at Wellington Court. In the meantime B. Tomkins, Jr., had acquired the Court House and Brook House Farms, King's-Pyon, in addition to the Blackhall, so that the father and son together occupied three and for a time four farms from 650 to 840 acres in extent. How closely they worked together is shown by his having his father's stock and farm on his death. The lives and work of father and son were so intimately blended together that no wonder after many years succeeding generations, without close inquiry, regarded their work as that of one man. The father worked for 47 years (from 1742 to 1789) and the son for about the same number (1769-1815), and they worked together for about thirty of the seventy-three years of their joint breeding career. This occupying of several farms with separate homesteads enabled them not only to keep more cattle, but to keep the different strains distinct in different places, enabling them to carry out their system of line breeding without necessarily using very near affinities.

It has been imagined that they inbred their cattle very closely, but this idea is not justified by facts. It is well known that they bred all the bulls they used, and that they kept several at each homestead, thus enabling them to cross in their own herds from selected variations in desired directions without close in-breeding. They likewise had a wide family circle breeding the same variety of cattle, which gave them a still wider range for the exercise of their judgment and skill.

When working alone we know that these two men achieved a most marvelous success. What must have been the power of their united judgments during the thirty years they worked together?

Eyton says, 1846: "The Misses Tomkins have been in possession of the same breed without a cross since the period of their father's death. They were in the habit of keeping four or five bulls and whatever is bred from their stock may be relied upon for the purity of the blood."

Here we have evidence to show that the Misses Tomkins continued to follow their father's and grandfather's system of breeding in the male line for over thirty years after his death on a single farm of less than 300 acres.

Eyton says further: "During the latter portion of Mr. Tomkins' life he used none but bulls bred by himself, and did not cross with any other stocks."

As Eyton was here treating the work of the two men as that of one he should have said: "During the latter portion of Messrs. Tomkins' lives they used none but bulls bred by themselves, and did not cross with any other stocks."

When B. Tomkins, Sr., practically ceased crossing from outside sources is uncertain, but most likely it was about the time he took Wellington Court, in 1758, as there were then herds of Tomkins cattle more or less pure-bred

TABLET OF BENJAMIN TOMKINS, SR., IN WELLINGTON CHURCH, HEREFORDSHIRE.

at the Hill, New House, Weobley's Field, Wooton, Court House, Canon Pyon, Wigmore Grange, and probably other places outside the family circle. How quickly successful he was is thus shown by the early and rapid spread of his cattle amongst members of the family.

Eyton says: "The bull which is often referred to as 'Silver Bull' he always considered as the first great improver of his stock. There is a prevailing opinion respecting this bull's name, that it was given to him because he was of a silver or grey color, but the fact is that he was a red bull, with a white face, and a little white on his back, and his dam was a cow called 'Silver.'"

Eyton, like Professor Low and others, did not know that he was treating the work of two men as that of one; referred this information to B. Tomkins, Jr., but it must have concerned his father, whose stock had become famous many years before the son began business in 1769. The father evidently at an early period of his career bred this bull from descendants of the Silver cow his father left him in his will in 1720.

Sinclair has shown that the red, with the white face, color markings for cattle were fashionable in the country before the death of Lord Scuddmore in 1671, and Richard Tomkins esteemed them fifty years after in 1720, since his favorite Silvers were thus marked, so that the taste for these colors dates back much further than has been generally supposed. Their system of breeding was essentially the same as that followed by Hereford breeders at the present time, only over a more prescribed area. In one respect it was the opposite, since they always bred the bulls used instead of purchasing them. In their case this was necessary, as there was no reliable source to procure them from outside their own herds, in fact no cattle so good as their own to improve them with were outside their own herds.

Professor Darwin says: "The power of man to accumulate the slight variations of our domesticated animals in a given direction by constant selection is very considerable. The improvement begins by crossing different types, and is afterward continued by constant selections from the varieties produced. When a cross is made the closest selection is more necessary than in ordinary cases between good animals of an established type or breed. To accumulate these slight differences, absolutely inappreciable to the ordinary observer, acquires an accuracy of eye, touch and judgment that not one in a thousand possesses. A man endowed with these qualifications, who devotes a lifetime to the work, will effect great improvements."

This work the two B. Tomkins were preeminently fitted for, as they carried it out with consummate skill and success, which the sale in 1819 demonstrated. They seemed to have intuitively grasped the physiological law enunciated by Darwin a hundred years after, "that given an equal amount of pureness of blood, the male animal possesses a greater amount of accumulated variation in a given direction than the female."

"These variations are at first artificial, but after accumulating them for a length of time they become typified, and constitute a distinct variety or breed."

It is remarkable that the elder B. Tomkins first observed the variation in the direction he desired in two females, and he and his son afterwards conserved it more particularly in the males.

Regret was at one time expressed that Tomkins did not exclusively adopt the red with white face colorings for their cattle, but considering that the cattle B. Tomkins, Sr., began with were a grey, a dark red with white spots on its face, and a red with white face: one starting with these animals differently marked, he and his son would have to subordinate color marks to the more essential qualities when developing a fresh type of animal from various sources. When selecting and blending the best materials from a limited number of animals, it would have been impossible, even if desired at that time, to make the places of the color spots on the body an all-important consideration. If they had bred exclusively from red with white face, mottle face or grey, they must have sacrificed some of their best animals and thus defeated their object. They knew the business too well to do that, and by continually crossing their differently marked cattle to develop and fix certain desired characteristics they kept these color marks on the body, liquid or movable, consequently when the old red with white face Herefords were crossed with the Tomkins cattle the color marks were easily made to conform to the originals, while the progeny retained the typically fixed, good qualities of the Tomkins breed, conserved through a long line of generations.

It is unfortunate that at the time when the Tomkins were systematically transforming the Herefords from rough, bony draught and dairy cattle into a superior beef-producing breed, that no written records were kept. In the absence of these the old writers took color markings as indicating what they chose to call breed. They spoke of the white-face breed, the mottle breed, the grey breed; and they took it for granted that breed and color necessarily went together and could not be separated. The universal acceptance of this great error led to endless, regrettable disputes amongst the old breeders. This misuse of the word "breed" was most misleading during the transition state of the Herefords, and all attempts to trace its history by color marks completely failed. Marshall, describing Bakewell's Longhorn in 1784, says: "Color is various, the Brindle, the Pinchbeck and the Pye are common. The lighter the color the better they seem to be esteemed, but this seems to be merely a matter of fashion." And when describing the points of a perfect Longhorn he says: "Any color that can be joined with the foregoing qualifications, it being perhaps of little (if any) essential import." Thus Bakewell, like Tomkins, disregarded color marks, and his improved Longhorns varied in this respect, but were not called in consequence the Brindle breed, the Pinchbeck breed and the Pye breed.

Instances are given in the first volume of the Herd Book where the sire and dam are white-faced and the offspring mottle-faced, and vice versa. In the phraseology of the day, the sire and dam would belong to the white-faced breed, and the offspring the mottle breed. This clearly illustrates the fallacy of taking color marking as a guide to the breed during the transition period of the Herefords.

Tomkins never line bred color markings, but rather used them together in every imaginable way. Their system was in the words of the poet:

> "White face, Pick face, Mottle face and Grey,
> Mingle, Mingle, Mingle, ye that mingle may."

The Herefords were then in a state of commingling with the Tomkins cattle, on whose bodies color marks had no fixed abode, so that at that time it was a matter of choice where these should be placed on the future Hereford. It could have been constituted a grey, a mottle face or a white-face breed. The overwhelming choice was to stick to the old red with white face markings, and although some old breeders resisted this for a time, they ultimately died or gave up the contest.

The red with white face markings left liquid by Tomkins have now through many years of selection carefully obliterated the spotted face and grey markings, and become typically fixed and the true index of breed, which they were not during the transition period.

Doubtless the Tomkins cattle would have spread much faster if B. Tomkins, Jr., had not been so extremely jealous of others obtaining his best blood. It is well known that he had many of his best bulls killed at home for the harvest men rather than others should have them, and many of his best cows were resold by the butchers for breeding purposes. The old butchers bore universal testimony that the Tomkins cattle were the most profitable butchers' cattle they killed.

Day of Credenhill, Bakerville of Weobley, Preece of the Shrewd, Davies of Canon Pyon, and others used to declare that for quality of meat, associated with smallness of offal, none they killed approached them. And Sinclair has shown in his history that all the old noted herds, without exception, that could be traced went back to what Hewer tersely called "Old Tomkins' Prime Cattle."

* * *

Beside the foregoing manuscript, prepared by Mr. Bustin, I wish to acknowledge here his great assistance in the preparation of the illustrations in this history. Without Mr. Bustin's help this great feature of the work would be most lamentably lacking. His skillful search has unearthed drawings and paintings that have

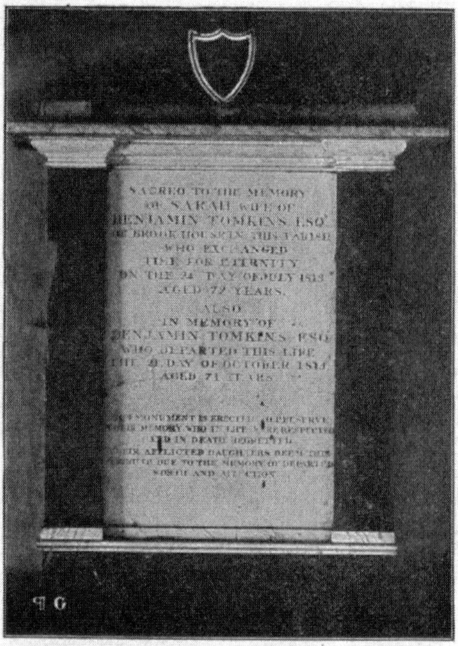

TABLET OF BENJAMIN TOMKINS, JR., KING'S-PYON CHURCH, HEREFORDSHIRE.

been hid for years; he visited various parts of England, securing photos of homesteads, farm views, ancient drawings, paintings and engravings, etc., etc., which are invaluable to the student of Hereford history. I wish to acknowledge also the kind co-operation of Mr. Geo. Leigh, of Aurora, Ill., to whom I am indebted for many photographs of English Hereford breeders, their homes and their cattle.

I would draw particular attention to the fact that the illustrations in this work are, as far as possible, reproductions of photographs from life. The successful photograph of live animals is only a recent accomplishment, and not as yet so successful in America as in England. This is one great work wherein England excels America, a condition which can hardly be expected to last. It is only in recent years that actual photographs of cattle from life has been the successful method of illustration in England. Therefore, the illustrations in this work of both English and American subjects, prior to 1895, are reproductions of lithographs, paintings, engravings, etc., all produced by hand. I thought at first that it would be best to have old drawings modernized by an up-to-date live stock artist, as has been done in some other prominent works on cattle, but after giving the matter much thought it seemed to me

W. H. BUSTIN, HEREFORD, ENG.

best to reproduce these old pictures exactly as they were made. They vary, as does all hand work, with the ideals of the artists drawing them. Messrs. Gauci, Page, Dewey, Burk, Hill, Palmer, Throop, etc., each had their ideal, so that were each of these men to portray the same animals those familiar with their work would have no difficulty in discovering from the picture who the artist was by the peculiar personal ideal invariably, and perhaps unconsciously, incorporated into the picture. Some of the ancient artists painted the forms of the animals they portrayed upon impossible stems, representing legs. If the readers of Miller's History will bear in mind these variations of the artists' ideals when examining the old drawings of cattle reproduced herein they will form a much more intelligent conception of the excellence of those old foundation animals by substituting in their mind's eye correct impressions of animal anatomy for the superfine limbs and heads portrayed by the artists. I have in the appendix added full page reproductions of photographs from life of choice specimen Herefords of different ages, being correct reproductions of actual photographs from life; every one knows that photographs have never yet been made to flatter animals, for, as a rule, they portray faults more plainly than virtues. These photographs, however, give the most correct ideas of anatomy attainable by any process known at this day. These illustrations have been selected carefully from photographs taken in England, and are specifically included in the appendix of this work to give correct ideas of Hereford form and character.

In closing I wish to say that there are personal references in this book, mainly of persons long since deceased, that I regret exceedingly to see again in print. Yet, to leave out these references, would be leaving out facts of history that would give the reader no conception of the trials and vexations to which the early supporters of the Hereford breed of cattle were subjected. No one coming freshly upon the scene to participate in the breeding of pure-bred cattle in these days of breed tolerance could understand the lengths to which jealousy, prejudice and selfish interests drove men in their opposition to Hereford cattle; in what was, in veriest truth, the "Battle of the Breeds." I have personally experienced something in this line myself. Several old show ring controversies are, in this work, again brought to light. Fraudulent entries and false ages are shown to have been prominent factors in past conflicts. Let no reader suppose that such things do not exist to-day. They are not so patent between the breeds, because the show ring contests between breeds has largely been done away with, but the fraudulent exhibitor is, if possible, more prevalent to-day than ever. He can be found in our own ranks, so much so that an exhibitor must take one of three courses, either one of which is equally unpleasant. I refer to the fact that ages are misrepresented (flagrantly, in some cases). Surgical operations are performed to change the appearance and eradicate defects of animals in a manner that should put the most unscrupulous horse farrier to shame, and there are, at times, as notorious manipulations of judging committees in the present day as ever occurred in the past. As I said before there are three courses open to the honest exhibitor who desires to bring his cattle before the public at the great shows. He must either (first) protest and prove these nefarious practices; (second) practice these unworthy methods himself, or (third) submit tamely and allow the unscrupulous exhibitor to win unmerited prizes and escape unscathed. The one redeeming feature of the show ring is that the unscrupulous exhibitors are a very small minority, so that whenever those who show fairly and honestly get together and protest in a body against crooked practices they can be overthrown, but as a rule up to this time exhibitors have preferred to allow these frauds to go unrebuked, because of the prominence of the parties committing them, or of a desire to keep peace regardless of price. Again, the class of exhibitors committing these depredations on the show ring usually last but a little while. They are, as it were, meteors, who come out and, to use their own language, "make a killing" in the prize ring for a year or two, and then disappear, only to be followed by some similar fraud upon whom their mantle invariably falls. As I said before, these unpleasant parts of the book are left as their author shaped them. Mr. Miller and my father had the habit of calling things by their real names, and both were accustomed to tell the truth regardless of who were hurt or benefited thereby, and therefore I have felt constrained to adopt the policy that was forced upon Pontius Pilate and say, "What is written is written." They could never in life forgive the garbling of their statements by the editor, and I could not be party to such action now that they are not here to protest for themselves.

This is Mr. Miller's work, and as such is submitted as the best work ever published on cattle. If every stockman in America will read this work and act upon its suggestions, in the light of its teachings, more will be accomplished in the profitable upbuilding of the beef interests of America in one decade hereafter than has heretofore been accomplished in a century.

T. F. B. Sotham.

Chillicothe, Mo.,
April 14th, 1902.

CHAPTER I.

FOUNDATION HEREFORD HERDS

In the year 1627, John Speed published a work on England, Wales, and Scotland, in which he says of Herefordshire, "the climate is most healthful and the soil so fertile for corn and cattle that no place in England yieldeth more or better conditioned." (¶ 1)

Starting from this data, it is fair to presume that the cattle of Herefordshire should improve, and that Mr. Benjamin Tomkins, who commenced the breeding of Herefords in the year 1742, should have found a class of cattle of great merit. (¶ 2) It is well here to give an account of the Tomkins family.

The Tomkins of Weobley were of considerable note and position in its neighborhood, prior to the civil war of Charles the First, but being enthusiastic Royalists, they suffered much, in consequence of that monarch's overthrow.

They were distinguished in music and painting, being patronized in both arts by royalty, and the leading members of the House; they were great and consistent politicians, for many generations, representing Leominster and Weobley in Parliament.

At successive periods during the seventeenth century, the branch from which the distinguished cattle breeder sprang was known as Tomkins of Garnestone, a considerable domain, situated immediately south of Weobley, which belonged to James Tomkins, Lord of Weobley, and M. P. for Leominster from 1623 to 1628, who was much esteemed as a country gentleman and noted debater in the House of Commons.

In the beginning of the eighteenth century, was one Richard Tomkins, of the New House, King's-Pyon parish (¶ 3), who spent his life there, and became a very successful farmer and breeder of work oxen. In his will in 1720, he bequeathed a yoke of oxen, called Spark and Merchant, to his son Richard, and a cow Silver and calf to his son Benjamin. Richard Tomkins died in 1723, leaving six sons and one daughter. Five of his sons established themselves as farmers in the immediate neighborhood. The fourth son, the first distinguished cattle breeder, "Benjamin Tomkins the elder," was born at the New House, King's-Pyon, in 1714, and commenced business at the Court House, Canon Pyon, about 1738. He married Anne Preece of Alton, in 1742, and subsequently moved to Wellington Court in 1758, where he died in 1789, leaving six children, four sons and two daughters. Of these four sons, Benjamin, who has been credited as the noted breeder and improver of the Hereford breed of cattle, was the second son of Benjamin of the Court House and Wellington Court, and from Richard of New House to Benjamin inclusive, there were ten of the sons and grandsons, who were all farmers and probably breeders of Hereford cattle.

Benjamin Tomkins, (¶ 4) the renowned breeder, was the second son of Benjamin Tomkins of Court House, Canon Pyon, where he was born in 1745 and commenced farming at Black Hall, (¶ 5) King's-Pyon, in 1766. He married in 1772, his cousin Sarah, second daughter of Richard Tomkins of the Grange, Wormsley. He occupied Black Hall until 1798 when he sub-let it to his nephew, George Tomkins, Jr., of Frogdon, and removed to Wellington Court, which he held as a bytake, from his father's death.

In 1812 he gave up Wellington Court and went to reside at his own place, Brook House, (¶ 6) King's-Pyon, where he died in 1815. From James Tomkins, Lord of Weobley, who was active in politics in 1623-8, to Richard, who commenced farming at New House, King's-Pyon, and died in 1723—nearly one hundred years—we are without a record.

Returning to Mr. Benjamin Tomkins, the younger, who commenced the improvement of the Herefords in 1766, we have very little information as to the course he pursued, except that his cattle obtained a very enviable reputation among breeders, and brought large prices from some of the best breeders during his time. At one time he took twenty cows to Herefordshire Agricultural Show and gave a challenge of £100 to any one who would show an equal number against him. His nephew, George Tomkins, after traveling over Herefordshire and other parts of England, among cattle breeders, when he came home, reported to his uncle that of all the cattle he had seen, there were none equal to his. Mr. Jno. Price, of Ryall, about the year 1804, became acquainted with the cattle of Mr. Benjamin Tomkins, from whom he bought a few cows, using them to bulls descended from Mr. Walker's stock. He first attempted to improve the Tomkins cattle by crossing them with the larger stock of Mr. Walker, with a view of increasing their size, but the result was so unfavorable that he put away all these crosses and returned to the pure Tomkins variety. Mr. Price continued to breed Herefords until 1841, his herd being solely of the Tomkins blood. So that, upwards of seventy years at least, this strain, first in possession of Benjamin Tomkins, and then in that of John Price, was bred continuously without an out-cross.

It would appear that Mr. Tomkins was entitled to the position of leader in the improvement of the breed, and for giving a fixed character both as to quality, color, and markings, and at the same time, the Hewers, William and John, were close seconds. It was perhaps to be expected that their friends would take sides, and a feeling of rivalry should grow up among them for the time being; but ultimately as these different lines expanded and new men took the places of the early rivals, the best of each were brought together.

Mr. John Price of Ryall and other eminent breeders acquired bulls and cows of the Tomkins breed and they soon spread widely over the country. In October, 1808, Tomkins had a large sale at the Court Farm, Wellington, which the auctioneer, Mr. William James, announced in these words: "For sale, the following valuable and much admired stock, the property of Benjamin Tomkins, who is going to decline breeding cattle; consisting of 20 capital cows and heifers, which have five calves now sucking, two four-year-old bulls, one ditto martin, nine three-year-old bullocks, six two-year-old ditto, two yearling heifers, one of which is heavy in calf, three two-year-old bulls, two ditto yearlings." No note of the prices or purchasers' names at this sale has been obtained, but we are able to give a private valuation of the stock at Wellington Court Farm, drawn up by George Tomkins in June, 1808,

SCENE IN HEREFORDSHIRE.—A FERTILE SOIL AND WELL-CONDITIONED CATTLE.

which will indicate the owner's estimate of their worth: "12 cows and calves at £40 ($200) each, £480 ($2,400); 12 oxen at £43 ($215) each, £516 ($2,580); 10 two-year-olds at £20 ($100) each, £200 ($1,000); 10 yearlings at £15 ($75) each, £150 ($750)." An average for old and young, steers and breeding stock of over £30 ($150) each.

Only a comparatively small number of the bulls bred by Benjamin Tomkins were entered in the Herd Book, and in few cases are particulars given of their breeding. Wellington (4) 160 (¶8) is registered simply as coming under the division of the mottle faces, and as having been bred by Mr. B. Tomkins. He passed into the possession of Mr. Price, and was purchased at his sale in 1816 by Mr. Jellicoe of Beighterton for £283 10s, ($1,400), being afterwards sold to Mr. Germaine. He was considered by Mr. Tomkins the best bull he ever bred, his Silver Bull (41) 432, excepted, and also the best stock getter. In Vol. 1 of the English Herd Book, there is a colored lithograph (reproduced herein) of this bull from a painting by Mr. Welles, representing a compact, straight animal, of fine size, with fine bone, mottle face, white dewlap, and white along the lower parts of the body. Another of Tomkins' bulls registered in Vol. 1, is Ben (96) 6703, of which the editor, Mr. Eyton, says that "Miss Tomkins informed him that Ben was by Sam (144) 6704, out of one of Mr. Tomkins' cows called Nancy." Sam (144) 6704, is without recorded pedigree, all that is said concerning him being that he was bred by B. Tomkins. Wild Bull (145) 3040, bred by Tomkins, was, on Miss Tomkins' authority, said to be by Silver Bull (41) 432, out of Tidy 340. Phoenix (55) 3035, a mottle face, out of Storrell 3039, bred by Mr. Tomkins and got by Wild Bull (145) 3040, was purchased at Miss Tomkins' sale in 1819 for 560 guineas (over $2,800) by Lord Talbot. Mr. Eyton has this remark as to his dam: "Storrell, Miss Tomkins informs me, was out of a mottle faced cow of the same name (Storrell 3041), by a Pigeon bull." The bull called Son of Prices 25 (84) 440, bred by Tomkins, was out of (Price's No. 25) 439, "who was out of a sister to the dam of Price's 23, or 'The Slit Teat cow,' by the Silver Bull (41) 432." Proctor's bull, (316) 376, was bred by B. Tomkins "out of his favorite cow, 'Old Pink.'" Voltaire (39A) 429, was a white faced bull bred by Tomkins, dam Price's No. 3. Wizard (59) 6699, was a mottle face of Tomkins' breeding by Ben (96) 6703, and was sold to Mr. Germaine for 300 guineas ($1,500). Wedgeman (166) was bred by Tomkins, but no pedigree is given in the Herd Book.

In the appendix to Vol. XI of the English Herd Book, Mr. E. F. Welles gave some interesting recollections of the stock of Mr. John Price, from which a very complete idea can be obtained of the character and appearance of the Tomkins cattle. It is, indeed, one of the most valuable statements that has been made on the subject. Mr. Welles says: "When Mr. John Price commenced cattle breeding, the character of bull most in esteem in the chief Midland districts was one having a throat with as little loose flesh as possible depending from it. This character was also introduced by some cattle breeders amongst Herefords. The celebrated Purslow bull, the property of the Haywoods of Clifton-on-Teme, had this character. Mr. Walker of Burton had also adopted it, and from him Mr. Price had a bull or two. Mr. B. Tomkins and other Hereford breeders had not been affected by this fashion, and Mr. Price, when he became acquainted with Mr. Tomkins' stock, relinquished it, preferring, and upon sounder principles, that character which better indicated the male animal, a considerable degree of throatiness not being objected to. This character belonged to Wellington (4) 160, the first bull, and I think, the only one bought by Mr. Price of Mr. B. Tomkins. This bull was very dark in color, with face and bosom both mottled and speckled. His dam, too, bought afterwards by Mr. Price (but did not breed with him), was also of the same color."

TYPICAL HEREFORD FAT OXEN OF THE OLD-FASHIONED SORT.

"The cows bought by Mr. Price of Mr. Tomkins were the following: First, a large cow with a speckled face, giving a blue appearance to it, with what may be termed an arched forehead or Roman nose, tips of horns blackish, body of lightish brown, dappled, under part of body and legs inclining to blackness, white along her back and well formed, but on rather high legs. Secondly, a cow commonly called 'the Mark-nosed cow'—a red cow with mottled face, square made, and on short legs, rich quality of flesh, with a soft and thick pile of hair moderately curled. This cow was unfortunate to Mr. Price as a breeder, the only produce I recollect out of her being the 'Marked faced bull,' alias 'Pyon' at his sale. Thirdly, a large yellow cow with white face, rather long headed, and not carrying much flesh. She was the dam of Voltaire (39A) 429, by one of Mr. Tomkins' bulls.

"Pigeon, by far the most remarkable cow he had of Mr. Tomkins—and her own character, as well as that of her descendants, will well warrant me in terming her the best—was a large cow, rather on high legs, somewhat shallow in the bosom, with very fine bone, neck rather light, head good but horn short; her color a speckled grey, the red parts being dark, growing still darker about her legs; hair rather short but soft, quality of flesh excellent, back and hind-quarters great, excepting thighs, which were rather light, but with good twist; her constitution hardy, and she was a regular

THE NEW HOUSE, KING'S-PYON, HEREFORDSHIRE.

and successful breeder. About this same time also, Mr. Price had another cow from Mr. B. Tomkins, which was called the Rough cow, from her coat being much curled; she was a middle sized cow, nothing remarkable in form, her color dark red, with white back, and she had the reputation of being of a family that were good ox breeders. Mr. Price had a bull from this cow called 'Rough bull' alias 'Original,' but he did not long retain any of his stock. There were sisters to him by other Tomkins bulls. Two more cows Mr. Price subsequently obtained from Mr. B. Tomkins—a half-sister to No. 25, and a daughter to Mr. Tomkins' famous 'Slit Teat Cow' No. 21. The former of these was a small cow out of a very true form, dark color, with white along her back; she was the dam of Lord Talbot's Woodcock (50) 654, sire of Mr. Price's Woodcock Pigeon 651. I am not aware that Mr. Price had any more cows from Mr. B. Tomkins, but he afterward obtained two cows of his blood —one called Damsel 371 from Mr. T. Tomkins, and another from Mr. Tomkins of Wormbridge, the former the dam of Woodman (10) 307 and the latter the dam of Diana 638. He also bought a few Tomkins bred cows from Mr. Jas. Price; among these was the dam of Peg Murphy 3559."

These notes, which furnish a complete picture of a large number of the Tomkins cattle, fully bear out what has been said as to their diversified colors. Mr. Price's selections comprised animals that were yellow with white face; speckled grey; dark red, with white back; red with mottle face; dark color with white along the back; and lightish brown dappled, with white along the back, etc. The only point in which there was an approach to uniformity as regards color was the white back. A few other notes as to Tomkins' cows are gleaned from the entries in the Herd Book. The "Slit Teat Cow" referred to by Mr. Welles was considered by Tomkins the best cow he ever had. Storrell 3039 by Wild Bull (145) 3040, was, as has already been mentioned, dam of Phoenix (55) 3035, sold to Lord Talbot for 560 guineas ($2,800). Old Rose was out of the dam of Silver Bull (41) 432, Old Lovely 657 was a daughter of the "Slit Teat Cow." All we know about others are their names, and in some cases those of their sires —Blowdy out of Old Pidgeon, the dam of Mr. Price's Pigeon 373; Margaret, by Silver Bull (41) 432; Stately, by Wizard (59) 6699; Blossom, by Phoenix (55) 3035; Old Lily, Nutty, etc.

After the death of B. Tomkins, in October, 1815, the herd, which had by this time been much reduced in numbers by private sales, was kept on by his daughters, the Misses Tomkins, until October, 1819, when part of it was sold. Through the courtesy of Mr. Haywood of Blackmere House we have been favored with a copy of the original sale bill, containing the prices and purchasers' names, marked by one who was present at the sale. The document has a historic importance, and is reproduced in full:

"A catalogue of the valuable stock of prime Herefordshire cattle, the property of the late

B. Tomkins of Wellington Court, which will be sold by auction without reserve, upon the premises at King's-Pyon, nine miles from Hereford, on Monday, 18th October, 1819, being the eve of the Herefordshire Agricultural Show, and two days previous to the great cattle fair at Hereford:

LOT.	NAME.	PURCHASER.	PRICE.
1.	Yearling heifer, Young Blowdy	Mr. W. West	56 14 0= $285
2.	Ditto, Young Fairmaid	Mr. Court	99 15 0= 500
3.	Two-year-old in-calf heifer, Young Blossom	Mr. G. Tomkins, for Lord Talbot	105 0 0= 525
4.	Ditto, Young Silver	Mr. John Tomkins	73 10 0= 370
5.	In-calf heifer, Duchess	Mr. G. Tomkins, for Lord Talbot	105 0 0= 525
6.	Ditto cow, Pigeon	Mr. W. West	159 12 0= 800
7.	Ditto, Stately	Mr. G. Tomkins, for Lord Talbot	52 10 0= 265
8.	Ditto, Silk	Mr. Lewis	70 7 0= 350
9.	Ditto, Beauty	Mr. Cooke	262 10 0= 1310
10.	Ditto, Silver	Mr. West	210 0 0= 1050
11.	Ditto, Cherry	Mr. G. Tomkins, for Lord Talbot	110 5 0= 550
12.	Ditto, Prettymaid	Mr. Lewis	99 15 0= 500
13.	Ditto, Piot	Mr. Turner	105 0 0= 525
14.	Ditto, Nancy	Mr. G. Tomkins, for Lord Talbot	252 5 0= 1260
15.	Ditto, Blowdy	Mr. Cooke	273 0 0= 1365
16.	Ditto, Fairmaid	Mr. West	65 2 0= 325
17.	Ditto, Tidy	Mr. Cooke	131 5 0= 855
18.	Ditto, Lovely	Mr. Cooke	53 11 0= 265
19.	Ditto, Storrell	Mr. G. Tomkins, for Lord Talbot	262 5 0= 1310
20.	Ditto, Pink	Mr. Edwards	141 15 0= 705
21.	Bull-calf off Ditto	Mr. Clarke	147 0 0= 735
22.	Ditto off Beauty	Mr. Crooke	215 5 0= 1075
23.	Fat Cow, Blossom	Mr. James	48 6 0= 240
24.	Pair of two-year-old steers (twins)	Mr. T. Cooke	47 5 0= 235
25.	Ditto	Mr. W. Cooke	48 6 0= 240
26.	Ditto	Mr. James Price	49 0 0= 245
27.	Single Bullock	Mr. James Price	20 0 0= 100
28.	Pair of yearling bullocks	Mr. Smith	24 0 0= 120
29.	Ditto	Mr. Patrick	27 0 0= 135
30.	Ditto	Mr. James Price	49 0 0= 245
31.	Ditto	Mr. Oliver	27 6 0= 135
32.	Ditto	Mr. Wedge	25 10 0= 130
33.	Pair of bullock calves	Mr. W. Cooke	20 10 0= 100
34.	Ditto	Mr. W. Cooke	16 0 0= 80
35.	Ditto	Mr. Wright	16 0 0= 80
36.			
37.	Heifer calf	Mr. G. Tomkins	30 0 0= 150
38.	Ditto	Mr. Cooke	26 0 0= 130
39.	Two-year-old bull of Pink	Mr. W. West	147 0 0= 735
40.	Ditto of Storrell	Mr. G. Tomkins, for Lord Talbot	588 0 0= 2940
41.	Four-year-old Ditto	Mr. Welles	162 15 0= 815
42.	Five-year-old Ditto	Mr. T. Cooke	173 5 0= 865

"N. B.—The above cattle are all of the pure breed, which have been so justly esteemed and admired by the most competent judges in every part of the kingdom where they have been introduced, and for which peculiar blood, the highest prices have been obtained, and particularly No. 23, which is considered to carry the greatest weight upon the smallest bone of any cow in the kingdom."

SUMMARY OF SALES.

	Total. £ s d	Average. £ s d=U. S. D.
17 cows	2249 2 0	149 18 9=$ 749.68
3 two-year-old heifers	283 10 0	94 10 0= 472.50
2 yearling heifers	156 9 0	78 4 6= 391.12
4 bulls	1071 0 0	267 15 0= 1338.75
2 bull calves	362 5 0	181 2 6= 905.62

28 Head; Total, £4172 6s 0d ($20,861.50); Average, £149 ($745.00).

It is interesting to compare the foregoing averages with those realized at the great Shorthorn (Durham) sales of the Brothers Colling, which took place about the same time. At Mr. Chas. Colling's sale at Ketton, in 1810, the average for 47 head was £151 8s ($757.00) (the bull Comet bringing 1,000 guineas or $5,000.00). At Mr. Robt. Colling's sale at Brampton in 1818, 61 head averaged £128 17s 10d ($644.35); and at his sale in 1820, 46 head averaged £49 8s 7d ($247.14).

A statement has recently been made to the effect that the Tomkins Herefords, if they had not from the effects of excessive in-and-in breeding fallen into disrepute before B. Tomkins' death, at least did so almost immediately afterwards. There is certainly no evidence of want of public appreciation in the sale list which has just been given. Moreover, Mr. John Price for many years retained without any mixture the blood of Tomkins, and also bred very closely, and yet when his herd was dispersed in 1841 (26 years after Tomkins' death) the average for 99 lots was £53 16s 4d ($270). But it is only necessary to glance at the composition of the foundation herds of Herefords as recorded in the Herd Book—those of Knight, Smythies, Yarworth, Hewer, Walker, Hoskyns, Perry, Jellico, Smith, Lord Talbot, Sir F. Lawley, etc., to see how largely the Tomkins blood was infused over the breed. There was scarcely one of the early herds that was not indebted to the Tomkins strains for part of its excellence, and if the results of injudicious in-and-in breeding had then, as is alleged, been so painfully apparent, the blood would have scarcely obtained such wide circulation. That some of Tomkins' cattle went into the possession of those who were not able to do them justice, and who failed in the attempt to carry out what they supposed was his system, is unquestionable; but that the herd retained unimpaired its high character when Benjamin Tomkins died, is proved by the results in 1819, and by the fact that John Price continued for 37 years to successfully breed on the Tomkins' lines without resorting to other blood.

After the sale in 1819 the Misses Tomkins remained at the Brook House Farm, and continued to breed Hereford cattle. They had a second sale in October, 1839, when it was announced that they were about to retire from business. The catalogue of this sale shows that their cattle still retained considerable reputation. Among the prices were £108 ($540) for the nine-year-old cow Pigeon, the purchaser being Mr. Gough; £50 ($250) for the eight-year-old cow Stately (Mr. Davenport); £56

BENJAMIN TOMKINS, Jr., 1745–1815.

($280) for the seven-year-old cow Diana (Mr. Galliers); £50 (or $250) for the nine-year-old cow Lovely (Mr. Galliers); £52 ($260) for the three-year-old heifer Countess (Mr. Jones); £51 ($255) for the two-year-old heifer Tidy (Mr. Galliers). A three-year-old bull No. 1, got by a bull from Old Pigeon by the same sire, made £82 ($410) (Mr. Griffiths). Among the other purchasers were Mr. Smythies, Mr. Yeld, Mr. Vevers, Mr. J. Moore, etc. The average for 48 animals was over £30 (or $150). The final sale was in October, 1854, one of the sisters having, in the interval, died. The entire herd, numbering 55 head, "descended from that peculiar blood which has for three-fourths of a century been the admiration of the county, and which have upon former occasions realized higher prices than any other breed of Herefords in the kingdom," was then dispersed. No catalogue of this sale seems to have been printed; the announcement having been made on a "broadside," specifying the numbers of the various classes of stock, and there is no note of prices. The auctioneer, however, stated that the animals were purely descended from the herd of the late Mr. Benjamin Tomkins, from whom the late Mr. Price of Ryall obtained that breed of cattle which, at different times, have been distributed through the United Kingdom at enormous prices, particulars of which will be found in "Eyton's Herd Book." From a note in Vol. 1 of the Herd Book it appears that it was the custom of the Misses Tomkins to give the same names to their cows through successive generations, and their bulls—of which they were in the habit of keeping four or five—were not distinguished by names but by numbers. From these causes no bulls or cows bred by them appear in the Herd Book. We are informed that for some years the Misses Tomkins had the advantage of Mr. George Tomkins' assistance, but when he gave up his farm in 1836 the herd does not seem to have been so carefully managed, though the system of close breeding was continued. Doubtless the Misses Tomkins parted with the best portion of their stock in 1819, and most of the remainder in 1839; and between that date and 1854, having only themselves and a bailiff to depend upon, it was only to have been expected that their herd should not continue to possess the special merits by which it was formerly characterized, and it may be from its decadence that the idea has arisen, that Benjamin Tomkins' stock had greatly deteriorated before his death.

Other members of the Tomkins family beside the Misses Tomkins engaged in the breeding of Herefords. Among them may be mentioned Mr. Richard Tomkins, Hyatt, Sarnesfield, a brother of the wife of Benjamin Tomkins. He was born in 1736 and died in 1819. After the death of this gentleman, his herd was sold in April, 1819. We have the sale list but it is unnecessary to print it in full. The prices were very good for the times. A pair of oxen named Summons and Merryman were sold for £80 ($400); another pair named Merchant and Lightfoot sold for £60 ($300); Mr. Westcar gave £50 ($250) for a pair of three-year-old bullocks, and £48 10s ($240) for another pair. Among the purchasers of breeding stock were the Rev. Mr. Smythies, and Mr. Jones, Breinton. Mr. Bray tells us that among Richard Tomkins' stock were a good many of the "hailed backed" variety, and several of the animals included in the sale are thus described in the catalogue. The Rev. J. R. Smythies purchased two "hail backed" heifers.

George Tomkins of Frogdon, born 1740, died 1797, brother of Benjamin Tomkins, the younger, had also a noted stock of Herefords, and he is generally believed to have been a remarkably good judge of stock.

George Tomkins, son of the gentleman just named, nephew of Benjamin Tomkins and a trusted friend of Lord Talbot, also bred Hereford cattle. Born in 1776, he occupied the farms of Wistaston and Frogdon. He gave up the former farm to his son-in-law, Thos. Galliers, in 1836, and then retired to the Green, Norton Canon. The portrait of a cow, bred by George Tomkins, was often pointed out by the mother and father of T. T. Galliers, Wistaston, as being a good representative of the Tomkins' "Silver-breed." This cow was purchased by Mr. Peploe, of Garnestone Castle, and was a favorite of his, being kept to a great age for breeding. He had her painted by Weaver in 1814, when she was eight years old, and the picture hung in Mr. Peploe's study during his lifetime, and during that of his successor, Captain Peploe. When the Rev. J. B. Webb-Peploe succeeded to the property he presented the oil painting to the late Mrs. Galliers on her requesting permission to have a photo taken of the portrait of her father's Silver cow. (¶ 9) This painting represents the Tomkins Silver variety of Herefords. Mr. T. A. Knight, of Downton, obtained some of his stock from Mr. Geo. Tomkins, who died in 1854, aged 79 years.

Other members of the Tomkins family who were breeders of the Hereford cattle were Richard Tomkins, of Dippers Moor, born 1757, died 1800; William Tomkins, of Wormbridge, born 1756, died 1821; and Thos. Tomkins, of Court House, born 1743.

Mr. T. C. Yeld (¶ 10) in an interesting article from which we shall quote, mentions Wigmore Grange sale, and says "that most of the purchasers secured several lots. They were cows, calves, and young heifers; the oxen, steers and bulls being sold in the following spring. The writer has seen a painting of one of the oxen, four of which, he has learned, from the family, sold for over £70 ($350) each.

"Old Mr. Tully also left three sons in farming business—one at Huntington, one at Clyro and one at Grafton; and these possessed by far the best of what would be called the white-faced Herefords, if I except Mr. Skyrme, of Stretton, but of whose stock I have no reliable account, except the opinion of Mr. T. A. Knight, which is certainly most favorable. In

BLACK HALL, KING'S-PYON, HEREFORDSHIRE.

giving an account of the Herefords of the last century, I have stated nothing but what is from correct sources.

"I now proceed to name the best herds at the commencement of the present century, and although Mr. Benjamin Tomkins was in the highest repute there were many who possessed equally good cattle. The late Mr. T. A. Knight in replying to my inquiry about the pedigree of the celebrated White Bull, writes as follows:

"'Sir: The account which you appear to have received respecting the bull from which you have bred is in every essential respect correct, but I did not give the calf to Mr. Turley. He bought it of me, and never paid me anything for it. The dam was bred by Mr. Skyrme, of Stretton, who at that time possessed, in my opinion, by far the best breed of cattle in the country, and which was Mr. Westcar's opinion. I reared several other bulls from the same cow, which were very excellent, and for one of them at eleven months old I refused 40 guineas. The sire of your bull descended from a mixture of the breed of Mr. Tully, of Huntington, and Mr. Isaac Martin, who possessed very excellent though small, stock. I do not think a better bred animal than that about which you have inquired ever existed in the county of Hereford. I never bred above two or three animals from Benjamin Tomkins' stock, which, I confess, I never liked. With good wishes, your obedient servant,

(Signed) T. A. Knight. (¶ 11)

To T. C. Yeld.

Downton, January 8th, 1836.'"

It may be pointed out that Mr. Yeld was evidently unaware of the fact that there were two breeders named Benjamin Tomkins. The associate of William Galliers, of Wigmore Grange, to whom he refers, was, as has already been explained, not Benjamin Tomkins, the younger, as he seems to have believed, but his father.

William Galliers (¶ 12a), of Wigmore Grange (¶ 12), was intimately associated with the elder Benjamin Tomkins in social and business relations, and was born in the year 1713, and died May 26th, 1779, and his herd passed to his son, John Galliers.

William Galliers, of Frogdon, was a son of William Galliers (¶ 13A), of Wigmore (¶ 12B) Grange, and hence a brother of John; was also a breeder of Herefords, and gained thirteen cups and two decanters before the Herefordshire Agricultural Society between the years 1802 and 1815. The Wigmore Grange herd was sold on October 15th, 1795. Prior to this date a portion of the herd had passed into the hands of William Galliers, Jr., who went to Oxhouse in 1765; to Eye in 1790; and to Frogdon in 1799.

The Wigmore sale, October 15th, 1795, comprised 82 head.

The two sons of William Galliers—William, born at Wigmore Grange (¶ 13) in 1744, who died at Oxhouse in 1832, aged 88 years; and John, born at Wigmore Grange in 1755, who died at Coxall in 1828—were both celebrated breeders. The prize list of the early shows of the Herefordshire Agricultural Society proves the character of the stock of William Galliers, and the sale list given indicates the estimation in which the herd, after it had passed into the hands of John Galliers, was held, although he does not seem to have long continued breeding Herefords after his removal to Coxall in 1795 (¶ 13B).

Miss Letitia Galliers, grand-daughter of

William Galliers, of Frogdon, remembers some animals of the mottled-face variety being at Oxhouse. She believes that at first a portion of the Galliers cattle were more or less mottle-faced, but they gradually assumed the red with white face markings, and by selection, they ultimately became wholly of that color. There can, in her opinion, be no doubt that her grandfather won his prizes with white-faced animals.

Some notes taken from a memorandum book belonging to Mr. William Galliers, of Frogdon, show that in 1775 an ox weighed 80 st. 4 lbs. (1,124 lbs.); while in 1787 an ox weighed 89 st. 11 lbs. (1,257 lbs.), and a cow weighed 84 st. 9 lbs. (1,185 lbs.). He seems to have sold his cattle by weight, at 4d per pound, off grass.

Benjamin Tomkins, the elder, began farming in 1738 and died in 1789.

William Galliers was born in 1713 (a year earlier than Benjamin Tomkins, the elder,) and died in 1779, ten years earlier than B. Tomkins, the elder; thus the two men were breeders for forty years or more. Mr. B. Tomkins, starting in 1766, was a breeder for 23 years by the side of his father, and for 13 years beside Mr. Galliers. John Galliers continued breeding from this date, after his father's death, until October 15th, 1795.

It will be seen from this date that the Tomkins family had been breeders of Hereford cattle for a century or more; and it is true, probably, from the time of Speed in 1627, that Herefordshire had many farmers who were breeders of cattle of a quality equal to the best that went into London market.

John Duncomb, Secretary of the Hereford Agricultural Society, and Historian of the county, says: "The cattle of Herefordshire have long been esteemed superior to most, if not all, other breeds in the Island. Those of Devonshire and Sussex approach the nearest to them in general appearance. A large size and athletic form and unusual neatness, characterize the true sort. The prevailing color is a reddish brown with white face. They are shod with iron in situations which frequently require their exertion on hard roads.

"The showing of oxen in thriving condition, at Michaelmas Fair, in Hereford, cannot be exceeded by any similar collection in England. On this occasion they are generally sold to the

BROOK HOUSE, KING'S-PYON, HEREFORDSHIRE.

principal graziers in the counties near the metropolis, and then perfected for the London market." I have introduced Mr. Duncomb at this point, as it is fair to presume that his statement may refer to the past as well as to his own time.

Mr. Henry Haywood (¶ 14), of Blakemere House, Hereford, informed Messrs. McDonald and Sinclair (editors of a history of Herefords, published in 1886) that in the division of John Haywood's property in 1713, he especially re-

WELLINGTON (4) 160, CALVED 1808, BRED BY B. TOMKINS.
(From an old lithograph.)

fers to his cattle and to one of his sons, and says further: "My father always told me that his great-grandfather (the said John Haywood) was considered a superior man of business, and was a breeder of Hereford cattle. My uncle, Joseph Smith, of Shellesley (who had always lived in that neighborhood), often mentioned this John Haywood as a leading man and breeder of Herefords."

The fact that the Tomkins family and the Haywoods were breeders, each in the eighteenth century, and probably much earlier, is sufficient evidence that the Herefordshire farmers were breeders of a superior class of cattle, and with such a foundation Mr. Benjamin Tomkins, Jr., commenced his work.

Mr. J. H. Campbell, of Charlton, Kent County, was a contributor to the "Annals of Agriculture," published by Arthur Young. He wrote two papers for the "Annals" treating on breeds of cattle and sheep, and mainly relating to the Herefords. Campbell had a controversy with Young as to the point that should characterize a model beef animal, and having been described as a warm advocate of the Herefords, he said: "I am so, because of long experience. If I am wrong, it is not for want of painstaking, or being thoroughly acquainted with several other breeds, and particularly those about which there has been much said, at least in print, as to which, after a long-continued trial (and in the outset of the trial, as confident in my expectation as anybody could have been of finding them better than the Herefords) in the end being of the opinion that in most cases they were greatly inferior to them."

Campbell's discussion with Young originated in a difference of opinion as to the merits of an ox, of the true Herefordshire breed, which the former had exhibited.

Campbell says that "the opinion of many who viewed this animal alive was that they never saw so much beef under a hide of the size, and upon so small proportion of bone." He also stated that he "knew, from experience through trials of various breeds, none that would fatten on less food and few that would not require more than the true Hereford breed. The difference in thriving, for the food given, between them and good specimens of other breeds, which he had fed along with them, did not require weight and scales to determine."

Mr. Campbell was a farmer in Charlton, Kent County, and a feeder of cattle for the butcher in London market, and commenced feeding cattle at or before 1779, probably before that time. The ox, a specimen of the true Herefordshire breed, over which the controversy was held, was slaughtered in 1779 and exhibited at Greenwich, on account of the fineness of his flesh, beauty of his shape, symmetry of his parts, fore and aft, the impartial distribution of his weight, and the regular fattening of all his parts. The ox was about seven years old, and the following are the figures of his size and weight: Live weight was 3,360 pounds; the forequarters weighed 1,016 pounds; the hindquarters weighed 896 pounds. The dressed weight of this ox was 1,912 pounds.

Mr. John Westcar, of Creslow, Buckinghamshire, an eminent grazier, identified himself with this breed. He regularly attended the Hereford fairs as early as 1779, and the high prices at which he sold bullocks in the London market doubtless convinced many of their adaptability for grazing purposes. We first note his selling fifteen oxen on September 17th, 1798, for a price in English currency equal to $4,637.00, an average of $243.00 each. The same year he aided in the organization of the Smithfield Cattle and Sheep Society, before which at their first meeting he took, with a Hereford, the championship for the best ox in the show (¶ 15).

Mr. George Dood wrote Rev. J. R. Smythies

that he had been permitted to examine Mr. Westcar's books, and made selections as of the Herefords which he had sold for £100 ($500) each, and he found between 1799 and 1811 twenty oxen sold for £2,123 ($10,615), or $531.00 each, and says that if he had selected such as sold for £80 ($400) each, the list would have been very largely increased.

Selecting from his sales from 1799 to 1811, cattle that had sold for $500.00 or over, there were twenty head that averaged $531.00, and the highest priced ox sold for $737.00—all selling to butchers.

In the year 1812 or 1813 he made a sale of fifty oxen at Smithfield for $250.00 each. These are the sales of which I find an account, though he fed and grazed 200 head or more each year, which found a market in London.

I have selected these two feeders, Westcar and Campbell, as coming nearer or contemporary to the Messrs. Tomkins' work, than any other. These men were graziers as early as 1799, and were experienced as graziers, feeders and sellers.

William Marshall, contemporary historian of Tomkins, Westcar, Campbell period, wrote in 1788, describing the cattle of the west of England, that the great writer Speed said in 1627, "that the Hereford breed of cattle, taking it all in all, may, without risk, I believe, be deemed the first breed of cattle in this Island." Here we have the fixed data of Speed in the year 1627, who was a historian, writing of England, Wales and Scotland.

Marshall was a native of Yorkshire, and journeyed all over the country, collecting facts illustrative of the various agricultural districts, and making inquiries as to the breeds of cattle, horses, and sheep, for facts to be used in a work published by him, entitled "Rural Economy of the West of England."

Marshall gives a description of the Hereford ox, as he found him in 1788, which it is well to quote here: "The general appearance, full of health and vigor, and wearing the marks of sufficient maturity, provincially oxenish, not steerish, or still in too growing a state to fat; the countenance open, cheerful, pleasant; the forehead broad, the eye full and lively; the horns bright, tapering and spreading; the head small; the neck long and tapering; the chest deep; the bosom broad and projecting forward; the shoulder bone thin and flat. No protuberance in bone, but full and mellow in flesh; the loins broad, the hips standing wide and level with the spine; the quarters long and wide at the nache; the rump with the general level of the back, not crooping or standing high and sharp above the quarters; the tail slender and neatly haired; the barrel round and roomy; the carcass throughout being deep and well spread; the ribs broad and standing close and flat on the outer surface, forming a smooth and even barrel; the hindmost large and full length, the bone small and snug, not prominent; the thigh clean and regularly tapering; the leg upright and short, the bone below the hough small; the cod and twist round and full; the flank large, the feet of a middle size; the flesh everywhere mellow, soft, and yielding to the touch, especially on the chine, the shoulder and the ribs; the hide mellow and supple, of a middle thickness and loose on the nache and huckle; the coat neatly haired, bright and silky, its color a middle red; with a bald face, the last being esteemed characteristic of the true Herefordshire breed."

We submit that this description, written over one hundred years ago, will pretty well answer for to-day.

SILVER COW, CALVED 1806, BRED BY GEORGE TOMKINS.
(From an old painting.)

Marshall also says that "At the Hereford fair on October 20th, 1788, we saw about a thousand head of cattle, chiefly of the Hereford breed. A large proportion of them were grown oxen, full of flesh. The most valuable collection I have met with," and then he adds: "Out of Smithfield, by much the finest show I have ever seen."

Mr. Fowler, on another page, gives an account of Mr. Westcar's visit to the Hereford October Fair with the Duke of Bedford and Lord Berners, to which we call attention.

As stated, Mr. Westcar took an active part in the organization of Smithfield Cattle and Sheep Society, afterwards changed to the "Smithfield Club," an account of which we have thought best to adopt, supplemented by items that I find in the "Annals of Agriculture." (¶ 16)

CHAPTER II.

Foundation Hereford Herds—Continued

JOHN PRICE, OF RYALL.

It was impossible to notice the career of Benjamin Tomkins, the younger, without making some reference to his greatest supporter and disciple, John Price. (¶ 17) Thanks to Mr. Price's habit of carefully recording his breeding transactions, and to the industry of his friend, Mr. Welles, we know almost exactly the character of the cows which he purchased from Mr. Tomkins; and his subsequent method of breeding is clearly narrated in the Herd Book entries, which were drawn up from his catalogues and notes. Mr. Price was scrupulous in his attention to pedigrees, and in his case, there is no occasion for regret at the absence of details.

John Price, the eminent breeder, was the eldest son of Job and Elizabeth Price, who occupied a farm at Earl's Croome, in Worcestershire (¶ 18), where he was born in 1776. The son of an industrious farmer, John Price was from an early age engaged in all the operations of the farm. Thus employed, he had little opportunity for receiving any other than a plain village school education. He was taught to read, to write, and the use of figures. Whatever disadvantage, however, he experienced from the want of a more extended education was amply compensated by the possession of great natural abilities—of a mind powerful and original in its conceptions and conclusions, and as soon as he commenced business on his own account, he let slip no opportunity of improving his education by reading and seeking the society of gentlemen of high respectability. Early in life he became a favorite with the Earl of Coventry. These facts are gleaned from an obituary notice that appeared in the "Farmers' Magazine" in 1845. Mrs. Pumfrey, Mr. Price's daughter, in a subsequent number of the journal, wrote: "All is true that you state of his humble birth; not that his parents were of mean grade or fortuneless; but farmers then lived and brought up their sons so differently to those of modern times. My father's transcendant and natural abilities and genius, however, surmounted every obstacle to improvement; by nature and habit he became a perfect gentleman, an ornament to any society, and this without any assumed polish. Humble and courteous even in his most palmy days, he was a favorite with all, the kind and assisting friend of many, his very faults leaning so much to the side of virtue as to disarm one of blame. Not only, as you say, was he an admitted, but an honored guest at Croome, for even during the visit of royal personages has the late Countess of Coventry insisted on my father being of their circle. I have known the late Earl of Coventry, with his brothers, to dine at my father's house five days of the week; the late Earl Plymouth, and many others, too numerous to name individually, none of whom need to blush in association with a man mentally superior to most. His fame as a breeder and judge of stock will not die for many an age; in which respect I have often been told since and before his death, he had no equal."

Mr. Price ultimately succeeded his father as tenant of Earl's Croome, and he early evinced a fondness for the live stock of the farm. The cattle he first possessed of any pretensions to good breeding were procured from Mr. Walker, of Burton. Mr. Welles states that with some of these he was induced to try crosses with the pure Gloucesters, an old breed famous for their milking properties, the improved specimens also making good carcasses of meat and producing good steers. An uncle, Mr. Barnes, of Corse Court, was in possession of an excellent herd of the Gloucester breed, and Mr. Price procured a few cows of him. Mr. Welles says he remembers a cow bred from one of these by a Hereford bull, making, when fed, an extraordinary animal—weighing upwards of 18 score per quarter (1,440 lbs.).

It was about the year 1804 that Mr. Price became acquainted with the cattle of Mr. B. Tomkins, from which he bought a few cows, using to them bulls descended from Mr. Walker's stock. Mr. Welles recollected the first bull so bred, out of the cow Pigeon, bought from B. Tomkins. But the cross did not suit and the animal was disposed of.

About 1811 Mr. Price gave up the farm at Earl's Croome and bought a small estate at Ryall, near Upton-upon-Severn. He also took a large field of pasture, a part of Croome demesne, of about 120 acres, which he held till his death. In a few years from this time he possessed himself of cows from Mr. B. Tomkins, and his herd began to attract considerable notice. Among the purchasers of the stock he was able to draft, being many of the nobility,

MR. T. C. YELD, OF THE BROOME.

including the Earl of Plymouth, Earl Talbot, and the honorable Mr. Germaine. In 1812 he gave a challenge, to be decided at the Lichfield Agricultural meeting, to show twenty of his cows in milk against twenty Longhorn cows for 100 guineas ($500.00). The challenge was accepted by Mr. Meek, and was decided in Mr. Price's favor. About this date he made a large speculation in purchasing land. The venture was not a success, and the estate had to be sold at great loss. A good stock of cattle and sheep which Mr. Price had collected also came to the hammer, and the prices showed that much judgment had been exercised in their breeding and selection.

Mr. Price then carried on his farming operations at Ryall, where he continued to reside, taking, however, more grass land of excellent quality at Mytton, near Tewkesbury. But previous to this he had obtained more Herefords of Tomkins blood and purchased the bull Wellington, and his dam, from Mr. Tomkins. Soon after 1816 Mr. Price left Ryall and took up his residence at Poole House, near Upton, still holding the land of which he had been tenant for so many years under Lord Coventry.

Mr. Price frequently expressed his views on the subject of breeding. He stated that among cattle the Highland Scot approached more nearly than any other animal to the standard of form, which he considered the true one. "This," he adds, "determined me in adopting them as my model. I was desirous of possessing a breed of cattle on a somewhat larger scale than the Scotch Kyloes, yet having the same symmetrical loggy forms, with similar coats and texture of flesh." In this opinion, Mr. Price only repeated that Mr. John Charge had heard Bakewell many years before state that from the West Highland heifer he thought the best breed of cattle might be produced.

In commencing to form a herd which should possess the form and qualities he thought most desired, Price, as has been indicated, fixed upon the stock of Benj. Tomkins, from whom he purchased a considerable number of cows and heifers and three bulls. These cattle were of smaller size than other herds he saw in Herefordshire, but had more of the good properties of the model he had in view than any others he could meet with. As we have seen, he first attempted to improve the Tomkins cattle by crossing them with the larger stock of Mr. Walker, with the view to increasing their size, but the result was so unfavorable that he put away all these crosses and returned to the pure Tomkins variety.

Mr. Price continued to breed Herefords until 1841, his herd being solely of Tomkins blood; so that for upwards of seventy years, at least, this strain, first in the possession of B. Tomkins, and then in that of John Price, was bred continuously without a fresh cross.

For a description of the various animals purchased from Tomkins by Mr. Price the reader is referred to the interesting notes of Mr. Welles printed on a preceding page. In reference to the statement that Price obtained the best animals that Tomkins possessed, Mr. Eyton says there was one old cow that must be excepted, a remarkably good breeder, which Tomkins always refused to sell, although Price offered him £250 ($1,250.00) for her. This remark suggests the idea of the sums Price paid for the animals he actually bought from the great breeder at Wellington Court. Mr. Welles expresses the opinion that Mr. Price had only one of the Tomkins bulls, the celebrated Wellington (4) 160. But in addition to that animal he owned Voltaire (39a) 429, a white-faced bull bred by Tomkins, and an unnamed bull of his breeding that appears in some of his pedigrees. Price seems to have followed Tomkins not only in his system of in-and-in breeding, but also in his disregard of color. It will have been noticed that the colors of the cows he acquired from Tomkins varied greatly. Then among the bulls, Wellington was a mottle-face; Voltaire a white face, and Victory (33) (¶ 19) calved in 1839, bred by Price, was chosen for illustration in the first volume of the Herd Book as a typical specimen of the grey variety;

THOMAS ANDREW KNIGHT, ESQ., PRESIDENT OF THE LONDON HORTICULTURAL SOCIETY, BORN 1759.
(From an old lithograph.)

while the portrait of Young Trueboy (32) 630 (¶ 20) is also given in Volume 2 as a specimen of the greys, although in the entry of the first volume he is stated to have been a mottle face.

One of the most remarkable cows owned by Mr. Price was Toby Pigeon 308 by Toby (5) 372, dam Pigeon, or Price's No. 6, 373, bred by B. Tomkins. It is stated in the entry of one of this cow's produce in Volume 1 of the Herd Book, that nearly the whole of Mr. Price's herd sold in 1841 were derived from her. At 19 years of age she had bred 19 calves, having taken the bull by chance when a calf, and at 3 and 4 years old she had twins. The following is a list of her progeny: Woodcock Pigeon 651, by Woodcock (50) 654; bull, Solon (92); bull died; Miss Woodman; bull, Young Woodman (12) 238; bull, Paris (19) 6657; bull, Plenipotentiary (23); cow, sold to Mr. Monkhouse; bull, Trusty (15) 643; cow; cow died young; ditto Burton Pigeon; bull, died; bull, Trueboy (14) 637; cow, Blue Pigeon 3697; cow, Stock Dove; cow, Nonesuch; bull, Washington (35).

Price frequently challenged admirers of other breeds to show their stock against his own, this, as we shall have occasion to point out, having been a favorite method of settling disputed points as to superiority, prior to the general acceptance of the more satisfactory arbitrament of the show ring. He attended one of Lord Althorpes' ram sales in Northamptonshire, and after the dinner gave a challenge to show one of his bulls against any Shorthorn. He succeeded in getting up a sweepstakes of £5 each, which he won with his bull Lundyfoot (16) 3560, which, according to the writer of the memoir in the "Farmers' Magazine," was allowed to be the completest animal any of the company ever saw. In 1839 he issued another challenge, of which Mr. Haywood, of Blakemere House, has furnished a copy. It is as follows:

"CHALLENGE! ! ! To all breeders of cattle in England. Mr. Price, of Poole House (¶ 21), Upton-upon-Severn, is willing to show a bull and 20 regular breeding in-calf cows, bred by himself, for any sum not exceeding £100 nor less than £25, to be shown before the last day of November next ensuing, against a bull and a like number of cows of any sort that have been bred by, and are now in the possession of, any breeder of cattle in the United Kingdom. The judges to decide on this occasion to be chosen by that noble patron of Agriculture and stickler for fair play, Earl Spencer, and his Lordship's friend, Sir Francis Lawley, Bart., or whom they may appoint. The stock to be viewed on the farms of their respective owners, and the judges to be paid by the losing party. N. B.—It is a well known fact that this herd has lived on worse and less food, owing to the dry summer, than any other herd of cattle in the county."

This challenge was not accepted but it led to a controversy between Mr. Thos. Bates (¶ 22), of Kirklevington, the well-known Shorthorn breeder, and Mr. Price. Mr. Bates, writing in 1840, said he had visited Herefordshire about fifty years previously, and was then, and continued still, an admirer of the best variety of the Hereford cattle. But he considered then and had for about 40 years been convinced, that "the very best Shorthorns, which were only a few, were capable of improving all other breeds of cattle in the United Kingdom, as well as the ordinary Shorthorns which were far from a good breed, and much inferior to the Herefords, Devons, and others." Mr. Bates added: "I have at present two red, twin, one-year-old bulls, out of the dam of the Duke of Northumberland, you may not think unworthy to be put to your herd of Herefords for one season, to give you an opportunity of testing the merits of this cross-bred. In my opinion they would prove an invaluable cross with the best Herefords; increase the growth of the Herefords, and at an earlier age be fit for the butcher, with a less consumption of food, and quality of beef unimpaired, and also give that breed an increased milking quality, both in quantity of milk and richness, yielding more butter." To this Mr. Price replied that he had inspected Lord Spencer's Shorthorn herd, and had never seen anything to shake his belief that Hereford cattle would pay more money for the food they consumed than any other breed with which he was acquainted. He said he had tried many crosses all of which signally had failed, where the object had been to obtain more size and weight by using large male animals with females of smaller dimensions.

Writing to the "Farmers' Magazine" in 1841, Mr. Price gave a description of his farm and the difficulties under which he labored, owing to shortness of keep, etc. He said: "The farm I have occupied since 1829 has not, at any time, much exceeded 150 acres, 20 of which are arable, totally unfit for the growth of turnips, and nearly 120 acres, part of the Croome Demesne (¶ 23), belonging to the Earl of Coventry, in one ground and rather below second-rate quality of land, greatly covered with ornamental timber, and neither buildings nor fold yards on my farm sufficient to hold 20 beasts. Yet, on this land I have usually kept 100 head of cattle, together with a flock of 150 sheep, 40 of which were rams, besides my cart and other horses. These are facts well known to the whole of my neighbors, who have always given me full credit for being the worst keeper of stock in England. I have seldom made use of oil cake, and on no occasion have I given corn or meal to any of my stock."

On October 17th and 18th, 1816, Mr. Price had an extensive sale at Ryall, which was thus announced: "The cattle stock are wholly descended from that of the justly celebrated one of the late Mr. B. Tomkins, of Wellington Court, in the county of Hereford, and are too well known to need any comment." The prices realized at this sale were very large, and the event forms such an important landmark in the history of the breed, that we give the list of prices.

WIGMORE GRANGE, HOME OF WILLIAM GALLIERS. 1713-1779.

The averages for the various classes were as follows:

	Average.	Total.
32 cows	$270	$8,650
13 three-year-old heifers ...	295	3,830
21 two-year-old heifers	285	5,995
10 yearling heifers	180	1,820
21 heifer calves	125	2,580
13 bulls	660	8,595
6 bull calves	360	2,170
116 head averaged $290.		$33,640

The highest-priced females brought respectively £252, £215, £189, £173, £120 and £110, or, in American money, $1,260, $1,075, $945, $865, $600, and $550 respectively.

The bulls bringing over $500 each were Waxy (3) 655, £341, or $1,705; Wellington (4) 160, £283 = $1,415; Ryall (45), £262 = $1,310; Original (40) 9779, £147 = $735; Warrior (44), £136 = $680; Moses (7) 426, £115 = $575; Leopold (1) 652, a calf, brought £126 = $730.

We have also a catalogue, with a few prices, of Mr. Price's sale at Mytton, Lodge Farm, near Tewkesbury, on March 21 and 22, 1820, but it is not necessary to reproduce it. A few high prices were realized. Mr. Barnes, £109 4s ($550) for heifer Thalia; Mr. Jellico gave £127 1s ($635) for the heifer Vesta. Mr. Price's final sale took place at Poole House on October 15th, 1841. Here the cow Toby Pigeon was sold in her 22d year to Sir F. Lawley for £14 ($70). The highest price was £166 ($830) for the bull Washington, Lord Talbot being the purchaser. Among the cows, Wood Pigeon made 150 guineas ($750), going to Mr. Byrd, Hampton Court, Hereford. The heifer Tuberose was taken by Lord Talbot at 100 guineas ($500) and Ceres at £115 ($575), by Sir F. Goodricke; Mr. Smith, Martly, gave £100 ($500) for the bull Tramp; Sir F. Lawley £140

WILLIAM GALLIERS, WIGMORE GRANGE.

($700) for Young Trueboy, and £100 ($500) for Victory; Mr. Samuel Peploe bought the bull Murphy DeLaney for 110 guineas ($550). The largest purchasers were Lord Talbot, Sir F. Lawley, Sir F. Goodricke, Captain Walters, Mr. S. Peploe, Garnestone, and Rev. J. R. Smythies, Lynch Court. Mr. Evans, Pendeford Hall, Stafford, took The Rejected for 110 guineas ($550). An average for 99 animals was £53 16s 4d ($270), and the total £5,328 ($26,640). Commenting on the results of the sale, Mr. Price said: "Although the average of my sale in 1816 is a little above the average price of my last sale in 1841, it will, I think, appear evident on taking into account the length of time (25 years) that has elapsed between the two sales, the great reduction which has taken place during that time (and since the sales of Messrs. Colling's Shorthorns) in the price of first-rate herds of cattle, and also of other herds of cattle, together with circumstances too well known to both landlords and tenants to need any comment from me, that the average of my last sale was much the best; thereby placing the herd on much higher ground compared with all others than they heretofore occupied, and I do sincerely hope that the hands these animals have fallen into, will take care that they keep their present high position. Should they not continue to do so, and lose caste, the fault will not be in the cattle."

At the Poole House sale it was resolved to present Mr. Price with his portrait, and a good picture was painted by Mr. Frederick Tatham. Mr. Price survived this sale only two years.

At his three sales of cattle, in 1813, 1816, and 1841, the proceeds amounted to no less than £16,690 ($83,450). A complete record of the sale in 1820 does not exist, but we should imagine that if the amount obtained at it could be added it would swell the total to £20,000 ($100,000).

In the article on Hereford cattle contributed to Morton's Cyclopedia of Agriculture, Mr. Welles has a few remarks on the character of Mr. Price's stock. He said he thought it must be generally admitted that unusual exertions had been made through great difficulties by an individual of an adjoining county, who had been the most zealous and (if high prices were the test) the most successful breeder of Herefords of that day (about 1830-40). And little as his opinions seemed to be in conformity with those of a large portion of the breeders of the county of Hereford, and though his great efforts to raise the character of the breed had been so little understood and appreciated on its native soil, he thought those who calmly and dispassionately examined the principles which guided him in the pursuit, must be convinced that there were many points on which he insisted as indispensable in the formation of a superior animal that could not safely be disregarded. Instances of failure might be adduced against him in some of his practice, but these often resulted with the most sagacious from the trial of new combinations; and Mr. Welles thought it very probable that the rising generation of breeders would find that a superior intellect brought to bear so exclusively on one subject had not been exercised in vain, and that time would dispel many of the prejudices existing in certain places against Mr. Price's breed.

Mentioning some of the exceptions that might fairly be taken to Mr. Price's system of breeding, he said one of the most prominent was a great disregard of the milking properties; and from his late practice of breeding from near affinities, this fault might be supposed to have been more permanently fixed in certain families. That it would not be desirable in a breed such as the Hereford to make too many sacrifices to the milking quality, he thought would be generally allowed; but there might be, he was convinced, a sufficient disposition to give a fair quantity of milk and the cow be equally good for any purpose required of her.

There were, however, many cows that from want of proper care of the udder after calving, and during the time that the grass was luxuriant, were rendered more or less incapable of a supply of milk afterwards, and he thought much inattention on that head was often the case in the stock he alluded to. Another objection that might be raised against Mr. Price's stock was the shortness and rather mean appearance of the horns, in many of his cows, not characteristic of Herefords in general, which had mainly risen from his attention being devoted to more important qualities; and as the family in which these were most concentrated was deficient in horns, he left them unimproved, thinking he might in the pursuit of a non-essential run the risk of losing a valuable property; still Mr. Welles believed the possession of good horns to be quite compatible with every other valuable requisite, and it was certainly a considerable advantage to the appearance of the animal.

These observations prove that a prejudice had arisen in the county against Price's cattle, which, on the evidence of even a favorable witness like Mr. Welles, was not without some justification. But the variety was very far, indeed, from being even at the close of Mr. Price's career, without substantial merit. As to the late appearance of Mr. Price's herd, we have the following interesting communication from Mr. George Smythies: "I had no intimate acquaintance with the Hereford herd of Mr. Price. I never saw it until after he had given up farming; when I knew the herd it was kept in Lord Coventry's park the greater part of the year and for a short time, in winter and spring, the cattle were tacked out in straw yards with anybody who would keep them. I

THOMAS TOMKINS GALLIERS (1902), SEATED IN AN ANCIENT CHAIR BEFORE TYPICAL OLD HEREFORDSHIRE WAIN (WAGON) HOUSE.

once saw the best lot of 14 two-year-old heifers I ever looked at in a yard where they got nothing but stubble—that is, the straw that remained after hand-reaping of wheat, only there was a little clover in it, the field it came from having been sowed with seeds. During the last few years that Mr. Price kept his Herefords he changed their forms a good deal. The Tomkins breed, which I believe he used exclusively, were very wide over their hips and narrow on their shoulders. This he altered, getting his cows much wider on the chine with less gaudy hips. These characteristics were particularly exemplified in Dove, bought by my father at the sale in 1841 for 77 guineas ($385), and by Tuberose, sold to Lord Talbot for 100 guineas ($500). Mr. Price's cattle were some of them red with white faces, some a beautiful roan, as was Dove (Dove was a smoky roan, differing from the roans as bred by Tully), others being white-backed with mottle faces. The bulls were brought up differently to what they are now, running, in almost wild state, with the cows, until they were fit for service, when most of them were let and kept from home as much as possible, Mr. Price having but little accommodations for them. Consequently they had a mean appearance as compared with the cows, which were magnificent animals."

WIGMORE GRANGE, SEAT OF THE GALLIERS FAMILY. (REAR VIEW.)

Cobbitt, in his "Rural Rides" (1830), writes from Tewkesbury: "I am here among the finest cattle and the finest sheep of the Leicester kind that I ever saw. My host, Mr. Price, is famed as a breeder of cattle and sheep. The cattle are of the Hereford kind, and the sheep surpassing any animals of the kind that I ever saw. The animals seem to be made for the soil and the soil for them. The sheep are chiefly of the Leicester breed, and the cattle of Hereford, white face and dark body, certainly the finest and most beautiful of all horn cattle."

The Earl of Coventry (¶ 24) says: "The fame of John Price's Herefords still lives in this neighborhood, and there are yet living people who speak of the noble herd with admiration, and describe them as being possessed of great scale and extraordinary constitution. They were accustomed to range the pastures summer and winter, and were almost always to be seen in the well-known 'Cubsmoor,' a large grass field of great repute among graziers. John Price had a bull which weighed 29 cwt.* (3,248 lbs.), and a bull calf 9 cwt. (1,008 lbs.) at nine months old" (¶ 25).

In the appendix to Vol. 1 of the Herd Book, a list is given of the principal breeders of the Tomkins and Price stock, from whom pedigrees had been received by Mr. Eyton. They were the Earl of Talbot, Ingestre; Sir F. Lawley, Bart.; Sir F. Goodricke, Studley Castle; Mr. G. Brake, The Manor Farm, East Tytherly; Mr. Shepherd, Eastwood House; Mr. Thos. Juckes, Tern Farm; Mr. N. Smith, Martly; Mr. Pratt, New Field; Mr. Gravenor, Wellington; the Rev. W. P. Hopton, Bishops Froome; Mr. J. Smith, Shellesley; Mr. T. P. Wight, Tedstone Park. Only a few of these were resident in the County of Hereford, and of course there are others who ought to have been included in the list—notably Mr. Smythies, Mr. Welles, and others.

Lord Talbot, as we have seen, was a liberal purchaser at the Tomkins' and Price's sales. There has been considerable uncertainty as to the reason why he gave up his herd. Mr. George Smythies, Marlow Lodge (son of the Rev. J. R. Smythies, Lynch Court), informs us that he was once at Ingestre about 1840, and naturally has not a very clear idea now of what he saw there, but he remembers he thought the land did not suit the Herefords. We are able to give in Lord Talbot's own words the explanation of the dispersion of the Ingestre herd, which quite confirms Mr. Smythies' impression. In a letter to Mr. Geo. Tomkins, Eccles Green, Norton Pyon, dated March 4, 1847, Lord Talbot fully states his reason for disposing of his herd, and as the communication has other interest, as showing the friendship existing between these two breeders, we print an extract from it:

"Dear George," wrote Lord Talbot, "Events of a very painful nature have occurred which have prevented our meeting, as we formerly did, in friendship and good fellowship. The

* Note: The English cwt. is 112 lbs.

remembrance of past times of this nature cannot but be most gratifying to me, and I feel not otherwise to you. * * * What you will not perhaps expect, I have to inform you, that I have resolved to give up being a breeder of Herefordshire cattle; not, be assured, from any want of partiality to the breed, but simply that I find my land, having been now more or less attended to, and constantly depastured, is becoming too rich for a breeding stock. Accidents have been so frequent with slipping calf —with the apoplexy; which over condition is sure to produce, and other causes of disappointment that, however painful the struggle, I have faced it and have advertised my breeding stock for unreserved sale. The die being cast, what is to be done in the future? I wish to feed Hereford oxen largely, which intention is perhaps fortified by the facility I have of sending up to Smithfield. The want of market, which formerly prevented my feeding these excellent cattle to the extent I wished, is now removed, and therefore I return with eagerness to the project of feeding instead of breeding Herefords." Lord Talbot proceeded to ask Mr. Tomkins' co-operation in obtaining suitable cattle for feeding, and concluded by inviting him to his sale, which took place on October 24th, 1838.

In addition to animals bred by Mr. Tomkins and Mr. Price, or descending from their herds, the sale included specimens from the herds of the Misses Tomkins, and Mr. George Tomkins. We have not a list of the prices, which, however, were not extraordinary; but the influence of the Ingestre stock still exists. Lord Talbot seems to have again collected a few pedigreed Herefords, as we find him purchasing at Mr. Price's sale in 1841.

Sir F. Lawley, as we have seen, also secured many of the Tomkins and Price cattle, and Mr. Duckham tells us that he has heard from old breeders that he had a very grand herd of heavy fleshed mottle faces. He had a sale in 1839 of which Mr. George Smythies gives us the catalogue. Mr. Smythies attended the sale, when several of the lots were purchased by his father, and described as having been very good animals. Mr. Smythies also supplies us with a priced catalogue of the sale of Sir F. Lawley's herd, which took place after his death. The prices were very low, the best being only 28 guineas ($140).

Lord Plymouth, Earl St. Germaines, and others, had at one time very good herds of this variety, but they had long since been dispersed and few traces of them now remain. In his Cirencester lecture Mr. Duckham mentions that in 1863 Mr. Smith, Shellesley, sent some well-fleshed animals of the mottle-faced sort to the Worcester show, but they were not successful. The last he says he remembers to have seen a winner was the heifer Superb 1824, exhibited by the Earl of Radnor at Salisbury, and then purchased for the royal herd, where she was put

WILLIAM GALLIERS, JR., OF KING'S-PYON, 1744-1832.

to the red with white face bull Brecon (918) 1810, and produced the heavy-fleshed bull Maximus (1650) 1817 (¶ 26), winner of the first prize at the Warwick and Battersea meetings of the Royal Agricultural Society of England. The marks on his face showed the transition from the mottle face to red with white face being larger than those with mottle face and fewer in number. Mr. Smith used many of Mr. Price's best bulls. In 1856 he received a letter stating that H. R. H. Prince Albert had been graciously pleased to patronize the Hereford breed and an appointment was asked by the representative of H. R. H. in order that Mr. Smith's celebrated herd might be inspected. Mr. Duckham also mentions Sir F. Goodricke, Captain Rayer, Captain Peploe, Mr. Drake, and Mr. Jellicoe as having been breeders of this variety.

SOME OF THE
SILVER CUPS
¶13B

CHAPTER III.

FOUNDATION HEREFORD HERDS—Continued

HEWER HEREFORDS.

(¶ 27) Indebted as we are and as all writers on Hereford history must be to Messrs. James Macdonald and James Sinclair for the concentration of facts compiled by them in their "History of Hereford Cattle" (1886), we cannot do better than, with this acknowledgment, quote them in this chapter almost exclusively.

The student of the Herd Book will find that nearly every valuable strain of the Herefords at the present day is full of Hewer blood. The influence of the Hewer cattle has, indeed, been remarkable. It is not merely that a few families that have become exceedingly valuable are of this line of descent, but that the modern character of the entire breed has to a large extent been determined by this variety, not alone as regards color markings, on which the Hewer impress has been very powerful, but on the more essential matters of shape and quality. This being the case, it is needless to say that a most important section of Hereford history is that relating to the proceedings of the Hewers. Here again, however, reliable information is not over plentiful.

It may be explained that the original idea of the founder of the Herd Book was to confine it to a record of the Tomkins-Price stock; but this manifestly would not have been a Herd Book of Hereford cattle, and the plan was so changed that all the varieties should be admitted. A Hereford Herd Book without the records of the Hewer cattle would certainly have been a curious production, comparable only to the performance of the play of "Hamlet" with the leading character omitted. This was evidently appreciated by Mr. Eyton, and so in Vol. 1, 96, bulls bred by William and John Hewer were entered; but Mr. Eyton was either unable to collect much information about the Hewer family and their herds, or he did not greatly trouble himself about the subject. All that he has to say regarding them is that Mr. John Hewer informed him "that the breed he now possesses has been in his family for many years. A great number of the principal breeders have had bulls from him. He at present possesses more bulls, most of which are let, than any other breeder in the county." Then if the pedigrees are closely examined, it will be found that so far as they are registered the Hewer cattle trace back to a bull called Silver (540) 358, as to whom the only facts vouchsafed are, that he was white-faced, was calved in 1797, and was bred by Mr. William Hewer, of Hardwick.

Obviously there was not within the covers of the Herd Book an adequate account of the Hewers and their cattle, nor had former writers on Herefords added any trustworthy information to these scanty details. An effort was therefore made to find whether all the records had perished, and if it were really impossible to get some light thrown on this branch of the history of the breed. Although Mr. John L. Hewer, Aston Ingham, Ross, has most cordially seconded our efforts, we regret that owing to papers having been mislaid and to the habit of the old breeders to look upon the sources and management of their herds as trade secrets, which must on no account be disclosed, we have not succeeded quite so well as could have been wished. Still, it is possible to remove much of the uncertainty and misapprehension in which the subject has been enveloped. The account of the Hewer family that follows is chiefly taken from communications furnished by Mr. John L. Hewer.

William Hewer, the father of John Hewer, was a native of Gloucestershire, being one of the Hewers of Northleach and was descended from William Hewer, so frequently mentioned in Pepy's Diary. He was born in 1757 and married a Monmouthshire lady—Miss Hughs, of Court Morgan, near Abergavenny, about the year 1787. In order to be near his wife's family, he went to live at the great Hardwick and Dobson's farms, remaining there for 28 years, and then took a farm at Llanlellen, about a mile from Hardwick. About the year 1825 there was a kind of panic amongst the banks, and one in which he had a large sum of money invested failed. William Hewer was so overcome by the disaster that, with his eldest son, William, he left the country for America, but he lived only about six months after he arrived, being quite heart-broken. He died in New York in November, 1825, and was interred at the cemetery of St. Mark's churchyard, Bowery, New York, on December 2d, being at the time of his decease 68 years of age.

John Hewer (¶ 28) was born on March 12th, 1787, and died September 28th, 1875. His son, Mr. John L. Hewer (¶ 28 A and B), never heard him say positively whether he was born at Kilkenny Farm, near Northleach, where his father resided before going to Monmouthshire, or at the Great Hardwick, Abergavenny. John Hewer, Sr., assisted his father at the Hardwick, and it was then he formed the idea of having the Herefords of uniform color and markings. During that time he had a few cattle of his own, and had the benefit of his father's experience. The statement that John Hewer went to his relatives in Gloucestershire in 1805 and remained with them for several years, is, we are assured, incorrect. He never left home, except on a visit, until about the year 1817, when he went to Purslow Hall, in Shropshire. Here he continued for several years. On his departure for America William Hewer left his wife and the younger portion of his family in England. They took a farm called The Grove, in Monmouthshire, and John Hewer managed it for his mother, and did very well until some misunderstanding took place between him and other members of the family; after which he went into Herefordshire, living first at Hill House, Aston Ingham. He subsequently proceeded to Moor House, about a mile from Hereford, and from that place to Brandon Cottage, where he had some land. In 1835 he occupied Hampton Lodge (¶ 29), near Hereford, and Litley Farm. He gave up the latter holding in 1839, the date of his first great sale. He was at Hampton Lodge until 1846; then at Lower Wilcroft, where he continued for two years. He was for two years at Palmer's Court, Holmer (¶ 30), whence, in 1850, he went to Vern House, Marden, where he settled down, having purchased it in 1855. He resided at Vern House (¶ 31) until 1875, when he sold it to Mr. H. Burr, of Aldermarston, and took Paradise Villa (¶ 32), Marden, where he died in the same year and was buried in the Holmer churchyard (¶ 33).

These are the salient biographical facts as to the two Hewers. We were naturally very anxious to ascertain, if possible, where William Hewer originally procured his Hereford cattle, he being a native of Gloucestershire, living in Monmouthshire, and never having resided in the county of Hereford. Mr. John L. Hewer says he cannot tell us where his grandfather obtained his stock, but he always understood from his father that his great-grandfather had a herd of Herefords and that William inherited them. Certain it is, says Mr. J. L. Hewer, he was a successful exhibitor at the Bath and West of England shows before the close of the eighteenth century. Mr. Thos. F. Plowman, Secretary of the Bath and West of England Society, has kindly searched the old records of that Society, and informs us that it was not until

MR. HENRY HAYWOOD, 1819-1902, WHOSE FAMILY BRED HEREFORDS FOR CENTURIES.

1794 that cattle were exhibited at its shows as stock, not as beasts of draught or burden, and no mention occurs of Herefords until 1799, when £5 5s ($26.25) was awarded to Mr. W. Smith for the best Hereford heifer. Mr. Plowman adds that he finds no further allusion to the breed until 1810, when £10 10s ($52.50) was awarded to Mr. Kemp for a fat cow of the Hereford breed. Any of the Hewer Herefords exhibited at the earlier meetings of the Bath and West of England Society must therefore have been draught oxen. In another letter,

Mr. J. L. Hewer says, in reference to the character of his father's and grandfather's cattle, that he believes that they were principally red, with white faces, and from what he has heard his father say, they must have been in possession of the family for some generations, as several of their relatives in Gloucestershire had the red with white faces before the nineteenth century came in. He has also heard Mr. John Hewer say that his father traced his best cattle back to the bull, called Silver (540) 358, calved

HEREFORD OX AT SEVEN YEARS. CHAMPION AT SMITHFIELD, 1799; BRED BY MR. TULLY.

in 1797, which impressed them with the red with white face character, and also with that massive, heavy flesh and full eye which distinguished all his late father's stock.

It is of course not improbable that the Hewer family in Gloucestershire had, during the last century, obtained from the best breeders in the county of Hereford some good specimens of the breed, of the old red with white face variety. Mr. Marshall has told us that the Gloucestershire graziers got their oxen from Herefordshire, and it is not likely that the transfer of cattle from the latter county would be confined to oxen. The Gloucestershire farmers would doubtless have secured a few of the cows that produced such excellent bullocks, and it may be assumed the Hewers were among those who did so. Besides, it is evident that the Herefords had penetrated, by the time of William Hewer's settlement there, into the county of Monmouth. There is nothing very definite in these theories as to the origin of the Hewer herds and hope of being able to discover a more precise explanation had almost been abandoned, when aid was received from an unexpected quarter. Going through the notes on herds contained in the appendix to the first volume of the Herd Book we came across a statement in the notice of the stock of Mr. Yarworth, New House, Brinsop, to the effect that in 1814 he sold to Mr. Hewer a bull calf by Trojan (192) 378, while at his sale at New House in 1820, the one-year-old bull Alpha, by Trojan, dam Red Rose, was purchased by Mr. Hewer, Northleach. The bull calf sold in 1814 went to Mr. W. Hewer, Great Hardwick, but that transaction having occurred a good many years after he had removed to Monmouthshire, the fact did not help to an explanation as to the original foundation of William Hewer's herd. But connected with this sale of a bull in 1814 is an incident that brought some welcome guidance. (¶ 34.)

In the year 1821 there was a furious newspaper controversy between Mr. William Hewer and Mr. Yarworth. It is a matter of regret to have to refer to this unfortunate affair, but it is desirable to explain the origin of the misunderstanding. It appears that Mr. Yarworth, before going to Brinsop, occupied the farm of Troy, near Monmouth. On leaving this farm in 1814 he had a sale described as of valuable Herefordshire cattle. William Hewer attended the sale and purchased stock to the amount of £145 ($725.00). In 1821 Hewer and Yarworth, probably as the result of show-yard rivalries, quarreled, and Yarworth then wrote to the Gloucester Journal (Feb. 4th, 1822) a letter addressed to Mr. W. Hewer, Llanlellen, near Abergavenny, in which he referred to the cause of the misunderstanding between them, and, as was the custom in those days, challenged him for 100 guineas ($500) to show 20, 15, or 10 of his heifers above three years old, of his own (Hewer's) breeding, and then in his own possession, against the same number the property of Mr. White, Upleadon, which were descended from Yarworth's bull Trojan. Yarworth then proceeded to write to William Hewer as follows: "I beg to inform you for the first time how the yearling bull, bull calf, etc., which you purchased at my sale at Troy in the year 1814 were bred, and from which bulls your stock since that time is descended. The bull calf was got by Trojan; his dam (which you bought) was got by the late Mr. William Smith's old bull; his grandam by a bull bought of Mr. Tully of the Haywood; his great-grandam by a bull bought of Mr. Howells of Hadrock, near Monmouth; his great-great-grandam by a cross-bred bull of little value, out of an old brindled Gloucestershire dairy cow, which was purchased by my father (she being an excellent milker) of my predecessor, Mr. Dew of Troy Farm, in the year 1797 for £8. The yearling bull was got by Trojan out of the grandam of the bull calf." Mr. Yarworth expressed surprise that Mr. Hewer had not asked for information as to the breeding of the stock at the time of the sale, and said: "Although they were only culls I then sold, I assure you there were only four of them but were well bred; three out of the four you made choice of —namely, the yearling bull, the cow, and her bull calf." Mr. Yarworth advised Mr. Hewer to part with his stock descended from the old brindled cow. He added: "When Mr. Bluck asked how your bull was bred, you answered that he was got by your old bull, which you bought of old Tomkins, and that you had let him that season to two gentlemen in Breconshire for 100 guineas ($500), but I am sorry you forgot it was the bull you bought at my sale in 1814, then a calf. The two heifers you sold at Tredegar your son said were out of the two old cows you bought of old Tomkins. Now in a letter I received from Miss Tomkins, of the Pyon, dated the 12th inst., she assures me that you never bought any stock of her, and her sisters, or to her knowledge of her father."

In reply to this attack on the character of his stock, William Hewer wrote a long letter, in the course of which he said: "Mr. Yarworth's bills of sale were headed thus: 'Particulars of the valuable Herefordshire cattle, the property of Mr. Jas. Yarworth, of Troy Farm, near Monmouth.' Mr. Yarworth, in reference to my purchasing a bull calf at such sale, observes that it was a most strange and unaccountable thing I did not inquire about the breed, and whether I did so or not I cannot charge my memory at this distance of time, but it is a matter of no importance to me, as I had stock far superior to any Mr. Yarworth possessed; and I added to them five cows and heifers by a purchase I had made of Messrs. Tully, of Huntington, previous to Mr. Yarworth's sale in Troy. And as to my old bull being the calf I purchased at Troy sale, it is sufficient, in contradiction, to state one fact, viz., my old bull was calved at least a year before Mr. Yarworth's came into existence. I deny having stated to Mr. Bluck that I ever purchased any stock of Tomkins, though I can prove I have descendants from some of Tomkins' prime cattle.... His (Mr. Yarworth's) advice to get rid of my stock descended from his brindled cow comes too late by many years, as I have none of them left." Mr. Hewer further expressed the opinion that Mr. Yarworth's stock sold at Troy were, as he himself had publicly described them, pure-bred Herefords, and that the story as to the brindled cow was an invention.

So out of this very unfortunate squabble two most important facts are gleaned—that William Hewer had at a very early period in his career obtained five cows and heifers from Tully, of Huntington, and that he had in his herd descendants from "some of Tomkins' prime cattle." That of course is not a full explanation of the foundation of the Hewer stock, but it shows some of its principal ingredients, and clearly proves its descent in part, at least, from the herds of Tomkins and Tully. It is thus apparent, as had always been supposed, although until the discovery of these documents we had no means of verifying the impression, that the Hewer cattle were not a new or mysterious element imported into the breed, but mainly the result of a most skillful compounding of the old strains.

In the Herd Book there are six bulls entered as having been bred by Mr. William Hewer, as follows:

Silver (540) 358 of whom all the information given is, as we have said, that he was red with white face, and calved in 1797; Old Wellington (507) 290 also red with white face, calved in 1801, by Silver (540) 358, dam Primrose, bred by William Hewer; Young Wellington (505) 294, red with white face, calved in 1812, by Old Wellington (507) 290, dam Silky 362 by Waxy (403) 356, grandam Silk 404 by Silver (540) 358; Old Favorite (442) 292, calved in 1819, by Young Wellington (505) 294, dam

DOWNTON CASTLE IN 1775, SEAT OF T. A. KNIGHT.
(From a water-color drawing.)

Cherry 360 by Wellington (507) 290, grandam Old Cherry 402 by Waxy (403) 356; Waxy (403) 356, calved in 1811, by Wellington (507) 290, dam Strawberry; Alpha the Second (457) 528, calved in 1814, by Young Wellington (505) 294, dam Silk 529, by Young Wellington (505) 294, grandam Silky 362, by Waxy (403) 356, great-grandam Silk 404 by Silver (540) 358.

No doubt some of these pedigrees which we have copied from the Herd Book are erroneous, and mistakes have also crept into the entries of several of Mr. John Hewer's cattle, arising partly from the evident desire of father and son to acknowledge no connection between their herds and other stocks.

MR. JOHN PRICE, 1776-1845.
(From an old lithograph.)

Both William and John Hewer were fond of giving animals the same name and not distinguishing them by numbers, but merely stating that they were old and young—an exceedingly confusing system of nomenclature. Mistakes of identity therefore occurred, particularly in the pre-Herd Book days, when records were not carefully kept. There is little doubt, however, that both father and son carried out a system of close breeding.

In several of Mr. John Hewer's catalogues the following note is made: "It may not be unworthy of remark that the celebrated and justly admired bull Old Sovereign (404) 221, (¶ 35), the sire of Cotmore (376) 150, the winner of the first premium for Herefords at the first meeting of the English Royal Agricultural Society, held at Oxford, July 17th, 1839, as well as the sire of Wormlow, the property of Sir H. Hoskins, Bart., Harewood and many others of equal merit, was bred by Mr. Hewer; his sire Favorite, grandsire Wellington, great-grandsire Old Wellington, dam Countess by Wellington, grandam Cherry by Waxy, great-grandam Old Cherry. Much has been said and written by most of the eminent breeders in the Kingdom against the practice of breeding in-and-in, as they termed it; but by referring to the above pedigree it will be seen that Old Sovereign was the offspring of an own brother and sister and acknowledged by the first judges the best bull ever bred in the county of Hereford, and the sire and grandsire of more prize cattle at Smithfield and elsewhere than any bull in the Kingdom. Old Sovereign was used by the following distinguished breeders, and died in his fifteenth year, viz., Robert Tench, Bromfield, Salop, three years; Lord Sherborne, Gloucestershire; Earl Ducie, ditto; Thomas Wells, Hamnet, ditto; Richard Kilmister, The Grove, ditto; Messrs. Hewer, Northleach, ditto; Sir H. Hoskins, Bart., Harewood; Thos. Jeffries, The Grove; Thos. Jeffries, Jr., Cotmore, ditto; Edmund Jeffries, The Grove; Richard Yeomans, Howton; John Turner, Noke; Messrs. Rogers, Sternsbach, and by the breeder." The other bulls bred by the Hewers were also let out in many of the leading herds, and their influence was thus widespread.

Particulars we have obtained as to weights and measurements on some of William Hewer's cattle will be interesting:

Weight of the bull Wellington (507) 290, 1 ton, 6 cwt. (2,912 lbs.) in the year 1815. The weight of the cow Silk 529, 1 ton, in the year 1820. Dimensions of the fat steer—length from the nose to the settings of the tail, 11 feet, 1 inch; girth, 8 feet and 10 inches; across the hips, 3 feet, 1 inch; weight, 1 ton and 6 cwt. (2,912 lbs.). The document from which these details are taken is dated April, 1822.

Another paper gives the dimensions of the bull Wellington, as follows: Length from the setting on of the tail to the end of the nose, 11 feet, 4 inches; girth, 11 feet, 3 inches; across the hips, 3 feet, 2 inches; length from the tail to the hip bone, 3 feet and 2 inches. The dimensions of the bull Alpha are thus given: Length from the setting on of the tail to the end of the nose, 10 feet and 11 inches; length from the tail to the hip bone, 2 feet and 9 inches; across the hips, 2 feet and 9 inches; girth, 9 feet and eight inches.

William Hewer conducted his operations as a breeder under serious difficulties, arising from the nature of the land he farmed. He himself says, "Notwithstanding the disadvantages incidental to half mountain land, and a sharp, gravelly soil, yielding almost sapless herbage, which I have encountered, I have frequently successfully opposed at various shows some of the first Herefordshire breeders."

John Hewer had, as we have seen, assisted his father at the Hardwick. He had then a few cattle of his own, and had the benefit of his father's advice in their management. It has already been mentioned that early in the present century, John Hewer went to Purslow in Shropshire. The gentleman who owned the Purslow estate, Mr. Browning, bought a number of cattle from William Hewer. He also obtained his consent to John Hewer going to manage the estate and cattle. There is some uncertainty as to the date when John Hewer removed to Purslow, Mr. John L. Hewer mentioning 1817, while others say it was a few years later. Mr. George Smythies says: "I learned that the whole herd of Mr. Hewer, The Hardwick, near Abergavenny, was bought by Mr. Browning, who had purchased the farm of Purslow, near Cravens Arms, Shropshire, and were taken there by Mr. John Hewer, who remained as manager of the cattle till the end of the year 1822, or the early part of 1823. Among the bulls taken to Purslow were the famous Wellington, Favorite, and Old Sovereign. This last bull was purchased by Mr. Tench, of Bromfield, Shropshire, and was given by him to his son-in-law, Mr. Edmund Jeffries, and was the founder of his good herd. In 1824 Mr. Browning was obliged to sell his stock and let the farm, the tenant taking the cattle. They were subsequently sold, and Mr. John Hewer bought some of them."

The more probable date of John Hewer's removal to Shropshire is, we think, 1817, a view of which is confirmed by the fact that Sovereign was calved at Purslow, in 1820. The change to Shropshire, it will thus be observed, did not cause an interruption of his connection with his father's cattle.

It was John Hewer who was responsible for the direction of Mr. Browning's herd, and it is not surprising to learn that it became the talk of the country, people going for miles to see it. Mr. Lloyd Roberts, Crofton Manor, was introduced to Mr. Hewer at this time and he says he never saw a grander lot of cattle in his life than those he brought to Purslow. It was also when he was at Purslow that the late Mr. Bowen, Crofton, became acquainted with Mr. Hewer, and he used bulls of Hewer blood for about 40 years.

When Mr. Hewer returned to Monmouth, it would appear that he took with him a number of the Herefords of his father's strains that had sojourned for a time in Shropshire. It is a very important fact that John Hewer never lost his control over the stock, in the breeding of which he had been closely associated with his father. In the words of Mr. John L. Hewer: "It was left to my late father (John Hewer) to finish what his father had begun, to produce a race of cattle which were beautiful to look at and good rent-paying animals of great scale and splendid quality, and he bred and let more bulls than any other man. There is not a single herd of Herefords in existence but what traces back to his stock.

"He was a great stickler for scale with quality, and some of the older bulls were immense animals, often weighing from 25 to over 30 cwt. (2,800 to 3,360 American lbs.). The General (1251) 1677 was 32½ cwt. (3,640 lbs.) when six years old. He was let for four seasons running for the sum of £84 ($420) per season. Governor (464) 87, the sire of General, was let for £100 ($500) per season, and I have heard my father say he let Favorite (442) 292 for £200

RYALL COURT, WORCESTERSHIRE, HOME OF JOHN PRICE.

($1,000) per season, and Defiance (416) 217 for £200 ($1,000). I have known him have 55 bulls let out at one time about the country, and he used to let bulls to go into Scotland. He occasionally sold bulls at long prices to go abroad as far back as 1835, and one Hampton (513) was sold for £500 ($2,500). He also sent a lot of cattle to Australia in the year 1840, and he saw afterwards in an Australian paper shown him by a friend, that one of his bulls was sold out there for 1,000 guineas ($5,000)."

About this time his friend, Wm. H. Sotham, bought several animals for export to America.

"My father had what he called his four favorite strains—Countess, Lofty, Red Rose, and Fanny, and those are the ones from which his cattle are principally descended. Red Rose 393 (¶ 35 A), by Chance (355) 289, from Rosebud 288, was his favorite cow. He kept her until she was twenty-three years of age, and I believe she bred him twenty calves. She had twins when she was seventeen years old—a bull and a heifer—by Governor (464) 87. The bull

VICTORY (33), BRED BY J. PRICE, CALVED 1839.

Grateful (1260) 1599 was sold to the late Mr. W. Stedman, Bedstone Hall. There were several cases of longevity amongst his bulls, especially Berrington (435) 3362, who was a stock-getter at twenty-one years old; Sovereign (404) 221, at fifteen years old, and a more recent case, Above All (2910) 3127, which was only fed in the winter of 1883 by Mr. Jones, Preston Boats, Salop, at seventeen years old."

It would be impossible to give here the names of all the famous animals bred by Mr. John Hewer during his career as a breeder, which practically extended from about 1803 to 1873, a period of nearly seventy years. Mr. Eyton mentions in Vol. 1 of the Herd Book that he was informed by Mr. Hewer that Pretty Maid, Primrose, Beauty, and Damsel, were the four cows from which his herd was originally descended.

Most of the following bulls bred by J. Hewer were extensively used in the county, and have left their impress on the breed: Sovereign (404) 221, Lottery (410) 185, Byron (440) 205, Hope (411) 282, Chance (355) 289, Defiance (416) 217, Prince Dangerous (362) 146, Lot (364) 846, Lottery 2d (408) 1413, Young Favorite (413) 350, Wonder (420) 451, Fitzfavorite (441) 366, Hamlet (512) 275, Original 1st (455) 219, Young Waxy (451) 301, Purslow (446), and Conqueror (412) 262. But the list could be largely extended, and the difficulty is to find a single animal of note in the present day that does not inherit Hewer blood.

The sum received for the letting of Sovereign (404) 221 was £640 18s ($3,205), for Lottery (410) 185, £710 ($3,550), for Lottery 2d (408) 1413 £645 11s ($3,225), and for Defiance (416) 217 £525 ($2,625). The produce of Countess the 2d 226, was valued at £455 14s ($2,280), and the produce of the cow Lofty 147 made £1,289 ($6,445). As illustrative of the weight of some of Mr. Hewer's cattle, it may be noted that the live weight of the cow Rosy 5469 was one ton and one cwt. 3 qrs. or 2,436 lbs.

Mr. Hewer gained the first prize at the R. A. S. E. Show at Oxford in 1839 for the best heifer (Lady Oxford 1414), and also the following year, at Cambridge, with Duchess of Cambridge. The widespread influence of the Hewer blood on the breed arose very much from Mr. Hewer's habit of letting out his bulls, as many as fifty-five having been, as Mr. J. L. Hewer has told us, let out at one time. In few herds was the Hewer influence more potent than in those of Mr. Thomas Jeffries of The Grove (who used Lottery, Sovereign and Byron), Mr. Yeomans of Howton, Mr. Yeomans of Moreton, Mr. Bowen of Crofton, Mr. Jones of Breinton; Mr. Turner, Noke, and Sir Hungerford Hoskyns. But in this respect again it is impracticable to make a full list without giving the names of nearly all the well-known breeders.

In Mr. Jeffries' hands, however, it will be shown, the Hewer blood was most successful. The celebrated Cotmore (376) 150, calved in 1836, bred by Mr. Jeffries, and winner of the first prize at the initial show of the Royal Agricultural Society at Oxford in 1839, was got by Mr. Hewer's Sovereign (404) 221, and his dam was by Lottery (410) 185. Then Chance (348) 119, the sire of the wonderful bull Sir David (349) 68, was from Victoria 186 by Lottery (410) 185, the grandam being Countess 264, by Old Sovereign (404) 221. Sir David's dam was also by Chance (348) 119. Instead of going into elaborate details now as to the influence of the Hewer cattle, it will be more convenient to allow the facts to come out when we refer especially to the various herds.

It has been somewhat positively stated that the whole of Mr. John Hewer's cattle, as well as those of his father's, were white-faced. On this point Mr. John L. Hewer writes us: "My grandfather's herd were not all pure white faces. He had some ticked-faced ones; also, I believe, a few a little mottled, and my father had one strain of the tick faces (¶ 36) which he prized very highly, the Lofty strain from which descended Governor (464) 87, one of the best sires of his day; he was ticked very much. Mr. Bowen, who bred mostly from Hewer stock, had several ticked-faced ones in his herd descended from Governor, of which he thought a great deal. They were heavily fleshed of nice quality, and with great aptitude to fatten; and, as I told you before, my father went in for great scale, with symmetry, weight and quality. There were one or two tick faces at his last sale, but they early wore out from continually crossing with the white faces, although now and then one would crop up.

"Lofty 147, I may add, was also the dam of Wonder (420) 451, used by the late Lord Berwick in his herd, also of a steer, winner of a first prize at the Rutland Show at Oakham in 1837, and first at Smithfield. Her produce realized £1,289 ($6,445). I remember her very well; she was a favorite with me when a boy."

The late Mr. E. Bowen, Crofton, wrote: "My first remembrance of the tick-faces in Mr. Hewer's herd was in 1838. I have heard him say that there was a particular strain that had these face markings, viz., Old Lofty 147, and the Lady Byron 218 (¶ 36A) families. In Governor (464) 87 the ticks were more blended than in any others. My Old Lady Wiseman 7723, by Cardinal Wiseman (1168) 2688, dam a Governor cow, was also ticked-faced; out of fifteen calves produced by her, there was only one ticked-faced. That was her last, named Leah, now in my possession. She is strongly marked with the ticks but she has produced me four calves, all of white-faces, so that I think with a continual crossing with white-faces the ticks will disappear."

Mr. Forester, Sherlowe, remarks: "Mr. Hewer's stock were all what is commonly called red with white faces. But this description is subject to a variety of which Mr. Hewer was rather fond, namely, a ticked-face—not what is usually termed a mottle face, that is, one with spots such as could be counted, but minute ticks of a bluish tint. The bulls Wonder (420) 451, and Governor (464) 87, had it and their dam, Lofty 147, by Original (455) 219, in a marked degree; also Above All (2910) 3127, and in a less degree his sire Abdel Kader (1837) 3135."

But, after all, the outstanding feature of Mr. Hewer's stock, as regards color, was their white faces. This uniformity was doubtless produced by careful selection in breeding. We may assume that the animals originally obtained by William Hewer possessed the white face and other markings now distinctly characteristic of the breed, and that in his case, and that of his son, stock that reverted back to the other shades were not (except in the strain of tick faces to which we have referred) retained for breeding purposes; size and quality also received great attention, and the impressive power of the Hewer sires in all these particulars was remarkable.

Mr. John L. Hewer wrote Messrs. McDonald and Sinclair: "It seems strange to you, no doubt, how my father managed to breed from his own stock for such a length of time, and so successfully to keep up the size and character of his herd. But that was his secret. He used to say that he had five different strains, which by judicious crossing kept up their stamina."

The influence of the Hewer strains is very clearly brought out in an analysis of the pedigree of the prize cow Queen of Hearts 1552, which was drawn up by the Earl of Southesk. It is seen from it how many of the best strains traced back to the bull Silver (540) 358 which Mr. Duckham, in the revised edition of Vol. 1 of the Herd Book, truly remarks: "Appears to have laid the foundation of William Hewer's eminence as a breeder."

Mr. John Hewer did not pamper his cattle. He kept them on plain fare and aimed at developing robust constitutions. It is believed, indeed, that his system of management was of a somewhat rough description. At any rate there were never any complaints as to his animals being delicate.

YOUNG TRUEBOY (32) 630, BRED BY JOHN PRICE, CALVED 1838.

Mr. John L. Hewer tells us that his father's system of management differed very little from that of other farmers in the district. He tried to have most of his calves dropped in the early summer. They ran with their dams in the pastures until they were weaned, and were then brought into the house and received a little cake, crushed oats, bran and chaff, with a few roots. The cows were wintered in the yards on straw—frequently tacked out on straw. The only difference he made was with his young bulls. As yearlings, he never liked to pamper them, but kept them in good growing condition, as he said they always lasted longer than animals that were pampered when young. He had some rough sheds put up in the orchards with thatch for covering. The bulls ran out in the orchards, and had those sheds for shelter from sun and rain. They had also an allowance of cotton cake and crushed oats daily.

The only fault that he had was that he was often overstocked, and one or two of his bulls were sometimes poor in condition. "But that could not be altered," remarks our correspondent, "as we frequently had as many as twenty-five to thirty bulls let out all over the Kingdom.

"POOLE HOUSE," UPTON-UPON-SEVERN, HOME OF JOHN PRICE.

The consequence was that some of them came home in a shocking plight, and it took some time to get them in condition again, although I must give most of our customers credit for sending them home in good condition—indeed, some of them took a pride in sending them home in better condition than they received them."

Consequent on his frequent removals and owing to other circumstances, Mr. Hewer had many sales of stock. The first of these was in 1839 at Hampton Lodge, when an average of £58 ($290) was obtained. The highest price was £346 10s ($1,733) for the four-year-old cow Lady Byron 218, got by Chance (355) 289, dam Fatfrumps 276. The purchaser was Mr. Williams, Bristol, who also bought the cow Red Rose 393, by Chance (355) 289, dam Rosebud 288, for £105 ($525), and the bulls Baron (418) 2860, for £120 ($600), Dangerous (419) 1699, for £252 ($1,260), and Lot (364) 846, for £267 15s ($1,340). Mr. Lumsden, Auchry, Aberdeenshire, purchased the bull Matchless (415) 2524 for £105 ($525), and the bull Wonder (420) 451 sold for the same price.

Another sale was held at Hampton Lodge in October, 1843, when the cow Lady, sire Chance (348) 119, dam Lady Byron 218, sold for 100 guineas ($500). A sale was held at Lower Wilcroft in October, 1846, and sales also took place at Vern House in 1855, 1861, and 1866.

At the Grove (Jeffries') sale in 1844, Byron (380) 190, calved in 1842 by Confidence (367) 255, dam Lady Byron 136 by Hewer's Byron (440) 205, was sold for £75 ($375), Lady Byron 136 going for £84 ($420) to Mr. Price. Confidence (367) 255, tracing to Hewer stock, and first at the Derby Show of the R. A. S. E., being sold for £100 ($500) to Mr. Smith.

Faugh-a-Ballagh (368) 5464, by Confidence (367) 255, Regulator (360) 174 by Sovereign (404) 221, dam by Lottery (410) 185 (the latter well known in connection with the Monaughty herd), and Hope (439) 324, by Byron (440) 205, from the same dam as Cotmore (376) 150, from which many of Mr. Carpenter's (Eardisland) winners were descended, may be mentioned as intimately connecting the Hewer stock to some of the best stock of the present day.

In connection with Herefords in America, the following bill of sale from the Hewers to W. H. Sotham is of interest.

The following are the pedigrees of the beasts sold by me this day, April 10th, 1840, to William H. Sotham, Perch Lake Farm, Jefferson County, New York, North America:

1. An eight-year-old cow, Lumpy, was sired by Nelson, dam by Panic, grandam by Alpha. Nelson's sire was by Trojan, dam Bloomy, grandam Old Bloomy.

2. A five-year-old cow, Gay, and bull calf, Sir George. Gay sired by Noble, which was sired by Sovereign, that sired Cotmore which won the prize at Oxford Royal Agricultural Society, 1839, and is admitted by all breeders to have sired more prize beasts than any other in the county of Hereford. Dam by Conqueror, grandam Spot by Alpha; Alpha by Trojan, the owner of which offered to show against any bull in England for a thousand pounds, and was not accepted.

3. Young Sir George, by son of Sir George that won the prize at Hereford for best aged bull.

4. Four-year-old cow Maria and calf, by Young Favorite, by a son of Alpha, dam by Noble.

5. Calf Matilda by Major, which won the prize at Hereford with his dam and sire, for

bred were The Sheriffs (356) 283,—by Sovereign, a prize bull by Gloucester and sold to Mr. Mason at The Grove sale in 1836 for £60 ($300); Portrait (372) 194,—by Lottery (410) 185,—sold in 1836 to Mr. Rogers for £52 ($260); Grove (370) 247 sold at the same sale for £80 ($400), and Conservative (270) sold for £70 ($350).

Mr. Thos. Jeffries is acknowledged to have been one of the most successful and skillful breeders of the Herefords. Beginning with the old Jeffries blood, he seems to have perceived that the best course for him to pursue was to infuse a large proportion of Hewer blood. He had on hire Mr. John Hewer's grand bulls Sovereign (404) 221, Lottery (410) 185, Byron (440) 205, and Fitzfavorite (441) 366. The cattle thus bred were of the very highest merit, being of large size, good form, splendid quality, and generally uniform in color markings. He did more than any other breeder to spread abroad the fame of the Hewer stock, and encouraged by his success many of the best breeders of the day imitated his example and crossed their stock with the Hewer bulls. Indeed, it is not too much to say that it is largely owing to Mr. Hewer, Mr. Yeomans, and Mr. Thomas Jeffries that the uniform color marking of the breed was established. It is not necessary here to go into much detail regarding the many impressive sires that were distributed over the country from The Grove herd.

Cotmore (376) 150, bred by T. Jeffries, calved in 1836 (got by Hewer's Sovereign (404) 221, when he was fifteen years old), dam by Lottery (410) 185, was considered to have been one of the finest Hereford bulls ever seen. Besides gaining first prizes at Hereford as a two-year-old, three-year-old, and later in the aged class, he was the first prize winner at the Oxford Show of the R. A. S. E. in 1839. His live weight was 35 old English cwt. (or 3,920 lbs.). Hope (439) 324 (¶ 38) from same dam as Cotmore, was a grand animal and impressive sire that left his mark on the breed. (¶ 38B)

Mr. Thos. Jeffries' services were not overlooked by his contemporaries. A subscription list, prefaced by the following notice, appeared in the Hereford papers in 1839: "Many admirers as well as breeders of Hereford cattle having viewed with feelings of pride the success of Mr. Thos. Jeffries of The Grove in obtaining at the first meeting of the English Agricultural Society, held at Oxford on Wednesday, the 17th day of July, 1839, a prize for exhibiting the best Hereford bull, desire to present

"CROOME COURT," WORCESTERSHIRE, SEAT OF THE EARL OF COVENTRY.

him with a piece of plate, as an expression of the highest estimation in which his services are held as a breeder of Herefords." (¶ 38a) A very handsome response was made, and Mr. Jeffries at a dinner at which he was entertained at Kington, presided over by Sir Robert Price, was presented with a magnificent service of plate. The service, along with a large number of cups, are in the possession of Mr. Henry Jeffries, of Guilford, who treasures them not only as evidence of the skill of his father and other members of the family in breeding Herefords, but also as a testimony of the esteem in which Mr. Jeffries was held by a wide circle of friends.

The most eminent of the early improvers who come in chronological order next to those already mentioned, may be appropriately introduced by continuing the account drawn up by the late Mr. T. C. Yeld of The Broome, from which a quotation was made in a preceding chapter:

"No one," Mr. Yeld says, "ever bred better cattle than the late Mr. T. A. Knight. There

THE RIGHT HONORABLE EARL OF COVENTRY.
(From a photograph taken in 1902.)

was no one who knew the principles of breeding cattle better, and he took great pains to try the various crosses, the only success being with Scotch heifers. His white bull, entered in the Herd Book as Snowball, or Knight's White Bull (246) 328, was used after Mr. Turley, by Mr. Rea of The Rock, and, I believe, by his son, Mr. Rea of Monaughty, by his son-in-law, Mr. Taylor of Eye, by Messrs. Hill and Trumper of Orleton, and afterwards by Mr. Yeld of The Broome."

Mr. Yeld tells us: "There were very few stocks at this time fit to breed bulls from besides those of Tomkins, Price, Galliers, Skyrme, Tully, Hewers, Jeffries, Knight, Mr. Proctor, Blackhall, Mr. Martin, Wistaston, Mr. Sheward, Little Dilwyn, Mr. Yarworth, Brinsop, in what may be termed the Pyon district; and in Pembridge district, Mr. Parry, Birley; Mr. Farrier, Luntly; Mr. J. Jones, Charbrook; Mr. Jones, The Lowe, Pembridge; Mr. Powel, Marston, and Mr. Turner, Aymestry. On the Hereford side was also, first, Mr. Weyman, Moreton, succeeded by Mr. Chute Hayton, Mr. Clarke, Lyde, Mr. Walwin, Sir John Cotterell, Col. Matthews, Belmont." Mr. Yeld adds: "I am speaking now of the first twenty years of the present century."

"All the above named herds," says Mr. Yeld, "possessed form and quality. There was another class of Herefords to appear to have been bred solely for working purposes, being large in size, with very heavy bone......There was a third class of what were termed Welsh Herefords, red and white-faced, but that carried no flesh and when grazed on the best land would never stretch.

"During the French revolutionary war, and up to 1821, the return to cash payments, everything sold high, and farmers could pay high rents, but with the winter and spring of 1820-21 Peel's Monetary Bill came into full force. Down went the manufacturers, down went the bank and down went the farmers. At this time scores were ruined by force of circumstances, and those farmers who had not real property to fall back upon were bound to go to the wall. I have seen whole streets filled with cattle in the years 1821-22-23, and no one asked what they were bought for.

"I well remember the stock of Mr. John Jones of Charbrook, sold in 1822. They were as good as anything I ever saw; the cows and heifers magnificent. Cows sold from £7 to £12 (or $35 to $60) each; most beautiful two-year-old heifers from £6 to £8 ($30 to $40) each. There was as good a cart team as it was possible to find; the highest price £11 ($55). At Hereford Fair in 1822 some very splendid barren cows, bred by Col. Matthews of Belmont, were bought by a neighbor of mine at £6, 7s, 6d ($34) each.

"At this time graziers found they could make no profit by feeding, rarely making more than £1 ($5) for summer profit over price, and farm-bred were The Sheriffs (356) 283,—by Sovereign, a prize bull by Gloucester and sold to Mr. Mason at The Grove sale in 1836 for £60 ($300); Portrait (372) 194,—by Lottery (410) 185,—sold in 1836 to Mr. Rogers for £52 ($260); Grove (370) 247 sold at the same sale for £80 ($400), and Conservative (270) sold for £70 ($350).

Mr. Thos. Jeffries is acknowledged to have been one of the most successful and skillful breeders of the Herefords. Beginning with the old Jeffries blood, he seems to have perceived that the best course for him to pursue was to infuse a large proportion of Hewer blood. He had on hire Mr. John Hewer's grand bulls Sovereign (404) 221, Lottery (410) 185, Byron (440) 205, and Fitzfavorite (441) 366. The cattle thus bred were of the very highest merit, being of large size, good form, splendid quality, and generally uniform in color markings. He did more than any other breeder to spread abroad the fame of the Hewer stock, and encouraged by his success many of the best breeders of the day imitated his example and crossed their stock with the Hewer bulls. Indeed, it is not too much to say that it is largely owing to Mr. Hewer, Mr. Yeomans, and Mr. Thomas Jeffries that the uniform color marking of the breed was established. It is not necessary here to go into much detail regarding the many impressive sires that were distributed over the country from The Grove herd.

Cotmore (376) 150, bred by T. Jeffries, calved in 1836 (got by Hewer's Sovereign (404) 221, when he was fifteen years old), dam by Lottery (410) 185, was considered to have been one of the finest Hereford bulls ever seen. Besides gaining first prizes at Hereford as a two-year-old, three-year-old, and later in the aged class, he was the first prize winner at the Oxford Show of the R. A. S. E. in 1839. His live weight was 35 old English cwt. (or 3,920 lbs.). Hope (439) 324 (¶ 38) from same dam as Cotmore, was a grand animal and impressive sire that left his mark on the breed. (¶ 38B)

Mr. Thos. Jeffries' services were not overlooked by his contemporaries. A subscription list, prefaced by the following notice, appeared in the Hereford papers in 1839: "Many admirers as well as breeders of Hereford cattle having viewed with feelings of pride the success of Mr. Thos. Jeffries of The Grove in obtaining at the first meeting of the English Agricultural Society, held at Oxford on Wednesday, the 17th day of July, 1839, a prize for exhibiting the best Hereford bull, desire to present

"CROOME COURT," WORCESTERSHIRE, SEAT OF THE EARL OF COVENTRY.

him with a piece of plate, as an expression of the highest estimation in which his services are held as a breeder of Herefords." (¶ 38a) A very handsome response was made, and Mr. Jeffries at a dinner at which he was entertained at Kington, presided over by Sir Robert Price, was presented with a magnificent service of plate. The service, along with a large number of cups, are in the possession of Mr. Henry Jeffries, of Guilford, who treasures them not only as evidence of the skill of his father and other members of the family in breeding Herefords, but also as a testimony of the esteem in which Mr. Jeffries was held by a wide circle of friends.

The most eminent of the early improvers who come in chronological order next to those already mentioned, may be appropriately introduced by continuing the account drawn up by the late Mr. T. C. Yeld of The Broome, from which a quotation was made in a preceding chapter:

"No one," Mr. Yeld says, "ever bred better cattle than the late Mr. T. A. Knight. There

THE RIGHT HONORABLE EARL OF COVENTRY.
(From a photograph taken in 1902.)

was no one who knew the principles of breeding cattle better, and he took great pains to try the various crosses, the only success being with Scotch heifers. His white bull, entered in the Herd Book as Snowball, or Knight's White Bull (246) 328, was used after Mr. Turley, by Mr. Rea of The Rock, and, I believe, by his son, Mr. Rea of Monaughty, by his son-in-law, Mr. Taylor of Eye, by Messrs. Hill and Trumper of Orleton, and afterwards by Mr. Yeld of The Broome."

Mr. Yeld tells us: "There were very few stocks at this time fit to breed bulls from besides those of Tomkins, Price, Galliers, Skyrme, Tully, Hewers, Jeffries, Knight, Mr. Proctor, Blackhall, Mr. Martin, Wistaston, Mr. Sheward, Little Dilwyn, Mr. Yarworth, Brinsop, in what may be termed the Pyon district; and in Pembridge district, Mr. Parry, Birley; Mr. Farrier, Luntly; Mr. J. Jones, Charbrook; Mr. Jones, The Lowe, Pembridge; Mr. Powel, Marston, and Mr. Turner, Aymestry. On the Hereford side was also, first, Mr. Weyman, Moreton, succeeded by Mr. Chute Hayton, Mr. Clarke, Lyde, Mr. Walwin, Sir John Cotterell, Col. Matthews, Belmont." Mr. Yeld adds: "I am speaking now of the first twenty years of the present century."

"All the above named herds," says Mr. Yeld, "possessed form and quality. There was another class of Herefords to appear to have been bred solely for working purposes, being large in size, with very heavy bone......There was a third class of what were termed Welsh Herefords, red and white-faced, but that carried no flesh and when grazed on the best land would never stretch.

"During the French revolutionary war, and up to 1821, the return to cash payments, everything sold high, and farmers could pay high rents, but with the winter and spring of 1820-21 Peel's Monetary Bill came into full force. Down went the manufacturers, down went the bank and down went the farmers. At this time scores were ruined by force of circumstances, and those farmers who had not real property to fall back upon were bound to go to the wall. I have seen whole streets filled with cattle in the years 1821-22-23, and no one asked what they were bought for.

"I well remember the stock of Mr. John Jones of Charbrook, sold in 1822. They were as good as anything I ever saw; the cows and heifers magnificent. Cows sold from £7 to £12 (or $35 to $60) each; most beautiful two-year-old heifers from £6 to £8 ($30 to $40) each. There was as good a cart team as it was possible to find; the highest price £11 ($55). At Hereford Fair in 1822 some very splendid barren cows, bred by Col. Matthews of Belmont, were bought by a neighbor of mine at £6, 7s, 6d ($34) each.

"At this time graziers found they could make no profit by feeding, rarely making more than £1 ($5) for summer profit over price, and farmers began to pay more attention to breeding.

"After 1820, among the very best breeders were Mr. Hayton, Mr. Smythies, Mr. J. Monkhouse (¶ 39), and especially Mr. John Turner (¶ 40) of Noke Court, who not only bred but managed his stock in a highly creditable manner, and his three-year-old steers were always greatly admired. I may also mention two gentlemen who never pushed themselves into notice, but who brought out some of the best steers I ever saw, viz., Mr. Richard Hill and Mr. Trumper, of Orleton. Besides those before named, there were many others possessing very excellent herds, including Mr. Davis, Ladycott; Mr. Davis, Oxhouse; Mr. Joseph Edwards, Kingsland, and Mr. W. Wheeler, Irving Park.

"At this time, 1825, several new stocks were creeping into notice and eventually took a leading place, viz., Mr. T. Roberts (¶ 41), of Ivingtonbury; Mr. James Bowen, of Monkland; Mr. Yeld, The Broome; Mr. John Morris, Stocktonbury; Mr. W. Bennett, Strettford; Mr. John Thomas and Mr. Vaughan, Cholstrey; also Mr. Wm. Parry, Mr. J. Williams, Kingsland; Mr. Samuel Peploe, and others. (¶ 42)

"Few people at this time had better stock or were better judges than the Rev. J. R. Smythies. He began about the year 1820, and, regardless of price, bought the best he could find. He bought the remainder of Mr. William Galliers' stock. After retiring from business, Mr. Galliers took the Lynch House, and a portion of the meadow land, and there took some of the very best of his herd, all of which were purchased by Mr. Smythies, among them the celebrated bull Cupid (198) 311 and the cows Venus and Browny, which were equal to anything ever bred in Herefordshire. Mr. Smythies also bought some of the finest of Mr. Yarworth's (of Brinsop) herd, among which were Countess and Larkspur. He also bought about the same time some of the very best of Mr. Sheward's (of Little Dilwyn) herd, which certainly was on the whole equal to any other . . . . . ."

In addition to the aid afforded them by Mr. Yeld's statement, Messrs. McDonald and Sinclair were enabled to estimate the position of the leading herds during the first twenty years of the nineteenth century by analyzing the prize lists of the shows of the Herefordshire Agricultural Society. These they compiled by the advertisements and reports contained in the Hereford Journal, the early records of the society not having been discovered, if, indeed, they are in existence. They found it necessary to qualify the record by mentioning, that "of course there were good herds whose owners did not exhibit," thus on only one occasion did Mr. Benjamin Tomkins send an animal for competition at the show. The list is, however, interesting and valuable evidence as to the relative position of the various herds.

The completion of the records mentioned relate to the shows commencing 1798 and concluding 1819, and from them it appears that the largest number of first prizes for breeding stock were won by Mr. Galliers, Frogdon, who,

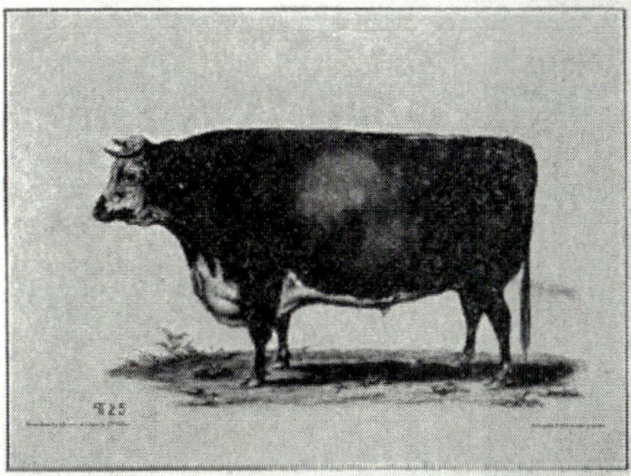

WOODSTOCK (24) 164, CALVED 1833, BRED BY J. PRICE.

as previously stated, secured thirteen—Messrs. Jeffries of The Grove and The Sheriffs won nineteen (nine falling to Mr. Jeffries, The Grove, and five to Mr. Jeffries, The Sheriffs); Mr. T. A. Knight followed with nine; Mr. Watkins, Brinsop; Mr. Yarworth, Troy, and Brinsop, seven; Mr. Walker, Burton, six; Mr. Tench, Bromfield, five; Mr. Walker, Wessington, five; Mr. Samuel Tully, Huntington, four; Mr. Moore, Wellington, four. Each of the following gained three first prizes: Mr. Jos. Tully, Baywood; Col. Matthews, Mr. Yeoman, Howton, and Mr. R. Wainright, Hereford. Those who gained two first prizes were Messrs. John Apperley, of Withington; Mr. Skyrme, of Stretton; Williams, of Thingehill; Rev. J. R. Smythies, of Lynch; Kedward, Westhyde; Williams, Brinsop; Smith, Gattertop; Deykin, Brierly; Weaver, Stretton; Hardwick, Wier; Jones, Fawley; Cooke, Wintercott. The winners of single first prizes were: Messrs. Croose of Sugwas, Smith of Sufton, Powell of Titley, Downes of Hinton, Clee of Downton, Downs of Ashford, Tomkins of Wellington; Croose, Ocle; Davies, The Rodd; Welles, Earl's Croome; J. G. Cotterell; Barnet of Ledbury; Lowe, Gattertop; Prichard, Eaton Mill; Edward, Dilwyn; Oakes, Lenthall; Downes, Mansell; Green, Stoke; Hughes, Marcle; Stevens, Cotmore; Hewer, Abergavenny; Woolaston, Lynch; Salwey, Ashley Moor; Proctor, Orleton; Harris, The Marsh; Wood, Buryhill; Mason, Wooferton; Hanbury, Shobdon; Stevens, Brinsop; Rev. W. Bayley, Womesley Grange; Bannet, Netherton; Turner, Bockleton; E. Jones, King's Caple; Preece, Leyecourt; Mrs. Berrow, The Green, Dewchurch; Symonds, Yatton; E. Walwyn Gravenor, The Parks; J. Purchas, Fownhope; Price, Norton Grounds; T. Barnaby, Brockhampton; Tomkins, Dippers Moor; Parry, Birley; Wood, Burghill, etc.

"All the gentlemen whose names have been given, and many more," say McDonald and Sinclair, "were noted breeders during the first twenty years of the present century. Their number demonstrates that the Hereford breed can claim a broad and solid foundation."

It would be manifestly impossible to give an adequate notice of these many herds. It is, indeed, inadvisable to attempt to do so, as many of them have not exercised a recognizable or known influence on the modern character of the breed. Our remarks will, therefore, be confined to those old herds that may be regarded as forming links with the present. As to the herds of Tomkins, Galliers, Tully, Skyrme, Hewer and Jeffries which are referred to in the list, all the material facts in our possession have already been given. It seems necessary to explain that the number of prizes won at the Herefordshire Show is not alone a reliable indication of the relative positions of the herds exhibited, inasmuch as some of them—notably those of Galliers, Tully, and Skyrme—were dispersed a considerable time before the meeting

MAXIMUS (1615) 1817, CALVED 1858, BRED BY H. R. H. THE PRINCE CONSORT.

in 1819, to which the list extends; while others were represented at the shows during the whole period.

Mr. Thomas Andrew Knight of Downton (¶ 43) was one of the most successful exhibitors at the early shows of the Hereford Agricultural Society. Mr. Yeld has told us of the estimation in which his herd was held by his contemporaries, and in a former chapter reference has been made to the investigations carried out by Mr. Knight as to the history of the breed. Born at Wormesley Grange on August 12th, 1759, Mr. Knight, after studying at Oxford, retired to his country seat, and devoted himself to the improvement of the leading industries of the county.

Prior to 1806 Mr. Knight had become well known as a practical agriculturist and as an improver of Hereford cattle. He was instrumental in founding the agricultural society of the county, and to the end of his life he was almost invariably present at its meetings. In the letter written by Mr. Knight in 1836, which Mr. Yeld preserved, there is an account of the breeding of the bull Snowball (246) 328, otherwise known as Knight's white bull. From this it appears that Mr. Knight's herd was descended from the stocks of Mr. Skyrme, Stretton; Mr. Tully, Huntington; Mr. Isaac Martin, and Mr. Benjamin Tomkins. Mr. Knight evidently preferred the first to all others, remarking that Mr. Skyrme "at the time possessed, in his opinion, by far the best breed of cattle in the county."

In an article on the Ashley Moor herd of Mr. Theophilus Salwey, which appeared in "Bell's Weekly Messenger" in 1873, Mr. Houseman remarked: "The reader, acquainted with more than the merest rudiments of Hereford history, need not be reminded that the animals of Mr. Salwey were descended from those of Mr. T. Andrew Knight. In describing the Ashley Moor stock, we therefore shall have occasion to make frequent allusion to the old Downton Castle tribes. And here a tribute is due to the memory of a public benefactor, who was far in advance of his generation in perception of the principles of animal and vegetable reproduction, and of hereditary recurrences, and ever ready to inform and encourage the seekers of knowledge. His views upon stock breeding are less widely known than his contributions to the stores of horticultural science; yet, upon the subject of which we now treat, he could speak as one who had made himself its master so far as patient observation, with long practical experience, could give an insight to its mysteries.

"Sometime about the commencement of the latter half of the eighteenth century, Mr. Knight had determined to form a herd that should be well adapted to a somewhat poor and uneven locality, and for that purpose he visited all the best herds of the county, and selected from them according to the best of his judgment, without much regard to size, but keeping in view symmetry and good quality. He soon discovered that some very noble cattle of the larger sorts were not suited to his purpose on account of their inability to stand and walk as he knew that animals should stand and walk. The setting of the legs, their shape and the way of using them were great points with him, and often did he repeat to eager listeners, who availed themselves of his counsel, the avowal of his strong aversion to 'lamb's knees and sickle hocks,' which he said were 'quite unfit to move upon Bringewood Chase' (near Downton Castle), where his Herefords were kept. The result of his antipathy necessarily was that animals characterized by the unpardonable similitudes were mercilessly weeded out, and after various sifting processes, the final selection fell upon a few animals from the herds of Mr. Tully of Huntington, near Hereford; Mr. Geo. Tomkins and Mr. Skyrme—the grey element of the stock (afterwards celebrated as 'the Knight's Greys') deriving their blood solely from the Tully strain, the dark red from the Tomkins, and the pale red from the Skyrme tribes.

"None of these varieties contributed remarkably large animals, but the Herefords thus retained were invariably very thick, and stood particularly well on their legs, so that they could easily move up and down the steep pastures they occasionally had to live upon. So fastidious was Mr. Knight upon this point that he would not choose a bull calf to rear for use as a sire until he had not only made him walk, but even trot—a practice which drew down upon him sometimes from his old neighboring farmers remarks of contemptuous merriment, which Mr. Knight most completely disregarded. The issue turned the laugh upon his side as the progeny of his 'trotting bulls' proved clever and free in their action to the very last stage of fattening."

"I well remember," says a correspondent who in early life knew that original thinker and successful breeder, "most of the leading points which Mr. Knight endeavored to obtain were the following: Broad nostrils, small from the nostrils to the eyes, and fine large eyes, broad bash (scope of forehead down to the line of the eyes); open and well developed horns, a little dipping in the first instance and then gradually rising; large measurements of girth was always a *sine qua non*, and likewise that the shoulder should not be an upright one, but well lying back from the neck, the blade being very oblique but lying open toward the chine. He would have his animals thick through the heart, with the forelegs going down straight like two pegs (the opposite of 'lamb's knees'), ribs broad and well arched, especially the last ribs at the adjoining of the quarters; the table-bones of the sirloin long, flat, and well developed, particularly the one adjoining the ribs, thus making a strong, well-formed back, and joining the quarters. Hips were always considered to be secondary in importance, though he never wished to see them prominent, but so formed and placed as to stand tolerably even with the sirloin and ribs. The catch (pen-ends, pin-ends, or fool's point) he wished to see well developed, with not the slightest prominence of frame between the catch and the hips (i. e., the packing of hind quarters) nor anything in that region which might come under the denomination of gaudy.

"Indeed," says the writer, "Mr. Knight's ambition was to see an animal as true in its formation and level as possible from the catch, all the way over the back, loin, chine, shoulders, and as far up the neck as possible; thighs true, deep and thick; purse full and very well spread over the abdominal region, with indeed a disposition to fatten all the way up to the brisket (lengthwise under the body); thick, mellow skin and long, soft hair."

CHARACTERISTIC HEREFORDSHIRE FARMYARD.

Mr. Welles has placed it on record that the variety called grey or roan would obviously arise from an intermixture of the red with those possessing a large proportion of white. They obtained their greatest celebrity from their being favorites with Mr. Andrew Knight, much of whose stock were of that color; one of the earliest being a white cow, from which he bred one or two celebrated bulls. "That he pursued his object with judgment as well as ardor," says Mr. Welles, "has been evinced by prizes having been awarded of late years to many descendants of his stock, among others the ox bred by Mr. Hill, Orleton, which obtained the gold medal at Smithfield in 1839, for the best beast in the yard."

Few of Mr. Knight's cattle have been registered. Of these Snowball (246) 328 has already been referred to. Lawton (223) was also bred by Mr. Knight, and used by Mr. Downes' Aston Ball in 1811. Stratford (264) 369 is entered as having been from Mr. Knight's celebrated white cow, and the Gatley bull (501) 3038 was of his breeding. Among those who obtained stock from him were Mr. Rea, Monaughty, Mr. Turner, Aymestery, and Mr. Salwey, Ashley Moor. Through all these herds the Knight blood is still represented. Mr. Salwey purchased from Mr. Knight four heifers, from which a valuable progeny descended (¶ 45).

JOHN L. HEWER, VERN HOUSE, MARDEN, HEREFORDSHIRE.

Mr. Boughton Knight (¶ 44), of Downton Castle, some years ago attempted to found a herd of the old Knight grey color, but finding the red with white faces were more easily disposed of, he abandoned the strain. Mr. J. A. Rolls, The Hendre, Monmouth, is one of the few gentlemen who now keeps the stock of the old color. In other herds possessing the Knight blood, the markings have not been retained, although the fine quality and true shapes, for which the sort was celebrated, were as conspicuous as ever, the heads being true to the old type.

Mr. Duncomb, Historian of the County, in drawing up his report on the agriculture of Hereford for the Board of Agriculture in 1805, relied almost exclusively on Mr. Knight's information for his description on the management of cattle. In his observation on the subject, Mr. Knight said some of the Herefordshire breeders had sacrificed the qualities of the cow for those of the ox. He does not value the cow according to the price which the grazier would give for it, but in proportion as it possesses that form and character which experience has taught him to be conducive to the excellence of the future ox. The cow of Herefordshire (1805) is very feminine in its character, light fleshed when in common condition, but is capable of extending itself universally in a short space of time when fattening. It may here be remarked that there is an extraordinary difference between the weight of a Herefordshire cow and the ox bred from her. Perhaps other sorts, eminent for producing fine oxen, are similarly distinguished, but it is a fact that the Herefordshire cow will not unfrequently be the mother of an ox of nearly three times her own weight.

Mr. Knight was convinced that the true function of the Hereford breed was to produce first-class beef. He did not believe in attempting to develop the milking properties of the race, considering that it was sufficient if a cow gave milk enough to keep its calf fat, and unless it could do so, it was disqualified from breeding a good ox. What he says descriptive of the cow bred in Herefordshire appears to mean simply that preference should be given to neat, compact animals. The system in Leicestershire of bestowing most of the attention on the improvement of the cow, and making her an excellent animal for the purpose of the grazier, was, in his view, unsound.

JOHN HEWER. BORN 1787, DIED 1875. THE GREATEST IMPROVER OF HEREFORD CATTLE.
(From an oil painting presented to him by his friends in 1861.)

CHAPTER V.

A Noted Feeder on Herefords as Beef Animals

As to the permanent record of the breed we find that in 1787 J. H. Campbell was a purchaser of Hereford cattle and a grazier in the County of Kent, near London, an account of which is given in the following pages with the correspondence that was published in the "Annals of Agriculture," an agricultural paper, from 1780 up to 1805. I have given this correspondence fully, not only to show what the breed was at that time, but to show also that Mr. Campbell was an intelligent breeder and feeder, and that his statement was: that the Hereford breed of cattle were the best and most economical feeders at that time; and in 1788 Mr. Marshall, a noted and intelligent writer on agriculture and live-stock subjects, after visiting Herefordshire and adjoining counties, pronounced the Hereford breed of cattle the finest in the United Kingdom.

Taking the testimony of Mr. Speed, given in 1627, and the testimony of Mr. Campbell, a grazier and feeder, and of Mr. Marshall, a noted writer on live-stock interests of the United Kingdom, twenty years from the time that Benjamin Tomkins commenced his work, it must be assumed that although the complete record is lacking between Lord Tomkins (1570 and 1640) and the time of Mr. Campbell and Mr. Marshall (1750 and 1820) the breed was in the hands of good breeders during that time. It is certain that by 1788, Mr. Westcar, of Creslow, Buckinghamshire, was one of a large number of appreciative purchasers of Hereford cattle for grazing and feeding, and that ten years later he was a party to the organization of the Smithfield Cattle Club and a successful exhibitor of Hereford cattle before that society from its first exhibition up to the time of his death in 1819.

The winnings of the Herefords before the Smithfield Club were in evidence as to the merits of the breed, and a very important feature in the history of the Hereford cattle, is the fact that Hereford breeders were farmers and not exhibitors, that they established weekly sale days (¶ 46) at Hereford City, and a yearly sale in October of each year, as far back as we have any record, and that those sale days have been continued up to the present time.

We have the fact, stated by Mr. Fowler, that on the annual sale day in October there have been brought to the Hereford market as high as 8,000 to 9,000 head, filling not only the market grounds but the streets of Herefordshire with Hereford cattle, and that during all the time from the first record we have to the present, that cattle feeders and graziers of Buckinghamshire and other counties near London have come to the Hereford market on these sale days and to the farmers and purchased their steers at prices much in advance of what like ages of any other breed have been sold for.

It has been claimed by breeders of the Shorthorn cattle that while the Herefords have made much larger gains before the Smithfield Society on oxen and steers, that the Shorthorn breeders have made larger gains by the exhibit of Shorthorn cows. This is explained by the fact that Hereford breeders were not exhibitors in the earlier years of this show. The show of Herefords being made by the graziers of other counties who bought their stocks in Herefordshire (¶ 47) (¶ 48).

We have followed Mr. Benjamin Tomkins and Mr. John Price in their work, and the Hewers in theirs; they were undoubtedly leaders, Tomkins and Price breeding more compact and smaller animals, and the Hewers breeding more for a larger scale and heavier weights, and between these two lines, Tomkins and Price, and Hewers, we know, there has been an army of breeders through Herefordshire and adjoining counties equally successful and intelligent in their works.

We have quoted from J. H. Campbell, of Charlton, in Kent, on breeds in a preceding chapter. The following is from a letter to the "Annals of Agriculture" published in London, dated Charlton, Jan. 15th, 1789:

"Sir: Enclosed I send, as you desire, a copy of the queries I received from a gentleman (who wished me to inform him of the method I used in feeding cattle on potatoes), with what I wrote in answer to them. I have added a few words which I should have put in my answer, but, being rather busily employed just at that time, and the season of the year requiring the answer should not be delayed, my answer was more hastily wrote than I wished, and I did not recollect all that I ought to have mentioned.

"You were pleased to think so favorably of my management of the stall-feeding business that you wished me to communicate in writing, with a view to its having a place in your Annals, an account of my method of proceeding throughout.

"If I conceived that I had really found out or was possessed of any means excelling what others knew or practiced, I should most readily and without delay (without vagging the public) give it into your hands, who, we know, would put it in a way doing the most good to all whom it might concern, and the community at large, but I am conscious that there is nothing in my

DARLING, FIRST BULL EVER BRED BY MR. J. L. HEWER.

way of carrying on that business that will not as well or better be learned from many other persons, indeed. The enclosed papers (though wrote only as answers to inquiries about feeding with potatoes only) tell you almost the whole of my proceedings. That you may not, however, suspect me of disinclination to satisfy you about it, and to show you there is nothing extra in my management, I just run over the course here.

"The sort of oxen I take for my stall feeding are such as should not go to market till there is the greatest call for prime beef, and for that time they should be completely fat; so that I do not propose any material or regular sale, until Candlemas, and sell so as to clear all the stalls some time in May. With this view, when I buy cattle in the spring, and on to August, I make no objection to good oxen (that is, mellow, supple-fleshed ones) for being as lean as poor keep and hard work can fairly make such; but from that time I look for having them in better and better order, till at the last buying for the year's feeding, at the end of October, I demand what may be called, full of flesh.

"I keep them at tolerable grass during the summer, mending the quality of that as the season draws later, and with good latter-math (to which about the beginning of November or how soon the hoary frosts hang much on the grass and the nights become long and cold) I add a small feed of inferior sort of hay in the mornings. The first of December (or sooner if severe rains or snow set in) I take them to the stalls, where at first I give them but small quantities of potatoes and a larger proportion of hay, but increase the potatoes and decrease the hay till they come to about three-quarters of a bushel of potatoes and about six pounds of hay per ox, one with another.

"When they have arrived at their full quantity of potatoes, I give them in addition some brewer's grains, beginning with about a half bushel to each, increasing till they will at last, in general, eat a whole bushel per day. To the grains after Christmas I begin to add either pollard or (if to be had proportionately cheap) pea or bean meal; of these, also, I begin with a quart and increase till they come to a peck or more of pollard, though seldom quite so much of meal except to such as are extra size and therefore fit to be made extra fat.

"Every other particular of my proceeding you will see by the enclosed paper.

"Many persons have expressed some wonder at the rapid progress they observe my cattle made in their fattening, and I believe with some, I get more commendation than my proper share, for the credit is justly due to the cattle, not to their master, who claims no other merit than having taken extra pains to make himself well acquainted with different kinds of cattle, giving a fair and thorough trial to several different breeds, thereby enabling himself to know what breed would be most to his purpose to attend to.

"Of this true Herefordshire breed (which Mr. Culley in his book on live stock says he is pretty clear is neither more or less than a mixture between the Welch and a bastard race of long horns), I may venture to say that (by the assistance of my good friend Mr. Samuel Pantall, of Warhamear, Hereford), I have some of the highest blood, a few of which I was very happy in the opportunity of showing you and Mr. Macro; I wished you two would have allowed me more of the pleasure of your company, and of bringing more of my own bred flock under your examination, which were unluckily most of them (cattle and sheep) at Mr. Cator's park, at Beckenham-place, and my working oxen out in teams.

"I will beg you to remember that I do not suppose the true Herefordshire cattle in respect to kindly disposition for feeding, or delicacy of flesh, to more than equal the true bred Sussex. But that they are yet more complete in their make, generally wider and fuller over the shoulders or fore chine, and the breast or brisket, also in the after part of the rump, which is much oftener narrow and shelly in the Sussex than the Hereford; the mouse-buttock or ham apt not to be so round.

"It is time to apologize for the tedious length of my letter and also for speaking so freely on some of the expressions above alluded to. Allow me to put you in mind that you have told us (who are of a different opinion from you in those points) that one person has 'established the superiority of his breed to all others beyond an idea of question or competition,' that the disposition to fatten, in that person's breed, is so much greater and beyond all others as to make a parallel absurd, which was surely rather unnecessarily treating other people's breed, other people's opinion, with more contempt than they deserved, so I hope to be more easily pardoned by you; and am, sir,

"Your most humble servant,

"J. H. CAMPBELL.

"To Arthur Young, Editor.

"P. S.—When I had the pleasure of your company here, I understood you meant to feed some cattle on bean-meal, and that you had not practiced it before; if so, it might not be known to you, or immediately occur, that chaffing hay and mixing it with the bean-meal, will keep the beasts' mouths clean, and stomachs from being clogged, and much promote their feeding, as I have heretofore experienced, which makes me take the liberty of mentioning it.

"I beg your pardon—More last words.

"When you set down the different articles of my stall-feeding, from which you cast up the daily expenses, I believe the grains were set down at a bushel per day and the meal at a peck; if so, you will remember I mentioned those quantities as the largest that the cattle were brought to eat toward the finishing of their fattening. It should be noticed also that some time lapses after their coming into the stalls before they have any grains, then on grains some time before any meal is added; then begin the meal at the proportion of a quart to the bushel of grains, and by degrees only to a peck toward the latter end. At this present time of writing they have but one-fourth of a bushel of grains (this, indeed, because the quantity of grains I get will not go to more per head of the number of beasts in the upper shed, and my people had given grains to all in that shed before I came home and I did not think it right to put any of them from it again), and with the one-fourth bushel of grains they have now one-fifth bushel of the sort of meal I mentioned to you.

"So, taking the average of daily cost of food from their coming into the stalls to their going out, I believe you will not find me much wrong in my reckoning one shilling (25 cents) per ox per day, and supposing the dung to pay for litter and attendance. My garden having required supply at different times in different carts, as I could spare them to carry it, and the very many matters I had to set to rights on my farm, etc., since I came here, prevented me, hitherto, from taking a true account of the quantity of dung made.

"As to the value per load (what four horses draw from London to this neighborhood) I have, since you were here, enquired of a sensible farmer, my near neighbor, who told me he has given and should always be glad to do it whenever he had occasion for more manure than was made by his stock, five shillings ($1.25) per

HAMPTON LODGE (NEAR HEREFORD), OCCUPIED BY JOHN HEWER, 1835-1846.

such load. Spit dung he allowed not so good as my stall dung, and when I said the bringing it, he added 7s 6d ($1.87) per load, he said, certainly it could not be called less, he rather thought it should be more. This is one instance that shows how necessary it is to take locality into consideration in valuing."

The following is extracted from another letter of J. H. Campbell to Editor Arthur Young:

"In your review of Sussex you say the Sussex are not so broad and heavy in the shoulder as the Herefords, but whether this is a fault will admit of argument. I wish you would give me what appeared to you as such, but I can conceive none. Mr. Ellman is plainly of my opinion by the rules given you. Speaking of the joints as particular in a Herefordshire ox, you say great breadth before; you ought to have added behind also. No ox, I am sure, will pass for a good one in Herefordshire which has not good hind quarters as well as good fore ones."

PALMER'S COURT, HOLMER, OCCUPIED BY JOHN HEWER, 1848-1850.

(¶ 49) COPY OF QUERIES ANENT FEEDING CATTLE ON POTATOES, with the answers which were written to them, published in the "Annals of Agriculture."

As Mr. Campbell's answers will not be exactly to the same terms he will beg to promise that he is satisfied beyond a doubt (though aware the contrary is the established custom) that the weight of an ox can be no rule to judge by as to the quantity of food he will require to make him fat, or how much of it he will consume in a day; nothing being easier than to choose two oxen (of even the same breed) of equal weight that would require very different quantities of the same food to make them fat, and would also consume very different quantities of it in a day, still greater difference if chosen from different breeds. A large ox, having more lean flesh, will, to be sure, allow, with propriety, of being carried to a greater degree of fatness than the ox that has not so much lean flesh to put with it. But the time and the quantity of any food required to make any beast fat does, in Mr. C.'s opinion, depend largely upon the thriving disposition (easily known by the view and handling) and not in the least on his weight.

Query 1st.—What quantity of potatoes on an average may be sufficient to fat an ox of any good weight, suppose 100 stone (1,400 pounds)?

Mr. C. is of opinion that about 100 bushels of potatoes, with a little hay added to or given between every meal of them, amounting on the whole, while consuming the 100 bushels of potatoes, to about 7 cwt., would be sufficient to complete the fattening of any ox that was a tolerable good thriver.

Query 2nd.—How many in a day, how often, any preparation or cutting?

Cattle generally take to the eating of potatoes as readily as any other food, but some will decline them for a few days, but if they take to them directly, they should be allowed but small quantities at first and increased by degrees to any quantity they will come to eat with appetite, always intermixing the dry food and regulating that by the effect the potatoes are observed to have; that is, if the cattle become very lax, to increase the proportion of dry food until that alters again. Mr. Campbell, being of opinion that the more an ox can be brought to eat (with appetite) in a day, the sooner he will become fat, and consequently the cheaper and with more profit, never puts them to allowance, but thinks, when feeding altogether on potatoes (with only a drying quantity of hay), the average daily consumption per ox would be about one bushel, though many could be brought to eat one bushel and a half, some even two bushels some days. With potatoes, as with every other sort of food, it will be supposed that any one ox will vary much in the quantity he eats on different days. Reckoning the potatoes and accompaniment of hay as one meal there ought to be at least five servings. Perhaps when the days are the shortest there may hardly be time for so many, but they should be multiplied as the days grow longer. Mr. C. always begins with a very little hay; small quantities at the intermediate meals, but ends at night with as much as they will be supposed inclined to eat, not having any potatoes with them at night, lest (though very unlikely to happen) any might stick in their throats.

Mr. C. never has any cut, except when an ox at first rejects them, a few to coax him to eat them. If they have much dirt sticking to them they are washed by putting them in a long trough under a pump or run of water and rattling them about with a birch broom or some such operation to clear them from gross dirt, but that can very seldom be necessary. They are stored in oblong pits in the dry ground, as nigh hand the place where they are to be used as may be, and with care not to let wet in and to cover the end soon when they are taken out; they keep quite good till grass comes in.

Query 3d.—Does Mr. Campbell give any stated quantity of hay, or as much as the ox will eat?

This is answered by what is said above.

Query 4th.—Is any corn or meal necessary, and at what expense, if so?

None necessary. An ox may be as completely, and to as high a degree, fattened on potatoes and hay, as it is possible he should be on any food; and beef so fed remarkably good. But a variety of any such sort of food as mentioned in this query, given in addition, does, in Mr. C.'s opinion, bring the ox on yet faster, and therefore when such food is to be had at a moderate price he should always incline for using it as, according to his way of thinking, it would tend to make the whole of the feeding come more profitable, and that is cheaper. Mr. Campbell, in his present situation, gets some brewer's grains, and to as many of his feeding oxen as the quantity he gets will serve he gives, to two about a bushel, among which is mixed about a peck of pollard or sometimes pea or bean-meal coarsely ground, whichever of these or such like matters happen at the time to be cheapest (he means a price cheapest in proportion to the feeding quality of the particular article).

Query 5th.—Is there anything particular in the management of the cattle thus fattened which the above queries do not include?

The greatest difficulty that Mr. C. finds in bringing the people who have the care of the cattle to the trying of his directions, is to break them of the practice of giving too much at a time. If this is done, it disgusts the beast, puts him off his appetite and wastes the food. If, when an ox has eaten up what was put before him, he craves, he should have more given him, but as near as may be, no more should be given than he will clear; some will, however, remain, and before they are served again, whatever is left should be taken away; and that never should be wasted. Mr. C. has always some milking cows or some other cattle tied near, to which such leavings of the feeding cattle are given.

Mr. C. places much dependence for the thriving of feeding cattle on their being kept clean. The part of their mangers in which the hay is given them, is (whenever fresh is put in) cleaned from dirt and dust, and the division for potatoes, or (if any) meal, or such like feed, cleaned up with a blunt-pointed trowel every morning after they have been served, and whilst they are feeding on their potatoes, their stalls are cleaned, all dung and wet litter taken away, and what tolerably dry, shaken up, and sufficient fresh litter strewed over; that invites them (as soon as they have satisfied their inclination for eating) to lie down; they seem to enjoy the clean bed much, and their resting, much. Mr. C. presumes it must promote their fattening. Mr. C. thinks that combing and carding their hides promotes their thriving more than equal to the very small portion of time it takes up, and orders that it be done now and then, whenever time for it may be spared from the other business of the attendants.

The following account of an ox fed by J. H. Campbell is dated 7th April, 1789:

An ox of the true Herefordshire breed fed at Charlton, near Greenwich, in Kent, was ex-

VERN HOUSE, HEREFORDSHIRE, PROPERTY OF JOHN HEWER, 1855-1875.

hibited at Greenwich, on account of his size and beauty. The above-named ox came in a lean state to Charlton, the beginning of July, 1787. That summer he was kept on coarse grass. In the latter end of the autumn, lattermath, but of very indifferent quality; first of December, put in the stall and had potatoes and hay; about Christmas had some grains, also; some time after that, pollard mixed with grass, thinking at that time to have sold him in the spring, 1788, but afterwards determined to keep him over to another season (expecting a good increase of weight on him, as he was then only six years old); I turned him out to grass the latter end of April, and he continued out until the latter end of October, generally in grass of a middling quality that many people thought he must fall away and said I should spoil him. The two last weeks of that time he had some coarse hay in the morning. At the end of October he was taken to the stall, and had potatoes and hay as before, and then grains after some little time, bean and pea-meal mixed, added among the grain, beginning with a small proportion and increasing by degrees to one-fourth of meal to the grains. The 4th of April, 1789, Mr. Adams, the salesman, sold the ox to Mr. Cowldry, at Greenwich, who took him to show at Smithfield, and afterwards brought him to Greenwich again. It was remarked that he walked as well and as easily as any ox commonly fat, and his appetite as good as a lean ox, readily eating any hay or potatoes given him, and as readily the straw thrown to him for litter.

PARADISE VILLA, MARDEN, NEAR HEREFORD, WHERE JOHN HEWER DIED IN 1875.

Measure.	Feet.	In.
Length from poll to end of rump	8	3
Height	5	2
Girt round the first rib	10	6
Plumbed to the outside of each first rib	3	3¾
Girt behind the shoulders	9	0
Plumbed across the hips	2	9¼
Ditto the shoulders	2	8½
Ditto the end rump, or tutts	1	4½
Round the leg below the knee	0	9
Round the hind leg below the hook	0	11

Weights: Live weight was 3,360 pounds.

The carcass dressed was: Fore-quarters, 1,016 lbs.; hind-quarters, 896; total fore and hind-quarters weighed 1,912 lbs. Fore-quarters weighed 120 lbs. more than the hind-quarters. Tallow weighed 228 lbs. The hide was not weighed, but Mr. Cowldry supposed that it weighed 120 lbs. The tongue weighed 12 lbs.; heart weighed 9 lbs.; the neck pieces weighed 20 lbs.; leg pieces weighed 18 lbs. Mr. Cowldry said, the blade was not thick or of more weight than one of a beast of only 640 lbs., but the entrails were much less than it were commonly in beasts of small size, and the liver was less than any he had ever met with in any full grown beast. Middle sirloin, 62 lbs., sirloin for roasting, 48 lbs.; decrease, 14 lbs. This ox was sold for £70 ($350).

The following is a letter on the breeds of cattle, by Mr. Campbell, in 1790:

"Charlton, Jan. 2d, 1790.

"To the Editor of the 'Annals of Agriculture:'

"Sir: When you favored me with a letter on receipt of mine, on the 29th of April last, you wrote me that you had seen the advertisement of my famous ox, which you said 'was not much to the purpose, in proving the butcher's opinion, if he did not keep an account of the product of all its parts; offal, hind and fore-quarters, tallow, etc., and that such a particular would, with the live weights, be valuable for the Annals.' In consequence of which I should have sent such particulars, but that Lord Sheffield informed me that he had sent them to you, telling me at the same time that you were then gone on a tour abroad. Hoping that you are returned well, and may now again be somewhat at leisure for such correspondence as I wish to trouble you with, I will beg leave for a few words in justification of the advertisement and the butcher's opinion you alluded to, and which you have before known to be mine also.

"The advertisement 'presumes that the exhibition of that ox would sufficiently prove it to be a mistaken notion and direction, that the belly, shoulder, and neck, should be light, and that if a beast has a disposition to fatten or to be heavy in those, it would be found a deduction from the more valuable parts.'

"You, and everybody conversant on these matters at all, know that any person, though but a tolerable judge, could find out, by view and handling a beast alive, whether it was defective, either in proportionable weights or fatness on any piece of the whole carcass. If, therefore, a beast is exhibited and submitted to such examination, which will be found by everybody who does so examine it to excel both in weight and fatness on these forbidden parts, and it is also found that so far from such excellence being at the expense of, and being found a deduction from, the more valuable parts, that those more valuable parts are also in the same beast, excellent both as to weighty valuable substance and fatness, I cannot conceive how it can be denied to be sufficient proof of the fallacy of the above quotation.

"When I had the pleasure of showing that very ox to you at this place I did not understand that you thought the beast defective in the weight or fatness, on any of those valuable parts, or on any part of the whole carcass. You will remember I had at that time declared to you how much I disliked such partial rules for breeding cattle; and I really then wished much that if you did not see convincing facts in that ox that you would have been so kind as to have shown me why you still adhered to that opinion and where (in that ox) I and all others whom I heard speak of him were mistaken.

"In my letter, which you inserted in No. 67, I took the liberty of observing that in your 62d number you (mentioning that the Sussex cattle were not so broad and heavy in the shoulder as the Herefordshire) said, 'whether this is a fault will admit an argument.' I expressed my wish that you had expounded the argument against its being a fault. You did not favor my wish with any notice, either in the Annals or in your letter. But as these particulars, by which the right and best shape and make to be aimed for in breeding cattle are, as you have often observed, of much consequence to public benefit, as well as to the particular interest of every individual who breeds cattle, many or few; I will still hope you will favor us with the argument to prove a light and narrow shoulder to be no fault. I still undertake to produce you, any day of the year, ocular and manual proof, that it may be weighty, broad and fat, also without being found a deduction from the more valuable parts.

"As by what you said of the advertisement above mentioned, I find my last year's weighty ox, which you viewed and handled here, did not make the impression on you I thought he must, and that I had conceived he did by your not then mentioning an objection to what I said. I will hope for the favor of your company here sometime this season, when I may have some of my feeding stock in a state of proof to that impartial equality all parts of a beef may be looked for, in respect to weight and fatness, as well as I shall have lean stock at home which may afford very intelligent signs of what, with the greatest confidence, might be expected, when coming to that proof.

"You mentioned in your 63d number of the Annals, when speaking of the stock of this place, that I was a 'warm advocate of the Herefordshire breed of cattle and sheep;' I am so from the result of long experience; if I am wrong it is not for the want of painstaking, or being thoroughly acquainted with several other breeds, and those particularly about which there has been most said, at least in the print, as to which, after a long continued trial (and in the outset of the trial as confident expectation as anybody could have of finding them better than the Herefords), in the end, being of opinion that in most respects they were very greatly inferior to them.

"As you have named me of council for the Herefordshire breed, you will allow me to entreat that an evidence, whose testimony (from the open, plain manner he believes it in and the candid, as well as intelligent manner in which he treats every matter of this kind on which he has written) is, I think, of much consequence, and I should presume must necessarily weigh much with the jury, may have access into court; I mean Mr. Marshall, whose 'Rural Economy of Gloucester,' never that I observed, found admittance, or was taken the smallest notice of, in your occasional reviewing of agricultural books in the Annals.

"Whereas, Mr. Culley's book on live stock stands forth very conspicuously in them, and is ushered in with such flattering marks of approbation, and so many very high compliments, as most certainly add much more weight to his evidence with the jury, than (with submission) it seems to me to deserve, and if admitted by them as recommended by you, would, indeed, completely upset the cause of the Herefords as a breed; for he makes the cattle a strange hodge-podge of Welch and some illegitimate, that he represents wandering about some two or three English counties, and the sheep only a degeneracy from the breed, which, in most respects, I cannot consider as other than one of the worst in the kingdom.

HOLMER CHURCHYARD, NEAR HEREFORD, WHERE JOHN HEWER IS BURIED.

"As Mr. Culley has brought so strong a charge against the Herefordshire breed of cattle and sheep, I will beg he may be confronted in your court with Mr. Marshall, evidence for the defendant, and that they may be placed in manner and form as understated:

Mr. Culley's book on Live Stock. (1794). Page 21. Mentioning the number of different breeds of cattle, and naming fixed breeds with intent afterwards, as he says, to point out the	Mr. Marshall, in Vol. 11 of his Rural Economy of Gloucestershire (1788-1798). Page 226, and between that and page 231: "The Herefordshire breed of cattle, taking it all in all, may,

perfection and imperfections of each kind, compare them with each other, and then offer his opinion with regard to which are best. He says: "As to Herefordshire brown cattle they are, I am pretty clear, neither more nor less than a mixture between the Welch and a bastard race of long horns that are everywhere to be met in Cheshire, Shropshire," etc.

Page 39. "It is true they draw a few oxen in Herefordshire."

Page 179. "The more milk, the less beef; whenever we attempt to get both we are sure to get neither."

Page 180. "If it had not been for Mr. Bakewell perhaps we might still have been groping in the dark; at least that great breeder was the first I know who pointed out the valuable parts and made those true distinctions, u n o b-served, I believe, before his day."

without risk, I believe, be deemed the first breed of cattle in this island." Of the Fair at Herefordshire, 20th of October, he says: "The most valuable collection of cattle I have met with out of Smithfield, and by much the finest show I have ever seen."

In Herefordshire working oxen are the particular object of breeding. "Besides their superiority as beasts of draught and their being eligible as dairy stock, they fat kindly, at an early age, the strongest proof of their excellency as fatting cattle. I have seen three-year-old heifers of this breed, to use a familiar phrase, as fat as mud; much fatter than any heifers of that age I have seen of any other breed, spayed heifers of Norfolk excepted. * * * *" "Viewing the Herefordshire breed of cattle in this light, which I believe to be the true one, how unfortunate for the rural affairs of these kingdoms has been the choice of the spirited breeders for midland counties."

Again, speaking highly in commendation of the improvement in the breeds of long-horned c a t t l e and mentioning some of the differences between them and the Herefordshire, he says he "hopes soon to have an opportunity of digesting his ideas respecting that breed, and, lest infatuated by the fairness of their form, I may, in their praise, be led beyond the truth, I have here compared their general nature with that of a breed (Hereford) which I consider as the first the island affords. By having a standard to refer to, I may be the better enabled to regulate my judgment."

"Charlton, Feb. 23, 1790.

"Sir: When I had wrote the enclosed (under date of Jan. 2d), I found you was not then returned to this kingdom, therefore laid it aside till after the receipt of your printed circular letter, which you favored me with. As answers to queries in that, from this situation, would be of no use, I will not trouble you on what you will be much better informed from other hands. I am, sir,

"Your most humble servant,

"J. H. CAMPBELL."

Of interest in connection with the Campbell ox, which was slaughtered in April, 1789, we give the dimensions of a Hereford cow, owned by the Duke of Bedford, some eight years later.

DIMENSIONS OF HEREFORDSHIRE COW

In the possession of the Duke of Bedford.
October, 1797.

		ft.	in.
Height of	Hind quarter	4	5
	Shoulder	4	3
	Knee	1	1
	Hock	1	7¾
From the ground to the	Dewlap	1	6½
	Brisket	1	7¼
	Chest	1	9¼
Length of the	Hind quarter from the rump to the extremity of the hip bone	2	2
	From poll to tail	6	7
	Face	1	6½
	Horn	1	2½
Round of	Chop	1	7¼
	Cheek and forehead	3	6
	Neck	3	0½
	Chest	6	9½
	Knee	1	0½
	Bone of the foreleg	0	7½
	Coronet of the fore foot	1	1
	Hock	1	4
	Hind leg bone	0	9
	Coronet of the hind foot	1	1
	Horn	0	6½
Breadth of the	Face across the eye bones	0	8½
	Hip	2	1

LOTTERY (410) 185, CALVED 1824, BRED BY J. HEWER.

CHAPTER VI.

"YOUATT" ON BRITISH CATTLE

One of the important events in the cattle interest, especially as regards the Herefords, was the issuing of what has been styled, "Youatt's History of British Cattle" (1835), and purported to be "a full history of the various races." The author, Mr. W. Youatt, is rated "a scholar, distinguished for the extent, variety and elegance of his attainments," "a veterinary surgeon." His work was published under the auspices of the "Society for the Diffusion of Useful Knowledge," of which the following were officers: Chairman, The Right Honorable the Lord Chancellor, F. R. S., member National Institute of France; Vice-Chairman, The Right Honorable the Lord J. Russell, M. P., Paymaster of Forces; Treasurer, Wm. Tooke, M. P., F. R. S.; Secretary, Thos. Coates, No. 59 Lincoln's Inn Fields, London. These committees were of the leading men of England in Parliament, in the church and in common business, but as with members of such societies in our day, they were not especially interested or proficient in agriculture. The north pole or a new star were of vastly more importance to these scientists. It was with the organization of the Smithfield Club in 1798 that we begin to make a connected and reliable cattle record.

That part of Hereford history easiest to establish is the pre-eminent fact that as early as 1766, B. Tomkins, Jr., evolved from the efforts of his ancestors a systematic work with a view to making a standard, by which, or to which, the Hereford breed should conform later.

Editor Youatt, in giving the history of the Herefords, says: "The Herefordshire white-faced breed, with the exception of a very few Alderney and Durham cows, have almost exclusive possession of this county. The Herefords are considerably larger than the North Devon. They are usually of a dull red and some of them are brown and even yellow. A few are brindled, but they are principally distinguished by their white faces, throats and bellies. In a few cases the white extends to the shoulders. The old Herefords were brown, or red-brown, with not a white spot upon them. It is only within the last fifty or sixty years that it has been the fashion to breed for white faces."

"Whatever may be thought of the change of color, the present breed is certainly far superior to the old one. The hide is considerably thicker than that of the Devon, and the beasts are more hardy. Compared with the Devon, they are shorter in the leg, and also in the carcass; higher and broader and heavier in the chine; wider and rounder across the hips and better covered with fat; the thigh fuller and more muscular, and the shoulders larger and coarser." He then refers to an ox that belonged to the Duke of Bedford, and he quotes Marshall's opinion as given elsewhere.

Youatt continues: "They fatten to a much greater weight than the Devons, and run to 50 or 70 score (or 1,000 to 1,400 lbs.). A tolerable cow will average from 35 to 50 score (700 to 1,000 lbs.). A cow belonging to the Duke of Bedford weighed more than 70 score (or more than 1,400 lbs.). An ox belonging to Mr. Westcar exceeded 110 score (2,100 lbs.).

"They are not now much used for husbandry, though their form adapts them for heavier work, and they have all the honesty and docility of the Devon ox, and greater strength, if not his activity.

"The Herefordshire ox fattens speedily at a very early age, and it is therefore more advantageous to the farmer, and perhaps to the county, that he should go to the market at three years old than to be kept longer to be employed as a beast of draught.

"There are few cattle more prized in the market than the genuine Hereford. The Devons and Herefords are both excellent breeders, and the prejudice of Devonshire and Herefordshire farmers being set aside, a cross of the two breeds will give the Devons more bulk and hardihood, and the Hereford a finer form and activity.

"They are evidently an aboriginal breed and descended from the same stock as the Devons. If it were not for the white face and somewhat large head and thicker neck, it would not at all times be easy to distinguish between a heavy Devon and a light Hereford. Their white faces may probably be traced to a cross with their distant relations, the Montgomeries."

In his notice of Sussex, he says: "Of the Hereford and Sussex, the Hereford beast has flesh upon the ribs and sirloin, the Sussex more upon the flank and inside. It cannot be denied that the Herefords have theirs in the best places, and it is on that account that the prize is so often adjudged to them at the cattle shows, and particularly at Smithfield."

Youatt again says: "In some parts of Glamorganshire the pure Herefords are cultivated in preference to any admixtures with the native breed. Mr. Bradley, near Cardiff, is partial to the Hereford and his stock does not yield to many in the neighborhood or the country generally."

Of Monmouth, Youatt says: "In the vale district, the farmers were content with the Glamorgans, and the better kind of hill cattle. Of late years, however, the Herefords have, in a manner, superseded both of these breeds."

Of Dorset, Youatt says: "Concerning the Hereford on the Dorset stock the points of superiority said to be gained over the Devon cross are larger size, more hardening, and a disposition to yield a greater quantity of milk."

Of the Devon, after speaking of some trials that had been made by the Duke of Bedford, in Devonshire, he says: "We are, however, compelled to add that the Duke of Bedford has to a considerable extent changed his breed at Woburn, and the Devons have in a great degree given away to the Herefords."

These are a portion of the records that Youatt gives, outside of the space given to what he terms the "History of Hereford Cattle."

We will now take up some of the items that he might have put to their credit, but did not. First, the prizes won at the Smithfield Club from 1799 to 1834, inclusive, are as follows: 88 premiums on Hereford oxen, amounting to $5,760.00 (£1,152); 13 premiums on Hereford cows, amounting to $750 (£150); total premiums on Herefords, $6,510.00 (£1,300).

As against 35 premiums on Shorthorn oxen, amounting to $3,175.00 (£635); 17 premiums on Shorthorn cows, amounting to $925.00 (£185); total premiums on Shorthorns amounting to $4,100.00 (£820). Showing 53 more premiums on oxen at Smithfield for the Herefords than for the Shorthorn.

SOVEREIGN (404) 221, CALVED 1820, BRED BY J. HEWER.
(From an old painting.)

We have called attention previously to the fact that J. H. Campbell, of Charlton, Kent, and Mr. Westcar, were buyers of Hereford cattle, at the Hereford Michaelmas fair in October, 1779, probably earlier and up to 1819. This is true at least of Mr. Westcar, and he won the first premium at the first Smithfield show in 1799 with a Hereford ox.

Mr. John Westcar lived at Creslow, Buckingham. There are reported sales of his at Smithfield at long prices. The first ox that Mr. Westcar exhibited, and which won first prize at the meeting of 1799, was bred by Mr. Tully, of Huntington, Herefordshire, and was a grey six-year-old, fed by Mr. Westcar; weight 2,198 pounds. He showed also a grass-fed ox the same year at 1,806 pounds.

Mr. Armesley says: "If memory serves me right, in the year 1812 or 1813, I saw sold in the Smithfield Christmas market 50 Hereford oxen belonging to the high-famed Mr. Westcar, living in the Vale of Aylesbury, that averaged 50 guineas ($250) each, making 2,500 guineas ($12,500). At the same time, Mr. Richard Kightley, of Castlethorp, Berks, sold 30 Hereford oxen that averaged 47 guineas ($235) each, making 1,410 guineas ($7,050). Mr. Kightley's oxen were sold by himself, and Mr. Westcar's were sold by Mr. Thomas Potter. The old and famous breed of Herefords are not so large in frame as they were fifty years back, and are more complete and mature at an earlier age. Nearly all the above high-priced, gigantic oxen had been worked, and had earned home money at the plow and other labor before they were fed. At that time it was well known that Mr. Westcar had a large close that kept the whole of the summer 200 large oxen, and with the 200 oxen he fed 300 ewes and their lambs, and all went to market the same year. It is impossible to overrate the value of the services rendered to the breed by the old graziers."

RED ROSE 393, BRED BY MR. JOHN HEWER.
(This was Mr. Hewer's favorite cow.)

Mr. Westcar is reported as saying that the heaviest Hereford he ever fed, dead weight, weighed 157 stone, 2 pounds (14 pounds to the stone, making 2,200 pounds). It is reported on examination of his books, that they show sales as follows:

In 1799, Dec. 16th, two oxen to Mr. Chapman for £200 ($1,000).

In 1800, Dec. 4th, one ox to Mr. Chapman for £127 ($635).

In 1800, Dec. 13th, one ox to Mr. Harrington for £100 ($500).

In 1801, Nov. 26th, six oxen to Mr. Giblett for £756 ($3,780).

In 1802, Nov. 26th, one ox to Mr. Giblett for £100 ($500).

In 1802, Nov. 30th, one ox to Mr. Chapman for £126 ($630).

In 1802, Dec. 4th, two oxen to Mr. Horwood for £200 ($1,000).

In 1803, Dec. 4th, one ox to Mr. Chapman for £100 ($500).

In 1803, Dec. 19th, one ox to Mr. Reynolds for £105 ($525).

In 1803, Dec. 19th, one ox to Mr. Giblett for £105 ($525).

In 1804, Dec. 5th, one ox to Mr. Giblett for £105 ($525).

In 1805, Dec. 4th, one ox to Mr. Giblett for £100 ($500).

In 1811 Nov. 28th, one ox to Mr. Chandler for £105 ($525).

We have the statement of the historian, Mr. John Speed, in 1627, that the Hereford cattle were the best-conditioned cattle in England.

Mr. Gillam says that the Tomkins family were breeders for a long time, and Mr. Haywood also confirms the same. In 1779 Mr. J. H. Campbell takes out one ox to slaughter, and shows quantity and quality that could not come from a mongrel race.

"THE TICK-FACE."
(Photograph from life, in 1901, of a cow bred by the late K. B. Armour, and descended from Hewer stock.)

In 1798 the Smithfield show brings together all the breeds of the United Kingdom, and before this tribunal the Herefords led the breeds of England, and England led the world.

Up to 1851 all breeds of cattle were shown in competition for the best beef animal. As near as can be ascertained from the records, the relative standing of the Herefords, Shorthorns, Scotch, and cross-breeds is shown by the following statement of the prizes won by the different breeds:

OXEN AND STEERS.

Herefords won 185 prizes.
Shorthorns won 82 prizes.
Devons won 44 prizes.
Scotch won 43 prizes.
Sussex won 9 prizes.
Longhorns won 4 prizes.
Cross-breeds won 8 prizes.

Total prizes, 185 won by Herefords; 190 won by all other breeds.

The Hereford oxen and steers winning within five prizes as many as all the other breeds combined.

Mr. Youatt published his so-called "History of British Cattle," and made no mention of these exhibits and the extraordinary standing of the Herefords, as to the prizes won, nor did he mention the more extraordinary sales they made.

The following is an extract of winnings as between the Herefords and Shorthorns, for a period of 36 years, ending with 1835, when Youatt wrote his book, and the tabulated statement herewith shows how the breeds stood:

HEREFORDS.				SHORTHORNS.			
Year.	No. of prizes.	Guineas.	Dollars.	Year.	No. of prizes.	Guineas.	Dollars.
1799	..	..	....	1799	..	..	....
1800	4	52	260	1800	..	..	....
1801	..	..	....	1801	..	..	....
1802	..	..	....	1802	..	..	....
1803	4	80	400	1803	..	..	....
1804	..	..	....	1804	..	..	....
1805	2	40	200	1805	..	..	....
1806	..	..	....	1806	..	..	....
1807	2	30	150	1807	..	..	....
1808	3	40	200	1808	..	..	....
1809	2	40	200	1809	..	..	....
1810	3	50	250	1810	1	20	100
1811	1	10	50	1811	1	20	100
1812	2	40	200	1812	..	..	....
1813	2	40	200	1813	1	20	100
1814	2	45	225	1814	2	40	200
1815	4	90	450	1815	1	20	100
1816	2	40	200	1816	1	20	100
1817	1	25	125	1817	..	..	....
1818	3	50	250	1818	..	..	....
1819	2	40	200	1819	1	25	125
1820	4	55	275	1820	3	60	300
1821	3	45	225	1821	..	..	....
1822	2	25	125	1822	..	..	....
1823	4	55	275	1823	1	20	100
1824	1	15	75	1824	2	40	200

HEREFORDS.				SHORTHORNS.			
Year.	No. of prizes.	Guineas.	Dollars.	Year.	No. of prizes.	Guineas.	Dollars.
1825	2	25	125	1825	3	50	250
1826	3	45	225	1826	2	30	150
1827	3	45	225	1827	1	20	100
1828	4	65	325	1828	1	10	50
1829	4	45	225	1829	2	40	200
1830	3	30	150	1830	3	55	275
1831	4	50	250	1831	3	40	200
1832	4	55	275	1832	1	20	100
1833	5	55	275	1833	2	35	175
1834	3	30	150	1834	3	40	200
1835	5	60	300	1835	2	30	150
Total	93	1412	7060	Total	37	655	3275

From Youatt is learned that throughout the whole of Gloucestershire the Herefords were preferred for working and for fattening. In the notice of the Sussex cattle, Youatt tells us that it cannot be denied that the Herefords carry their fat on the best places, "and it is on this account that the prize is so often adjudged to them at the cattle show at Smithfield;" an observation that proves that Youatt was well aware of the success of the Herefords, although he did not mention the fact in his account of the breed.

He also quotes a remark by Arthur Young, editor of the "Annals of Agriculture," that "both in quality of flesh, thriving disposition, etc., both the Sussex and Devons exceeded the Staffordshire Longhorns and the Herefords left them far behind."

According to Youatt the Hereford had at that time spread over much of the kingdom. In Dorset some farmers were, with every probability of success, engrafting the Hereford on the Dorsetshire stock. Three points of superiority were said to be gained by the Hereford over the Devon cross—a larger size, more hardiness, and a disposition to yield a greater quantity of better milk.

In Somerset some of the dairy cows were red with a white face, which marked the Hereford cross. In some parts of Glamorganshire the pure Herefords were cultivated in preference to any mixture with the native breed.

"In Radnorshire," says Youatt, "they have principally had recourse to the Hereford as a cross with their own cattle, and though they have thus produced a beast large and too capable of yielding beef to be perfected on their poor land, they have obtained one that will thrive and pay otherwise, and that will find a ready market."

Of the Montgomeryshire cattle, he says: "The native cattle bear considerable resemblance to the Devons, but in the grazing districts they are generally abandoned for the Herefords, which are found to be suitable to the soil and climate, and are better feeders. Con-

LADY BYRON 218 (CALVED 1833), BRED BY JOHN HEWER, SOLD FOR £346 10s, or $1,730. A "TICK-FACE."

siderable attention is here paid to the dairy, and particularly to the production of cheese, which is little inferior to the Cheshire. About nine months feeding with grass, hay and turnips, will add about three score pounds weight to each of their quarters."

In Breconshire, recourse had been had to the Devons and Herefords, with evident advantage in favor of the Hereford both for work and grazing. The cattle on the side of Brecon that was nearest to Herefordshire were in a particular manner becoming very strongly mixed with the Herefords.

A cross with the Herefords had been with evident advantage attempted by the graziers in Carmarthenshire.

In Bedford, the Duke of Bedford had given an impetus to the rearing of Herefords. In Hampshire the Norman crossed with the Hereford was not injured as a milker while she was improved in size, and disposition to fatten. In Worcestershire the Herefords and Shorthorns were struggling for superiority on the grazing ground.

When writing of the Monmouthshire cattle Youatt said: "The Herefords will never find their way into the dairy; they belong to the graziers and butchers," to which he appends a foot-note by Mr. Walker, of Burton in Warwickshire, who tells him that this is too strongly expressed, it being his opinion that "they want nothing but management to bring them into the dairy; being so admirably adapted for the grazier, their milk is quite neglected. The Herefordshire farmers want early calves, and their cows and heifers calve between the middle of December and February after living entirely on dry meal, and usually by the time the grass comes they are nearly or quite dry, but if the Hereford heifer calve for the first time at grass and about the middle of May, she might become a good milker. Some of the cows will, under the present management, yield from ten to twelve quarts of milk at one time, and their

COTMORE (376) 150, CALVED 1836, BRED BY T. JEFFRIES.

Weight, 1 ton 16 cwt., English, equal to 3,920 American pounds. Champion and acknowledged the greatest bull ever produced up to his day. Note.—The above engraving is reproduced from an old painting, on the face of which all this matter is written, including the following:

"Let each succeeding race employ your care,
Distinguishing which to slaughter, which to spare;
Mark well the lineage from purest make,
And from pure blood its just proportions take."

"Robert Hewer, one of the most celebrated herdsmen, fed upwards of 200 winners in different parts of the kingdom."

milk is superior to that of any other cow except the Alderney. The quantity of milk given by a cow will greatly depend on her treatment with her first calf. If she has not proper feed to swell the milk veins at first starting she will never afterwards make a good milker. The Hereford cow seldom has a fair chance here. I speak from experience," he says, "for I have had much to do with the Herefords for several years, and have had many good milking cows of that breed," to which Mr. Youatt says, "these are very important observations, and although we are not sufficiently convinced to alter what we have written, and what almost universal experience and belief confirm, the remarks of Mr. Walker deserve serious attention."

Mr. Youatt, in making up the history of the Hereford cattle, quotes from Marshall his description of the points on which to judge the Hereford breed, but does not quote Marshall when he says: "The Herefordshire breed of cattle, taking it all in all, may, without risk, I believe, be deemed the first breed of cattle in this island. Their frame is altogether athletic, with limbs in most cases sufficiently clean for the purpose of traveling (¶ 52).

"Their form as beasts of draught is nearly complete. Besides their superiority as beasts of draught and their being eligible as dairy stock, being in this respect similar to Gloucestershire, the females fatten kindly at an early age,—the strongest proof of their excellency as fattening cattle. I have seen three-year-old heifers of this breed, to use a familiar phrase, as fat as mud; much fatter than any heifers of that age that I have seen of any other breed, the spayed heifer of Norfolk excepted.

"Viewing the Herefordshire cattle in this light, which I believe to be the true one, how unfortunate for the rural affairs of the kingdom has been the choice of the spirited breeders of the midland counties," who had selected the longhorn variety for their use. Nor does he quote Marshall when Marshall says that "at the Hereford fair, in October, 1788, I saw about 1,000 head of cattle, chiefly of this breed, a large proportion of them of grown oxen, full of flesh, sold for, or were worth, at the selling prices of the day, from £12 ($60) to £17 ($85) an ox, the most valuable collection I have met with out of Smithfield market, and by much the finest show I have anywhere seen."

These statements, as made by Marshall, were accessible to Mr. Youatt, and should have been used.

Neither does Youatt refer to the J. H. Campbell Hereford ox, shown and slaughtered at Greenwich in 1789, whose live weight was 3,360 pounds and dressed weight 1,912 pounds (for the four quarters), and was sold to the butchers for $350 (£70). Nor does he mention the correspondence that grew out of the exhibition and slaughter of the animal as contained in the "Annals of Agriculture" in Vols. 11, 12 and 13.

Neither does he refer to the formation of the Smithfield Club, nor to the fact that from 1799 to 1835, inclusive, the Herefords won 93 premiums, amounting to $7,060, against the Shorthorns winning 37 premiums and $3,275.

Neither does he refer to the sales of fat cattle made at Smithfield during the years from 1779 to 1835, and more particularly of the especially well-known sale by Mr. Westcar on Sept. 17th, 1798, of fifteen oxen at an average of $243, and of the still better known fact that from 1799 to 1811 Mr. Westcar had sold at Smithfield twenty head of Hereford oxen at an average

COTMORE, LYONSHALL, HEREFORDSHIRE, HOME OF MESSRS. JEFFRIES.

of $531 as butchers' beasts, or that Mr. Westcar made a sale of fifty Hereford oxen in 1812 or 1813 for $250 each.

These are facts that were matters of record in London and accessible to Mr. Youatt, as was the fact that at the London market Herefords were selling at 5 shillings ($1.25) per hundred pounds of dressed weight higher than the Shorthorns (¶ 53).

We have made copious quotations from Youatt's "History of British Cattle" to reiterate the fact that he had information that would have given the Herefords an entirely different position among British cattle if the facts had been accredited to the breed in their history. I repeat that in quoting from Marshall, the eminent agricultural historian, he omitted important facts. He recognized that the Herefords were successful at the Smithfield Club from its organization to the time he was writing, but fails to give any particulars as to the winnings before that society, or the sales that were made in the London market, which were accessible to him.

Beside the foregoing facts which were in his possession and used in the description of other breeds, there were the following leading events in the history of the Hereford cattle that were, or might have been, accessible to him:

First, Mr. John Speed, of London,* in his history (1627), said of the Herefords, that they were the best conditioned cattle in England. Again, he fails to notice the work of Mr. Benjamin Tomkins, commenced in 1766, and continued for about fifty years, a successful breeder of Hereford cattle and of wide reputation. A public sale made of his cattle averaging nearly $750 per head, and attracting the attention of the kingdom. He also fails to notice Mr. John Price, of Ryall, who followed Mr. Tomkins, and was a leading Hereford breeder

THE SHERIFFS, LYONSHALL, HEREFORDSHIRE, OCCUPIED BY THE JEFFRIES FAMILY.

at the time he wrote what purports to be a history of Hereford cattle, selling cattle of his breeding at extraordinary prices.

He also fails to notice the work of the Hewers, who were at the time Hereford breeders of national reputation, having bulls on hire in many counties.

(¶ 54) He also fails to notice the weekly sales of Hereford cattle at Hereford, and also the annual October sales held in Hereford, which sales were of national reputation; and the fact that graziers of cattle in Buckingham, Kent, and other counties near London, visited these fairs and made purchases of cattle to be fitted for the London market, and for exhibition at the Smithfield show. While he quotes from Mr. Marshall a description and character of the Herefordshire ox as he had seen them in Herefordshire, he fails to notice that Mr. Marshall says: "The Herefordshire breed of cattle, taking it all in all, may, without risk, be deemed the first breed of cattle in England, their superiority as beasts of draught, and their being eligible as dairy stock." All of which, and more of the same import, appears in the same article from which Mr. Youatt quotes in his history.

Neither does he take any notice of the formation of the Smithfield Cattle and Sheep Society, which was afterwards changed to the Smithfield Club, or of the winnings of the Hereford cattle at the first and subsequent exhibitions of the society, all of which were a matter of record in London on the books of the society.

Having slighted the leading British breed of cattle (the Herefords) in every conceivable way, Mr. Youatt deliberately delivered the preparation of the Shorthorn section of his work into the hands of a Shorthorn enthusiast, and in embodying it as part of his work, says, "for every portion of the text in this excellent account of the Shorthorns we are indebted to the Rev. Henry Berry, than whom there are few more zealous breeders of cattle, while there is no better judge of them."

In giving a description of the Herefords he presents two cuts, one of a Hereford ox, and the other a cow, in store condition, while for Shorthorns and other breeds he gives cuts showing the animals in full flesh.

We are thus particular in criticizing Mr. Youatt for the reason that his description of the Hereford cattle has been quoted from that time to the present day as a correct description, and his work has been used to the disadvantage of the Hereford breed. It would seem that not only the Shorthorns were described by Mr. Berry, but that Mr. Youatt took Mr. Berry's prejudiced and selfish view of the Herefords, or that he allowed the Shorthorn fanciers to revise and abridge the Hereford history; certain it is that he was unfair in writing the description, and unjust (amounting to caricature) in the cuts that he used to illustrate the character of the Hereford breed (¶ 50).

When the true character of Youatt's work began to be realized there was an urgent demand for an unbiased history. The Herefords have ever been in the main owned by tenant farmers, who had more care for producing a superior breed of profitable cattle than in advertising them. There being no Herefordshire

*England, Wales, and Scotland described (1627).

writer to prepare the work, the Scotch editors whose experience in preparing the history of their favorites, the polled Aberdeen-Angus cattle, qualified them for such work, took up the matter. They had no trouble in securing many of the facts that, though easily accessible to Youatt, were ignored by him, a portion of which being so suitable to this work, will close this chapter (¶ 51).

Benjamin Tomkins, the younger, drove 20 of his cows to Hereford on the day of the agricultural show, and offered 100 guineas to anyone who would show an equal number superior to them, but the challenge was not accepted.

In 1810, Mr. Meek, of Lichfield, gave a challenge to show his Longhorn bull against any Hereford for 100 guineas. His challenge was accepted by Mr. J. W. Walker, Burton Court, who sent his bull Crickneck (175) 305 to Lichfield, but when he got there it appeared that Mr. Meek had made himself acquainted with the superiority of Mr. Walker's bull, and rather than submit to defeat, allowed judgment to go by default in favor of the Hereford.

Mr. Haywood, of Blakemere, and Mr. Hill, Felhampton Court, have portraits of the Hereford bull Prizefighter, bred by Mr. Samuel Haywood, Clifton-on-Teme. The inscription on the painting states that he was "shown at Shifnal, 29th December, 1800, by Mr. Tench, of Bromfield, against Mr. Knowles, of Nailston, Leicester (whose nomination would doubtless have been Longhorn), to decide a bet of 100 guineas; determined in favor of the Herefordshire by Mr. Pestear, Somersetshire."

Mr. Price, of Ryall, issued two challenges. In 1812 he gave a challenge to be decided at Lichfield Agricultural Meeting, to show 20 of his cows in milk against 20 Longhorn cows for 100 guineas, which was accepted by Mr. Meek, and was decided in Mr. Price's favor. In 1839 Mr. Price sent forth a public challenge to show 20 cows and a bull of his own breeding against the same number of any other person's breeding, or any breed, open to all England, but no one came forward to accept it.

Mr. Weyman, of Stockton, challenged all England with his bull Stockton (237) 167 for 500 guineas, which was accepted by one of the Tomkins family, who, however, afterwards withdrew, leaving Mr. Weyman the victor.

Mr. Turner, of Aymestry, in 1803 offered to show a six-year-old Herefordshire ox, of his own breeding, against any breeder in the county of Hereford, or any adjoining county, for 100 guineas, for "weight and least coarse meat."

Mr. Walker, of Burton, was offered 60 guineas ($300) by Mr. T. Day, the exhibitor of the Durham ox, for a four-year-old working ox.

In 1825, at the Smithfield Club, there was a sweepstakes between three Herefords belonging to the Duke of Bedford and three Durhams belonging to the Right Hon. Chas. Arbuthnot, which was won by the Herefords.

The preparation of an account of what the Herefords have done in the show yards especially engaged the attention of Mr. William Housman, whose intimate acquaintance with

HOPE (439) 324, CALVED 1836, BRED BY T. JEFFRIES.

the history of the breed is well known. In order, however, to trace the progress of the breed it is desirable to briefly refer here to some of their early achievements at shows. The Smithfield Club and the Agricultural Society of Hereford were founded about the same time. The Smithfield Club was instituted under the title of the Smithfield Cattle and Sheep Society, at a meeting held in December, 1798, the name by which it is now so widely and favorably known having been adopted in 1802. The history of the club has been written by the late Sir Brandreth T. Gibbs. From this valuable little book we learn many interesting details regarding the position taken by the Herefords at the early shows.

At the first show, held in 1799, Mr. Westcar's first prize bullock, a Hereford, sold for 100 guineas ($500), was 8 ft. 11 ins. long, 6 ft. 7 ins. high, and 10 ft. 4 ins. girth. In his Cirencester lecture Mr. Duckham said that from the dimensions given upon a colored print which he then exhibited together with the names of the feeder and purchaser, all corresponding with the figures given by Sir Brandreth T. Gibbs, he had no doubt it was intended to represent this prize animal of Mr. Westcar's. If so, says Mr. Duckham, he was bred by Mr.

Tully, Huntington, near Hereford; his weight was 247 stone (3,458 lbs.), and he bore the distinctive marks of the red with white face, with the exception of the white stripe which now extends along the back, and just over the shoulders, being shown in the picture as far as the hip bones, and it also differed from the modern markings by the lower part of the legs being red instead of white. We may add that the drawing in Garrard's book removes any doubt as to the identity of this animal, which was unquestionably bred by Mr. Tully.

At the show in 1799 it is also stated that Mr. Grace, of Buckinghamshire, exhibited a Hereford ox 7 ft. high, which weighed upwards of 260 stone (3,640 lbs.) and measured in girth 12 ft. 4 ins. Among the other winners at this inaugural show of the Smithfield Club were the Duke of Bedford and Mr. John Ellman, both of whom were breeders of Herefords; the latter had the prize for the best ox fattened with grass and hay only, in the shortest time from the yoke.

It is thus apparent that the Herefords constituted a very important feature of the inaugural show of the Smithfield Club.

Among other curious details mentioned by Sir Brandreth T. Gibbs, in connection with the meeting in 1812, were the following: "A Hereford with a red ring round the eye," and a "smooth-coated Hereford." Mr. Duckham points out that these apparently trivial circumstances go far to prove that at that time it was expected that Herefords should have white faces and rough coats.

From the establishment of the Smithfield Club to the year 1851 all the different breeds and cross-breeds of cattle were exhibited at its show in competition with each other, except during the period from 1807 to 1815, when there was a classification of breeds, separate classes being assigned for Herefords, Longhorns, Shorthorns, Sussex, or Kent, Devons, and mixed breeds. Some of the records of the early shows are incomplete, but Sir Brandreth T. Gibbs states that during the time the breeds competed together—that is, before the new classification in 1852—the general prizes were thus distributed:

The Hereford oxen won 185 prizes...	$13,790
The Hereford cows won 22 prizes.....	1,155
	$14,945
The Shorthorn oxen won 82 prizes....	$ 6,995
The Shorthorn cows won 92 prizes....	5,665
	$12,660

The result, says Sir Brandreth T. Gibbs, shows that the number and amount of general money prizes was vastly in favor of the Herefords, their principal winnings being in the oxen and steer classes. The Shorthorns owed the fact of their approaching the Herefords in total amount of winnings to the success of the Shorthorn cows.

The Agricultural Society of Hereford, we are glad to say, under the name of the Herefordshire Agricultural Society, conducted with

LADY GROVE, CALVED 1838, AND CALF. FOIG-A-BALLAGH, BRED BY T. JEFFRIES.

vigor and success, was established in 1797, its object being "to carry the breed of cattle and sheep as to fleece and carcass to the greatest points of perfection."

The old minute books of the society have not been found, and for notices of the early shows we are indebted to contemporary newspapers. At the show of June, 1799, the first prize for best bull not exceeding 20 months old was awarded to John Apperley, Withington; and that for the best bull not exceeding three years and seven months old, to Samuel Tully, Huntington.

At the show in June, 1800, it is reported that there was an exhibition of cattle of very superior form and beauty. "Great praise," says the Hereford Journal of that date, "is due to those gentlemen who so warmly patronize this institution, which, whilst it ultimately promotes the interest and advantage of the farmers and breeders, will extend and perpetuate the fame of the county for a species of stock already in reputation with competent judges."

The first prize for best bull was awarded to Mr. Croose, Sugwas, for a bull bred by Mr. Jones, Fawley, for best yearling bull to Joseph Tully, Haywood, and for best heifer to Mr. Skyrme, of Stretton.

At the show in March, 1801, it was announced that the following gentlemen and farmers intended exhibiting bulls: Mr. Weyman, Moreton; C. Bodenham, Rotherwas; H. Moore, Wellington; J. G. Cotterell, Garmons; E. Waring, Lyonshall; T. A. Knight, Wormesley Grange (2); S. Tully, Huntington (2). Thirteen bulls were presented for the two premiums, and the most successful exhibitors were Mr. Smith, of Sufton, and Mr. H. Moore, Wellington.

Another show was held in June, 1801, when the prize winners were Mr. J. Tully, Haywood, and Mr. Williams, Thingehill. The report on the show in March, 1802, states that it seemed to be the general opinion that so many fine animals of the sort were never seen together before. The successful exhibitors were the two Messrs. Tully, of Huntington and Haywood; Mr. Powell, of Titley; Mr. Galliers, of King's Pyon, and Mr. Apperley, of Withington. The bulls of Lord Essex, Colonel Cotterell, Mr. Croose, of Ocle; Mr. Lowe, of Gatterton; Mr. Lewis, of Burghill, and Mr. Moore, of Bartonsham, were also, it is added, much admired.

Certificates were received from nearly all the candidates that their animals had not been fed with corn or straw imperfectly threshed, during the last six months; and in future, says the chronicler, this is to be made an express condition, without which no animal is to be entitled to a premium.

At the first show of the Royal Agricultural Society at Oxford, in 1839, the Herefords made a very good appearance, and the Rev. J. R. Smythies, of Lynch Court, won the first prize for the best cow in milk, "which," in the opinion of the judges, "was best calculated for dairy purposes." Mr. T. Jeffries, of The Grove, was first for Hereford bulls with Cotmore (376) 150, full of Hewer blood. Mr. Walker, North-

HOPE (439) 324, CALVED 1836, BRED BY T. JEFFRIES. (From an old painting.)

leach, was first for cows; Mr. J. Hewer first for yearling bulls, Mr. J. Walker, Burton, first for bull calves, and Mr. E. West first for heifer calves.

The favor in which the breed was held by graziers, the numerous challenges, most of them resulting successfully for the Herefords, the victories at Smithfield, and the spirit exhibited at the meetings of the County Society—all tend to show that the breed was constantly improving; that the farmers of Herefordshire were proud of their cattle, and that their merits were attracting widespread attention.

Narrow in its treatment of the Herefords and partisan to the Shorthorns as was the original Youatt's work, the American revised and abridged edition was narrower and more partisan and fraudulent, eliminating most of the little that Youatt had admitted. The American edition, edited by Ambrose Stevens, issued in 1851, and purporting to be Youatt's work, was purely a Shorthorn advertisement. The almost criminal fraud against the Herefords being thus introduced in the preface:

"In presenting an edition of Youatt to the American public, the American editor may justly say, that of all the treatises on cattle, none is so valuable as his. Mr. Youatt was a man of rare ability; a scholar, distinguished for the extent, variety and elegance of his attainments; for his power of research, historical, and scientific; for the brilliancy of his style; and as a veterinary surgeon of profound knowledge, in both the science and practice of his art, and of devotion to its pursuit. Scarcely any man of all the world was so happily fitted as he to produce a great historical and medical work on cattle.

"And while he was so peculiarly qualified to write such a work, the circumstances that originated it were eminently the ones to insure him success in the undertaking. An association existed in England, under the name of 'The Society for the Diffusion of Useful Knowledge.' Men of eminence in every variety of learning were its members; the publication of practical treatises in all departments of useful knowledge, its object. Appreciating the ability of Mr. Youatt to give the world a valuable work on the history, breeds, management, and diseases of cattle, this society enlisted him in its production.

"In preparing this treatise for publication, the American editor has abridged it of the history of local and inferior breeds of cattle in England, in which the American farmer and amateur has no interest. There is not a page in the whole but has been carefully considered, and where it required its matter advanced to the present state of knowledge on the subject."

J. A. MONKHOUSE, THE BLIND VETERAN OF THE STOWE.

CHAPTER VII.

THE COUNTY OF HEREFORD—HEREFORDSHIRE

It is of interest to know about the section of England where the best breed of beef cattle of the present day originated. Probably no more authentic or reliable information can be obtained than the account as given in the "Encyclopedia Britannica."

It will be seen in the parliamentary returns of 1873 that some of the well-known thoroughbred Hereford cattle breeders were large land owners. The Hon. J. H. Arkwright (who bred our famous Sir Richard 2d (9702) 4978 and Mr. Sotham's recently imported sensational Improver (19206) 94020) having over ten thousand acres; Sir A. R. Boughton Knight over ten thousand acres, and Lord Bateman (¶ 55), who bred the first cattle imported by Mr. F. W. Stone, nearly seven thousand acres.

"The English county of Hereford is on the south border of Wales. Its greatest length from Ludford by Ludlow to the Doward Hills near Monmouth is 38 miles; its greatest breadth from Cradley to Clifford, near Hay, is 35 miles. The country is well watered with numerous rivers and is pre-eminently a grazing district. The climate is variable, owing to the damp and fogs, which moisten the earth and account for its great verdure as well as its large proportion of timber, not only in parks and on landed estates, but almost on every hedgerow.

"The surface of the country is undulating in long ridges (¶ 56), as if by subterranean ripples. Ash and oak coppices and larch plantations clothe its hillsides (¶ 59) and crests. Its low lands are studded with pear and apple orchards, of such productiveness that Herefordshire sometimes, as well as Kent, is called the garden of England.

"Herefordshire is also famous as a breeding county, for its cattle of bright red hue, with mottled or white faces and sleek and silky coats. The Herefords are a stalwart and healthy breed, and put on more meat and fat at an early age in proportion to food consumed than almost any other variety. They produce the finest beef, and are more cheaply fed, than Devons or Durhams, with which they are advantageously crossed.

"Breeders' names from this county are famous at the national cattle shows, and the number, size and quality of the stock are seen in their supply of the metropolitan and other markets. Prize Herefords are constantly exported to the colonies.

"Agricultural horses of good quality are bred in the north and saddle and coach horses may be met with at the fairs, especially Orleton, Brampton, Bryan, and Huntington."

LIVE STOCK.

Horses	21,206
Cattle	77,402
Sheep	340,741
Pigs	24,169

"According to the parliamentary returns of 1873, the county was divided among 13,731 proprietors, owning a total area of 506,559 acres, with a rental of £924,640 ($4,623,200). Of the proprietors 9,085 (66 per cent) held less than one acre; 2,478 (18 per cent) held between one and ten acres. (¶ 61)

"The owners of the largest holdings are J. H. Arkwright (¶ 57), Hampton Court, Leominster, 10,559 acres; A. R. Boughton Knight, Downton Castle, 10,081 acres; R. D. Harley, Brampton-Bryan, 9,869 acres; Sir Geo. H. Cornewall, Moccas, 6,946 acres; Lord Bateman, Shobdon Court (¶ 58), 6,815 acres; Earl Somers, Eastmor Castle, 6,668 acres."

CHAPTER VIII.

The Smithfield Club; National Show

The records of the Smithfield Club are matters of history, that record the first authentic victories of the Hereford cattle. The character of this great national society, the nature and importance of its exhibits, are important, as for over one hundred years they have been the court of last resort for exhibitors of the meat producing animals of Great Britain. Cattle of all breeds and crosses were shown in one general class, as fat cattle, before this society until the year 1852, when the several breeds were separated and shown by themselves.

"The Annals of Agriculture" contain an account of the formation of this club, by the Smithfield Cattle and Sheep Society, by common agreement, December 17th, 1798, being the market day before Christmas. The Duke of Bedford in the chair.

"We, the underwritten, do hereby agree to institute a society open to all subscribers, to the premiums hereafter mentioned, and subject to such conditions as should be agreed upon by a committee to be named this evening. The subscriptions, 10s 6d ($2.62), each to be paid immediately." (¶ 62)

Thirty-seven persons subscribed and paid their initiation fees, agreeing that Mr. Arthur Young be empowered to receive other subscribers, and that each candidate for a prize should give Mr. Young one month's previous notice; then Mr. Young should inform the candidate of the time and place of showing. That it is the intention of the present committee to give premiums to the best beasts fed on grass, hay, turnips or cabbage; and to the best fed on corn or oil cakes. The meeting then adjourned to the Duke of Bedford's Sheep Shearing at Woburn, the 17th of June next.

Expenses of the first show in 1799 were £101 ($505). Receipts were £100 ($500).

The following extracts from the minutes of the Smithfield Club from its organization show the splendid work of this most powerful of factors in the encouragement of British breeds of live stock.

1798. At a meeting of the agriculturists held December 17th, 1798 (being Smithfield's great market day), the "Smithfield Cattle and Sheep Society" was founded by mutual consent. Present: Francis, Duke of Bedford, in the chair; Mr. J. W. Wilkes, of Measham, the original proposer, and twenty-nine others, including Lord Somerville, John Hennet, the Earl of Winchelsea, John Westcar, Richard Astley, John Ellman, Arthur Young, etc.; subscriptions 10s 6d each. Eight other names were added later in the day, including Sir Joseph Banks.

A committee was appointed, who decided to offer premiums for the best beast above a stated weight, and fed on grass, hay, turnips, or cabbages; also for the best beast fed on corn or oil cake; for the best sheep fed on hay, grass, turnips, or cabbages; or for the best sheep fed on corn or cake.

Arthur Young, Esq., was requested to receive subscriptions.

1799. The first show was held at Wooton's livery stables (The Dolphin Yard), Smithfield. The three days' admission money to this show amounted to £40, 3s ($200.75). The first dinner of the subscribers took place at the Crown and Anchor Tavern, on the Friday previous to the Christmas market. The following account is given of some of the animals:

"The largest sheep were of the true old Gloucester breed (Cotswolds), bred by Mr. Haines and grazed by Mr. Poulton—6 ft. 5½ in. in girth, 27 in. across the back, 22 in. over the shoulders and stood only 26 in. high.

"A Hereford bullock fed by Mr. Grace, of Buckinghamshire, 7 ft. high, weighed upward of 260 stone (3,024 lbs.) and measured in girth 12 ft. 4 in. Mr. Westcar's champion bullock (Hereford), which sold for 100 guineas, was 8 ft. 11 in. long, 6 ft. 7 in. high and 10 ft. 4 in. around the girth." [We have given a picture of this, the Tully ox, in a preceding chapter. T. L. M.]

Among the winners with Herefords were Mr. Westcar, the Duke of Bedford, Mr. Edmonds, and Mr. John Ellman; the latter "for the best ox fattened with grass and hay only in the shortest time from the yoke."

A committee of management of fifteen members was appointed.

1800. The pieces of plate offered this year as prizes were divided between vegetable-fed and cake and corn-fed cattle, under 150 stone (2,100 lbs.), and above 80 stone (1,200 lbs.), also above 150 stone (2,100 lbs.), and varied in value from 20 guineas ($100) down to 10 guineas ($50). The pieces of plate offered as prizes for sheep were for wethers fed the same as the cattle, and were of the value of 12 guineas ($60), and 8 guineas ($40). The judges had to select the two best animals, each class to remain for public exhibition. (¶ 63)

The owners of prize animals were expected to agree with the butchers that the "judges have full power to inspect the killing, for the purpose of ascertaining the weight of each animal, distinguishing the fore and hind quarters, the tallow, hide, pelt, or offal, including blood, etc., or by some other satisfactory mode, to ascertain the comparative live and dead weights," and, having ascertained such points, the judges were then to decide which was to be placed the first, and which the second best in each class, and report accordingly.

The exhibitors were requested also to furnish an account of the "breed, age, time of fattening, sort of food, and the time the animals had eaten cake or corn," etc. Early maturity in sheep was to be particularly considered.

The butchers having objected to the conditions respecting the weights, and having declared that they would not buy the animals under such an engagement, it was resolved that this be not insisted on, but that the proprietors and purchasers be requested to furnish such information to the judges as may enable them and the public to decide as accurately as possible on the respective points of the prize beasts and sheep.

The judges were appointed by a committee, and were to consist of three graziers and three butchers, and the instructions to the judges were, "to look to the quality of the meat, proportion of valuable meat, proportion of meat to offal, and time of feeding, and not to consider certificates satisfactory unless explicit as to the mode of feeding for some time back."

The proprietors of the yard received the money of the admission of the public, and paid the society 60 guineas ($300). The proprietors had also to furnish hay, straw, etc.

THE COURT OF NOKE, PEMBRIDGE, HEREFORDSHIRE, SEAT OF J. TURNER.

The cattle prizes were won by Herefords, Sussex, and Longhorns.

At the meeting of the committee December 13, 1800, the Duke of Bedford first made the proposition of a new constitution of the society, and "that it be formed into a permanent club, to consist of fifty members, viz., the present committee, together with Mr. Wilkes of Measham, the original proposer of the society, and thirty-four to be elected by ballot. Annual subscription 1 guinea ($5)."

THE COURT OF NOKE IN 1902, OCCUPIED BY MR. EDWARD FARR.

A guinea was paid this year for each successful candidate's servant (nine in number).

Distinct prizes were proposed for oxen, and for cows and heifers.

It was considered "that if the cattle shown had not been worked, then early ripeness was a merit equally as in sheep."

Two oxen were exhibited this year (1800) by his Majesty the King.

1801. The prizes offered this year varied in amount from 30 guineas down to 10 ($150 to $50). There was a separate class for cows and heifers.

Exhibitors were required to state particularly the condition of the flesh of the animal at the time of putting to fatten.

March 30, 1801. The following were elected officers: President, the Duke of Bedford; Secretary, Arthur Young, Esq.; Stewards of the Show, Mr. Bennet and Mr. Ellman.

It was determined to have five judges for the entire show, and that each member of the club might suggest in writing the names of such persons as he thought qualified to be judges. A selection was made by the committee, and the president and stewards ultimately appointed the judges. It was recommended that in future the state of the flesh of the animals when put to fatten be attested in the certificates by two respectable parties. The number of members of the club was now extended to sixty-five.

1802. It was required that beasts shall have had no cake, except in the year 1802, and that the whole of the food consumed from October 1 to November 30 be certified.

Heifers were excluded from competing. Only cows that had had three calves and had calved either in 1801 or 1802 were qualified. The time when dried of milk was to be certified.

The title of the Smithfield Cattle and Sheep Club last occurs in the minutes December 8, 1802, and from that time it is styled the "Smithfield Club."

Francis, Duke of Bedford, the president, died. John Drake, of Bedford, was elected president December 8, 1802. The number of members of the club was extended to one hundred.

Placards were placed over each animal, stating breed, age, etc.

A guinea was paid this year to the exhibitors' servants in charge of first-prize animals.

1803. *Resolved,* That it be a condition of the prizes to be offered next year (1804) that no beast shall have been put to fatten previous to the 1st of January.

Several animals were disqualified, in consequence of the certificates not giving the requisite details as to the mode of feeding.

1804. Several prizes were withheld, in consequence of no returns of dead-weights having been sent in. The prizes were not paid until the returns of dead-weights were produced.

It was resolved to appoint a person in future to attend at the killing and weighing of live stock, who shall report the results to the club. Mr. King, Jr., appointed.

The number of members increased to 120. The show was held in the Swan Yard.

1805. *Resolved,* That the number of the members of the club be unlimited. Election to be by ballot, and one-third present to exclude.

A Devon ox was driven 126 miles to the show. The show was held at Dixon's Repository.

1806. Arthur Young, Esq., resigned the secretaryship, not being able to attend at the time the meetings were held.

The offices of treasurer and secretary were made distinct. Mr. John Farey was appointed secretary, with a salary, and "liberty of publishing the proceedings and papers of the club for his own benefit, as Mr. Arthur Young had heretofore done."

Mr. Paul Giblett was appointed treasurer. The show was held this year at Sadler's Yard, Goswell street; H. R. H. the Duke of York gained a prize for a pig. Lord William Russel was elected chairman during the Duke of Bedford's absence abroad.

Lord Somerville and Mr. H. King, Jr., were appointed inspectors to attend the weighing of prize animals.

It was determined to offer for next year (1807) seven prizes for fat oxen or steers of the weight of 120 stone (1,680 lbs.) and upwards, viz.:

Class I, Herefords, 20 guineas ($100); Class II, Longhorns, 20 guineas ($100); Class III, Shorthorns, 20 guineas ($100); Class IV, Sussex or Kent, 20 guineas ($100); Class V, Devons, 20 guineas ($100); Class VI, mixed breeds, 20 guineas ($100). Also an additional prize for the best ox or steer in the foregoing classes, £10 ($50). The following conditions were added: "The animals exhibited for the above premiums must have worked at least two years ending the 1st of January, 1807, and must not have been put to fatten previous to the 5th of April, and the whole of the food consumed from the 1st of October to the 30th of November must be certified. The time of putting to feed, and the state of the flesh, must be certified, under the attestation of two respectable witnesses."

1807. There were no exhibitors for the Longhorn, Shorthorn, and Sussex prizes. The prize in the class for Devons was not adjudged, for want of sufficient merit. There were no exhibitors for the cow prize.

Resolved, That in future there be three judges instead of five.

Also that the secretary's salary be 30 guineas. The club's dinner was held in the Free Mason's Hall.

1808. The proprietor of Sadler's Yard, Goswell street, paid the club only £50 ($250), he finding provender as usual. The club also entered into the same agreement with Mr. Sadler for holding Lord Somerville's show. Again this year there was no exhibitor for the Longhorn prize. The only competitor for the Shorthorn prize was disqualified. There was no exhibitor in mixed breeds. The Duke of Bedford very liberally paid upwards of £68, charged for three years' use of rooms at Freemasons' Tavern for meetings, etc. *Resolved,* That no person who has gained a prize in either of the first five classes be entitled to show a beast in the same class next year.

1809. For the Shorthorn and Sussex prizes there were no exhibitors. The judges were not to be informed of the names of the owners of the animals, but were to adjudicate by numbers placed over each. *Resolved,* That from next year (1810) a prize be offered for pure Merino sheep.

1810. The conditions respecting Longhorn and Shorthorn oxen having been worked were dispensed with. It was decided that in 1811, Hereford, Sussex, or Kent, and Devon oxen or steers, be shown in pairs or yokes, of the same age and one person's breed, the premiums to be equally divided between the breeder and grazier. A portable weighing machine for cattle was hired from Mr. Shepherd, of Woburn.

1811. There were no exhibitors for the prizes for yokes or pairs of oxen in the Hereford or Devon classes. The arrears of subscription amounted to 468 guineas.

1812. Members were required to subscribe to the "signature book," engaging to pay subscriptions (¶ 64).

Some curious descriptions of the colors, etc., of the animals occurred:

"A Hereford with a red ring around the eye;" "a red, frosty-faced Sussex;" "a red and white Devon;" "a smooth-coated Hereford."

1813. The Duke of Bedford resigned the office of president when about to proceed to the continent. The Marquis of Tavistock was requested to accept this office, but declined. A committee of management was elected, consisting of the stewards and thirteen members. Thos. Coke, Esq., was requested to become president, but declined. *Resolved,* That in future the prizes be in plate instead of money. The number of show days was reduced from three to two. The subscriptions and arrears due to the club amounted to £1,086 15s ($5,434). The number of members was 272.

IVINGTONBURY, SEAT OF MR. T. ROBERTS, WHO BRED SIR THOMAS.

1814. The Duke of Bedford offered 100 guineas ($500) annually for additional prizes. His Grace afterwards increased this sum to 125 guineas ($625) in order that a five-guinea ($25) medal might be given to the breeders of the animals in each of the five proposed classes. These were subsequently styled the "Bedfordian" plate and medals. The Duke of Bedford was requested to continue president; Lord Somerville, Sir John S. Sebright, Bart., C. Callis Western, Esq., M. P., afterwards Lord Western, Thos. Mellish, Esq., were elected vice-presidents. Mr. Thomas, of Bond street, was consulted respecting a die with a profile of the Duke of Bedford, for the Bedfordian medals, and subsequently a die was engraved by Mr. Jos. Porter of Fleet street, from the original model for a bust by Mr. Nollekens. Mr. Thomas undertook to have suitable articles of plate on view for the successful candidates to choose from. It was ordered that the certificates be delivered to the secretary eight days before the show.

EDWARD FARR, PRESENT OCCUPANT (1902) OF THE COURT OF NOKE.

1815. The Duke of Bedford was again elected president. Mr. Farey was elected treasurer as well as secretary, with 40 guineas salary. The rule requiring the dead-weights of animals was rescinded. *Resolved*, That no animal once shown be exhibited again except as extra stock.

1816. Arthur Young, Esq., resigned being a member of the club. The finances were in so bad a state that the president proposed that the club should not offer any prizes next year.

1817. The classification of cattle according to breed was discontinued. No prizes were offered this year out of the club's funds, only the Bedfordian plate and medals, value 25 guineas. Nevertheless, the judges reported favorably of the show. The Duke of Bedford suggested "whether the ends for which the club was associated were not sufficiently answered;" but in case the club should judge otherwise, he expressed his readiness to continue the Bedfordian premiums. Sir John Sinclair having expressed his anxious hope that the club should continue its useful exertions in this time of agricultural depression, it was resolved:

"That it is the opinion of this meeting that great advantages have accrued to the landed interests and the community in general from the exertions of this club, which have tended materially to increase the supply of animal food of superior quality to meet our greatly increased population and consumption. That the late exhibitions, and the present one in particular, show that the improvements in live stock are yet in successful progress as to the essential points of disposition to fatten, early maturity, and consequent cheapness of production, and that further and greater benefits may be rendered to the community by the continuance of these exertions; under these impressions it is the decided opinion of this meeting that the club should continue and receive the utmost support from its members."

1818. The prizes offered out of the club's funds amounted to £50 without classification as to breed of cattle, but with conditions as to weight only. The Bedfordian premiums were also offered as before.

Resolved, That one steward retire each year. That the judges be taken in rotation from the list of names which shall be determined on at the meetings of the club.

That in future there be three judges of cattle, and three of sheep and pigs, to be chosen from the following lists, viz.:

Experienced breeders and graziers of cattle; experienced breeders and graziers of long wools; experienced breeders and graziers of short wools; experienced butchers in or near London.

That the stewards supply, as heretofore, any vacancy that may occur by non-attendance.

1819. That no exhibitor be allowed next year to gain a prize in the same class in which he has gained one this year, nor in future to win in the same class any two consecutive years (¶ 65).

That the sheep be shorn before the judges award the prizes.

A class for cows was re-instituted, the same as in the year 1815.

The secretary reported the death of Lord Somerville, a vice-president of the club, and one of its original members.

N. B.—From the prize sheet for 1819 it appears that an exhibitor was allowed to enter two beasts in the same class, but where two prizes were offered he was only allowed to gain one.

1820. Mr. Sadler was allowed to charge non-members for the standing room for implements.

Sir Joseph Banks, one of the original members and promoters, died.

1821. December 6, 1821. A letter was received from the Duke of Bedford, stating his determination to withdraw from the club, and to discontinue the Bedfordian prizes, and expressing his opinion that "the advantages which on the first formation of the institution were held out to the public have been amply realized and that any further incentive to improve the breeds of cattle, sheep, and swine is become wholly useless.

"The only object was to increase the animal production of the kingdom and this object has been fully attained. The markets of the metropolis and throughout the kingdom are abundantly supplied. The best and most profitable breeds of cattle and sheep have been brought into notice, and have made rapid and extraordinary progress in the estimation of the breeder and grazier."

It was, however, unanimously resolved by the meeting of the members after considerable deliberation that "the club ought to continue and receive the utmost support from its members."

Sir John Sebright, Bart., was requested to become president, but declined.

Resolved, That in future an exhibitor be entitled to win in the same class two consecutive years, provided he be both breeder and feeder.

Lord Strathaven, a vice-president, died.

1822. The judges declared this to be one of the best shows they ever witnessed. The club was without a president.

1823. The Duchess of Rutland became a member.

Resolved, That in future the prizes be in cash or plate the same as previous to 1814.

That in future the restrictions as to an exhibitor not winning in the same class two consecutive years, unless he be both breeder and feeder, be abolished.

1824. *Resolved,* That in future there be two classes of çows, viz., one for fattened dairy cows, which have calved in their full time twice at least, and have given fair proportion of milk. The other class for cows or heifers (not spayed) which may not be eligible for the above class.

From the prize sheet for this year, it appears that exhibitors sending two beasts for the same class had to select which of the two should compete for the prize.

1825. Viscount Althorp (afterwards Earl Spencer) was elected the president of the club. There was a sweepstakes between three Herefords, belonging to the Duke of Bedford, and three Durhams, belonging to the Rt. Hon. Charles Arbuthnot, won by the Herefords.

Resolved, That "no motion having for its

DOWNTON CASTLE.
(From a photograph taken in 1902.)

object either the dissolution of the club, or materially altering its constitution," should be taken into consideration until a copy has been sent to each member, with a statement of the position of the club's funds, etc., and a second meeting being appointed, at least twenty-one days from the first, for discussing and deciding upon it.

It was ordered to be mentioned in the prize sheet, that "next year the prizes will be distributed to the successful candidates at the annual dinner."

1826. Mr. John Farey died, having been secretary twenty years.

Resolved, "That no officer receiving pay from the club shall be made a member of the club."

Mr. William Farey was appointed secretary. (¶ 66)

Resolved, That in future the ears of one-year-old sheep be marked to prevent their being shown again.

That in future no animals be shown without the exhibitors certifying that they have had them in their possession six months at least.

MR. A. J. R. B. KNIGHT, OF DOWNTON CASTLE, 1902.

Resolved, That the butcher's returns of the dead weights be again required.

1827. *Resolved,* That there be three stewards.

That there be three judges for the whole show.

That each steward be expected to procure a breeder of cattle or sheep, or a grazier, who will be willing to act as judge of the show.

1828. *Resolved,* That the prizes be in pounds instead of guineas.

That in future the officers of the club shall not be allowed to answer in their official capacities any inquiries respecting the proceedings of the club, except such as are entered and written in the minute book.

That the judges be paid their traveling expenses, fixed at £5 ($25) for 1829.

The president proposed sweepstakes for oxen and steers, and cows and heifers, to be decided by the club's judges next year.

1829. *Resolved,* That silver medals be given to the breeders of the first prize animals in Classes 1, 2 and 3 this year, and for the future to the breeders of first-prize animals in each class.

Mr. Kitelee placed at the disposal of the club the £10 prize awarded to him, to be offered in a gold medal for the best beast that may be exhibited in any of the classes at the club's next show.

Resolved, That silver medals be given in future on extra stock for the best beast, the best long-wooled sheep, the best short-wooled sheep, and the best pig.

1830. *Resolved,* That in future a gold medal be given to the breeder of the best beast in any of the classes, and also a gold medal to the breeder of the best pen of sheep in any of the classes. The officers and leading members of the club decided to raise subscriptions of a guinea each, in order to present a piece of plate to Viscount Althorp, the president, in testimony of his valuable services in raising the club to its present eminence.

1831. An extra gold medal (£5) was voted for the second best animal in Class 1, there being no second prize offered.

A gold medal (£10) was offered for any ox not gaining the prize in Class 1, but which shall, in the opinion of the judges, possess extraordinary merit.

Resolved, That stewards in future audit the club's accounts, and, if possible, previous to the Friday's meeting.

That there be separate gold medals in future for long-wooled sheep and for short-wooled sheep.

Sir John Sebright, Bart., offered to give a gold medal in 1832 for the best pen of three-year-old Southdowns.

At the dinner a candelabrum, value 200 guineas ($1,000), was presented to Viscount Althorp.

1832. Richard Astley, Esq., the then father

of the club, and one of its most active supporters, died; also Thos. Mellish, Esq., a vice-president.

His Grace the Duke of Richmond was elected a vice-president.

Resolved, That in future the breeders' certificates be required.

That any member who shall become 10 guineas in arrears be excluded from the list of members.

1833. *Resolved,* That the extra gold medal in Class 1 be discontinued.

1834. The Duke of Richmond won a match made with Lord Huntingfield for the five best shearling Southdown wethers.

Messrs. Hoars were appointed the bankers of the club, the secretary to draw upon them.

1835. The Marquis of Exeter won a match with Earl Spencer for the best freemartin heifer.

Mr. Ellman won a match with Thomas Coke, Esq., for Southdowns.

Resolved, That a silver medal be given to the butcher who shall purchase animals to the largest amount out of the classes, upon his duly producing to the secretary the certificates of purchase. (¶ 67)

N. B.—From the prize sheet for 1835 it appears that only one beast, one pen of three sheep, and one pen of three pigs belonging to the same person could be exhibited in each class.

1836. Humphrey Gibbs, Esq. (afterwards Humphrey Brandreth), was elected honorary secretary.

1837. A committee was appointed to see if better premises could be obtained for the club's show.

Resolved, That any member of the club who has paid up all arrears of his subscription, or any new member on his admission, may compound for all future annual subscriptions by the payment of 10 guineas.

The Duke of Richmond having called attention to the importance of the dead-weight returns, the Hon. Secretary offered to be present at the weighing, which offer was accepted with thanks.

1838. This was the last year of the show being held at Goswell street.

1839. The show was held at the Baker street Bazaar for the first time.

Resolved, That in consequence of the great increase of stock shown of late years, there be for the future three judges of cattle and long-wooled sheep, and three judges of short-wooled sheep, and pigs.

That three additional stewards be appointed for the purpose of selecting the judge of short-wooled sheep, and pigs.

That new classes of Scotch and Welsh cattle be formed.

1840. His Royal Highness, Prince Albert, visited the show.

A die for the medals, with a profile of Earl Spencer, president, was ordered to be engraved by Wm. Wyon, R. A.

1841. His Royal Highness, Prince Albert, was elected a member of the club.

BROCKSWOOD (485), CALVED IN 1843, BRED BY J. RICKETS.
(From an old lithograph.)

Resolved, That in future no animals fed on milk during twelve months previous to the show, except pigs, be qualified.

The president having expressed a wish that the meeting would give its opinion as to the age that drew distinction between a cow and a heifer, it was found to be the opinion of the meeting that the term heifer applied until the animal was four years old, and after that it should be considered a cow. Also, that the term steer applied until the animal was four years old; after that it should be considered an ox.

That the butchers' medal be offered for the future the same as in 1835.

His Royal Highness, the Duke of Cambridge, was elected a member of the club.

1842. A new class for cross-bred sheep (long and short-wooled cross) was formed for 1843.

On the motion of the president, Earl Spencer, it was

Resolved, That in future the club's gold medals be given to the feeders, in place of the breeders.

That members who have paid twenty annual subscriptions be allowed to compound for five guineas.

"That no animal, the property of or bred by

any person who has been expelled from any agricultural society, or otherwise disqualified from exhibiting stock at any agricultural show in consequence of having been proved to have exhibited stock with a false certificate, shall be exhibited at the shows of the Smithfield Club, provided: that if the person so disqualified or expelled shall have taken any legal proceedings in consequence of such expulsion or disqualification and a court of law shall have decided in his favor, the case may be submitted to the club, who shall have the right to determine whether the above rule shall apply to his case."

1843. Humphrey Gibbs (H. Brandreth) resigned the office of honorary secretary.

B. T. Brandreth Gibbs elected honorary secretary.

Her Majesty, the Queen, and Prince Albert honored the club by visiting the show December 13.

Earl Spencer, the Duke of Richmond, and B. T. Brandreth Gibbs were elected trustees.

HEREFORD HIGH TOWN IN 1850, SHOWING THE OLD MARKET HOUSE.
(From an old print.)

Resolved, That in future a separate gold medal be offered for the best cow or heifer in the classes.

That no member who is more than one year in arrears of his annual subscription shall enjoy any of the privileges of the members of the Smithfield Club.

Lord Weston, a vice-president, died.

1845. The Earl Spencer died, having been president of the club twenty years. (¶ 68)

His Grace, the Duke of Richmond, K. G., was elected president.

On the motion of Philip Pusey, Esq., P. M., the following resolution was unanimously passed:

"That we desire to record our lasting gratitude to the late Earl Spencer for his long exertions in the service of the Smithfield Club, which were neither interrupted by the discharge of more arduous duties nor damped by the retirement from public life or by the advance of years; and also to express our affectionate veneration for his manly and noble character."

Resolved, That for the future, in the sheep classes which are restricted to weights, live-weights be substituted for dead-weights.

Lord Portman was elected a trustee, in the room of the late Earl Spencer.

A class for Scotch, Welsh and Irish cattle was re-established for next year. It was determined to offer a gold medal for the best pen of pigs in the classes.

Sir John S. Sebright, Bart., vice-president, died.

Earl Spencer elected vice-president.

1846. J. M. Cripps, Esq., a vice-president, died.

Resolved, That in future all restrictions as to feeding in the oxen and steer classes be done away with.

That in future non-members pay a fee of a guinea to the funds of the club to entitle them to exhibit.

That in future the judges be not made acquainted with the kind of food on which the animals have been fed.

That disqualified animals be removed from the yard before the public exhibition.

That in future the cross-bred sheep be judged by the short-wooled judges.

That in future no return of dead-weights be required from the exhibitors or the butchers.

1848. *Resolved,* That the restrictions on the feeding of sheep be done away with.

That the judges be not made acquainted with the traveling of the animals to the show.

That a new class be formed for short-wooled sheep, not Southdowns.

1849. *Resolved,* That in future exhibitors of stock be admitted to a private view of the show, viz., between eight and ten o'clock in the evening of the day of adjudication, provided the judges have signed their awards.

Thomas Gibbs, Esq., the father of the club, died.

1850. Her Majesty the Queen, and his Royal Highness, the Prince Albert, again honored the club by visiting the show.

Full power was delegated to the president to admit the royal family of this or other countries at whatever time he may consider expedient.

Resolved, "That, the club being anxious to evince and perpetuate its feelings of respect toward the Duke of Richmond, the president de-

HEREFORD OCTOBER FAIR.
Scene in Broad Street, October, 1881.

HEREFORD OCTOBER FAIR, 1901.
An institution centuries old, held in the new cattle market, Hereford.

sires that his Grace's profile should appear on the reverse side of the club's medals, and that his Grace be requested to sit to W. Wyon, Esq., R. A., of her Majesty's Mint, for the preparation of a die."

1851. *Resolved*, That in future the cattle be classified according to breed.

That all notices of motion be delivered to the honorary secretary on or before the last day on which certificates are to be received.

That a copy be sent by the secretary to each member stating the meeting at which such motion is to be considered, etc.

1852. The cattle were classified into distinct breeds, viz., Herefords, Devons, Shorthorns, Scotch, Welsh or Irish; other pure breeds, cross or mixed breeds.

A committee reported its recommendation to add poultry to the show, but it was subsequently found that the space would not admit of it.

1853. *Resolved*, That it shall be incumbent on the exhibitors and breeders of animals exhibited at the Smithfield Club's shows to prove the correctness of their certificates, if called upon by the stewards to do so.

Separate classes were formed for Welsh cattle. (¶ 69)

1854. His Royal Highness, Prince Albert, his Royal Highness, the Prince of Wales, and his Royal Highness, Prince Alfred, honored the show by visiting it.

Resolved, That in future the ages of animals be calculated to a fixed date, viz., the first of December.

That the judges give in a reserve number in each class, in case of prize animal being disqualified.

That in future no person be eligible to be elected a steward unless he shall have been a member of the club three years.

That in future steam engines, etc., may be exhibited down stairs.

1855. *Resolved*, That for the future, in extra stock long-wools, and also short-wools, there be a silver medal for the best wether sheep, and another for the best ewe.

That in future members who shall have paid their subscription for the current year be admitted to the private view by a special card.

That a silver medal be awarded to the breeder of each first-prize animal in the Scotch and Welsh classes, provided he has furnished the breeder's certificates.

That the outgoing stewards shall nominate

¶ 49

THE CELEBRATED FEEDER OF PRIZE STEERS, RICHARD SHIRLEY, AND FAMILY, OF BAUCOTT.
(House built in 1600.)

their successors to the honorary secretary previously to the annual notice paper being sent to the members, prior to the annual meeting; and that no member shall be considered eligible to be appointed unless he shall have been three years a member of the club.

A testimonial of plate was presented to B. T. Brandreth Gibbs, honorary secretary, at the annual dinner of the club.

Philip Pusey, Esq., a vice-president, died.

1856. The amount of prizes offered to the club was still further increased, and additional

¶ 50

YOUATT'S TYPICAL HEREFORD COW.
(Drawn from imagination. Reproduced from "Youatt on Cattle.")

separate classes added for Sussex, Norfolk, or Suffolk-polled, Longhorned, Scotch-polled and Irish cattle.

Resolved, That in future the award be not read at the dinner, except the portion relating to the gold medals.

That in future the certificates be lodged with the honorary secretary on a fixed day, viz., always November 1, except the first fall on Sunday, and then to be on Monday, the 2d.

1857. The thanks of the club were voted to the honorary secretary for his "History of the Club," and ordered to be recorded on the minutes.

That in future the club's accounts and statements thereof be made up the first of December; that the statements show the balance carried forward from year to year; and that the stewards audit the accounts up to the first of December annually; and that the statement be that of cash actually received and paid.

Earl Spencer, a vice-president, died.

1858. The Right Honorable Lord Feversham elected a vice-president. On the motion of Mr. John Giblett, a committee was appointed to inquire into the practicability of procuring a better and more commodious place for the Smithfield Club to hold its annual exhibitions.

That there be three additional judges, viz., three for cattle, three for long-wooled sheep and pigs, and three for short-wooled sheep and cross-bred sheep. Those for cattle and long-wooled sheep and pigs to be nominated by the stewards of cattle and long-wools.

That in future all the pigs shall have their dentition examined by a competent authority, previously to the judges making their awards, and if the dentition shall satisfactorily indicate that the age of any pig had been incorrectly returned in the certificate, the stewards shall disqualify such pig and report their having done so to the first meeting of the club, and that such disqualification shall be final and without appeal.

That there be two silver medals for extra stock cattle, viz., one for steers or oxen, and one for heifers or cows.

That the judge's award be not read in extenso, but laid on the table.

That no alteration be made in any of the implement stands, except under the actual directions of the stewards or secretary, both as to the articles to be exhibited and the arrangement of their stands.

1859. His Royal Highness, the Price Consort, visited the show.

The show yard committee reported on the site, etc., that had been offered for the club's show, including a proposition from the Crystal Palace at Sydenham. (¶ 70)

Mr. John Giblett having suggested the formation of a company, the show yard committee recommended, "That, if a responsible company be formed and adequate terms offered to the club, the club should lease their exhibitions for a term of not exceeding twenty-one years." The committee was empowered to further consider the subject and report again in May, 1860.

That a tabular statement of the amounts offered in prizes and the number of entries in the different classes during the last three years be printed and furnished to the members.

That no article (except agricultural books) exhibited in the implement galleries be allowed to be removed during the time the show is open to the public.

The Right Honorable Lord Walsingham elected a vice-president.

C. T. Tower, Esq. (the father of the club), elected a vice-president.

1860. Preliminary prospectus of the proposed Agricultural Hall Company issued, with Mr. Jonas Webb as chairman.

The show yard committee reported their recommendation in favor of the site known as "Dixon's lair," at Islington:

"That if a responsible company be formed who will enter into an agreement to erect a suitable building on this site to the satisfaction of the club, also to pay the club £1,000 ($5,000) per annum, and to enter into an arrangement on similar terms to those now made with Mr. Boulnois, the club shall lease their exhibition for a term of not exceeding twenty-one years, commencing 1862."

Several meetings were held on the subject of the removal of the show to the proposed new Agricultural Hall.

On the 17th of July the following resolution was carried: "That the report of the sub-committee appointed at the general meeting of December 9, 1859, to inquire into the practicability of providing a more commodious place for holding the annual exhibitions of the Smithfield Club, having been adopted at the special general meeting of the 22d of May, 1860, and the report of the legal arrangements committee, appointed on the same 22d of May, to conclude the terms of an agreement with the agricultural meeting held on the 6th of June, 1860, this meeting does in the fullest manner confirm those proceedings."

Power was given to the committee, or any two, to sign the agreement, and a copy of it ordered to be entered on the minutes.

His Grace, the Duke of Richmond, K. G., president of the club, died.

Resolved, That the meeting desires to record its deep regret at the irreparable loss the club has sustained by the decease of the late president, His Grace, the Duke of Richmond, K. G., who has, during a number of years, given the greatest attention to further its objects and promote its prosperity.

The Earl of Yarborough, a vice-president, died.

Her Majesty, the Queen, and his Royal Highness, the Prince Consort, visited the show, December 10, 1860.

Her Majesty, the Empress of the French, visited the show.

Resolved, That the president be elected for the term of one year, and that the said president be not eligible for re-election for the term of three years, and that this be the rule of the club for the future.

That Right Honorable Lord Berners, elected president for 1861, be the first of the annual presidents.

Lord Portman and Mr. Brandreth Gibbs resigned their offices of trustees, and the thanks of the club voted.

Charles Barnett, Esq., Thos. Greetham, Esq., and Samuel Druce, Esq., elected trustees.

Resolved, That all exhibitors of cross-bred animals shall be required to specify the exact nature of the cross—that is, the breed of sire and dam respectively, and whether the animal exhibited is the result of a first or more remote cross.

Prize sheet rearranged and fresh divisions and classes added, and prizes to amount of nearly £800 ($4,000) extra.

The following were added to the list of vice-presidents: Lord Berwick, the Honorable Col. Hood, the Earl of Leicester, the Duke of Richmond.

Lord Berwick, a vice-president, died.

1861. Lord Tredegar elected a vice-president.

Implement committee appointed to arrange as to allotment of space and determine the rules and regulations for the implement department.

Resolved, That the honorary secretary be requested to make the same arrangements as to catalogues that he has been in the habit of making with Boulnois, and that the entire arrangement shall rest with the honorary secretary of the club, who is to fix the maximum prices at which the catalogues are to be sold by the Agricultural Hall Company. (¶ 71)

¶51

YOUATT'S TYPICAL SHORTHORN COW.
(Drawn to flatter Rev. H. Berry. Reproduced from "Youatt on Cattle.")

1862. This show held at the Agricultural Hall for the first time.

His Royal Highness, the Prince of Wales, visited the show, accompanied by his Royal Highness, the Crown Prince of Prussia. His Royal Highness, the Duke of Brabant (now king of the Belgians), and his Royal Highness, the Prince Louis of Hesse; also his Royal Highness, the Duke of Cambridge; her Royal Highness, the Duchess of Cambridge, and her Royal Highness, the Princess Mary of Cambridge.

Prize sheet revised and £600 ($3,000) added to the prizes.

Resolved, That an implement catalogue be published on the same terms as the live stock catalogue.

That in future any member of the club who has duly served his three years as steward of

¶ 52

YOUATT'S IDEA OF A HEREFORD WORKING-OX. (Drawn from imagination. Reproduced from "Youatt on Cattle.")

the yard shall not be eligible for reappointment for the next six years.

That both the nomination and election of the new stewards be in the business of the general meeting, and that no retiring steward have the especial right of nominating his successor.

On the motion of Major-General, the Honorable A. N. Hood, it was resolved:

1st, That the management of the Smithfield Club be vested in a council consisting of the President, Vice-Presidents, Trustee, Honorable Secretary and twenty-four members.

2d, That one-third of the twenty-four members shall go out annually by rotation, and not be re-eligible for one year.

3d, That the council shall prepare rules and regulations for the management of the club, and shall submit them for consideration and adoption at a special general meeting of the club in May, 1863.

His Royal Highness, the Prince of Wales, was elected a member of the club.

Two silver cups, value £40, for cattle; three silver cups, £20, for sheep, offered in lieu of gold medals, also separate silver medals for ewes in extra stock.

1863. On March 5th, the council of the club held its first meeting and considered the draft of by-laws as preliminarily prepared by the honorary secretary.

Resolved, That the honorary secretary be empowered to appoint an assistant secretary, at a salary of 50 guineas per annum; that the appointment, removal and control of such assistant secretary shall rest entirely with the honorary secretary.

Resolved, That one month shall be added to the ages of sheep, viz., to be under twenty-three months and under thirty-five months, instead of twenty-two months and under thirty-four months.

That a silver cup, value £10, be offered in lieu of the butcher's medal.

By-laws of the club finally discussed and agreed to.

The private view arranged to take place on the Monday afternoon of the show, and the public to be admitted at 5s ($1.25) each.

Resolved, That every member of the council be invited to send the names of fitting persons for judges in each particular class.

That the council shall select the judges.

That the council shall have power to add names to the list.

That the members of council sending in any names shall first ascertain that the respective parties are willing to act as judges.

That the names be delivered to the honorary secretary on or before the 1st of November.

Resolved, That there be fifteen judges, viz., six for cattle, six for sheep and three for pigs.

Resolved, That all the judges of cattle shall be joined to decide the adjudication of the silver cups, and in case of equality of votes, the stewards shall call in a judge for umpire out of one of the other divisions.

That the six judges of sheep shall be joined to decide the last cup named on the prize sheet, and in case of equality of votes, the stewards shall call in an umpire, as in the case of cattle.

A special divine service given for the first time by the Vicar of Islington for the men in charge of live stock.

The first report of the council laid before the general meeting.

Two stewards of implements appointed.

Humphrey Brandreth, Esq. (formerly honorary secretary), elected a vice-president.

1864. *Resolved,* That members of the club and exhibitors be admitted to the galleries during the judging.

That a framed diploma and a sovereign be given to the man (to be named by the exhibitor) who has had charge of the stock winning first prize in each class. The ages in the class for steers were rearranged. (See prize sheet.)

The marking of the sheeps' ears was discontinued.

Resolved, That all protests against animals exhibited at the club's show must be delivered in before six o'clock P. M. on the Tuesday of the show, and that no protests be received after that time.

Fines were instituted for the non-exhibition of animals that had been entered.

Resolved, That the club will not, in any case, or under any circumstances, hold itself responsible for any loss, damage, or mis-delivery of live stock or article exhibited at the club's shows.

Rosettes placed over winning animals.

Resolved, That for the information of the agricultural and such other journals as may wish to publish the awards the same evening, the winning numbers be posted on a placard in the gallery from time to time as the judges proceed.

That the reporters of the press generally shall obtain the complete award by applying at the honorary secretary's office at the hall at 3 o'clock, when a clerk will be in attendance to read over the numbers, so that the reporters may mark their catalogues.

Veterinary Inspector's fee raised to £15.

Judge's fee raised to £7.

Freemartins to be allowed to compete in the heifer classes.

The Marquis of Huntley, vice-president, died.

Humphrey Brandreth, Esq., vice-president, died.

1865. *Resolved,* That the stewards be paid a fee of £10 each.

That a silver cup be substituted for the gold medal hitherto given for the best pen of pigs.

That a gold medal be given in lieu of a silver medal for the breeder of the best ox or steer, also to the breeder of the best cow or heifer in any of the classes.

That there be three butcher's cups—one for beasts, one for sheep, and one for pigs.

The date of the show was put a week later than usual in consequence of the cattle plague necessitating early slaughter after the show.

Various stringent rules were enacted in order to guard against the plague, animals exhibited at any show within a month being excluded; veterinary certificates being required to be sent with the animals, conveyances being required to be disinfected; animals to be examined by a veterinary inspector before being admitted to the show; constant attendance of veterinary inspector night and day, etc.

Resolved, That this year the show shall close on the Thursday evening instead of the Friday.

Resolved, That the stewards of live stock and implements, not already members of the council, be so ex-officio, during their terms of stewardship.

Resolved, That it is the opinion of the general meeting of the Smithfield Club, held this 12th day of December, 1865, that it is the duty of the government, under the formidable visitation by which this country has been afflicted, to issue such orders for the regulation of the cattle trade in Great Britain as may be necessary to check the extension of the cattle plague, so that the practice may be uniform throughout the country; also that this meeting concurs in the spirit of the recommendations made to the Privy Council by the Council of the Royal Agricultural Society of England in reference to the cattle plague, and begs to press on the government the extreme importance and urgency of its taking immediate steps to insure uniformity of action throughout the country.

1866. *Resolved,* That the condition disqualifying spayed heifers be struck out.

That the butchers' cups be discontinued.

That the rule excluding animals exhibited elsewhere within a month be rescinded for this year.

That all the other rules and regulations in reference to cattle plague be enforced as last year, with the exception of that altering the date of the show.

That no steps be taken by the Council of the Smithfield Club to obtain a relaxation of the orders of Privy Council in reference to the cattle plague.

¶53

YOUATT'S HEREFORD FEEDING OX.
(Drawn from imagination. Reproduced from "Youatt on Cattle.")

That a deputation consisting of the president and honorary secretary shall wait on the Privy Council to advocate the following recommendations:

(¶ 73) 1st. That no fair or market for store-stock should be opened before the 1st of April.

2d. That the greatest care should be taken that the regulations as to quarantine be rigidly carried out.

3d. That permanent lairs be established at the ports of debarkation.

1867. *Resolved,* That a list of the winners of the gold medals and silver cups at the past shows of the Club be printed at the end of the annual Prize Sheets, the same to be a reprint and continuation of the tabular statement in the appendix to the honorary secretary's "History of the Smithfield Club."

That a list of the members of the club be printed at the end of the Prize Sheets.

That none but the official placards respecting the food on which the animals have been fed be allowed to be placed over the heads of the animals in the hall.

¶ 54 -

YOUATT'S ORIGINAL DURHAM COW.
(Drawn from imagination. Reproduced from "Youatt on Cattle.")

Lord Feversham, vice-president, and C. T. Tower, Esq., vice-president and father of the club, died; also Thos. Greetham, a trustee, died.

The club entered into arrangements with the Agricultural Hall Company respecting various alterations in the hall, by which on the one hand the club had to pay the Hall Company £1,000 ($5,000) in consideration of the enlargement of the galleries; building of a dining room, in which the club's annual dinners are to be held; store room for animal's food, and construction of a new building for the pigs, and other improvements.

On motion of the President (Major-General Hood),

Resolved, That £15 be given for a report on the animals exhibited at the club's show.

The Duke of Marlborough and the Earl of Powis elected vice-presidents.

The president was requested to lay the following resolutions before the Privy Council as recommendations from the Council of the Smithfield Club:

That all foreign stock be slaughtered at the place of landing.

That sheep be included in the present quarantine regulations for store stock.

Mr. H. H. Dixon was appointed to write the report of the present year's show.

Resolved, That in future each member of the Council shall not nominate more than one person to each division of judges, and that the conditions requiring that the nominators shall have first ascertained their willingness to act be remitted.

Her Majesty the Queen exhibited for the first time in her own name.

His Royal Highness, the Prince of Wales, exhibited for the first time.

A donation of £25 ($125) was received from Her Majesty the Queen in lieu of that of £5 ($25) hitherto annually paid by Major-General, the Hon. A. N. Hood.

SMITHFIELD CLASSIFICATIONS.

(¶ 74) The following are the Smithfield Show classifications made on cattle from the organization of the society to 1835:

(Note.—A stone was reckoned at 14 lbs. for live weight and 8 lbs. for dressed or dead weight, "sinking the offal." In live weight 2 stone equaled a quarter or 28 lbs. Four quarters made one hundred weight (cwt.) or 112 lbs.)

1799. Oxen or steers. Class for beasts fed on oil-cake or corn, class for beasts fed on grass feed, etc., class for beasts under 24 score (480 lbs. dressed or 840 lbs. live weight).

1800. Oxen or steers. Class for beasts under 150 stone (2,100 lbs.), above 80 (1,120 lbs.), grass fed, etc. Class for beasts under 150 stone (2,100 lbs.) above 80 (1,120) fed on oil-cake, etc. Class for beasts above 150 stone (2,100 lbs.), grass fed. Class for beasts above 150 stone (2,100 lbs.), cake or corn fed.

1805. Oxen or steers. Class for 160 stone (3,240 lbs.) or upwards, not to have cake or corn before the 5th of April, 1805. Class for 100 stone (1,400 lbs.), not to have had cake or corn. Class for under 100 stone (1,400 lbs.), not to have had cake or corn. Class for oxen or cows that have gained the greatest weight from the 1st of April or later, to 1st December, grass fed.

Cows. Class for fat cows, must have had three calves at least.

1810. Oxen or steers. Class for any age, classified separately, as according to breed, as follows: Herefords, Longhorns, Shorthorns, Sussex or Kent, Devons. Class for any breed, 5 years, above 100 stone (1,400 lbs.). Class for any breed, 5 years, under 100 stone (1,400 lbs.). Cows. Class for fat cows, must have had three calves.

1815. Oxen or steers. Class for any age, classified separately, according to breed, as follows: Herefords, Sussex or Kent, Devons, Longhorns, Shorthorns. Class for any breed under 5 years, 100 stone (1,400 lbs.) or upwards. Class for any breed under 100 stone (1,400 lbs.). Bedfordian plate for oxen, 110 stone (1,540 lbs.), or upwards. Bedfordian plate for oxen under 110 stone (1,540 lbs.).

Cows. Class for fat cows that have had at least three calves.

1820. Oxen or steers. Class for any breed under 36 months. Class for any breed, 160 stone (2,244 lbs.), not exceeding 5 years. Class for any breed above 120 stone (1,680 lbs.) and under 160 (2,240 lbs.), not exceeding 5 years. Class for any breed under 120 stone (1,680 lbs.), not exceeding 4 years. Bedfordian plate for 130 stone (1,820 lbs.) and upwards. Bedfordian plate for under 130 stone (1,820 lbs.).

Cows. Class for any breed of cows that have calved twice.

1825. Oxen or steers. Class for any breed under 36 months. Class for any breed 160 stone (2,240 lbs.) or upwards. Class for any breed under 160 stone (2,240 lbs.) and above 110 (1,540 lbs.). Class for any breed under 110 stone (1,540 lbs.).

None of the above to have had cake, etc., previous to the September twelve-months preceding.

Cows. Class for cows that have calved twice at least. Class for cows not spayed.

1830. Oxen or steers. Class for any breed under six years. Class for any breed or age, 160 stone (2,240 lbs.) or upwards, that have had no cake, corn, etc., before August 1st, 1830. Class for any breed under 160 stone (2,240 lbs.) and above 120 stone (1,680 lbs.), no cake, etc., as above. Class for any breed under 120 stone (1,680 lbs.), no cake, etc., as above.

Cows. Class for cows that have calved twice at least. Class for cows of any age, must have calved once in years 1828-29, and not been dried last time previous to November 1st, 1829. Class for cows or heifers not eligible as above, freemartins and spayed heifers not qualified.

1835. Oxen or steers. Class for any breed under 5 years. Class for any breed under 6 years, 90 stone (1,260 lbs.) and upwards, no cake, etc., previous to August 1st, 1835. Class for any breed under 5 years, under 90 stone (1,260 lbs.) and above 70 stone (980 lbs), no cake, etc., as above. Class for any breed not more than 4 years, 3 months, under 70 stone (980), no cake, etc., as above.

Cows. Class for cows under 5 years. Class for cows 5 years and upwards. Freemartins and heifers not qualified.

LORD BATEMAN, 1826-1901, LORD LIEUTENANT OF HEREFORDSHIRE, 1852-1901, THE CELEBRATED BREEDER OF HEREFORD CATTLE.

CHAPTER IX.

Contemporary Reports of Smithfield Club Matters

We find that considerable light can be shed on the cattle history of the early years of the nineteenth century by giving what was published in the agricultural publications of that day.

The Smithfield Cattle and Sheep Society, gathered in London, Dec. 13, 1799, appears to have been a meeting for general business, and the following classification was published:

Class 1. Beasts fed on grass, hay, turnips, cabbages, or other vegetables, under the weight of 150 stone but above 80 stone, a piece of plate not exceeding £15 ($75) for the best, and a piece of plate not exceeding £10 ($50) for the second best.

Class 2. Beasts fed on oil-cake, corn, or any other food except grass, hay and vegetables, same premium.

Class 3. Beasts fed as Class 1, above the weight of 150 stone (2,100 lbs.), a piece of plate not exceeding in value £20 ($100) for the best, and a piece of plate not exceeding in value £15 ($75) for the second best.

Class 4. Beasts fed as Class 2, and the same weight as Class 3, same premiums as Class 3.

"That the cattle be brought, on the following or any other conditions that the committee may think proper, to a place fixed upon by the committee on Thursday preceding the Christmas market day for the purpose of being accepted by the judges on the day succeeding, who will be directed to select the two best from each class.

"Those selected to remain at the place fixed upon by the judges, for the purpose of their being exhibited to public view, and that they be exhibited accordingly on Saturday, Monday and Tuesday immediately following the day of selection, when they shall be returned; but it will be expected that the proprietors shall agree with whomsoever they may sell them to, that the judges shall have full power to inspect the killing for the purpose of ascertaining the weight of each animal slaughtered, distinguishing the fore and hind-quarters, the tallow, hide, offal, including blood, etc.; or by some other satisfactory mode of ascertaining the comparative live and dead weight, and having ascertained such points the judges are then to decide on which is the first and which the second best in each class, and make their report to the committee.

"*Resolved,* That one of the conditions be, that each candidate shall at the time of showing, produce to the judge a paper, signed by himself, containing an account of the breed, age, time of fattening, sort of food, and time they have eaten cake or corn, etc.; also the name of the breeder in case where it can be known.

"*Resolved,* That no candidate shall be entitled to two prizes in the same class.

"*Resolved,* That the committee be instructed to direct the judges to take particularly into their consideration age of the sheep, the society being of the opinion that early maturity is a merit.

"*Resolved,* That the committee be instructed to advertise the premiums in the following papers once: County Chronicle, Bath, Hereford, Lewes, Leicester, Stamford, York, Northampton, Cambridge Intelligencer, Oxford, Canterbury, Sherburn, Ipswich, Reading and Warwick, on the first day of publication in the month of October, 1800.

"*Resolved,* That it be left to the committee to fix a price to be paid for admission to see the different prize cattle and sheep.

"*Resolved,* That the committee be instructed to dispose of the surplus arising from the subscribers, shows, etc., in such manner as they think fit, rendering such disposal public, and if there is still a surplus after their object is accomplished to pay it over to the committee for the year ensuing.

"*Resolved,* That the committee be instructed to have a dinner provided on the Christmas market day, subject to such regulations as they may deem proper.

"*Resolved,* That the committee be instructed to draw up a report of the whole of their proceedings, including their receipts and expenditure, at as early a period as they conveniently can, and to make the same public in the cheapest manner."

The first premium equal to $75 for the best beast in the show and another equal to $50 for the best ox fed on oil-cake, was awarded to Mr. Westcar, on a Hereford. A prize equal to $25 was awarded to the Duke of Bedford's Hereford for the most complete beast under the weight of 48 score (960 lbs., dressed, or 1,680 lbs. live weight), fed on oil-cake and grain. The Duke also gained $75 for the best beast fattened on grass and hay only. A prize of £10 ($50) was awarded to Mr. John Edmonds, of Welford, for the second best ox fed on grass and hay only, and £5 ($25) to Mr. Ellman, of Glynd, for the best ox fattened on grass and hay only in the shortest time from the yoke.

The following are sample certificates presented with cattle exhibited. The Duke of Bedford's reads:

"Dec. 12, 1800.

"This is to certify that my two oxen were purchased of the breeders, lean from the yoke, on the 15th of March, 1800; and the Herefordshire ox was bred by Samuel Patrick, of Middleton, near Ledbury, and the Shropshire ox was bred by Thos. Bishop, of Moor, near Ludlow; both were five years old, and have been fattened with grass and hay only, and have not eaten any hay before the 17th of November last."

Another reads:

"This is to certify that the two Sussex shown by me for the prizes given by the Smithfield Society were bred by Mr. John Ellman, of Glynd, Sussex; were eight years old last spring and worked constantly until the last week in May, 1799, and have been fed on grass and hay only until the 14th of December, 1799, by Mr. John Ellman; since that time by Henry King, making together nineteen months from the yoke.

"Witness our hands this 12th day of December, 1800.

JOHN ELLMAN,
HENRY KING."

At the Christmas market at Smithfield, Dec. 21, 1800, Messrs. Hixcock and Farrow made a show of beasts that was never equaled or exceeded in the kingdom. Their largest bullock, a real Herefordshire one, was fed by Mr. Grace, of Buckinghamshire, and on account of his very extraordinary bulk and fatness, was conveyed to London by water. He was seven feet high,

MR. WM. PRICE ON HIS FAVORITE HORSE AT "THE VERN," ONCE OCCUPIED BY JOHN HEWER.
"The surface of the country is undulating in long ridges."

weighed upwards of 260 stone (3,640 lbs.) and measured in the girth twelve feet, four inches. The other killed by them was a real Glamorganshire, grazed by Mr. Woodman, of Buckingham, and weighed 220 stone (3,080 lbs.), and was the fattest ever seen on the ribs and sirloin.

A prize Hereford ox was purchased at Smithfield in 1800 by Mr. Chapman for $500, and exhibited to public inspection in the Fleet market on Wednesday. This fine animal was fed by Mr. Westcar, of Buckinghamshire, and weighed nearly 300 stone (4,200 lbs.), was eight feet eleven inches long, six feet seven inches high and ten feet four inches around the girt. He carried the first prize at the Smithfield Show of Cattle.

SMITHFIELD CATTLE SHOW, 1801.

Dec. 12, the judges met and examined certificates of the exhibitors who had cattle to show. Mr. Westcar, for two oxen fed on cake, six years old, one bred by Mr. Tully, the other one by Mr. Holman; have eaten cakes since last September, at Ledbury, March 17, 1799; also, two others fed on grass and turnips, bought at Hereford, October, 1799, bred by Mr. Williams, of Thinghill.

Class 1. Several cattle were exhibited by Mr. Westcar who took the first prize on a Hereford ox, and the Duke of Bedford second prize on a Hereford ox.

Class 2. The character of the cattle in this class was not sufficient in the opinion of the committee to be awarded a premium.

Class 3. Mr. Westcar's white Hereford took first premium.

1802. The following is a report of the judges of the Smithfield Cattle and Sheep Club, Dec. 8, 1802:

We, the judges, appointed by the Smithfield Club to examine and report the merits of the cattle, sheep and pigs shown for the prize of this year, having received and read the certificate of the several candidates and duly considered the instructions this day received from the club, do adjudge:

Class 1. First prize in this class to Mr. Westcar, second prize to the Duke of Bedford, for Hereford oxen.

Class 2. The first prize to Mr. Westcar for the Hereford ox; second prize to Mr. Ladds.

Class 3. Only the Duke of Bedford's French ox shown—if a prize must be given in this class the Duke of Bedford is entitled to it.

"SAY WHEN." MR. J. H. ARKWRIGHT AND SON TROUT FISHING AT HAMPTON COURT.

Class 4. For cows, the first prize to the Duke of Bedford.

The following certificate shows that Hereford oxen of the Tomkins sort were strong competitors at the early shows:

Class 1. This is to certify that the Duke of Bedford's Hereford ox just shown for a prize offered by the Smithfield Society (winner of second prize in Class 1) was six years old when put to fattening, was bred by Mr. Tomkins in the County of Hereford, was put to fattening the 1st day of May, 1801, being then respecting flesh very poor, having been till then at hard work; has been fed on cakes, turnips and hay.

Signed EDMUND CARTWRIGHT,
JOHN CLAYTON.

Class 2. The following is the certified account of the food given to Mr. Westcar's grass-fed Hereford ox:

1802.	Tankard turnips per day.	Hay per day.	Hay taken up.
Oct. 1st to 8th......	108 lbs.	21 lbs.	½ lb.
Oct. 8th to 15th......	108 lbs.	21 lbs.	½ lb.
Oct. 15th to 22d......	108 lbs.	21 lbs.	½ lb.
	Swedish Turnips.		
Oct. 22d to 29th.....	94 lbs.	20 lbs.	½ lb.
Oct. 29th to Nov. 5th	94 lbs.	20 lbs.	½ lb.
Nov. 5th to 12th....	94 lbs.	20 lbs.	¾ lb.
Nov. 12th to 19th....	94 lbs.	20 lbs.	¾ lb.
Nov. 19th to 30th....	94 lbs.	20 lbs.	¾ lb.
Total	5934 lbs.	1221 lbs.	36¼ lbs.

We, whose names are hereunto subscribed, do certify that the ox Mr. Westcar shows for the grass-fed prize, was purchased by him of Mr. Williams, of Thinghill, March 20, 1801, directly from work, and in store condition.

THOS. HEDGES.
ROBERT BYNG.

An account of the food given to the cake-fed ox, Mr. Westcar's:

1802.	Cakes per day.	Turnips per day.	Hay per day.	Hay taken up.
Oct. 1st to 8th......	7 lbs.	24 lbs.	20 lbs.	1 lb.
Oct. 8th to 15th......	7 lbs.	24 lbs.	20 lbs.	1 lb.
Oct. 15th to 22d....	8 lbs.	17 lbs.	19 lbs.	1 lb.
Oct. 22d to 29th.....	8 lbs.	17 lbs.	19 lbs.	1 lb.
Oct. 29th to Nov. 5th	9 lbs.	10 lbs.	18 lbs.	1¼ lbs.
Nov. 5th to 12th....	9 lbs.	6 lbs.	18 lbs.	1¼ lbs.
Nov. 12th to 19th...	9½ lbs.	2 lbs.	17 lbs.	1½ lbs.
Nov. 19th to 30th...	10 lbs.	2 lbs.	17 lbs.	1½ lbs.
Total	512½ lbs.	722 lbs.	1036 lbs.	72½ lbs.

We, whose names are hereunto subscribed, do certify, that the ox of Mr. Westcar shown for the cake-fed prize, was purchased by him of Mr. Tully, of Huntington, near Hereford, March 30th, 1801, directly from work in store condition, and that the said ox never had any cake except in the year 1802.

ROBERT BYNG,
THOMAS HEDGES.

The dressed beef of Mr. Westcar's cake-fed Hereford ox (1802), bred by Mr. Tully, was as follows:

	Stone.	Lbs.		
Fore-quarter	72	1	= 577	pounds
Hind-quarter	65	2	= 442	"
One side	137	3	=1019	pounds
W't of whole body	274	6	=2038	pounds

Mr. P. Giblett, the celebrated London butcher, gives the following particulars of Mr. Westcar's grass-fed prize ox of 1802:

	Stone.	Lbs.		
Carcass	225	6	=1,806	pounds.
Fat	28	6	= 230	"
Hide	15	2	= 122	"
Liver	1	5	= 13	"
Entrails, not emp'd	22	7	= 183	"
Pluck	1	6	= 14	"
Head	5	3	= 43	"
Feet	4	1	= 33	"
Tongue	1	2¾	= 10¾	"
	306	6¾	=2,454¾	"

SHOBDON COURT, SEAT OF LORD BATEMAN.

Smithfield Show, Dec. 13, 1803. Certificate Admitted. Class 1. This is to certify that the two oxen I showed for the prize offered by the Smithfield Society were bred by Mr. Tully, of Huntington, near Hereford, of whom I purchased them the 12th of April, 1802, directly from work and in store condition. And I also certify, the said oxen never ate any corn of any description while in my possession, nor any oil-cake, except in the present year, 1803.

JOHN WESTCAR.

Creslow, Dec. 3, 1803.

Witnesses to the above:
THOMAS HEDGES.
JOHN ROADS.

Particulars of the food eaten by the large ox: 570½ cakes and 919 pounds hay.

Particulars of food eaten by the smaller ox: 315 cakes, 1,302 pounds turnips, 1,005 pounds hay.

Mr. Grace, two oxen cake-fed and two grass-fed, the grass-fed five years old, and the cake-fed seven years old, bred by Mr. Farmer, of Weobly, Herefordshire, and by Mr. Holmes, of Hereford.

THOMAS GRACE.

Premiums awarded as follows: To Mr. Westcar, first premium of $60 for beast under 150 stone (2,100 lbs.), fed with grass and hay, on a Herefordshire ox.

To Mr. Edmonds, first premium of $75 for beast above 150 stone (2,100 lbs.), on a Hereford ox.

To Mr. Westcar, first premium of $75 for best beast above 150 stone, fed with cake and corn, being a Hereford ox.

To Mr. Grace, for second premium of $50 for Hereford ox in the same class.

The dressed weight of one of Mr. Westcar's oxen was as follows: The four quarters weighed 1,952 pounds, one sirloin and rump weighed 240 pounds, six of his fore ribs weighed 176 pounds, buttock 128 pounds, his leg 20 pounds, his head 28 pounds, his tongue 14 pounds; bought by Mr. Chapman, of Fleet market.

Two oxen were shown by His Majesty at Smithfield, 1802. One of them was taken from work Oct. 17th, 1799; he was seven years old, and worked three and one-half years; he was fed on hay and grass only, except on a few potatoes for a few weeks prior to his going to London; the other was five years old, and worked nearly two years, and was taken from work Sept. 12th, 1799, was fed with grass and hay only, except on a few potatoes for about five weeks; both were bought for the King by Passey, in Herefordshire.

SMITHFIELD SOCIETY, 1804.

Certificates of Mr. Westcar's brown ox, oil-cake fed:

Carcass weight	1,674	pounds.
Fat	204	"
	1,878	pounds.

Mr. Westcar's dark brown grass-fed ox:

Carcass weight	1,626	pounds.
Fat	193	"
	1,819	pounds.

A SCENE AT "THE WHITTERN," KINGTON, PROPERTY OF MR. R. GREEN.
"Ash and oak coppices clothe its hillsides."

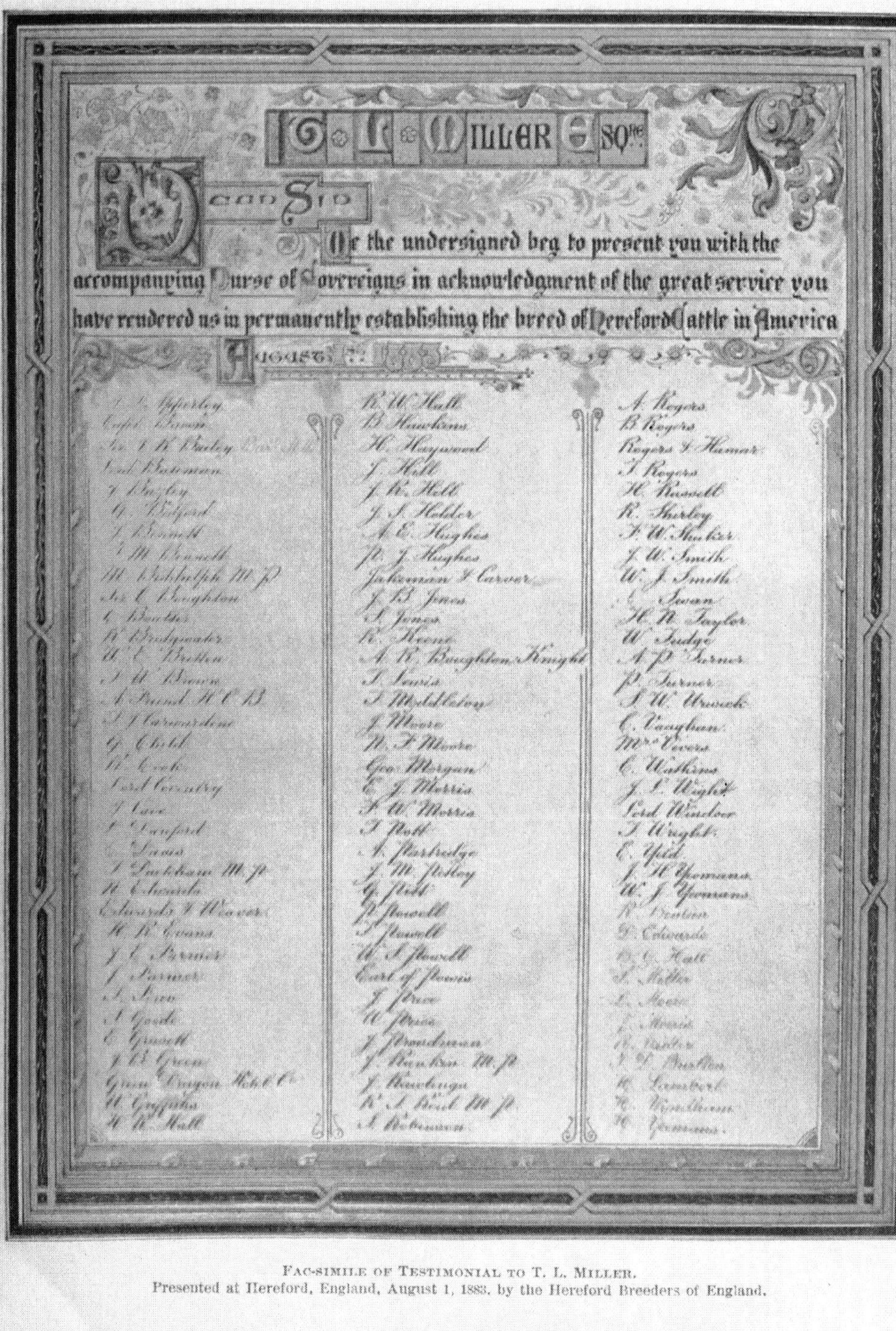

T. L. Miller Esqre.

Dear Sir

We the undersigned beg to present you with the accompanying Purse of Sovereigns in acknowledgment of the great service you have rendered us in permanently establishing the breed of Hereford Cattle in America

August 1st 1883

[illegible] Apperley	R. W. Hall	A. Rogers
Capt. [illegible]	B. Hawkins	B. Rogers
Sir J. R. Bailey Bart. M.P.	H. Haywood	Rogers & Hamar
Lord Bateman	J. Hill	T. Rogers
J. Bayley	J. R. Hill	H. Russell
G. Bedford	J. J. Holder	R. Shirley
J. Bennett	A. E. Hughes	T. W. Stalker
J. M. Bennett	P. J. Hughes	J. W. Smith
W. Biddulph M.P.	Jakeman & Carver	W. J. Smith
Sir C. Boughton	J. B. Jones	[illegible] Swan
C. Bowlter	T. Jones	H. R. Taylor
R. Bridgwater	R. Keene	W. Tudge
W. E. Britten	A. R. Boughton Knight	A. P. Turner
J. W. Brown	T. Lewis	P. Turner
A. Brand K.C.B.	T. Middleton	T. W. Urwick
J. J. Carwardine	J. Moore	C. Vaughan
G. Child	W. F. Moore	Mrs. Vevers
R. Cook	Geo. Morgan	E. Watkins
Lord Coventry	E. J. Morris	J. L. Wright
J. Cave	T. W. Morris	Lord Windsor
F. Danford	T. Nott	T. Wright
[illegible] Davis	A. Partridge	E. Yeld
J. Duckham M.P.	J. M. Pistley	J. H. Yeomans
H. Edwards	G. Pitt	W. J. Yeomans
Edwards & Weaver	P. Powell	R. Denton
H. R. Evans	T. Powell	D. Edwards
J. E. Farmer	W. T. Powell	G. G. Hall
J. Farmer	Earl of Powis	J. Miller
J. Fenn	J. Price	E. Moore
J. Goode	W. Price	J. Morris
E. Gravell	J. Pondeman	R. Butler
J. B. Green	J. Rankin M.P.	T. D. Burlton
Green Dragon Hotel Co.	J. Rawlings	W. Lambert
W. Griffiths	R. T. Reid M.P.	R. Wintham
H. R. Hall	A. Robinson	H. Yeomans.

FAC-SIMILE OF TESTIMONIAL TO T. L. MILLER.
Presented at Hereford, England, August 1, 1883, by the Hereford Breeders of England.

SMITHFIELD SOCIETY, 1805.

Mr. Westcar's prize ox, fed on oil-cake:

Carcass weight	1,988	pounds.
Fat	216	"
	2,204	pounds.

Duke of Bedford's grass-fed ox:

Carcass weight	1,061	pounds.
Fat	192	"
	1,253	pounds.

Mr. Westcar's grass-fed ox:

Carcass weight	1,439	pounds.
Fat	163	"
	1,602	pounds.

The Fourth Sheep Shearing Gathering of the Duke of Bedford commenced on Monday, June the 18th, 1800. A large attendance was on the ground and the meeting occupied four days; the time was given almost entirely to the shearing exhibition, sale and renting of sheep on Monday. There dined at the Duke's table 160 persons; in the steward's room, 60, making 220 who dined on Monday, and about the same proportion, 245, on Tuesday; 244 on Wednesday, and 138 on Thursday.

The Duke offered and awarded large premiums to those who expended the largest sums in the purchase of pure-bred sheep in the county of Bedford.

On the Wednesday, after dinner, Mr. Westcar, of Creslow, Bucks, informed the Duke, in the hearing of the whole company, that Mr. Tully, of Herefordshire, and his friends would produce twenty Hereford oxen at Woburn this time twelvemonth, fresh from the yoke, for one hundred guineas, against twenty of any one breed in England, the Duke of Bedford to appoint the judges. This offer of a bet was instantly accepted by Sir Thomas Carr, of Bedingham, in Sussex, who offered to produce twenty Sussex oxen against them.

A friend of Sir Thomas Carr's here interfered, requesting explanations, which, in the opinion of very many, did not appear necessary for a plain bet explicitly accepted.

A conversation ensued upon fattening the oxen, and various other circumstances, on which Mr. Westcar (who declared that Mr. Tully would bet on any fair conditions) retired in order to offer two propositions in favor of Sir Thomas Carr, that if he liked the second better than that he had already accepted, he might be indulged with it. This proposed the following extension of the bet, should Mr. Thomas Carr wish it, which, before reading, was explained to be distinct from the first proposition, and not annexed to it as a necessary condition.

"And that such oxen may be fatted and produced at the Christmas following at Smithfield, for a second hundred guineas, value to decide the superiority."

Here a fresh debate ensued. Mr. Ellman, of Glynd (Sir Thomas Carr's friend), proposed an explanation by adding these words: "Disposition to fatten and quality of flesh to decide the superiority *without size* being a chief object of consideration."

SCENE ON THE RIVER ARROW, COURT HOUSE. FARM OF JOHN PRICE.

"Its low lands are often called the 'Garden of England.'"

This Mr. Westcar rejected, observing that if size was thrown out of the question, twenty runts might be shown against Herefords of two hundred stone (2,800 lbs.).

Propositions and explanations being multiplied and mixed with conversation, the original acceptance of the bet slipped from attention, and the whole was eluded, but not without the Sussex breed suffering somewhat in the reputation, as it was deemed all escape on that side.

One bet, however, was clinched. Mr. Tench, of Broomfield, near Ludlow, in Salop, offered to show a Hereford bull against any bull in England for one hundred guineas, which was accepted by Mr. Knowles, of Nailstone, in Leicestershire, immediately entered, viz.:

"Mr. Tench bets Mr. Knowles one hundred guineas that he shows a Hereford bull against any Leicester bull to be produced at Shifnall, the Monday fortnight after the great show at

Smithfield. The Duke of Bedford to appoint judges.

ROBERT TENCH.
SAMUEL KNOWLES."

The above trial came off and the Hereford won.

The editor of the "Annals of Agriculture," writing in 1800 of a great friend of the Hereford and his sheep shearing show, says: "I might expatiate on the husbandry of the Duke of Bedford and the uncommon improvement he has made even since the last sheep show. The show house for the tups is admirably contrived, but these and various other articles highly interesting I reserve for that register of observations made at Woburn which has been drawn up on a former occasion, and which yet waits for some drawing not completed.

"The meeting passed off to the satisfaction of all present, and it was an animating circumstance to see the lovers of agriculture assembled together from countries so remote, from the most distant provinces of Ireland, from Germany, and from Switzerland. Several came expressly for this purpose about five hundred miles. It shows the spirit with which agriculture is at present prosecuted; it marks the genius of the age; it presages (may the providence of the Almighty permit) the future prosperity of this flourishing empire.

"To see a prince of the royal blood and many great lords sit down to the same table and partake of the conversation of the farmer and the breeder; to see all animated with the spirit of improvement, and listening with delight to the favored topic of the plough, is a spectacle worthy of Britain, and in her blest isle alone to be beheld: *Esto perpetua.*

"The conversation throughout the meeting was entirely agricultural, as it has been on every former occasion, opinions of stock, of cultivation, discussed, facts related, ideas sported, questions debated, bets proposed, and emulation active and promoted.

"He little knows the secret springs that move the public good who does not see the excellent effects that must flow from prejudices being worn away by the attrition of contrary sentiments, by exhibitions of superior stock being examined and compared, and by the sphere of rural knowledge being thus extended.

"The Duke announced premiums for the year 1801. May the new century open auspiciously to the plough; may the spirit of this sheep shearing improve the flocks of Britain; may her

HEREFORDSHIRE PEASANTRY. CHEAP LABOR FOR THE REARING OF LARGE CROPS.

fields smile again with ample harvests; her wastes by a general enclosure covered with cultivation; her farmer rich; her poor well fed and happy, and may we all, by reverence of that being from whom all blessings flow, endeavor to deserve them."

THE DUKE OF BEDFORD'S SHEEP SHEARING IN 1801.

On Monday 255 people attended, on Tuesday 259, on Wednesday 232, on Thursday 132. The sale and rental of sheep made a very satisfactory progress. Liberal premiums were paid to farmers of the county for the best cultivated farms and for the largest improvement of the sheep, and on different classes of farm machinery.

The Duke of Bedford died in the spring of 1802. The Sheep Shearing Shows at Woburn bearing his name were continued in 1803 by his son.

Those who attended this meeting, hitherto so bright and cheerful, animated as it was by the enlivening presence of a nobleman so greatly beloved and respected, looked around on every scene with heavy eye and sorrow in their hearts. There hung a gloom over the whole business, which would not dissipate. The succeeding Duke had given orders for conducting everything exactly as on former occasions. Lord Somerville presided at the dinners, and the general arrangement of business was conducted by his lordship, and Mr. Coke, with the assistance of the Rev. Mr. Cartwright.

The company was very numerous all the three days and on Tuesday equal to the appearance on any former occasion. In addition to the sale of sheep there was a sale of Hereford cows and heifers, including two bulls, ten head, all of the Herefordshire breed. The sale amounted to £974 ($4,870), an average of $487 a head. Premiums were awarded and paid to the farmers for the best conducted farms and the largest improvement in live stock.

HEREFORD OX, CHAMPION AT SMITHFIELD, 1816. (Bred by S. & C. Haywood, Worcestershire.)

We quote Mr. Young's "Annals of Agriculture" (Vol. 35, p. 91) to show the aims of Herefordshire agriculturists in founding their county Society, which, being the oldest and most flourishing in England, is another proof of the intelligence of the Herefordshire farmers:

"Rules and orders of the Herefordshire Agricultural Society, with an account of premiums annually offered for the encouragement of agriculture and industry, lists of members and subscribers, and directions for the field culture for the early Lancashire Dwarf Potato.

"Such is the heading in view of establishing an Agricultural Society in Herefordshire.

"The advantages which have already arisen to the public from the establishment of Agricultural Societies in the various parts of the kingdom, first gave use to the idea of a similar institution in the county of Hereford. The alacrity of persons of fortune, and of many of the most respectable land-holders, in support of the measure, affords a well-grounded confidence that the exertions of this Society will not be ineffectual.

"To point out the utility of such an institution it can only be necessary to state what are the principal objects of its attention; these are to excite by premiums and otherwise, a general spirit of emulation amongst breeders and practical farmers.

"To encourage industry and fidelity among servants employed in husbandry.

"To reward laborers who shall bring up, or have already brought up, the greatest number of legitimate children, without any or with the smallest relief from their respective parishes.

"To promote the knowledge of agriculture by encouraging experiments on those subjects which are of the most importance to it, and by distributing rewards to such persons as shall produce the best and most abundant crops of grain and grass, in proportion to the quality of land they occupy.

"To encourage the improvement of waste and other lands by enclosing, draining and manuring in the most cheap and effectual manner.

"To ascertain from actual experiment that course of crops on either light or heavy soils, which shall prove most profitable, and leave the lands in the best state.

"To make generally known in this county the most successful modes of husbandry adopted in others.

"To promote all improvements in the several implements now used by the farmer here, and to introduce such new ones as experience has proven to be valuable elsewhere.

"To improve the breed of horses of the cart kind and to carry our cattle and sheep to the greatest point of perfection.

"To ascertain and make public the best means of raising and protecting orchards, of

CHAMPION HEREFORD OX, SMITHFIELD, 1837, AT 4 YEARS AND 10 MONTHS.
(Bred by J. Hewer.)

propagating the best fruits; and the most easy, certain and efficacious manner of proceeding in all the stages of manufacturing their produce into cider and jelly.

"In short, to recommend and bring into practice all the means of facilitating labor, of exciting and rewarding industry, and of receiving at the least expense the greatest quantity and the most approved quality of animal and vegetable food.

"These objects will readily be allowed to be of no small importance, and the spirit with which they may be promoted, and the extent to which they may be carried must depend much upon the liberality of subscriptions. And although the Society looks with confidence to general support, they hesitate not to say they most particularly invite the aid and concurrence of practical farmers."

One of the battle grounds where Hereford cattle have won many honors is the Bath and West of England Society's yearly shows. The yearly accounts of this society are meager.

An account of their 1799 show, taken from Mr. Young's "Annals of Agriculture" (V. 32, p. 244), states that a Hereford heifer won a champion prize at that meeting.

At the meeting of the Bath and West of England Society, 1797, John Billingsley, Vice-President, in the Chair; Lord Summordor was chosen President for the year ensuing.

The exhibition of cattle, sheep and swine for the premium and bounties were considerable in number and generally valuable in qualities.

The premiums awarded were £3.3s. ($15.75) each.

To Mr. Whipley for raising twenty children.

To Thos. Lucas for bringing up in like manner eight children.

To Wm. Spencer for bringing up nine children.

CHAMPION HEREFORD OX, SMITHFIELD, 1838. (Bred by H. Chamberlain, Leicestershire.)

To John Hooker for bringing up eight children.

To John Bartlett for faithful service in one family for sixty-five years.

To John Thomas for living in one family thirty years.

To James Batten for like service for twenty-nine years.

To Joseph Budgell for like service for twenty-seven years.

To Elizabeth Noyes for like service for forty-five years.

To Rebecca Hunt for like service for twenty-nine years.

To Mary Batten for like service for twenty-nine years.

To Benjamin Reynolds as shepherd in one family during sixty years.

The exhibition and sale of cattle and sheep was held on the next day. No awards were reported. It is stated there was a respectable show. It may be seen here that other societies did not cultivate the cattle interest with Herefordshire intelligence.

We have thought best, even at the risk of being tiresome, to give and continue the details of the organization and continuance of the Smithfield Club, and the awards to the Hereford cattle up to the time that Youatt wrote the history termed the History of British Cattle. We have also given an account of the Duke of Bedford's sheep shearing reports and the awards that he made and paid to those gatherings for the improvement of agriculture and live stock of the farmers of his county, thus showing his interests in live stock improvement and proving that it was not for want of care or investigation that the Duke accepted and adopted the Hereford breed of cattle at Woburn as the most valuable for farm purposes. We do this because Mr. Youatt, while stating the fact that the Herefords had been adopted by the Duke of Bedford, failed, except in one instance, to show any of the numerous experiments that he had made, and these the American editors left out entirely.

From 1839 the breeders of Shorthorn cattle in this country and in England quoted this pretended history of Hereford cattle by Youatt and abridged it to discredit the Hereford breed, and advance the interest of the Shorthorn breed of cattle.

CHAMPION HEREFORD OX, SMITHFIELD, 1839. (Bred by R. Hill, Orlton; exhibited by the Earl of Warwick.)

CHAPTER X.

TWENTY YEARS OF HEREFORD BREEDING—1799 TO 1819 (¶ 74A)

The Herefordshire Agricultural Society, naturally, had the largest exhibits of Hereford cattle, in the early days of the breed, and all information that can be preserved of its doings will be of interest to Hereford cattle breeders.

We give a summary of the breeders of Hereford cattle who were prize takers at the shows of the Herefordshire Agricultural Society, from 1799 to 1819. The record says: The list has not been put into the present shape without a good deal of trouble, for the minute books of the society have gone astray, and the record has had to be made up by a diligent search of newspaper files, the chief source of information being the back numbers of the Hereford Journal. Mr. T. Tomkins Galliers of Wistaston has gone over these files for us and has extracted the notes which constitute a useful chapter in the early history of Herefords.

Herefordshire Agricultural Society—established 1798.

We find no account of live stock shown until June meeting, 1799, when, for best bull, not over twenty months, John Apperley, Withington, gains the premium; and for best bull not over three years, Samuel Tully, of Huntington, £5, 5s ($26.25).

1800. Premiums were £5, 5s ($26.25) cups; second prizes, £3, 3s ($15.75) plate; general prize decanter stands. At June meeting, 1800: for best bred bull, Mr. Croose, Sugwas (this bull was bred by Mr. Jones of Fawley); for best yearling bull, Joseph Tully, Haywood; for best heifer, Mr. Skyrme, of Stretton.

1801. March meeting. Bull, three years, seven months, Mr. Smith of Sufton; yearling bull, Mr. Moore, Wellington, Wooton. At this meeting it was suggested to offer more prizes for stock.

June meeting. Three-year-old heifer, J. Tully, Haywood; yearling heifer, Mr. Williams, Thinghill.

1802. March meeting. Notice is given that the following gentlemen will show bulls in the several classes at the coming meeting. Class all ages, M. Crosse, of Ocle; Mr. Tanner, Hampton Court; Mr. Vorce, Warham; Mr. Goode, Dunswater; Mr. Watkins, Brinsop; Mr. Powell, Titley. Class three years old: Mr. Tully, Huntington; T. G. Cotterell, Garmons; Mr. Tully, Haywood; W. Galliers, King's Pyon. Class yearling bulls: Mr. Low, Gattertop; Mr. Tully, Haywood; T. A. Knight, Elton; Mr. Apperley, Withington; Mr. Tully, Huntington. The prizes were awarded as follows: Best bull, any age, Mr. Powell, Titley; best three-year-old, Mr. Tully, Huntington; second prize, W. Galliers, King's Pyon; best yearling bull, J. Tully, Haywood; second prize, Mr. Apperley, Withington. June meeting, best heifer, under sixteen months, W. Downes, Hinton; best heifer, under four years, Mr. Skyrme, Stretton. No stock mentioned at October meeting.

1803. March meeting. Best bull all ages, E. Jones, Fawley; best three-year-old, T. A. Knight; second prize, Mr. Williams, Thinghill; best yearling bull, T. G. Cotterell; second prize, Mr. Jones, Breinton. June meeting: Best heifer, T. A. Knight; second best, Mr. Jeffries, Lyonshall. At this show the committee who awarded the premiums for cattle expressed the opinion to the public that the heifers exhibited (but not entered for premiums) by Messrs. Tomkins, E. Jones, Andrew Knight, and Joseph Tully, were such as would have done credit to the first breeders in England.

1804. March meeting. Best aged bull, Mr. Barnet, of Ledbury; best two-year-old bull, S. Tully, Huntington; best yearling bull, T. A. Knight. Leominster meeting (June). Yearling bull, Mr. Lowe, Gattertop; three-year-old bull, Mr. Fencott, The Broome; three-year-old heifer, Mr. Williams, Brinsop; yearling heifer, Mr. Prichard, Eaton Mill. October meeting. Best three-year-old heifer, Edward Walwyn; best two-year-old heifer, T. A. Knight. Only two stock prizes given.

1805. March meeting. Best aged bull, Mr. Jeffries, Pembridge; best three-year-old bull, Mr. Galliers, King's Pyon; best yearling bull, Mr. Yeomans, Howton. June meeting. Mr. Tully gained prize for working oxen; Mr.

Knight for yearling heifer. These are the only two prizes mentioned, but there must have been more given. Leominster meeting—June. Best yearling heifer, Mr. Williams, Thinghill; best three-year-old heifer, T. Clee, Downton. October meeting. Best two-year-old heifer, Mr. Stevens, Cotmore.

1806. March meeting. Aged bull, Mr. Jeffries, The Sheriffs; three-year-old bull, Mr. Watkins, Brinsop; yearling bull, Mr. Weaver, Stretton. The above premiums were awarded by a committee of the following gentlemen: Mr. Apperley, Mr. Cheese, Mr. Edwards, Mr. Jeffries, E. Jones, Mr. Redward, Mr. Knight, Mr. Tench, Mr. Watkins and Mr. Williams. This is the first notice of how the prizes were decided. Leominster meeting, June 20. Best yearling heifer, Mr. Watkins, Brinsop; three-year-old heifer, Mr. Deykin, Brierley; two-year-old heifer, Mr. Woolaston, Lynch; three-year-old bull, Mr. Salway, Ashley Moor; aged bull, Mr. Proctor, Orleton; yearling bull, Mr. Downes, Ashford. (¶ 74 B) Hereford meeting, June 30. Best working ox, T. A. Knight; best yearling heifer, Mr. Tully, Huntington. Only two stock prizes given. October meeting. We failed to find an advertisement of the awards of this show, but in the general news of the paper of October 29 we find the following: "Mr. Tomkins, Wellington, gained the premium for best two-year-old heifer. The stock shown was very fine and never surpassed on any former occasion." This is the only notice we find of Mr. B. Tomkins, Jr., showing. We suppose he was offended at something that took place, and never competed again.

1807. Spring meeting. Best aged bull, W. Galliers, King's Pyon; best three-year-old bull, Mr. Hewer, Abergavenny; best two-year-old bull, Mr. Weaver, Stretton; best yearling bull, Mr. Yeomans, Howton. Leominster meeting. Best yearling heifer, Mr. Kedward, Westhide; best three-year-old heifer, Mr. Williams, Brinsop; best yearling bull, Mr. Green, Stoke. Hereford, June 30. Best yearling heifer, Mr. Hughes of Marcle; best working ox, Mr. Dawes of Mensell. October meeting. Best three-year-old heifer, T. A. Knight; best two-year-old heifer, T. A. Knight. A notice is inserted to the effect that Mr. Knight declines accepting the premiums since he has gained so many, so William Galliers, King's Pyon, being next best, takes two.

1812. Candlemas. Best yearling bull, B. Wainwright, Hereford; best two-year-old bull, Mrs. Berrow, The Green, Dewchurch; best

CHAMPION HEREFORD OX, SMITHFIELD, 1841. (Bred by Mr. Mason, of Tarrington; fed by Mr. Senior.)

three-year-old bull, Mr. Watkins, Brinsop; best aged bull, W. Galliers, King's Pyon. Leominster meeting. Best aged bull, W. Walker, Burton; best two-year-old bull, J. Walker, Wesington; best yearling bull, H. Moore, Wellingtons, Wooton; best three-year-old heifer, Mr. Tench, Bromfield; best two-year-old heifer, Mr. Tench, Bromfield; best yearling heifer, Mr. Watkins, Brinsop. Hereford, June. Best yearling heifer, Colonel Matthews, Belmont. October meeting. Best two-year-old heifer, Mr. Yarworth.

1813. Candlemas. Best yearling bull, Watkins, Brinsop; best two-year-old bull, not awarded; best three-year-old bull, Mr. Galliers, King's Pyon; best aged bull, Mr. Pugh, Thinghill. Leominster meeting, June. Best yearling bull, T. Jeffries, Pembridge; best aged bull, J. Wainwright. June meeting. Best yearling bull, J. Purchas, Fownhope; best two-year-old bull, Mr. Fluck, Moreton; best yearling heifer, Col. Matthews; best two-year-old heifer, T. Jeffries, Grove. October meeting. No advertisement of meeting, but in general news it is stated that Mr. Welles, Earl's Croome, in Worcestershire, and Mr. Yarworth of Brinsop, took prizes for cattle.

1815. Candlemas. Best yearling bull, Mr. Price, Norton Grounds, Gloucestershire; best two-year-old bull, Mr. Yarworth, Brinsop; best three-year-old bull, T. Barnaby, Brockhampton; best aged bull, Mr. Yarworth, Brinsop. Hereford June meeting. Best yearling heifer,

CHAMPION HEREFORD OX, SMITHFIELD, 1846. (Exhibited by John Hudson, of Norfolk.)

bull, Mr. Symonds of Tatton; two-year-old bull, Mr. Walker, Burton; best yearling heifer, Mr. Jeffries, Grove; best two-year-old heifer, Watkins, Knightwick (late of Brinsop). At the Leominster meeting it was proposed to discontinue the shows there owing to the low state of the society's funds. Hereford June meeting. Nostock prizes. October meeting. Best two-year-old heifer, C. Walwyn; best three-year-old heifer, James Yarworth.

1814. Candlemas meeting. Best yearling bull, Mr. Grovenor, the Parks; best two-year-old heifer, not awarded; best three-year-old John Morris, Marsh. October meeting. Best two-year-old heifer, Mr. Tench, Bromfield; best yearling heifer, Mr. Walker, Burton. N. B.—All bulls to be shown in future at Candlemas, and heifers at October meeting.

1816. Candlemas. The committee for deciding the merits of cattle report that the aged bull exhibited this day by Mr. Yarworth of Brinsop is the finest animal ever shown before this society; the dam of this bull is now in the possession of Mr. Price of Morton Grounds, Worcestershire. Best yearling bull, Mr. Tomkins, Dippers Moor; best three-year-old bull,

Mr. Jeffries, Grove; best aged bull, Mr. Parry of Birley, near Stretford. (Yarworth's bull disqualified, having taken the prize last year.) June meeting. No stock shown. No account of other shows this year.

1817. Candlemas. The meeting only noticed in general news. Yarworth, Wainwright, and Wood of Burghill, successful competitors. June meeting at Leominster. Best yearling heifer, Mr. Jeffries, Grove; best two-year-old heifer, Mr. Tench, Bromfield; best two-year-old bull, Mr. Smith, Gattertop; best three-year-old bullock, Mr. Jeffries, Grove. October meeting. Short notice. Mr. Walker of Burton and Mr. Eckley of Tillington, got premiums, but it does not state for what.

1818. Candlemas. Best aged bull, Mr. Smith, Gattertop; best yearling bull, Mr. Cooke, Wintercott. Leominster meeting. Best three-year-old bull, T. Jeffries, Grove; best yearling and two-year-old heifers, Mr. Walker of Burton. October meeting. Best yearling heifers, Mr. Smythies, Lynch; best two-year-old heifer, Mr. Welles, Earl's Croome.

1819. Candlemas. Best aged bull, Mr. Dawes, The Rodd; best yearling bull, Mr. Yeomans, Howton. Leominster meeting. Best pair working oxen, Mr. Walker, Wesington; best yearling heifer, Mr. Jeffries, Grove; best yearling bull (¶ 74 C) Mr. Preece, Comberton; best three-year-old bull, Mr. Cooke, Wintercott. October meeting. Best yearling heifer, Col. Matthews; best two-year-old heifer, Mr. Smythies, Lynch.

These records, uninteresting in themselves, give an insight into the exhibits and exhibitors of Herefords in a past century and show a classification not equalled at the time by any other county show, and proving again the thorough establishment of the Hereford breed at that early day (¶ 74 D).

CHAMPION HEREFORD OX, SMITHFIELD, 1846.
(Bred by Mr. Thomas, of Cholstrey, Herefordshire; exhibited by the Earl of Warwick.)

CHAPTER XI.

Early Hereford History in America

HEREFORD-SHORTHORN CONTROVERSY, FROM 1834 TO 1841.

In 1834 Kentucky, Ohio, Indiana, Illinois and Michigan were organizing agricultural societies; and Wisconsin, then a state one year, was discussing the propriety of such a move. An agricultural convention was held at Albany, N. Y., in which the different interests of farming were discussed, and the necessity of state aid urged. Among those prominent in this movement were H. S. Randall, L. F. Allen, F. Rotch, R. L. Allen, J. J. Vail, Jesse Buel and C. N. Rement.

The American Institute held its eleventh annual fair in October, 1838, and among the managers were Jesse Buel and C. N. Bement of Albany. At this date there were active efforts for the establishing of agricultural societies, but our investigations will be confined mainly to New York.

It is probable that at this time the Shorthorns had a stronger hold in Kentucky than elsewhere. A sale was advertised at Powelton, near Philadelphia, of Mr. Whittaker's cattle, an eminent English breeder. A sale was held at Bloomfield, N. Y., at which the highest-priced animals were sold to Gen. Dudley of Kentucky. At a sale held at Paris, Ky., eleven head sold for $8,157; and ten animals of mixed blood for $2,580; the prices at the Whittaker sale referred to above ranged from $360 to $540. H. Clay, Jr., sold this year his cow Princess for $2,000. At a sale in Cincinnati ninety-one head were sold, averaging $305, one four-year-old bull selling for $1,450. These prices will indicate the standing of Durham cattle at this time.

The New York State Agricultural Society met the first Tuesday in February, 1839. Buel, Allen, Vail, Van Berger and Spencer were a committee to report names of offices, and among the officers were Jesse Buel, corresponding secretary, and C. N. Bement, treasurer.

In 1839 the Royal Agricultural Society of England was organized and received a charter from the queen. A society had existed previously, known as the English Agricultural Society, and the Royal absorbed it. At this first show a premium was offered for the best cow calculated for the dairy. The first was won by a Hereford, the second by a Durham cow, and this class, we think, was dropped from that time.

In 1841 the New York State Society held their first fair at Syracuse. These movements for the establishment of the Royal in England and New York Fair at Albany, were made by those in the interest of the Shorthorns as was the movement of the writing of the "History of British Cattle," by Youatt, and each of these movements, if they had been written and planned for the advancement of the Shorthorn interest, and so given out, would have been legitimate and proper—commendable even; but when the Shorthorn men took the machinery and charter of the Society for the Diffusion of Useful Knowledge, to write up the Shorthorn and write down other breeds, it became dishonest, and so when they, under the sanction of the crown of England, established the Royal for the advancement of agriculture and the improvement of live stock, gave preference to the Shorthorn race of cattle unfairly, it was dishonest.

We have made these points—the writing of the "History of British Cattle" by Youatt, the establishing of the Royal Agricultural Society of England, and the New York State Agricultural Society—because to these influences, more than any other, the Shorthorns owe their standing.

Previous to the writing of the "History of British Cattle" the Duke of Bedford had made very careful experiments in grazing and feeding of Herefords and Shorthorns, and these experiments, widely published, resulted in the adoption of the Herefords by the Duke of Bedford on the score of economy; the details of those experiments were before the writer of that history, but were not used because not satisfactory to the Shorthorn breeders. We have shown that the Herefords and Shorthorns were in competition before the Smithfield for thirty-six years; and the Herefords had taken ninety-three premiums equal to $7,060 on oxen, while the Shorthorns had taken only thirty-seven premiums, amounting to $3,275, and these facts were not noticed by the writer of that history.

We have shown that in the London market, at the time that history was written, the Hereford beef was selling at a half penny (one cent) to a penny (two cents) a pound more than Shorthorn beef, and that the writer of this history did not note this fact. We have shown that he quoted the sales of Shorthorns by Mr. Arrowsmith between 1801 and 1808, ranging per head from $80 to $175, and that he neglected to quote the sales of Herefords made by Mr. Westcar between 1799 and 1811, ranging from $500 to $737 (¶ 74 E). From these and similar facts we have charged that the history was written in the interest of Shorthorn breeders and we shall be supported in this view by all impartial men. We have not the data to enable us to dissect the action of the Royal. We have given one case in which the Herefords won at their first meeting, and that, when the Shorthorns claimed the greatest strength; and shown that, from that time, they did not bring their cattle into competition with Herefords as milkers.

In later years there has been no question as to the bias of the Royal in favor of the Shorthorns, and whatever breeders in this country may claim, English breeders will hardly question our statement. As to the New York Society, we shall present conclusive testimony to show that it was managed entirely in the Shorthorn interest. We might refer to other societies, as the state societies of Ohio, Kentucky, Indiana, Michigan, Iowa and Illinois, and many others, but for the present we propose to leave them out.

In 1835 the only reliable experiment, that of Smithfield, showed the Herefords to be the best cattle. Reliable records of trials of the Herefords and Shorthorns in every instance showed the former to be far in advance of the latter. The London market showed the value of the Hereford to be 10 per cent over the Shorthorn, and these facts were ignored by the writers of the "History of British Cattle." From 1836 to 1840 inclusive, before the Smithfield Society, the Hereford bullocks took thirty-seven prem-

FIRST PRIZE HEREFORD OX AT SMITHFIELD, 1846.
(Bred by T. Roberts, of Ivingtonbury, Herefordshire; exhibited by Mr. Trinder.)

iums equal to $1,875, the Shorthorns taking eighteen premiums, or $740.

With these basis facts established, we turn to the importation of Mr. W. H. Sotham, and the correspondence and controversies growing out of that importation, and these we shall quote freely. The following extract is from a letter of Mr. W. H. Sotham, of date June 1, 1840, to Messrs. Gaylord & Tucker, publishers of the "Cultivator," V. 7, pp. 113, 114:

"Of our cattle I shall not say more than that they are of the Hereford breed, and from the same breeder who agrees with me that the Herefords are decidedly the best and most profitable when taken in the aggregate. The Smithfield show will acknowledge this, as the Herefords take top price against all others.

"Other instances are in their favor. The oxen are excellent workers, the best feeders, and when in market fetch one-half penny per pound more than the Durhams. The fat and the lean is so well interlarded. This assertion is backed by most of the London butchers, of whom I made inquiry, and I send you a letter for insertion from Mr. Guerrier to me, one of the best salesmen in London, who presented me with an engraving (¶ 63) of the prize ox in 1837, the property of Mr. John Thomas Smith, Portland, Lincolnshire. This ox obtained the first prize of 15 sovereigns at the Rutland Agricultural Society's show at Oakham. Also, at Smithfield Club show, the first prize of 20 sovereigns in class first. Age of this ox, four years and two months; weight 122 stone (1,708 lbs.), bred by Mr. John Hewer of Hereford. It is my opinion that Herefords are better milkers than generally represented. It is not clear to me that they are inferior to the Shorthorn or Durham. When our cows come to grass I will endeavor to give the quantity. Appearance on the vessel are much in their favor. I will not say more on this subject until we have some for sale; these are intended for our own use.

CHAMPION HEREFORD OX, SMITHFIELD, 1848. (Bred by H. R. H. Prince Albert.)

"Should it meet the views of improving, intelligent, enterprising farmers, such as know how to lay out their money judgmatically, we have no objection to enter into a trade with Mr. Hewer, to take the whole of his extra stock yearly, who says no other person shall have them for exportation. Any other breed of the first order, or any kind of animal England can produce, I will endeavor to procure by the purchaser representing it to me, and the highest price he will give in New York, or any kind of field or garden seeds. But I must admonish the purchasers to show a little spirit, not to be afraid of their shadow in a good cause; the best things in this world cannot be bought at low prices; the expense and risk of shipping is very heavy, and not a very pleasant business for a sea-sick sailor."

The following is the letter referred to, from Mr. Guerrier to Mr. Sotham, dated London and West Smithfield, 17th April, 1840:

Dear Sir: As I could not conveniently, during the busy engagements of our market, reply to your inquiry respecting the breeds of Durham and Hereford cattle, I take this opportunity to state that never, during twenty years' experience as a salesman of cattle of all breeds in this market (Smithfield), although I have tried time without number, when I have had some of the best descriptions of Durhams to sell, could I succeed in persuading my best customers at the west end of London to purchase Durhams when I had any well-bred Herefords to part with; they one and all stated in cutting up the beef they find in the Herefords so much more roasting beef to that of boiling. To satisfy your mind still further, just cast your eye over the particulars (which I send for your perusal) of our last Christmas show of cattle. There you will perceive the Hereford takes the top prize. Attend our Smithfield show and you would be more than ever assured of the importance of the best breed of Herefords before that of Durhams. Last year I had Durhams 15 stone per ox (200 lbs. live weight) heavier than Herefords, but could not realize so much by 4d per stone of 8 lbs. (dead weight) as I could for the Herefords——Wishing you success in your undertaking,

I am, sir, yours respectfully,

WILLIAM GUERRIER.

To W. H. Sotham.

N. B.—I need not observe to you the remarks I have made regarding Herefords are not because I am at all prejudiced against Durhams, Scots, or other breeds, but having with my father grazed all breeds for many years, and as a salesman, having for twenty years past, during which time I have annually sold from 5,000 to 10,000 cattle, consisting of all breeds, from Ireland, Scotland, as also in this kingdom, never found any breed of cattle more profitable than the said Herefords, if well bred.

From the same volume of the "Cultivator" we quote (p. 104) the following editorial matter:

"One of the most important importations of cattle and sheep that has ever taken place in this country has just been made by the Honorable Erastus Corning of this city and Wm. H. Sotham of Jefferson County. It consists, as will be seen in the list given in another part of

HEREFORD STEER, 2 YEARS 11 MONTHS OLD, CHAMPION AT BIRMINGHAM AND SMITHFIELD, 1853.
(Bred by T. Carter, Dodmore, near Ludlow, Herefordshire; fed by Mr. Heath of Norfolk.)

this paper, of twelve cows, calves and heifers and twenty-five sheep. The cattle are of the Hereford breed from Herefordshire, and the very best animals that could be selected. The sheep are of large size, being the Cotswold, cross with the Bakewell, and probably as fine animals of the kind as ever imported. No one can avoid being struck with the extraordinary size of the cows, their fine forms, their muscular development, denoting strength and power, and showing the basis of the reputation which the Herefords formerly had for working cattle, and now have for feeding. The expense of the importation was nearly $8,000.

HEREFORD OX, 4 YEARS OLD, CHAMPION AT SMITHFIELD, 1863.
(Bred by T. L. Meire, Shropshire; fed by Mr. Heath of Norfolk.)

"The attention of cattle breeders has within a few years been much directed in England to the improved Herefords, and principally in consequence of the numerous prizes which these cattle have taken at the great cattle shows of Smithfield, and lately at the fairs of the English Agricultural Society. Although they have not in general reached the great weight of some of the improved Durhams, yet the rapidity with which they take on flesh, the superior excellence of the beef and their early maturity render them great favorites with the English feeders, and the Michaelmas fair of Hereford is one of the finest shows of the kingdom.

"We copy the following from the Encyclopedia published by the Society for the Diffusion of Useful Knowledge, Vol. XII, article, "Herefordshire," as an accurate account of the general qualities of the breed:

"'The prevalent breed of cattle is that for which this country is justly noted; their color is red with white or mottled faces, and frequently white along the back and about the legs. *Good milkers are occasionally found among the cows, and it is possible that a race might be reared from this stock that would be useful for the pail. But dairy farming is never practiced here*, and the milk of the cows which are kept only for breeding is given to the calves.

"'It was formerly the custom to work oxen at three or four years old, and to feed and send them to market at five; but there is now a complete change of system. The oxen are no longer worked, *but are commonly fed when they are two years old and sent to market before they are three. Their early maturity and the readiness with which they fatten make them suitable for this system of farming.* Graziers from the south and middle of England drive a large number of this popular stock from the Hereford Candlemas and October fairs.

"'*The Hereford ox fattens more rapidly than the Devon and in proportion to the quantity of food consumed lays on a greater weight of flesh than a Durham ox.* The result of a trial of this kind may be seen at page 34 of Youatt's "Cattle." That the flesh of the Hereford is of finer quality than the Durhams is proved by the superior price per stone which it obtains in the Smithfield market.'

"Comparatively few of the Herefords have as yet been introduced into this country. Mr. Bement of this city has a bull and a cow of this breed which are fine animals. The Honorable W. C. Rives of Virginia, and the Honorable Henry Clay of Kentucky have made importations of these animals and they have been in their possession so long that either of these gentlemen could speak fully of their value as compared with the Shorthorns, in the points of feeding, milk, and endurance of our climate, and we think at this time, when attention is turned to the subject, they will confer a great benefit on the American public by stating the result of their experience and their opinions with regard to these cattle. We need not say we should be happy to be the medium of presenting such history and opinions to the public.

"We have presented these remarks in the hope of eliciting from some who are qualified for the task, a discussion of the comparative merits of the Herefords and the Shorthorns and their adaptation to our country for the purposes of feeding and the dairy. Both are valuable breeds; the question to be decided is, which, in all respects, is the most proper for us?"

It will be noticed in the foregoing that the "Cultivator" quotes the encyclopedia published by the Society for the Diffusion of Useful Knowledge, Vol. XII, article, "Herefordshire." It should be compared with the same society's Youatt book. We call attention to that portion of the "Cultivator's" quotations in italics (which are our own). Will the Shorthorn men recognize them as facts? In the same volume, p. 158, a correspondent of the "Cultivator" says as follows:

"Among the recent importations for improvement that have taken place in this state is that of Hereford cattle and Cotswold sheep, by Messrs. Corning & Sotham of Albany. Attracted by a letter of Mr. Sotham that appeared in the July number of the 'Cultivator,' when down last month, I made an inspection of these

HEREFORD OX AT 4 YEARS; CHAMPION AT SMITHFIELD, 1868.

superb animals that gave me a very different opinion than I have heretofore entertained of these breeds. All other Herefords that I had previously examined, in comparison with these, though noble in appearance, had large heads, thick necks, narrow hips and thin loins, compared with the best Shorthorns, but these nearly approach them now in all such particulars, especially in the great width of the hip bones, showing a capacity, when well fed, to place their meat in those parts where it is most valuable, and I cannot but coincide in the remark of one of our most distinguished breeders of Durhams, 'that on the right soil they would give the Shorthorns enough to do to maintain their native dairy cows, they could be called even fair milkers. It is apparent, therefore, that in the general purposes of improving our native stock the Herefords cannot be rivals to the Durhams at least till they are further advanced to good milkers, which will then make them but in fact another race of improved Shorthorns. Yet, if the Herefords yield to the Durhams at the dairy, in the yoke they must be far superior to any other of the ox kind, for they have nearly the quick step, the fine bone, the sinew and muscle of the Devon, with a much greater weight and size. I could not but admire the great length and rotundity of the barrel, the smooth, powerful structure of their frames, and

HEREFORD OX, 2 YEARS OLD: CHAMPION AT SMITHFIELD, 1882.
(Bred by Mr. F. Platt.)

present high position.' But how are they enabled to accomplish this? Why, only by approaching Shorthorn perfection in these particulars. Yet, at present they are only the graziers' and butchers' stock; for, though Mr. Sotham talks of their good milking qualities, I must confess that though I eyed them sharply, and handled them closely, I was not favored with any such discoveries in their veins, nor did the appearance of their udders make up at all for this deficiency, and I should require some proof of the facts before I could be convinced that even in comparison with our good clean, elastic limbs. In fact, as workers, they seem to me to be that happy medium on the race of oxen that I am so desirous of seeing cultivated more generally in horses; neither the light mettlesome racer on the one hand, nor the slow, fleshy cart-horse on the other, but the superior and more happily mixed general utilitarian. To those who are breeding working oxen, or stock expressly for the butcher, I would strongly recommend these Herefords. I should think them particularly well adapted to the rich interior of the Western states, where cattle must be driven a great distance to market. In that case the blood of the Devons, to which the Herefords are so nearly allied, could not but tell, like that of the thoroughbred racer on the course."

In the same volume of the "Cultivator," p. 28, there is an essay on cattle by Henry S. Randall. Of the Herefords he says as follows:

"The Hereford ox is supposed to be descended from the same stock with the Devon, but is larger, heavier in the bone, usually of a darker red or brown color, with a white face, throat and belly. They are shorter-legged than the

HEREFORD CATHEDRAL.

Devon, hardier and kindlier feeders, but less docile in temper, and even worse milkers. Indeed, a Hereford cow is rarely seen in an English dairy. Their hardihood and great muscular power give them the first rank among working cattle. This, together with their superior grazing qualities, has led to their introduction into the United States, by the Hon. H. Clay of Kentucky and several other individuals. But it is probable that their deficiency in milking properties will always prevent their very general adoption, either as a cross or in a pure state."

This, as well as what is further stated, is selected mainly from Youatt (by Berry), and the essay has the evidence in itself as being for the purpose of bringing forward the Shorthorns. Mr. C. W. Bement notices Mr. Randall's reference to Herefords in the same volume, page 125:

"I was much gratified on perusing the excellent 'Essay on Cattle' in the February number of the 'Cultivator,' from your talented correspondent Henry S. Randall, Esq. His short, pithy history of several varieties is well calculated to arrest the attention of farmers, and guide them in the selection of that breed best adapted to the different sections and purposes, for which they may be wanted, whether for the dairy, yoke or shambles. For instance, where the climate is mild, and a full and rich bite of grass at hand, and the dairy and beef the object, I would by all means recommend the Durhams; but where the climate is cold, seasons short, land rough and hilly, with a short and sweet bite of grass, with labor and beef the object, I would recommend the Hereford or Devon; and at the Northwest, in Missouri, Illinois and Wisconsin, where only beef is wanted, and where they have to be driven any great distance to market, from what I have seen and can learn, the Hereford certainly would be preferable, being hardy in constitution, good travelers, of great size, will fat at an early age, and will make more pounds of beef with the quantity of food consumed, and when better known in market will command the highest price."

Mr. R. L. Allen says, as follows, on p. 112:

"The best specimens I have seen are the Herefords recently imported by Mr. Sotham, and now in the neighborhood of Albany; but as he promises a description in your journal, we may all hope to know more about a breed that has for a long time assuredly been held in high estimation abroad."

Mr. J. H. Hepburn, on page 102, says:

"Another matter of surprise, not only to me but to many others who derive their book knowledge of these matters from your paper, is the effort now apparently making both in England and America to elevate a different breed of cattle over the heads of the Durhams. I have particularly examined the account of the recently imported Herefords in the few last numbers of the 'Cultivator,' and read the appended recommendations. I have never seen a sample of the Herefords, but have been familiar with their history *as recorded by* Mr. Youatt in his work on 'British Cattle;' and taking the text as laid down by him minutely, the recent discoveries of excellencies in these cattle, calculated to place them before the Durhams or Shorthorns, have been matters of considerable surprise. This, to some of the advocates of the Herefords, may sound strange, but the strangeness of the observation will disappear, *if they examine fully the chapters devoted by* Mr. Youatt *to the different breeds of cattle,* in what may be now, strictly speaking, called their native country. If I understand the work referred to, *the engravings in it are calculated for correct representations of the living animals; and, independent of the writings, they alone will be sufficient to satisfy any breeder that some of the allegations made of the superiori-*

ties of the recent importations of Herefords, if they resemble their progenitors, cannot be correct. If these statements are correct, and the appearance and test of the animals will prove it, then improvement has been extended to them. *If they have now properties that the breed in the time of* YOUATT'S *writings had not* [but five years previous. T. L. M.], it is a very important question to know how they have acquired those properties. MR. YOUATT says: 'The Hereford cow is apparently a very inferior animal. Not only is she no milker, but her form has been sacrificed by the breeder.' These observations or these assertions of positive facts, for such we must take them to be *when from a standard work*, do not read well with the recent assertions of Mr. Sotham and Mr. Bement—the first of whom holds out the idea that they are equal to the Shorthorn or Durham, and the latter that they are very good milkers and large. The latter gentleman, however, states that he has understood their qualities for milk have been 'recently improved.' How have they been improved in their quality for milk? Not, certainly, by breeding among themselves, for the trite and true axiom is settled, I believe, that 'like begets like.' If, then, the Herefords in the time of YOUATT, and for years before that, were no 'milkers,' how has the present improvement in that quality been effected? It must have been by the aid of some other breed, celebrated for their possession of that quality, and by whose aid, also, the form of the Hereford cow has been so materially improved, for Mr. Sotham says his are fine looking animals, and so says Mr. R. L. Allen, if I recollect right, who states that he saw them near Albany.

"Now, Messrs. Editors, may it not be possible that we are at the commencement of another 'stock mania,' by which John Bull is about to realize thousands from the farmers of America by selling them a compound breed of beautiful cattle, the essential qualities of which have been derived and that very recently, from the Shorthorns, that we have been making heavy importations of, for years back? Let any candid man answer the question for his own satisfaction. See what the Herefords were; hear what they are now, and then say if there is any impropriety in charging their admitted 'recent improvement' in points in which Shorthorns excel, *to an admixture with them.* If such is the fact, and we have a number of the finest specimens of the Shorthorns, cannot we, by judicious crossings and attention to these matters breed, for ourselves, if I may be allowed the expression, a breed of cattle without expending enormous sums to pay our trans-Atlantic neighbors for doing work that we ought now to do for ourselves? The state of the times is such as to call loudly upon every man in every station of society to do his duty to himself and to his country; and I state it boldly, without fear of contradiction, that there is now abundant material in our own country to retain, by judicious breeding, the purity of the full-bred Durham, and to commingle their perfections with the stock of our own country, in such a manner as to produce a race of animals equal to that of any other country under the sun, for dairy qualities as well as for the butcher."

The reader of to-day will see the Shorthorn advocate quoting Youatt (Berry). At page 161 is the following letter written by Mr. Sotham, from Portsmouth:

"Messrs. Gaylord & Tucker: I am so far on my journey with the best lot of stock ever seen together. They consist of the following [sheep omitted. T. L. M.]:

"One Hereford cow [Spot 1074—alias Matchless, T. L. M.] that won the first prize at Oxford, 1839, against all England, and a young bull [Young Prize 1070 (2333) T. L. M.] from her eleven months old.

"Two five-year-old Hereford heifers. These heifers are in calf by Dangerous 1619 (419),

CITY OF HEREFORD, CATHEDRAL AND WYE BRIDGE.

a yearling bull that is to be shown against all England next year.

"One half-bred between the Hereford and Durham to show the cross, which I think is an excellent one, probably better than the pure-bred of either and from what I saw of Mr. Cother's stock, of Middle Aston, it might be extended much further than is generally supposed, for his fourth cross was equal to the first—not the least sign of degeneration. Of this I will say more when I have more time, for it is now precious. * * *

"We shall show a number of our cattle at the show at Niblo's, in October, where we invite the owners of the best cattle in America to appear against us (with the best of feeling), for it is opposition and competition that spurs us on to superiority, but prejudice must be put out of the question. Let reality be our helmsman and perseverance our motto, and then our country can be equal in stock to any on the globe. It all depends on the people. * * *

"Mr. Hewer has numerous backers, if he will allow it, to show from one sheep to a hundred against any person in the world, either ewes or rams. He has been very careless about showing his sheep and cattle, having met with a ready sale without it.

"I am, dear sirs, yours sincerely,

"WM. HY. SOTHAM.

"Portsmouth, Aug. 24, 1840."

We find the following letter from Mr. Sotham at page 176. We quote from this:

"Messrs. Gaylord & Tucker: * * * In perusing your valuable paper I noticed an article from Mr. Randall on cattle, which differed widely from my opinion in many instances. I cannot say I agree with him or his quotations on Herefords. I am sorry to dispute a person who has so much zeal for his country, but I know his good sense will hark back to a fault if caught on a bad scent. When he catches me running the same course I will hail his correction with pleasure and profit from his good intention.

HEREFORDSHIRE FARMYARD SCENE.
(Taken at Mr. Newton Moore's, Sutton.)

"He says: 'They are larger boned, usually of a darker red, or browner color than the Devon and even worse milkers; indeed, a Hereford cow is rarely seen in an English dairy.' Probably he might have made these assertions twenty or thirty years ago without fear of contradiction, but no practical man of the present day will allow them to pass with impunity, unnoticed. The Herefords, like everything else that is looking towards perfection, have met many unjust accusations, and I must say that breed, with the Cotswold sheep, have had to contend against prejudice and abuse in England more than any other breeds ever exhibited; but they have fought their up-hill course with great credit; their good qualities have triumphed even over the most prejudicial, and have won a permanent standing in the estimation of the first breeders that will not be easily forestalled. The Hereford oxen have taken the first prize in each class at the Smithfield shows for the last two years. * * *

"I must say to Mr. Randall that the pure Herefords are not larger and heavier in their bone; there is as much good breeding shown in their limbs as any breed in existence. The working oxen are as good and as docile as any, and I think I can forward a Hereford cow that will fill the pail as high as most Durhams or Devons, and if Mr. Randall will take an ocular survey of English dairies he will find in them more Herefords and crosses from them than any other breed, though the best breeds of cattle are far between, even in England. This condemnation has arisen more from theory and hearsay than practice. The best proof of this is to refer Mr. Randall to Class 5 of the Royal Agricultural Society at Oxford, 1839, and he will there find that the Hereford cow won the first prize against all England, in the opinion of the judges, as the best calculated for dairy purposes; the Durham obtained the second. This is a quotation that cannot be disputed, and one much calculated to retrieve the injured character of the Herefords as milkers—though an animal that can be fattened with facility is sufficient recommendation, as the steer, the ox and the cow must ultimately come to the shambles. A pure-bred Hereford bull will cross well with any breed, and I shall be much disappointed if they do not afford more actual benefit to the United States than any other breed ever imported. It was this impression that induced me to give them the decided preference. The female Durham is thougnt to be a better cross with other breeds than the bull. Mr. Randall, in extolling the Durhams, quotes from the 'Farmer's Series' the following:

"'In early maturity they have confessedly no rivals, being ready for the butcher for from two to four years earlier than the other English breeds.' I should imagine his author meant Herefords when he advanced this; if not, practice and experience will admit them to have at least one year in advance of all others; beyond this would look too much like fiction, for it must be deemed unprofitable to keep steers over three years, unless for use of the yoke. Herefords decidedly hold the first place in England for early maturity and a tendency to the secretion of fat. They often go to market at two years old.

"In my opinion, good hips and rumps with expanded chest in cow or bull are very prominent points. A kind chop, a straight chine well lined with good quality of flesh, backed up by good round sides, straight with the shoulder, are valuable acquisitions when you cannot get perfection. There are more cattle fail behind the shoulders than any other point, and when this is the case they are apt to show much paunch, which, to me, is a very great objection; reason seems to say: exorbitant belly, consumes much food, very unprofitable, and are not so apt to fatten; this I have frequently noticed minutely, and invariably found it so. Flesh hides a multiplicity of faults, but will never hide this, which I consider a very important one.

"There is also much to be learnt in feeding cattle, and it is very essential to discover the daily consumption of each beast so as to give them just as much as they will eat, leaving a clear manger to sleep over; if they have hay before them to blow on it weakens the appetite almost to satiety. If a beast is cloyed with any kind of food he does not relish it again for many days. Cattle, when feeding, require much water, and it is very essential. A person who has a thorough knowledge of feeding, sees, immediately he enters the stable, whether his cattle have been regularly and sufficiently fed and watered; if they are at all restless, they are not satisfied, and it is a certain omen of something wrong. They should not be disturbed more than absolutely necessary; the more quiet they are kept, the better they thrive. Much has been said on feeding raw potatoes to cattle. I think them valuable, given in a limited degree. The generality of farmers give too many, which loosens the bowels, without aiding the body. They can be well supplied with meal once a day, so as not to affect the quietness of the stomach. The Swede (better known in America as rutabaga) may be fed more extensively. They suit both the palate and constitution, and are a very profitable root to the farmer. England would be lost without

"THE WOODLEYS," WOOTON, OXFORDSHIRE, ENG.
(Estate of the Sotham family; birthplace of Wm. H. Sotham.)

it, as it is fed to advantage to many kinds of animals.

"The breeders of stock and tillers of soil ought to be intimately united, or the anticipated improvement in husbandry will end in disappointment. Could we establish agricultural societies and combine in friendship, the more we met together and the oftener we discussed subjects connected with agricultural matters, the greater would be the advantages that would ensue. I hope this feeling will be established in Jefferson County and that we shall be able to meet together to awaken the farming interest and promote the public good.

"I am, dear sirs, yours most sincerely,

WM. H. SOTHAM.

"Perch Lake Farm, Jefferson County, N. Y., 1840.

"N. B.—I have met with the last three numbers of your valuable paper since writing the above, and will answer all reference to my communication in your next."

It is most notable that Mr. Sotham advocated a system in 1840 akin to the all-prevailing "Agricultural Institute" of to-day. He was half a century ahead of his time.

In same volume, pp. 193-4, we quote H. S. Randall in full:

"Messrs. Gaylord & Tucker: In the August number of the 'Cultivator' Mr. Bement of Albany expresses the opinion that the statement made by me in a previous number, 'that the Durham would lose in milking properties by a cross with the Devon or Hereford' was incorrect. Mr. Bement's skill as a breeder and his candor as a gentleman, to both of which I am happy to bear ample testimony, entitle his opinion to much respect. The subject, too, derives additional interest from the recent importation of Herefords and the introduction of several valuable herds of Devons into our state.

"The essay of mine from which Mr. Bement quotes was designed rather to correct several popular errors than to furnish a text-book for the scientific breeder. But brief and imperfect as was the account it contained of the Herefords, much subsequent examination of both English writers and living English breeders has only confirmed me in the correctness of the positions therein assumed, particularly of the one controverted by Mr. B. His evidence, founded on the results of his own experience, is good as far as it goes, but a few instances does not establish a rule. The entire preponderance of testimony, both of writers and breeders in England, is against the milking proper-

"THE WOODLEYS," OXFORDSHIRE. (Birthplace of Wm. H. Sotham. Rear view.)

ties both of the Devon and the Hereford. *The work on* BRITISH CATTLE, *BY* MR. YOUATT, *is acknowledged the standard work of the day on the subject on which it treats.* Mr. Bement relies on it as his authority in favor of the Herefords. On the same page from which he quotes, I find the following (one would think) conclusive statements: 'The Herefords are far worse milkers than the Devons. This is so generally acknowledged that while there are many dairies of Devon cows in various parts of the country, none of which, however, are very profitable to their owners, a dairy of Herefords is rarely found.' Again, 'the Hereford cow is apparently a very inferior animal. Not only is she no milker but even her form has been sacrificed by the breeder.' Our friend of the Three Hills Farm, is, I presume, too good a lawyer to attempt to invalidate his own witness. If such are facts, how are we to suppose that the Shorthorn, concededly the first breed in England or America as milkers, will suffer no deterioration in this property by a cross with a breed so decidedly inferior? Whatever results individual experiments have led to, such a position would be at variance with every established maxim of breeding.

"I desire to be distinctly understood in my remarks as taking no ground against either the Devon or the Hereford, except in the single point involved in this controversy. The stylish and highbred Devon has always been a decided favorite with me. *Of the Herefords, I know little personally, but am prepared from the English publications of the day* as well as the testimony of American breeders, to believe that in positions suited to them they may be regarded as a valuable acquisition to our American breeds of cattle. On the large and little cultivated prairie farms of the western states, where early maturity and milking qualities are of little consequence and where they must be driven great distances to market, the Herefords can have few equals. The Hon. Henry Clay, of Kentucky, was the first importer of this breed into the United States, and the following extracts from a letter which I received from him, bearing date Sept. 21, will, I doubt not, be read with much interest:

"'I first imported, upwards of twenty years ago, two pairs of the Hereford reds, and bred from crosses between them until I was induced to discontinue in consequence of an apprehension that I should breed in too far, which in some instances I found to be the case. I could not obtain conveniently crosses from other females of the same race.'

"Mr. Clay afterwards remarks: 'My opinion is that the Herefords make better work cattle, are hardier, and will, upon being fattened, take themselves to market better than their rivals. They are also fair milkers. On the other hand, the Durhams, I think, have the advantage in earlier maturity, in beauty, and in the quality of milk which they will yield. They will also attain greater weight and size even.

"'The choice between the two races should be regulated somewhat by circumstances. If one has rich, long and luxuriant grasses, affording a good bite, and has not too far to drive to market, he had better breed the Durhams; otherwise, the Herefords.'

HEREFORD COW, "MATCHLESS," ALIAS "SPOT." (V. 5, P. 113) 1074.
(Bred by J. Turner, Noke Court, Herefordshire; imported by Wm. H. Sotham. First prize R. A. S. E. show at Oxford, 1839; dam of the first prize yearling bull at the same show.)

"After some remarks in relation to the origin of the two races, he continues: 'The Herefords resemble the Devons—the race of New England cattle, and a fine race it is. But the Herefords have the advantage over them of greater size, greater length, more power consequently for draft, and are, I think, quite as quick in the step and as good at the pail.'

"'I have thus,' he says, 'expressed my opinion; but I must add that here in Kentucky, the Durhams are generally preferred to all other races. Our grasses are rich and abundant, and our blue grass especially (a name improperly given, for it is a green sward) is an object of great admiration. The Durhams are much more generally distributed than the Herefords, there being none of the latter, within my knowledge, but what have sprung from my importation.'

"Mr. Clay's opinion cannot be regarded otherwise than as strong testimony in favor of the Herefords, though I presume his general statements are to be understood to apply as much to his mixed as to his pure-bred animals. That mixture was with the Shorthorn, and it is not impossible that this drop of alien blood has had a significant bearing on their 'fair' milking properties. Nor is it to be presumed that this cross has resulted in very serious deterioration to the Herefords in other respects.

"Since writing the above the 'Cultivator' of November has come to hand, containing a communication from Mr. Sotham on the subject of the Herefords, in which he takes the same position with Mr. Bement in relation to their milking properties, and also denies the correctness of other portions of my description of them, in the essay already alluded to. Mr. Sotham has doubtless imported some very valuable animals. I have not had the pleasure of seeing them but have conversed with several intelligent breed-

MR. RUST'S GRADE HEREFORD OX, WEIGHT 3,700 LBS.
(The sensation of the first N. Y. S. F., 1841.)

ers who have examined them closely, and who speak of them favorably. One gentleman writes me: 'They have nearly the size and breadth of loin of the Durham, but are coarser in the head and are not so handsome.'

"This is certainly a very different description from that given by MR. YOUATT. How are we to account for this discrepancy? It is but six years since MR. YOUATT wrote, and his remarks, therefore, cannot be supposed to apply to the breed as they existed 'twenty or thirty years ago,' as suggested by Mr. Sotham. It is well known, too, that his great work on 'British Cattle' was published under the auspices of the Society for the Diffusion of Useful Knowledge, and that he was aided in it by the first breeders in England, who are equally responsible with himself for the correctness of his statements. Writing with little individual interest or bias of his own, it is perfectly apparent through his whole work that it is his aim to present every breed possessing any charms, in its most favorable coloring. The quotations made by me are neither isolated nor garbled ones. The history of the struggle between the Durhams and the Herefords, as well as the Devons, Lancashires, etc., is the same throughout nearly all the most fertile districts of England. In almost every one the star of the Shorthorns has risen to the ascendant. It strikes me as idle to talk of 'unjust accusations,' 'prejudice and abuse,' as directed towards any particular breed. The contest has been a warm one—many severe things have been said on both sides, but how are we to suppose that the Herefords have been more 'abused' or called on to encounter more 'prejudice' than their rivals? One would naturally infer precisely the contrary. The Herefords are an indigenous breed—or, at all events, they separated from the parent stock (the Devon) at a period 'whereof the memory of man runneth not to the contrary.' On the other hand, it is little less than a century since a little handful of cattle, in the hands of three or four breeders, on the banks of the Tees, were attaining that superiority which has since asserted itself beyond intervening oceans, on the then scarcely discovered Ohio. 'Prejudice' is ordinarily enlisted on the side of antiquity and opposition to innovation. The Herefords were a known and favorite breed long before the improved family of Shorthorns had their origin. The Shorthorns were the innovators—innovators in size, shape, and last, but not least, in popular estimation, color. Is it not they instead of their opponents which have been made the particular victims of 'abuse' and 'prejudice?' I confess I have always so regarded it. I will cite one specimen of unfairness and 'abuse' which has been frequently resorted to against them. It is this: The advocates of other breeds in making their pretended experiments between improved Shorthorns and their own favorite breed, in feeding properties, etc., have repeatedly selected the unimproved Shorthorn (known as Lincolns, Teeswater, Holderness, etc.) to make the trial with, and then publish the result to the world as a fair experiment. But enough of this.

"Mr. Sotham differs as widely from Mr. Clay as from MR. YOUATT. Mr. Clay says: 'The Herefords resemble the Devons, the race of New England cattle.' Do the Devons or New England cattle approximate in size and breadth of loin to the Durham? A single instance of this kind might well be deemed an extraordinary one.

"I will not refer to Marshall, Lawrence, Culley, Loudan, or the other old English writers, because it may be objected that the breed has changed since their day [discretion the better part here. T. L. M.] It may be remarked, however, that they speak of the Herefords in the same general terms with Mr. Youatt. [Not Marshall. T. L. M.] I might cite the opinions of many eminent English [Shorthorn. T. L. M.] breeders residing in this country, in corroboration of their opinions, but shall not do it on this occasion as it would occupy much space in an otherwise sufficiently extended article.

"I will now ask if the concurrent testimony of all the standard English writers of cattle, ancient and recent, corroborated, too, by statements of many eminent breeders, do not prove that the Herefords have been—have been, too, if we may rely on the authority of MR. YOUATT, unchanged until within six years. If it is pretended that any so great and radical a change has been effected during that period, does it not devolve on him who asserts it, to show and to prove when and where, and how that change has been effected?

"If a few specimens of animals called by name are relied on alone to prove it, the question arises, what certainty have we that they truly represent the breed, or that they are pure blooded and unadulterated animals. The Shorthorn cross, for instance, engrafted on the Hereford might work wonders if the color was attended to. It would offer a very convenient solution to the physiological enigma of converting a small, shapeless and milkless cow into a stately, broad-hipped, deep-milking animal, all within the space of six years. I do not pretend to state that such a cross has been made, because I know nothing about it. Mr. Sotham doubtless ascertained the character of the men he dealt with. That Hereford oxen, and sometimes Hereford cows, have recently received many prizes from English Agricultural Societies, I have no disposition to deny or conceal. The oxen in such cases are exhibited in reference to their ripeness for the hands of the butcher. The Herefords are conceded on all hands to be a breed that feed kindly, and that they should occasionally receive prizes is much less surprising than would be the fact that they did not. But if a Hereford cow has triumphed over the Durham as a dairy cow, it is certainly a new thing under the sun, provided the Durhams were properly represented. I presume that no breeder of Herefords in this country will be found willing to challenge the breeders of Shorthorns to such a contest.

"But suppose we concede the point that a few English breeders have approximated the Hereford to the Shorthorn in size, shape, early maturity, etc., the question immediately presents itself, *cui bono?* If the Durham already possesses certain properties in the greatest attainable degree, what is the practical benefit of forming a new, or remodeling an old variety, only to attain the same properties? Is it not better to avail ourselves of the skill and industry of those who have gone before us, commencing where they left off, than it is to spend our whole lives in trying to overtake them by a road of our own? What should we say of the student of mathematics, who, disdaining to avail himself of the labors and discoveries of Newton and La Place, should begin *de novo* with the nine digits and attempt to build up a mathematical system of his own?

"But it is contended that the Hereford, if made equal to the Durham in its peculiar points of value, will superadd to them that hardihood and muscular power peculiar to the old Herefords. This cannot be. It is contrary to physiological laws, which every man's observation has recognized. The wild boar of the desert, or the pencil immortalized 'Land-pike' of Mr. Allen, is a better traveler and possesses more muscular power than the quiet and fat-secreting Chinese hog. The untamed Argall, that subsists among the rocks and glaciers of the Alps, is an animal of greater endurance and muscular energy than its descendant, the

WOODBINE.
(Calved 1850. Bred by Wm. H. Sotham; property of H. Bowen, Jr., Summit, N. Y. First prize N. Y. S. F., 1853.)

Leicester sheep. That course of breeding which modeled the Chinese hog, the new Leicester sheep or the Durham ox, has, and inevitably must deprive the animal from which they descended of that fleetness, strength and endurance of the rigors of climate, which was necessary for their protection before they were subjected to the dominion of man.

"The Hereford, if converted into a Durham, will cease to be a Hereford. Marshall, in giving his eulogistic description of the Hereford ox for labor, did not write of a quiet, sleep-loving animal, which would become a lump of fat at two years old. Mr. Clay, in referring to the same point, said nothing of animals possessing the size and broad loin of the Durham, and one year's earlier maturity. He spoke of a race 'resembling the New England cattle.'

"But after all, I am not quite convinced that the Herefords have been converted into Durhams. If Messrs. Walker, Hewer, and other breeders in Gloucestershire, had possessed such animals six years since, *it is astonishing that they should have escaped the notice of* MR. YOUATT—if they have been created, so to speak, since, it is a little short of miraculous. On the bleak highlands of Gloucester no breed has been cultivated with any very marked success. In the vale of Berkeley (a name so cherished by all lovers of good cheese) the prevailing breed is a compound of nearly everything, the old Gloucester, Hereford, Devon, Durham, Leicester, Suffolk, Dun, North Wilts, etc. MR. YOUATT says expressly: 'There are (in the vale of Berkeley) no Herefords for the pail, a few Devons, some Suffolks, a few North Wilts, and the rest Gloucesters, with various crosses.' And not a word does he say of a race of Herefords in this county equaling the Devons in size and form, 'filling the pail as high as most of them,' of one year's earlier maturity, etc., etc.

WILLIAM MILLER (UNCLE WILLIE), STORM LAKE, IA. (One of the Millers at Markham who bred Mr. Sotham's kind of Shorthorns.)

"Now, who shall decide when doctors disagree? MR. YOUATT has certainly given as much attention to the comparison of the English breeds as any other individual—is thought in England to be about as well qualified to arrive at a correct decision as any other individual, and, above all, had no personal interest in the result. This last consideration is of no little import. 'It is according to our gifts,' as the Pathfinder would say, that every man's goose should have a most swan-like appearance to himself. I do not doubt the propriety of Mr. Sotham's motives, or the sincerity of his convictions. He doubtless believes as he would have us believe. He deserves high credit for his enterprise for introducing so valuable a stock of cattle, but when he calls upon us to give up opinions supported by all the best English authorities—opinions until now unquestioned, we must demand something besides the authority of an interested witness.

"Yours truly,

"HENRY S. RANDALL.

"Cortland Village, Nov. 16, 1840."

The reader will here recognize the reasons for placing prominently heretofore the fact of Youatt's "History of British Cattle" being written by Berry, a Shorthorn breeder, and why we have been so particular to show the record of the Smithfield Show; the experiments of the Duke of Bedford; the sales of Mr. Westcar, and the value of Hereford beef on the London market.

In replying to Mr. Hepburn, Mr. Sotham wrote the publishers of the "Cultivator," as follows:

"Messrs. Gaylord & Tucker: Anxious as I may appear to defend the Herefords against all unjust attacks, I hope not to depreciate any other breed by it. However high I may value their good qualities, there are other breeds that are commendable. I think the right sort of Durhams hard animals to be beaten, and will never be intentionally depreciated in value from my remarks. If I advance my opinion to society, I am subject to public criticism, and am ready and willing to meet it. I neither aim at display nor ever expect to derive any pecuniary benefit from it; whenever I am in error, I hope to be corrected, as what I advise is from my own observation and practice, and I trust founded on facts. When proved so, all I ask is to have them confirmed. Should they tend to promote the public good, my hopes are realized, the obligation canceled, and I am satisfied. I will now endeavor to answer the objections brought against the Herefords by Mr. Hepburn.

"Mr. H., I see, has read Youatt, and quoted a passage from him that he would do better to reperuse. How far Youatt's practical knowledge extends is to me a matter of doubt, especially when he says, 'the Hereford cow is an inferior animal; not only is she no milker, but even her form has been sacrificed by the breeder.' I refer Mr. H. back to his own quotation, and he will perceive that Mr. Y. acknowledges the breeder to have tended to form his cow, to breed a good ox. I would ask Mr. H. if a good ox can be bred from an inferior animal, or if any breed of any kind can be kept up with inferior dams? There seems to me to be something wrong in the remark of Mr. Y. that does not read well to a practical man. I will also refer him to a sale of Herefords twenty years since that will show that they stood high at that time in the estimation of breeders. A sale of Hereford cattle took place at Wellington, Herefordshire, Oct. 18th, 1820, the property and extra stock of the Misses Tomkins, at which 29 lots, consisting of 16 cows, 5 heifers, 2 two-year-old bulls, six calves, were sold for £4,709 7s; the highest price was a two-year-old bull, which sold for £588, or $2,910. The average price of the 29 animals was £162 7s, nearly $800 each. This sale can be referred to by any Hereford or Durham breeder of note in England as a bona fide sale. If Mr. Youatt had been at this sale prior to writing his views of Herefords it might have been couched in a different position, as I cannot allow the good judgment of English breeders to be so disgraced as to give such prices for inferior animals. I cannot think Mr. Youatt gleaned his information from practical men, or where were the advocates of all other breeds when the owner of Trojan offered to show him against all England for a thousand guineas? This was more than twenty years since. I can refer to date and all particulars if necessary. If he had been an inferior animal it certainly would have been accepted, or could any inferior animal breed a bull of this description? Let England produce a better bull of any breed than old Cotmore, or Major, of the present day; then I will be convinced that Herefords can be beaten.

"I might trace the origin of the Durhams from the ancients without any proof that it was that breed only, deserved from it the ascendancy. I might pursue their advancement and find them supported by the hands of wealthy, popular men, who nursed them with the most tender care, who spared no expense in effecting their improvement, and gave them the advantage of their art, skill and study. In the meantime, a better breed may spring up, go through a less flattering process, and come out triumphantly, and I think all will allow that Durhams have had the decided advantage over Herefords in this respect.

"I am sorry that my remarks do not read

CARDINAL WISEMAN, 1202.
(Calved 1850; bred by Rev. J. R. Smythies, Lynch Court, Herefordshire; imported by Corning & Sotham, Albany, N. Y. First prize N. Y. S. F., 1853.)

well to Mr. H. after reading a standard work. I should like him to explain to me the true derivation of that standard, and establish its firm foundation. My idea is that the practical man that makes the improvement is the truest standard. A man may be proficient in disease, and effect many miraculous cures, which I have no doubt Mr. Youatt has done, and is as highly respected as any man in England, but we are all liable to mistakes, and are all open to correction. I am sorry that my efforts are called forth against him. Whatever the Herefords may have been, they are all they are represented to be, beautiful, noble animals. And I think other breeds will find that they are something to contend against, to maintain the top standing, and, as our worthy and esteemed friend A. B. Allen, Esq., says, that good and gentlemanly judgment is showing itself in Ohio, that is all I ask to judge, and will cheerfully abide by the decision. True judgment, and a right spirit in agriculture, is all that America requires to make her what she ought to be, the head of all nations. It was that only that kept England up, and raised her to what she is.

"I cannot agree with Mr. H. when he tries to deter men from importing. My idea is that we should get the best animals that can be obtained and of every breed that are likely to render service, leaving prejudice out of the question, and as many of them as the purse will allow. A good start is half the battle, and now is the time to make that start. Those who wish to obtain glory and honor in warfare must not come to the field when the battle is won, but must share some part of the burden. Money cannot pay a man for the care, anxiety, and risk he is exposed to in importing stock. It is a task that I shall be glad to see others undertake, as I have done my share of the duty. I must now do justice to those we have, which is much the pleasantest part; there is no art, no science, no study so pleasing, so substantially gratifying to the mind of man, as to fully develop the good points of animals, in his own superior skill and management. It may be made the theme of usefulness. My opinion of securing good stock, with economy in view, is

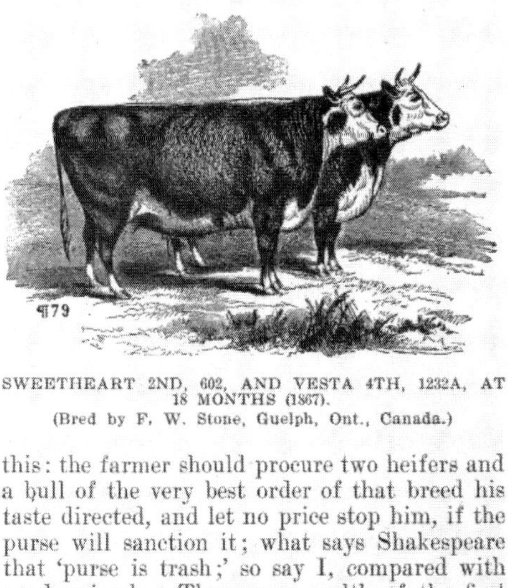

¶79

SWEETHEART 2ND, 602, AND VESTA 4TH, 1232A, AT 18 MONTHS (1867).
(Bred by F. W. Stone, Guelph, Ont., Canada.)

this: the farmer should procure two heifers and a bull of the very best order of that breed his taste directed, and let no price stop him, if the purse will sanction it; what says Shakespeare that 'purse is trash;' so say I, compared with good animals. They were wealth of the first people; why not continue that of the present?

"With any ordinary luck the progeny of two females will soon extend while the male improves the inferior, and adds value to the whole; this consideration will sustain high prices, and pure animals cannot be imported without paying high for them.

"I am further of opinion that pure breeders for sires that can be depended upon, should be upheld in high prices, so as to enable them to keep their stock without a stain, as I think the most important part of breeding lies here. The progeny always degenerates if the sire has the lightest cross; it is not so with the dam; the stock will improve if the sire is well selected.

"Mr. H. talks of showing spirit in selecting a breed of our own. I should like to hear his suggestions in commencing the breed, without having the best animals to resort to; even their crossing with a bad breed will take a full century to make anything out of it, if founded on the best judgment. I will allow there are a limited supply of the best Durhams, but not a tenth part there should be.

"No person can go to England and fetch the best animals, unless he is favored in freight, for less than $500 each.

"I see our friend, Mr. A. B. Allen, says that Ohio is the home of the lordly Shorthorns. I hail their prominent name, and as ours are principally of the feminine gender, will say York State is the home of the ladylike Herefords. As ladies are considered the first race of animals, I hope the Herefords will maintain it. I am, dear sirs, yours sincerely,

"Wm. H. Sotham.

"Perch Lake Farm, Jefferson Co., Nov. 25, 1840."

The "Cultivator," on page 16, Vol. 8, gives a correspondence that was published in the "Farmer's Magazine" of December, 1840, growing out of a challenge given by John Price, a Hereford breeder of England, to show a bull and twenty breeding cows of the Hereford breed against a similar number of any other breed. This challenge called out Mr. Bates, the noted Shorthorn breeder of England, and in the correspondence Mr. Bates says: "But I consider now, and have for about forty years been convinced, that the very best Shorthorns, of which there are only a few, are capable of improving all other breeds of cattle in the United Kingdom as well as the ordinary Shorthorns, which are far from a good breed and inferior to Herefords, Devons and others."

On page 19, same volume, Mr. Sanford Howard steps in and gives his testimony as follows:

"Messrs. Editors 'Cultivator': I recollect noticing in your paper some time since a request that those who have any knowledge of the Hereford cattle would give their opinion of its relative merits compared with the improved Shorthorn and other varieties. In Mr. Bement's communication, published in the August number, he says he thinks there has been no importation of Herefords excepting by Mr. Clay, Messrs. Corning & Sotham, and himself. This is a mistake. In the year 1825, if my memory serves me right, the Massachusetts Society for Promoting Agriculture received as a present from Admiral Sir Isaac Coffin, of the Royal Navy, a bull and a cow of the true Hereford breed, selected either by himself or his agent in

England. There came also with them, as a present from the same beneficent gentleman, an improved Shorthorn cow (Annabella), a bull of that breed (Admiral) had before been sent.

"The Hereford cattle were kept for one or two years in the section of the state where I then resided, on the farm of John Prince, Esq., of Roxbury.

"The cow never had a calf after coming to this country, and, it being supposed that she would never breed, she was slaughtered, but I have been informed that on killing her she was found to be with calf. As to the quality of her beef, I cannot speak, but presume Mr. Prince can give information on this or any other point relating to these cattle. The cow was certainly considered one of the most extraordinary animals for shape and size ever exhibited at the Brighton (Mass.) Cattle Shows.

"I believe the bull was not patronized to a very great extent while at Roxbury, it not being a stock-growing neighborhood, and he was subsequently taken into the interior of the state, where, if I am not mistaken, he was sold. I saw many of this bull's progeny in the vicinity of where he was first kept, and have owned some of them. They were generally highly esteemed. They made prodigiously powerful and active draught cattle—there was a majesty in their gait, and an elasticity and quickness of movement which I never saw equaled, and which, together with their beautiful mahogany color, and strong constitution, made them decided favorites with the Yankee teamsters.

"For dairy qualities, the progeny of this bull was, as far as my observation extended, an improvement on the stock with which he was crossed. I am aware that the Hereford has not the reputation of a milking race, and, though this may be generally correct, it is by no means certain that some grade of that blood may not be superior milkers. I will mention an instance in support of that idea. Several years ago some cattle of Bakewell's celebrated breed of Longhorns, called Dishleys, or improved Liecesters, were imported to different parts of this country. They were famous for fattening, but not for dairy qualities, yet the half-bloods were generally good milkers, and some of them were uncommonly superior in this respect. I know not how to account for this fact, unless on the ground of the constitution of the cross being so much strengthened that they were able to resist the inclemencies of the weather and to digest their food more perfectly and make the most of it.

"I will here remark that I knew many and owned several of the progeny of the Improved Shorthorn bull, Admiral, before mentioned as having been presented to the Massachusetts So-

HEBE 469, BONNY LASS 679, BRED BY LORD BATEMAN, AND GRACEFUL 545, BRED BY LORD BERWICK.
(Imported by F. W. Stone and foundation of his Hebe, Bonny Lass and Graceful families.)

ciety by Admiral Coffin, and I have no hesitation in saying that for the ordinary uses to which cattle are applied to the northern section of our country, I consider the stock of the Hereford bull above alluded to, decidedly preferable.

"From 1830 to 1837 I resided in the State of Maine, and my business was the breeding of various kinds of stock. I had never seen any full bred improved Shorthorns which appeared to possess sufficient hardiness of constitution to adapt them to so rigorous a climate and the

EMPEROR AT 2 YEARS (1867).
(Bred by F. W. Stone, Guelph, Canada. First prize N. Y. S. F., 1867.)

hard labor of the yoke to which oxen are there subjected. Accordingly, on commencing operations, I purchased of Hon. John Wells, of Boston, a bull of a cross between the Hereford and the Improved Shorthorn, with a slight dash of the Bakewell. This bull was the easiest animal to fatten that I ever saw—and was of very perfect symmetry, vigorous and active, and very heavy in proportion to the bone. His weight at six years old, after having been wintered on the coarsest fodder that the farm afforded, was 2,000 pounds.

"With this bull I bred some selected cows of various grades of different families, my object being to manufacture and establish a breed better adapted than any other, to the soil, climate and purpose of the section of country for which they were intended.

"As my stock arrived at an age to have the qualities tested, they gradually got into favor; but as the passions of too many was for stock of enormous size, and mine had been bred wholly with regard to useful properties, I had to wait until the superiority could be proved before their merits were generally admitted. I, however, took several prizes on different descriptions of cattle at cattle shows of the Kennebec County Agricultural Society. On leaving Maine, I sold some of the best stock which I had bred, to Mr. J. W. Haines, of Hallowell. I am informed that it is now considered preferable to any stock ever known in that section. Mr. Haines carries several of the highest prizes at the Kennebec shows; he took the first on milch cows, with one which was bred by myself —the first on bulls, and the first on heifers of the same stock.

"The recent importations of Herefords by Messrs. Corning & Sotham, of your city, must, by all accounts, be very valuable. A gentleman from Boston, and a good judge of stock, lately passed through here on his return from a tour of the West. He had seen the Herefords above spoken of, and also had seen the fine herd of Durhams belonging to Mr. Sullivant, near Columbus, Ohio, as well as much other fine stock. He pronounced the Herefords superior in fineness of bone and symmetry to anything he had ever met with.

"Cannot some of your distant readers be gratified by a sight of some of these fine Herefords in the 'Cultivator'? The portrait of Mr. Bement's bull, Dallimore, is certainly one of the best figures I have ever seen. His death must be a great public as well as individual loss.

"The remarks of Mr. A. B. Allen in the October number of the 'Cultivator' on the working and fattening properties of the Herefords, I have no doubt are correct; neither have I any doubt that wherever strong constitutions are required, and oxen are wanted for the yoke, the Herefords will be preferred to the Improved Shorthorns, and perhaps to all other breeds.

"Mr. Allen expresses some surprise that the lately imported Herefords have wide loins, and says the Herefords he had before seen were narrow at these points. I cannot think the Herefords are generally deficient in this particular. Youatt, in his work on cattle, published under the direction of the British Society for the Diffusion of Useful Knowledge, speaking of the preference given by the Duke of Bedford to the Herefords, after repeated trials with nearly all the other breeds in England, after mentioning some objections to the Herefords generally, says of the Duke's cattle: 'They retain all the length of quarter, and much of the wideness and roundness of hip and fullness of thigh, which have ever been esteemed the peculiar excellencies of the Herefords. A few of them might, in their fore-quarters, be mistaken for Devonshires, but with a broadness of chine and weight behind, which the Devons have rarely attained.' (Page 211.)

"But in choosing the breed, the grand object

THOS. DUCKHAM, M. P., 1816-1902, BAYSHAM COURT, ROSS, HEREFORDSHIRE.
To whose painstaking efforts the greatest credit for the maintenance and accuracy of the Hereford Herd Book is due.

would be to obtain the breed best calculated for the particular situation for which it was designed. Each has its peculiar excellencies and defects, and the one which may be best for some locations and purposes, may be worst for others. The great error in public opinion of the present day in regard to all kinds of stock is the passion for great size. The only correct medium is, I think, that of your correspondent R. in the August number, 'that the only desirable size is where we find the greatest weight in the smallest relative compass.'

"Sanford Howard.

"Zanesville, Ohio, Nov. 7, 1840."

Wm. Hy. Sotham thus wrote the "Cultivator":

"Messrs. Gaylord & Tucker: In looking over the back numbers of the 'London Farmers' Magazine,' I find in that of September, 1837, a portrait of Sir George, and as he is the sire and grandsire of a portion of our Herefords, I copy the accompanying account:

" 'The subject of the plate is a Hereford bull, called "Sir George" [* (405) 9999 T. L. M.], the property of John Hewer, Esq., of Hampton Lodge, near Hereford. In giving some description of him to our readers we cannot do better than to quote Mr. Hewer's own account of him. "Sir George," now ten years old, was got by Waxy, Waxy by Wellington, his dam Beauty by Old Wellington, the sire of Wellington. They are of the old original blood, and I believe confined to myself. Sir George has won five premiums.'

"I send you this account to show you a portrait of the original breed; and it is my intention to answer, as briefly as possible, the remarks of Mr. Randall and Mr. Hepburn. The latter says: 'I have selected my weakest adversary, confident of an easy victory.' I know not why, when each derives his information from the same source, both making many quotations from theory, the language of one may be rather more classical than the other, each grounding their hope of success on the name of a standard work, that standard not written by a breeder. Mr. R. launches out with similes from other writers, and then says: 'But, after all, I am not convinced that the Herefords have been converted into Durhams,' and I see from the extract of the Hon. Henry Clay's letter, that his cattle must have been selected from such breeders as Mr. Youatt describes, not from the best; therefore I perceive nothing formidable in any remark from either that would cause me fearful thoughts of an overthrow. If I cannot gain the victory justly, I shall not feel entitled to the reward.

"Now, gentlemen, I am quite willing to take you individually or collectively; you may quote theory back to the fall of Adam if you choose, and breed cattle from that instruction, and if you do not possess a fallen race I shall be mistaken in judgment.

"If we are to enter into a detailed controversy, we must suggest some means of bringing it before the public for inspection and decision. The 'Cultivator' (according to your account) is pressed with too much important matter to admit of long articles; this can be agreed upon, and winter evenings are best for farmers' study.

"I do not see anything more to be noticed in Mr. R.'s article, further than this: He presses me to challenge the Shorthorn breeders for milking. I am not disposed to give challenges, but as he has called it forth, I will show a milking cow in May next against him or Mr. Hepburn for a silver cup, value $15; mine shall be a pure Hereford, my opponent's a pure Shorthorn. I shall not extend it unless it is urged upon me. Mr. H. says: 'According to my own showing, many a good ox has been bred from an inferior animal.' I should like to know when this assertion was made to me. Have I acknowledged this Hereford cow to be inferior,

WALFORD (871) 47, CALVED ABOUT 1844.
(Bred by T. Longmore. Walford won many prizes and his blood has a powerful influence in the improvement of the breed.)

because Youatt has? Never. This must be a wild thought; probably I have seen as many of the best oxen in this country as Mr. H., or any other person. In the fall of 1838 I purchased myself, for barreling, nearly five thousand head of cattle, in the western country in the course of two months, and was engaged in that business four years previous, when from four to six

---

* Note. Where, in this volume, in early communications the herd book numbers are given for Herefords, they have been inserted by the author for proper identification, the herd books not being in existence at the time. As far as possible the American as well as the English (in parenthesis) numbers are given. T. L. M.

FREDERICK WILLIAM STONE, GUELPH, ONT., CANADA.
Canada's greatest importer of pure-bred live stock.

thousand were slaughtered annually, selecting the best for the New York market, and the whole time I did not see an ox that was bordering on goodness (for what I call a show beast); good breeding must be looked to for such an animal, and requires as much skill as it does to breed a bull or cow. I should have been much surprised at this remark from a breeder, had I not known he was led astray by the study of theory alone. Mr. H. may have seen larger oxen, but I doubt whether he ever saw the first quality. I shall leave the breeders of our cattle to answer the unwarrantable attack he has made on them. I shall only tell him that were they disposed to steal 'crosses,' represent them pure when not so, they could not show them for

ERASTUS CORNING, ALBANY, N. Y.

a premium; a true certificate of their pedigree is demanded. If they are entered as pure Herefords, they must prove so, or the beast is disqualified, the owner erased from the list and never allowed to show an animal again. Matchless [Mr. Sotham's cow, entered in the Herd Books as Spot 1070. T. L. M.] was entered as a pure Hereford, won the first prize, and I defy any man to prove there is a drop of Durham blood in her veins, or any other of our animals that I say are Herefords. I hope this is plain enough for Mr. Hepburn. We have a half-bred Durham heifer, as I have before stated, and a man that can distinguish a mule from a Spanish ass, can see this cross, though I think her a superior animal. I refer these gentlemen to a late number of the 'Farmers' Magazine,' to peruse the letters of Mr. Price and Mr. Bates; there they may profit from practical knowledge, and see what the Herefords were forty years ago in their estimation.

"I was honored by a letter from the Earl of Warwick a few days since, which I send you, and trust his lordship will not feel displeased at his name appearing in your valuable paper. Much information may be gleaned from it.

"I think I may say, with confidence, we have just received from R. Lovel, Esq., Edgecott Lodge, one of the finest Shorthorn heifers, coming three years old, that ever crossed the water. I will send you her likeness and pedigree at some future period; she is a pure herd book animal.

"I could say much on Mr. R. and Mr. H.'s articles, but will leave it to consult your space, or some other means, to do it. Their ideas of arriving into notice as breeders, in six or eight years, are quite ridiculous unless at an enormous expense to start with.

"Yours sincerely,

"Wm. Hy. Sotham.

"Perch Lake Farm, Jefferson Co., N. Y., Feb., 1841."

The following is the letter from the Earl of Warwick, mentioned by Mr. Sotham:

"7 Carlton Gardens, Dec. 28, 1840.

"Sir: I received your letter of the 13th of November, and accompanied by a 'Cultivator' paper; your wishing an answer induces me to send one, but not breeding myself any Herefordshire cattle, I fear I have little satisfactory to communicate; or any breeding stock on my farm to show you should you visit this country. I consider they can breed better stock in the counties where the whole attention of the farm is devoted to that object. I therefore buy in my stock, poor, between two and three years old; fat them, and feed them off for the butcher as I can get them ready on grass for summer keep; Swede turnips (on which I mostly depend) and hay for the winter keep, with some finish only, of oil-cake when necessary. To do which it would be immaterial to me what breed I bought, or where, depending on what would grow fat fastest as good butcher's beasts, the butcher being my customer, and I have no doubt the Herefords are the best breed for such objects, and notwithstanding well-bred ones are bought very dear (a proof of their estimation), I think they pay for their keep, turning into money fastest for the food they eat, and less liable to casualties from the thriving disposition

"THE WOODLEYS" (SOTHAM ESTATE), WOOTON, OXFORDSHIRE, ENG.
(View from the fields.)

of their constitutions to do well, and lay on flesh while growing. I send into Herefordshire and buy in lean, young stock of the best breeders, not trusting to fairs; I know how all are bred.

"The ox I won the first prize with at the Smithfield Show last year (¶ 65) (and the best beast of the class shown, ninety competitors) I so bought as a three-year-old (with many more) for £17 ($85), lean. He ran with the other steers, and was put up with them to fat, and I never thought of sending him to the Smithfield till two months before the show, as I never sent a beast there before. After the show I sold him to a London butcher to kill for £70 ($350). There is a wretched print of him in the London shops. The painter who did his picture for me is Mr. Davis, animal painter to the Queen, and lives at Chelsea. He has made a very good colored engraving, and one of which I shall be very happy to procure for you, if you will have the goodness to write to me when you arrive next summer in England.

"We breed some pretty good Leicester sheep in Warwickshire. My bailiff will be happy to show you anything I may have, if you are likely to be that way. I only breed sheep—we consider nothing like the Leicester for long wool, and Southdowns for short. A Leicester tup has been known to let for the season for £1,000. My House in London is 7 Carlton Gardens.

"Your obedient servant,

"Warwick.

"W. H. Sotham, Esq."

"Messrs. Gaylord & Tucker: Since writing you, I learn from Mr. Thornton (the person who has charge of our stock) that the heifer Spot, by 'Sir George,' coming three years old, calved on the 18th of January, is now giving from five to six quarts of milk daily over what her calf sucks; he is a fine bull, thriving very fast, is her first calf and lies loose in the stable with her; her feed has been cut straw and hay of about equal quantities, mixed with about a peck of bran. Gay is the dam of Spot, whose pedigree may be seen in the July number of the 'Cultivator.'

"Yours sincerely,

"Wm. Hy. Sotham.

"Albany, Feb. 20, 1841."

We submit that Mr. Sotham conducted this controversy in an able and gentlemanly manner. We next find Mr. H. S. Randall coming to the attack in an article entitled "The Shorthorns and Herefords."

"Messrs. Editors: It is with deep concern

¶ 87

TOMB OF THE PARENTS OF WM. H. SOTHAM, IN WOOTON CHURCHYARD, OXFORDSHIRE.

that I perceive by a communication in your March number that Mr. Sotham has lost some of that comity of feeling with which he professed to enter into the discussion of the relative merits of the Shorthorns and the Herefords, more especially that he has, without pausing to consider the pain and injury it would inflict, spoken in terms falling little short of utter disrespect of the breeding skill, and even the knowledge of what constitutes the valuable points of cattle, possessed by Mr. YOUATT, Hon. Henry Clay, Mr. Hepburn, and lately, my humble self. No man, I will venture to say, intimately acquainted with Mr. Sotham, values him more highly than I do, and I have ever believed that he was by no means properly appreciated by a large portion of our breeders.

"I well recollect with what profound pleasure I hailed Mr. Sotham's communication from England (published in your October number), that he was on his way to this country, 'with the best lot of stock ever seen together,' and his modest challenge to the owners of the best cattle in America, to meet him at the cattle show at Niblo's in October. It is an affair of more magnitude than every one may suppose for one man to buy up 'the best lot of stock ever seen,' even in the small Island of England, and I could not help fancying the deep mortification, the unavailing regrets of such men as Earl Spencer, Mr. Bates, and other English breeders, when they ascertained the fact and found that the broad Atlantic rolled between the aforesaid 'lot' and England. What deep and lasting gratitude then should every American have felt towards so great a public benefactor. But alas! all know that the 'ingratitude of republics' has passed into an adage. The fair day arrived at Niblo's, and with it arrived Mr. Sotham, with the Hereford cow (with the aliases) that 'won the first prize at Oxford, 1839, against all England;' the cart mare and cart colt, 'allowed by the best judges to be as good as England could produce;' the twenty-four rams that 'could not be beaten even in England;' and lastly, 'the pigs of various descriptions, the best that could be procured in England.' Mark the astounding sequel. The cow that 'beat England' was beaten by an American-bred Durham, though rumor says that Mr. Sotham actually condescended to wait personally on the committee, after they had retired to their room for consultation, for the generous purpose of imparting his superior knowledge of what constituted the true excellencies of cattle, and to inform them in what low estimation the

WOOTON CHURCH, OXFORDSHIRE, ENGLAND. (The cross marks location of the Sotham tomb.)

cow that subsequently proved to be the winner would be held in England. Whether any of the four-and-twenty rams—like the fiddlers of old, 'all in a row,' received premiums, I am unable to say, though I learn that some of Mr. Sotham's sheep were beaten by Mr. Clift's. And to show at what an utterly low ebb the taste of our country is in such respects, I will state that I am credibly informed that those splendid steeds, as well as those pigs, whose loss it is feared England will never be able to repair, were actually laughed at by two-thirds of the ignorant, impudent Yankees present.

"But enough of this. It only proves that men in advance of their age are rarely appreciated by it. Galileo found it so; Copernicus found

A TABLET IN WOOTON CHURCH.

it so; Capt. Symes found it so, and Mr. Henry Sotham, if he finds it so, should neither be grieved nor disappointed. It is your empty, swaggering, conceited fellow, who always proclaim their own 'best,' who are most successful in these degenerate days; for the modest and unassuming there is but little chance. The simple fact that the Committee of the American Institute decided against Mr. Sotham's Herefords, proves nothing. That they were 'the best of the lot' at Niblo's, or that 'were ever seen,' we have the most indisputable authority—the same which the Marshal Montmorenci had, that the Dauphin was a brave man, the Dauphin told him so himself. Will any man deny that this was

> "'Confirmation strong
> As proof of holy writ.'

"But I must say I think it was hardly magnanimous of Mr. Sotham, after seducing Messrs. Clay, Hepburn, myself and others into this controversy by honeyed assurances of dealing gently and lenient with us, to suddenly, without a word of warning, convert a merely friendly passage of arms into deadly strife. It might have evinced considerable nerve on the part of Fitz James to say to a party of wild Gael, 'Come one, come all.' But Mr. Sotham, when he says he is willing to stand a brush with Messrs. Youatt, Clay, Hepburn, etc., 'individually or collectively,' well knows that he utters a safe challenge. True, Mr. Youatt is concededly the first writer in England on cattle, Mr. Clay is a clever man in the Senate, and one of the first breeders of the various kinds of improved cattle in the Western States. Mr. Hepburn certainly writes like an intelligent man—but which of these men ever 'purchased five thousand cattle in two months,' or belonged to a concern 'which slaughtered from four to six thousand annually for four years?' If there be truth in the sage old apothegm that, 'He who kills fat cattle must himself be fat,' does it not follow by a parity of reasoning that he who buys and slaughters cattle must be an adept in the science of breeding them? Cannot your butcher, who wields the knife and cleaver, manufacture these implements better than your mere blacksmith who, perhaps, never cut up a beef in his life? We doubt whether this last process was ever performed by Mr. Youatt, unless in the way of dissection; and as for the Kentucky Senator, confess,

> 'An' thou lovest me, Hal,'

that there's many a man within half a dozen miles of Ashland, who has bought more, killed more, barreled more and ate more beef than thyself, and argal, knows better how to breed it. The fierce old Hepburns, of East Lothian, were drovers and butchers both in a border foray, but we doubt whether their peaceable Pennsylvania descendant has ever drove or slaughtered 5,000 cattle in his life. If not, what should he presume to know about breeding fine cattle?

"By the way, we should like to know what was the 'head and front' of this unfortunate man's 'offending' that he should be selected as the especial victim of Sotham's ire—used up—as little of him left as the famous Kilkenny cats. Was it for suggesting that the recently imported Herefords were indebted to a Shorthorn cross for their improved points? Mr. Hepburn undoubtedly considered this the greatest compliment he could pay them. Mr. Sotham, in the October number of the 'Cultivator,' page 161, in enumerating 'the best lot of stock ever seen together,' says, 'one-half blood between the Hereford and Durham, to show the cross, which I think an excellent one, probably better than the pure breed of either, and from what I saw of Mr. Cother's, of Middle Aston, it may be extended much further than is generally supposed, for his fourth cross was equal to the first —not the least sign of degeneration.' Can words be more explicit? And what did Mr. Hepburn do but 'follow in the footsteps of his illustrious predecessor?' Why, Mr. Sotham is as difficult to please (the best have their foibles) as the drunkard, who, reeling home at midnight, declared 'if he found his wife up, he would whip her for not going to bed when she ought to have done; if he found her abed, he would whip her for not sitting up and waiting for him.' Disagree with Mr. Sotham and you incur the fate of Clay and Youatt; agree with him and the shade of the mangled Pennsylvanian arises warningly before you. Mr. Hepburn only surmised (in common with many others) that Mr. Sotham has been a more fortunate man than he himself supposed—that instead of mere Herefords he had actually obtained that cross between the Herefords and the Durhams, which we have Mr. Sotham's own authority for supposing 'better than the pure breed of either.' Why should Mr. Sotham object to having it believed that his cattle have received that last finishing touch, which, by his own showing, will make perfection more perfect. Is it modesty—the fear of claiming for his own more than they deserve? Yet, he may be mistaken against himself. He says of the cow that was beaten at Niblo's, 'I defy any man to prove that there is a drop of Durham blood in her veins.' I presume there is not, but, supposing there was, and a good many of them, who could testify to it, except the breeder? I assert that my pig is pure land-pike, and now grant, for the sake of argument, there are several drops of the 'alligator' in him, how is Mr. Hewer or Mr. Sotham to ascertain the fact if I see fit to conceal it?

"Finally, Mr. Sotham challenges Mr. Hepburn or myself to exhibit a pure Shorthorn against one of his pure Herefords, as a milking cow in May next, for a silver cup, value $15.

SIR CHARLES (3434) 543.
(Bred by F. W. Stone, purchased by T. L. Miller, 1872, for $1,000 gold. From drawing by E. H. Dewey.)

Now did Mr. Sotham ever hear of a shrewd Yankee by the name of Ezekiel Peabody, who, having tried in vain to dispose of a horse for $40, rode him one bright morning briskly into the town of C. (where several sportsmen resided), and offered to wager $15 that the aforesaid horse could out-run anything in C. for a mile; and how that he finally consented to sell 'Connecticut Eclipse' to young 'Squire W. for $150, before the race came off? But would not Ezekiel have betrayed a great want of discretion had he suffered his motive for making the bet to leak out in the presence of a third person who was under no obligation to keep it from the other parties? I will merely add that this 'silver cup' challenge came to my ears long before it was openly made to me. *Verbum sap.*

JOHN R. PAGE'S CONCEPTION OF SIR CHARLES (3434) 543.

"In reply to that challenge I may be permitted to say that I never have paraded my own stock before the public. I have discussed the merits of the Shorthorns and Herefords as families without converting my communication into advertisements setting forth the merits of my own animals, as the vendors of patent medicines recount the wonderful virtues of their nostrums. My herd, always small, has been rendered still more so by recent sales, and circumstances have induced me to turn much of my attention to crosses between the Shorthorn and other varieties. But if Mr. Sotham is really in earnest—if he wishes to risk $50 or $100 on the milking properties of a pure Shorthorn and a pure Hereford, by the proper tests (say the actual measurement of milk—or, rather, the pounds of butter made during one week of each month from the time of calving to the time of exhibition—the matter to be decided at the annual fair of the State Agricultural Society—I hereby distinctly inform him he shall be met. Satisfactory proof to be had, of course, of the milk and butter produced by each cow, and the bet to be forfeited in case either party should 'sell out.' If Mr. Sotham accepts this challenge he will signify it in your next number, to enable the proper steps to be taken.

"HENRY S. RANDALL.

"Cortland Village, March, 1841."

We find in this controversy Hepburn charging fraud in the breeding, and when obliged to admit the merit of Mr. Sotham's cattle, then claiming the credit due to Shorthorn crosses. Mr. Randall there leaves the argument and the merits of the breeds and proceeds to personal abuse. These two men are representative Shorthorn men from the time that Berry entered upon the Hereford-Shorthorn controversy, from 1820 to 1830 down to the present time.

The Randalls, Hepburns, Allens, Pages, Stevens, Nichols, Matthews, Andersons, Judys, Sanders, Rusts, and the entire list are of one class, and made use of the same measures, and substantially the same language. They came before the stockmen of the world in 1834 with the livery of that august society termed the "Society for the Diffusion of Useful Knowledge," officered by the nobility of England, and used as their scribe a learned and scholarly student, and asked the world to witness the impartial history written by this scholar and endorsed by the nobility of England.

And then follow such men as Randall, Stevens, Page, Allen, Nichols, Anderson and Matthews, asking the stockmen to believe that history, and accept their claims on that title. If a political party had come into power on such fraudulent claims they would be buried beyond a resurrection when their crime was brought to light. If a lawyer, in the interest of the client, should have committed such a fraud in his practice as Berry did in the interest of himself and his associate Shorthorn breeders, he would have been prohibited from practice.

What Berry did under the cloak of the "Society for the Diffusion of Useful Knowledge," Randall, Allen and Stevens and their associates did in organizing the State Board of Agriculture of New York, and they used it as Berry used the English Society, and with this machinery broke the Hereford interest.

Let us look at this. Mr. Sotham, in 1840, brought to Albany a herd of Hereford cattle. Mr. Erastus Corning, a wealthy citizen of that city, bought an interest in them, and would have probably furnished capital to any extent needed but Henry S. Randall used the fraud of Berry, and copying after Berry organized the New York State Fair, and used it to defeat the Herefords, and the result was that Mr. Sotham and his enterprise was financially ruined, and on these ruins Mr. Randall and his associates were enabled to perpetuate the fraud that Berry inaugurated, and the State Agricultural Societies have been brought under this Shorthorn influence and each and all have been made tributary and obedient to the Shorthorn behest, and every man that has dared to question their right has been broken. Thus, the Shorthorn breeders again, in my time, have endeavored to do with this movement. This was witnessed by one of the most fraudulent conspiracies ever conducted, that was pushed during the year 1881 with all the venom and malice that could be devised. But it was met and hurled back upon the originators.

Let us see how Mr. Sotham met this abuse. We find his answer in the "Cultivator," on page 83:

"Messrs. Gaylord & Tucker: I have read with much amusement the highly facetious letter of Mr. Henry S. Randall, in your last 'Cultivator,' and laughed at its contents almost as much as he says 'the impudent Yankees' did at my steeds and pigs. My mirth, however, was mingled with a feeling of regret that I had incurred his wrath, and that his good sense (of which I had previously a high opinion) had allowed him to stoop to ridicule and ribaldry as a 'dernier resort,' certain signs of a weak argument.

"I know nothing of his friend, 'Mr. Ezekiel Peabody,' and therefore give Mr. Randall all the credit he can gain by his acquaintance; and

THE BATES SHORTHORN IDEAL, IMPORTED DUKE OF AIRDRIE (12730)
(From a drawing by J. R. Page.)

I leave all such passages, dictated as they are, by anything but good breeding and gentlemanly discretion, to pass for what they are worth, which I think will be very little, less even than the shadow of one of the Kilkenny cats, so wittily served up in the 'Olla Podrida' with which we are regaled. Such insinuations are unworthy of Mr. R., and are beneath my notice. I will merely answer those parts of his letter which appear businesslike.

"In the first place, I am aware I spoke highly and confidently regarding my purchases; but it was not bravado. I knew the prejudices I had to contend against, the true worth of my animals, and was willing to back my own opinion. I see by your last paper that I have raised up opponents who will throw out every obstacle against me, but 'Truth is mighty and will prevail,' and they shall not drive me from my purpose. I did not intend to say anything regarding the judges at Niblo's fair, but only wished for an opportunity to test their judgment. Mr. R.'s article has given it to me. I will show 'Matchless' against any cow in the United States, of any breed, as the most valuable animal, for a cup of the same value as was there given, adding the expenses of the judges, who shall be selected in the following manner: I will name mine, my opponent his, these two selecting a third; and to prove that I am not actuated by any exclusive feeling I will show a three-year-old Durham heifer lately imported against the one produced by my opponent, whether winner or loser, on the same terms. I will also show a Berkshire boar and sow now in the possession of Mr. John Thomas, of Albany, nine months old, which were a part of the identical stock so cruelly 'laughed at,' against anything in this state of the same breed, excepting Mr. Allen's herd, of Buffalo (as I think him the best judge to decide), for a prize similar to the one given by the Institute. I will go still further; Mr. Clift has the two ewes shown at the fair now in his possession, and I presume both have lived together and fared alike since then; if so, they shall be exhibited at Mr. C.'s farm; I will accompany Mr. R. there, who shall be the judge, without knowing to whom either originally belonged, and I will venture a similar cup to the one Mr. C. gained on his decision.

"As regards the milking dispute I shall not go beyond the proposition I have made, but am ready to fulfill that 'in earnest.' My cow shall be pure Hereford, but I could not agree to forfeit if not in my possession. She shall be one of the 'twenty-four' I imported last summer. It is not my wish to make any sort of gambling contest of this business, but to give an excellent breed of cattle a chance of being appreciated as they ought to be, and to silence their traducers,

A TYPICAL HEREFORD OF 1840, COTMORE, WEIGHT 3,920 LBS., AT 9 YEARS OLD.
(From an old painting.)

and I refer the public to Mr. R.'s various epistles to say whether or not they have made it incumbent on me so to do, or whether I had any alternative but to reply, or to submit in silence to his injurious misrepresentations. Should my offers be taken and the triumph be mine, I shall value the prizes as mementoes worth preserving; should I fail, a second trial may be more successful, as I am resolved on perseverance.

"Mr. Randall says I have 'seduced' himself and others into this controversy. Let him refer back to the 'Cultivator' if his memory fails. He will soon see who commenced it, and I will leave it to your readers to say which of us has most violated that amenity of feeling which ought always to accompany us, however eager we may be to impress our own opinions on the minds of others, or controvert those of our opponents. I ask Mr. R. what accusation can be more opprobrious than that of 'stealing?' And has not Mr. Hepburn endeavored to fix this charge on the breeders of our cattle? I also again ask him to cross his pure Berkshire with the 'land-pike' he speaks of, exhibit the produce to proper judges, with his certificate of pedigree and purity, and if he is not detected by them, the said produce and his own conscience, he may then venture on the 'alligator,' and I will believe that 'stolen' crosses may be concealed. I say now what I have said before, that a cross with a Hereford and Durham is an excellent one, probably better than the pure breed of either, but cannot be brought under the observation of a judge without immediate detection.

"I must now call your attention to Mr. Youatt's text book, to which Mr. Randall and Mr. Hepburn seem to have pinned their faith with an obstinacy which regards anything that may differ in the smallest degree with its dogmas, as an innovation and a heresy not to be tolerated. Mr. Youatt is unquestionably a man of great talent and judgment; he, however, is but a man and is liable to errors like others; while, therefore, his opinions have all the weight which they so justly deserve, with practical men in his own country, these men do not, like Messrs. Randall and Hepburn, make an agricultural Pope of him, or consider his book (excellent though it be) as the agricultural gospel. Science and improvement are too much on the advance to rest long on any particular point. But to return to the text book: On the 11th page he says, speaking of Devons: 'They have long been celebrated for a breed of cattle beautiful in the highest degree; and in activity at work and aptness to fatten unrivaled.' In speaking of the Herefords, page 32, he says: 'They are even more kindly feeders than the Devons, and will live and grow fat where a Devon will scarcely live.' And, further, in the same page: 'The Devons will acquire bulk and hardihood, and the Herefords a finer form and activity.' These are his exact words. Gentlemen, reflect on these conflicting statements, and reconcile these contradictions if you can. When Mr. R.'s classical knowledge, backed by Mr. H.'s 'intelligence,' have shed their light on them and converted such language into reason and proof, I shall then believe there is real and sound advantage in studying the classics; but till then I shall be content to make my quotations in humble dog Latin; to '*gallus meusego et nunquam animus*,' notwithstanding Mr. R.'s terrific range of literary acquirements from the 'Kilkenny cats' and his 'friend Peabody,' to 'Galileo' and 'Copernicus,' his high sportive and illuminating prose, his poetic taste, and 'slaughtering' satire. But enough of this. I will turn to another attack.

"When I sent you the 'Earl's letter' for publication I did not expect my motives to be impugned, or my character calumniated by an illiberal suspicion; my object was to show that the men of the highest rank in England know the value of good stock equally well as the practical breeder; and the advantages to be derived from a careful selection. The latter was a 'genuine,' off-hand, businesslike communication, in which forms and phrases were evidently less thought of than the sound information conveyed in it; and I very much regret that it did not please the fastidious taste of your several correspondents. Doubtless some of them thought that an 'Earl' should convey his information on the science of breeding cattle in a most labored and elegant pastoral composition. For the criticisms on myself, I care not; I write with a view only of promoting the farmer's welfare and the cause of truth. I am not so easily pulled down, therefore, by any error in etymology and syntax. I have taken my stand and propose to maintain it. Now, Mr. Editors, I will be very much obliged if you would tell me the names of your private correspondents, or publish their communications, so that I may distinguish my friends from my enemies.

"I send you a letter from Mr. Turner, Court of Noke, Herefordshire, a gentleman well known in the agricultural world, so that you may again state that you have seen the 'original' from which my extracts below are taken.

"'It is a well-known fact that there is no other breed that can compete with the Herefords as regards a profitable stock. For instance, the memorable old Mr. Westcar, the Buckinghamshire grazier, at the dinner of the Christmas Cattle Show, some years ago, when the first prize in the first class was awarded to a Durham ox, the owner in the heat of the moment rose and offered to bet a wager that he won the same prize on the following year with an ox of the same breed. Mr. W. silenced him with this public challenge. He would go to the next October fair at Hereford and purchase one hundred oxen of the Hereford breed, which he would feed and show against the same number of all the breeds in England, for one hundred guineas per head, or two thousand guineas, which offer no one dare accept.' This was done to show the rich men who then possessed the Durhams, that he was ready to support the Herefords with his purse. 'He then told the company that during his experience as a grazier he had fed and sold twenty Herefordshire oxen to the butcher at an average of one hundred guineas each, and he would defy all the breeders of Shorthorns together to say they had done the like. As regards my own breed of Herefords, I can boast of having bred the last ox sold to the slaughterer for one hundred pounds, which was in 1827, since that several steers under four years old, from sixty to seventy pounds. I have much pleasure in saying my herd has not degenerated, as will appear by the show at Hereford last week. All the prize cattle were either bred by, or descended from, bulls of my own breed. Mr. Perry purchased the "Goldfinder" of me when a yearling.'

THOMAS BOOTH,
The Great English Shorthorn Breeder.

"'As regards the Hereford breed for dairy purposes I can speak from experience that when well kept few will answer better. I can give an instance of a prime cow of my father's producing *thirteen pounds of butter per week* when allowed hay and cabbages the whole of the winter; but the principle of the Herefords being the stock, little attention is paid to the dairy. We all know to keep up cows to their milk requires the most nutritious food, which is seldom allowed to cows in this country. We have experienced a very severe winter, and the epidemic so prevalent, I fear good stock will not be very plentiful this spring. The present prices are very satisfactory to the breeder. The fair at Hereford was very pleasing to do business at, a complete clearance of all good stock, particularly bulls, of which there was a most splendid exhibition.'

"I now conclude with one remark for the satisfaction of Mr. R., that 'Matchless' was bred by Mr. Turner, purchased by Mr. Hewer for Mr. James Walker, and the only reason I altered her name was I had one named 'Spot' in my first importation.

"I will give my views on sheep breeding at some future period. I think nothing is a better test for profitable animals than those that get the greatest weight of animal food with the least vegetables.

"Wm. Hy. Sotham.

"Perch Lake Farm, April, 1841."

We now give a letter from Mr. Wm. Kingham, which may be found in the "Cultivator," on page 116:

"Messrs. Editors of the 'Cultivator': I have been induced to take up my pen by the perusal of a controversy between Messrs. Randall, Hepburn and Sotham. I have been acquainted with the Hereford cattle for the last thirty years, as a grazier, a dairyman and butcher. My father used to graze about fifty and dairy about seventy cows, of different breeds, Longhorns, Shorthorns and Herefords, in Oxfordshire, England. For feeding, the Herefords are not surpassed by any breed I have ever seen; for the dairy, I do not say they are invariably good milkers, though they are, many of them, very good. I never kept an account of the produce of a single cow, but one of my neighbors says he knew a Hereford cow that made sixteen pounds of butter per week. As a butcher, which business I worked at in London, and afterwards followed in Berkshire for eight years, I say the Herefords cut the best stall of beef I ever put a knife in. Mr. Hepburn takes Youatt for his author, and by his description is led to believe the Herefords to have been a trifling breed as lately as six years ago. I know them to have been as good thirty years since as now, and whenever they have been shown against the Shorthorns have oftener gained the prize than their antagonists. Mr. Youatt must have been prejudiced in favor of the Shorthorns, was ignorant of the qualities of the Herefords, or got his information from some one no better informed than himself. Mr. Hepburn very ingeniously endeavors to impose on his understanding by supposing the Herefords of Mr. Sotham to be the produce of a stolen cross with the Durhams. I have no doubt but Mr. H.'s experience, if he lives long, will convince him that his supposition was erroneous, and that he was misleading the judgment of the community and doing injustice to the breeder and importer by supposing them guilty of imposition. I should say, as a breeder, the Herefords need no such cross; but I should say as a breeder, that many of the Durhams, some of even Herd Book pedigrees, would be materially improved by a cross with the Herefords, as having a tendency to lengthen the rump and widen the hip of those whose edge or round bones are wider than their hips, the rump short and low, the tail high, and their skin as tight over their rump as if stretched over it with a pair of pincers. Such animals have a thick, heavy thigh, the thick, coarse buttock, supported by a large bone, coarse leg, the sides, as Culley* describes, being one layer of black flesh across another, the shoulder bones large, the points projecting. Such beasts are sure to be bad handlers, never get very thin, and never get fat, will get fleshy, but when it is on is no better than bull beef. With such animals a cross from the Herefords would be a great improvement. I have handled many Durhams with high pedigree, with all the above objections. I never yet put my hand on a Hereford that was not a good handler. My opinion of the Herefords is that they are the nicest breed of cattle, taken for all purposes, that walk the earth, and would have had some in Ohio long ago if my means had been equal to my wishes.

"I remain, gentlemen, yours respectfully,

"William Kingham.

"Springfield, Clark Co., O., March 11, 1841."

We now give a letter from Mr. Wm. Cother, of Oxfordshire, England, to Mr. Sotham, under date of Middle Aston, Oxfordshire, Feb. 1, 1841, which may be found on page 132 of the "Cultivator" for 1841:

"Mr. Wm. H. Sotham: I am happy to inform you that our ram season closed very satisfactory for the breeders of Cotswolds. Their superiority is acknowledged by the extraordinary demand and the high prices given for them, which is very easily accounted for by

MARCHIONESS, BRED BY E. CORNING, JR., ALBANY, N. Y.
(First prize, N. Y. S. F., 1867. From a drawing by Page.)

their being much better sucklers, maturing earlier, producing more lean, and heavier fleeces than the Leicesters. Their fame is now spread far and wide, and I expect in a very few years that nearly every long-wooled flock in England will be 'alloyed' by Cotswold blood.

"The heifer you call Eliza is not by Young Sovereign, but by Favorite, Jr., a son of Fitz Favorite. Fitz Favorite was by Favorite, the sire of Old Sovereign, and the latter was the sire of more prize beasts than any other bull of his day, and was a remarkable instance of successful in-and-in breeding, being the produce of a mother and son, the pedigrees of which can be given, if required, for a period of more than forty years. And I would add that the dams of each of the bulls I have named were pure Herefords, the pedigrees of which can be given with equal accuracy. The dam of Favorite, Jr., I knew well, and it is my firm opinion that I never saw ten so good.

"This, I should hope, would be sufficient to satisfy the most skeptical as to the purity of her pedigree on the male side. Her dam was bought of Messrs. Brown and Lion, the great North-country cattle dealers, and was purchased by them at Darlington, in the County of Durham. It was fortunate that you did not have her sister, as she has cast her calf.

"The whole of my three years and nine months old steers by Favorite, Jr., grazed on my inferior land and finished with £3 worth of oil cake each, have made this Christmas £41 apiece, a price which I am of opinion where very few Shorthorns or Herefords have arrived at, in the same time, and under similar treatment. These were the second cross from Shorthorns, with a pure Hereford bull each time, a cross I do not by any means recommend you to adopt. You are in possession of some of the very best Hereford blood this country can produce (a few of the heifers are perhaps a little too much in-and-in bred), and all you now require is a couple of first-rate bulls, with the addition of a few heifers of different blood to make proper crosses with, to start you a first-class Hereford bull breeder, and depend upon it that your neighbors will ultimately discern that they can live harder, work better, feed equally quick if not more so, and produce a stall of meat superior in quality, with a less quantity of coarse, than the Shorthorns, and milk as well as the highest-bred animals of other kinds. In short, I believe they will pay more money for the food they consume than any other breed, in which opinion many of the most eminent graziers agree, some of whom reside in Shorthorn districts and travel nearly a hundred miles to Hereford fairs to buy oxen—a distance unparalleled by graziers in search of cattle of any other kind in this country.

"I advise you to breed pure bulls, and let others cross for the shambles, and in so doing I think each will benefit his country as well as himself.

"Hard things, indeed, may justly be said of some cattle wearing white faces, and with equal truth it may be observed that there are many very bad ones of beautiful roan and spotted color, with Shorthorns; such are frequently good milkers, and so are the Herefords which are of low breed and bad form. I imagine that Mr. Youatt and others who have designated the Hereford cow 'an inferior animal,' could not have done so from actual observation, for it so happens that in symmetry of form, with substance and quality combined, a more beautiful animal (of her species) cannot be found in Britain. Nor can it be very pleasing to owners

TROMP, BRED BY W. H. SOTHAM.
(Property of A. Ayrault, Geneseo, N. Y. First prize, N. Y. S. F., 1851. From a drawing by Forbes.)

of such superb animals as are very many of the Hereford cows, to have them so dominated in what is called 'A Standard Work on British Cattle.' Nor do I think such an opinion (libel) ought to be so uncontradicted, given as it must have been (one would suppose) from information and not from ocular demonstration.

"However 'astonishing' it may appear to Mr. Randall that the Herefords in the Gloucester Hills 'should have escaped the notice of Mr. Youatt six years since,' they have located there long before that period, and in many instances cows and heifers have been known to weigh from fifteen to nineteen scores [300 to 380 lbs. T. L. M.] per quarter [or 1,200 to 1,520 pounds to the carcass. T. L. M.], when dead, exclusive of hide and tallow, and the oxen from twenty to twenty-eight scores [400 to 560 lbs. per quarter, equal to from 1,600 to 2,240 lbs. per

WILLIAM H. SOTHAM, IN HIS 80TH YEAR (1801-1884.)
Fifty years the Champion of Herefords in America.

carcass. These carcass-weights would make the live-weights, according to the fixed English system, 2,100 to 2,660 lbs. for cows and 2,200 to 3,920 lbs. for oxen. T. L. M.]

"A Hereford steer and heifer, both bred in the parish of Northleach, Gloucestershire, fed in the County of Wilts and slaughtered at Oxford this Christmas, the former under four years old, weighed nearly eighteen scores per quarter [1,440 lbs. dressed, equal to 2,520 lbs. live weight. T. L. M.] and the latter three years and six months old, more than seventeen scores per quarter [1,360 lbs. carcass, equal to 2,380 lbs. live weight. T. L. M.]

"I now take my leave of the 'lady-like' females, for the purpose of pursuing their 'lordly' sons of the Smithfield Club Show, where the tug of war is annually kept up between the two contending breeds. Of their extraordinary fame there, let the annals of the club testify; but I would observe, by the way, that they are by no means well represented there, in proof of which a great number of graziers of high reputation, viz., Messrs. Rowland, Lidbrook, Terry, Hewett, Manning, the three Pains, Bull, and many others equally well noted, who are purchasers of a large quantity of the very best steers Herefordshire produces, seldom, if ever, exhibit an animal; the reason of which, as some of them have stated to me, is 'that winning a prize entails a certain loss,' while the breeding and feeding of them is almost neglected by wealthy owners of the soil.

"Not so with Shorthorns. They are reared, fed and shown under the fostering care of the Marquis of Exeter, Lord Spencer and Brownlow, Sir Charles Knightly, and other opulent men to whom expense is not an object. In making these observations, do not suppose that I wish to detract anything from their merits. On the contrary, I consider them a credit to their noble owners. But I cannot pursue this subject further without transcribing the opinion of an eminent breeder, Mr. Bates of Kirkleavington, Yarm, Yorkshire, whose cattle bore away nearly all the Shorthorn prizes from Oxford. He says, at page 426, 'Farmer's Magazine' for December, 1840: 'I visited Hereford about fifty years ago, and was then and continue still an admirer of the best variety of cattle (Herefords). But I consider now and have for above forty years been convinced that the very best Shorthorns, which are only a few, are capable of improving all other breeds of cattle in the United Kingdom, as well as the ordinary Shorthorns, which are far from a good breed, and inferior to the Herefords, Devons, and others.' And so would any moderate judge of stock conclude from taking a survey of Smithfield market at Christmas, where and when some of the meat of nearly every kind are pitched, the Herefords reigning paramount to any other breed, in numbers and quality combined, making more money per head than a like number of any other variety.

"Should the position I have taken be doubted by any of your American opponents I would say to such, come and see, and do not be satisfied with a view of a few inferior of their kind, but go home to the best breeders' houses, where they will be met with a hospitable reception and a hearty welcome, and will find such animals as are worthy of a place in a herd book; but in the absence of their names and pedigrees in print their own good qualities will be found sufficient passport.

"It may be asked what reason can a man find for resorting to a cross breed who so extols the

LUTHER TUCKER, SR.
(America's first great Agricultural Editor.)

Herefords? The question is solved in a few words. I was resolved to breed rather a large size, and it being difficult to procure large, well-bred Hereford cows, except at very high prices, and, not having a long purse, I preferred as good Shorthorns without pedigrees as I could procure, rather than Herefords under the like disadvantage, considering, with the 'alloy' in my mind's eye, that by so doing I should procure a rent-paying, though not a bull-breeding stock, and I have much reason to be satisfied with the steps I took at that time, since which I have added some well-bred Herefords to it, and am now in possession of one bull by Cotmore, the Oxford pet, and two others, embracing the blood of old Trojan and Old Sovereign in a high degree, which, I think, would be likely to do good in America, as a very near relative of the two latter has done in Scotland, a son of which won a prize in London this Christmas.

"I should deem it impossible to get up a

Hereford herd book here. The breeders are so satisfied of the superiority of their own breed that they are at perfect ease on the subject. But I advise you to have a well authenticated pedigree with every beast you import, and commence a herd book of your own.

"Mr. Wm. Hewer's Major, the sire of most of your heifers, won a prize at Farringdon, beating a number of Shorthorns, and at Cirencester he did the same. Major and a daughter of Sir George and his (Major's) son won the prize for bull, cow, and offspring, a daughter of Old Sovereign winning another prize as the best breeding cow. And a bull calf, ten months old, by Major, won another prize as the best under two years old, and here also they came in competition with Shorthorns. Will it be asserted again that 'on the bleak highlands of Gloucester no breed has been cultivated with any very marked success'?

"I say it has been for twenty years, and can prove it, and I say more: no man has a right to

CASSIUS M. CLAY, WHITE HALL, LEXINGTON, KY.

injure another by an assertion he cannot prove. These may be considered hard words, but they are just ones, used only for the purpose of correction, and not intended to give offense.

"My three fat ewes, which you saw, were killed at Oxford last month by Messrs. Greening, Alden, and Barr, and weighed respectively 228, 185, and 184 pounds.

"Your friends and acquaintances are well, many of whom speak anxiously of your success, which I sincerely hope may exceed your most sanguine expectations, nor do I doubt it, for I consider you have fixed on the breeds of cattle and sheep well suited to the rigor of your climate; recollect Devons are natives of a much warmer country than the Herefords.

"Yours truly, Wm. Cother."

It will be remembered that Mr. Sotham, in support of his claim for the Herefords as milkers, gave in proof that the year before (1839) at the first show of the Royal Society of England the Hereford cow of Mr. Smythies took the first prize for the best cow in milk. On page 57 of the "Cultivator," Vol. 8, may be found the following testimony, under the caption of "The Hereford Prize Cow:"

"We have a letter from our friend, Joseph Cope, of East Bradford, Pa., in relation to the statement made by Mr. Sotham that 'a Hereford cow won the first prize, as the best cow for dairy purposes of any breed at the great show of the Royal Agricultural Society at Oxford in 1839.' Mr. Cope doubts this fact, and to show that such was not the case he gives a detailed list of all the awards of prizes to cattle at that show, quoting from a newspaper and pamphlet account in his possession. We have compared Mr. C.'s account with that published in the 'Farmer's Magazine,' the organ of the society, and find it all correct with the exception of so much as relates to the fifth class, in which there is no notice of the animal which took first prize. The first prize of fifteen sovereigns, 'for the best cow in milk, which shall, in the opinion of the judges, be best calculated for dairy purposes,' was awarded 'to the Rev. J. R. Smythies, of Lynch Court, Hereford, for his Hereford cow, aged nine years, six months.' The second prize was awarded 'to Joseph Badcock, of Pyrton, for his Durham cow, aged fourteen years and two months.'"

With this array of testimony in favor of the Herefords as brought before the public by Mr. Sotham in 1840-41, the breeders of to-day will be surprised that they did not succeed in 1840, and then become the dominant breed for this country as they are now in the West and are bound to become in the East as well.

We find on page 125 of the "Albany Cultivator," for 1840, a letter from C. N. Bement, and take the following extracts from the same:

"In Missouri, Illinois, and Wisconsin, where only beef is wanted, and where they have to be driven great distances to a market, from what I have seen and can learn, the Hereford certainly would be preferable, being hardy in constitution, good travelers of great size, will fat at an early age and will make more pounds of beef with the quantity of food consumed, and when better known in market will command the highest price.

"Three Hereford and three Durham cattle were put in the stalls to be fattened on the 3d of November. The weight of the Herefords was then 33 cwt. and that of the Durhams 38 cwt. and 14 lbs. Between that period and the 30th of March, when all were sold at Smithfield, the Durhams had consumed 12,755 pounds weight more of turnips and 1,714 pounds more of hay than the Herefords, but the Durhams, notwithstanding the large size when put to fatten, and the greatly larger quantity of food consumed, sold for only twenty shillings more per head than the Herefords, and such, I believe, will ever be the results of similar trials, when one class of animals has been properly fed and the other overfed, the merits of the breed equal.

"At the meeting of the Smithfield Club in December, 1839, the Herefords took the first four premiums in Classes 1 and 2; in Class 3 the second prize and second prize in Class 7. In the prizes for extra stock, the Herefords took the only prize. In the list of commendation, the Hereford stands 13, Durham 8, Sussex 1, North Devon 1.

"I have examined the Hereford cattle lately imported by our spirited and enterprising citizens, Erastus Corning, Esq., and Wm. H. Sotham, as noticed in the last 'Cultivator.' I must confess I was greatly disappointed in the size and general form of the cows, for they were apparently as large as the Durhams and possessing the broad loin, large, capacious bodies, deep, broad and projecting brisket, but with a coarser head and neck, which, to those familiar with the Durham, would appear oxy, if I may be allowed the expression. The shape and size of their udders would indicate fair milkers. Mr. Thornton, the person who has charge of them, informed me they gave on an average full a middling quantity of milk."

R. A. ALEXANDER, LEXINGTON, KY.

CHAPTER XII.

More Early American Hereford History

It is reported that W. C. Rives of Virginia imported Herefords, the date and number of which I have no account.

The Hon. Henry Clay of Kentucky imported two pair of Herefords in 1817.

Admiral Coffin, of the Royal Navy, imported and presented to the Massachusetts Society for Promoting Agriculture a pair of Herefords. The cow, not breeding, was slaughtered, and the bull sent to North Hampton, Mass., where he was kept for several years, and his produce highly esteemed as work cattle, as butcher's beasts and in the dairy.

In 1839 Mr. W. H. Sotham made his first importation of Hereford cattle. On his arrival in New York he sold an interest in them to Hon. Erastus Corning of Albany, N. Y.

Mr. Sotham was born on the 25th of January, 1801, in the village of Wooton (¶ 73a), Whitechurch, Oxfordshire, England. His father owned his own estate of 200 acres, which descended to the oldest son. William Henry was kept at school until fourteen years old, when he took a working place on the farm for two years as under teamster, having a team of four horses, and a boy to drive. From that time he was in charge of the sheep, the farm carrying 150 Cotswold ewes. He then had charge of the colts, breaking and training them; thus for several years having charge of the several departments upon the farm, in the handling of the stock and marketing of the same until 1832, when he came to America with large expectations, obtained from descriptions received.

Stopping a short time in New York he thence went to Buffalo by the New York Canal, from there to Cleveland, Ohio. He here met Mr. Henry Coit, of Euclid, near Cleveland, who owned a large tract of land at Liverpool, Medina County, about twenty-five miles from Cleveland. He went with Mr. Coit to look at the land, finding a comfortable brick house with a small farm cleared. Mr. Sotham made an engagement to take charge of the farm and sales of land. He remained here for two years, having accumulated a considerable stock of cattle and horses. He took the cattle to Cleveland and sold them, and the horses he took to New York.

After selling the horses and paying the proceeds over to a brother of Mr. Coit's, Mr. Sotham returned to England. Before making the importation of Herefords in 1840 he had been to England four times, having brought over in that time a few sheep.

In 1840 he imported twenty-one Hereford cows and heifers, a two-year-old Hereford bull and two Shorthorn cows. One of the Shorthorn cows, bred by Sir Charles Knightly, died on the passage. The other Shorthorn cow, "Venus," bred by R. Lovell of Edgecott, was sold to Mr. Thomas Hillhouse for $1,000.

He again returned to England and brought out ten Hereford heifers and a bull, six first-prize Cotswold ewes, shown at the Royal fair in 1839, and seventeen Cotswold rams.

A third importation was made in 1843, when he took the whole steerage of the ship Hendrick Hudson, which he filled with Hereford heifers, one bull and several calves. He met a heavy and severe storm off the bank of Newfoundland. The severity of the storm made it necessary to close the hatches for several days, and the result was the suffocation and consequent loss of the entire shipment.

During this time the Hon. Erastus Corning was interested with Mr. Sotham in his importations. Mr. Corning had large business interests and was influential in the political world. Recognizing these facts, the Shorthorn breeders made special efforts to detach Mr. C. from the Hereford interest, and were finally successful. They went still further and endeavored to prevent the stock from going to Mr. Sotham at all, but on this point they failed to carry out their plans. Mr. Corning met their endeavors in this particular way by saying that Mr. Sotham should have such as he wished and sold to him at favorable prices and terms.

We wish here to call particular attention to the chapter which precedes this, and gives the details of the Hereford-Shorthorn controversy as waged in the agricultural papers and at the different shows and fairs from 1834 to 1841; the New York Agricultural Society being founded for the special purpose of promoting the Shorthorn interest.

We give in the chapter on Fairs and Shows a full account of the formation of the New York State Fair, as a matter of reference.

It will be seen that the Shorthorn breeders used every means in their power to defeat Mr. Sotham. We will follow that controversy from 1841, in which year the New York State Society organized their first show. It will be shown that that society was under the control of the same parties that met Mr. S. and his Herefords in 1840, with the correspondence we have given.

The writers of that day speak of the Sotham importation of Herefords in the highest terms. A. B. Allen says of them: "that they gave him a different opinion of the breed from what he had before."

The committee of judges at the first fair held by the New York State Agricultural Society recommended a special premium to the Hereford cow "Matchless" (¶ 74F) and further spoke of the Herefords belonging to Messrs. Corning and Sotham as entitled to a class, and recommended that a class should be made and special premiums should be awarded them, which, so far as we are enabled to find, was not done.

Mr. T. C. Peters, writing from London January 6th, 1842, says: "One of the best Hereford bulls I have seen, indeed one of the best I ever saw of any breed, is going out to Albany by the packet ship Hendrick Hudson." Major was purchased by Mr. Sotham in person.

We have in our review of the Hereford-Shorthorn interest from 1834 to 1841, brought Mr. Sotham's connection with that controversy fully before our readers. We have shown that he brought in support of the Hereford claims an array of testimony that was a complete establishment of his claims; that it was done in a manner creditable to him as a man and a breeder. We have shown that his opponents admitted that his stock was excellent, of the best quality; that it was better than the history of Youatt would warrant them in expecting to see, and they claimed it must have a cross of Shorthorn to give character. Other and disinterested parties came to Mr. Sotham's aid to bear testimony to the merits of the Herefords, and their testimony was not only for the time then under controversy, but covered from a quarter to a half century previous, and the only testimony that the Shorthorn men who opposed Mr. Sotham brought forward in favor of the Shorthorns and in opposition to Herefords was Youatt's History of British Cattle. This history, we have shown by Youatt himself, so far as the Shorthorn breed was concerned, was written by Rev. Henry Berry, a Shorthorn breeder.

We have shown from Youatt's history of the Herefordshire cattle that extensive experiments had been made by the Duke of Bedford in the feeding and grazing of Shorthorns and Herefords that had resulted in displacing the Short-

"WHITE HALL," NEAR LEXINGTON, KY.
(Residence of Cassius M. Clay.)

horns and establishing the Herefords on that estate, and we quote Youatt as saying that the reason why the date for these experiments was not given was because they were not satisfactory to the patrons of the Shorthorns. We have shown that Youatt, in the history of several of the local breeds, credits the Herefords as giving character to those breeds in improving the feeding, grazing and dairy qualities, that, had they been given in the history, would have placed the Herefords in the first position at the time he wrote.

We state distinctly and confidently that Mr. Sotham produced a class of cattle in 1840 that finally established the claims he made for superiority, that he brought forward testimony, clear and convincing, that these qualities had been in the breed for half a century, and that the success of the Herefords at that time was defeated by a combination of Bates' Shorthorn breeders, and we have shown that the New York State Fair was organized by these men in 1841 at Syracuse. At this fair Mr. Rust showed a fat ox with a mottled face (¶ 75), which took the first premium, for which the Shorthorn men claimed the credit. Mr. Sanford, one of the leading writers of that day, and a breeder of large experience, met this claim by showing that the Hereford blood was the controlling element in his breeding, and that he owed his excellence to the Hereford blood. He says:

"While on my late trip to the East I saw this ox of Mr. Rust's. He is truly a most superb animal. His portrait in the 'Transactions' does not do him full justice. He is finer in the neck and every way more finished than the picture represents. He has, both in shape and color, all the leading characteristics of a Hereford; his shoulders are well set, his chine full, back short, loin and hips very wide, rump long, legs clean and sinewy, and he is considerably heavier than any other animal I ever saw of so little bone and offal. At the time I saw him Mr. Rust thought his weight could not be less than 3,700 pounds, and it has been ascertained by repeated weighings that his gain was at least three pounds per day. Notwithstanding his immense weight, he was from the justness of his proportions very active. When lying down, he would get up as quick as a suckling calf.

"I saw the man who said he raised this ox and the history which he gave of him was that the bull which sired him was 'part Hereford.' In this both he and Mr. Rust agreed. I cannot see why this statement need be doubted, for, according to an account which Mr. Bement has published, some Herefords were introduced into this part of the country several years ago. But history and tradition out of the question, it appears to me there would be as much propriety in taking an animal which would show all the principal points in shape and color of an improved Shorthorn as a specimen of the 'native stock' as there is in taking this ox as such. An example of this kind would probably be regarded by the advocates of the Shorthorns as not altogether fair."

At the close of this fair the judges made the following report: "Your committee further report that a new and beautiful race of cattle was presented for their examination, the Herefords imported by a distinguished breeder of cattle, residing in Albany County, which they take pleasure in recommending to the attention of those who desire to improve their stock. Your committee recommend a special premium of twenty dollars for the Hereford cow 'Matchless,' as we consider her a very superior animal, and they would also suggest the propriety of offering and awarding premiums for the best blooded animals of each individual breed, Improved Shorthorn Durhams, Herefords and Devons, at their next agricultural meeting, in addition to premiums offered for the best animals of any breed." (Report of Com., page 49, Transactions N. Y. Agl. S., 1841.)

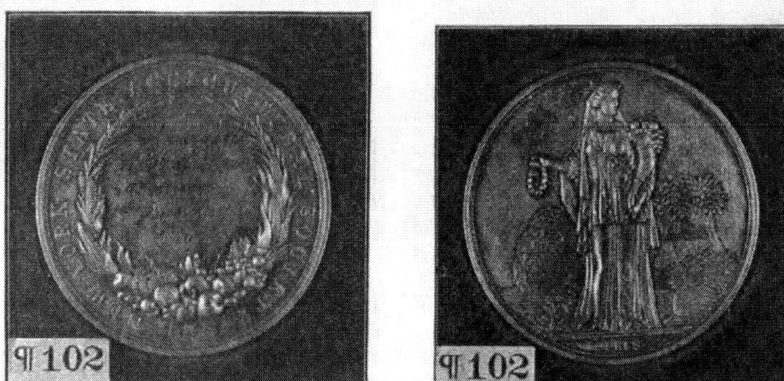

MEDAL OF THE N. Y. S. A. S., AWARDED TO WM. H. SOTHAM, 1856.

In 1842 this fair was held at Albany, and we quote from the judges' report, published in March following, six months after the fair. Report of cows, heifers, etc.:

"In publishing the account of the State Fair in the November number of the 'Cultivator,' the following report was omitted: 'The committee of the New York State Agricultural Society on Cows, Heifers and Heifer Calves, comprising Classes V, VI, VII and VIII, respectfully report:

"'That they have attended to the duties of their appointment, in which they experienced the embarrassment usual on such occasions, from finding the animals numerous while the premiums were few, but adopted the rule that a majority in number of the committee should of course in all cases be decisive.

"'Your committee found it no easy matter to decide among so many fat calves as were shown in Class VIII, their respective merits and faults being alike covered and hidden by flesh, so much so that they were really better adapted for the inspection of a committee of butchers than of breeders.

"'The practice, now too common, of fattening breeding animals for exhibition is not only wholly without utility, but is so bad and injurious in every point of view that it ought to be discountenanced.

"'It will not fail to be noticed that all the foregoing premiums are given to animals of the valuable breed known as "Durham Shorthorns," against which kind there were on this occasion no other breed shown in competition, except Herefords, of which there was a beautiful and very creditable exhibition, consisting of a portion of the herd of Messrs. Corning and Sotham, some individuals of which this committee would highly recommend, especially as being good specimens of that important quality, good handling, always essential to excellence.

"'Your committee (of which a portion if not a majority is composed of what might be called "Shorthorn men," either by preference or interest, as Shorthorn breeders), from motives of delicacy, not to say generosity, did not deem themselves called on to decide between these two rival breeds and against the Herefords, which would have been, for the most part, and in effect, their decision, if made on this occasion.

"'In England, the home of both breeds, where beef is the first and almost governing consideration, the Herefords as a breed, it is well known, have long maintained a sharp and often successful competition with the Shorthorns for feeding purposes, especially as a grazing stock, while it is claimed and now generally conceded by well-informed, dispassionate persons in England, that the well-bred Shorthorns have the merit of earlier maturity and are also entitled to the preference for stall feeding and more especially and decidedly so for dairy purposes, in which the Shorthorns and their crosses are believed to excel all other breeds and that the pure-bred males of this breed are capable of improving all other breeds of cattle, certainly a most important consideration, and especially so in this and all the northern portions of the United States.

"'It is understood that the Herefords have not yet been sufficiently tried in this country as milkers, in the absence of which there seems to prevail at present an unfavorable impression of them as dairy stock, which impression, it is hoped, may be soon done away, if, as their friends claim, the Herefords are really a superior milking breed. Some of their crosses with native live stock, now existing in Massachusetts, descended from an importation of Herefords made many years since, by Admiral Coffin, are understood to have proved excellent milkers.

"'It is, besides, claimed for the Herefords that they will make good working cattle, being strong and active, which is not doubted. It is also conceded that the quality of the Hereford beef is excellent. Therefore, taking no more than a fair view of their case, the Herefords must in all probability prove a highly valuable stock in those portions of this country where the grazing cattle for beef is a primary object.

"'Under these circumstances your committee would ask in behalf of the newly-imported Herefords a fair chance—and that they be allowed after coming from on shipboard to get well upon their feet before they "enter the lists" against the now well-established Shorthorns. If the Herefords are cherished and encouraged for a time it is to be hoped that the now favorite Durhams may by-and-by, in future competition, find in them "foemen worthy of their steel."

JOHN MERRYMAN, OF "THE HAYFIELDS," COCKEYSVILLE, MD.

"'If the Herefords were distributed in more hands so as to give room for competition among themselves, your committee would suggest the expediency of hereafter offering premiums for them in a class distinct from other breeds. Also for North Devons, a highly useful and most valuable breed, especially on light soils and in hilly districts of country.

"'Your committee cannot, in justice, close their report without remarking that the want of information as to how the animals had been fed, also as to the milking qualities of the cows, and occasionally as to pedigrees, was much felt by the committee, who in the absence of this needed information, were in many instances left to grope their way in the dark to a decision, of course in some cases by no means satisfactory even to themselves. Nor did we find persons in attendance to lead out the animals for a more full and careful examination, especially as to their style of carriage or movement, which, it is needless to remark, is, as well as form and handling, an essential element of any intelligent opinion or critical decision on their merits. All of which is respectfully submitted.

"'David C. Collins, Ch'n.
Hartford, Conn.
I. S. Hitchcock,
W. K. Townsend,
Chas. Broom,
R. L. Allen,
Committee.

"'Albany, Sept. 29, 1842.'"

We give this quotation in full. The committee state that they are Shorthorn men. We have clearly shown that Mr. Sotham had a contest single-handed against the State of New York. These Bates men had taken that society and organized it and ran it in their interest. The report, it will be noticed, states that all of the premiums were given to Shorthorns, and that there was no competition except the Herefords. With Shorthorn judges it would not be expected to be otherwise.

Mr. Sotham exhibited his Herefords at the New York State Fair for a number of years, as he did at other fairs in Ohio, Kentucky, Missouri and Pennsylvania. He sold stock to go to Massachusetts, New Hampshire, Maine, Vermont, Connecticut, Ohio, Kentucky and Tennessee.

We have found descendants of his stock in all parts of the country from Maine to the Rocky Mountains. Whatever proportion of blood remained they made friends of their owners, and the influence of his work we have met everywhere and have been benefited thereby.

We met a Mr. Kelly, formerly of Rutland, Vt., who said he had owned Herefords forty years ago, and they were the best dairy cows he ever owned, and the young cattle were always ready for the butcher.

We met Judge Downing of Denver, who has

HEREFORDS. PROPERTY OF T. F. B. SOTHAM.

Herefords from Mr. Corning, and they were used as mountain teams at three years old, from grass, made 1,600 pound bullocks.

A Mr. Gird took a half-blood Hereford to Colorado and he gave character and reputation to the cattle of his neighborhood, and Judge Wilcox, whose herd run with the descendants of this bull, said the calves from the quarter-blood were as good at two years old as the others were at three. We might repeat these influences almost without limit.

About 1852 Mr. Sotham sold a number to Mr. H. Bowen, Jr., Summit, N. Y. (¶ 76), and four to Mr. Remington of Auburn, N. Y. They bred for several years and were exhibitors at the New York State Fair as late as 1860. We met Mr. Bowen at the Centennial Exhibition at Philadelphia, where he introduced himself. We recognized him at once from what we had known of his connection with the Herefords. He spoke in the highest terms of the Herefords as the best cattle he has ever had or known, and to the question why he did not continue to keep them, his reply was, the Shorthorn interest was too strong for him; he could not make sales at paying prices.

Mr. Sotham sold to Mr. Ayrault of Geneseo, N. Y., and to Mr. Murray of Mount Morris, and about the same time he sold to the Hon. John Merryman of Cockeysville, Md., and Mr. Merryman continued an active and successful breeder from that time and the family are still breeding Herefords.

That those he desired to reach should the better hear both sides, Mr. Sotham began the publication of his "Hereford Journal," a copy of the first number of which we herewith present in full, as again illustrating his aims and the opposition met with:

MEDAL OF THE N. Y. A. S., AWARDED TO WM. H. SOTHAM, 1859.

HEREFORD JOURNAL.

By W. H. Sotham. Vol. 1, No. 1.
Owego, Tioga County, N. Y., April, 1857.

AMERICAN JUDGES OF CATTLE.

Owego, Tioga Co., N. Y., Jan. 21, 1857.

To the Editor of the "Mark Lane Express" (London):

Dear Sir: As it is a very stormy, snowy day, and a very severe "cold blow," I will show what kind of a "judge" Mr. Chapman proves to be, and I am exceedingly sorry to say there have been numerous judges of this kind at our state shows, by which more second and third-class Shorthorn bulls have gloried in their triumph with first and second prizes over those of first class. The following is my letter to Mr. Chapman:

Sir: As you say you treat every person "gentlemanly," allow me to ask you as a gentleman whether you put your hand on all or any of my cattle during the time they were in the ring? You were closely observed by many, who feel confident in saying your eye was your judge, and that you did not put your hand upon either of them properly. I have a right as an exhibitor to ask this question, and you as a gentleman have a right to answer it.

I am, sir, yours, etc.,
W. H. Sotham.

The following is Mr. Chapman's reply:

Mt. Pleasant Farm, Clockville,
Madison Co., N. Y., Dec. 22, '56.

Sir: In yours of the 15th inst., just received, you ask if I put my hand on either of your animals while in the ring? I answer, No, sir—but for you to suppose that "my eye alone was my judge" is simply ridiculous. I think that the animals "in the ring" would not handle materially different from animals in their stalls. And I endeavored to give all animals (cattle) of any merit a good examination during the three or four days previous to their appearance in the ring. For you to feel so sore about the decision of four such men as composed that committee, setting aside myself, the fifth, I think, will lead you to injure rather than benefit your herd. Respectfully yours,

S. P. Chapman.

Wm. H. Sotham, Esq.

My reply to the above:

Owego, Dec. 26, '56.

I received yours this morning, and in reply must ask you one more question, which you are bound as a gentleman and chairman of that committee to answer. Did you put your hand on either of my animals while in their stalls at Philadelphia? Both my young men (students of mine), my herdsman and myself all know you; either one or the other, and more frequently two of us, were with the cattle; all four will say you never entered "either stall" in our presence. When I ask your advice you are at liberty to give it; I will risk my herd myself, and I fear no man. My object is to know whether you did your duty as chairman, and you ought to know, as a breeder, that is my right. If Mr. Tainter is as rich as Golconda he is "no judge of cattle." What you can say or do will not injure my herd; it will stand the test under sound judgment.

I am, sir, yours, etc.,

Wm. H. Sotham.

S. P. Chapman, Esq.

I was perfectly satisfied Mr. Chapman did not put his hand upon my cattle, and I waited three weeks for his reply to this, in hopes of getting something more in Mr. Chapman's own "hand"-writing; but I suppose he found out his true predicament and declined to expose himself further. Therefore I wrote him the following letter:

Owego, Jan. 16, 1857.

Sir: As you did not answer my last letter, I suppose you silently confess all I said were facts you could not contradict; but as you pro-

CITY OF HEREFORD AND THE RIVER WYE.

fess to treat all men "gentlemanly," why refuse to answer the questions therein contained more than the other?

All seem to correspond. I have written another letter to the "Mark Lane Express," in which I shall copy both yours and my letters. I also send you a copy of my letter from the "Mark Lane Express," so that you may answer for yourself as you deem proper. I have also sent a copy to each of your associates (one of whom now owns the four cows I there exhibited), so that you may bring them to your aid. I know the person who purchased them handled them thoroughly, both in and out of the ring, and I also know his opinion of them. On the other hand, I must confess I never saw or heard a plainer case of prejudice or ignorance on cattle than yours. Or, if you mean to infer by saying "I endeavored to give all the animals (cattle) of any merit a good examination during the three or four days previous to their appearance in the ring" that mine were of no merit, which evidently shows your object, to escape the accusation, proves wilful in the superlative degree; therefore you can place yourself on which horn of the dilemma you think best. I am prepared to meet you on either horn of the dilemma, or both, or any of your associates who choose to take up this all-important subject. I consider it "gentlemanly" to send you a copy of everything I write, therefore I shall do so, also your associates on the sweepstakes committee.

I am, sir, your humble servant,

W. H. SOTHAM.

S. P. Chapman, Esq.

Now, Mr. Editor, I send you the "Ohio Farmer" containing Mr. Cassius M. Clay's answer to my former letter, published September 15 in your paper, in which he says, "I made a mistake in calling his brother's, Brutus J. Clay's, herd his. I therefore transfer the remarks I made on that herd to where it belongs, and will truthfully maintain all I did say."

LEONORA, "THE INCOMPARABLE," BRED BY MRS. S. EDWARDS, WINTERCOTT, LEOMINSTER, HEREFORD. (From a painting.)

Mr. Cassius M. Clay says, "the best judges I have heard speak on the subject regard the Herefords as not a pure breed; and if they were I have all the more no fancy for them." Is this not hearsay and prejudice of the worst kind and carrying his fancy to schoolboy extreme without investigation? Can any man say more against any breed unfoundedly? Still, he says, "Shorthorn breeders do not attempt to pull other breeds down to build up their own."

I would ask Mr. Clay, as he professes to be an old breeder and dealer, and also your more experienced breeders, whether Shorthorns with long, silky coats are thoroughbred animals; or whether the origin of that coat does not belong to the Scots? I am perfectly satisfied in my own mind that that coat came from Scotland, clandestinely bred into Durhams, for it was never seen in their originals; so say all the original breeders.

Again he says, "I do not know what Mr. S. means by proof. The cow of mine which weighed 2,020 pounds" (I suppose Mr. Clay

WINTER DE COTE (4253) 3204.
(Bred by T. Edwards. Champion of England, 1871 to 1875, inclusive. Sire of Leonora, Beatrice, etc. From a painting.)

here means live weight, much the best way of selling Shorthorns), "was only stuffed six months, and as good judges as Mr. S. offered me $140 for her." He (Mr. C.) challenges any cow in America to beat this. I could have fed twenty cows of my own breeding that would have beaten it fairly; all raised on a light soil; much more, had they been raised on the rich grazing land in Kentucky, with a close adherence to the breeders' abundant corn cribs in winter. My cattle never could enjoy such luxuries of "stuffing" as Mr. Clay says his "can bear." My purse was always too light and too much embarrassed to afford this extravagance. So, Mr. Clay may boast of stuffing, but I wish he would produce his balance-sheet with that "stuffing."

In 1844 and 1845 I took a Hereford cow and a half-bred Hereford and Shorthorn cow to Boston. I sold to Mr. Bennett (the noted salesman of Brighton) both cows. The former for $150—she weighed 2,313 lbs., when put on the railroad scales at Albany, and when weighed again at Brighton she weighed 2,247 lbs. The half-bred Shorthorn weighed somewhat lighter and fetched a little less money, but was an excellent cow. I did not keep the dead weight of either. The Hereford was milked once a day until August 1st; afterwards grazed as long as there was a good bite of grass and then put up to feed. She was in good store condition while milking and came to her stall good average American beef. In the last of March I started to Boston with her. The butcher who killed her took the meat round Boston for public inspection on a sunny day and spoilt the whole of it; it had been hanging in his stall long enough for this purpose previous to this, and for good keeping, and this was the reason I could not get her dead weight. I never saw a cow that would sink less offal. She was perfect in her symmetry and her quality of meat proved it could not be excelled.

Mr. Clay further says: "As to the Herefords I have nothing to say for or against them, except they have never in public opinion risen to the rank of contending for the supremacy."

Can Mr. Clay, professing to be an old breeder and dealer, plead ignorance for this assertion? Does he know nothing of Smithfield and Birmingham shows? "I guess not." I shall leave this to you, Mr. Editor, as you are well and impartially "posted" in this matter; and you know how frequently the Herefords have beaten the Shorthorns while each breed were contending for the "supremacy" and the former always met the latter under adverse circumstances. I see by one of your papers "that a snug little family party of Shorthorn breeders had previously controlled the Smithfield Club," but latterly the "march of progress" had taken its seat amongst them; the pressure without called loudly for justice, to break up this combined clique, and has now succeeded. We shall now be able to see which breed deserves the "supremacy."

The home of the Shorthorns in England is as rich land as Kentucky, which gives them every advantage—forces them on in early life. Still the Herefords have always beaten them in early maturity, when brought under a fair trial. No better proof of this could be shown than in the decision of Mr. White of Upleaden and the Rev. H. Berry, producing the yearling Herefords and Shorthorn heifers at Sir Charles Morgan's shows at Tradegar—Mr. Berry giving the "challenge" and Mr. White beating him as a yearling. Mr. Berry, not being satisfied, challenged to show Mr. White the second year, the heifer gaining the greatest weight to be the winner; Mr. White again accepted the challenge and beat him in weight 112 pounds. I refer you to this statement in one of the late Rev. J. R.

Smythies' letters in your paper a short time previous to his death, and who was present at both decisions.

The home of the Herefords in England is light sterile soil, and they are generally bred on light soil here. Still, we beat the Shorthorns, both here and there, under impartial and proper judges, and with this just treatment will ultimately rank first of all breeds.

You see plainly by the prejudice advanced by Mr. Clay and Mr. Chapman what Hereford breeders have to contend with here; and there are many such writers as "C. M. C."—fresh ones springing up every day like mushrooms—assuming that a breeder can be made in a day by purchasing "herd-book cattle" and the next week copy from some one else an article to teach old breeders. This you will see by the "Ohio Farmer," which I send you with this.

Such "judges" as Mr. Chapman have been frequently put on the committee at our State fairs. When "Halton" took the first prize at Rochester in the foreign class, one of the "judges" was a schoolmaster. This I know to be true, for I was manager of the whole cattle department that year and laughed heartily while they were "judging" of "Belted Will," and "Halton" (both Shorthorns), the former being by far the best bull. They call "Halton" an enormous size, therefore "Halton" was the best. Whether this was the "schoolmaster abroad" I know not. From such causes as these, Mr. Editor, springs my spirit of controversy; and in such defense you cannot blame me. Whatever you see amiss in my letter tell it to the people; such conduct will never offend me. I write my views on a subject as plainly as I am capable, not being a "classical man," and solicit information from the soundest source. If I deviate from the truth intentionally, trim me with a severe pen and I will retire from the field disgusted with my own actions. I am the only breeder who will write in defense of Herefords against Shorthorn boasting, and I am determined to stand my ground against unlimited numbers of those who call themselves "breeders" and who are constantly puffing their cattle in this country either by tongue or pen.

Since my last letter to you I have traveled through much of Canada, and from casual observation in the townships of Markham and Pickering, near Toronto, was impressed most favorably with the quality of the soil, more especially for turnips. It is mostly inhabited by

GRATEFUL (4623) 2572.
(Bred by A. Rogers. Champion of England, 1876-1880. From a painting.)

Scotch. The crop of swedes, carrots, etc., etc., grown by George, William (¶ 97) and John Miller in these two townships proved to me that they had not forgotten their old Scotch customs. Their crops of swedes were equal to anything I ever saw in England under a similar season, and cultivated in a very workmanlike manner, perfectly free from weeds and perfect, straight drills.

Messrs. George and William Miller have twelve head of Shorthorns, which they imported from Scotland. Ten of these heifers and a bull were the best lot I ever saw come from one man's herd; they were uniform in symmetry, first quality, size and compactness, "straight bottoms and large, straight tops," on short legs,

NORTH HEREFORDSHIRE HOUNDS.
(From a photograph.)

free from hollow crops and large paunches, most of them long, beautiful "silky coats." My "hand" was upon the whole of them; and after "a good examination" I pronounced the breeder of those Shorthorns worthy of the name, although I never before heard it, or ever saw it in print. He would not sacrifice his herd for the purpose of ruining them for "high shows." I consider this a wise man and should very much like to know how he bred them. I feel very much inclined to say Scotland and Scotch coats was the making of that herd. After examining another bull that came out at the same time I found him second-class. I then began to think Shorthorns could not be bred uniformly in first-class quality, imagining this a descent from it, but ultimately discovered that this bull was not bred by the same person. He has searched far and wide for similar quality to his own; not succeeding, he sent the best he could. Here lies the principal evil in breeding. Had the Messrs. Miller used this second-class bull, in what class would the offspring of these first-class heifers have ranked? Would they not have gone back to their sire generally? But the Messrs. Miller concluded not to use him. I advised them to search the country through for a first class, as their means were ample; recommended them very strongly to get "Balco," imported by L. G. Morris, or "New Year's Day," imported by the Ohio Company—the two best bulls of that breed I ever put my hand upon. Such Shorthorns as these, in a breeder's hands, are very hard to be beaten by any breed, and when I see them compare generally with this herd, or a similar one, my pen shall cease to complain of want of true uniformity in that breed.

On the other hand, suppose Messrs. Miller used the one they had, or "Halton," "Meteor," "Marquis of Carabus," "Locomotive" (which Mr. Cassius M. Clay says "beat the world"), and many others that I could name, most of which have gained first prizes at our State shows, how quickly this herd would go back into second and third class, or mingling all these qualities with it. Here the very secret lies: "Judges" have not sufficiently studied all three classes—given prizes more to large size than any other object; and purchasers have been influenced by the endorsement of these Societies. So long as this principle is allowed to govern, thus long shall we possess all qualities in Shorthorns, and more frequently third class. I will call on all proper breeders to deny this if they can. Mr. George Miller lived in a mud house built by his own hands. I was much amused by the description he gave me of the operation while building it. A man chopped straw with an ax, mixed it with the mud and served him with this material while he laid up the walls with a dung-fork, hewed them straight with a broad-axe when sufficiently dry. It was afterwards covered with lime mortar and small stone, having the appearance of a stone house, costing him 113 days for one man building.

I am, dear sir, yours sincerely,
WM. HY. SOTHAM.

A CHALLENGE!—W. H. SOTHAM.
HERD BOOKS.

Written for the "Ohio Farmer," by Cassius M. Clay.

Whitehall, Madison Co., Ky., Dec. 18, '56.

My Dear Brown: I was not a little astonished at the article of a Mr. Sotham, which you copied from an English paper. I make a few remarks in response to his assertions about my stock, not so much to repel his untruths as to "vindicate history" for its own sake.

Mr. Sotham seems to think that I ventured

upon asserting that the "Shorthorn" was the "best breed of cattle" without "*proof.*" Now, every intelligent agricultural writer and reader ought to know that the word of a man is not more worthy of credit because they may go before a magistrate and make oath to the same. Each one gives his opinion, no doubt biased in some extent, by habit and the amiable weakness of self-interest. But, after all, a man's ability and honesty will be very properly estimated by the public, as Mr. S. will no doubt find out after awhile.

The imputation against me as one of those who are getting up an "excitement" about Shorthorns need not be proven to you to be untrue, when you remember that I only write the few articles in the "Ohio Farmer" by your request. I am the first breeder of Shorthorns in Madison County, and a breeder and dealer for more than twenty years, and yet I never before wrote a line on the subject. For Mr. Sotham's consolation I will say that my herd is very small, and I have a desire to increase it, rather than sell, even for "exorbitant" prices! His remarks about my cattle at the National fair at Springfield are rather singular, as I had only one-half of the bull Locomotive (who took the first premium in his ring against the world) and *no other cattle at all.* If through mistake he applied his remarks to me instead of my brother, B. J. Clay, I need only say that he took more premiums than any other man upon the same number, and many thought he was entitled to the premium for the "best herd," which was not given, I learn, because some other parties were not willing for additional judges to be called in.

The breeders of Shorthorns do not deem it necessary to attack other breeds in order to elevate their own. In giving my views upon the "best breeds" I was willing to admit that the Devons and Ayrshires were the best breeds in some localities and under certain circumstances of climate and food.

Mr. S. complains of our "stuffing" our "cattle." Ours are the cattle that will bear "stuffing" and "pay" well for the stuffing. We always avowed that they were heavy feeders and would not recommend them where "heavy feeding" was not desirable, and yet as I suggested, the Boston "Cultivator" asserts that those fine ones in Maine and New Hampshire are crosses of the Shorthorn, Devon and Longhorn. As to the Herefords, I have nothing to say for or against them, except that they have never, in public opinion, risen to the rank of *contending for the supremacy,* and with Mr. S. for an advocate I think the day is far distant when an "excitement" in their behalf will be gotten up! The best judges I have heard speak on the subject regard the Herefords as not a *pure* breed; and if they were I must say that I have all the more no fancy for them. I know not what Mr. S. means by "*proofs.*" The cow of mine which weighed 2,020 pounds was only "stuffed" six months; and the butchers at Albany, N. Y., who are, I presume, as good judges of fine beef as Mr. S., offered $140 for her, which was refused by her owners. Now, this cow was by no means the largest or finest of the breed, and never was much "stuffed," and yet I venture to assert there is no "Hereford" in America which will weigh as much or bring as much money at the block! The above facts, if not "proofs" can be proved by certificates, if we have to go outside of the rules of gentlemen in this discussion!

What though I admitted that some families or strains of Durhams or Shorthorns were "delicate," do not facts aver the same of all breeds of all animals? I desire to elicit truth and promote the general good. If I was merely a defender of Shorthorns, I should say *buy* any of long pedigree, without regard to quality, beauty, form or constitution; but as I profess to enlighten others, as far as my knowledge and experience go, I wish them to use their own judgment; and I give them hints upon which I have founded my own.

I don't know how Mr. S. could dare to venture upon the assertion that the many engraved likenesses of Shorthorns in the United States were simply copies of the same animal; when

YOKE OF HALF-BRED HEREFORD STEERS, 2 YEARS OLD, OUT OF DEVON DAMS.
(Favorites of Wm. H. Sotham.)

there are so many thousand living witnesses in disproof of his allegation! and yet his *admission* of a common type and common peculiarities, when so proved, are in direct establishment of the *purity* of the breed, and their great excellence in reproducing their like with so much certainty. So far as you and my friend Allen are concerned, you can answer for yourselves.

In conclusion, I'll give you one "proof" of my belief in facts. I will, at the next National fair, to be adjudged by the Society's judges of sweepstake cattle, show two heifers, bred and now owned by me, of the "short-horned" breed, one year old last November, against any two heifers, bred and owned by any one breeder of "Herefords" in America, the loser to pay the expenses of going and returning from the fair. Or, in case there are no heifers of the same age or near the same, I will show two 'shorthorn females' of any named age of my own and my

YOUNG HEREFORDS, BRED BY T. F. B. SOTHAM, CHILLICOTHE, MO.

brother B. J. Clay's herd against any two 'Herefords' of any two breeders in America, owned and bred by the same parties.

Your obedient servant,

C. M. CLAY.

* * *

Herd Books. P. S.—Any attempt to create jealousy between American and English breeders, or controversy about which is the best herd book, the American or the English, is ridiculous. Herd books are simply general and convenient *registers* of pedigrees, and their merit depends upon the fidelity and ability with which they are edited. Both the English and American are good authority, and both *needed* for *convenience;* and any refusal to put pedigrees in either is based upon narrow views of the thing to be answered by "herd books;" for they neither give nor take away credit due to the authenticity of pedigrees, further than that the editors are presumed to be well versed in such things; and that errors are more liable there to be corrected, and frauds to be exposed.

C. M. C.

THOMAS BROWN'S EDITORIAL.

WILLIAM H. SOTHAM.

Mr. Sotham has sent us the copy of the "Mark Lane Express" in which *his* notice and critique of the National Agricultural Exhibition, of Philadelphia, is contained. With the above came also a letter, over his name, concerning "Shorthorns," containing a vast amount of spleen and emphatic, under-lined words and sentences. We are requested to publish it or send it back. Now, we don't think it is worth the attention of the public, and not worth preservation by Mr. Sotham; we shall therefore gently drop it beyond that "bourne" whence no letter returns.

Since writing the foregoing, we have received from the same source two other long letters for publication, made frightful by italics and *threats;* but they have gone the way of the other!—*Ohio Farmer.*

MR. KEARY'S ESSAY.

Mr. Keary, in 1849, wrote an essay on breeds of cattle, favorable to Shorthorns, for the Royal Agricultural Society; the Council of which then comprised Shorthorn breeders, who had much influence over that society; much in the same way as they have had over the New York State Agricultural Society. I can vouch for the truth, that Mr. Lewis F. Allen and Mr. Francis Rotch, Sen., and a few other such Shorthorn breeders, have nominated more "judges," more presidents and more vice-presidents, more members of ex-committee, than all the rest of New York State people put together. The two former have always been very officious in these matters. This every one will admit who knows anything of the society; and I think much to its injury. Mr. Allen, being author of the Herd Book, cannot back out of what he has said so exultingly favorable to Shorthorns, although turned to a Devon breeder.

Mr. Rotch was not satisfied in helping nominate "judges" but assumed the responsibility of teaching them the "true points of excellence" in the different breeds; which I consider the most absurd stuff ever penned by a breeder, although the quality he advocated, if cellular substance could be called quality, exactly suited to his own herd, giving the preference to soft, flabby flesh, characteristic of the very herd he then possessed. This was endorsed and adopted by the New York State Society, and would have been permanently posted on the books, had not a few of us strongly opposed. These "points of excellence" died a natural death, very composedly, for I have not heard of them since their first appearance at the show at Saratoga, which I believe was the end of them.

I may be called "pugnacious," "fire and tow," and many hard names for saying what I do, but who can speak and write patiently in defence of Herefords with such men to encounter? I can be as courteous as any man with reasonable men to deal with; but every impartial man must admit that I have had a certain clique of Shorthorn men to oppose, who were determined to drive me out of market with the Herefords.

There are some kinds of men in the world whom the truth cannot reach, and such men are most apt to accuse others of untruth unfoundedly—notwithstanding this, I shall speak the truth boldly, and fear no man. The time will come when we shall have just and proper judges of men, as well as of cattle.

When Mr. Keary wrote his "essay" for the R. A. S., he found his opponent, Mr. Smythies, a straightforward, just man, who wrote the truth, and Mr. Keary found it went home to him. I did not keep Mr. Keary's letters in M. L. E., but publish the following challenges Mr. Smythies gave to Mr. Keary, and his last letter in reply to him, which will show about the whole of the discussion.

EDITORIAL FROM "MARK LANE EXPRESS" (LONDON).

Day by day we are coming to a more distinct classification as to a more becoming recognition of our several breeds of stock. Without exactly undertaking to assert which is really the best, we now give to almost every variety a fair opportunity of displaying its merits and attractions. We have for some time been gradually approaching this, but never so directly nor so decisively as during the last Smithfield Show week. The admiring public is to be puzzled no longer, but to go methodically through every class or kind of animal it ever heard of. It is no longer Shorthorn, Hereford, and Devon only; but as equally defined, Sussex, Welsh, Scotch, and any other high-bred cattle that can prove to a local habitation and a name. All this is very good. We not only encourage our breeders and enlighten our visitors, but we even ease the duties and lessen the responsibilities of our judges. Years back, the upright judge went into the yard instructed to say at once which was the best beast there—to pick him out valiantly from all sorts and sizes, thoroughbred or mongrel, no matter which! He owned, perhaps, to some little sympathy with the Durham, or to some slight antipathy to the Devon, and he decreed and got abused accordingly. Now, however, he can pronounce on a Shorthorn simply as a Shorthorn, without any of those invidious comparisons which so often ere this have brought him to grief. To be sure there is the gold medal still, but then a man who takes the first honors of his school will always look with some little philosophy on any little "mistake" his friends may fall into.

The labours, then, of our judges are considerably facilitated, while their decisions are like to be freed from much of that angry discussion which has too often attended the publication of the awards. And yet, strange to say, there never was more difficulty in making out an efficient corps for such duties than there is just at present. Crabbe, who, whatever his merits as a poet, always wrote with wondrous truth and fidelity, thus describes the man we are looking out for:

"He was of those whose skill assigns the prize
For creatures fed in pens, and stalls and sties;
And who in places where improvers meet
To fill the land with fatness had a seat;
Who plans encourage, and who journals keep,
And talk with lords about a breed of sheep."

We will not venture to say how many years it is since this was penned; but this we may say, that the lines are far more applicable now than they possibly could have been when originally composed. Where the poet had one or

LONGHORN BULL, REPRODUCED FROM YOUATT'S BOOK ON CATTLE.

two such models in his eye, we have them in scores and hundreds. He might perhaps have pointed to a Bakewell or an Ellman. We turn at once to the list of the Royal Agricultural Society—some six or seven thousand strong—and "tick off" name after name of men who sit in places where improvers meet, who plan, encourage, journals keep, and talk with lords about a breed of sheep, or of horses, or of cattle, or of pigs. The Society would indeed seem to have been born to "make" such men; and as, no doubt, to a great extent it has done. Yet now it has made them it cannot use them. At this moment the Royal Agricultural Society has nothing more trying to contend with than the appointment of its judges. Exhibitors must not be judges; members of the council have something else to do during this busy week, and so we return to Mr. A and Mr. B., who have very often obliged us before, and will probably be kind enough to oblige us again.

That a certain sort of exclusiveness has had something to do in creating this difficulty there

"SEVENTEEN STEER," SOLD AT CINCINNATI, 1841. (From an old print.)

can be but little question. As with the Smithfield Club, gentlemen have duly gone the round of their duties, and then like Bloomfield on his visit to Vauxhall,

"Why, then, they go round them again!"

It has been Mr. A out and Mr. B in, and Mr. C. re-elected, until at last, should we ever get through our A, B, C we are fairly at our wit's end. We don't know what to do. We have been preparing no one else for the place, and now that it is vacant, we have nobody ready to take it. The only thing, of course, is an advertisement in the papers, which we insert here, duly free: "Wanted, a few good judges of stock, etc., etc., for the ensuing meeting of the Royal Agricultural Society of England. Apply by letter or in person, at 12, Hanover-square."

Once more do we ask, do the members generally of the Society know their own rights and privileges? Are they aware, that according to the rules and regulations of their own body, they are directly requested to send in the names of any of their acquaintances whom they feel are competent to assist in any way in the proceedings? Could they tell us the names of a few gentlemen qualified to act as judges at the great national meetings, but who never yet have acted in such a capacity? If so, as no doubt very many of them can, let them oblige, not us so much as the Members of Council, and send such names in on their nomination. Never mind if they are yet untried in so large a field. We have been working a little too much by line and rule as it is, and if we want a precedent of any kind here, it is that every man must have a beginning. So far we appear to have been selecting men as judges and stewards before. Let us now extend the classes a little, as we do with the stock. Let it no longer be all Shorthorn, Hereford, and Devon, but let us have grace enough to name a good man for the office, on the very excellent showing, as we take it, that he never held it before. There is precedent even here, if we must have it. One or two of the very best judges ever enlisted—for stock or implements, we will not care which—never acted at all until within the last two or three years.

The selection, of course, must rest with the Council, and it is only right they should have something good to select from. However much or little encouraged, so far, it is a duty the members of the Society generally owe to themselves to assist in making this list out. We will not go quite so far as to say that every member reads the "Mark Lane Express." If he does not, however, there is the more necessity for his being directly invited to appreciate a little more demonstratively the privileges he enjoys. Why, there is not a man amongst us who does not meet, every day of his life, some capital judge of a horse, or Down, or an ox. The Council of the Society requires the services of this gentleman. It may be a gratifying and well-merited distinction to him, and it will be an essential advantage to them. Send up his name, then, by all means—and if you know of another as good, don't hesitate to send his, too. It is impossible to have too many to pick from.

We have said that this selection is made by the Council; that is, by as many Members of Council as choose to attend any meeting appointed for that purpose. It is sometimes asserted that a large meeting cannot get through its business anything like so efficiently as some two or three members of it would. But this, on the contrary, is especially the business, if not of a large, of a full meeting of Council. Depute it to two or three, and you will have the old prejudices and precedents stronger than ever. Go carefully through the list in open Council, and if you want a name here or there, you will be far more likely to get it from the experience of two or three and twenty than from two or three only. Let us never forget the united strength of a bundle of sticks.

MR. SMYTHIES' CHALLENGE.

In 1849 Mr. Smythies gave the following challenge: "I will show 100 Hereford beasts, which were the property of Sir Francis Lawley, Bart., on the 1st of January, 1849, and the same number which were the property of Mr. Aston, of Lynch Court, on the same day, against an equal number, the breeders of Shorthorns or Devons in any part of Great Britain, on the same day, for one hundred sovereigns. I am willing to leave the decision to the three judges at the last Smithfield Show, two of whom are unknown to me, even by sight."

At the same time Mr. S. made the following offer: "I am ready to place four Hereford calves, on the 1st of May next, in the hands of any respectable grazier in the midland counties, against four Shorthorns, and four Devons; no calf to be more than four months old on that day; the twelve calves to be turned to grass together, to have nothing but grass till the 20th of October following, then to be put in stalls and to be fed as the grazier thinks proper, but the food to be weighed in each lot, till the following May, when they shall be again turned to grass till the following 1st of October; then to be again taken into the stalls, and the food weighed as before; the whole to be shown as extra stock at the Smithfield Show, at the Bazaar, and after the show to be slaughtered, the four beasts that pay the best to be the winners."

HEREFORDS VS. SHORTHORNS AND DEVONS.

To the Editor of the "Mark Lane Express":

Sir: It was not my intention to have troubled you with any more letters on the subject of Mr. Keary's essay, but his misinterpretation of my last letter is too gross to allow it to pass unnoticed. As to what he means by his assertion that my letter can have no weight with *practical* men, I do not understand. Having occupied fifteen hundred acres of land for forty years, and having purchased every beast on the estate myself, and every animal upon it having been bred under my own immediate direction, and the whole management of the estate having been conducted by me, I think I am as much a practical man as Mr. Keary. He accuses me of having hastily arrived at an er-

OX WEIGHING 3,500 LBS., RAISED IN SANGAMON CO., ILL., 1834. BRED FROM "SEVENTEEN BLOOD."
(From an old print.)

roneous conclusion, in having accused him of having drawn a comparison between Shorthorns, Devons and Herefords unfavorable to the latter. Yet, his essay speaks for itself. Does he not, in every particular, claim the preference for his two favorite breeds; while I, on the contrary, declare that they are both inferior? and I have offered to test their several merits and to back my opinion. Mr. Keary says: "My knowledge of Herefords is questioned because I have omitted to notice the grey ones, a pet sort of Mr. Smythies. I cannot call to

"SEVENTEEN STEER," JOHN SHERMAN.
(Bred by J. D. Gillette, Elkhart, Ill. Champion Chicago Fat Stock Show, 1878.)

mind having often seen many of these extraordinary greys, and Mr. Smythies admits himself that after a lengthened inquiry and a great deal of trouble, he could only purchase four." Four what? Not four greys, but four of the pure blood of Mr. Tully's of Huntingdon. I think no man could mistake the meaning of my letter who did not do so willfully. If I had only desired to purchase grey ones, I might, I suppose, have got four hundred in the time. Some of our best breeders had nearly all greys; Mr. Jones, of Brierton, who used to produce as good a lot of steers as most men, had all grey ones for many years; and Mr. Ricketts, of Sarnesfield, whose herd was a very good one, had nearly all greys, and many other breeders in the county had a good many. As an excuse for omitting the name of Sir Francis Lawley as a breeder of Herefords, he said he did not profess to mention the names of all the breeders of Herefords that may be; but still, it appears singular to me that he should have omitted the name of a breeder who not only has the best herd in existence at the present day, but the best herd of any breed. Did he do so with respect to either of the other breeds? Mr. Keary denies that he states in his essay that the young Shorthorns have rather a liberal allowance of cake, and that he only asserts that of the Herefords, and requests that I will refer to it again. I have done so and copy the following sentence from the "Journal": "From the time of rearing, little or no difference may be said to exist in the treatment of the young stock, between the Shorthorns and Herefords. Warm and well sheltered paddocks, with hovels or yards with open sheds, form, in both cases, their winter quarters, in which they are supplied with hay, roots, and, generally speaking, a rather liberal allowance of cake, or other artificial food."

Now, sir, I confess I was stupid enough to consider, from this statement, that Shorthorns were so treated. I will leave your readers to determine whether or not I had a right to come to such a conclusion from the foregoing sentence. I denied in my letter that Herefords were so treated—at least, that mine were, and I asserted this on my word of honor as a gentleman. Mr. Keary then says: "To practical men such statements are somewhat startling," or, in other words, What Mr. Smythies has stated is false. Never having heard Mr. Keary's name till I saw it at the end of this essay, I have no means of knowing in what society he may have been brought up, but I beg to inform him that the men with whom I have passed my life hold such an assertion as I have made above as sacred as the most solemn oath. It is very difficult to prove a negative, but the doubt which Mr. Keary has attempted to throw on my veracity renders the endeavor to do so imperative. If Mr. Keary will do me the favor to accompany me into Herefordshire, I will bring before him men who have worked on my farm for many years, some ever since I occupied it. I will produce two of my bailiffs who lived with me nine or ten years each, and one twenty-three years. He shall also see the men who looked after my cattle; he shall also see a variety of men who have worked on the farm at various periods, and who were discharged, and who are not likely to make any false statements in my favor; and if he can find any one man who can prove that he ever saw an oil cake on my estate during the time I occupied it, excepting the year mentioned in my former letter, or ever saw any lot of steers of mine eating cake, corn, seeds, meal, or any artificial food whatever, I will present him with a hundred sovereigns and pay all his expenses. I beg also to refer him to Mr. Rusbridger, the Duke of Richmond's bailiff; to Mr. Rowland, of Creslow, one of the largest graziers in England; to Mr. Senier, of Broughton House, near Aylesbury; to Mr. Druce, of Ensham; to Mr.

William Trinder, of Wantage; to Mr. Bailey, near Wolverton, and to a variety of other graziers who have bought my steers, whether, when they put them into the stalls to feed, they took to cake as if they had been accustomed to it.

I think, sir, I have offered such proof of the truth of my statement as must satisfy the mind of any unprejudiced man, and if I have not satisfied Mr. Keary, which I think not unlikely, for

> "A man convinced against his will,
> Holds the same opinion still,"

I shall be glad to give any further proof that the nature of the case admits of, if he will be so good as to point it out.

Mr. Keary goes on to say that he has learned that the practice of treating young cattle as he has described was extensively, if not universally, used. The best steers at Hereford fair were usually found in the show yard. When I was a member of the Herefordshire Agricultural Society, all animals exhibited for their prizes were restricted from having cake or corn, and no person could become a candidate for a prize without first signing a certificate that the animal exhibited by him had not eaten cake, corn, seeds, meal, or any artificial food whatsoever. One thing is certain, that if those persons who have been in the habit of showing their stock there have told Mr. Keary that their steers were fed as he says they were, they must have stated what was not correct in one instance, for they cannot both be true. I have obtained many prizes there, and I always had to sign such a certificate.

Mr. K. seems very indignant at my accusing him of knowing but little of Hereford cattle. I was willing to attribute his unfair comparison to ignorance from which, indeed, I really thought it did arise; but if he wishes to have it considered willful misrepresentation, I can have no objection; he is welcome to place himself upon which horn of the dilemma he likes best.

Another striking instance of Mr. K.'s perversion of the meaning of my letter I cannot pass over. He says that though I bred seventy a year for a long period, I never had but one good milking cow!! Does the sentence in my letter to which he alludes convey any such meaning? It is this: "I have seen Hereford cows milk well, and had one myself that made eleven pounds of butter a week for three months." Does that imply that I had but one good milker? Was that any reason to infer that I might not have had a hundred good milkers? It is true that from this sentence it might be inferred that I had only one that would make eleven pounds of butter a week, but I might have had any quantity that would have made seven or eight; but to exemplify the argument I was using, I selected the strongest instance that had occurred in my herd.

He also adds that I admit it takes several months longer to make up a Hereford than it does a Shorthorn. I fancy he is the only man who read my letter that did not perceive that that was a sarcasm on his assertion, which I clearly showed my disbelief of by offering to place four Herefords in competition with four Shorthorns, to be slaughtered at two years old.

I will only allude to one more instance of gross perversion of the meaning of my letter. Mr. Keary says, "With Mr. Smythies' challenge I can have nothing to do." In this wise decision he shows his creed. The prudence of the Yorkshireman is quite apparent in refusing to accept a challenge, which he knows he must lose if he accepts. But the way in which he evades it is most to be noticed. He says it is impracticable; he knows no Shorthorn breeder who breeds a hundred steer calves in a year. Nor I, either, nor of any other breed, and I would ask, is there any other man in the United Kingdom who has read my letter, besides Mr. Keary, who imagined that I had challenged him to show a hundred steers bred by one man in one year? What are the words of my letter? "I will show one hundred Hereford *beasts* that were the property of Sir Francis Lawley, on the 1st of January, 1849, and the same number that were the property of Mr. Aston, Lynch Court, on the same day, against the same number that were the property of any two breeders of Shorthorns or Devons on the same day in any part of Great Britain." Is it not clear that I meant the general breeding stock upon their farms? He then adds, "His next challenge is equally difficult to carry out impartially and fairly." What, sir, is there no grazier in the midland counties to be found with sufficient honesty to try this experiment? I could name twenty myself who would do it. I am the

WM. POWELL, CHANNING, TEX., FORMERLY OF BEECHER, ILL.

advocate of truth, and therefore court investigation; I have nothing to conceal, and therefore have no occasion to put false interpretations upon passages as clear as the sun at noon, nor to make use of subterfuges of any sort. As to the boasted superiority of the Shorthorns, let us see what the past says. If Mr. Keary will search the records of the Smithfield Club, I think he will find the Herefords have taken away more prizes than his favorite Shorthorns, and five times as many as the Devons. Mr. Westcar sold twenty Hereford oxen for two thousand one hundred and fifty odd pounds. I

"QUEEN OF ATHENS" AND CALF, "MY MARYLAND."
(Bred by John Merryman, Cockeysville, Md.)

once stated this fact in a letter in the "Farmers' Journal," which statement was ridiculed in a letter the following week, written by Mr. Henry Berry [Youatt's authority. T. L. M.], who was a great advocate for the supremacy of Shorthorns, as a perfect impossibility. A few days after his letter appeared I received one from a gentleman staying at Creslow, containing an extract from Mr. Westcar's books, giving the date of the sale, the name of the butcher they were sold to, and the sum paid for them, amounting, together, to the sum I have mentioned. Six of them were sold in one deal to Mr. Giblet, of Bond street, for six hundred pounds. Have twenty Shorthorn bullocks ever sold for two-thirds of the money? So much for the past. As to the future, I am ready and willing to test the respective merits of these three breeds in any way Mr. Keary will point out. I only wish it to be clearly ascertained which are the most useful beasts. I am convinced in my own mind that no animal of any description can be fatted on the common vegetable produce of a farm so soon, or brought to such perfection on grass, hay and roots only, as a Hereford, and this appears to me to be the great desideratum at the present day, that we should be able to convert the vegetable produce of our farms into animal matter at the greatest advantage, and that we should not have to go to the foreigner to buy oil cake to feed our beasts when we have enough at home to feed them with if we select the proper animal. I know it is a notion amongst Shorthorn breeders that oxen cannot be fatted without corn or cake. I once showed a fat cow at Bath, and won the prize with her, and sold her to Mr. Hale, the celebrated butcher in that town, who told me she was the fattest animal he had ever seen. I had given in a certificate that she had been fed on grass, hay and roots only. Some people who saw her thought it impossible she could have been so fed, and somebody, just before the judges entered the yard, threw some oil cake into the manger before her, in hopes the judges might see her eating it. This I was told by one of the judges when he came out of the yard, and he added, she evidently had never seen such a thing before, for she took no notice of it; we tried her competitors with it, and they ate it up in a minute. So convinced am I of the decided superiority of Hereford cattle for feeding purposes over any others, that I am most anxious to see them brought fairly into competition, and whenever they are so, honestly and honorably. If the "Journal" committee of the Royal English Agricultural Society wish to act fairly, they ought to publish my letters, that the poison and the antidote may travel together, and let the world determine who is right. I have done nothing underhand, or in secret; I have referred you by name to many most respectable men. If I have stated one syllable that is not true, I have afforded you every means of detecting me; if, therefore, you fail to do so, I have a right to claim from my agricultural brethren an implicit reliance on the truth of my statements. Regretting that I have occupied so much of your valuable paper,

I remain your obedient servant,

J. R. SMYTHIES, (¶ 78)

Grey Friars, Colchester, Feb. 16.

* * *

I wrote to Mr. Thomas Brown, editor of the "Ohio Farmer," to find out the *clandestine man* who signed himself "Agricola," who wrote on "in-and-in breeding," etc., as I considered the remarks he made on that subject doing much injury to the public. His assertions on other points seemed to me truly ridiculous, and that a *common sense* editor would refuse to publish such stuff. His teachings on breeding, his Shorthorn puffing, and his writing anonymously, condemned him in my estimation. I never knew a man (who is a man) ashamed to own his own productions. The following was Mr. Brown's reply:

Cleveland, April 23, 1856.

My Dear Sir:

I do not think there is any special danger of my being seriously *used* by Mr. Allen. I know your alarm is honest, and your cautions well meant, and I therefore thank you for your letter. "Agricola" is *not* Mr. Allen, and I don't know that Allen ever met him. He is one of the "oldest breeders in America," a D. D. and LL. D., whose fame is as wide as the earth. He is now president of a flourishing theological college and professor of Biblical Literature—and is none other than the Rev. Robert J. Breckenridge, of Lexington, Ky. Are you satisfied on this point? Keep it to yourself.

I have just received a good article from Aston on Herefords. I mean to do you and your Herefords full justice, but I must be permitted to take my own time and way to do it. I am obliged to you for any advice, and reproof, even, given in kindness. I desire, too, to continue on friendly relations with you.

I am, very truly,

Your friend,

THOS. BROWN.

I wrote Mr. Brown that this was the first I had ever heard of the Rev. Robt. J. Breckenridge, and that it was high time he declined to give instructions to breeders. If Mr. Brown's *extravagant* encomium of the man had gone still further, and he had said this famed Mr. Breckenridge had been known beyond "the earth," I should have believed it quite as readily as many other assertions made by him and his correspondents on Shorthorns; but I do and always shall believe that neither of these assumed writers on breeds of cattle knew how much mischief he was doing the community by advising "sire to daughter and continue it." One of these noted writers and breeders, to my knowledge, has bred blind calves, calves void of symmetry and quality, calves minus their legitimate power of reproducing, calves with long pedigrees, and who has not gone near so far into the "in-and-in" system as he recommends others to breed, and at the time he talks of "Colling's successful in-and-in breeding in Shorthorns; the long Scotch coat was produced in them, the origin of which no man can dispute who knows both breeds." Here is a mystery that ought to be solved and who is more capable of doing it than the man *"whose fame is as wide as the earth?"*—a man anxious for notoriety as a breeder and dictator, a man careful and kind to his "flock and herd," a man who advocates truth, purity, morality, honesty, justice, one who loves his country. I say again, there can be no better man to solve this mystery than this intelligent and learned divine, upheld and supported by his one-sided editor and contemporary, who is so well versed in the improvement of his country, and in the *protection* of his pet correspondents. This same one-sided editor speaks of my letters being *"made frightful by underlining." Mark* the letter above, that his readers may "learn and inwardly digest." This is a wide world to live in, and the people are diversified. Some feel "warm when they have seen the fire," and are satisfied with the comfort such a blessing bestows. Others, with more extravagant ideas, heedlessly clench the flames, and if a man will put his hand in the fire he must expect to get scorched. Editors may, in their ignorance, madness or self-conceit, commit truth to the flames, but can never *destroy;* it will, phœnix-like, rise again from its ashes, to teach such vain editors decency and justice. One-sided editors will frequently add brighter coloring to glaring *untruths* from a favored correspondent, while his opponent is repulsed with editorial power, in forbidding entrance to his columns what he *knows* to be true. But truth will out—nothing can suppress it. The editor of the "Ohio Farmer" will sooner or later find out that "two heads are better than one, if one is a sheep's head."

"DOLLY VARDEN" (V. 9, p. 279) 5.
(Bred by J. Morris, Madley, Eng.)

I did not see Mr. Aston's *"good* article on Herefords," of which Mr. Brown speaks, nor do I believe it ever appeared. It was in answer to C. M. Clay, therefore, I suppose it was too "powerful" for Shorthorn breeders, and was committed *"gently"* to the flames, from whence, he thinks, such articles never return.

I have been accused of "untruth" unfoundedly by an editor before this, and my accuser had fairly been convicted of perjury; nothing saved him from a prisoner's cell but *money,* untruths, and a flaw in the indictment; but this same editor is now reformed, and, I see, strongly advocating morality. I glory in his wisdom, and therefore freely forgive him. If I am "pugnacious," unclassical, severe, and forbidding, as my opponents choose to characterize me, I can forgive. Nor do I consider publishing Mr. Brown's article a breach of trust, after his conduct to me.

* * *

EDITORS AND JUDGES.

There is nothing more important to the agricultural community than just and good judges. Nor is there any true evidence of a man's being a good judge of cattle because he lives in a large, expensive house, and gives high prices for imported stock. Such men, if reported rich, are very apt to be looked up to for this department—many of which are very fearful of soiling their hands and their clothes, frequently handle cattle with their gloves on; some of kid and others of a different material. Such men generally consider such labor beneath their dignity, while practical men leave not a single point unexamined, making their hands and their eyes guide them to a just decision. Such men will not be swayed from the true points of an animal by prejudice, favor or aristocratic influential men, but will give the animal its due without reference to its owner. All societies will be a curse to the country until such men are nominated.

In the present course of things, a costly animal, say a "five thousand dollar one," must be the best, because he costs so much, when good, unbiased judgment would probably pronounce him in second or third class.

There is no lack of display in portraits and publications, and these enormous prices quickly go the rounds. Editors of agricultural papers are generally too anxious to "obtain first" such exciting articles, and are very jealous of others who "copy from them without credit." Neither do they forget to flatter and feed the vanity of those extravagant purchasers, many of whom never see the animals until they arrive home, spurring them on to this destructive mania, and when such prices are given without judgment or discussion, it is a ruinous principle.

Most judges are apt to select animals for prizes from their "important" puffed character, made public by editors, who had probably been well paid for their puffing rather than trusting to their own judgment, and frequently listen to these dictates as settled facts without consideration. The owner, too, is an important man, and frequently money only has made him

HEREFORDS ON THE PLAINS OF COLORADO.
(From a photograph taken in the spring, after a hard winter.)

so. Editors have puffed his cattle—puffed his management as a breeder—courting an equality to his exalted society, and to obtain a golden fee, he knowing at the same time the true credit belonged to his agents. By this means some editors are made notorious, their papers popular, their judgment consulted, and frequently put on as judges of cattle, some of which find difficulty in distinguishing the different breeds, without "guessing" and inquiring. Still they can write about these breeds with as much confidence as the best judges.

These editors may call me "pugnacious," but they know my statements are true, and therefore they may poke their editorials at me to suit themselves, if they are only sufficiently manly to send me a copy and allow me to answer them; if not, it is no better principle than stabbing a man in the dark. I think it will benefit the community if they are "reproved in kindness."

There are many editors truly good fellows, if they did not assume to be judges of cattle. Many remarks they make on this subject are truly ridiculous.

It is very gratifying to receive "visits" from many of them if the *more vain* would not tell the public how kindly they had been treated by great men, in what kind of carriage they had been drawn to such breeders' mansions; how upright and stately they sit in that carriage, ruminating over the "big" words that should compose their description and flattery; not forgetting to tell such a man what advantage and favor it was to have such puffs in their columns.

Some editors are even weak enough to tell what they had for dinner, and even add to this the enormous size of the beautiful Tom Turkey presented to them to carry home for another "luxurious meal." My readers who have read agricultural papers must admit all this to be true, and they must further admit that a moderate breeder who understands his business practically, lives in a moderate house, with moderate ideas, and possessing substantial judgment, is very seldom noticed, but more frequently avoided and unjustly repulsed. To be the guest of an honest, plain farmer would not be etiquette, if he lived in a plain house. Such editors as the "Ohio Farmer" would deem him beneath his dignity and unworthy of an editorial remark if he had not a fine pair of mules or a splendid pair of horses to ride behind, seated in an aristocratic carriage for the purpose of taking his observations; a Tom Turkey, however fat, however "enormously" heavy, would scarcely be considered a present, while that of the reported great man's would be puffed and stuffed with "enormous extravagance."

Such is the way of the world, and we are compelled to succumb to it as long as such editors as that of the "Ohio Farmer" "has his own way," and is allowed to burn up a proper answer. I suppose he is lord of all he surveys when he beholds the whole of his paper, and therefore has power to forbid.

There are many agricultural editors who require more substantial judgment; more independence of character; more sound investigation into the numerous articles they publish, and although some of them are willing to treat both sides of the question fairly, every one must admit there is a vast deal of cattle trash published only for the want of better judgment to guide them.

A LETTER TO THE EDITOR OF THE "OHIO FARMER."

Owego, N. Y., Jan. 17, '57.

Thos. Brown, Esq.:

Sir: A friend has sent me your paper today, containing Mr. C. M. Clay's letter, from which I see you have previously copied mine from the "Mark Lane Express." If you publish the *whole* of that letter, I am satisfied.

If you publish my letters *just as I write them* I will meet Mr. Clay or any other person on that subject. "All I ask is a fair field and no favor." It is very easy to say and to print that another tells an "untruth," but it is more difficult to prove it, nor can I believe it *gentlemanly* to say so without proof. When Mr. Clay or any other person proves an "untruth" in any of my letters they are at liberty to condemn me as one of "low degree," "low bred," "third-class quality," and "no breeder." I will then retire from the field disgusted with my own actions.

HON. J. W. PROWERS, WEST LAS ANIMAS, COLO.

On the other hand, I am perfectly willing to admit all that Mr. Clay says as a "gentleman," but I cannot agree with him in opinion, nor in his *judgment* on cattle, however long he may have been a dealer or a breeder.

There is but one true standard of quality, symmetry, weight, etc., and he who differs from that is no judge.

I honor the name of "Clay" as highly as any other man, and I regret that I made the mistake of calling Mr. Brutus Clay's Mr. Cassius M. Clay's. I fully understood that it was C. M. C. that took the prizes at Springfield. I therefore transfer the remarks I made upon that herd to where they belong, and will fully maintain all I have said in answer to C. M. C. When we have *both* had a fair trial, I will leave to the public, whose just opinion has most weight in the scale of true judgment.

HEREFORDS BRED BY THE REYNOLDS LAND & CATTLE CO., CHANNING, TEX.

My desire is to test this matter *honestly*, and fear no man.

If you publish this, I will accept Messrs. C. and B.'s challenge in a *fair way* in another communication. If not, I shall answer him in other papers.

I send you another letter from the "Mark Lane Express," for your perusal and publication, *if you choose*.

I am, sir, Your humble servant,
WM. HY. SOTHAM.

MR. SOTHAM'S ANSWER AND ACCEPTANCE OF C. M. CLAY'S CHALLENGE.

Owego, Tioga Co., N. Y., Jan. 25, 1857.

Mr. Editor: In reply to Mr. Cassius M. Clay's letter. He says, "The word of a man is no more worthy of credit because he may go before a magistrate and make oath to the same." I fully endorse this, Mr. Editor, and will say still further: a man who is over-anxious to take such a course is less likely to be believed and trusted.

I take it for granted (inside the rules of gentlemen) that Mr. Clay had no self-interest in Shorthorns, but has more of it in Mr. Brown's invitation to write for his paper. The half of "Locomotive," and the whole of that "robust family of Shorthorns"; the former "beating the world," and the latter as good as can be found in it, can have no "exciting interest" to their owner, nor can there be the least sign of "elevation" towards these valuable animals by such expressions from Mr. Clay. He having "*self*-possession," "*self*-control," and an "amiable disposition"—rests composed and contented with the increase as it surrounds him—(enviable position this!) He needs no demand, has no desire to attract a man in the "habit" of giving "high prices," or taking advantage of his "amiable weakness." Mr. Clay has no such intentions, his only "habit," object, and *self*-interest is in writing for the *Ohio Farmer*, and Mr. C. seems to think "that a Mr. Sotham will 'find out' these profitable advantages in time, when the public become more familiar with his character." I must say, Mr. Clay, that I have no desire for this public "estimate," and rather your "*self*-interest" would rest where it is. Neither have I any desire to excite high, artificial prices in Herefords. I always sold at reasonable ones, and it is still my wish to do so.

Now, Mr. Clay, I see plainly by your letters, you consider yourself a judge of cattle. You have studied it practically, therefore you must know that the herd that takes most prizes is not always the best; neither are the judges who award those prizes. Excellent proof of this was shown at Springfield, which every practical judge must candidly admit. I met a Shorthorn breeder, whom I am well aware ranks "inside the rules of gentlemen," and who openly confessed that Shorthorns there were miserably lacking in quality, nor did he attempt to hide the remarks from some of his own herd. A white heifer belonging to Mr. Duncan was decidedly the best female in that class and the best in her own. There was another red and white heifer that compared well with her, still these two were beaten by one as hard and as tight as flesh could be forced upon her—third-class of the lowest kind. The white heifer went home prizeless, and, in my opinion, reflecting much animal or "amiable weakness" on the majority of the judges. Hence, my minor report. I have too much regard for my associates to believe for a moment there was any self-interest in this decision, for I feel assured they came inside " the rules of gentlemen."

Mr. Clay says that breeders of Shorthorns do not deem it necessary to attack other breeds in order to elevate their own, and further says, "the best judges I have heard speak upon the subject regard the Herefords as not a pure breed, and if they were I have all the less fancy for them." This remarkable injustice; this hearsay; this fancy; this self-contradiction, is it not untruth? Mr. Clay, I will not have it so; neither will I deem it self-interest, if I am considered by the public "to step outside the rules of gentlemen." I cannot help it—I must say it is your amiable weakness to vindicate history. Notwithstanding this, your opinion is the same opinion still. Your robust family is the same and may still be the best in the world under the same opinion. Durhams may still be the best breed of cattle, and Locomotive, who took the first premium against the world, is still Locomotive if his weight has kept such steam pressure from his becoming too exaltingly elevated. So the world goes—we are all entitled to an opinion, and many of them will be remembered as long as they remain in print.

Again, Mr. Clay says: "As to the Herefords, I have nothing to say, for or against them, except they have never in public opinion risen to the rank of contending for the supremacy." Can such an assertion as this come from the pen of an old breeder and dealer?

They have beaten the Shorthorns five times out of eight, when each were contending for supremacy, under the most adverse circumstances, and influential nobility on the other side. I refer, Mr. Clay, to the annals of the Smithfield and Birmingham Clubs from their beginning for an endorsement of this fact. I also refer him to the report of the "Mark Lane Express" of December 15th, of this Christmas Show at Birmingham; he will there find the Herefords stood far superior to any other breed classes at that exhibition. Mr. Clay did not mean to pull down the Herefords by these remarks, I know he did not. He had previously said it was not necessary, and he would have studied the facts in the case before he had ventured to attempt it, though I cannot help thinking there is a little amiable weakness in this mistake to vindicate history.

As to the animal likenesses. Did Mr. Clay ever see one of Mr. Page's bull portraits on paper show hollow crops? On the contrary, did Mr. Clay ever see the original except New Year's Day and Balco with full crops? There may be more in these two questions than Mr. C. imagines, and I would advise him to consider before he answers them in the name of an old breeder. I have something further to say on this important subject hereafter.

I have now to acknowledge the proof of Mr. C.'s cow weighing 2,020 lbs., and also his own assertions that no Hereford in America could weigh as heavy, or realize as much as $140. I have bred more than twenty cows that would beat her, myself, under the same circumstances, and to confirm "my opinion" I sold to Mr. Bennett, the noted salesman at Brighton (who almost every dealer knows), a Hereford cow that weighed on the railroad scales at Albany 2,313 lbs., and when in Brighton 2,267 lbs.; Mr. B. gave me $150 cash for her; I also sold him a half-bred Hereford and Shorthorn cow, a very superior animal, she weighed somewhat lighter, and sold for less money. The Hereford was milked once a day until August

¶123

¶123

MEDAL OF THE CENTENNIAL AT PHILADELPHIA, 1876, AWARDED TO T. L. MILLER.

1st, when she became dry. I kept her on a good bite of grass, on a light sandy soil, as long as the grass lasted, then put her up to feed. She was *economically fed (and not "stuffed")* until the last of March, when I started to Boston with her.

This was the extent of her feeding.

I will now for a finality, propose to meet Mr. Clay's "challenge" in a fair way; I have two heifer calves, "Prudence," calved August 29, 1856; "Woodlark," calved September 30, 1856. These are all I have left this season. They ran with their dams as long as any grass; when the cows came to winter quarters, they were allowed

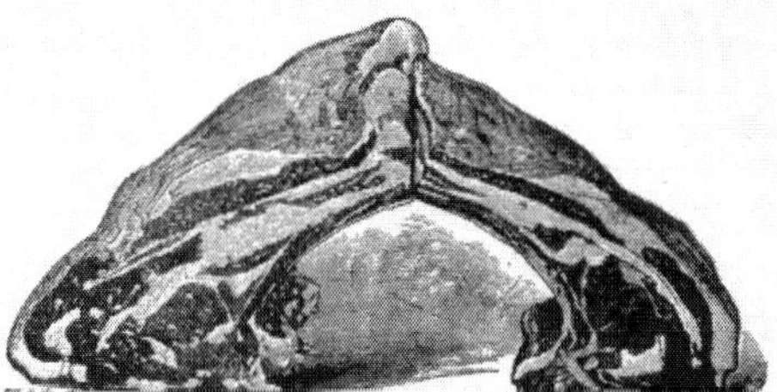

¶124

HEREFORD. SHORTHORN.

(The thickness of Hereford and Shorthorn roasts compared. Actual photograph of roasts from Hereford and Shorthorn bullocks dressed at Chicago Fat Stock Show, 1879, showed Hereford 33 1-3 per cent thicker than the Shorthorn.)

to suck their dams once a day; about the 20th of December they were taken away, and now live on oat chaff and cut cornstalks, mixed with about a pint of oatmeal each per day, mixed with their chaff, one common sized rutabaga per day, cut up between them regular, and are in growing store order, as I do not believe in forcing calves; as spring and warm weather approaches I shall increase the roots, and keep to about the same quantity of oatmeal. These calves shall be turned to grass in the spring, until next fall, when I will bring them with me to Kentucky, at the National Show, place them in any honest Kentucky grazier's hands, against any two heifer calves *now* owned by Cassius M. Clay, or Brutus J. Clay, of a similar age; if they have not any exactly the same age, a few months difference must be allowed for accordingly. The four calves shall be weighed when delivered to the receiver, the feed to be weighed the whole year, and at the end of it all four of them again weighed. The following year shall be pursued with the same treatment; coming in at three, again at four years old. They shall be milked both seasons, each as long as she will hold out in milking under the just and economical management of their receiver; the butter regularly weighed, and disposed of by him, he reporting quality and giving a just account at different times, as he deems right. At the end of this trial they will be five years old; they shall be fed for one year, or two, as agreed upon, the two heifers that make the most butter and most beef, for the food consumed, to be the winners. The quality of beef at the "*block*" to be taken into consideration. Each pair of heifers to be *charged with what they eat*, of each kind of feed, all living on the same, and allowing them a good and sufficient grazier's quantity through the whole trial, the losing heifers to be forfeited to receiver, to pay expenses. Pedigrees of heifers to be given at the time of delivery. There are plenty of men in Kentucky that will act fairly between us.

One more question, Mr. Clay, and I have done for this time. Did you ever see a "*Durham,*" "Shorthorn," or "Teeswater" with a "*long, silky coat*" that you would "*venture*" *as a breeder* to pronounce "*thoroughbred?*" I will "venture" to assert that kind of coat is descended from the Scots. The original "*Durhams*" have no claim to it, or ever possessed it; so say all the old breeders in England with whom I am acquainted, and they are not a few. Most of them speak from knowledge descended from their ancestors. Every effort "*Shorthorn*" men have made to contradict this charge has created a *stronger* desire in me to believe it and that it was *bred into them clandestinely*. I have every reason to believe the noted Hubback was *half Scotch*, no "*proof*" has ever been shown to the contrary, but the more breeders try to *hide* this plausible "history" of him, the more likely to be *true*.

I hope Mr. Clay's next letter will contain more *practical teaching*, and I heartily wish him success, though we differ widely in "*opinion.*"

I am, etc.,

WM. HY. SOTHAM.

WM. HY. SOTHAM'S LETTER TO B. P. JOHNSON, ESQ.

Secretary New York Agr. Society.

Owego, N. Y., Aug. 15, 1855.

B. P. Johnson, Esq.:

Sir: As you have undertaken to write an uncalled-for article in your "New York State Journal" of that Society, placing the Shorthorns predominate in your opinion, in value, weight, early maturity, etc., I think you have done great injustice to the Society. No such body has the right to endorse the opinion of any man, without his producing the weight of each breed satisfactorily, and prices sold for to the butchers, also the facts of their early maturity. You, as editor of that "Journal," ought to be extremely cautious how you advance your opinion, extolling any one breed over another; it is a ruinous, selfish principle for such a Society to adopt. Such opinions adopted by the Society are puffs unfounded, and very injurious and destructive to all connected with them.

The Hereford breed is the only one that has taken first prizes at Smithfield and Birmingham Shows at two years old. They have done this three times, proof of which you can find by referring to the record of each club. The Shorthorns have never won a prize younger than three, and if you refer back to the records you will not only find one at that age; they are generally four and upwards.

You are the first man who dare risk his reputation in print that Shorthorns are uniform in quality. I will defy you to refer me to a just breeder of Shorthorns who will admit this. They *know better,* but prefer being silent on the subject. It is those who are ignorant of the facts, and wish to make themselves notorious who advance such opinions, and it is by this abominable system that the Shorthorns have obtained their assured notoriety, and called "first class."

Societies, especially those of the Empire State, ought to be impartial and give all breeds an equal chance. If members of that Society are expected to look up to its superior officers for example it is very important that those officers are upright and just men.

Now, Mr. Johnson, I ask you (as I know you to be an honest and just man) whether you know a herd of Shorthorns uniform in quality? No breed on earth varies more in this respect. I have seen first, second and third quality in one herd, and that of *puffed notoriety.* I have never seen a herd of Shorthorns (and I have seen a great many, both in England and this country) without two qualities, and those frequently of second and third.

As you have decided for the Society that Shorthorns are *best,* and Herefords are *second,* it is high time we show outside the gate, and challenge to show the Shorthorns inside, as we are there forbidden to show by the remarks advanced by the secretary of that Society, nor can Herefords ever succumb to Shorthorns, when we have shamefully beaten them in England under the most adverse circumstances and by the best judges.

I send you an extract from Mr. R. Rowland's letter to me, dated March 13, 1855, giving weights of Herefords and prices sold to the butchers. When you can honorably beat them, you are at liberty to state all the particulars in your "Journal."

Hereford breeders challenge you to do so, and until you can do so truthfully we shall consider our breed triumphant in reality.

Richard Rowland, Esq., and successor to the well-known Mr. Westcar, writes me as follows:

"I will add the prices of some of Mr. Westcar's oxen. The weights I do not know, except in a few cases. I was an exhibitor for some years after I came here and took three prizes with Hereford oxen, which weighed from 290 to 270 stone, eight pounds to the stone—these were the four quarters—[that means dead weight, or 4,000 to 3,800 lbs. live weight, T. L. M.], and that is about the weight Mr. Westcar's large oxen reached. I have shown for many years, but have now left off. Shorthorn men in the aristocracy always ruled at Smithfield, and Hereford feeders are very tenacious of feeding there on that account, but the prize Shorthorns that win there (as you have seen) go to Birmingham, and get fairly beaten by Herefords where no favor is shown to breeds of any kind; they are there governed by true merit."

Now, Mr. Johnson, if you will give a prize for the six heaviest breeding cows of any breed from one breeder, fed on hay, roots and pasture only, I will be an exhibitor for that prize; the scales will then be the only judge wanted. I will also show the same six cows for quality—also for butter, separately. Now, sir, you have a chance of proving all you have said so highly favorable to Shorthorns, and you must admit that the course you have taken in puffing them in our "State Journal" will tend to bring the Society into disrepute. Every judge the Society selects reads these puffs and many of them are much guided by them in their decisions.

JOHN D. GILLETTE, ELKHART, ILL.

Your remarks on Mr. L. C. Morris's bulls, Balco and Marquis of Carabus, extolling both bulls to the highest pitch of eminence (in your opinion) in the State Society's "Journal," a paper destined to be the organ of the Society, is also very ruinous. Both these bulls were, in the opinion of the Secretary, "hard animals to beat," both were equally praised for their various good qualities. While Balco was an excellent first-class animal, and the Marquis of Carabus a very inferior second-class—very flabby flesh and very thin skin—this, every man who is really a judge of cattle, will admit.

Some time before this, I wrote a letter to the Secretary, condemning Mr. Rotch's "points of excellence," showing the Society the fallacy of countenancing such stuff, and told them it was a direct insult to good judges. These letters were then read to the ex-committee, and then

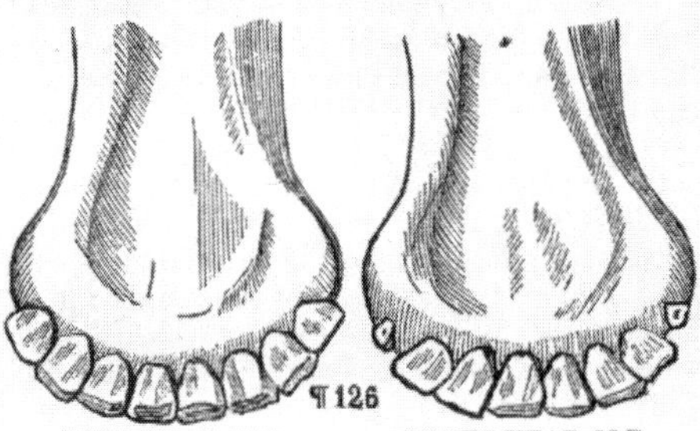

FOUR-YEAR-OLD. THREE-YEAR-OLD.

(Ages of cattle compared by teeth.)

"laid upon the table," which was the last of them. The influence of Shorthorn men prohibited anything of this kind from having its proper force.

With the above letter I gave the following memorandum of the price of twenty fat oxen sold by the late Mr. Westcar, Bucks, England, taken from his books by his nephew, Richard Rowland, Esq.

Sold in Fleet Market, London:

Dec. 16, 1800,	2 oxen to Chapman	£200
Dec. 4, 1800,	1 ox to Chapman	147
Dec. 15, 1800,	1 ox to Harrington	100
Nov. 26, 1801,	6 oxen to Giblett & Co.	630
Dec. 31, 1801,	1 ox to Chapman	126
Dec. 31, 1801,	2 oxen to Harwood	200
Dec. 4, 1803,	1 ox to Chapman	100
Dec. 19, 1803,	1 ox to Reynolds	105
Dec. 19, 1803,	1 ox to Giblett	105
Dec. 5, 1804,	1 ox to Giblett	105
Dec. 4, 1805,	1 ox to Giblett	100
Nov. 28, 1811,	1 ox to Chandler	105
		£2123

Averaging £106 3s, or $513.04 each.

These proofs were entirely unnoticed—and there has been a time since when you could scarcely pick up a New York State agricultural journal that did not contain a puff for Shorthorn cattle. I ask the members of the Society whether this is right? I have not shown an animal at our State Society Show, except at Elmyra, where I was bound to sustain our southern tier of counties, and which will ultimately show what they can really accomplish when put to the test.

When I found the Secretary of the Society would not publish my letters I wrote to the President, whose letter I have now, and will probably appear at some future time. I asked him why my letters could not be published in the "Journal." He said nothing could appear there unless connected with the Society. I did not see why my opinion on cattle, as a member, had not as much right in that "Journal" as that of the Secretary, and I thought that "the antidote ought to go with the poison." "But no!" all was of no avail, and here it ended. I ask my readers to look at this in its true light, for the next "Quarterly" may bring something more for the members to investigate.

WM. HY. SOTHAM.

PEDIGREE.

My idea of pedigree is to refer back to those breeders who have universally bred first-class cattle. Uniform in quality, symmetry and size, more especially the two former. If like produces like, of which I have no doubt, when the breeder understands his business, is sufficiently versed in the three distinct qualities, and never allows a second or third-class beast to enter his herd. From this process and care in breeding spring all genuine herds. If a breeder has gained just celebrity in breeding uniformly best animals, long ago, and the offspring of that breeder follow his example, inheriting the true judgment of their sire, and never deviate from the true course laid down to them, such animals can be perpetuated. From such a parent tree branches extend their influence, the *old stock* is genuine, many of the branches are genuine, while others should have been cast off and committed to the flames before it had contaminated the original and substantial standard.

A pedigree from the *old stock*, conveyed through the best branches, where no dark stain has entered, and where sober thoughtfulness has perpetrated that true delineation of character, in a long continuance of good, uniform breeding, is worth much, and such pedigrees only should be recorded in a Herd Book. I contend that the composer of that book should know, before he enters an animal in its pages, whether it inherits quality and symmetry; if not, it cannot be genuine. The principal point in a herd is uniformity of first quality, and should that herd possess first, second and third, does it deserve recording in the Herd Book? I decidedly say no. Such a herd ought not to be countenanced by judge or breeder, and how many coarse third-class and flabby second-class bulls there are in the Shorthorn Herd Book, with long pedigrees. There is no class of cattle in the world so uneven in this important particular; therefore, what good is their pedigree, which is intended to perpetuate uniformity in good qualities? The Herefords and Devons are generally uniform in first quality, if descended from proper breeders.

It is no criterion that a man is a genuine breeder because his name is puffed in almost every paper by editors and their puffers. I would rather risk a pedigree from a man whose name is scarcely known at all. A good judge can seldom be deceived in the true points of an animal. And in these exciting times a large, coarse Shorthorn bull in second or third class, with a long pedigree, is a curse to buyer, seller and the country.

The late Mr. Bates, for instance, was puffed as a breeder in the most extravagant manner; and whose herd can show more unevenness than his? He has sent to this country more coarse second and third-class bulls than any other breeder. His name and pedigree have created a mania which I feel much inclined to say is bordering much on lunacy. And as I am bound to tell the truth, I think a pedigree from such exciting causes, descended from such bulls as the late Mr. Bates', are worthless. I would much rather trust my hand and eye, guided by sober judgment, than such pedigrees, for there must be something wrong in a herd possessing three qualities.

I think a pedigree from a man who advocates in-and-in breeding ought to be avoided. I have tried this ruinous experiment to my own satisfaction, and when I come to a "deadlock," as many breeders have, I will mix with another breed of cattle where I shall not lose my quality rather than engender disease.

W. H. S.

I owe an apology to my readers for the hurried way in which this paper is got up. The next quarterly will explain all.

(END OF PAPER.)

After Mr. Sotham's dissolution of partnership with Mr. Corning he moved to Black Rock, near Buffalo, N. Y., and from thence to Genesee Flats and to Owego, and finally to Islip, Long Island, all in the State of New York, continuing to breed Herefords.

In 1850 Mr. John Humphries and Mr. Aston of Elyria, Ohio, imported the Hereford bull John Bull (3885) E. H. B. and 464 of the American Hereford Record, bred by Mr. E. Price of Pembridge, Herefordshire, Eng., by Goldfinder 2d (959) 474, bred by John Perry, Machowarne; by Wichend (1118) 486, bred by Mr. Perry; by Monkland 2d (1012) 498, bred by Mr. Perry; by Monkland (552) 504, bred by Mr. W. Perry of Cholstry; by Lion (335) 519, bred by Mr. W. Perry; by a bull of Mr. Jeffries; and on his dam's side tracing to Sir David (349) 68. Mr. Humphries also imported Victoria 478, bred by Mr. Bowen of Markland, Leominster, tracing to Old Court (306) 60.

T. Aston imported the cow Duchess 15, bred by Mr. Bowen of Markland. For continuance of her breeding see Vol. 1 of the Herd Book. Also imported Curly (801) 14, bred by Mr. T. Roberts, Ivingtonbury, Leominster, Eng., tracing directly to the stock of Mr. Benj. Tomkins. From these herds H. and N. Abbie of Elyria, H. Chappel, Thos. Clark, Thomas Cox, A. Dyke, H. S. Kline, N. G. Porter, William Richardson, H. T. Smith, W. W. Aldrich and G. W. Byers, all of Ohio, had stock from the produce of these importations.

Mr. Frederick William Stone of Guelph, Can.,

"SEVENTEEN" SHORTHORN STEER, McMULLIN.
(Champion Chicago Fat Stock Show, 1881-2, bred by John D. Gillette.)

imported in 1860 or 1861 quite a number of Hereford cattle. (¶ 79-80-81) Among the bulls imported by Mr. Stone was Sailor (2200) 12, by Severn (1382) 24, and he by Walford (871) 47. (¶ 82) Mr. Stone's importation was largely of Lord Berwick's herd of Shropshire, and Lord Bateman's of Shobdon, Leominster, Eng. Mr. Stone also bred Shorthorns, and while the Herefords were his favorites, having both breeds, he was not a decided advocate of one more than another, but distributed much very valuable Hereford blood.

Mr. Taft of Williston, Vt., writing to the "Breeder's Journal," August 7, 1882, says:

"The Albany 'Cultivator' in August, 1846, contained the following notice of some Herefords that were brought to this town:

"'Rev. L. G. Bingham of Williston, Vt., has lately purchased of Messrs. Corning and Sotham some fine Hereford cattle. The lot consisted of the imported cow "Aston Beauty," two yearling heifers, a yearling bull and heifer calf. They were animals of excellent qualities, and we think will prove particularly valuable to that section of the country. Their vigorous constitutions will adapt them to the climate, and on the sweet pastures of the hills and mountains they will easily and quickly thrive and fatten, while in any fair trials in the yoke or for the production of butter they will not be 'found wanting.'

MR. J. H. ARKWRIGHT, HAMPTON COURT, HEREFORDSHIRE, ENG.

"The cattle more than justified all that was said of them in the above extract. They produced splendid oxen, were not wanting in the production of butter, and one of the best drovers in the Boston cattle market told me he was never cheated by buying a Hereford 'in the lump.' But the man who brought the cattle here failed in his extensive enterprises and left the State, and they were not bred after that; but the grades were here and traces of the blood, with its excellent characteristics, remained in this vicinity until lately. There is now but one full-blooded Hereford in Vermont, 'High Chief 2d,' recorded in the English Hereford Herd Book No. 5966, bred by Mr. Hawes in Maine, and now owned by me; and there are very few in New England.

"Yours truly,

"R. S. TAFT."

Commenting on which, the editor of the "Journal" says:

"There are quite a number of Herefords in Maine, New Hampshire, Massachusetts and Connecticut which we know of. A herd was started in Oxford, Me., by Mr. Holmes."—Ed.

THE RUST OR SYRACUSE OX created a sensation in his day, being proven by Mr. Sotham and others to have Hereford blood. This ox was owned and fed by Mr. N. P. Rust, Syracuse, to whom was awarded the first premium of the New York State Agricultural Society for the best fat animal exhibited at the Albany Fair, in 1842. This ox was eight years old; his live weight February 19, 1841, 2,360 pounds; on the 18th of July, 1842, it was 3,400 pounds, and when exhibited at the State Fair in Albany September 28, 1842, it is said to have weighed 4,200 pounds, which would be a gain of three pounds per day for nineteen months. At this weight he retained his activity and appetite, and continued to take on flesh as fast as ever. If it is said that the weights and gains are too large, we have only to say that they are given by Shorthorn men, when claimed by them as a Shorthorn grade. Mr. Rust made a certificate as to the feeding of his ox, as follows:

He has been fed nineteen months on cornmeal, from twelve to sixteen quarts a day, and during the winter he was fed a bushel of potatoes or rutabagas each day. During the summer he was fed four quarts of oil meal in addition to his cornmeal. Mr. Sanford Howard, one of the careful, practical, painstaking writers of that time, gave his opinion that the ox owed his excellence to Hereford blood and was a typical Hereford.

(¶ 84) CONTEMPORARY WITH THE RUST OX, a gigantic Hereford ox was being exhibited in England, of which the following account we find in the "Chamber of Agriculture Journal" of November 14, 1881. It must be remembered that the English ton is 2,240 pounds, which would make the steer the immense weight of 4,480 pounds:

"The records of Hereford cattle are not devoid of information respecting gigantic oxen. Some forty-five years ago one animal gained some notoriety by the name of 'Wettleton Ox,' on account of his immense scale, deep flesh and wonderful symmetry. The ox was exhibited upon

Ludlow race course, and also at many of the principal towns in England. To give some idea of his immense size we quote an old poster calling attention to the animal when it was exhibited in Ludlow race course. It is as follows: 'Stupendous Ox. Now Exhibited on this race course. The celebrated Wettleton Ox, bred and fed by Mr. Sheppard of Wettleton, the proprietor. He is of the pure Hereford breed, stands eighteen hands high, girts twelve feet, three feet six inches over the first rib; three feet three inches across the hips, and weighs upwards of two tons (4,480 pounds).' At that period the exhibition of the ox created great attention, and no animal approaching it in scale has ever before been exhibited in that part of the kingdom."

Mr. Sotham never lost interest in the Herefords even after he had given up breeding.

Writing to the "National Live Stock Journal," May 12th, 1871, Mr. Sotham says, under the caption of

"HEREFORDS VS. SHORTHORNS."

To the Editor of the "Journal":

Much has been said in comparison of Herefords with Shorthorns, and much more can be said; and as the Herefords are but little known in the West I think it only fair to bring their true merits before the people, and as I do not own a single animal of the breed I cannot now be accused of interested motives. Much has been said in favor of Shorthorns, and extraordinary prices have been published to the world, given by men of money, for none other could purchase them. There are but few men capable of breeding Shorthorns, and all who know anything about them are aware that they are a made-up breed, hence their want of unity in breeding. Tell me, ye Duchess men, ye followers of Bates, did you ever see uniformity of breeding in the herd of Thomas Bates? A more uneven herd I never beheld, either at home in their glory, with their admirers around them, or divided in America by those gentlemen who strongly advocate the Duchess tribe of Bates' breeding of thin skins and *soft handling.* What is more remarkable, these advocates both in England and this country were generally men who knew nothing about breeding; were literary men who had acquired a fortune by some lucrative business, became possessed of it by heirship, or leaped into it by some "lucky" speculation. These men gave high prices because they had the money, not because they had the judgment to select for themselves; because the

THE RIGHT HON. EARL OF COVENTRY, CROOME COURT, WORCESTERSHIRE.

Duchess stock had become fashionable in America, and men who did not know their true value paid fabulous prices for them. Still, they soon began to find out that it cost more to keep up appearances than the animals could realize, and they retired from the Duchess contest as soon as propriety would allow, or with as little injury to the cause as possible—Morris & Becar retired in favor of Thorn,—Thorn retired in favor of Sheldon,—Sheldon retired in favor of Walcott & Campbell—all Duchess men. Vail retired from a Duchess to a Devon breeder. Lewis F. Allen, author of "American Shorthorn Herd Book," followed Mr. Vail's example and went

MR. S. W. URWICK, HEREFORD, ENG.
(For twenty years Secretary of the English Herd Book Society.)

from Duchess to Devon. After spending much more than he got from them in hawking them about from place to place with flattering advertisements he finally disposed of the tail-end in Illinois, after a winter's feeding on corn to lay on the flesh. He flew to Devons for relief. Then came Stevens in literary force, with arguments strong in favor of Sherwood and himself. They imported freely from Bates. What became of them, all know. Then came Chapman, with all his Duchesses, full of puffs, portraits and pedigrees. This ended in smoke, with Halton at the head of the herd, who sold at the sale for about what his owner advertised as the price of a cow. Then came Page, with his flattering portraits made of straight lines as if all done with a ruler, accompanied with constant puffs of the Duchess tribe. But where are Mr. Page's Shorthorns? They can be mainly seen on paper and there he gets his profit. Flattery is not lost upon his admirers, and these pictures please fancy men of money, and attract novices by their sameness in straight lines. I could enumerate others to the end of a *very long chapter,* but space in the "Journal" is too valuable. Let me ask the Duchess men the cause of all this? Because, in another communication, and at a more convenient season, I shall show that Bates could not be compared to Booth in breeding Shorthorns, and I shall endeavor to show that Shorthorns well bred by scientific breeders, such as Booth and his *true* followers, are equal to any cattle in existence. Notwithstanding this, Hereford breeders have pursued the even tenor of their ways, bred some very superior animals and have won more prizes than Shorthorns at the Smithfield and Birmingham shows in England, more especially when they have come in competition with each other. The Herefords, being a race, they breed more uniformly, and the breeders, not being led away by fictitious prices or sham auctions, as the Shorthorn breeders in England and this country were, they stick to good breeding, and, being content with reasonably remunerative prices, made money and extended the breed all over England, Wales and Australia; and now the West Indies and Scotland are beginning to encourage them and they are increasing strongly among the farmers of Ireland. You seldom hear of sales of Herefords where they are thoroughly known, because the demand at home, at reasonable and remunerative prices, is greater than the supply; while Shorthorn breeders, waiting for their extraordinary prices, look in vain for moneyed men, who are a long way between, but who pay well to have their name conspicuous in print, injure the Shorthorn cause by bidding far beyond the value, vainly seeking that fame which can only be realized by fancy men with money.

As the Shorthorns have many strong advocates among these men of money, let me show you what Herefords have done against their strong influences. You have seldom heard of a Hereford sale, unless when the head of the family had gone to his last home and his estate had to be divided; while the Shorthorn sales are everlastingly in print, under the sheriff or, to attract novices, with under-bidders to spur them on, aided by strong and frequent drinks, to keep up notoriety. This cannot be denied.

In the early part of the Smithfield shows,

when Shorthorns and Herefords contended against each other, the Herefords almost invariably took the first prizes. Mr. Westcar of Creslow, Buckinghamshire, took the first prize with a Hereford twenty years in succession, but subsequently retired from the contest, disgusted with the trickery of Shorthorn breeders who, at length, insisted that Herefords and Shorthorns should be put in separate classes—because repeated failures drove them to it. Notwithstanding this, the Herefords took more prizes in steers and oxen from that alteration than Shorthorns, while the cow class in Shorthorns took the majority of prizes because Hereford breeders would not sacrifice their best animals for that purpose. Shorthorn breeders forced their females from birth for show and appearance, thus producing barrenness, hence their show cows for the gold medal. This being fashionable, a large majority of Shorthorn cows over Hereford could always be found in the show yard, enabling them to win more gold medals under the forcing system.

Again, Shorthorn breeders were constantly boasting of early maturity over Herefords, until they were obliged to succumb. In 1862 Mr. Heath won the gold medal with a Hereford steer two years old, while there were thirteen Shorthorns from three to five years old in the same class. Soon after this, I believe in 1854, Mr. Shirley showed another two-year-old Hereford, winning the gold medal under precisely the same circumstances. The following year he did the same. I may be mistaken in the years of Mr. Shirley's triumphs, but these are facts that cannot be denied, for they are on record, and they silenced the traducers of Herefords, and you heard of no more bragging in print of the early maturity of Shorthorns. Even Allen, Stevens, Tucker and Tom Brown had to knock under and fly to some other plea for Shorthorns.

Then the great milking qualities were brought before the public. A prize was given for that quality at the Royal Show at Oxford, in 1839, open to all England and all breeds. There was great competition, Shorthorns predominating. Mr. J. R. Smythies' Hereford cow took first prize for the best milker, a Shorthorn second. This prize was discarded afterward, because Shorthorn men had control and they feared a repetition. This somewhat silenced the bragging for milking Shorthorns and encouraged the dairy men in favor of Herefords. Mr. Smythies, in a communication to the "Mark Lane Express," February 5, 1849, discussing the merits of Herefords as milkers, says: "I have seen Hereford cows milk well, and had one myself which made eleven pounds of butter per week for three months;" but he observes that "beef is much more profitable than butter in this section, and on that account it is not an object with Hereford breeders to have good milkers." This is undoubtedly the fact, and it explains why Hereford breeders have not paid more attention to the milking properties of their cattle. Mr. Duckham (¶ 83), in his lecture, says of Herefords: "In the dairy counties, where the milking qualities of the cow are well attended to, the most satisfactory results are realized." He quotes from a letter from Mr. Reed, to whose experience with the Herefords as workers reference had been made, the statement, "that they have been used for dairy purposes for nearly half a century upon the farm and that he believes they yield a larger return than could be obtained from any other breed upon a similar class of land." Mr. Duckham also quotes from a letter of Mr. James of Mappowder, Dorsetshire, whose dairy herd of Herefords has been formed thirty years. He states that the stock has been much improved since he obtained it, and that Hereford dairies are becoming very common in that country. He adds: "In proof that they are good for milk, we let near a hundred cows to dairy people, and if I buy one of any other breed to fill up

MR. J. H. ARKWRIGHT.
(First President of the English Herd Book Society; on his favorite hunter, "Bagpipes.")

the dairy they always grumble and would rather have one of our own bred heifers. We let our cows at so much a year, finding land and making the hay." Mr. D. also quotes from a Cornwall correspondent, who says that according to his experience "the Herefords are good milkers and he is convinced that when the cows are deficient in their yield of milk it does not arise from any constitutional defect, but rather from mismanagement in rearing, or a deficiency of the constituents essential to the production of milk in their food. My cow Patience," he continues, "bred by Mr. J. Y. Cook, Moreton House, Hereford, has this summer given fourteen pounds of butter per week, and Blossom, bred by the late Mr. Longmore, Salop, gave twenty-two quarts of milk, yielding two and one-half pounds of butter per day, equal to seventeen and one-half pounds of butter per week." The same correspondent says: "I consider the Herefords are particularly adapted to this humid, fickle climate, where Devons become small and delicate and Shorthorns grow long and coarse." This statement of butter, seventeen and one-half pounds, is somewhat extraordinary, but I know Mr. Duckham to be a gentleman of veracity, who would not state anything to the public but what he knows to be true.

WM. H. SOTHAM.

Detroit, Nov. 20, 1870.

C. M. CULBERTSON, NEWMAN, ILL.
(First President American Hereford Cattle Breeders' Association.)

CHAPTER XIII.

Mr. Wm. H. Sotham's History of the Herefords

We have quoted from the standard publications of the day the early correspondence of Mr. W. H. Sotham and his opponents to show his and their methods. We have proven beyond doubt that the Herefords were during all this time the thriftiest graziers and the most economical producers of the best beef of all the British breeds. Before taking up the later history of the Herefords we feel it but justly his due to more fully present a history of Mr. Sotham's connection with the breed. Mr. Sotham fought a good fight, actuated by the highest motives; his was a practical mind that, looking through the fog of fashion, fad, prejudice and self-interest, saw nothing but the beef-making qualities of beef-bred animals. Prime beef at minimum cost was to Mr. Sotham the desideratum in cattle, and no animal had value in his eyes that did not meet the requirements of our motto: "Economy of production and value of product." It is given to few men to have such a rich and varied experience. His sterling honesty and love of truth added to this experience founded a character that could not patiently brook opposition based on inexperience, subterfuge, mediocrity and self-interest.

Mr. Sotham gave his life-work to the Hereford breed of cattle, without at any time much hope of financial reward. He came naturally to look upon the Hereford breed as a ward under his fostering protection, and upon their thrift and perfect quality as inherent good traits of faithful adopted children. And undaunted by vicious opposition he continued his championship half a century, till, at the ripe age of eighty-three, he died in 1884 at Chicago.

We quote herewith as a suitable place for its preservation, from Mr. Sotham's own pen without comment his "History of the Herefords," written for the "Chicago Drover's Journal" in 1881, two years before his death.

PART I.

Stonington Park, Ill., Oct. 2.—Referring to the stock on exhibition at the recent New York State Fair, the "Country Gentleman" of the 23d of September says: "The twenty-six Herefords are owned by two exhibitors, Erastus Corning of Albany, N. Y., and Burleigh and Bodwell of Fairfield Centre, Me. Mr. Corning's cattle were all bred by himself with the exception of one imported bull, Comus (4457) 6665, and it is complimentary to the skillful management under which this well-known herd has always been conducted that it defeated the Maine white-faces, including, as they did, a number of cattle recently brought over and regarded as among the best in England. The Hereford show, all things considered, was certainly the best for years."

I think this pretty good proof of what I told more than once, that the Herefords were as good when I imported them to the United States as they are now. In 1839 I bought the first-prize cow at Oxford Royal Agricultural Show. The bull, cow and offspring that took first prize at Tredegar and Cirencester, England, and several others about as good in my first importation. Mr. Erastus Corning, Jr., has only brought one female into this herd since we first imported them, and she never bred—a beautiful animal—and every means were tried to obtain an offspring without success. All the improvement that has been made on the herd to my knowledge has been from the bulls Mr. Erastus Corning, Jr., has imported, and they have been good. I am not at all surprised to hear of his success over the boasted herd of Mr. Burleigh, as I fully believe that herd and others sold from it was imported on speculation rather than for good breeding purposes; and I am fearful that there will be too much of this, to the injury of the Herefords. I had supposed that Hereford breeders had profited by the bad example set them by Shorthorn speculators on pedigree.

The late Hon. Erastus Corning (¶ 85) was a true nobleman at heart, and a sincere friend. Whatever he undertook was based on sound principle, and, being just in all his dealings, remarkably liberal in his views, coupled to an enterprising spirit, he was a true benefactor to his country. He did everything in his power, while obtaining a huge fortune, for its benefit; though not a speech-maker, while Senator of his own State, and at Washington, his sound judgment was the foundation of many that were made. Many who were popular for their best speeches based them on the foundation of his soundness. This I know ocularly and demonstratively, having been present at such consultations many times to witness his superiority. Unfortunately for the Herefords, he was a strong politician and, although he delighted in im-

T. E. MILLER, BEECHER, ILL.
(First Secretary American Hereford Cattle Breeders' Association.)

provements, he had but little judgment in cattle.

When I first went to England for Herefords I had very limited capital, but great faith in them as being the best breed for this country. I felt that I understood their true merit when in England, and some of their best breeders were my most intimate friends. Having but little capital, I felt that it was impossible for me to get any from my mother country.

The fall of 1839 I had just finished buying nearly 4,000 head of cattle in droves for Mr. Ebenezer Wilson, as they were making their way from the West to Albany. I paid a portion down on each drove to be then delivered at the price stated per hundred at his slaughter house for barreling purposes. After contracting for many droves I returned to Albany, and as the droves came in we selected some of the best and (¶ 86) I took them to the Bull's Head (N. Y.) for market. Mr. Wilson barreled a little over six thousand head that season; he sold his hides for cash, his tallow for cash, his feet, horns, and bones also. His beef as soon as in the barrel was sent to New York to his factor and shipper, of whom he drew at sight for the proportion per barrel agreed upon, he holding the beef as security. The whole of his capital was invested in his beef, and the prospect being favorable for higher prices he held on to it.

I had frequently spoken of the Herefords in high praise to Mr. Wilson and he, being highly pleased with my description, proposed that if I would go to England and get credit for a good lot that when I arrived in New York he would meet me there with money enough to pay the whole investment, in which he said he should have ample means from his returns in the spring. I told him that I would do this, if I could do it with safety. He assured me that I could, and with this promise I started.

I purchased twenty-two head, paid what little money I had of my own, and promised to pay all on my return to New York. Among them were those I have above stated. When I arrived I found that barreled beef had gone down and much depressed, that much of his had soured and was unsalable, as he had involved himself so much that he was compelled to fail; but he went to the Hon. Erastus Corning, told him in what way he was situated with me, that he knew my judgment was good and that the cattle would be superior.

On this information Mr. Corning sent Mr. Watts Sherman, then cashier of the Albany City Bank, of which Mr. C. was president, to examine the cattle on board. This gentleman was highly pleased with them, examined my bills and prices for them, and drew a draft on Mr. Corning for the whole amount. I took them to his farm and, they being about the first Herefords ever imported, they raised considerable excitement; but their *true merit* was but little known.

The Shorthorns had full sway, were owned by rich men who determined to support them at all hazards and who had the controlling power over the executive committee of the N. Y. State Agricultural Society, and when they (the Herefords) came to be shown in the fair by their side (the Shorthorns), could not avoid seeing the Hereford superiority for beef-making, hence became jealous and full of prejudice, put on Shorthorn men as judges, who would not report anything in their favor, but tried to make a bad impression, considered me an innovator, and were determined to keep them in the background; would not make a class for all breeds, although I kept constantly urging it, on paper, and at their meetings. They often insinuated that the Herefords were far below the Shorthorns, and a portion of them puffed and portrayed Bates and his tribes beyond control; he must have the ascendancy, above everything. This brought me into a controversy with many Shorthorn men, with whom I had to fight a hard battle, to which the editors of the "Country Gentleman and Cultivator" must confess, and their paper will prove.

No person could be more kind to me than Mr. Corning, and my endeavor was to reciprocate. He was much pleased with the cattle and the same year gave me a draft on Baring Brothers, London, to go and buy another lot. They were similar to the first purchase, and being so successful in bringing them out I did not insure. Took the whole steerage and filled it with cattle and feed. Just before we got to the banks of Newfoundland a heavy storm came upon us, shattered our bulwarks and swept off our galley. This storm blew us back 300 miles; the hatches were closed, and the waves sent their foaming white-caps and heavy spray over us in quick rapidity; no air could be conveyed into the steerage, consequently the cattle broke loose, were jumbled up together and died in their suffocation and confusion. Such a spectacle I never before beheld, as each animal was drawn up to be deposited in its watery grave. I had become attached to several of the beautiful animals that thus suffered and perished, as I fed and nursed them on the voyage; I thought of the kindness and liberality shown me by Mr. Corning and felt that he would blame me for not insuring.

When the calm came I began to reckon on the loss, and found that gentleman's would not be less than $8,000. My embarrassment was most trying, although I was not ashamed to meet Mr. Corning, but his loss, under the kindness he had shown me, weighed heavily. When we met and he heard my explanation not a murmur did I hear from him. He was so well versed in this world's affairs, and the disasters belonging, that he saw the situation at once. He never hinted the loss to me.

After this he became so beset by Shorthorn breeders, who did everything they could to discourage him, and they being men of capital and influence, he listened. They tried to make him believe that the Herefords did not amount to anything, that I was only a braggadocio, that there were no other cattle in the world like Shorthorns, and the fictitious prices they sold for made him believe there was something in it; still, they could not get him to adopt them. Between politics and his other extensive business he found he could no longer stand the worry of the special pleading of Shorthorn breeders, of their abuses toward me and the disparagement of the Herefords. He resolved to become clear of it.

I never was so kindly treated by any person in the world as I was by the late Hon. Erastus Corning and Mrs. Corning; had I been a near relative they could not have treated me more kindly. They did much to lead me into the best society in Albany. I fully appreciated all they did, and exerted my utmost to reciprocate. I frequently went to his house to spend the evening, and in one of those events he said to me: "Sotham, I know your strong faith in the Herefords, and the strong prejudice that Shorthorn breeders have against you and them. I like the Herefords and believe all you say of them; but I cannot stand the constant pleading of that body of men to turn my attention to their favorite breed. I know nothing of either

ADAMS EARL, "SHADELAND," LAFAYETTE, IND.
(First Treasurer American Hereford Cattle Breeders' Association.)

breed, practically; therefore, if you have faith in taking the most part of them at half value, and will get an endorser for the amount, you shall take them. You may make the papers to suit your time. I will keep three heifers, the cow Victoria and a bull, to show that I am still a believer in them. You may take the bulk of the herd, and if you do I sincerely hope you will be successful." This seemed to me a most reasonable, kind and valuable offer for me, but I felt that the little capital I had in them was gone in our wreck at sea, and how to get a farm to support them upon was an important consideration. However, I secured an endorser and a farm at Black Rock, near Buffalo, and took the herd there.

J. M. STUDEBAKER, SOUTH BEND, IND.
(Member Organization Committee, American Hereford Cattle Breeders' Association.)

* * *

PART II.

From my early boyhood I had an exceedingly high opinion of the Hereford cattle, and have since that time been a strong and staunch advocate for them, and think I can fully support all I have said. The uniformity of their character, the superiority of their flesh, combined with rich milking and substance of body, induced me to patronize them to the best of my ability.

My frequent visits to the herds of Messrs. Hewer of Northleach, Gloucestershire, England, in an early day, gave me an opportunity of seeing some good ones bred by them. Mr. Wm. Hewer, Sr., Wm. Hewer, Jr., and Joseph Hewer, the father and two sons, did all in their power to outvie with each other, and each tried his utmost to get at the pinnacle of good breeding. They were as earnest in excelling each other as opposite breeders. All were practical men of good, common sense. Their herds originated from that well-known breeder, Mr. John Hewer of Herefordshire. If either procured a very superior bull, each derived the benefit and each bred him at pleasure.

They studied together the improvement each animal made, and vice versa the defects, and by their combined good judgment none excelled them in good breeding. This prosperous course of rivalry was pleasant and instructive, and rendered superior aid in arriving at superior judgment, without which no man can become a proper breeder. The Messrs. Hewer's Hereford cattle and Cotswold sheep obtained as high a stand as any in England, and by which all became wealthy, derived from their practical knowledge of a superior animal and the coupling of male and female. (¶ 88)

Much care and attention are required in the effect of improvement, and can only be obtained gradually. Skill in the advancement is inherent, which cannot be learnt by lessons or lectures or by professional theoretical novices. The result must be practically satisfactory to enable you to pursue with confidence; without this you cannot succeed. He who trusts to the opinion of others will never make a breeder. Results from his own experience must be his guide, and when a breeder arrives at the highest point of excellence, his name spread far and wide, it is a very difficult matter to keep there. Prosperity is apt to make men careless and consequential, which is almost certain to create degeneration, and when this takes place the downward strides are long and rapid.

The Messrs. Hewer made vast improvements in the Hereford cattle and Cotswold sheep. The senior held his fame and his untarnished reputation until his death; Wm. Hewer, Jr., until he retired; but Joseph was taken away in his early career, by jumping into the water to save a favorite ram when he was saturated with perspiration in his haste to arrive. He, in his usefulness, left a fine herd of Herefords and Cotswolds to be divided. The Hereford cattle and Cotswold sheep of my first importation were derived from the Messrs. Hewer, with the exception of the first-prize Hereford cow I purchased from Mr. James Walker, with her yearling bull, which took first prize at the show of the Royal at Oxford in 1839, and bred by Mr. Turner.

I showed the kindness I received from the late Hon. Erastus Corning of Albany. This gentleman was a pattern to rich men. He was a princely honest man, and was always ready to encourage true enterprise. His object was to do good, and his thorough knowledge of business and the world gave him the advantage over most men. There was no man who ever knew him thoroughly but that loved and revered him. He was moderate in all his views, kind, even to a fault, and no man worthy of support ever called upon him in vain. He was a true promoter of deserving enterprise. I only wish there were more such men at the head of our Government now. It would then have a solid foundation. His mind was based upon a sound and solid principle, and being just in all his dealings, he advised others into the same course. With such a mind as this, Mr. Corning could not help seeing the true value of the Hereford cattle when they appeared on his farm, and after paying all the charges upon them, advanced me more money to go to England again for others of a similar character, as they were admired by all who saw them. (¶ 89)

The trio, Lewis F. Allen, Ambrose Stevens and John R. Page, were the scribes for the "Bates mania," and to denounce the Herefords, and, although neither of them had any practical knowledge of stock, they had an unbounded conceitedness connected with their brass to teach men more practically informed how to breed.

Then there were Thos. Brown of the "Ohio Farmer," Francis Rotch of Bates fame, George Vail, an importer and constant puffer of Bates and his tribes, in connection with S. P. Chapman, neither of whom really knew anything more of Herefords or Shorthorns than a cast-iron soldier. All wrote and re-wrote, but neither knew what they were writing about. They might know enough to distinguish a heifer from a steer. All these scribes went to Erastus Corning with high praise of Bates and claiming that Herefords had no character. Although Mr. Corning felt differently, he was disgusted with their familiarity and constant harangue, that, with his other important business, was a great annoyance to him, and on that account he made me the generous offer which I explained.

Now, let me show you what became of all these scribes and deceivers. Lewis F. Allen, editor of the Shorthorn Herd Book, became a hawker of his Shorthorns; tried public sales at great expense, without effect; then sent them to Illinois to be fed on corn for many months to make a better appearance, so that he could dispose of the whole. All who purchased know what trouble there was in obtaining correct pedigrees—in the Red Ladies more particularly. After the disposal of these he went to Devons. While in Shorthorns and their grades his diary presents a most laughable tale, with which I am familiar, and may present at some future day.

Ambrose Stevens, the bosom and confidential friend of the editor of the Herd Book, who kept two Shorthorn cows for him, the only stock of the kind he then possessed, bred to Allen's bulls, but bred nothing of character. Allen, finding they were no profit to him, wished me to take them until Stevens could find a place for them. I put them to my Hereford bull Major, and one of them produced a heifer calf of true Hereford character, except she had a "sweet head" so

¶ 136

GEO. F. MORGAN, LINWOOD, KAN.
(Member Organization Committee American Hereford Cattle Breeders' Association.)

puffed up by Bates. I saw this heifer at Batavia, N. Y., when she was two years old, and among the "grand importation" made by Ambrose Stevens, Esq., from Mr. Stevenson, the noted breeder of the Princess tribe, Ambrose could not help but admit that she was the best animal in the lot. She was of pure Hereford character except in head and horns, which were short, small and crumpled. I tried to buy her, but he would not set a price, as he valued her "sweet head" very highly. I never learnt what subsequently became of her or her progeny.

I now hear but little of John R. Page; his light seems to be "hid under a bushel." I hear nothing of his herd, and his notorious flattering portraits on paper seem to have lost patronage. John was once a shining light; I should not be surprised to see him swaggering amongst the white-faces with as much pomp as he did at Shorthorn sales, if he can make it prove as agree-

THOS. CLARK, BEECHER, ILL.
(Member Organization Committee American Hereford Cattle Breeders' Association.)

able to Hereford breeders as he did to those of Shorthorns. John puffed Morris & Becar, who retired in favor of Jonathan Thorn; he retired in favor of Sheldon, who retired in favor of Walcott & Campbell—all Duchess men. Vail retired to S. P. Chapman and became a Devon breeder. Chapman failed when in prime of life—when he considered himself just in his glory, while advertising Halton (purchased by Mr. Vail) at $20 per cow, which was more than the brute was worth. I should like to hear the first man say that he ever saw a good one from him. All Page's puffs and portraits of him at the head of the herd ended in smoke. Halton sold at the sale at about the price advertised for a cow to be bred to him.

Francis Rotch was another Bates puffer and a fancy pet scribe of the "Albany Cultivator and Country Gentleman," the organ for the Shorthorns, which was solely under these scribes and breeders. If a Hereford breeder advocated their breed he was either "strongly prejudiced against Shorthorns" or was no judge of them or of Herefords. They knew "on which side their bread was buttered," and they took advantage of it. Rotch's prestige in Shorthorns failed before he died; his herd became extinct; there is nothing left of his work as a pleasant memorial, and this same "Country Gentleman" has ever since shunned the Herefords, though conscious of how much they had abused them.

I name some of these gentlemen and their organ because they were the principal ones who were constantly worrying Hon. Erastus Corning by condemning Herefords and speaking in high praise of the Shorthorns. Thirty to forty influential men of money against one individual, who had but little means to defend himself against such men, glorying in their power. It was that overbearing power that brought them to a sense of their weakness, when put into the balance scale of profit and loss.

It is difficult to say how much they lost. I do most earnestly wish that good and just man, the late Hon. Erastus Corning, was here now to witness the change and realize the true character of those men in their present state, who did all in their power to influence him in their well-known deceit, none of which can again visit him for the purpose of deceiving him in his calm and unalloyed resting place. He is now receiving his just reward for the good he has done on earth.

* * *

PART III.

These Shorthorn men were fully aware they had something to contend with in the Herefords, and exerted their utmost to keep them in the background. Criticised their white faces and bellies, the long horns of the cows and the large horns of the bulls, their thick hides, not knowing that the two latter were the best signs of constitution and good quality, of which the most fashionable Shorthorns were deficient. Bates obtained a name for breeding superior cattle from in-and-in families, at the same time produced his best cattle from "outs in the dark," deceiving his followers, thus condemning himself and destroying his reputation for "pure breeding." Although the truth will out, nothing can suppress it, the Bates mania became so strong that his disciples were not sufficient judges or observers to detect this fraud, or they were determined to fully endorse it.

The facts were so plain to a practical breeder, and, when coming before the public, though startling, the more they were stirred, the more plainly the proof appeared. The way Mr. Matthews sifted Mr. Bates' pedigrees in the "National Live Stock Journal" and stating undeniable facts of their mixed-up alloys were sufficient proof of his intentional misleadings. Judge T. C. Jones and J. H. Sanders, publishing these articles without comment, were at the same time insane on Bates and his "top crosses," neither of them being capable of detecting which alloy had the advantage.

The mania cry was "pure Bates," "absolutely pure," and men went headlong into this "purity" like maniacs released from an asylum, proof of which was so palpable at the New York Mills sale, that "he who runs could read." Bates and his clique consisted of the men I have named; the tongues and pens of those who had but little money were freely exercised by favor of those who had.

Mr. John R. Page was a special pleader. He made in-and-in pedigrees pure, assisted by L. F. Allen and Ambrose Stevens; to make this more sure he sketched very flattering portraits of pet animals, and Lewis F. Allen placed them in the Herd Book, which was sufficient to create an excitement. John R. Page had just the tools to do it. (¶ 90) (¶ 91) His pencil and ruler could draw straight lines out of an original crookedness. He had a faultless *art* of making crooked side-lines straight, could make high hips low, coarse bone fine, smoothen rough shoulders, transfer thin necks into prominent neck veins; "sweet heads" was a specialty with him, as he invariably carries that pattern in his eye, and his brain was always addled with it. He always patronizes "up-standing style," consequently could not shorten the legs to change that character in the fashionable Dukes, but he made their bone finer and much out of proportion. John could not make a picture in Shorthorns without excessive flattery no matter how uneven the original was; the one on paper was all straight lines, and thus they appeared in the Herd Book and sale catalogue, which were John's principal advertisers, assisted by Lewis F. Allen's and Ambrose Stevens' tongues and pens. Examine all his pictures there and you will find a straight furrow along the back of those so-called breeding animals, as if made up of blubber and over-ripe for Christmas show.

Here let me ask any practical man who has seen the original Dukes (¶ 92) whether they ever saw a full neck vein, a smooth shoulder point, a straight under-line or full crops on either of them—all strong signs of constitution and quality? Then look in the Herd Book and see how John R. Page has straightened them and blended each together, so that the picture on paper appeared ideal. Can any reasonable man see such transactions with such proof before him in any other light but that of deception? But it fully corresponds with the "ins and outs" of the Bates pedigrees, and thus all went hand-in-hand. I am exceedingly sorry that the *best* Shorthorns should be abused by novices which brought a curse upon them.

The next position was that John should be the Bates auctioneer. I saw him at his first appearance on the rostrum sell the noted herd of Mr. Haines of Elizabethtown, New Jersey. His audience was looking for a strong "opening speech" on the Dukes, and "pure breeding," but were much disappointed. On going to the scaffold, raised for his exaltation, all eyes were upon him, expecting great things to come to pass in

¶ 138

BEN HERSHEY, MUSCATINE, IA.
(Member Organization Committee American Hereford Cattle Breeders' Association.)

the Shorthorn world, and from the tongue of him who professed to be "the Herd Book in breeches,"—the infallible man of Bates.

No sooner had he shuffled himself into position, he stood erect as if studying attitude. In this state of mind he resembled an automaton. He then moved gracefully, flourishing his right arm to the assemblage, then, pointing his fore finger in vacancy, his phiz looked beseechingly. This was done in silence, to command attention. Then came the opening: "Gentlemen," said he, with a short pause, "you are all aware of the great importance of the breeding of Bates' Shorthorns, and how exceedingly popular I have made them in this country and in England, and I have the honor of being selected above all other men as the sole auctioneer for that excellent tribe, the breeders of them knowing how thoroughly I am posted in their pedigrees; having every one committed to memory, I can vouch for their being correct. As there are many in this herd that have top crosses in Bates I shall expect you to bid very spiritedly. This is all I have to say." He then looked gravely at his audience, but his gravity was not that of a cynic, for I suppose he felt like the ass when amongst the monkeys, that they were all "making faces at him."

¶139
W. H. TODD, VERMILLION, O.

"Now, Mr. Haines, have the first animal brought before me."

"There, gentlemen," said John, "is a most beautiful animal, one of the most fashionable pedigrees (which he read) that a fancy man can desire; she has four top crosses in Bates, by Dukes, in-bred to Duchesses, and one in Oxfords. Now, gentlemen, give me a bid." Silence ruled for a while. He calmly repeated, "Come, gentlemen, give me a bid." The audience looked at John and John looked at the audience. "I have seen animals not so good as that, not so high in Bates, sold for five thousand dollars. Give me a bid, gentlemen; she is to be sold. Shall I say a thousand dollars for you, Mr. L. F. Allen?" who shook his head. "You, Mr. F. Rotch?" who looked over his nose on to the ground. "What do you say, friend Stevens?" who wriggled in his boots, as if he wanted to bid, had he the purse to endorse him. A gentleman from Rahway offered him $100. "Did you say one hundred?" said John; "I suppose you meant one thousand—the very lowest I expected to be the first bid, but as Mr. Haines will not allow any under bidders, and means to sell, I suppose I must take it." S. P. Chapman offered $125; there she stood and John looked as if he was struck dumb; twisted his curled mustache round his fore-finger, hung his lower lip, looked solemn, mumbled out in his confusion, "Gentlemen, I am surprised at such a Bates cow as that going at $125; going, going—remember, I shall knock them down quickly." One dollar advances were afterwards made, until she reached $133—"going, going, gentlemen; I cannot dwell—gone." John looked like a mummy rising in a muddle when the next animal came out, and the next sale went on as in the beginning; he looking incidentally like a "live auctioneer," until the sale was postponed.

Luther Tucker, Sr., was the "chiel takin' notes" for the "Country Gentleman," who expected John to be a shining light, and sustain the flattering advertisement he had given him, and more fully support the Bates mania. Both felt gloomy and forlorn in their unexpected disappointment. To make a little amend, John sketched a flattering picture of the Haines unsold bull, so high up in Bates, to be "transferred on stone," to appear in the "Country Gentleman" at the editor's earliest convenience. When it appeared, John's zeal for the Bates mania cheeringly revived, and he again considered himself the leading star of the Bates clan; his cheek added an additional shade of brass, and he again felt in his prodigality that he was born a wonder, was ready for the second sale, and that he was armed and equipped to again urge on the Bates fiction with the strictest propriety. I am sure that all who attended the Haines sale will vouch for the truth of my statement.

About this time I was preparing a herd of seven of the Herefords for the State Show at Poughkeepsie, N. Y., the first time I ever fed for show, and they were in fine store condition. The Shorthorns were always pampered for this purpose. The Shorthorn breeders controlled that society, demanded high prizes for them, which were more in the aggregate than all the other breeds put together.

The above gentlemen I have named were the leading stars to make premiums and appoint judges, and they combined together to shut out all opposition to them. I made a plea to the committee for the privilege of showing against the Shorthorns. They insisted that I had no right to expect to be heard; that the Herefords were so far inferior to Shorthorns that they could not listen to such a preposterous proposition, notwithstanding my stalls were crowded with admirers and with the novelty of seeing the "new breed" that the Shorthorns seemed slighted by the visitors. I was treated as an innovator, a braggadocio, or an adventurer unworthy of notice by the clique; men I have mentioned looked over the other's shoulders and withdrew with a puff; a few good judges pronounced them a superior breed of cattle. Luther Tucker, Sr., was quite taken with them, and wrote a splendid short article praising them highly, for which he received many upbraidings from his pet Shorthorn correspondents. As there was no Hereford Herd Book when I imported them, this was held as a plea for their not being worthy of showing against the Shorthorns, with all their alloys, and many made a laughing stock of themselves by abusing their thick, mellow hides, and turned their white faces into ridicule. Then their excuse was that it was impossible for judges to decide between two breeds. I asked them if there was more than one proper standard for a good and profitable beefing animal, and whether there was not to be found unprejudiced men of good, sound, common sense, capable of judging, impartially, a good animal regardless of the breed. I was aware that flesh governed almost all judges, but if the Shorthorn men were determined to pamper, and state societies allowed them to do so, Hereford breeders must do the same. Shorthorn men would not accept of this. Neither would they accept of a challenge. When I offered to show four cows and a bull against them on the show ground, on my own account, for $100, they evaded it, well knowing they would be beaten under good and impartial judges, as I consider three of these cows were as good as any of the late importations. One was the first prize cow at the Oxford Royal, two others first and second at Tredegar, and the bull first prize with dam and their offspring. Although they had not been pampered, they showed evenness of flesh, with substance, symmetry and quality; they did not require pampering to hide their faults in their coarseness, or require long legs, long necks, high crests and high hips, to make them stylish or fashionable, but I OFFERED TO WEIGH THEM ON THE SCALES AT THE SAME AGES. They considered this "impudence" in me; they declined to accept. They knew they were beaten and kept aloof, instead of embracing the opportunity. I called upon the officers of the State Agricultural Society in the following letter to the Albany "Cultivator," September number, pages 250-53:

"I did not intend to have said anything more in favor of Herefords, as I had made up my mind to let them take their chance till their real value should be proved, but as certain individuals are continually boasting of particular *tribes* of Shorthorns, in your paper, I am anxious to see the Herefords brought into fair competition with them. I think the State Agricultural Societies should do something to bring the different breeds to a fair trial. I am ready to stand a brush with any breed and in any way the society will point out. All I ask is a fair field and no favor. My idea is that some of each breed should be placed in the hands of an honest, disinterested person, to try the experiment, and the society should pay the expenses. An accurate account kept of the weight and kind of food consumed; the beef, butter, or other products should also be weighed and disposed of, and the cattle that yield the greatest return from the weight taken at commencement for cost of food, etc., should be determined the best. I hope the Executive Committee will take this matter into consideration and propose an honest trial."

This, and many other such trials, I offered these boasting men of Bates, but not one dared to take me up. They were aware that discretion was the best part of valor, in the position in which they were placed. The tongues and the pens of these Shorthorn men before named had given them a widespread notoriety, and their money gave them a partial command of the press. Most of the agricultural editors and proprietors were poor, and money to them was tempting. They puffed and praised, where no praise was due, but in reality censure might have been more properly administered. Lewis F. Allen strained every nerve to bring the Shorthorns, more especially Bates, into notoriety; but he took great precaution to keep degrading and more important facts in the dark, as Bates did in his pedigrees; took great care to keep the best qualities of the Herefords out of sight, and exhibit fancied weak points conspicuously.

G. S. BURLEIGH, VASSALBORO, ME.

I refer you to Mr. Allen's fraudulent book on cattle. Never were more infamous intentional mistakes printed on paper. He read Youatt, and grounded his artificial knowledge of Herefords on this unfounded authority.

Every one of common sense who read his work was aware that his object was to make money out of the Shorthorn breeders who paid him for his puffing of Shorthorns and condemning Herefords (¶ 93), by which he had to stretch his conscience, but his self-contradictions in the latter partially destroyed the effect which he intended to convey. This is how all such men lose their prestige. Lewis F. Allen, when he wrote his book on cattle, had no practical knowledge of breeding. Had he studied it impartially he would have found while praising the Bates so highly for purity of blood they had more of the alloys, even to "the old black cow," than any other breed; although he pretended to have the most horrid abhorrence of the Galloway cross, he was aware of the improvement, and procured it in his repeated mysteries. Still, he stuck to it that not a particle of it was in his herd, though Mr. Matthews, of Virginia, proved conclusively from his own statements in "Bell on Bates," and other authority, that such a statement was utterly false. Lewis F. Allen, who was constantly quoting Bell, etc., must have been aware of this fact, but he overlooked all imperfections in Bates and his breeding.

¶ 141

W. S. VAN NATTA, "HICKORY GROVE," FOWLER, IND.
(First Chairman Herd Book Committee American Hereford Cattle Breeders' Association.)

All who have ever read Bell on Bates will corroborate Mr. Matthews' articles in the "National Live Stock Journal." Allen was, like Youatt, determined to uphold the Shorthorns with extreme flattery, at all hazards, although he felt too poor to support them; he patronized the Devons because they were cheaper, better and more profitable than the noted herd of Thos. Bates, who disliked Booth, because he beat him in almost every instance when they came fairly into competition together.

Bates was continually condemning Booth (¶ 94) for his Galloway cross, to which Mr. Booth openly confessed. Bates contended that his "pure bloods" were far above Booth's "alloys," which ought to have been taken into consideration in the showing. Here lies a serious myth. Could it be possible that Bates did not know how his cattle were bred at the time he made the statement; that all his families possessed it? which has since been proved, by his own conviction, which his weakness could no longer hide, for his general positiveness was one of the certain marks of his weak judgment. There he allowed his fancies to rule over reason, and thus they ruled over him. When Bates showed against Booth he was very desirous of having such committees favorably disposed toward him, and if he did not succeed in this, his rage deprived him of his reason, and made him a laughing stock, forgetting the old adage, "that reason governs the wise man and cudgels the fool." It was hard work for his friends to make Thos. Bates believe this, and no man living can believe contrary to his convictions, or doubt when he is convinced; if he affects to do otherwise he deceives himself.

Thos. Bates proceeded, flattered by his followers, some of which were never more happy than when sheltering themselves behind intrigue, proof of which has been ample in the sales of Shorthorns in this country. I say much of Bates and his followers because I think they have done much injury to Shorthorns in this country and in England, and it was their combined influence intended to injure me and the Herefords.

There are no better men in the world than Shorthorn breeders, and if so many of them had not been led away by the farcical Bates, the Shorthorns would have been in high repute. I always liked a good Shorthorn cow, if not pure; and I shall be highly pleased to see the best contending against the Herefords under competent judges, who will reward merit where it is justly due. It is profitable to have two breeds that can contend against each other; it is stimulating to success.

* * *

PART IV.

All who have read the description of Lewis F. Allen's comparison of the Herefords and the Shorthorns in his book on cattle must have been satisfied of his extreme prejudice against the former. It was certainly distinctly plain to all who read it that it must have been written to give a very unfavorable impression of the abused Herefords. [Allen did not mention the name of Sotham in his book, purporting to be a history of cattle breeds entitled "American Cattle," referring to Mr. Sotham as "an Englishman." T. L. M.] He tried to make it appear that those I sent to the East from Albany had no reputation, while, in fact, the steers bred from the bulls I sold to Mr. Bingham, of Vermont, and others, stood higher in the Boston market than any other, and were much sought after by the butchers, although only half-bred.

I refer you to a letter from Mr. Gregory, in the Albany "Cultivator" of 1851, page 305. He says: "A car-load of two-year-old Hereford steers, on ordinary keeping, astonished the Boston buyers and butchers, and sold higher per pound than any other." What could L. F. Allen have been thinking of when he tried to injure those Herefords which he said the Englishman (meaning W. H. Sotham) sent to Vermont and Maine? He knew when he wrote that pretended history that he was deviating from the truth. I sold several to go to Maine, where they succeeded admirably.

He knew full well the full history of the importation of Mr. Corning and myself; frequently sympathized with me for the loss we had sustained at sea, but I always mistrusted his sympathy was feigned for a selfish purpose, and so it ultimately proved. His object was to obtain the Herefords, if possible, and applied to Mr. Corning to see if he could purchase them, by giving him time, but Mr. C. informed him that if the Herefords were sold that I should have the advantage, and at half the price he would sell them to anyone else.

Mr. L. F. Allen did not dream of the statements he made, for he knew the whole particulars. He knew I was dependent on Mr. Corning's generosity. He came to see me at Albany and went to see the Herefords (¶ 95). Mr. Corning told him that I should have them as above stated. Mr. Allen had been in the Legislature, was a prolific speaker at every meeting connected with agriculture, generally chief spokesman, and was never more happy than when upon his legs and his tongue going.

Under the circumstances I confided in what he told me. "You buy the Herefords," said he, "take them up to Black Rock. My brother will sell you his cows, eighteen in number, his cans, and half of the team and wagon, which draw his and my milk to Buffalo. The milk business is good, you can soon pay for the cows from their milk, and the Herefords will sell there. The timothy grows as high as the fence, the shady pastures are always good; you cannot help but make money, and you and I can send up our milk together, as my brother and I have done." This struck me as an excellent opening. I made the purchase and took possession — sent up my milk with L. F. Allen; but I soon found out that the demand for it was not so good as represented, the price lower, and, for the first time, I began to suspect that the Allens were getting out of it on that account; but I got on in the best way I could. My milk paid my expenses and a little over. In about a year I sold to R. L. Allen two Hereford heifer calves, to be sent to Cuba, the price of which about half paid for the Allen cows. He purchased at the same time about six Shorthorns of L. F. Allen. They were all shipped together to New York; from there by R. L. Allen to Cuba. The whole of the Shorthorns died on the voyage, and the two Herefords were the only ones that landed safely. This Mr. A. B. Allen published in the "Agriculturist," when he was its editor.

W. E. BRITTEN, HEREFORD, ENG.
(A large exporter of Herefords to America.)

After trying the milk business one year, in connection with the Herefords, I found the object of Mr. L. F. Allen getting me there was to get his brother as well out of it as he could, and give him a chance to get some of the Herefords, of which I sold him three—a bull, a cow and a heifer calf. He took these three to Albany, with the whole of his Shorthorns, with one of the most flattering advertisements in the Albany "Cultivator and Country Gentleman" that was ever put into print, and a comment made by the editor of the high value of this herd of the editor of the "Shorthorn Herd Book." All of the Shorthorns that were sold at the sale were two, purchased by Mr. Geo. Vail, of Troy, whose sale was soon to follow, and the cause of his buying, as he told me subsequently, was to encourage L. F. Allen, and a few weeks after they were included in his sale. The Herefords were the only ones for which he received the cash. Mr. Erastus Corning, Jr., seeing Rarity among such a miserable lot of Shorthorns of the "Allen display," which those who came there expected to see from the flaming advertisement, bought her for $100. And Baron DeLonegdale, of Kingston, Canada, purchased the bull and heifer.

I must deviate from my subject here to tell you that L. F. Allen was not only a very careless breeder, but a most miserable feeder, to which I was witness. I did not see a spring while I was at Black Rock but many of his cattle had to be lifted up by the tail in their weakness. This was the case the same spring he sent his Shorthorns to Albany for sale, hence their miserable appearance. The best Shorthorns are an excellent breed, but it is just such novices as these, under the false pretense of breeders, who have destroyed their reputation. The clan of Bates men, of which I have before named, have been a direct curse to the Shorthorn breed. J. H. Sanders, and Judge T. C. Jones, who were supposed by many to be "somewhat sound," proved themselves to have been almost insane. The latter does not know which horn of the dilemma to hang his hat upon, "Booth or Bates." Here it was where he got himself while recently in England. He is now in a quandary as to which side he had better take, and Sanders is in the same situation as to discover the best way to advise him.

J. H. BURLEIGH, MECHANICSVILLE, IA. (Forty years a Hereford breeder.)

It is these novices I have described, and others of the same sort, that have injured the Shorthorns. The breed itself is a good one when in the hands of good and practical men; and it is proper to have two such breeds as Herefords and Shorthorns to contend against each other. If I was put on as judge of the two breeds I would give it to the best animal and to the best of my judgment, notwithstanding some men suppose I am prejudiced against Shorthorns. I may have a dislike to the novices I have named that have abused the Shorthorns, but I value the best of the breed; and I have frequently been told by some of the best breeders of Shorthorns that I have done them much more good than those flattering scribes who had not brains enough to sustain what they wrote.

I found that the supply of milk in Buffalo was greater than the demand, and when reduced to two cents per quart there was no profit. At this critical point Lewis F. Allen and his farm manager fell out and a dispute arose between them that could not be settled without a lawsuit; at the same time the Grand Island farm milk had obtained a bad reputation, and as my milk was sent up with it, and the sour milk returned was divided in proportion to the number of gallons of each, I supposed there was something more than common the matter, as I had so much coming to my share.

Whenever I went to Buffalo the customers always told me that my cans, which had a spot of solder on each, to distinguish them from L. F. Allen's, were always sweet and good [naturally attributed to the richness of the Hereford milk. T. L. M.], while those without the spot were always not only inferior but often sour before half was sold. The lawsuit came on between Allen and his man, and I was present through the whole trial. Mr. John Townsend, his foreman, produced his witnesses to prove that he had done his duty faithfully, and Mr. Allen produced his to defeat their testimony. After the former had brought all in, the witnesses of the latter came. The first on the stand swore as follows: "Do you work for Mr. Allen?" "I do." "In what capacity?" "I milk part of the cows, and help take charge of the milk." "What did you do with the milk after you put it into the cans?" "We stood the cans in the river up to their necks; the night's milk went to Buffalo in the morning, and the morning in the evening; we put ice in the cans to cool it." "What else did you do?" "We took off the cream." "By what orders did you take off the cream?" "By Mr. Allen's." "Did he ever go with you to do this?" "Yes." "Did he say that it was all right?" "Sometimes he told us we did not take off enough, and showed us how to do it." "What did you do with the cream?" "Mrs. Townsend made butter." "What did you do with the butter?" "Part of it was kept for the use of the house on the farm, and the other was sent to Mr. Allen's house." The second witness was called, whose name was Edward. He was the second person who milked and helped take care of the milk; he principally corroborated the former's statement. Then came the man who drove the team and delivered the milk to the customers, who was a German. "Do you take the milk of Mr. Allen to market?" "I do." "Was there any complaint of his milk among the customers?" "There was. Mr. Allen told them that his man did not send the milk from the farm according to orders." "Did you take up the milk of Mr. Sotham at the same time?" "I did." "Was there any complaint about his milk?" "No; all the customers wanted the cans with the spot on it, as all sold before it became sour." "What did you do with the sour milk?" "It came back and I measured and put some in both cans, to be divided according to the agreement made by Mr. S. and Mr. A. Mr. Allen collected the money every week and paid Mr. S. his share, deducting the number of gallons of sour according to the number of gallons taken up." This settled my determination to quit the milk business, as I could rely on the just testimony given, and the witnesses are now living in the neighborhood of Black Rock and Buffalo, who will vouch for the facts just as I have stated them. I then began to look out for other quarters for the Herefords, but my means were limited, the times were very hard, I had gotten into debt, and I had not the ready cash, without sacrificing my Herefords to meet them or to procure suitable fresh quarters.

The late banker, Hon. Allen Ayrault, of Geneseo, N. Y., came to see the Herefords, having two splendid farms of his own, and command of the late Mr. Spence's, near Geneseo. He probably was one of the best judges of cattle in that part of the country, and made more money by feeding the best than any man in that part of the country. Some of the most prominent graziers in the United States are located in this valley. Such gentlemen as the late Mr. Jas. Wadsworth, Dr. Fitzhughs, Mr. C. Jones, the late Mr. Wm. Wadsworth, the Messrs. Budlong, Judge Sibley, and other wealthy graziers, took very active parts, all of whom thoroughly understood their business and who owned some of the best land for grazing of any men in the Union.

The Genesee Flats are as well known through the United States as any other part of it. Although a very circuitous, muddy stream winds its way through this fertile valley, it was the means by which it was enriched. When the floods came the sediment remained as a fertilizer, and the water did not remain long enough to do very much damage, and the occupants were generally prepared for it. I knew the country well and felt it would be just the country for the Herefords.

Mr. Ayrault was highly pleased with the cattle. He said he had previously seen them at State Fairs, but he said that the Shorthorns had got such a strong hold, were so strongly supported by rich men, that Herefords could have no chance. He viewed them very closely in their pastures, and said, "what do you feed these cattle beside what they get here?" I told him the grass was all they had. He seemed to doubt me, but did not say so. "To tell you the truth, Mr. Ayrault, I have not the means to buy forcing feed for them, and if I had, I do not believe they require it. They, like myself, are willing to work for their living, and they will live upon the roughest kind of feed, which you see growing in these pastures."

He proposed a loan and again looked them through very steadfastly, and said: "Money is exceedingly tight now, and chattel security very treacherous. I will make a proposition to you. There is a very nice cow;" pointing to my best three-year-old. "If you will give me that cow, and drive all the others over to Geneseo, I will rent you the best flat land there is in the valley, a portion of the late Major Spence's splendid farm, at a very reasonable rate. Remember, it is a very hard matter to get money now. You may not get such a chance again." I asked him if he would not take some other cow than the one he had fixed upon; that she was my best cow and was named (Anne) after my wife, and that I considered her as good a cow as there was in America. He said, "I am better 'fixed' to put her in condition than you are. I think it will be an advantage to you. It is worth more money than you would ask for her in these close times." The Genesee Flats were very tempting, and I thought such grass would so much improve the Herefords that I accepted the offer. Mr. Ayrault drew a check for the money, and I made arrangements to leave Black Rock, and more especially L. F. Allen and the milk business.

WALTER M. MORGAN,
IRVING, KAN.
(Forty years a breeder of Herefords.)

PART V.

It was a hard blow to part with my favorite. Not only was she the best cow I had, but I believe she was as good a one as I ever saw. Nothing could have induced me to part with her had I the means to carry me through. I went back with him to look at the pastures, and found them all that he had represented, and rented a house in the village of Piffard. I felt satisfied that the Herefords would get fat in those pastures "without cake or corn."

I got the Herefords together and drove them myself to Geneseo, glad to leave Allen and the milk business, with the privilege of his abusing the Herefords and myself to his heart's content, and that he, Ambrose Stevens, and John R. Page, might enjoy themselves making pedigrees to suit their own inclinations, and that the latter might make all crooked lines straight as conveniently as he could stretch out his conscience to flatter the picture, beyond the original, to its fullest extent. These were my thoughts as I traveled slowly behind my cattle to their new home, and wondered how thinking men could be led away by such vain pretenders, who in reality could not discover quality in the live animal, and were too indolent to follow them to the shambles.

WILLIAM A. MORGAN,
IRVING, KAN.,
of the firm of W. A. Morgan & Son.

Without the practical knowledge of "handling" (by "scientifics" called "the touch") no man can become a judge or successful breeder. It is very difficult to discover real worth from tinsel, and it is certain that wrong will always bring its own punishment, therefore I will leave these worthies in their flatteries and fiction in their full enjoyment and proceed on my way to a better harbor, where prejudice is not so strong and judgment more matured. What struck me more forcibly was that this trio should be so conceitedly vain as to attempt to teach others so much better informed than themselves.

It has been very unfortunate for Shorthorn breeders to have such men to lead them into visionary scheming and recklessness beyond control. Excitement led them on to speculation, pride stimulated them to worship pedigree, and in their moneyed power rode the hobby with whip and spur, not having the judgment to discover the true valuable points to constitute a perfect beefing animal, which leaves their faith intact, and their knowledge of quality only a sham. If Shorthorns are still held up I must do my best to meet them.

As I traveled through the country I was asked all manner of questions, most of them too tedious to be answered. I turned my cattle into their pastures, which were luxuriant, and the following morning Mr. Ayrault drove in with his splendid pair of dappled greys. The cattle had filled themselves splendidly, some of them resting themselves from their journey, and sleeping in the sun. "Mr. Sotham, I like your cattle very much; they are just what we want on our rich flats. There are many rich men here, and I shall have much pleasure in driving them here to see them; most of them visit me frequently. I will send my man for my cow to-morrow. She is a good one, is she not?" "I should like you to show me a better amongst all your rich men," was my answer. I delivered the cow to his man, who drove her to his farm, and in the midst of a heavy shower put her into a luxuriant clover pasture, where she bloated and died, and when opened had a heifer calf, within six weeks of calving. If some Shorthorn men had had such a truly valuable cow as this die, they would have said they had lost $10,000. When the news came to me, if I had lost one of my children I could not have felt more dejected. I thought fate was against me. After they had been there about a month I never saw cows improve faster or fatter calves by their sides—not one but was first-class beef or veal. The heifers not sucking were thriving too fast.

Mr. Ayrault drove into the pastures, surveyed the cattle with scrutinizing eye. "Well, sir," said he, "I have been reading your articles in the Albany "Cultivator and Country Gentleman" very carefully. Your reply to Mr. Henry S. Randall is practically magnificent. I think neither he nor Judge Hepburn will trouble you again. That letter is a grand help to the Herefords, and I suppose all you say is true." "Mr. Ayrault," I said, "I always endeavor to write the truth, without which no man is capable of entering into a controversy. Mr. H. S. Randall is a prolific and classical writer, Judge Hepburn a rich and prominent man on the bench, but my opinion is that neither of them has sufficient knowledge of Herefords to write against them. I know nothing of the classics, and but little of grammar, and advance my opinions in my own way, and in plain language. Have you seen the letter of Wm. Kingham?" "Yes, I have. You could not have had a better endorser. Although he is not a classical writer, he certainly is a very practical man, and seems to write facts." I will here copy the letter.

[We omit the letter of Mr. Kingham, which we have reproduced heretofore in Chapter XI. —T. L. M.]

Mr. Ayrault asked me if I would not sell him my best bull for $100. "If you will do so I will have his picture taken and my name shall be put under him as the owner, which will go far to help you, and you know as you are situated you cannot afford to pay for that picture." I asked him what I should do for a bull to show at the State Fair at the head of my herd. "Your young bull is good enough," said he, "and my name appearing in the paper under the picture (¶ 96) of Tromp will do you great service. I am so well known amongst cattlemen and have great influence with some of them. I lend them money to buy the cattle they take to New York market. Remember, the cow I got for lending you the money, died, from which I had no benefit; you must sell me the bull lower on that account." "Had your man known enough not to have turned her into that wet, young clover the cow would have been still living, which you must be aware." This he admitted.

I consented to let him have the bull, though I thought him worth double the money. Tromp was put in the "Cultivator," but his picture did not do him justice. The State Show at Rochester came. The prizes were as they always had been, half in value and half in number, that for Shorthorns. Mr. Ayrault beat me in bulls, but I gained most of the others in Herefords, but they were not sufficient to pay the expenses. Many who had a strong passion for red had a forcible effecting prejudice against any other color; condemned their white faces and long horns. The Shorthorn men were exceedingly jealous of them as rivals. L. F. Allen was there and blew his horn loud and long. The majority of the Executive Committee were Shorthorn men, or influenced by them; many had been so from the commencement of the New York State Agricultural Society. It developed that the object of my money-lending friend was to get these cattle of me by degrees at very low prices. His constant plea was that chattel mortgages were very precarious, and that I must consider it a very great favor to have money lent upon one. I sold him two heifers to apply on the debt, for $60 each, about half as much as they were really worth. My calves were not in the mortgage, so I sold a bull and a heifer calf to a gentleman for $200, and the gentleman drove them off. I went to Mr. Ayrault and paid $125 of it on my debt. "Why, you had no right to sell anything without first consulting me. I shall send my man and have them driven back again." His lawyer lived next door, whom he consulted, who subsequently informed me that he informed him that I had a perfect right to sell those calves, and apply the money as I thought proper. "Now, sir, if you sell another calf of either sex out of that herd I will have them all sold."

This transaction made quite a stir in Geneseo and the neighborhood. Now, my money-lending friend was known by the familiar name of "Old Slikey." Numerous men came to caution me against "Slikey's" tyranny, and told me of many he had ruined by lending money and taking advantage of them. Beware of "Old Slikey"

¶ 146

CHAS. B. STUART, LAFAYETTE, IND.
(Framer of the American Hereford Cattle Breeders' Association's Rules and By-Laws.)

was constantly brought before me. "He will catch you unawares, as is his custom." I felt that I was in a very precarious situation. I realized this to be the case, for he foreclosed. the sale was advertised, and before I knew of it, the bills were out and the sale was to take place ten days from the date of the bill.

My case was soon noised abroad, and many came to see me, knowing that I was fully in the clutches of "Old Slikey, the money-shaver," and I received much sympathy. Amongst these gentlemen was Mr. Murray, of Mount Morris, but a few miles from Geneseo, and who well knew the shaving principle of "Old Slikey." I told him my situation and showed him the cattle. Although he was a perfect gentleman in every particular, he was no judge of cattle, but he admired them very much. He inquired how much was the indebtedness, which I showed him. He offered to lend me the money. Mr. Murray was a large capitalist and a true philanthropist. We talked the matter all over and concluded that it would be much better to let "Old Slikey" sell under mortgage, and to call my friends together, which I did. The day came and I found that I was surrounded by friends who were all strangers to "Slikey." He drove up to the crowd with his pair of dappled greys and addressed his auctioneer. Their faces were familiar to each other. "Go on with the sale," said "Slikey." The first cow was put up. The money-lender was the first bidder—$50. The biddings were spirited, and she was knocked down to Mr. Murray at $165. "Slikey" looked around with a little surprise, and seeing strangers he could not recognize he did not know what to make of it. The next cow came; the banker offered $50, she ran up to $162, and was knocked down to Mr. John Lapham, of Penn Yan, N. Y. The next came a very good cow, and known to be a very great favorite of "Slikey's," who put her in at $75, then bid up to $100. There she stood for a while; Mr. Murray bid $150. Hon. E. Casner, of Penn Yan, bought her for $175. "Slikey" looked around with astonishment. Although, when he alighted from his carriage he looked as *slick* as "Beau Brummel" ever did look, and as he viewed the crowd, he found he had some substantial men around him.

EDWIN PHELPS, PONTIAC, MICH.

Another cow was ordered in, with a young heifer calf; a wag inquired if the calf was in the mortgage. "Slikey" looked at him, and said, "What business have you to make any remarks, sir; I know my business." "So do I," said the wag, "and if the cow and calf is to be sold separately, and I buy the calf, I want to know who I am to pay my money to." All understood this thrust at "Old Slikey," and the laugh was loud and hearty. He stood erect, as if before the glass after fixing his toilet. At length he said, "The cow and calf will be sold together." The wag said, "Will Mr. Sotham agree to that without being consulted?" "Slikey" was all confusion. He began to feel that the public knew the secrets of his heart. The cow was of the first-class, and "Slikey" often wanted me to sell her to him before she calved, at $100. There was but little choice in ten of them. The cow and calf were put up together. "Slikey" put them in at $100. Mr. Murray offered $200; Mr. Casner $210; Mr. Lapham $220; "Old Slikey" $225; Robert Rome, the well-known cattle buyer of Geneseo, $230; "Slikey" $240; Mr. Murray $250, and she was knocked down to him among loud cheers from the crowd, and this was the highest price I ever sold a Hereford.

At this juncture Mr. Murray and Mr. Casner came to me and asked if I wanted to have any more sold. I told them that I could not help myself. "You go tell the banker that you will have no more sold," which I did. He looked almost black in the face. "What right have you to stop the sale? there is not much over half enough sold to pay your indebtedness of mortgages and rent. I am surprised at your impudence, sir." There were lots of listeners around to hear what was going to be done. He looked at me as contemptuously as he well knew how. "Go on with the sale, Mr. Auctioneer; but little over half of the indebtedness is yet reached." Mr. Murray and Mr. Casner stepped up to him and told him to make up the balance of Mr. Sotham's indebtedness, and they would pay it. The "shaving banker" knew not which way to look; all eyes were upon him.

* * *

PART VI.

I have not said much of Mr. Francis Rotch, of Butternuts, Otsego County, a retired banker, and a follower of Bates, a pet scribe of the Albany "Cultivator and Country Gentleman." He was also somewhat of an artist—sketched many animals for the paper—though they were not so highly flattered as those of John R. Page. He was about generalissimo of these papers and took the junior editor, Luther Tucker, Jr., under his care. He made much of him, and he frequently visited him at his farm. He advised his father to send "young Luther" to England,

and the "Bates ring" promised to do much for him.

Luther Tucker, Jr., was then but a stripling, but being the son of a very worthy father, whose character for integrity and truth was well known, was an advantage to him. He managed through Mr. Bates and others, to whom he had introductory letters, to obtain a good showing off. He became acquainted with the "Druid," a gentleman who wrote much for the "Mark Lane Express." At that time Luther wrote some communication for the paper, copying as near as his ability would admit, the Druid style; tried to imitate his style, and in his deep study impaired his constitution for a while, which compelled him to give it up. Ultimately the young man began to think himself out of his "leading strings:" he could adopt a style and language of his own. He praised Bates highly for favors received, gave him a puff in the "Cultivator," and returned to America to take the position as junior. A short time had elapsed before he paid Mr. Rotch a visit, who at that time had a pure Bates bull, which he contemplated sending to England to catch the enthusiastic followers of Bates, whose mania was nearly at full height. The junior editor in his youthful state was in ecstasy when flattered by such a man as Francis Rotch, Esq., of Butternuts, whom he supposed stood so high in the Bates ring and was very wealthy. "But what went he out for to see? A reed shaken with the wind," or a Bates bull he had so partially extolled in England. The truth was evident. They went to the stable and the bull was ordered out. About this time the strongest objection to the Dukes were their coarseness and sluggishness, even by the most imaginative of Bates men. "Now," says Mr. Rotch, "I want to show you the activity of this bull so you can explain to the public that the sluggishness of Shorthorns is unfounded."

The bull was brought out. He described him as follows: "He came out of his stall with his head and his tail up." As he was led to and fro Mr. Rotch said, "Did you ever see such activity in a bull of his great size?" "No, never," said the junior editor. "He can move like a race horse, his action is extremely good; he has such a beautiful high crest he cannot help but move actively, more especially with his large size, large bones, and large, long legs; he is extremely active."

"See what a loin he has," said Mr. Rotch.

"And look at his high hips," said the other; "that in horses is an indication of speed."

"Look what a cupboard he carries; that is what we want to make beef," said Mr. Rotch.

"Then look what a slender waist he has; more like a Duchess than a Duke," said the editor.

Although this may not be exactly the same language as was published in the "Cultivator," it is so near that it would be folly for anyone to dispute it, and I can refer to the original. After it appeared I wrote a criticism of this ridiculous, supposed-puff for Rotch, but it was refused publication. I sent it to the "Mark Lane Express," where it was published. Wm. Carr, who wrote that interesting book on the Booth Shorthorns, embraced the opportunity and commented upon it, part in poetry and part in prose, in which Mr. Rotch's bull was eulogized as coming to England (¶ 97), and the junior editor of the "Cultivator" was coming with him to ride him to fox-hounds to show his activity as a Bates Duke. The poem was a laughable one, which I am sorry to say was burned up with my furniture in the Chicago fire.

This bull that Mr. Rotch anticipated sending to England was one of the coarsest Dukes of those exciting Bates times. Here let me say, that there was no man in this country I more highly esteemed than Luther Tucker, Sr. He was a purely honest man, and I believe a sincere Christian; but the Bates mania overpowered him. He had his family to support and it was his duty to do it. I never could properly blame the senior editor. The Bates mania afforded him a profit, and he encouraged it. He was not to blame. The Shorthorns were very valuable cattle, but the Bates mania was a curse to them, which all who read must have learned, and those who adopted it found it a curse to the community. After the New York State show at Poughkeepsie the following editorial appeared from the pen of that worthy gentleman, Luther Tucker, Sr., editor of the "Albany Cultivator and Country Gentleman." I copy verbatim from the "Cultivator and Country Gentleman," page 312, October number, 1844: "Herefords. —The only specimens in this class were eleven head from the capital herd of Messrs. Corning & Sotham, Albany. These were splendid animals. The two-year-old bulls and bull calf, which were all of the masculine gender exhib-

H. C. BURLEIGH,
VASSALBORO, ME.

ited, were good. They had fine limbs, very spacious chests, round bodies, etc. Several of the cows were very extraordinary. 'Perfection' is one of the most massive cows of her age to be found anywhere, and 'Aston Beauty' and 'Victoria' 1075, for beauty and finish, can scarcely be surpassed, if equaled. It is but justice to say that no animals on the ground excited more praise than these."

I called upon this worthy and impartial editor (¶ 98) and thanked him for the high compliment he had paid our cattle, and told him that I thought he had told the truth. I asked him why the Society would not allow me to show against the Shorthorns. He told me that "it was impossible for anyone to say less of them, for their superiority was apparent, but the Shorthorns had become so strongly established in this country that it was almost impossible to contend against them. Your letters have been very pointed, although I publish them. The Shorthorn breeders thought they were very severe, and my son, who you know has just returned from England, and is now a partner, is averse to having them published. He speaks so highly of Mr. Bates and the hospitality he met with at his house, that he is highly impressed in favor of the Bates tribes, and you know how high they stand in this country now."

TOM. C. PONTING, MOWEAQUA, ILL.
(The first man to drive Texas cattle to New York.)

Whether the son had any influence with the father I had not heard, but there has never been an editorial in favor of Herefords since that time.

After my article was refused publication in the "Cultivator," I met the junior editor, but he looked cross-eyed at me. I began to joke him about the Rotch Duke, but he made himself scarce. A Shorthorn breeder, not of the Bates clan, was present. I said I intended to have asked him if he was going to take the Bates clique with him to England to see the bull take his fences in a style becoming his action, as it was admitted by all of that clan that they were "all stylish; that their heads and tails were always up," and legs long enough to get over the ground. No editor could be more conspicuously situated than to be thus placed, for all the Dukes, Duchesses, Lords and Ladies in England would be there to witness such a transaction, and the example would be great for all such editors to follow.

The poem of Wm. Carr, and the comic way he described "view halloo" and "bull bellow," was one of the most laughable productions of the kind that was ever written. It stirred up the wrath of all belonging to the Bates mania, as the poem was founded upon my article.

Auctioneers were more in demand, the ring became more excited, sales more frequent, counterfeit prices were obtained to make the Bates mania popular. Pedigree was all in all. "Pedigree, oh, pedigree, thou art my darling, we praise thee, we worship thee, we give thanks to thee with sincere affection. Bates, the great 'I Am' is pure, he has no 'alloy,' he detests it, and we, his devoted followers, endorse him." Such were the views and actions of this devoted tribe.

Thus went on the battle. Shorthorn breeders of Bates tribes were seen traveling all over the country, to consult each other, as the Herefords must be kept down at all hazards, regardless of their true merit. The ring was continued and none admitted unless in secret. Bates Shorthorns must go up above any other tribe, and the Herefords must be put down; there was no other alternative. Expense was no object; the time had come for the Bates men to be up and stirring; every man must be at his post when necessary. Every flunky belonging to the ring must do his duty. The editor of the "Shorthorn Herd Book" must add flattering notices to the preface of his book; Ambrose Stevens' wits must be stirred up to aid him; Page's pencil and brush must be kept going, and, if possible, in more flattering style; nothing must be left undone that can be done.

Such was just the state of things when the unexpected panic came. The Bates men dropped off financially, one after another. Imaginary rich men failed, on whom not the least suspicion rested, except from the maniac prices given, which, in the minds of prudent men, was a caution to prevent them from entering into this mad extreme of speculation so contrary to reason and prudence. Being supported by men of supposed moneyed influence was the principal cause of its long continuance. Even now some of the aristocracy of England continue the farce, but they, like their fat calves, decline as the "milk fails." Those who laid great stress on the traveling exhibition of the "Bijou" and the "White Heifer," knew nothing of the rules of good breeding, or they might have known at once that both of these animals were very injurious patterns by which novices in breeding wished to guide them. Every description of them has been of extreme coarseness of frame and softness of flesh, and a paunch purposely made to consume expensive provender without profit.

Mark how the brains of L. F. Allen and Ambrose Stevens used this supposed elevation of the Shorthorns—this portable caravansary, containing the show ox and white heifer as an example of exaltation in "American Shorthorn Herd Book" (page 5 or 6 of Vol. II). What would either of these gentlemen say now should they see an exact picture of either, without flattery, transferred to the "American Shorthorn Herd Book" to compare with the flattering ones of John R. Page?

When all of these flattering gentlemen speak of these "wonderful animals" they think "they struck ile," at the same time the Messrs. Collings were taking a cross with the Galloways to reduce the Durhams' coarseness and improve their quality. Both were overgrown, overfed and forced to the extreme to create wonder, by which they did much to the injury of good breeding. Notice what stress Mr. L. F. Allen made on the craft of the Messrs. "Collings" when they started this wonderful caravansary. Barnum never succeeded so well. This, his strongest plea for Shorthorns, in the second volume of the "American Shorthorn Herd Book." There were hundreds of Herefords that outweighed them of far more compact character, supported with less feed and expense, less bone and offal, that were of far superior quality and held to their original character, which had been established for over a century without, like the Shorthorn, changing their pedigree into hodgepodge confusion, of which all the hidden secrets are coming to light.

Old Father Time does much when he undertakes to search into the secrets of men's hearts. Mr. Allen also says in the same volume that Mr. Collings was a close man. Are we to infer from this that he kept the stock bull in the dark and the show bull in the sunshine? What will he say now, as he discovers those secrets are exposed? Probably Collings will "be in a fix" when he is compelled to reveal those secrets he kept in that closet.

There has been more mystery in breeding Shorthorns than in any other course of breeding, but why should there be such mystery, unless these breeders like darkness better than light? The deceit is far more injurious. It is now generally believed that the Shorthorns are a mixture of Durhams; the white cattle with red noses and red ears of Chillingham Park; the Dishley; the Devons; the Galloways; the West Highland, and last, though not least, the Herefords; at least in the Seventeens, which are now generally conceded to be the best of the American Shorthorn tribes.

I have no objection whatever to this mixture; such experiments are beneficial, and are the art of good breeding, had they not been kept in the dark; but the deception kills the merit due to it. Dukes, Duchesses, Lords and Ladies were superfluous names to exalt, and a sham to destroy "good breeding." They were fascinating to the moneyed man, who had more of it than brains, and who was extremely anxious for a conspicuous title. The white-faces hold to their color and keep up their character for symmetry, substance and quality, principally derived from good breeding. Their breeders have not studied pedigree so exclusively; they have not pinned their faith entirely to it; know quality by the hand, symmetry by the eye, and, being aware of the superiority and purity of their own breed over all others, have jealously exerted every effort to keep it pure. Here let me quote a letter written to me by Wm. Cother, Middle Aston, Oxfordshire, England, which I had published in the Albany "Cultivator," page 132, August number, 1841, soon after my first importation. [This letter we omit as it is produced in full in Chapter XI, to which the reader is referred. T. L. M.]

A. A. CRANE, HOUSTON, TEX.

I always considered Mr. Cother one of the

most practical men and as good a judge of any kind of animal as I know, and he stood as high as a Cotswold breeder as any in England, being a son-in-law of Mr. Wm. Hewer, Sr., of whom he secured a part of his Herefords and Cotswolds; he was a school-fellow of mine, and we traveled much together in younger days to the different sales of Cotswolds in Gloucestershire and in Herefordshire to examine the Herefords. I need not say anything in favor of his substantial mind, as the letter is sufficient proof. When I imported the Herefords there was no Hereford Herd Book, but I had full and correct pedigrees of all I imported and kept them strictly so, but the great fire at Chicago, 1871, destroyed the whole of my records, with all of my furniture.

* * *

PART VII.

Soon after the New York Mills sale of Shorthorns Mr. Cassius M. Clay published a challenge in the "Country Gentleman" that he and his brother Brutus would show a herd of one bull and seven Shorthorn cows and two yearling heifers against any two breeders of Herefords in the United States. My anxiety was great to accept that challenge, but I had not the means to meet it.

Mr. Frederick Pumpelly of Owego, N. Y., of whom I rented my farm, saw the challenge and came to see me. "Why," said he, "you ought to accept that. Your cattle look well enough to show anywhere." I told him I had not the means to meet the $200 deposit, or the money to get them in order fit for show. "Call upon me for all the money you want for expenses, and I will deposit $200 in the bank at Louisville, Ky., where he proposes to meet you at the National Show." I immediately accepted the challenge.

F. P. CRANE, CHICAGO, ILL., of firm of A. A. Crane & Son.

My cattle, having lately come from the flats, were in excellent store condition. I fed them corn and oats ground together, in addition to their pastures, to harden them for the journey. I took one aged bull, seven cows, two yearling heifers, three bull calves and one yearling bull, supposing I might sell all of the latter.

I had at that time never met Mr. Cassius M. Clay. (¶ 99) I had the Herefords nicely fixed in their stalls on the show ground and left my man and a young friend of mine from Boston, who accompanied me to witness the contest, and who felt as much interested as if they were his own cattle. I went to the Gault House and engaged a room for the week, and then entered my cattle in the different classes of the Hereford department.

The late Mr. Robert Alexander's (¶ 100) Shorthorns were there, loaded in flesh to extreme. I learnt subsequently that the two of the highest character (Bates) had not had a calf for two years, and as they walked the blubber shook under their thin hides as if in a jelly bag. Thinks I to myself, "I am in for it now. I might as well have put my head into a hornet's nest as to have come to Kentucky, the home of the Shorthorns."

I had shown the Herefords at the national shows previously, and had become familiarly acquainted with Col. Wilder of Boston, who was president of that society. The Colonel used to always call me "Billy;" that was before I was old enough to be called "Uncle Billy." I went to his room at the Gault House; he took me by the hand and said: "Well, Billy, I am mighty glad to see you. I suppose you have brought the Herefords here to meet Brutus and Cassius M. Clay?" I said I had, but I thought it almost an impossibility for me to have justice here in the midst of so much opposition. "Never mind, Billy," said the Colonel, "you shall have justice to the utmost, as far as I am concerned; and the Kentucky breeders are honorable men. Keep up your former courage and all will come out right." "I am aware of the honor and integrity of the Shorthorn breeders," I answered, "but a case like this is out of the common way. There is much interest at stake, and I am here alone to contend against so many rich men, whose influence is great. What can I do among so many?" "Keep up your natural courage, Billy," said the Colonel, "and you will go through."

In the morning I left for the show grounds, and I saw a gentleman sitting in a buggy in front of my stalls. As I was looking through the stock this gentleman descended from his buggy and came to me. "Is this your stock, sir?" said he. "It is, sir, when it is paid for," I remarked. "Is your name Sotham?" "That is my name, sir." "My name is C. M. Clay." We shook hands and congratulated each other on the meeting.

Mr. Clay called me aside and began to explain. "Since I wrote that challenge my circumstances are differently shaped. I cannot meet you. I have only brought one heifer here and my brother declines." "Well, Mr. Clay," I replied, "I did not expect this; but I know you to be a gentleman, and I will endeavor to meet you as such. No man knows better what it is to be in a difficulty than myself. I sympathize with you heartily, and will consider the trial at an end."

There was a crowd around my stalls to see the white-faces; among them Mrs. Dr. Watts of Chillicothe and her daughter. The Doctor's reputation is well known as a Shorthorn breeder. I had met the whole family at their house at Chillicothe previously, consequently I walked with them through the stalls of the Shorthorns; but I found that Mr. Alexander had been politely showing them through his previously, and Miss Watts, being not only a belle, but a very sensible young lady, he paid her much attention, and, being a bachelor, it was very appropriate.

Mr. C. M. Clay and I walked together to the amphitheater, where most of the breeders had assembled. There we met Mr. Alexander and Mr. Clay introduced me. "Well, Clay," said the former, "I suppose you and Sotham are to settle your differences in opinion to-day." "No," said Mr. Clay, "we have amicably settled the difference." "Indeed," said Mr. Alexander, "how came that so? I suppose you are afraid of each other." "No, Mr. Alexander," said I, "that is not the case. I will leave Mr. Clay to explain." He repeated, "You are afraid of each other." With this repetition my blood began to rise. I then said, "In my lot I have brought two yearling heifers here to show against Mr. Clay. I will place these two heifers in any honest, competent man's hands in Kentucky against any two yearlings in your herd you choose to placé against them. They shall be fed alike in weight and quality of food, and those who shall pay the most for food consumed at two and a half years old shall be the winners." (¶ 101)

Mr. Alexander turned round with rather a forbidding look, and said: "If you come here to fight, I can fight." I replied that I did not come here to fight, but that I could fight, and I had the determination to meet him on that score. The atmosphere began to feel warm, and quite an excitement ensued. Mr. Brutus Clay remarked to me that I was just the man for Kentuckians! The difficulty must be settled, so he said, "I propose that we all go to Alexander's camp, take a drink, and bury the hatchet." This was agreed upon. On our way we encountered Mrs. and Miss Watts; the latter said: "I think, Mr. Alexander, that you and Mr. Sotham had better compromise." He said we were on our way for that purpose. We seated ourselves on the rustic seats; the display of the numerous silver cups, all trophies of Mr. Alexander's success in the Kentucky show rings, were distributed, and bottles of old bourbon, the contents of which were transferred to the cups, in which we drank each other's health in friendly terms.

¶ 152

DR. ORLANDO BUSH, SHELDON, ILL.
(Ex-President American Hereford Cattle Breeders' Association, the friend of all Hereford breeders.)

We adjourned to Mr. C. M. Clay's tent for lunch, which was amply displayed on the grass in true picnic order, characteristic of true Kentuckian style at agricultural fairs. Those not acquainted with Mrs. Clay and her daughters were introduced; here we again met Mrs. and Miss Watts, and a more happyand a more merry party never met together. We separated to look after our own interests in the show ring.

I took all the premiums in Herefords, as I had no competition, which paid my expenses. I drew the two hundred dollars in a draft on New York, payable to Mr. Pumpelly, and had it enclosed to him. Sold my yearling bull to the well-known Hon. Robert Wickliff of Lexington for $100, which was then considered a high price, and my bull calves at $80 each;

they went to Tennessee, all of which gave me great satisfaction.

The next week was the show at Lexington. Although there were no premiums offered, the society agreed to give me first-prize cups for bull and cow if I would exhibit, which I agreed to do. The Hon. R. Wickliff, who took much interest in me and my cattle at Louisville, told me to make his house my home while I remained in Lexington, and Mr. Clay gave me a general invitation to their picnic lunch on the show grounds, and I never met with kinder treatment in my life from all parties.

Mr. Wickliff, although partially blind, ordered his carriage and took me out to his valuable estate, directly opposite to Hon. Henry Clay. Although he had no thoroughbred cattle, he had some fine grades with which he intended to breed to the Hereford bull he purchased of me. I never heard the result of this cross. I then shipped my stock to Cincinnati, to the Ohio State Fair. There was no class for Herefords there, and the Shorthorn classes were strong from Ohio and Kentucky. The society gave me a special prize on each animal of bull, cow, heifer and calf, about half as much as given for Shorthorns, but about as much as paid my expenses at the show; altogether, I returned home pretty well satisfied.

THOS. FOSTER, FLINT, MICH.

I presented the cups I received at Lexington to Mr. F. Pumpelly, with which he was highly delighted, told me he had received the draft safely and said that nothing could have given him greater satisfaction than to know that he had the power in helping me, and that he felt proud of the triumph. (¶ 102)

Soon after my return I sold ten head of my Herefords to Mr. John Merryman (¶ 103) of Cockeysville, Md., for $1,000; and they were a very nice lot. Mr. Merryman is a nobleman in heart and soul, and was just the person to join me in the fight for supremacy against Shorthorns, but they had such a strong hold that it seemed to be an impossibility; their moneyed influence was so powerful over agricultural societies, the press and judges, that made it very discouraging. Still, this $1,000 and my sales in Kentucky helped me much.

I sold Mr. George Clark of Otsego County, N. Y., four cows. He was a large land owner, having several farms. I had previously sold him bulls, and he bred some very fine half and three-quarter-bred steers, one large lot of twenty-five he sold to Mr. Van Alstyne of Kinderhook, N. Y., who fed them, and when they went to the New York market made quite a sensation. They sold readily at half a cent more per pound than anything present, notwithstanding what has been lately said in a Kentucky journal to the contrary; this was true for I was present at the sale. Mr. Van Alstyne and many others will vouch for the facts. This was another of Ambrose Stevens' unfounded thrusts on the Herefords, which, had it been made to me instead of T. L. Miller, I would have replied, as I intend to let him fight his own battle, all I ask of him is due credit for my former articles. Mr. Miller is just in his glory, and he may become the "father of the Herefords," which, if rightly achieved, may be honor sufficient to gratify the ambition of the best of mankind, who are engaged in the true improvement of stock.

Rich and influential men are apt to have their hobbies and preferences, and the Shorthorns had their advantage in this respect, and, being fully aware of the superiority of the Herefords, found that they had much to do to obtain the ascendancy. They combined together as one man in England, where the origin of the well-known feigned sales originated, where animals were transferred from one herd to another at fictitious prices, to present to the world an artificial value, becoming men of wealth and character, which, like all such transactions, ended in failure.

Let me here refer you to a small portion of the text of Mr. Youatt's book. In the eleventh page he says, speaking of Devons: "They have been long celebrated for a breed of cattle beautiful in the highest degree, and in activity at work and aptness to fatten unrivaled." In speaking of the Herefords, page 32, he says: "They are even more kindly feeders than the Devons; will live and grow fat where a Devon will scarcely live." Further in the same page: "The Devons will acquire bulk and hardihood and the Herefords a finer form and activity." These, Mr. Editor, are Youatt's exact words. Let me ask the reader to reflect on these conflicting, glaring statements and reconcile themselves to these contradictions, if they can. Many such passages can be found in Youatt on the breeds of cattle, which convinced me that he was no authority on either breed. His object was to endeavor to please all, at the risk of his own reputation.

Let me here refer you to a letter written by Hon. Henry S. Randall, in the Albany "Cultivator," April number, 1841, and my reply to him in the following number. [These letters appear in Chap. XI. T. L. M.] These two letters will give some insight into the prejudice I had to meet against Shorthorns, and here I will copy the letter published at the same time, which I received from Mr. Turner, Court of Noke, Herefordshire, England, who bred the first-prize cow at the Royal Agricultural Show at Oxford, 1839, which I purchased of Mr. James Walker. [This letter appears also in Chapter XI. T. L. M.]

Some time after Mr. Turner's letter appeared in the "Cultivator" Mr. Richard Rowland, a nephew of Mr. Westcar, whom I know well in England, and who inherited his property, wrote me a very pleasant letter, giving me the names of the butchers, the prices each paid, copied from Mr. Westcar's books, which corresponded exactly with what Mr. Turner wrote me. T. L. Miller used this letter, as he did Mr. William Gurrier's, having the old books to refer to; every one of my letters were marked so as he could refer to them.

Notwithstanding the proofs I had given of the Herefords, the Shorthorn men I have named had great prejudice against them, knew their value, and were jealous, and they felt satisfied that with my light purse they could drive me to the wall. They run their cattle up to fictitious prices—far beyond their value—for the purpose of running the Herefords out of the market. A lot of speculators formed themselves into a ring (all were underbidders for each other), with the three flunkies I have named to do the drudgery, to catch novices with more money than brains who came to purchase and who were easily caught in the snare laid for them.

Many changed hands understandingly, and were transferred from auction to auction, frequently without any money or notes being transferred. I know of one instance where four Shorthorns, sent to a sale at Toronto, Canada, from Kentucky, were knocked down there at high, fancy prices up in the thousands. About four months after I saw the same animals at the owner's farm, who sent them there, and they remained until they were sold by the sheriff. Pedigree, regardless of the animal, was the chief attraction. (¶ 104)

* * *

PART VIII.

It was with much pleasure I recently revisited the breeding farm of Erastus Corning, Esq., the only son and heir of Hon. Erastus Corning, and who, I know, inherits most of his late father's noble qualities. The cattle stables were familiar to me. I was present at their erection, and some of the Herefords called to mind an inheritance of the old sort. Old Victoria, whose likeness Mr. Corning retains in a painting by Van Zant, was fed at the age of nineteen, after raising sixteen calves. Some of her progeny are still here; two of them took first and second prizes over Mr. Burleigh's imported ones at the New York State Fair last fall, but I think Mr. Corning has one better cow than either, and his young stock are very promising.

The bull he imported to cross with the old stock was bred by Mr. Turner, Herefordshire, England, from which he has bred some fine animals. I should have liked him better had he inherited a thicker and more mellow hide, although Mr. Turner has bred some good ones. The cow Matchless I purchased at the Royal Agricultural Show at Oxford, England, 1839, winner of first prize, was bred by him; he also bred many winners. He is now breeding for thicker hides, under a close observance to mellowness. Matchless formerly occupied one of the stalls, and there were four others equally good, Victoria, Perfection, Pretty Maid and Gay, in the first importation, all but Matchless bred from the original stock of Mr. John Hewer of Herefordshire, to which all the best recent importations can be traced, and I suppose there is not a herd of the present day but that contains a portion of his blood, the demand for Mr. Hewer's bulls being so great that he let them out by the year.

WALLACE LIBBEY,
OTTAWA, ILL.

In 1856 Mr. Corning imported three heifers from one of the best breeders in Herefordshire, which the high character of the stock proved. Although he has not kept a register of their pedigrees, no other breed has been crossed in, and although only kept in good store condition they are as good and well-bred as any herd. The good example he has always set in showing his cattle in natural shape is an excellent one, and the judges had the practical soundness to see their situation against pampered ones, hence his success in the show ring. I hope we are fast coming to this legitimate state of things; the country would profit much by it. (¶ 105)

Mr. Corning showed a fat heifer at the New York State Fair last fall against a dozen Shorthorns. She took first prize and was considered by all who saw her a perfect beauty. At Christmas she was killed, and some weeks since an appropriate poem was written of her by a butcher boy who admired her attractive appearance; the poem was copied into the "Drover's Journal." The butcher who killed her, Mr. J. Battersby, told me that she was the best carcass of beef he ever cut up; not only was she of most excellent quality, but steaks cut from her neck vein were beautifully marbled and fit to serve his first-class customers—better even than the choice steak of many other animals.

A. H. BULLIS,
WINNEBAGO CITY, MINN.

How many times such truths have been told to me by other butchers, under similar circumstances, and of equal standing as Mr. Battersby, whose father was one of the prominent butchers in Albany for the past forty years. No man killed better meat, for which he had a high reputation, and his customers were of the highest class. His son is following his good example, and I value his testimony, so fully corroborating that of many others of a like practical soundness.

* * *

PART IX.

Before going further into the Herefords, I must give you an additional insight into what I had to go through with in the Bates mania, of which, as I have told you before, Lewis F. Allen, Ambrose Stevens (who were called twin brothers) and John R. Page were the leading proselytes to that injurious imposition that so much injured the Shorthorn cause. I pronounced this trio "the three flunkies" to induce men of means to join the hue and cry of fancy and fashion that had taken possession of all who belonged to it.

The first wrote a book on the different breeds of cattle, to extol the Shorthorns, and did not only overstretch his ability in the task, but made gross misrepresentations, one must suppose purposely, to mislead. All who had patience to get at all interested in the work, and read his history of the Shorthorns and Herefords, could see his aim to effect high favor to the former, and create a panic against the latter, both of which he grossly and, I think, intentionally, misrepresented. The case was so plain to every unprejudiced reader that my attention was called to it by several gentlemen, among whom were some of the best Shorthorn breeders. I had an intimate knowledge of the writer's character, so I did not look into his book until my friends strongly advised me, and then the comparison between these two valuable breeds was all I had the patience to investigate. Fancy and fashion are capable of leading even the best men astray, and Lewis F. Allen did everything he was capable of doing to promote both of these delusions.

The second man was an adventurer precisely of the same calibre as the first; they were called "twin brothers," as they constantly coupled their visionary brains together, to support the fancy and fashion adopted to boost the "Bates mania." This was their hobby and they expected to reap their reward from the profit made by the Bates clan, but, like all such theories and profitless scheming, the bubble burst, which all who read can prove. He was the man who rewrote the fictitious Shorthorn compilation called history, to defeat the facts published by Rev. Henry Berry, but his misrepresentations soon found him out, and the Rev. Henry Berry's unpleasant truths now stand as firmly as if this misjudged prodigal had not so thoughtlessly interfered.

Still further this notorious Batesite brought out from the Bates herd of such notoriety the bull Duke of Cambridge. Stevens, after a long and familiar stay at Mr. Bates' house, had persuaded Bates that he (Stevens) was the great "I am" of the Shorthorn fraternity; thus prepossessed, Bates presented him with this "Noble Duke of Cambridge" as a memento of his kindness in so strongly supporting the Bates cause in America, the urgency for continuance of which Bates had strenuously instilled into the anxious mind of this supposed exalted breeder.

The bull arrived here with others from another breeder, of which I shall hereafter give a description through Mr. Stevens' letter to the "Cultivator" at the time of their arrival here. The Noble Duke of Cambridge was taken to the herd of Col. Sherwood of Auburn, and the owner charged him a very high price for half of him. Subsequently Stevens wrote asking Mr. Bates to state in a letter to him that he (Stevens) had given Bates an enormous price, naming the sum. At this Mr. Bates became indignant; he felt that a secret gift for puffing was not so profitable to him as represented, consequently wrote Stevens in gentle language that he was sorry to find he (Stevens) was somewhat the reverse of a "great I am;" told his housekeeper if he ever came to his house again to forbid his crossing the threshold. I refer your readers to "Bell on Bates" to corroborate these facts.

Now comes the third chap in the Bates mania who, in his vanity, supposed himself to be the head of all, having been selected by James O. Sheldon of Geneva, N. Y., to take a portion of his Bates tribe to England. This gentleman, being so elated with his flattering pictures in catalogues and Herd Books, placed Page on equal terms with the other great "I am," and at that time the Bates mania was at its zenith. Page was so fortunate in his pretenses of the knowledge of pedigree, fancy, fashion, upstanding style, thin hides, beautiful, soft touch, sweet heads, high hips, extended paunch and grand thighs, so peculiarly essential to denote the Bates tribe, that he had precisely what these fanciful, fashionable supporters of Bates wanted on the other side of that "big pond." The only thing against his personal appearance was his natural swagger, for which he had no checkrein. He caught the rich commoner and dipped a little into the nobility, who were at that period anxiously striving to gull each other at every sale, to find out which could build the most expensive castle in the air, consequently Page returned home highly elated.

I do not describe this trio with the idea of injuring them or the Shorthorns; far from it; but I do say that it was this kind of men that did that cause more injury than any other class; their tongues and their pens were "too fast" for their brains, and this short sketch will plainly show how the Bates mania was created, and I challenge any man to deny with truth the facts as I have stated them. I have no malice against this trio, but I had pity for them in their weakness, which I always predicted would end in the disappointment it did. This trio was quite successful with fancy and fashionable men, their smooth tongues, in the plenitude of their politeness made excessive flattery plausible to them, but when they met together to flatter each other upon the glorious impression they were making upon the Bates mania, Satan stepped out without interfering.

Bates, with all his faults, showed much cunning. When Mr. Price challenged to show twenty Hereford breeding cows and a bull for £100 against any breeder or breed in Great Britain, no man ever showed greater weakness in judgment than Mr. Bates when he accepted it. He felt that he should be sure to be beaten unless the judges were favorable toward him, and he would only have the choice of one of them. Influenced by this conviction, after depositing the forfeit, he began to realize his critical situation, fully aware that if he did not accept that the horns of his cattle must be drawn in much shorter.

JOHN GOSLING, KANSAS CITY, MO.
(America's greatest expert judge.)

Alarmed at his situation, Mr. Bates sent his confidential man Friday disguised in his smock frock on Sunday, while Mr. Price was in church, to examine the cattle he had to contend with. Friday went into the stable under the pretense that he wanted to hire to Mr. Price as a stockman. After looking them through he left, without leaving his name. Soon after Friday returned home Mr. Bates became exceedingly nervous; raised the most frivolous quibbles and excuses amongst his neighbors and friends; his schemes to enable him to back out made him querulous. He wrote to the "Mark Lane Express" that he had a great aversion to gambling; advanced this as a plea, notwithstanding he had been showing for money exactly on the same principle.

Bates' friends became alarmed; rode over to Kirklevington to inquire if that plea of gambling was the only reason; others wrote with anxious inquiries. Bates, in his perplexity, seemed to be wandering about the farm like a hen by candle light, in search of his man Friday, to

GRADE HEREFORD STEER, CONQUEROR, AT 27 MONTHS.
(Bred by T. L. Miller; the rightful champion of 1879-80.)

caution him to be sure and keep his secret; but, unfortunately, Friday could not keep from saying that Price's white-faces was a most excellent herd, and that his "master" would have very hard work to beat him. Friday was asked many questions by all comers to Kirklevington. At length one Shorthorn breeder came who had been into Herefordshire to buy steers for feeding. "Why, Friday," said he, "what were you doing in Hereford a few days ago? I saw you riding through that city and tried to hail you, but you seemed in a hurry as you put the spurs to your horse. Had anything gone wrong? You seemed to have your India rubber rolled up, as if you had been a long journey."

This breeder, in opposition to Mr. Bates, began to be a little quizzical, as he had strong suspicion of Friday's visit to Herefordshire, after reading Bates' letter in the "Mark Lane Express." "I hear you have been calling on Mr. Price. I understand his herd is a very superior one. How did you like it?" Friday was as much confused as Bates. It was the truth, but how did he know it? The Shorthorn breeder tried to look Friday in the face, but he looked every way but his. He was determined to sift him a little further: "Now, Friday, I am as good at keeping a secret as Mr. Bates. Tell me how you liked Price's herd of Herefords? I heard you were there." "Who told you?" "What made you take your smock frock along, rolled up in front of you, to look conspicuously? That was not a garment you usually wear on Sundays. There is some mystery which you ought to explain to me, as I am as much interested as Bates; I do not want to see the Shorthorns defeated. Why did he deposit the forfeit and then withdraw? There is something wrong." After this it was generally known what was the object of Friday's visit to Herefordshire, and for a time Friday was made the market talk, and one of Bates' myths was made public, much against his secret wishes. Friday had kept the West Highland cross confidentially in secret, but this one, so far from home, was too much for him.

I have obtained these facts about as I state them from a source that can be relied upon, and I was familiar with the transaction through the "Mark Lane Express."

In his letter to that paper, Mr. Bates wished Mr. Price to call upon him. None but a man of extreme vanity could have expected this. Mr. Price, being aware of Bates' prominent peculiarity of bribing the judges with insinuations, was the principal reason of his making the challenge, as he was determined that such a transaction should not be allowed on either side.

Mr. Bates wrote to Mr. Price, saying he declined to meet him on account of gambling. This, and more frivolous excuses, he made in his letter to the "Mark Lane Express." In these letters he wished Mr. Price to call upon him, and he would return the call. Here was a chance for more quibbles on etiquette, but Mr. Bates, being a confirmed bachelor, I suppose did not study that part of his ceremonies, as a quibble for delay and a chance to back out. How they were to settle between themselves which was the best herd was only a schoolboy transaction.

Of course each would declare his own the best, and neither they nor the public would have been satisfied. At the same time Mr. Bates said that he had a pair of twin bull calves of the Duchess tribe, he thought one of which would improve Mr. Price's Herefords. Mr. Price replied, "If you want to back out, tell me plainly; then I will listen, but I will not have any more cavil. The three judges having been agreed upon, can go to Kirklevington, and I will accompany them; they can return to my place with your escort, then neither herd can be removed. I feel very reluctant in taking your proposition as a direct insult, but circumstances indicate it. I must say that I was much surprised at such a proposition from one who professes so much to be a pure breeder. I assure you that no such cross as that can enter my herd."

Mr. Price was indignant, and well he might be. Although Mr. Price might think that such a cross on the other side might have been made with impunity in addition to Bates' other outs, he had more sense than to advise it, and etiquette forbade him taking that liberty. When the news came to America, there was much consternation. The Bates organs modified it to the best of their ability. The advocates of the Bates mania, supposed to be moral men, thought the plea of gambling sufficient, notwithstanding more than half of them would buy pools or bet on a horse race. The trio of Bates flunkies—Allen, Stevens, and Page—were silent for a time, but the other part of the clan, like Bates, made many excuses, and as the mania their nods as if they were familiar acquaintances. Although John had on his best clothes, there was something more wanting. It is very difficult for a plebeian in England to know how to act among aristocracy. No matter what clothes he wears, he is easily detected under any display of "independence"; however, the duke, lord or baronet in this "independent country" has no privilege, where all men are equal. Here John had the advantage over them in "independence"; but dukes and duchess titles in cattle were highly honored here, their money value far exceeding those whose titles were inherited under the law of primogeniture, as they knew not the value of it.

Reckless in their "independence," the English squandered it; consequently John had an excellent opportunity to show himself, and his tongue had no curb to check it from having its full scope. John, fully aware the transactions in England amongst Bates men would justify

AN AMERICAN SHOW-RING. UNDER THE MINNESOTA STATE FAIR TENT.

was previously up to fever heat, and the advocates in their glory, it was a difficult matter for them to form combined action, as the indiscretion of Bates was so glaringly ridiculous. John R. Page, as he made himself very conspicuous at the Shorthorn sales, when praising a Bates pedigree so highly, looked slyly and cautiously over his shoulder to see if Mr. Price or any Hereford breeder was behind him. In this connection I must give you more experience of this "herd book in breeches" as an auctioneer. All who knew John knew the frivolous airs he put on to assume consequence, his call to the "New York Mills sale" was supposed to be of such an exalted nature, and as he had to meet some of the nobility and aristocracy of England, he was compelled to put on as much polish as consistent with his high calling, and to receive his case in action and attitude, felt independent, his tongue moved as rapidly as was considered discreet for him to make it.

After being introduced to the foreigners, some of whom we knew, John mounted the rostrum, and when so elevated, the loftiness was of a very exciting nature. He stood erect in silence; not one of the nobility ventured to say, "Is that a dagger I see before me?" "His stature will not compare with our old veteran Stafford in England," said one. His attitude was one foot forward, the other its rear guard, but he was evidently confused; he pointed his forefinger towards heaven, and the canopy above was clear. He stammered a little and then said that he was exceedingly glad to see so many wealthy and influential gentlemen around him from both countries, "and you are aware that I have to offer you to-day some of the most fashionable blood descended from Mr. Bates' true and unadulterated original Dukes and Duchesses, well known the wide world over, and you all well know that I am the gentleman who sold Mr. James O. Sheldon's cattle of precisely the same families at Windsor, at the very gate of your Queen, and as there are many exalted Bates breeders among you I shall expect you to bid spiritedly."

Here was John Bull and Jonathan pitted against each other, regardless of consequences.

AT A COUNTRY FAIR IN MISSOURI.

After the first bid John's flattering tongue was of no use to him—nods were rapid and the order of the day. Fanciful John Bull was so extremely anxious to bid, and Jonathan outnumbering them, nodded conspicuously or gave tongue. The auctioneer became so confused he did not know which bid to take first; he almost fancied that it was he that was up for sale instead of the Duchess. An automaton would have answered the same purpose as John, amid such bidders. The representatives of both nations acted more like lunatics released from an asylum than breeders of discretion, seeking proper and profitable animals, as an example to benefit mankind.

When the Duchess sold for forty thousand dollars, the shouts from the crowd were loud and long, neither John Bull nor Jonathan could command themselves; hats were thrown in the air, men were crazed in delight; Jonathan and John Bull shook hands with each other, and with such a grip that all supposed that the two nations would never be divided, and that Thos. Bates would be king over all. One would suppose that Bedlam had been let loose, or that the lunatic asylum at Utica was on fire, and that the scattered inmates were attracted, by the hideous noises, to take a part.

When Page asked for a bid on the next animal, he was overcome; his voice failed, and when articulation came, a half dozen bids had been offered. "Go on," said John, and they did go on like men on the high road to ruin. When the sale was ended, some of them supposed they had been dreaming; others boasted of their bargains, and felt as proud as "Lucifer." After the Englishman for whom the Duchess was said to be purchased was informed of the transactions he preferred to pay $10,000 forfeit rather than $30,000 more and take the Duchess, which offer was accepted, and a rich New Yorker agreed to take her, but she died before she was delivered. The majority of the Duchesses were either hopelessly barren or followed the fate of the Duchess, who was buried in utter silence, and in the dark with her thin skin on, in honor of Thos. Bates, Esq., Kirklevington, Eng. I am not posted as to her having a marble tomb erected over her to denote her sacred memory. I could tell you much more; I have said so much on Bates and his followers to show how much this clan set against the Herefords. Now I intend to show what I and my Herefords went through after this bubble burst. All I have told you about it has been true, which I will challenge any man of truth to deny.

* * *

PART X.

You saw by the letter from Mr. Wm. Cother, Esq., that there was no Herd Book for Herefords when I imported; but the breeders from whom I purchased kept a reliable record of their own. After the present Herd Book was established, some of the best breeders were very careless about entering their cattle. Being so well satisfied with the pedigrees kept by themselves, and feeling confident of their practical judgment in selecting the sires to maintain the superiority of their herds, they felt perfectly safe in keeping the sound doctrine of self-preservation.

The Hereford breeders were principally practical men, not easily led away by novices or theory. The profits of their cattle kept them independent. Any one conversant with the Hereford fairs (¶ 106), and seeing the immense number of white-faces that came in on these occasions (the money that changed hands there, aside from that paid to the breeders of steers, at their homes), and sent directly to those graziers spoken of in Mr. Cother's letter, could not but see their true value.

The graziers, knowing the truth of the superiority of Herefords for grazing, purchased them. Amongst them were Shorthorn breeders to my knowledge. I will give you an instance of one of these of much prominence, which came under my observation. My brother-in-law, Mr. Marmaduke Matthews, lived a close neighbor to the well-known rich commoner, Mr. Langston of Sarsdon, Oxfordshire (whose only daughter and heiress married Earl Ducie), the well-known purchaser of the Bates Shorthorns, at Earl Ducie's sale, that were subsequently purchased by Mr. Jonathan Thorn and Mr. L. G. Morris, brought to this country, and with which all breeders of Shorthorns are familiar.

While staying with my brother-in-law he proposed to drive me over to see Mr. Langston's Shorthorns, and as I had frequently met Mr. Langston in the hunting field and once sold him one of the best hunters he ever had in his stables, at a high figure, and which proved a great favorite with him, I readily accepted. We drove to the house of Mr. Savage, who was Mr. Langston's steward, which situation affords him a splendid home, and one of his best farms, which he rented. There was another farm he managed for Mr. L., which surrounded Mr. L.'s mansion and pleasure grounds. Mr. Savage and myself had previously met at fairs and markets and were pretty well known to each other; therefore etiquette was not called into question, and an introduction was unnecessary.

We had a good, hearty shake, and I was tendered a hearty welcome. "I will send over to the house and tell Mr. L. you are here, as I know he will be glad to see you; he has the old hunter you sold him, and although now sixteen years old he is about as good as ever. He often speaks of the daring way you used to ride across the country here, and that you were as close to the tail of the hounds as any of them."

Here I must digress a little, and may be considered a little egotistical, but never mind. Probably there was not a more daring rider in the whole of the Duke of Beaufort's hunt, and at the age of twenty-five I never knew what fear was. I sold many horses to the nobility and gentry, and I was a fair judge of a hunter, but did not spare them when under me and the saddle. Bullfinch, brook, wall, or awa-awa never came amiss to me. The horse had to take them if there was the least prospect. By this daring I was as well known to the Duke of Beaufort's hunt as any man. We had just got through our "bread, cheese and ale" when Mr. L. appeared, seated on the old horse. He held out his hand and gave me a hearty shake, which was the grip of the true English sportsman. "Do you know the old horse?" said he. "I do, and it calls to mind the happy days through that winter in which I rode him." "Ah," said he, "you were enough for any of them then, and the old horse was master of your weight. My careful riding has saved him until now, and he is good yet."

We took a survey through the Shorthorns, which were all of the Bates tribes, a draft in the beginning from the Earl of Ducie. Although then considered in the height of fashion and petted by fancy men, I must say I never saw a worse herd. They seemed to me longer and coarser in the legs than any of the tribes I had seen in America, or even what I had seen at Kirklevington. I asked Mr. Savage if there was any propriety in keeping such stock. This brought out a long argument between myself and Mr. Savage. He held up for style and grandeur, and I went for compactness, substance and quality. I condemned their thin hides and blubbery handling, at which Mr. Savage was a little inclined to take offense. Mr. Langston laughed, and Mr. Matthews said but little, as he was breeding unpedigreed Shorthorns to a Hereford bull, raising steers and sending them to market at two years old, by which he made a handsome profit. There were lots of young bulls. If I had been a Shorthorn breeder I would not have taken either as a gift. "Who do you get to buy these bulls, Mr. Savage?" said I. "Oh! We let our tenants have them to improve their stock." At this period I

PRACTICING FOR THE SHOW.

saw a lot of white-faces in a field at a little distance. "What are you doing with those white-faces?" I said. "Oh, they are some two-year-old Hereford steers I bought at Hereford fair to feed." We took a survey through them; they were a very nice lot; probably not quite so good as those of graziers who picked them up of the breeders before the fair, but they were even, and showed good breeding. They were in a good bite of grass; some of them were stretched out in the sun sleeping, others standing chewing their cuds contentedly. If we disturbed one from his lair he would twist his tail round and stretch himself, which satisfied Mr. Savage that they were thriving finely. One of them seemed contented, and with a mild look in his eye allowed us to handle him. His hide was thick and mellow, his flesh firm and ripe; I called Mr. Savage's attention to this steer's compactness, symmetry, quality and weight. "Have you a Shorthorn in your herd that can compare with him? Is there an animal amongst them with such a constitution? Have these and your Shorthorns all lived alike?" He said there was but little difference in their pastures. "If Shorthorns are so superior to Herefords, why did you not go to Darlington fair or market and buy high-grade Shorthorn steers and spayed heifers, which could be bought in any quantity, and at less money than Herefords? There seems to be a myth among Shorthorn breeders that I cannot understand. I think you are more attached to fashion, fancy and pedigree than you are to a rent-paying animal." "I tell you, Sotham," said Mr. Savage, "it is of no use disguising the fact. Those Hereford steers will go to market a month hence, without cake or corn, but Shorthorns want to be finished on cake, and I am free to say they fetch the higher price over the Shorthorns. If Shorthorns make greater weight, the extra weight does not compensate for food consumed." "I am satisfied, Mr. Savage," said I, "so let us change the subject," which was converted to prospects of things in America. Mr. Matthews and myself dined with Mr. Savage, and spent the afternoon smoking our long clay pipes, drinking our port and sherry in moderation, and in the evening sat down to a game at whist, which we kept up until a late hour.

Probably it is one of the greatest sights in the world to see at any Hereford fair so many superior cattle with white faces, and so uniform in color, symmetry, substance and quality combined, as to make that breed superior to any other. I was never more amazed than at my first visit. It confirmed my belief and a fixture of me, in favor of the Herefords, probably to the end of my life. I at once felt satisfied I was on the right track in cattle breeding, however the Americans might be prejudiced against them.

Notwithstanding they were shut out of state and county shows and driven under the lash

A "LINE-UP" AT WEAVERGRACE, FARM OF T. F. B. SOTHAM.

of influenced judges, they could not always be robbed of their good name and real merit; the reputation of the Herefords was too well established to be forever injured by a moneyed power or to allow prejudice to long rule over them. Under such a pressure, however, it was impossible for Herefords to win prizes, unless their superiority was so palpable that conscience, though seared, was compelled to give way.

Soon after this, the Earl of Warwick won the first prize at Smithfield with a Hereford ox of which the editor of the "Mark Lane Express" spoke very highly. I wrote to the Earl of Warwick for a little information of the Herefords and received the following letter:

Warwick Castle, Eng.

Sir: I buy my stock in poor, at two and three years old; fat them for the butcher; fat them on grass in the summer and on hay and turnips in winter, finishing them on oil cake when necessary. To do which it would be immaterial to me which breed I bought or where, depending on what would grow or fat fastest, as good butchers' beasts—the butcher being my customer—and I have no doubt the Herefords are the best breeds for such objects. Notwithstanding well bred ones are bought very dear (a proof of their estimation) I think they pay best for their keep, turning into money faster for the food they eat, less liable to casualties, from the thriving disposition of their constitutions to do well and lay on flesh while growing. I send into Herefordshire and buy lean, young stock, of the best breeders, not trusting to the fairs. I know how all are bred. The ox I won the first prize with at the Smithfield show last year, 1840, and the best beast of any class shown by the ninety competitors, I bought as a three-year-old, with many more at seventeen pounds each, lean. He ran with the other steers, and was put up with them to fat, and I never thought of sending him to Smithfield until two months before the show, never having sent a beast there before. After the show I sold him to a London butcher to kill, for seventy pounds. There is a wretched print of him in the London shops. The painter who did his picture for me is Mr. Davis, animal painter to the Queen, and lives at Chelsea. He made a very good colored engraving, one of which I should be happy to procure for you if you will have the goodness to write to me when

¶ 162

THE CATTLE RING AT THE DETROIT INTERNATIONAL FAIR AND EXPOSITION. T. F. B. SOTHAM, SECRETARY.

you arrive in England next summer. My house in London is 7 Carlton Gardens.

(Signed) WARWICK.

I called at Warwick Castle in 1842, and, after sending in my card, the Earl ordered his valet to show me in. As soon as he remembered my letter to him, and the prize ox, he ordered his man to bring up two saddle horses and he would show me the steers feeding in the park. The pasture was excellent on the old sod. They were a capital lot of two-year-olds, he purchased in Herefordshire in the spring. We talked of the prospects of the Herefords in America, and other different subjects, and he gave me a note to Mr. Davis, to give me one of the engravings of the ox, which I brought

STOCK BARN AT T. L. MILLER'S "HIGHLAND FARM," BEECHER, ILL.

here and presented to the late Hon. Erastus Corning, and, if I mistake not, it still hangs up in the residence in Albany now occupied by his widow, whose age is 87 years.

This letter to me from so prominent a peer in England made quite a stir. Mr. Francis Rotch felt quite indignant. He wrote a classical criticism upon it, in which he hinted that the letter could not be genuine. Luther Tucker, Sr., called upon me, and told me what Mr. Rotch had written. He did not publish it, but wrote him that the letter was certainly from the Earl of Warwick; that his "coat of arms" was on the seal, and there was nothing relating to the letter that could be construed into deceit. I immediately sat down and wrote Mr. Rotch as severe a letter as was ever written to another. I knew the circumstances for which he left New England, and I did not scruple in asking him if he supposed my character was like unto his, obliged to leave home for evil deeds done in the body. I told Mr. Tucker what I had done, who knew the circumstances of the late banker's sinning, notwithstanding I told him "that he who was without sin, let him cast the first stone," still I felt certain that I should never be guilty of forgery and for this broad insinuation I should chastise him the first time we met.

A few days after, Mr. Tucker came to me and said there was a gentleman in his office who wished to see me very particularly. Mr. Tucker had shown him the original letter, which he pronounced genuine, and Mr. Tucker, when he told me who it was, asked me to keep my temper. "Why, Mr. Tucker, I think you know me well enough that I can always forgive a man who acknowledges himself in error." Mr. Rotch then walked in out of the back room, and made me a very satisfactory apology, but there was much said about this letter in the "Cultivator," which can be seen on investigation, in 1841.

Here is another substantial evidence in favor of Herefords, and I have never yet seen a single instance where Herefords have been fairly tried but what they were appreciated. The following is an extract from a letter to me from Mr. Fisher Hobbs, Marks Hall, Essex, England, one of the most prominent agriculturists of his day: "When I commenced farming and breeding on my own account, in addition to my stewardship to Lord Western, I purchased a Shorthorn herd, probably but few better in England. I kept them and had to nurse them tenderly for five years. Notwithstanding this, I had become almost wedded to them. The nobility, with whom I much associated, patronized them, and I followed their example. I purchased a small herd of Herefords, because they sold cheap, and to my utter astonishment they could live and grow fat, while the Shorthorns grew poor, all faring alike. Although much against my wishes, I sold off the Shorthorns and took to Herefords, and have kept to them with much faith."

I sent the whole of this letter to the Albany "Cultivator" just before the beginning of the Bates mania, and when the junior editor was a boy playing marbles in the street. Mr. Luther Tucker, Sr., published it and was much pleased with it. Mr. Hobbs won the first prize for the best cow-in-milk, with a Hereford, against numerous Shorthorns, and kept to the Herefords until his death. Wherever similar trials have been impartially made, the Herefords have invariably been successful.

Mr. Fisher Hobbs and Lord Western were the originators of the Essex hogs, the latter imported the Neapolitan hog and the former crossed them with the "old Essex" breed, hence the improved breed.

When in England, after my second lot of Herefords, I called upon Mr. Hobbs, at Essex

Hall, and was received most kindly. We rode over his farm, viewed the cattle, which consisted of Herefords, and discussed the demand for his Essex hogs, which at that time was very great. I found in Mr. Hobbs one of the most substantial men I ever met, and never spent a more pleasant time than while under his roof. Many of the nobility looked up to him for advice. No man stood higher as a steward or as an agriculturist, and he was a leading man at the principal meetings and exhibitions of the country. I purchased a pair of his pigs, and had them sent to London to be shipped with my Herefords.

The testimony of such men is worth preserving, while those who write for the sake of seeing their names in print copy from others, no better informed than themselves, put on airs unbecoming men of judgment, do a great deal of injury, much more than they have any idea. How many writers of the past have been practical men? Youatt is yet often quoted by editors, novices in breeding, such as J. H. Sanders, Judge T. C. Jones and many others I have named, who have pinned their faith to him, when all practical men who have read Youatt can plainly see that he knew little more about cattle than cattle knew of him, and it was sheer affectation for him even to attempt so important an undertaking. He was educated as a veterinary surgeon, which profession required a classical education, and upon which acquired education such men are apt to place the foundation of their knowledge so that when they write, they study the classical part more fully than the more practical.

So it was with Youatt. He looked more to his orthography, etymology, syntax and prosody than he did to the good points and character of animals, or the true science in breeding them. He never studied good breeding practically, but gained his information at market ordinaries of those interested, and his friends were principally amongst the Shorthorn breeders. It was mostly hearsay without proper foundation. Any practical man can discover this through the whole of his work, which I have read closely, and have frequently quoted critically, which can be plainly seen in my former letters. He decidedly shows that he listened attentively to the dictates of Shorthorn breeders, because he knew they were rich, and by their riches had run themselves into a rapid current of notoriety. Such men have done great injury to the progress of good breeding. There are too many who have followed this example, feeling themselves exalted because they have written a book, no matter whether mankind had received any benefit from it, or it had inflicted a great and palpable injury.

All L. F. Allen's works are borrowed from such men as Youatt, while the portion emanating from his own brain rests upon a very shallow foundation, or smacks strongly toward his own interest. You might easily suppose that L. F. Allen was the founder and self-preserver of all the breeds in the universe, but it was this consummate vanity that prompted him to borrow the plumage of others in forming the bulk of his books. Shorthorn men themselves saw the exaggerations of Shorthorns and his desire to depreciate the Herefords. All knew his motives and his strong prejudices, and some of them made allowances for some of his idiosyncrasies. Although L. F. Allen was exceedingly anxious to become a noted breeder of Shorthorns, which was at one time partially supposed, on paper, he never bred or owned a good animal of the Shorthorn breed in his life. He was a man of great notoriety, without the necessary essentials of true merit to back him. I cannot but believe that his book on the different breeds of cattle has done much more harm to the community than he can ever repair. He supposed in his vanity that I was so far inferior to him in position, amongst the cattle men, that there was no paper in this

A MISSOURI COW AND HER TRIPLETS.

country whose editor would allow me to reply to him. I am very anxious to have him reply to my strictures upon him, backed by the other flunkies I have named, so intimately connected with him. I tell them all that the editors and proprietors of the "Drover's Journal" are honest in their purposes, and will publish anything they choose to write. My suggestions to them are, not to change a word, but give them their full force, and that they will find the "Drover's Journal" impartial toward either breeder or breed, and for which they have been rewarded for their honesty and independence in the success they have achieved in the field of journalism. May they still become more prosperous is my sincere wish.

My object is to write the truth, and if any of my former opponents think me in error, I am open to criticism, I care not in what shape it comes. If I cannot vindicate myself I will cry *peccavi*, and acknowledge the fault. All I ask is a fair field and no favor.

* * *

PART XI.

The Bates ring is now broken up, the Bates mania defunct; titles are looked upon with suspicion. Men who supported them did so to their hearts' content, but would now gladly return them to the country in which they were created.

GRADE STEER CALVES IN AN IOWA FEED LOT.

The wise men at the East have to be fed with roast beef by the practical men at the West; therefore, they have to study the best and most economical way of making it to their own advantage. They have found out that the Hereford has a strong constitution, has a good, thick, mellow hide to protect it, under which good quality invariably rests. Their meat being well interlarded is put on in the right places; this gained in their *activity* is unchangeable, which enables them to take it to market in good shape.

The Bates tribes under their thin hides and soft handling, so vehemently supported by their breeders, cannot stand the hardship they have to endure, especially when weaned from their nursing and pampering. The character of their soft flesh is bred in them, under the absurd doctrine Thomas Bates was constantly preaching to his followers. This flabby flesh wasted away in their coarseness from the exertion which compelled them to travel for their food, and what they did put on was carried to market in bad shape, which made them only fit for stockers to be finished amidst plenty, where they could be indulged in idleness, then the most profitable parts were deficient, not having been bred there; their pedigree must have been a fraud. So goes the world and the people in it are subject to many changes. L. F. Allen and his herd book, J. R. Page and his flattering portraits, even with the assistance of Ambrose Stevens, can never compensate for the injury they have done to Shorthorns. * * *

* * *

PART XII.

* * * When about in my 25th year, I used to go to visit my uncle in London, who lived in White Lion street, a close neighbor to Mr. William Gurrier, who was salesman for my father; and being fond of stock, I used to get up in the morning between 1 and 2 o'clock to ride with Mr. Gurrier to market, as Smithfield began at this time by lamp light. He kindly took me through the cattle as he handled them in their pens, frequently asking me which was the best bullock. If I differed with his judgment he explained it to me. I was always on hand at my uncle's door when he called for me. Mr. Giblett was one of the best customers, and we frequently went to Bond street to view the stall of meat killed by this noted purveyor. We had not much chance of judging between the Herefords and Shorthorns, as he seldom killed any of the latter, which were at that period very coarse and of inferior quality. Mr. Giblett's principal purchases were Herefords and Scots. Here we had a good chance to compare our judgments of alive and dead, and I again say here that no man can be a successful breeder of cattle of any breed unless he can discover true quality in the live animal. To do this, his hand must be his guide.

The more Galloway and Highland blood they introduced into the crossed Durhams the more they improved them, hence, the name of "Improved Shorthorn." When the passion for upstanding style in Dukes and Duchesses became fashionable and fanciful, this improvement was destroyed, and good breeding was lost sight of. Mr. Giblett's establishment and his reputation for the best stall of meat was extended far into the country, he killing nothing but Herefords and Scots. It was a great gratification to me to see the display of meat in this well-known and extensive establishment, and I was a frequent visitor when on a visit to my London friends. Everything was kept in the neatest order, he would not allow inferior carcasses of beef, mutton or veal to enter it. It was a pattern to all such establishments in that metropolis. There was not a breeder or farmer of any note in England, when visiting the city, but took a pleasant survey of it before they left. I frequently talked to Mr. Giblett and his son about the Herefords. They always spoke of them in the highest terms, and as being the most profitable beasts for the butcher. I inquired of several of the best butchers in several of the market towns in the country; all told me the same tale, but many of them said that they bought more Scots and Welsh cattle on account of the smallness of the joints, and because they could buy them cheaper than they could Herefords before they reached London.

Go to Oxford and you find that all the first-class butchers say that they buy none but first-class Herefords to supply the colleges with beef, and here much of it is consumed. Go and examine the extensive meadows rented by the principal butchers in that city, containing some of the best grazing land in the world; there you will see none but the best Hereford steers and oxen, all belonging to butchers, placed in the different lots, as a supply when wanted.

It is highly gratifying to the lover of good stock and well-fed beef to go and examine the carcasses behind these white-faces, every meadow containing from twenty to forty head. I remember Mr. A. B. Allen, when in England, then editor of the "American Agriculturist," being struck with this display of Herefords, could not refrain from writing a paragraph of high praise as he viewed them from the terrace of Christ Church College.

This was the principal market to which the noted and most worthy supporter of Herefords, Mr. Westcar of Creslow, Buckinghamshire, sent his well-fed Herefords. Mr. Richard Rowland, his successor, taking all things into consideration, found this his best market; he still holds to the Herefords as most profitable. Having become, like his uncle, thoroughly disgusted with the way the shows were conducted, he declined ever again showing an animal, notwithstanding he feeds the very best. In a letter to me, which I published in the Albany "Cultivator" of 1842, he said he never would allow another animal of his under such treatment as the Herefords had at their shows.

* * *

TEXAS 2-YEAR-OLD STEERS IN AN OHIO PASTURE.
(Champions at Chicago Live Stock Exposition, 1901. Property of D. W. Black, Lyndon, O.)

PART XIII.

As many of the Herefords come, when calved, with spots on their faces, let me tell those concerned that it is no disparagement to the breed. I have seen some of the best with mottled faces. Tomkins was a prominent breeder of the best, and preferred the mottled face, and many are descended to his favorite mottled faced cow, which, at that day, was considered as good as any, and those that run back to the Tomkins sort in this respect are generally good animals.

The defeat the Bates Shorthorns have experienced at the Bath and West of England

A BUNCH OF MISSOURI YEARLING HEIFERS.
(Bred by T. F. B. Sotham.)

show, by the Herefords, has had a tendency to check their breeders' vanity. I will here give extracts from the leading papers of England.

The following is from the "Mark Lane Express" of June 17th, 1878, on the Royal Show of that year:

"We consider the Shorthorn cow class to be a disgrace to that breed, and therefore to the breeders. We are quite unable to discover the 'grandeur' and 'superb character,' and the 'magnificent character' of these old crocks, which some of the Shorthorn fanciers appear to have the faculty of discovering, and do not hesitate to record them as being just a rough lot of cows. If their blood is of the bluest, their carcasses are of the ugliest, and are not worth anything beyond contractor's price when they come at last to the shambles. We can't help thinking that to a really unprejudiced mind there must be an evidence of a something outside agriculture, and quite useless to the rent-paying farmer, in this Shorthorn 'fancy.' We are sensible of the improvement that has been effected already in the rank and file of our cattle throughout the country by the use of this Shorthorn blood. We do not wish to detract one iota from its legitimate merits, but simply to point out wherein it becomes sometimes a matter of ridicule to non-believers. We see prizes awarded systematically which are not calculated to improve the production of either meat or milk in their descendants, and we are told that there is some marvelous power and virtue stowed up in their veins, and that although their bodies—the casket—are unsightly, their blood—the jewel—is pure, potent and almost priceless. Well, we simply do not believe it as they put it. We are perfectly well aware that any 'terribly in-bred weed,' a wretch to look at, but having an exceptional pedigree, will, if matched with mongrel-bred stock, produce a result which is far, very far, in advance of the mongrel-bred dams. So would any absolutely purely bred animal. Therefore, we think that farmers, those who have the production of beef and milk in view, have a right to expect something which is calculated to effect their object in a direct manner; no breeder of bullocks would give herd room, much more a high price, for the bluest blood bull which did not carry a frame the character of which it was desirable to transmit. When a lot of highly bred, but not correctly fashionable young bulls, are to be bought for about thirty pounds apiece, and here and there one that has been bred correctly to fashion fetches three thousand pounds and would not be worth one shilling more to the food producer, then we think we are justified in saying that rent-paying farmers have already drawn the line between business and fancy by refusing to give more than a business price for a fancy article. We should be glad to see every young Shorthorn bull now in the breeders' hands sold to tenant farmers, who would use them for meat or milk, as their requirements might decide, and are firm believers in the general usefulness and superior adaptability of the breed to any other; but we feel it a duty to point out the ridiculous position the Shorthorn breeders seem content to occupy at our great shows, by the mixed qualities and low status of many of the animals exhibited, which gives opportunity for pointing the finger of scorn, and leads to such questions as 'when is the bottom of this Shorthorn humbug likely to drop out?' We hear a great deal about the 'alloy,' and if correctly informed, it means an infusion of Scotch blood of some kind, and to our mind the very thing these Shorthorns are now needing is another infusion—a strong one of some *alloy* which will give them the thickness of flesh, the wealth of hair, and the butchers' form they so seldom possess, and then we should be prepared to expect great things from the renovated blood."

The same paper, speaking of the Herefords at this same show, says:

"They are not so numerous as the more fashionable breed, but the quality throughout is excellent. In the aged bull class there are five animals of which the Hereford men need not be ashamed. The heifers in milk numbered only three, but two of them were such animals as it was worth while coming to Oxford on purpose to see. Mrs. Sarah Edwards of Wintercott took first and second, leaving Mr. Lutly the reserve, but Mrs. Sarah Edwards' Leonora (¶ 107) is one of the most perfect animals that has been shown for years. It was first last year as a yearling at Liverpool, and will likely be first wherever it goes. The Champion prize given by the Oxfordshire Agricultural Society was also awarded to this heifer, as the best female horned animal in the yard. The champion heifer, Beatrice (¶ 108), is also very handsome, and took second to Leonora's first at the Royal last year, as it did last week at Oxford. Mrs. Edwards may well be proud of such stock as that. If Leonora had been a Grand Duchess, Shorthorn, a poem would have been composed in her honor, and translated in several languages by this time, but no Shorthorn that we have ever seen was cast in such a mold as this Hereford heifer."

Of the award of the champion prize, the same authority gives the following account:

"One of the most attractive features of the opening judging was the awarding of the champion prizes given by the Oxford Agricultural Association, and a great deal of interest centered in it. For the best Shorthorn animal in the yard the award was given by the Shorthorn judges, of whom there were but two, Mr. Drewery and Mr. Trindall, and the contest was between Mr. Linton's Sir Arthur Ingram and Mr. Marche's heifer Diana, and after some little time the bull took the prize card. To our minds the white bull calf Prince Victor was a better Shorthorn than either of them. He would be a bold man who would stake his reputation on a calf, as it may not realize the expectation formed of it, but taking the animals as they stood simply as merit actually possessed at the time, we thougnt the calf the best, nor were we singular in that opinion. Then the whole conclaves of judges came into the ring to decide which was the best male horned animal in the yard, and here the Shorthorn men were hopelessly in the minority, cornered in a manner that does not often happen in a show yard. In vain Mr. Drewery contrasted the strong points of Sir Arthur Ingram with the weak ones of Mr. Roger's Hereford bull Grateful (¶ 109); it was all to no purpose; the rest could not get away from the Hereford, whose wonderful rib and forehand was too much for the Shorthorn, and the show of hands showed an overwhelming majority for the Hereford. There is no doubt but that, had the Shorthorn judges been of sufficient numerical strength, they would not only have prevented this Hereford triumph, but also that which followed when the best female horned animal had to be decided upon, for it went sadly against the grain to award even such a heifer as Mrs. Sarah Edwards' Leonora the championship over the Shorthorn Diana, but it had to be done, and we think there were few outsiders who were not thoroughly satisfied. These champion awards were an unquestionable streak of lean for the 'fancy,' and we may depend on their not allowing such a thing to occur again if they can help it. This, coming after Paris, is about as much as they will be able to bear with patience. We should be glad to see sweepstakes judging of this kind more frequently."

The Shorthorn men, gaining full control of the Royal Agricultural Society and the Smithfield Club, forbade the contest between the Shorthorns and Herefords thereafter, well knowing the consequence, should it again occur under impartial and sound judgment. The editorial in the "Mark Lane Express" instituting comparison is true and was written by an impartial scribe, shows plainly how others see it. The "Mark Lane Express" supported the Shorthorns above any other breed, until they saw their error. The interest in that paper was supported by the aristocracy of the country.

NEBRASKA RANGE CALVES IN THE KANSAS CITY STOCK YARDS.
(En route to Eastern feed lots.)

I will conclude with a short sketch of Captain Morgan's visit to Mr. Hewer's in Gloucestershire. As I had crossed the ocean with him in the old Philadelphia and Hendrick Hudson seven times and return, we became thoroughly acquainted, were never more happy than when we could get a good joke on each other. The Captain's laugh was so hearty that it cheered everyone who heard it, and this being my third trip for cattle, I engaged the steerage in New York to bring them in. Going out Captain Morgan agreed to go with me into the country, to see some of my friends.

We went into Oxfordshire and tarried all night at the house of an old acquaintance, who was exceedingly glad to see us, to talk about the old country and the new, and, being early spring, the young rooks were fully fledged and ready to depart from their nests, a rook-shooting party was expected the next day. The host, the Captain and myself went out to view the young ones, perched by the side of their nests ready to take their flight when confident of the strength of their wings. The host asked the Captain if he was a good shot. "I used to be a good shot with a rifle when young, and I think I could hit one of those chaps on the highest tree with one." The host fetched his gun and loaded both barrels; the Captain brought down his bird. "Try again," said the host; he did, and was equally successful. "You may as well kill enough for a rook pie for dinner to-morrow, before the shooters come in the afternoon."

RANGE CALVES BOUGHT BY T. F. B. SOTHAM FOR EASTERN FEEDERS.

"Why, do you eat crow in England?" said the Captain, and he gave the American definition, with one of his hearty laughs.

"We consider young rook quite a favorite dish with us, and I do not think you will despise it to-morrow, after tasting it." So the Captain went on shooting until he killed a dozen, only missing two shots.

"Why, you will do to compete with the best of them to-morrow." The uproarious noise the old and young rooks made in the rookery attracted a great deal of the Captain's attention; it amused him much. We strolled round the farm until tea time, looking at the Cotswolds, all of which were very good.

After, the neighbors called in, to whom I was principally known, and the Captain soon became acquainted, as he prided himself on being a "full blooded Connecticut Yankee," and knew how to ask a few questions. We smoked our long pipes, and drank our ale, or brandy and water, as tastes differed. The Captain told some good yarns about land and sea that amused the whole party. He almost made them believe that roast pigs were in every shop window, with a fork stuck in their backs, inviting emigrants to come in and eat them free of cost. One old gentleman said he thought it must be a great country, but that he thought he was too old to emigrate.

The next day at 1 o'clock we sat down to the rook pie. The Captain could not help thinking of crow. He said to me, "Do you *eat crow?*" "Certainly," said I, "and I bet you praise the crow as much as you did the gun with which you killed them." The top crust was short and flaky, the under crust savory; the crow tender and of very nice flavor. The Captain looked at me and said *"crow pie,"* as soon as he tasted it. I said, "How do you like it?" "It is good, so far, anyway." All were helped, and all enjoyed it. The Captain began to think it excellent.

"I challenge you to a glass of wine," said the host. "Which will you have—port or sherry?" and after he washed the first serving down, he was ready to be served to a second. Who would have thought a Yankee could have supposed crow so good as this? The Captain said he never before enjoyed a dinner so much. The host loaned the Captain his gun, and nearly every one of his shots brought a "crow." The guests thought the Captain a dead shot, and there was about a wagon-load of young rooks in a pile from all the guns. All were anxious for a rook pie, and each took his share, the best shots having no more than those less successful. The Captain asked the hostess how she made that "crow pie." She told him that she took out the backbone, which was considered bitter, and jointed the rest, which was finer than even partridge or pigeon, and the rooks being young were always tender and of fine flavor. The Captain said he wished they could convert the crows into rooks in America.

It was now time to leave for Northleach, Gloucestershire, to visit Mr. William Hewer. William and myself being intimate friends from school-fellows, I was as much at home as if I had been in my own house. William knew we were coming and met us at the station. I said to him that "this is a Yankee Captain from New England, and I advise you to look out for him." "All right," said he, "and we will make a Hereford man of him." We got there just in time for supper, which was a cold leg of mutton, bread, cheese and water-cresses. The Captain said that he had eaten so much "crow pie" that he did not feel like eating any supper. The Captain liked water-cresses fresh from the spring, and that, with the Stilton cheese, refreshed his appetite, and we supped together in merry mood, notwithstanding we had eaten *crow*.

After supper we took our long pipes, and Miss Hewer brought in some mulled elder wine for a night cap. The next morning after breakfast William ordered his hunter, a mare fourteen years old I had sold him before I left England, as good a hunter as was ever under a saddle, and as good a pony that he rode around the farm, as ever was crossed by a man, and who knew as much as half of them. The greyhounds came jumping round, and we were off for a course. Hares were plentiful. "Which will you take, the old mare or the bay? Sotham can ride either, and I will ride the pony," but the Captain chose the pony. We joked the Captain on his seat in the saddle, and asked him if we should tie his legs. The first field produced a hare, and the greyhounds killed her before she got out of it. The pony became a little excited, but the Captain stuck to the saddle, and was highly delighted with the course, the first he had even seen. The next field William saw a hare in her form. "Now, Captain," said William, "go start for that tree and you will see her, whip her up; a hare will always lie very close when the dogs are round. We will give her plenty of law." The Captain could not see the hare in her form, but he supposed he must be near her, so he popped his whip. Up jumped the hare and the pony jumped from under the Captain. He lay upon his back and the pony kicked up his heels at his leisure. The course lay in the adjoining field. Our horses took the wall in good style (¶ 110). The Captain was soon on the top of it in great excitement. It was a capital course and our mutual friend was highly delighted; the wall gave him the advantage of seeing the whole of it. We afterwards killed a brace more hares and returned home to eat a coursed hare that had been hanging in the larder until ripe. It was fine, with the port wine gravy and currant jelly. "I tell you, Sotham," said the Captain, "this is about as good as crow. Why," said the Captain, as we joined in a glass of wine, "you English fellows live well. A fine boiled leg of Cotswold mutton and turnips, a roast hare, and an English plum pudding."

The Captain was as much at home as if he had known Mr. Hewer his whole life, as they were two hearty, good fellows, well met.

The next day we proposed to go and look at the cattle and Cotswolds that were to be shipped. As it was only half a mile to where the stock were, we agreed to walk and take the greyhounds with us. We walked about twenty yards apart. "So-ho!" said William, "there she sits just before you, Captain; put her up." The hare being so near the color of the ground, and she crouching into it so closely, the Captain could not see her until she jumped up. It was a good, strong hare, and the turnings were

A ROW OF STOCK BULLS, AT SHADELAND FARM, LAFAYETTE, IND.

numerous. The hare got a little law, and went straight for the Captain. A hare, when pursued by greyhounds, never sees anything before her, as her eyes are always back upon her pursuers. The Captain took off his new beaver, and as she was coming straight to him, put it between his legs to catch her. She went right straight into it, knocking the crown partly out, and the greyhounds being so close, came in contact with the Captain, nearly upsetting him. The Captain, gaining his equilibrium, gave one of his most hearty laughs, which echoed over the hills and through the vales. Such a burst of laughter issued from this, our trio, that will be remembered to our latter end. The Captain was highly delighted, and at her death, ran to take the hare from the dogs, forgetting his hat, which lay at a short distance from him.

"By Jupiter," said the Captain, "did you ever see anything like that? It will do for me to tell the Connecticut Yankees." We examined the hat, and the crown was about half circled. "Never mind," said the Captain, "I suppose there is a hat store in Northleach."

We then went to see the cargo destined for America. The Herefords were in one yard, the Cotswold rams, seventeen in number, in one pen, the ewes, eight, in another. I never saw a man enjoy anything more than the Captain did in viewing this stock. "Remember, Mr. Hewer," said he, "I am a Connecticut Yankee, and like to ask questions. How do you manage to raise such stock as this, and keep them in such fine condition? Why, the cattle look all alike, and the sheep all alike; how can you tell them apart?"

"I know every sheep in my flock by its face, as you know your children. The cattle are also familiar to me from calves. I like all of them as I do my family. I do not like to sell any of

SOME SHADELAND MATRONS.
(Herd of the late Adams Earl, Lafayette, Ind.)

my best, but as my old friend and I have been attached to each other from school boys together, I could not do less than let him have some of them. He bought the Cotswold rams at my sale and I think they are a good lot. Five of the ewes took first prize at the Royal Show at Oxford, 1839. I should be very sorry to have my culls go to America under such circumstances."

"There is something in that," said the Captain, and we all agreed it was an amicable trade, and no man could feel more proper pride than the Captain that he was going to convey such stock to his country.

As the Captain had to leave in the 4 o'clock coach, we made for home, and on our way had another course. This made the fifth hare. "Now," said the Captain, "Mr. Charles Dickens, of Pickwick fame, is going out in my vessel this trip, and if you will say which day you will be in London, I will invite him to dine with us on the vessel." So we agreed to be there three days before sailing day. The five hares were packed in a flag basket, and checked to Captain Morgan's quarters in London, and the Captain departed, saying that he never had such a pleasant time in his life.

At the entrance of the dock we met the Captain. As soon as he caught sight of us he hailed us with his new Northleach hat in hand, and with a laugh bearing no deceit, a most hearty grip was exchanged. On our way along the dock (¶ 111) there was one of the greatest old hags that could possibly be beheld. The Captain gave me a flat-handed slap on the back. "Do you see that beautiful female?" said he. "What would you do if there was no other in the world?" "Well, Captain," said I, "I am at a loss to decide in haste. What would you do?" "I would wash her up and take her."

We entered the cabin of the Hendrick Hudson, full of merriment on our past actions. "We dine at five," said the Captain, "and Mr. Dickens will be on hand."

Mr. Griswold, then the senior proprietor of the firm of the London Packets, was in London, and was invited to meet us. About half an hour before dinner we all met on deck. I had met Mr. Griswold before, in New York, and after a formal introduction to Mr. Dickens, the Captain began to give a sketch of his visit in the country, and his merry way of telling it was capital. "I ate crow there," said he, "and it was magnificent;" he explained why it was so. Mr. G. understood what was meant by "eating crow" in America, but Mr. Dickens did not until he heard the Captain's explanation.

The bell rang for dinner in the cabin; there was a nice boiled leg of mutton and turnips. To this we did justice. Then came one of the coursed hares. Had her head been put in a crouching position instead of up, she would have resembled one sitting in her form. "So-ho," said the Captain, "shall I cut her up, Mr. Hewer?" "I suppose that is what you mean to do," said he, "and that will be the end of her."

"Is that the hare that went through the Captain's hat?" said Mr. Griswold. The port, sherry and champagne was ordered in to wash her down, and the steward had cooked her to a charm; the port wine sauce and the currant jelly to encourage the appetite were all that could be desired.

The Captain's peculiar history of his visit was highly interesting, as he told it with such glee. When we came upon deck the tide had come in, and the vessel was far above the dock. Mr. Hewer said, "How are you going to get the cattle into this vessel?" The Captain looked at me as much as to say, "we have the joke on Hewer now." He called the first mate. "Mr. Starks," said he, "take Mr. Hewer down below and show him where you are going to cut the hole by the side of the ship for the cattle to enter." The sailors, with tools in hand, could not keep straight countenances, and Mr. Hewer began to suspect, and with one of his most hearty laughs the Captain said, "So-ho, Mr. Hewer, you are floating now." "Sold," said William. "I'll pay the fine," so he handed out a half sovereign to the mate to treat the sailors.

Mr. Dickens invited all to dine with him the next day, which we did, and had a very merry time. All through the trip, not one of us exceeded what Englishmen call "market merry," which is just enough to become pleasant to each other.

When the cattle came in Mr. Hewer came to the vessel to see them loaded, and as the bull Major was dangling in the air in a sling, between heaven and the decks, a tear came into his eye. When he was landed in the steerage he went down and caressed his old favorite for the last time, and so with all the others. He always disliked to see any of his favorites go away.

The vessel was hauled out into the river, Mr. Hewer returned home, and the Captain and myself took train to Portsmouth. Here the Captain was informed that Mr. Dickens could not leave until the next packet, so we were deprived of his company.

* * *

The old gentleman (Mr. Sotham) has gone to his rest, but not until he saw the triumph of the Herefords, and the breeding of Shorthorns started safely on the road to improvement, by rational breeding for practical results. The Scotch, which but recently was rated plebeian by Shorthorn breeders, and therefore utterly unfashionable, bids fair to revolutionize Shorthorn breeding, and is destined to do the breed great good, if the breeders do not overdo it and replace the "Bates mania" with a "craze for Scotch."

Mr. Sotham would rejoice to see the present state of the Hereford breed of cattle. Coming into their own has not addled the brains of Hereford breeders; they remain aloof from fads and fashion in pedigrees; they insist ever upon a superior individual with a good pedigree, and for such animals as embody a large degree of perfection in this desired combination, splendid prices are readily paid. Like the veteran champion and father of the breed, all influential Hereford breeders ever bear in mind that the end of every Hereford is the block, and they deprecate any Hereford that would fail to give a good account of itself in the butchers' hands.

Mr. W. H. Sotham was a half century ahead of his time; he made every sacrifice for the Hereford breed, because he knew their true value. His work must be deemed successful, for he blazed the path that is essential to pioneer work in every important movement. He laid the foundation upon which the Hereford structure rests (¶ 112) and he laid that foundation so soundly that, built up upon the same principle of truth and merit, it will endure so long as beef cattle are bred.

OUT FOR AN AIRING.
(Scene at Shadeland Farm, Lafayette, Ind.)

CHAPTER XIV.

Early Cattle in Kentucky—The "Seventeens"

The following letter, written by Lewis Sanders and published in March, 1849, in the "Cultivator," gives a full and exhaustive account of the early cattle interests of Kentucky:

The first emigration to Kentucky—"the dark and bloody ground," the hunting grounds of the Southern and of the Northern Indians, with the view of permanent occupancy, of holding the country at all hazards, by men determined to overcome the tomahawk and scalping knife by the use of the rifle, took place in 1775-6. The country then belonged to Virginia. A large proportion of the settlers were from that state, next from Pennsylvania, then North Carolina, Maryland, New Jersey, etc. It is presumed that the emigrants brought with them domestic animals, such as were then in use. H. Marshall, speaking of General Ben Logan, in his history of Kentucky, Vol. I, says: "In the fall of the year 1775, Colonel Logan removed his cattle and the remainder of his slaves to his camp (near where Danville now stands). Horses and cattle subsisted in the summer on the range, consisting of a great variety of nutritive grasses, including the buffalo, clover and pea vines, luxuriant beyond description, and in the winter in the cane brakes."

It seems to me that the general characteristics of the cattle of the United States at the commencement of the present century were very similar to those of Devonshire, Dorsetshire and Somersetshire, in England, as represented in prints of cattle in those counties in the last century. I have observed the cattle of Virginia, Maryland, Pennsylvania, New Jersey, New York, and the New England States; they seem to have had a common origin.

The first improvement of cattle in Kentucky was made by Mr. Matthew Patton and his family, to whom the country is much indebted, for the introduction of several animals. An historical account is given by Dr. S. B. Martin, a highly respected and spirited agriculturist of Clarke County, in this state, which is herewith forwarded as a part of this communication. Judge Beatty, in his very valuable "Essays on Practical Agriculture" (a book I recommend to all beginners to own), treats on this subject. These two papers combine all the evidence it is thought that can now be obtained relative to the Patton cattle. I have heard it intimated that the introduction of the Patton cattle increased the weight of the four-year-old bullocks 25 to 30 per cent., besides improving the quantity and quality of the milk. This was a great gain.

The next marked improvement in the breed of cattle was brought about by the importation of some animals direct from England in 1817. At that period and for many years previous, I lived in Lexington. My pursuits were otherwise directed than to agriculture, but I had early imbibed a fondness for fine stock, particularly horses and cattle. I admired good fruits and gave some attention to their culture. For several years I was in receipt of a variety of English publications on agricultural subjects and agricultural improvements, from which I got a glance of what was going on, in some respects, in the old country. It astonished me greatly to see the enormous prices paid for particular breeds. First, the Longhorns brought to a high state of perfection by the justly celebrated Bakewell, Princep, Munday and Fowler. Towards the close of the last century they were at the height of their popularity. Mr. Princep refused 500 guineas ($2,500) for a two-year-old bull of his breed. He was offered £100 each for twenty dairy cows. He refused to let his best bulls go to his neighbors' cows for thirty guineas ($150) the cow. At this period (1789) the circulating medium was gold. The bank did not suspend specie payments until 1797. Mr. Fowler refused 500 guineas ($2,500) for ten bull calves of the same breed, and let his bulls go out for the season (April 1st to August 1st) for from £60 to £80 ($300 to $400).

Much time was required, combining capital, skill and untiring perseverance, to bring this breed to such a high state of perfection. Notwithstanding all this, it was suffered to run out, almost to disappear, in the course of a few years. About the time the Longhorns were held in such high estimation commenced the improvement of the Shorthorns. Skillful breeders, with Charles Colling at their head, brought this breed to a very high state of perfection. Their value was at its height in 1810. In this year a public sale took place. The list of animals sold and the very high prices paid for each has been often published. Countess, out of Lady, four years old, brought four hundred guineas ($2,000); Comet, six years old, brought 1,000 guineas ($5,000). He was bought by four farmers.

It seemed to me that if four farmers were willing to pay $5,000 for a bull, there was a value in that breed that we were unapprised of, and that I would endeavor to procure it. I made up an order for six bulls and six cows. My views were then more inclined for good milking than for a beef breed. The weight of the authorities, given by the writers on the subject of cattle at the close of the last, and the commencement of the present century was in favor of the Holderness breed as the best for milking and the Teeswater and Durham as having the handsomest and most perfect forms. I settled on these breeds. In frequent conversations with Captain William Smith about the contemplated importation, he strongly urged me to include the Longhorns. He had witnessed the marked improvement made by the use of old Mr. Patton's first Longhorned bull, and he was extremely anxious to have a bull of that breed. I had some respect for him as a man, and confiding in his judgment, two pairs of Longhorns were added to the list. The order was forwarded in the fall of the year 1816 to Buchanan, Smith & Co., Liverpool, with instructions to cause selections to be made of the best young animals for breeders. All to be two years old in the following spring.

First a bull and heifer of the Holderness breed, to be procured from that district in Yorkshire. Next two bulls and two heifers of the Teeswater breed, to be procured on the river Tees, in the county of Durham. Then a bull and heifer of the Durham breed and two bulls and two heifers of the Longhorn breed, no limit as to price. If the money sent was not sufficient to put that number on board ship, they were to be reduced so as to have the best animals that could be had for breeders.

Buchanan, Smith & Co. employed Mr. Etches of Liverpool to go into the different districts to make the selections and purchases, and he seems to have executed the orders with much ability. The following is the invoice: Cattle shipped on board the Mohawk for Baltimore, consigned to Messrs. Rollins & McBlair, merchants there:

1. A bull from Mr. Clement, Winston, on the river Tees, got by Mr. Constable's bull, brother to Comet.

2. A bull of the Holderness breed, of Mr. Scott, out of the cow that gave thirty-four quarts of milk per day—large breed.

3. A bull from Mr. Reed, Westholm, by his own old bull.

4. A bull of the Holderness breed from Mr. Humphreys, got by Mr. Wase's bull, of Ingleton.

5. A bull of the Longhorn breed, from Mr. Jackson Kendall, out of a cow that won the premium.

6. A bull from the Longhorn breed, from Mr. Ewartson, of Crosby Hall—is of a very fat breed.

SOME SHADELAND YEARLINGS.

7. A heifer from Mr. Wilson, Staindrop, Durham breed.

8, 9, 10. Three heifers from Mr. Shipman, on the river Tees—his own breed.

11, 12. Two heifers of the Longhorn breed, from Mr. Ewartson, Crosby Hall—of Westmore.

The Mohawk arrived in Baltimore in May, 1817. The cattle were safely landed, in good condition; great pains had been taken in procuring comfortable accommodations for them in the ship, and an experienced herdsman was employed to feed and take care of them on the voyage. On arrival they were taken in charge by my friend, Mr. John Rollins, who caused them to be put into the pasture and particularly cared for.

After the cattle had been shipped and before their arrival in Baltimore, I sold to Captain William Smith, one-third of the concern, and to Dr. William H. Tegarden another third; reserving to myself one-third only. A suitable agent was sent to Baltimore for them and they would be brought to Kentucky at the joint risk and expense of the three parties. On their arrival at Lexington they were divided. There fell to my lot: No. 1, which I named Tecumseh; No. 2, named Sam Martin; No. 8, named Mrs. Motte; No. 10, named Georgiana.

Captain Smith's lot: Bull No. 5, which he named Bright; No. 7, cow, which he named

YOUNG BULLS AT SHADELAND STOCK FARM.

The Durham Cow; No. 9, cow, named Teeswater Cow.

Dr. Tegarden's lot: Bull, No. 4, which he named Comet; No. 6, which he named Rising Sun; No. 12, Longhorn Cow.

No. 10 died in Maryland; No. 3, bull, became lame on the travel out to Kentucky and was left on the way; he was afterwards received and sold by the company to Captain Fowler, who sold him to General Fletcher of Bath County, Ky., where he died.

When the division took place, Captain Smith evinced great anxiety to own the largest Longhorn bull. Dr. Tegarden preferred No. 4, and, as neither of them were my favorites, I cheerfully yielded, and in consequence they gave me the choice of the cows. I selected one of the Teeswater heifers and named her Mrs. Motte. It was a very pleasing occurrence to have each highly gratified with receiving the very animals he preferred.

The narrative of a pertinent coincident will not, I think, be deemed ill-placed:

Mr. H. Clay being in England in 1816, having always had a fondness for other fine stock, concluded to send home some fine cattle. At this time the Herefords were great favorites at Smithfield. Either from Mr. Clay's own taste, or from the recommendation of others, he selected that stock, purchased a cow, a young bull, and heifer of that breed, and sent them to Liverpool to be shipped to the United States. It so happened that they were put on board the Mohawk, the same ship with my cattle, and they arrived together at Baltimore, where they were placed in the same pasture, and the agent that was sent for my cattle brought out Mr. Clay's to Kentucky.

Although Mr. C. and myself at that period resided in the same city, and had always been personal and political friends from the time of his coming to Kentucky, in 1789, till March, 1825, and our social and personal relations have been unchanged for fifty years, yet, neither Mr. C. or myself had the slightest knowledge or intimation of the intention or views of the other in regard to importing foreign cattle.

Mr. Clay at one time had a good stock of horses. He bred the dam of Woodpecker, one of our best race horses, and he proved to be a good stallion. His flock of sheep were celebrated for the fineness of their fleece.

Having introduced the Herefords, I might as well finish them.

At this time, 1817, Mr. Isaac Cunningham owned the largest and best grass farm in Kentucky—the identical farm settled by old Mr. Matthew Patton, the father of the Patton family, who introduced the Patton cattle. Mr. Cunningham was wealthy, had a good stock of Patton cows, and had been in the habit of selling his young ones for breeders. Mr. Clay's good judgment led him to place the Herefords in the hands of Mr. Cunningham; notwithstanding all these advantages, the Herefords made no impressions. In a very few years they were unknown as a breed in Kentucky, and at this day a part blooded one is rarely to be met with.

As to the Longhorns (¶ 113), although there were two bulls imported, the breed was nearly run out. Captain Smith kept them for a while, but he died soon after they were introduced; his stock was neglected. The Rising Sun left a good stock in Clarke and Bourbon Counties, and for a while they were very popular with the feeders in these counties, but they have gradually yielded to the Shorthorns. A mixture of Longhorn blood in a remote degree is deemed by many feeders of great value (and that is my opinion). The hide is thick, the hair is long and very closely set; they are of very hardy constitution, well adapting them to our mode of feeding. Cattle are not housed or sheltered, but fed out in the fields, taking the weather as it comes. The Shorthorns have thin hides, fine, short hair, and do not stand exposure to the weather so well.

The importation of 1817 (alluding to which it seems that the Longhorns are to be omitted) gradually gained favor with the breeders and feeders. The young ones were very much sought after throughout Kentucky and parts of Ohio, and were all sold for breeders. The Tecumseh and Sam Martin were the principal instruments used in effecting this great improvement. Mrs. Motte, the Durham cow, and the Teeswater cow were excellent breeders. The Durham cow was equal to the best milk cow I ever saw. Napoleon was her best bull calf. Mrs. Motte was the neatest, the finest animal of the importation.

A year or two previous to 1831 I observed that my young cattle were not up to the mark of improvement that I wished to see progressing, but were rather falling back. The only remedy that I then thought, and still believe necessary, to arrest this downward tendency and to give a fair prospect of improvement, was the introduction of remote blood.

Col. John Hare Powell of Philadelphia imported a number of animals of the improved Shorthorn breed, several years subsequent to 1817. He ordered his selections from the best herds in England, with great particularity as to pedigree from the milking qualities, and without stint as to price.

My attention was directed to this stock to procure a cross on the Shorthorns of 1817.

In the spring of the year 1831 I procured of Mr. Barnitz of York, Pennsylvania, a young bull and three young cows of Col. Powell's stock. In several points their forms were better than those of 1817. The cross was very beneficial to me.

Some few years afterwards David Sutton of Lexington introduced several animals of Mr. Powell's stock.

Then other gentlemen imported cattle from Philadelphia, and from other parts of the United States and from England, so that we had a number of bulls and cows from the best known breeds in England and in the United States. From this basis intelligent gentlemen, with abundant capital and great skill, have continued to improve by judicious crossing until we have arrived at a high state of perfection, as to form and early disposition to take on fat, points most desired of all others by the grazier and the feeder.

Notwithstanding that Col. Powell's stock were drawn from the best milking families in England, their descendants did not prove with us to be as good milkers as the stock in 1817, nor were they so healthy.

The dairy is but a secondary consideration with a Kentucky farmer—beef is more profitable, and, as the great object of all pursuits is money, the one putting most in the purse will be pursued. For a dairy of cows where there is a demand, selling milk is most profitable, next cheese, if the climate suits; last, making butter. Cheese can be made here as well as anywhere else, but it costs too much labor to save it. Some writers say that it ought not to be relied on as a business south of 40 degrees. Butter could be made of the best quality, and in quantities, but it seems that the farmers prefer taking only as much milk from their cows as supplies their families with milk and butter, giving the remainder to the calves. From these considerations it would seem that the breed of cattle bringing most money from the butcher at two and three years old will give the preference with the grazier and the feeder, they using nine-tenths of the cattle bred in the State.

It will be seen from what has been stated that great attention has been given to the breeding of cattle in this State for more than fifty years, and the course pursued has been to procure the best breeds to cross with, so that we now have an excellent breed for the grazier and feeder—forms approaching near and nearer to perfection, and an aptitude to take on fat at an early age, but in obtaining these grand objects, perfect form and early maturity, so much desired by the grazier and the feeder, we have sacrificed mainly the milking qualities.

IMPORTED BULL DIPLOMAT (18328) 81537.
(Sold for $8,000; property of C. A. Jamison, Hamlet, Ind.)

Whatever be the breed, there are certain conformations which are indispensable to the thriving and valuable ox or cow. If there is one part of the frame the form of which, more than another, renders the animal valuable, it is the chest. There must be room enough for the heart to beat and the lungs to play, or sufficient blood for the purposes of nutriment and of strength will not be circulated—nor will it thoroughly undergo that vital change which is essential to the proper discharge of every function. Look, therefore, first of all, to the wide and deep girth around the heart and lungs—we must have both. The proportion in which the one or the other is preponderate may depend on the service we require from the animal; we can excuse a slight degree of flatness of the

THE FOUNDATION OF AN OHIO HERD.

sides, for he will be lighter in the forehand and more active, but the grazier must have breadth as well as depth. And not only about the heart and lungs, but over the whole of the ribs must we have both length and roundness—the hooped as well as the deep barrel is essential. There must be room for the capacious paunch, room for the materials from which the blood is to be provided. The beast should also be ribbed home. There should be little space between the ribs and the hips. This seems to be indispensable in the ox, as it regards a good, healthy constitution and a propensity to fatten; but a largeness and drooping of the belly is excusable in a cow, or, rather, notwithstanding it diminishes the beauty of the animal, it leaves room for the udder, and if it is also accompanied by swelling milk veins it generally indicates her value in the dairy.

The introduction of the Patton stock into Kentucky effected as much benefit to us in the improvement of our cattle in a little more than twenty years as was effected in England in more than sixty years.

A printed report of a select committee of the House of Commons, in 1795, stated that cattle and sheep had increased on an average, in size and weight, about a fourth since 1732.

The average weight of cattle slaughtered for the London market in 1830 was 656 pounds. (McCullough's Dictionary of Com.) [Undoubtedly dressed weight of 82 stone, which would be equivalent to 1,148 pounds live weight. T. L. M.]

At Liverpool, about the same period, 600 Irish beasts averaged 720 pounds; 140 English beasts averaged 730 pounds; 60 Scotch beasts averaged 810 pounds. [These being dressed weights would make the live weights 1,148 pounds, 1,277 pounds and 1,417 pounds respectively. T. L. M.]

It would seem that our improved breeds exceed these weights. Twenty fat cows were sold in the early part of this month by one drover at Cincinnati, the average weight of which was over 1,000 pounds the four quarters [or by the English system of calculation 1,750 pounds live weight. T. L. M.] These cows were Kentucky bred. All but three had produced calves.

I expected to receive authentic data to state the average age and weight of the four quarters of cattle slaughtered at Louisville and Cincinnati, for three periods. Though promised, the paper has not yet come to hand.

In 1833 I took to New Orleans three bullocks, produced by a cross of the cows of the Patton and Miller stock by bulls of the importation of 1817.

No. 1, red, six years old, live weight, 3,448 pounds; No. 2, red, six years old, live weight, 3,274 pounds; No. 3, brindle, four years old, live weight, 2,868 pounds.

I sold these three animals together at auction for the sum of $925. I was at the New York State Agricultural Exhibiton at Saratoga in September, 1847. I very attentively examined the cattle stock there shown.

The oxen were better than are generally to be met with in Kentucky, all others not so good.

The Ayrshire cattle may be classed with our half-blood Durhams, from common cows. We can derive no benefit from a cross of Devon blood. The diminutive size and ill forms of the Alderneys would exclude them from our pastures. Our climate is favorable for breeding and rearing cattle. They are free from any marked disease. I have never known an epidemic among them.

It is the custom with some farmers as soon as the corn is in the roasting ear to cut it up, giving stalk and all to the hogs. The hogs masticate the stalk—suck and swallow all the juice, throwing out the remaining fibrous matter, which soon becomes dry. Cattle are very

BACHELOR (Grade).
CONQUEROR (Grade).
BARNEY (Grade).

SUCCESS (5231) 2.
MAID OF ORLEANS 1146.

KANSAS (Grade).
WILL (Pure).
WASHINGTON (Pure).

(FAC-SIMILE OF MR. MILLER'S CELEBRATED CHROMO.)

This Lithograph in colors, freely distributed, was one of Mr. Miller's effective methods of drawing attention to Herefords.

fond of this refuse stuff, but when taken in quantities it causes a derangement of the manifolds, for which no remedy has as yet been discovered. At first the animal becomes restless, and is feverish; soon after it begins to rub its head down and up a post, or anything it can rub against, manifesting the greatest pain and misery. It continues rubbing until it dies. I have seen several so affected with it and after rubbing commenced I knew of none to be cured. Upon opening the animal it is found that the manifold is entirely deranged, dry and hard, mortification having in some instances already commenced. The only remedy is to keep your cattle from the place where green corn stalks have been fed to hogs.

Cattle of Ohio and Indiana are not so healthy as are the cattle of Kentucky. I was told by a Cincinnati butcher who supplied with beef a portion of the Jews of the city that he was compelled to procure his cattle for these people from Kentucky. The priest sticks the animal, which is dressed in his presence by the butcher. Upon opening the animal if any imperfection of the intestines is visible, such as blisters on the liver, etc., the priest remarks: "This one may do for the Christians, but will not do for the Jews—you must bring up another." The cattle of Kentucky have no blemish; the intestines are in a perfectly healthy condition; so, we only can supply the Cincinnati Jews with beef.

I was informed by Dr. Watts of Chillicothe, a gentleman of intelligence and great enterprise, who feeds and grazes on a large scale, that he would pay five per cent more for Kentucky raised cattle for either purpose than he would for Ohio or Indiana cattle. He considered the risk of life this per cent in favor of the cattle of Kentucky.

There are three epochs in the history of Kentucky cattle; first, the introduction of the Patton cattle, say in the year 1790, and some years afterwards the Miller stock of the like. These are generally diffused throughout the State, improving our stock twenty-five to thirty per cent in a period of twenty-five years.

Second, the importation of 1817, which gave us finer forms and an aptitude to take on fat at an earlier age, adding twenty-five to thirty per cent upon the Patton improvement, in a period of less than twenty years.

Third, the numerous importations made into Kentucky and into Ohio, from 1831 to 1836, from which has arisen the superior breed. To keep up this breed as it now is requires sound judgment and unceasing vigilance, or a decline must follow.

I recommend to the breeders in Kentucky to import at least half a dozen bulls from the Netherlands, Holland, or Northern Germany, at once, and renew such an importation every five or six years, for twenty years, rather than to draw their bulls from the best stock to be found in England. I do not think it is desirable to have a very large breed, but form and early maturity are not for a moment to be lost sight of. A skillful breeder endeavors to shape the animal so as to carry most flesh on the valuable points, to have the loin and hind quarters much the heaviest, as these parts bring to the butcher the most money.

LEWIS SANDERS.

Grass Hills, Ky., December, 1848.

Dr. Martin, to whom Mr. Sanders refers, gives the following information:

Dear Sir: Your letter of the 25th ult. is just received, and I will try to answer your inquiries. Your first question is, What breed, cross or variety (of cattle) has been found most profitable in your region for beef; and what for the dairy? The improved Shorthorns and

THE EXPOSITION BUILDING, CHICAGO.
(Home of the American Fat Stock Show.)

their crosses are most profitable for beef. They are of large size and fatten easily at any age, so as to come to early maturity, and they carry a large portion of their flesh upon the best parts and their beef is of an excellent quality. They pay better for food consumed than any other cattle that I have fattened or grazed.

In regard to the milking qualities of the improved Shorthorns there appears to be much diversity of opinion, some contending that they are the best milkers had in the country, and others that they are worthless. The truth is, that some tribes of Shorthorns are remarkable for the quantity of milk they give and other tribes are equally so for their small yield.

I purchased two cows at Col. Powell's sale in 1836. One of them, a cow of the Daisy tribe, was a steady milker, giving from twenty-eight to thirty-two quarts of milk daily. The other was scarcely able to raise her calf. And the qualities of each have been transmitted to their descendants for several generations. The cows that

I imported from England were all fine milkers, and so are their descendants. The cows of these milking tribes are generally thin whilst giving milk, but fatten very quickly when dry. The steers of the milking tribes are equal and generally superior as grazier's stock to the others. Mine has been superior, which I attributed to having been better nourished by their mothers.

Second. Which of the breeds imported in 1817, the Longhorns or Shorthorns, have succeeded best?

There was a close contest, for many years, between the Longhorns and Shorthorns and Herefords. Each had their advocates and each produced a stock that was a great improvement as grazing stock upon the native and "Patton stock" (as the old unimproved Shorthorns introduced by Mr. Patton were called). This contest was kept up until about 1830, when the advocates of the Shorthorns became most numerous. The Longhorns and *Herefords were gradually bred to Shorthorn bulls, until the pure breed of the former are nearly extinct.* (¶ 114)

JOHN P. REYNOLDS, CHICAGO.

Third. How do the Longhorns of that importation (1817) or their descendants compare with the Patton Longhorns?

Mr. Patton was one of the original importers in 1783 of two breeds of cattle. They were then called the milk and beef breed. The milk breed was Shorthorns. The beef breed had longer horns; but *I have always supposed they were the unimproved Herefords.* I am not aware that there were ever brought into Kentucky any of the full-bred beef breeds, so that my opinion that they were Herefords is based upon the appearance of the half-bloods that I have seen. Mr. Patton brought to Kentucky the full-bred milk breed and half-blood cows of the beef breed, and Mr. Smith brought also a bull which was half beef and half milk breed, called Buzzard. Mr. Patton's Shorthorns were very fine animals. They were fine-boned, heavy-fleshed and came early to maturity and fattened kindly and were extraordinary milkers. They were much larger than cattle that we had in the State previously. Mr. Patton brought only one cow of this breed, and she had no female descendants. The produce of these fine cattle were very much injured by breeding them to bulls which were descendants of the beef breed, such as Inskeeps, Brindle and Smith's Buzzard. These cattle produced large, coarse, big-jointed stock that came slowly to maturity, difficult to fatten, and when fully grown were of enormous dimensions. This was the state of things in 1817 when your importation of Shorthorns and Longhorns was made. I remember well examining the Longhorn bull Rising Sun, soon after Messrs. Cunningham & Co. bought him, and I then thought him the finest animal of the ox kind I had ever seen. His stock was very fine —vastly superior to the coarse stock above described. I sold a cow (got by Rising Sun) to a butcher who paid me for a thousand pounds, net meat [about 1,750 lbs. live weight. T. L. M.], a very unusual size for a cow in those days.

Fourth. How do the Shorthorns imported in 1817, or their descendants, compare with those that have since been introduced, including those of the Ohio Importing Co.?

The Shorthorns of 1817 were fine-boned, heavy-fleshed animals that came early to maturity and fattened much easier than the Patton stock (especially after the latter had been mixed with the beef breed). They fattened mostly on the outside, so that they always showed their fat to the best advantage. Their flesh was rather inclined to hardness, which was a considerable drawback upon their excellence.

The best of the improved Shorthorns, introduced within the last twenty years, have all the good qualities that the stock of 1817 had, and have these additional advantages: Their flesh is soft (tender), and they throw a portion of their fat among the lean so as to marble it. The beef is of a better quality and they take on fat much easier. They are as forward at three years old as the stock of 1817 were at four, or as the Patton stock were at six. But the later importations have had greatly the advantage of the stock of 1817 in having the improvement made by the latter to start with. Some of the finest animals I have ever seen fattened were a mixture of the two breeds. I think there was some of the importation of 1817 that did not have that hardness of flesh, but they soon became so mixed in their descendants that it was a general characteristic. (¶ 115)

Fifth. If you were now to choose a stock for general grazing purposes in your State, what breed or breeds would you select from?

I should have no hesitation in preferring the improved Shorthorns to every other kind of stock that I have ever seen, for the grazing in this region of country.

Sixth. Give as full a description as you can of the qualities of each breed, as they have been developed with you, embracing remarks on comparative size, form, activity, hardiness and tendency to disease of the different breeds.

The original breed of cattle in Kentucky strongly resembled the old unimproved Devonshire cattle. They were small, thin and difficult to fatten—cow weighing when fat from three to four hundred pounds. [This undoubtedly means dressed weight, the corresponding live weight being 525 to 700 lbs. T. L. M.] These cows were good milkers, giving a moderate quantity of rich milk. I do not know that they were subject to any other disease but the hollow horn, a disease brought on by poor keep in winter, so that the pith of the horn is frozen. It was cured by boring a hole in the horn.

The introduction of the Patton stock in 1785 and subsequently made a considerable improvement in these cattle. Cows of the Patton cross would weigh when fat from 6 to 7 cwt. [Meaning dressed weight, or 1,050 to 1,225 lbs. alive. T. L. M.] There was such a general disposition to increase the size that the coarse-jointed, large-boned animals were selected and saved as breeders generally, from 1785 to 1817, and the consequence was at the latter period the Patton stock (as all these cattle were called) were very coarse. The size of some of these cattle was enormous; but they did not weigh, net, near equal to their size. The graziers at that period did not like to attempt to fatten cattle until they were four years old. The importation of 1817 improved the coarse cattle very much, increasing their disposition to fatten. They came earlier to maturity, were gentler, better disposed and had much less offal. Whatever reputation the Shorthorns acquired in Kentucky prior to 1830 was owing to the importation of Shorthorns, and they had great reputation. (¶ 116)

The improved Shorthorns introduced within the last twenty years has been a great improvement upon those imported in 1817, and those of Mr. James Prentice of Lexington in 1818. At the last cattle show that we had in Winchester I showed a three-year-old steer, a mixture of the stock of 1817 and the improved Shorthorns since introduced; and the judges put his weight at 750 pounds. Shortly after the fair I sold this steer to Mr. Brinegar, who took him to New Orleans, and when butchered he weighed 1,242 pounds. I mention this circumstance to show how much more weight is contained in the same bulk; for if this bull had been of the Patton stock his bulk would have given him about the weight the judges laid him at. A few months before I had sold to a butcher in Lexington a steer two years and eight months old that weighed 1,025 pounds. I sold a heifer six years old to B. Roberts, that weighed when driven to Cincinnati 1,487 pounds. Last year I sold to Mr. Horn a five-year-old heifer that weighed 1,116 pounds. Both of these were mixtures of the stock of 1817 and later importations, and the last was uncommonly small for her weight. I regret that this last was not weighed before she was slaughtered that I might know the difference between her gross and net weight. However, I can give you the gross and net weight of a four-year-old steer sold to the same gentleman. His gross weight was 2,000 pounds and his net weight 1,280 pounds. All these net weights are exclusive of hide and inside tallow, taken out with entrails.

As regards the diseases of all the above, they are very few, if bred from healthy stock. The most formidable disease of the improved Shorthorns with me has been the milk fever. I lost two of my imported cows and one that I purchased at Col. Powell's sale with it. It chiefly attacks cows that are fat and have their calves in very warm weather—the attack being in a few days after calving. I never knew any but fine milkers to have it, and not until they have had several calves. The udder becomes very large, hard and hot. They soon appear to lose the use of their hind legs so that they cannot stand. I have cured some by large bleeding and purging freely with Epsom salts. But prevention is far better, which may be generally accomplished by preventing the cows from having calves in warm weather. Healthy parents generally produce healthy offspring in this region. You are aware that cattle in most of the adjoining States are diseased, particularly in the liver. These cattle produce sickly progeny, which seldom look as well as stock from more healthy parents. And I have noticed calves from them to be very subject to bowel complaints.

¶ 191

H. D. EMERY, CHICAGO.

Seventh. What breed of cattle is best for driving long distances?

This question is more difficult for me to answer than any of the others, as I have very little experience in driving cattle; but I am told by persons who have been engaged in this business that the improved Shorthorns, when fattened young, do not stand long journeys well. I should suppose from their make that the *Herefords would be the best travelers.* The improved Shorthorns make excellent oxen, as they never get overburdened with flesh while they have plenty of hard work to do. The breed is more gentle and docile than any others that we have had. Yours respectfully,

SAM'L D. MARTIN.

Near Colbyville, Ky., Dec. 4, 1848.

We present these complete and authoritative statements to show that while the "seventeen" importations included Herefords and that these were not by any means first-class specimens of the Hereford breed, yet they had enough of the naturally inherent thrift of the breed, so that the "seventeens" owed considerable (to say the least) to the Hereford for the thrift and quality they became particularly noted for. The "seventeens," as a family of Shorthorns, were never fashionable because of this Hereford alloy. But at the same time no Shorthorn family surpassed, if any equaled, them in real merit.

Evidence of this Hereford blood crops out frequently in the descendants of the "seventeen" importations to this day, as witness the steer John Sherman, whose portrait we give. (¶ 116) Mr. Gillette, who bred this steer, reports the dam a "white-faced cow" of unknown breeding, descended, as are all his herd (Shorthorns), from the earlier Kentucky stocks. The so-called "seventeen" families being given the greatest credit for excellence of the Gillette herd.

The presence of this old grazing blood undoubtedly has had a most favorable influence in making the reputation of "Gillette Shorthorns" for superiority above the pure-bred Shorthorn for range purposes.

WILLIAM WATSON
(UNCLE WILLIE).
(Father of the Angus cattle in America.)

CHAPTER XV.

Revival of Hereford Interest in America

From 1850 to 1871, although Hereford cattle in America were making friends for themselves in their immediate vicinity and had very staunch friends in all who bred and handled them, they were not so widely known as their merits deserved.

This, possibly, was owing to the tremendous upheaval in the political world of America, terminating in the Civil War that occupied the entire attention of the people from April, 1861, to 1865. There were very few families either North or South but what sent their contribution of father, son or brother to the armies. The price of all produce of the farm was high. The price of meat was correspondingly high. The farmers and stock raisers all made money with such stock as they had, and did not feel that urgent need of improvement that came later.

In 1871 I determined to spend the balance of my life upon the farm if we could find a way to make it profitable; and in view of this determination we gave two years to careful investigation as to the best methods and practices for obtaining this result, and this investigation brought us to the adoption of the Hereford breed of cattle. We were familiar with Mr. Sotham's fight for Hereford cattle through the files of the Albany "Cultivator," etc., which we had preserved, and these cattle were again urged upon our notice by Mr. William Powell (¶ 117), who was then acting as our foreman. As a result of our investigations early in February, 1872, we made our first purchase of six or eight head of Hereford cattle.

In March or April following we bought three more at the sale of W. W. Aldrich, Elyria, Ohio; two or three months afterwards we bought of Dyke and Creed, in Ohio, some twelve or thirteen head. Soon after we purchased six heifers from John Humphries, Elyria, Ohio, and about the same time we bought for $1,000 gold the Hereford bull Sir Charles (3434) 543 from F. W. Stone of Canada.

From time to time as opportunity presented itself, we purchased from others; from D. K. Shaw of Chautauqua his entire herd, and later the herd of H. C. Burleigh of Maine, excepting one cow and calf; from H. Woodward's estate in Kansas, the entire herd, and quite a number at different times from Mr. Parsons of Pittsfield, Ohio; several from Mr. Thomas Clark, then of Elyria, Ohio; several from the Hon. John Merryman, Cockeysville, Md. (¶ 118); and in 1873 imported from England Dolly Varden (Vol. 9, p. 279) 5, and her bull calf Success (5031) 2.

When we had become satisfied of the value of the breed we undertook to make it known through the advertising columns of the agricultural and live stock journals of the country. We found a very warm opposition from the Shorthorn interest as against their introduction, and ceased to make any strenuous efforts to introduce them among farmers in the States.

In the meantime quite an extensive correspondence had grown up with the ranchmen at the West. There seemed to be a demand from that quarter that promised success, but it proved difficult to bring out of that correspondence any results.

The plains of Colorado (¶ 120) were, in 1874, still the home of the buffalo, but cattle were being rapidly brought north from the great breeding grounds in Texas to stock them with beef animals. Denver, at that time, was the center for cattle men, they coming here for their supplies and making it their headquarters when not off on the range.

We took our first shipment of Herefords that went away from home to be sold, to Denver, Col. This consisted of five Hereford bulls. Three of these were sold in the spring of 1874 to Mr. Geo. Zweck of Longmont, Col. They were Plato (4843) 590, Duke of Beaufort (4527) 744 and Hervey (4644) 815. The first was four years old, the second was two years old, the third one year old. These bulls were put upon the range. Plato, the four-year-old, had been a show bull, and kept in high condition. He held this condition when on the range and continued a vigorous stock-getter for eight years. The other two bulls were reported eleven years later as still in fine condition and yet in service in Wyoming, in the herd belonging to

Mr. Zweck and his nephew. Mr. Zweck paid in 1874 for the three bulls named $1,250. Having had eight and eleven years' service, we would like to ask ranchmen whether such purchase would seem to them to have been a good investment? It is simply a representative one that may be repeated with Hereford bulls in

A CARTOON OF 1882, BREEDERS' JOURNAL.
("The Shorthorn committee discover the dam of Mr. Miller's grade Hereford steer Kansas.")

the experience and practice of any ranchman. Long prices may be, as in this case, the best investment.

Two other bulls taken at this time we sold to Mr. Powell at Canon City, Col., for $900. We followed this shipment by another in the fall of the same year that were sold to the honorable P. P. Wilcox, Geo. F. Lord and others.

In 1875 we took a carload of Hereford bulls to Denver, and they were sold to Mr. John Hitson at Deer Trail, Col., about fifty miles east of Denver, on the Bijou. The grass was fine there, the country not being overstocked. These bulls, although in high condition when turned out on the ranges, not only held it, but next spring after running out all winter, never seeing a shed or a barn, or any feed except what they got from the plains themselves, had rounded out, and thickened up, and had such splendid rough, curly coats of hair that they had proved their fitness for range conditions.

We took two carloads of Hereford bulls to West Las Animas, on the Arkansas River, Col., in the spring of 1876, all thoroughbreds and good ones. After five months of work, we sold them to cattle men as far west as Pueblo, on the river, and south to the Cimarron River in New Mexico. The Hall Bros., John W. Prowers (¶ 121), Abe Cronk, the Reynolds Bros. (¶ 122) and the Jones Bros. being among the purchasers. These shipments were followed by others up to 1877, or until a demand was created that reached back to the States for grade Hereford bulls. And this demand from the plains for grades made a demand for the purebreds to cross upon Shorthorns and other cows to produce bulls for this trade; and from that time there has been a steady and increasing demand for Herefords that has more than kept pace with the supply.

While at work among the great cattle herds of the West to create a demand for the Hereford cattle by taking the cattle out to these gentlemen and showing them what superior beef animals they were, we were not idle in exhibiting our cattle at the various shows in the East, and doing all that lay in our power to secure fair play for the Herefords.

This was a gigantic undertaking, as the fairs were, as we shall show, under the control of Shorthorn men. The judges were all appointed by them. We trust we shall be pardoned if it shall appear to any of our readers that we have gone too much into detail in showing up the frauds that were practiced; also the opposition that was encountered by the Hereford cattle and their breeders.

It was a matter of great expense to fit show herds, and a further great expense to ship them over the country for the purpose of educating the people as to the great merit of the breed. This was particularly so in 1876 when we took our show herd to the World's Fair, the Centennial Exposition, held at Philadelphia. This trip cost $1,200 for expenses only, we being there two weeks. There were no money prizes, but we have a bronze medal (¶ 123) in our possession that was awarded to us as first prize, the cattle being spoken of in the following language:

"Their exceedingly fine character, form and quality entitle us to consider them to be first-class specimens of the Hereford breed and worthy of our highest commendation." Success (5031) 2 was at the head of this sweepstake herd. This herd at the Northern Ohio Fair, held at Cleveland, Ohio, won first sweepstakes for best bull and four cows or heifers, owned and bred by exhibitor ($150), and first sweepstakes for bull, Success (5031) 2, and five of his get, $75. This was in competition with the best Shorthorns in the United States.

We visited Lafayette, Ind., with our herd in 1877, at which time there was no class for Herefords, they not being recognized in their premium lists. The society, however, awarded the following certificate:

"Office of the Secretary of the Tippecanoe County Agricultural Association, Lafayette, Ind., Oct. 9th, 1877. On Saturday, the 8th day of September, 1877, at a meeting of the Directors of this Association, on the fair grounds, the following proceedings were had, to-wit: On motion, the following preamble and resolutions were unanimously adopted by the Board:

"*Whereas,* Mr. T. L. Miller, Will County, Illinois, breeder of thoroughbred Hereford cattle, has exhibited his herd of thoroughbred Herefords at the annual fair for the Tippecanoe County Agricultural Association for 1877, at which there had been no ring or class of individual entries provided in which he could show his stock, which, of necessity, compelled him to exhibit at great disadvantage, and

"*Whereas,* The said herd were very generally admired by the visitors attending the fair, including gentlemen conceded to be experts in the matter of breeding and handling beef cattle, for their great uniformity of marking and their general feeding and fattening qualities; be it therefore

"*Resolved,* That the thanks of this Association are due and are hereby extended to Mr. Miller for affording the visitors at said fair, in so substantial a manner, with a pleasing variety in the show of thoroughbred cattle, and that he has the best wishes of the Association for his future success in an enterprise which has the laudable purpose of improving the breeds of beef cattle.

"And the secretary is ordered to make out a certificate of this action of the Board and forward same to Mr. Miller.

"(Signed) DANIEL ROYSE, Secretary."

Among those who examined the Herefords here were Messrs. C. M. Culbertson, Moses Fowler, W. S. Van Natta, Adams Earl, A. D. Raub, Robt. W. Sample, Chas. B. Stuart and others, whose attention for the first time was then drawn to the great value of Hereford cattle, and all of whom afterward became breeders of Herefords, in most instances getting their start from our herd.

Two weeks later Mr. Culbertson bought five head that were in this exhibit. Five years later the show of Herefords at Lafayette in 1882 was very gratifying and encouraging to the Hereford interest, and it seems hardly possible that such a change could be wrought in five years, and thus, what had at the time seemed a somewhat discouraging exhibit, really proved one of the most successful ever made by an exhibitor.

In the year 1879 Mr. C. M. Culbertson sent Mr. Geo. Morgan (who was then my superintendent) to England to purchase as good a herd of cattle as could be found. The result was, if we remember rightly, the importation of thirteen head. In 1880 we imported 109 head from England, and Mr. Morgan imported for Mr. Culbertson and Mr. Hershey of Muscatine, Ia., something over 200 head. From that time the importations have increased, in some years amounting to over 1,000 head. Public sales were held in this country at which the higher range of prices has been from four hundred to fifteen hundred and odd dollars per head.

Our beginning with Herefords, like Mr. Sotham's, was up-hill work, and had we depended upon the farmers' trade of the agricultural States we should, like him, have been overpowered by the hostile Shorthorn interest, but on the plains we were enabled to promptly show results, so plain by comparison, that the Hereford was victorious from his first introduction. But we are convinced that the Herefords are the best "general purpose" or farmers' cattle, and believe Hereford breeders should never be content till their merits are well known by every farming community of the whole country. On the range, when known, they have the field alone without a rival, but until every bovine has assimilated the true Hereford thrift his all-

SHORTHORN COW CONQUEROR. A BREEDERS' JOURNAL CARTOON OF 1882.

("While the Shorthorn breeders claim the virtue of grade Hereford steers due to Shorthorn dams, the farmer looking on decides that the credit is due to Hereford bull.")

conquering work of improvement must be vigorously continued.

It is easier for some Hereford breeders to "let well enough alone;" to be satisfied with existing conditions; to curry favor with the opposition, and at the same time reap where they have not sown from the active sowing of others; and while the herd of an active advocate of Herefords should not be patronized more than these, unless he really has a breed of Herefords equally good or superior, if equal, the great body of Hereford breeders always will, as they always have, yield a loyal support to those whose cattle, pen and voice are ever doing honest, intelligent and meritorious service for the Hereford cause, and therefore for the betterment of the chief product of American agriculture,—good beef.

CHAPTER XVI.

An Incident of an Early Chicago Fat Stock Show

The Hereford breeders considered that if they could only have a fair contest with other breeds on the butcher's block they would win a victory over all other breeds that could not be gainsaid. They estimated that a reliable and accurate record of every part of the slaughtered animal would be kept, and facts as to the percentage of dressed meat to the live weight, and also the weight, both live and dressed, would be something that could not be talked down or hid.

We are very glad that these records of the Chicago Fat Stock Show of 1879 are available. At that show a Hereford, a Devon and Shorthorn were dressed in competition. We present here an engraving (¶ 124) taken from a photograph of the rib roast cuts from the Hereford and Shorthorn steers. Both show well-marbled meat, but neither show as well as they would have, had the weather been cooler. Neither of the beeves ought to have been cut; the meat of either not being as good for being cut so warm as it would have been had it had time to cool. The Hereford suffered the most in this respect, from being the thicker-fleshed—nearly, or quite, 40 per cent. Both show marbling and well-marbled. Another point is that the beef was too fat. This is perhaps true. True for economy in making and economy in use; but the breeder who came before this show or Smithfield with lean steers would have gone away with little credit and no honors.

If these steers were made over-fat, there was a point where they could have stopped, and their meat shows them to have had good feeding qualities.

Thus much in common; but it is fair to give somewhat of a comparison of the two. The Hereford steer was one of fourteen steers, all that there was of the lot, and was from a common native cow and by a Hereford bull.

The Shorthorn steer was one of sixty, selected from 600, and from a herd that has had an American and English reputation for years as the best steer herd in America, if not in the world, and probably no breeder in the two countries stood higher as a breeder of steers than did Mr. John D. Gillette. (¶ 125)

Probably no man understood better the character and merits of a bullock than did Mr. Gillette. There was no man that was more familiar with the individuals of his herd than was Mr. G. Under all of these advantages the Shorthorns came to this test with the prestige of being from the most noted herd in this country. A selection of one from sixty and sixty from six hundred the one of six hundred selected by the most experienced breeder in America.

It must therefore be conceded that the Shorthorn came to this test with all the advantages that an animal of the breed could claim or wish.

Mr. G. entered two steers for dressing—Snowflake and Drake, both three and under four years, and selected from these the steer Drake—said to be three years old May 15, 1879, out of a three-quarter Shorthorn cow and by a thoroughbred Shorthorn bull.

The Hereford steer was three years old May 15th, 1879, and it is well to state that the exact age of the two steers was not a matter of record, but were claimed to be three years old, and dropped as near as could be ascertained in May, and called the 15th.

On the record the two steers stood as to age alike.

For some reason the Gillette steer was not measured, and therefore the measurements of the Hereford are not used, as there is no means of comparison on these points.

The age alike, the gross weight of the Hereford 1,963 pounds, and the gross weight of the Shorthorn 1,795 pounds. Stating these facts, we will give the official statement of the society, made up by the secretary:

LOT 9. DRESSED BULLOCKS.

Not less than two entries will be considered. The bullocks to be killed, dressed and weighed under directions of the awarding committee.

The dressed carcasses to remain the property of the exhibitor. Bullock dressing the largest percentage of meat in proportion to the live weight, $50.

No.	EXHIBITOR.	Age in days	Weight Nov. 14, 1879	Av. gain per day since birth.	BREED.	NAME OF STEER.
1	T. L. Miller, Beecher	1274	1963	1.56	Grade Hereford	"Barney."
2	L. F. Ross, Avon	1254	1614	1.31	Grade Devon	"Jim Lockwood."
3	J. D. Gillette, Elkhart	1274	1795	1.41	Grade Shorthorn	"Drake."
	Average	1267	1790	1.42		

Premium: $50, to grade Hereford steer Barney, exhibited by T. L. Miller of Beecher, Will County, Illinois.

REPORT OF COMMITTEE.

NAME OF STEER.	BREED.	Live weight Nov. 14	Dressed weight	Percentage net of gross weight	Left forequarter	Right forequarter	Left hindquarter	Right hindquarter	Head	Hide	Paunch, guts, liver, heart, feet, lights	Rough tallow, gut-fat, straight guts, caul, etc.	Contents of stomach, blood, shrinkage
"Barney"	Grade Hereford	1963	1317	67.09	371	354	305	287	55	106	113	178	194
"Jim Lockwood"	Grade Devon	1614	1055	65.36	277	275	256	247	49	95	95	145	175
"Drake"	Grade Shorthorn	1795	1179	65.73	308	303	285	283	47	90	97	155	227
	Average	1790	1183	66.06	318	310	282	272	50	97	101	159	198

The superintendent of the Cattle Department made up on all classes a report of the reasons as given by the committee for their judgment. And such a report was commenced on this class and started off by stating the Hereford steer to be the oldest by three or four months. We have sought from the records of the society a copy of his report on dressed steers, but are informed that no report was filed.

Learning of his claim, or that of the committee as to age, the Hereford men called the superintendent's attention to the heads of the two steers, which hung near by. The Shorthorn with a full four-year-old mouth and the Hereford with a three-year-old mouth.

To explain more satisfactorily as to the ages by the teeth, we had engravings made of a three and four-year-old mouth. (¶ 126)

It will be perceived that the three-year-old shows for the corner teeth the two calf-teeth; these were gone in the Hereford steer, one of the cavities still empty, and the other showed the new tooth just above the gum.

The four-year-old teeth represent the mouth of the Shorthorn (Drake) that was dressed.

This is perhaps enough to say on this subject here, as we have treated the subject elsewhere, but there are some few points of comparison that the report does not bring out. The steers being of the same age, the Hereford had made 1.56 lbs. per day. The Shorthorn had made 1.41 lbs. per day, a difference of .15 in favor of the Hereford. *The reader, from the teeth statement, may add to this as per his judgment.*

The Hereford head, paunch, stomach, etc., weighed 362 pounds.

The Shorthorn head, paunch, stomach, etc., weighed 371 pounds, or, for the Hereford 18.4 per cent; for the Shorthorn 20.8 per cent. The Shorthorn 2.4 per cent more waste offal than the Hereford.

The Hereford shows hide and tallow, 284 pounds; the Shorthorn shows hide and tallow, 245 pounds, giving the Hereford 14.46 per cent valuable offal and the Shorthorn 13.65 per cent valuable offal, being .81 per cent more for the Hereford; and of net to gross weight 1.36 per cent in favor of the Hereford.

Thus, in every point of comparison the Hereford shows the best record.

The reader must not forget that the Hereford was one of eight three-year-olds, and that five of the eight were selected to show in St. Louis in October and contracted then to be delivered in December. It was intended to take the best five, and hence five of the eight were drawn before this selection was made.

That the Shorthorn was one of six hundred.

And another point to keep in mind is the two sets of teeth.

With these facts accessible to each reporter of this test, the impression attempted to be carried out was that the Hereford won by main strength; that is, because of greater net weight, seeking erroneously to convey that if the quality should have been taken into account the victory would have been with the Shorthorns. This attempt to detract from the value of the award we endeavored to make fruitless. It might have been stated that had the superintendent had the selection of a committee to pass on the merits of the steers, it would have gone as the awards on the three-year-olds, the one-year-olds and the cows.

A BREEDERS' JOURNAL CARTOON OF 1885.

CHAPTER XVII.

Opposition Encountered by Hereford Exhibitors, 1877-8-9

It must be remembered that during the years immediately preceding the first strictly beef show that was inaugurated by the Chicago Fat Stock Show in 1879, the Hereford cattle were handicapped at the different State and district fairs, by having no class provided for them, or, if there was, small and insignificant premiums were offered for them in comparison with those offered for the Shorthorns; and at the best it was a one-sided affair and it was intended to be so. Shorthorn breeders or their friends were made judges in each competition and the programme was made up for them to win.

We are aware that the charges of manipulating judges for the purpose of carrying any interest is quite a serious charge. We would not make it did we not believe it to be true; and with the fullest evidence to support the charge; or if we believed the practice would be discontinued without it. We will not be tedious in bringing up a large number of cases, but will touch on a few of the prominent ones.

First, we will name what occurred at Ottawa in 1877. The Hereford exhibitors asked for an impartial and able committee when they should come in competition with the Shorthorns. This appeal was made to the president and several of the vice-presidents. It was granted that the Hereford breeders had a right to an impartial committee. And it is possible that the superintendent had instructions on this point; at any rate, he gave the Hereford exhibitors the assurance that they should have his best endeavor to get an impartial committee or judges. He advised the Hereford exhibitors that he had finally selected a referee to complete the committee, who he believed would be impartial and competent. That referee was Mr. J. H. Spear, one of the leading Shorthorn men of the State. It is well, however, to state that both the superintendent and the marshal of the ring who officiously assisted him were and had been prominent Shorthorn breeders for years; and the marshal had been the salesman for the Shorthorn herds of the entire West.

Again, at Freeport in 1878, the Hereford exhibitors asked of the board that they would take special pains to see that fair and impartial judges should be selected for herd and sweepstakes premiums, and proposed that President Gilham and ex-President Reynolds should select these judges. This was agreed to. They were so selected and the Herefords in the first contest took the first premium. The Shorthorn breeders made a row and would not submit their claims to this committee for further contest, whereupon the aforesaid Chief Marshal was delegated to form and select the committee, and did select them in the interest of the Shorthorns, when the Hereford exhibitors withdrew from competition.

Again, at the fat stock show in 1879, when the Herefords and Shorthorns came into competition, the Hereford four-year-old steer, belonging to T. L. Miller, took the first premium over all the breeds. This was the signal for another row, the result of which was a partisan contest.

We have before called attention to this exhibit, and we now repeat some of the figures to show the character of the animals in competition, and the challenge given to dress the bullocks and cows from the Hereford exhibits against the winning Shorthorn steers.

Our State Board had for many years, perhaps always, brought all the beef breeds into competition at the State Fair until this year they followed the practice of the Northern Ohio, Michigan and other State societies in excluding competition between breeds. This and other instances positively indicate that when Shorthorn breeders cannot control the decisions they use their influence to prevent competition. It is perhaps only another evidence that the Shorthorn men had influence enough to say what should be and what should not be at fairs and fat stock shows.

In the first class of steers in the Chicago Fat Stock Show in 1879, four years old and over, the winning steers were a grade Hereford first and a grade Shorthorn second, the former belonging to T. L. Miller and the latter to J. D. Gillette.

For steers three and under four there was a large number of entries. The steers taking the awards were first a Shorthorn and second a Devon:

Premium.	Name of Animal.	Date of Birth.	Age in days.	Wt. in lbs.	Av. gain per day.
1.	T. Stevens, G. S. H.	Apr. 25, '76	1294	1986	1.53
2.	Jim Lockwood, D'n.	June 4, '76	1284	1649	1.28
0.	Barney, G. Hereford	May 15, '76	1275	1991	1.56

There were other Hereford steers before the judges, perhaps better than this, but this is well enough.

For steers, two and under three years, both awards went to Shorthorns; there was only one Hereford entry. We give below the age, weight and gain per day of the winning steers and of the Hereford steer:

Premium.	Name of Animal.	Date of Birth.	Age in days.	Wt. in lbs.	Av. gain per day.
1.	Victoria Duke, G.S.H.	Apr. 22, '77	932	1532	1.64
2.	P. Cooper, G. S. H.	Dec. 14, '76	1059	1534	1.44
0.	Alex., G. Hereford	Aug. 15, '77	820	1474	1.80

This record needs no comment.

For steers one and under two years old there were fourteen entries—three or four of those were Herefords. The first and second premiums went to Shorthorns. We give the following figures in reference to the winning steers and one of the Herefords:

Premium.	Name of Animal.	Date of Birth.	Age in days.	Wt. in lbs.	Av. gain per day.
1.	McMullen, G. S. H.	Mar. 5. '76	605	1196	1.97
2.	C. S. Reed, G. S. H.	May 15, '78	644	1300	2.38
0.	Putnam, G. Hereford	July 12, '78	483	1152	2.39

This review of the four classes of grade steers where the Herefords and Shorthorns were in competition with the tables showing ages, weights and average gain per day from birth, is in each class in favor of the Herefords. The only question remaining open is the one of quality. To test this Mr. Miller, the owner and exhibitor of the Hereford steers, made a proposition to test these awards by dressing the bullocks as follows:

Chicago, Ill., Nov. 14, '79.

Hon. J. R. Scott, President of the State Board of Agriculture of Illinois:

The object of your Board is to determine, among other things, the comparative value of the different breeds of beef animals. Among other ways, expert judges of beef animals are called upon to pass upon the merits of such animals as are brought before you. This will determine the merits—excepting the errors of judgment and prejudices in favor of different breeds.

To correct such errors and prejudices, if any exist, I will submit my cattle that have come in competition with other breeds to the direction of your Board, to be slaughtered as follows:

My grade Hereford steer "Ben," four years old, which took the first (1st) premium in Lot 5, against the four-year-old steer that took first (1st) premium in Lot 6.

My three (3) year-old steer Barney against the first (1st) premium steer in Lot 5.

My yearling steer Putnam against the first (1st) premium yearling steer in Lot 5.

My yearling steer General against the yearling steer that took first premium in Lot 6.

My cow Jennie against the cow that took first premium in Lot 6.

The best of each of these animals to be determined by the quality of the meat of each animal, and the greatest amount of dressed meat to the gross weight of each.

Very respectfully yours,

T. L. Miller.

In reply to this proposition the following endorsement was made, and the papers returned to me:

Illinois State Board of Agriculture,
Chicago, Nov. 15, 1879.

Respectfully returned with the statement that the matter contained herein has been duly considered by the Board, and that the Board has decided not to depart from the published programme for the Fat Stock Show in this request at this late date.

Jos. R. Scott, President.

The following entries were made for the premiums for dressed bullocks:

Name of Animal.	Date of Birth.	Age in days.	Wt. in lbs.	Av. gain per day since birth.
Boynton, S. H.	Dec. 13, '77	697	1338	1.92
Snowflake, G. S. H.	June 1, '76	1257	1978	1.57
Drake, G. S. H.	May 15, '76	1274	1799	1.41
Barney, G. Hereford	May 15, '76	1274	1991	1.56
Putnam, G. Hereford	July 12, '76	483	1152	2.39
Jim Lockwood, Devon	June 4, '76	1254	1619	1.31

The above entries were made by parties as follows: Boynton, by J. M. Brown & Sons, Berlin, Ill.; Snowflake and Drake, by J. D. Gillette, Elkhart, Ill.; Barney and Putnam, by T.

L. Miller, Beecher, Ill.; Jim Lockwood, by L. F. Ross, Avon, Ill. Besides, our Hereford cow Jennie was entered for dressing. We offered the owner of the cow taking the premium to dress his cow against ours, which was declined. We then offered ten cents a pound for the cow. This was refused. The dressing would have been a fitting and proper test by which to have tried these verdicts.

In the above table there is evidence to show in the three-year-old classes that the Hereford was certainly a better steer than the Devon, and the dressing of these steers proved this conclusively, as between the first premium steer and the Hereford the difference is so light that only the dressing would determine the fact—the difference is, however, in favor of the Hereford.

In the two-year-old class the Hereford shows .16 pounds per day from birth the largest gain; and he was certainly the ripest and smoothest steer of the lot.

In the one-year-old class the Hereford shows a trifle the largest gain per day, and for quality and thickness of flesh the Shorthorns could not compare with him, and he would have dressed ten pounds to the hundred more than either the first or the second premium steer.

The large exhibits of Hereford cattle that we have had made at many of the prominent fairs were found to be a desirable feature by the several managers, and we had many cordial invitations to show over the West. We were promised fair play and a classification for the Herefords. To these invitations we gave in August, 1880, the following reply:

"Your favor inviting me to exhibit at your fair this fall is at hand. While I should be glad to exhibit on your grounds, if I were exhibiting at all, I have to reply that I determined last fall that I would not exhibit breeding stock in the future, and for these reasons: that the condition in which stock must be put, if success in taking premiums is secured, endangers the breeding quality of the animals exhibited, and, beyond this, is wasteful and illegitimate, and the fact that this custom has been followed for years is no reason why it should be continued.

"There are none that realize these facts more fully than the exhibitors themselves. I followed the exhibitions for years, with some success as an exhibitor. I have had large numbers on exhibition and in good condition. If it has not resulted in as much loss of good breeding animals as some others, it has been because the feeding has not been pushed to such an extent as others have done, and the feeding has been conducted with good judgment and my breed of cattle has more constitution to stand the strain.

"Our showing brought the Herefords into notice and gave them a credit they might not have obtained without it. We followed this showing, not that I might prove my cattle better than other Herefords, but that we might prove the Herefords better than any other breed, and this showing of breeding stock has been the only opportunity that we have had of bringing them in competition with other breeds. We were obliged to show under the great disadvantage of having Shorthorn breeders make our rules and our judges. We, however, won substantial honors, even under these conditions, over the Shorthorns, their breeders themselves being our judges.

"This is not all. Nearly all the societies that have been prominent at the West exclude this competition as between the Herefords and Shorthorns. Ohio, Michigan, Indiana, Illinois, Wisconsin, Iowa and St. Louis offer no competition between breeds.

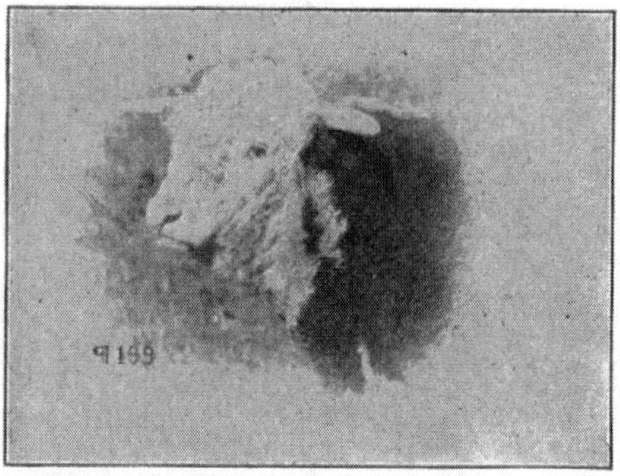

A TYPICAL BULL'S HEAD, CORRECTOR 48976.
(Photograph from life.)

"The Minneapolis Fair Association, conducted by Col. King, does offer open competition between breeds, and he offers liberal premiums, and I should be glad to go there and give that association such help as my herd and flock could, to make his fair the greatest success of any show this fall, and I will give all the aid I can to see that the Herefords are well represented there, and should he another year give the competition he offers now to breeding stock and fat steers I will agree that the Herefords shall have a representation that will be creditable to his show and to the breed.

"The societies that I have named as having shut off competition as to breeds are run by Shorthorn breeders and in the interest of Shorthorn cattle. Some of the officers of these societies say I cannot afford to keep my cattle off their show grounds. Perhaps not. The future will determine this. There is one thing about this: these societies cannot afford to run them in the interest of the Shorthorn breed of cattle. The people are inquiring for the best breed and they are looking for the proper tests of merit as between the different breeds, and they would look to the State and district societies for a solution of this question. They have done so in the past, supposing that societies taking the names of our great commonwealths to designate their special organization, conducted by men selected from the different congressional districts, and holding office under official seal, in the name of the people, for the purpose of selecting the best kind, and the best of the best kind: I say, it has been supposed that men acting thus under the great seals of the different States, and ostensibly for the public

THOS. ASTON, ELYRIA, O.

benefit, were doing what they professed, and not for their individual benefit; and when they, the people, shall recognize that all this machinery has been used to advance the Shorthorn interest, there will be as much use for them as there is for the old ruined castles and monasteries of the Old World.

"I have shown under these managements, under this order. I have known when on these grounds that there was as much chance of winning against Shorthorns as there was to be struck by lightning; still, I have gone on, hoping there might be a time in the future when this partialism should pass away. But this hope has never passed. When these men could no longer carry their ends, they say they will not permit competition as between breeds. Having decided on this, there is no further object in visiting these show grounds.

"This work of Shorthorn breeders was fairly illustrated at Ottawa (Illinois State Fair) when Mr. ——— was superintendent of the cattle department. I was making a very good show, and I asked of the president and ex-president and several of the vice-presidents that they would see that fair and impartial men were selected as judges to pass on the herd and sweepstakes premiums. I claimed I had a right to this. It was admitted, and Mr. ——— admitted the right, and after great protestations of trying, said to me he thought he had a good committee, and the last men selected had been selected after a very careful and laborious search, a search that had been so burdensome that he was obliged to get the aid of that other impartial and disinterested vice-president and official, Col. ———, and the combined efforts of these officials—one the cattle superintendent, the other the marshal of the ring, and, after a laborious and painstaking search of the State they found Mr. Spears, another disinterested and impartial man, to place as one of the judges to pass on the merits of the Herefords and Shorthorns. Mr. ——— (the superintendent) is and was a prominent Shorthorn breeder. Col. ——— is and was all this and, besides, was the salesman that sold all the Shorthorns of the great West; and the judge selected was one of the leading, and perhaps at his best, the leading Shorthorn breeder in Illinois.

"These are the kind of men that have run the cattle department of the great Board of Agriculture of the State of Illinois. The farmers of Illinois and the West wish to know the best breeds for a given purpose, and they will find a way by which to determine it. I am not prepared to say that the showing on the fair grounds of Illinois has not been a benefit to the Herefords. I think it has; but there may be ways and means by which the Herefords can be brought before the public. We may find one of these ways, and find it by bringing breeding stock in breeding condition and fat stock in fat condition.

"The steer is the legitimate product of any beef breed. I will endeavor to show to a reasonable extent this product, and will take the liberty of using the public shows or find some other way. For the present I will not show breeding stock in fat stock condition, and will not, if I can avoid it, show Herefords against Shorthorns with Shorthorn men for judges. A prominent Shorthorn advocate calls my attention to the premiums offered at Minneapolis—$825 in class and for best herd—and asks if Herefords are not in full force what will be the reason? Well, for myself, I have decided not to show breeding stock; but it is somewhat singular that this Shorthorn advocate is anxious for the reputation of the Herefords.

"Very respectfully,

"T. L. MILLER."

FIRST WESTERN SHOWING FAT STEERS.—The end and profit of all beef herds of cattle is the butcher's block; but previous to the determined effort made by Hereford cattle breeders to show to the public that their cattle were by far the best beef cattle, both in economy of production and dressed carcasses, no effort had been made at the several State and Agricultural Shows, to bring out a beef exhibit in the way of fat steers.

The St. Louis (Mo.) Fair always had, in those days, much the best attended and successful show in the West. That fair in 1879 made a move in the right direction, by offering a large and remunerative premium for the five best fat steers. This was a new feature in their fair and the premium was awarded to five Hereford steers, shipped from within forty miles of Chicago and owned by the writer.

PURE HEREFORD COW JENNY.
(Champion Chicago Fat Stock Show, 1878, bred by T. L. Miller.)

CHAPTER XVIII.

Impartial Judges Needed

It had been remarked upon several occasions, "the enormity of Miller, that insignificant Hereford cattle breeder", charging a conspiracy upon a large number of Shorthorn men, members of State Agricultural Associations and directors of Fat Stock Shows, to suppress and defraud the Hereford cattle of their rightful honors. To meet this, as well as to show what Herefords have had to contend with, we have gone quite fully into the facts, and presented the evidence tending to prove our position and say what we think that it proves, thus giving all a means of judging whether we were right or not.

That impartial judges were needed, and also men of intelligence and fairness, to award the prizes at the several fairs, was so conspicuous a fact that the leading daily paper at Chicago, "The Tribune," had the following to say, after the Fat Stock Show in November, 1881:

"Nothing is plainer to the average observer than the necessity for improvement in the matter of selecting judges to act in the various cattle rings at the Chicago Fat Stock Show, and unless radical changes for the better are made before another year rolls round, the great show will lose caste, and become, in the eyes of the people, simply a place where a certain class of breeders are given preference over all others, without reference to merit.

"The members of the State Board are gentlemen of excellent reputation, who, of course, would not countenance anything which had the appearance of unfairness.

"Several of the committee awards of last week were the subject of unfavorable comment, and the grounds for complaint and adverse criticism were perfectly well grounded. It is to be deplored that charges of unfairness were made under any circumstances, but in the cases referred to, there was a palpable lack of discretion, at least, on the part of the management in allowing the same set of judges to act in more than one sweepstake ring. The judges could take but one course, and that was to simply endorse their own decisions, as to do otherwise would have amounted virtually to stultification.

"In several cases it was unnecessary to take the cattle into the ring to undergo the farce of an examination by the judges, and it would have been just as well to have tied the blue ribbon onto the winning cattle while in their stalls.

"The force of this proposition may be readily shown in the case of Mr. Gillette's steer McMullen. (¶ 127) This animal, which, by the way, is fairly entitled to rank among the very best of high grade stock, was shown in a ring of twenty-one steers, composed largely of Shorthorns, for the prize offered for the best steer in the show three years old and under four. It may be presumed that the judges were governed entirely by the question of merit, and after a critical examination the prize was awarded to McMullen. This occurred on Wednesday. On the following day in the sweepstakes ring for the best steer or cow in the show the same animals were entered and the judges were called upon to act. The decision of the men who had already decided in a ring composed of identically the same cattle was of course a foregone conclusion, and the examination of the cattle under the circumstances closely resembled a farce. The judges were handicapped by their own previous action, and were absolutely compelled to again award the premium to the steer McMullen.

"When it is remembered that this animal was overlooked entirely in a ring where the Hereford men were victorious, and did not even receive second or third place at that time, it looks very much to a man up a tree as though a change was necessary. It is not proposed to criticise the men, but the idea of allowing the same judges to be placed in a position to pass upon their own previous judgment is a great mistake, to say the least.

"In the interest of harmony among exhibitors of the different breeds of cattle, it is hoped that some better method of selecting judges will be adopted ere another year has passed. It has been openly charged that the Shorthorn ring, so called, has secured control of the show, and that other breeders are denied certain rights and privileges to which they are entitled. 'The Tribune' does not believe that any ring exists who can control the management of these shows, but it does insist upon a better system of judging, to the end that exact and impartial justice may be the rule hereafter.

"In this connection the following plan is suggested, by which, at least, disinterested men might be secured for these trying and important positions: the Board should select the best butchers from eastern cities and pay them for their services, and thus avoid the scandal that yearly crops out because judges and exhibitors are from the same localities. Having good judges, the rule should be, that no man who owns an animal shall be allowed to hold it while in the ring, or to enter the ring under any pretext while the judges are at work. The adoption of some such rules as the above, or something of that character guaranteeing a better state of things, must be adopted at once, or the greatest of American shows will become a byword and reproach."

Those who read our great daily papers know that they are too apt to leave the beef interests and agricultural necessities of our stock breeders and farmers out in the cold, by saying nothing about them and leaving them severely alone. But this matter of showing beef cattle, and having justice done to the deserving ones, was of such moment that "The Tribune" in January, 1882, again came out in the following editorial:

"During the past year much has been said and written upon the subject of choosing judges to act at fairs, and the matter of these selections has become one which must attract still greater attention before another season. There has grown up in the minds of farmers, a pretty well grounded opinion that in nearly all exhibition rings there is a lack of judgment displayed, as well in the selection of judges as in the method in which their awards are given, and last year's experience has not had a tendency to change their views very materially. During the next three months many State Boards of Agriculture will hold their annual meetings, at which arrangements for this year's fairs will be perfected, and it is certain that no single subject in connection with their work demands more thoughtful consideration than that of an improved method of selecting and working judges. There will of course be many suggestions offered, all having in view the employment of men whose integrity cannot be questioned, and who, from experience, are competent to pass intelligent judgment and discover merit where it exists, without regard to who may, or may not, be pleased or benefited. It has been openly charged, particularly by cattle exhibitors, that no one but a Shorthorn breeder or feeder can hope for justice at the hands of Western fair associations, and breeders of other families claim that at these fairs the judges are invariably chosen from localities where Shorthorn cattle are raised almost exclusively. Protests against this manner of choosing awarding committees have been repeatedly made, especially by Hereford breeders, not a few of whom assert that the whole system of selecting judges, particularly in the Western states, is controlled by what they are not backward in terming the 'Shorthorn ring.'

"If such a combination exists, it is high time that measures were adopted by which its influence and power can be broken. The struggle for supremacy between the different breeds of beef cattle has become an exciting one, and the

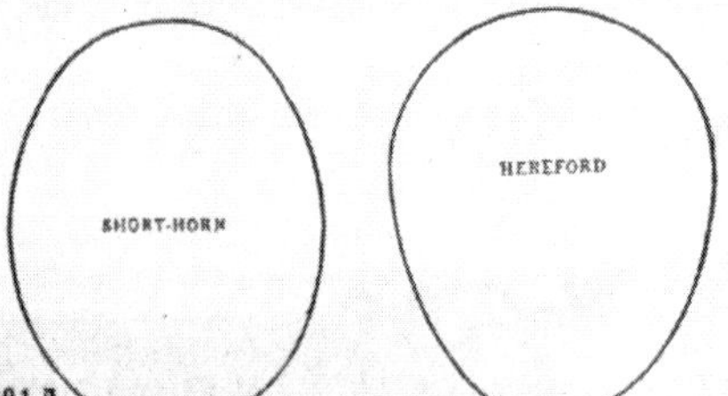

¶201A

HEART GIRTH OF HEREFORDS AND SHORTHORNS COMPARED.

consumers are deeply interested in the outcome. The several associations owe it to themselves, and the public to whom they look for patronage and support, to adopt such rules and regulations governing the award of premiums as will guarantee perfect fairness between contestants. There should be an honest effort made to stop the wholesale charge of unfairness that has acted for years past as an injury to the exhibitions, and which has become a matter of adverse comment, both at home and abroad. Breeders of fine cattle cannot afford to have their interests jeopardized through the actions of incompetent or biased judges, in whose hands they are compelled to place them. 'The Tribune,' during the week of the recent Fat Stock Show, November, 1881, took occasion to criticise the method then in practice, and unhesitatingly stamped several of the awards as entirely wrong. Thoughtful and conscientious breeders in all parts of the country have heartily endorsed the opinions given at that time, and nothing is more certain than that a radical change is necessary or the usefulness of this particular show will cease. The practice of selecting local judges cannot be successfully defended. Assuming that judges chosen promiscuously from districts from which exhibitors may come are perfectly fair, yet, there remains cause of complaint. The men thus chosen, in many cases, being friends or acquaintances of exhibitors, that fact of itself creates a feeling of distrust which may or may not be warranted, and leaves the door wide open for criticisms which would not be possible if the judges were selected from remote localities. This course would entail quite an expense, no doubt, but associations can better afford to close their doors without a profit than to countenance any system by which their usefulness and integrity can be impeached to the slightest extent.

"It is said that the Illinois State Board, under whose auspices the Fat Stock Show is conducted, will, at their annual meeting, which occurs this month, take measures to quiet the dissensions that have heretofore existed by adopting some plan that will guarantee to exhibitors the utmost fairness, based upon the test of merit, and that alone. In this the Board will be unanimously endorsed, and other Western fair associations should adopt similar measures."

After the appearance of the above article in "The Tribune," at the meeting of the Illinois State Board, the president, Mr. Scott, called the attention of the Board, in his address, to the necessity of procuring experts to serve as committeemen in making awards.

It must be remembered that two-thirds of the Board were Shorthorn breeders, and they had friends outside, and that every move that was made in the Board was known to these outsiders. That President Scott was surrounded with these men, and that they were determined to hold the Shorthorns in position; and, though vastly outnumbered, there were men in the Board that were desirous of having its methods all that they should be.

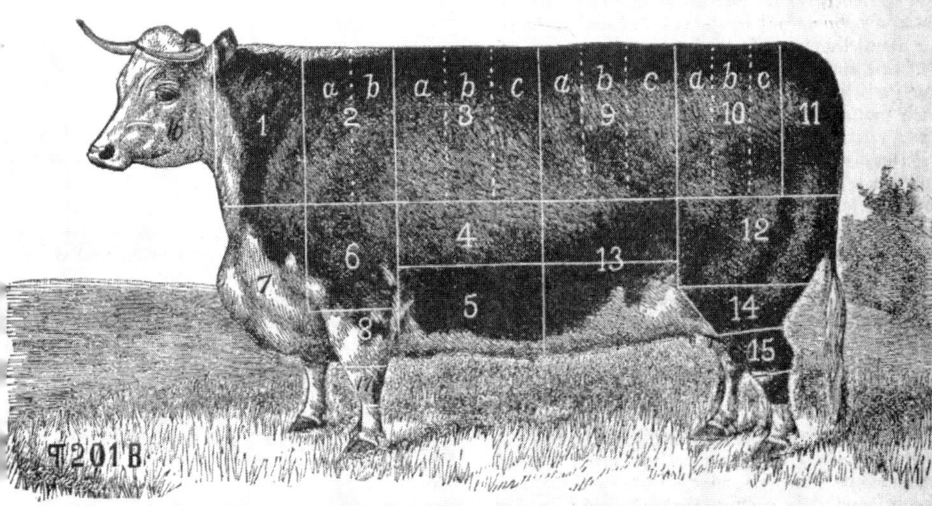

DIAGRAM OF BUTCHER'S CUTS.

EXPLANATION.—1, neck; 2, chuck; 3, ribs; 4 and 5, plates; 6, shoulder-clod; 7, brisket; 8, shank; 9, short-loin, or porterhouse; 10, sirloin; 11, rump; 12, round; 13, flank; 14, lower-round.

CHAPTER XIX.

American Hereford Record

The Hereford cattle had become so numerous in America by 1879 that it became a matter of necessity to have a Herd Book in which to record their pedigrees.

The English Herd Book was commenced by Mr. Eyton; he published the first and second volumes, recording only bulls, numbered from 1 to 901. Mr. Powell followed him with the first part of Volume III, up to number 1,137. At this point Mr. Thomas Duckham of Baysham Court, Ross, took up the work, and published the Herd Book from 1,138 of bulls up to the close of Volume IX. Eyton and Powell in their work recorded only bulls. Mr. Duckham, starting at part second, in Volume III, recorded cows as well as bulls.

At the close of Volume IX the breeders of Hereford cattle in England formed the society, termed the Hereford Herd Book Society, bought of Mr. Duckham, on March 5, 1878, the copyright and continued the publication. Mr. Duckham commenced the publication of the Herd Book in 1857, and was at that time a tenant farmer at Baysham Court, in the town of Ross, in Herefordshire. He was a prominent and skillful breeder of Hereford cattle.

The first president of the Hereford Herd Book Society was J. H. Arkwright (¶ 128), Hampton Court, Herefordshire. Vice-president was the Earl of Coventry (¶ 129), Croome Court, Worcester. The council consisted of twenty-four prominent and well-known breeders. The editing committee: Sir J. Russell Bailey, Mr. T. Duckham, Mr. H. Haywood and Mr. J. Hill. The secretary, S. W. Urwick (¶ 130), Leinthal, Ludlow.

Mr. Thomas Duckham was prominent in keeping the Herefords before the public, through the press, and was the means of distributing large numbers of them to Australia and other parts of the world. There is no man to whom the Hereford breeders, and through them the stock breeders of the world, are more indebted. In 1876, at our great Centennial Exhibition at Philadelphia, the managers of that exhibit sent to the English Government for a man competent to act as judge on the Hereford breed of cattle that should be exhibited at that show. The Duke of Richmond, then at the head of the Agricultural Department, selected Mr. Duckham for that position, and he visited this country, accredited from that government.

(¶ 131) Taking the English Hereford Herd Book as the foundation, I, in 1879, commenced the compilation of the American Hereford Record. I realized that in any thoroughbred race of cattle the cow was as important in securing the purity of the breed as the bull, and that she should be equally well identified in the pedigree of any animal. To secure this identification we could not see any better way than to give each cow a number as well as the bull.

To avoid the great repetition common to other herd books, that was a consequence of giving the entire pedigree with every entry, each entry in our book consisted of only sire and dam, with their Herd Book numbers, the breeder's and owner's names, and date of birth.

The Hereford "Times," England, commenting on this, says: "As everyone knows, Americans never like following a beaten track, and this volume is certainly most unique in its arrangement, and presents a very different appearance from anything that we have ever seen."

The first volume came from our press at Beecher, Ill., in 1880, and we thought that, considering the great service which had been done the Hereford breed of cattle by the Hon. Thomas Duckham, that he was entitled to the place of honor in the frontispiece of the first volume.

There were also twenty-one illustrations of prominent Hereford cattle. Volume II was issued in 1882, and carried the number of entries up to number 6,415, the frontispiece being Mr. Benjamin Tomkins, the oldest known Hereford breeder, that being followed by the earliest advocate of Herefords in America, Mr. William H. Sotham, and following in this honorable company, our own portrait.

As was natural and right, after the formation of the American Hereford Cattle Breeders' Association, the arduous work of publishing the Herd Book was assumed by them, they purchasing from us the copyright and volumes on hand. To say that the work and expense of establishing the American Hereford Record was very large is stating the case entirely within the bounds of truth.

There was a change in the form of the American Hereford Record, beginning with Volume VIII, arbitrarily undertaken, as we believe, by the Executive Committee, without proper thought and reference to the combined wisdom of the Association, and we have watched with interest the spirited and intelligent efforts of Mr. T. F. B. Sotham to have the form we instituted returned into use.

We therefore derive much satisfaction from the information recently conveyed to us in a letter (1889) from Mr. Sotham, as follows: "You will doubtless be pleased to know that my repeated attacks on the error of the Executive Committee in changing your form of record to the 'old fashioned,' 'out of date' style of other books, and common to other breeds, is about to bear fruit. Though laughed at for my pains, and called a 'pedigree crank' because I insisted that the tabulated pedigree is the only simple form, easy of comprehension, I have had victory enough in seeing even the 'pooh-poohing' members of the Executive Committee adopt this form; forced to do so because of the custom everywhere prevailing among Hereford breeders. Your victory is coming, too, for, much against his will, the leading Executive Committeeman has just told me that owing to the increasing number of Herefords he is obliged to go back to the original form established by you, and forms have been ordered for preparing the copy for Volume XXI after your original and unique style." Mr. Sotham adds: "It is decidedly unpleasant to set up an opinion in opposition to this willful and too powerful Executive Committee, but I have the true interests of the Hereford cattle at heart, and being sure, first, of being right, I have gone ahead in my dear old father's belief that 'the right must in the end prevail.'"

We would only add that when we established the American Hereford Record we were as sure as we are now that the Hereford will ultimately be the world's prevailing cattle, and in founding the American Herd Book we wanted that foundation simple and solid. We foresaw the increase of the breed and had we remained in charge of the Record we should have begun with Volume III to put more entries on a page. The only reason we had for limiting the entries (9) on a page in the Volumes I and II, got out by us, was, that with our unique form of entry, more would limit the pages and make too thin a volume.

We trust that young Mr. Sotham's prediction of a return with Volume XXI to the original form may be verified. We believe such a return inevitable; and further, we believe that the day will come when the unwarranted change of form will be generally condemned and a reprint demanded that will practically eliminate Volumes VIII to XX, inclusive, from the libraries of Hereford breeders and prove the indefensible work of the committee on these volumes a waste of the Association's money.

A BREEDERS' JOURNAL CARTOON OF 1883.
(Mr. Miller informs President Scott of the fraudulent entries.)

CHAPTER XX.

American Hereford Cattle Breeders' Association

In pursuance to a general desire for united action of the Hereford Cattle Breeders of the United States, a call was made for a meeting. This was fixed for the 22d of June, 1881. Upon that day there was a meeting of the Hereford breeders at the Grand Pacific Hotel, Chicago, and Mr. C. M. Culbertson (¶ 132) was nominated by Mr. T. L. Miller, Beecher, Ill., for temporary chairman, when, upon taking his seat, Mr. T. E. Miller (¶ 133) was chosen secretary. The Chairman then appointed a Business Committee of the following gentlemen: Adams Earl (¶ 134), Lafayette, Ind.; J. M. Studebaker (¶ 135), South Bend, Ind.; George F. Morgan (¶ 136), Carmago, Ill.; Thomas Clark (¶ 137), Beecher, Ill.; R. W. Sample, Lafayette, Ind.; B. Hershey (¶ 138), Muscatine, Ia.; N. Abbe, Elyria, Ohio, for the purpose of putting the business of the Association in shape and bringing it before them.

They presented the following plan of organization:

This Society shall be known as the American Hereford Cattle Breeders' Association.

The officers shall be a President and nine Vice-presidents, Secretary, Treasurer, and an auditing committee of three, and nine directors.

The term of office of President, Vice-president, Secretary, Treasurer and Auditing Committee shall be for one year, or until their successors are elected and qualified. The Directors shall be elected for one, two and three years, respectively, three for each year.

Any breeder of Hereford cattle in good standing can become a member by handing his name to the Secretary, and paying the sum of $20 for individual or firm.

The officers shall be chosen every year, by a majority vote of the members present, or by proxy, at a meeting called by the President for that purpose. The President shall have the power to call a meeting of all members, or of the Auditing Committee, at any time that he thinks proper.

Two of the Auditing Committee shall form a quorum to do business.

The duty of the Secretary shall be to keep an account of all the doings of the Society. The Treasurer shall keep all funds paid into the Society, and pay the same out on the order of the Auditing Committee.

In case the American Hereford Record is approved and endorsed by the Society it shall be the duty of the Auditing Committee to have the "American Hereford Record" thoroughly examined by a competent person or persons, and if any serious or important error is found, to take note of the same, so that it may be corrected by insertion in the second volume, soon to come out. All records and other matter that go into the second volume of the "American Hereford Record" must first be examined and approved of by the Auditing Committee.

The object of the Society is to promote and improve the beef cattle of the country, by the introduction of the Hereford strain of cattle, more generally throughout the beef-producing region, claiming, as we think, justly, that where the Herefords have been fairly tried they have proved themselves far superior and more profitable as beef cattle than any other breed known. We claim them as being better graziers, winter better on rough feed, mature earlier and bring better prices than any other known breed of cattle, and in proof of this we ask all who have given them a fair trial with any other strain of cattle to come and testify.

The Committee presented an address to the Society. Some points made in it were as follows: "We want it distinctly understood that we protest now and shall at all fairs protest, as is our right, against any man judging as to merit between different breeds of cattle who is interested in the breeding of any of the breeds competing, either directly or indirectly. We think it would be unfair, as if C. H. McCormick was to be made judge of the best make of mowing machines, or Mr. Studebaker as a judge of the best wagon or carriage at a fair. We approve of fat stock shows. It is the great educator, for on the block all beef has finally to be tested. We heard old butchers say at the last fall show that they had been butchers for twenty to thirty years, but had learned more in two days at the fat stock show than they had in the past thirty years. Of course they must have been Bourbons all that time. We do not want such for judges; we want butchers that try to learn something every day. Our ambition should be, not how or by what tricks we can obtain premiums, but rather how to merit them, then if beaten unfairly, it is sometimes worth more than a victory, where the spectators see that palpable injustice has been done.

"Having examined the American Hereford Record, we have no hesitation in saying that we approve it for its simplicity and brevity, and so far as we have examined it, find it correct and reliable, and in connection with this, we must say that T. L. Miller deserves much praise from the Hereford breeders for not only getting up the Record, but also for the long and steady fights he has made, almost alone, in championing the Hereford cause, until he brought them to the front to stay; for, once to the front, they will take care of themselves."

The following officers were elected: President, C. M. Culbertson, Chicago, Ill.; Secretary, T. E. Miller, Beecher, Ill.; Treasurer, Adams Earl, Lafayette, Ind.; Vice-presidents, A. H. Swan, Cheyenne, Wyo.; W. H. Todd (¶ 139), Vermillion, Ohio; William Hamilton, Flint, Mich.; R. W. Sample, Lafayette, Ind.; G. S. Burleigh (¶ 140), Mechanicsville, Iowa; J. M. Studebaker, South Bend, Ind.; William H. Sotham, Chicago, Ill.; A. H. Seabury, New Bedford, Mass.; A. D. Raub, Earl Park, Ind.; N. Abbe, Elyria, Ohio; Mr. Lee of Lee & Reynolds, Camp Supply, Indian Territory.

A BREEDERS' JOURNAL CARTOON OF 1883.
(President Scott stands on the rule.)

Auditing Committee: William Powell, Beecher, Ill.; Thomas Clark, Beecher, Ill.; T. L. Miller, Beecher, Ill.

One-Year Directors: H. Norris, Aurora, Ill.; E. R. Price, Chicago, Ill.; C. Gudgell, Pleasant Hill, Mo.

Two-Year Directors: Joseph Frank, Chicago, Ill.; George F. Morgan, Camargo, Ill.; W. S. VanNatta (¶ 141), Fowler, Ind.

Three-Year Directors: Thomas Clark, Beecher, Ill.; W. E. Campbell, Caldwell, Kas.; T. L. Miller, Beecher, Ill.

Mr. Earl, Chairman of the Business Committee, recommended the adoption of the American Hereford Herd Book, as gotten up in good shape, and giving correct pedigree of Hereford cattle in all its arrangements. It was adopted by a unanimous vote.

Mr. Miller handed the President a copy of the American Hereford Record to be presented to Mr. W. H. Sotham, the old champion and original Hereford advocate. (¶ 142) It was presented him in the name of the Association. (¶ 143)

It was moved and adopted that the Association meet annually for general purposes of business connected with this Association during the week of the Fat Stock Show at Chicago, Ill. (¶ 144)

The second meeting of the Association (¶ 145) was held during the Chicago Fat Stock Show, Friday, November 11, 1881. The same officers were re-elected and general work planned. The meeting was very successful, and the enthusiasm that it produced among the breeders of Herefords contributed largely to the success and larger show of Hereford steers that were made the following year.

The third meeting of this Association was held on November 17, 1882, in the club room of the Sherman House, Chicago, Ill. The meeting was called to order by the President, C. M. Culbertson, of Chicago. In his opening address the President stated that, as the most of the Hereford breeders had not arrived in Chicago, this meeting would be merely a preliminary one for the purpose of talking over business matters that would be brought up before the regular annual meeting, to be held November 21st.

The matter of the American Hereford Cattle Breeders' Association purchasing, as a society, from Mr. T. L. Miller, of Beecher, Ill., the American Hereford Record, was introduced. The President appointed a committee of three gentlemen, consisting of Mr. W. S. VanNatta, Mr. R. W. Sample and Mr. J. M. Studebaker, to wait on T. L. Miller and ascertain from him at what price and on what terms the Herd Book, with its copyright, good will and volumes, could be purchased, and to also present a financial plan by which the money necessary could be raised.

It was moved and carried that T. E. Miller be appointed a committee of one to submit to the next meeting of the Association rules to govern the admission of cattle to the third volume of the Herd Book. Mr. J. M. Studebaker gave notice that he would present to the next meeting a plan for a testimonial to Mr. William Henry Sotham for his long and efficient service in the Hereford cause.

The meeting then adjourned to Tuesday evening, November 21st, to be held in the same place.

The regular annual meeting of the American Hereford Cattle Breeders' Association was held November 21st, at the Sherman House, Chicago, Ill. The meeting was called to order by the President, C. M. Culbertson.

The Secretary read the minutes of the preceding meeting. The first business transacted was the election of officers for the ensuing year, which resulted in the election of Mr. C. M. Culbertson, of Chicago, President; C. K. Parmelee, of Wolcott, Ind., Vice-president; C. B. Stuart (¶ 146), Lafayette, Ind., Treasurer. The three Directors whose time had expired were re-elected for three years, viz.: H. Norris, of Aurora, Ill.; E. R. Rice, of Chicago, Ill.; C. Gudgell, of Pleasant Hill, Mo.

The report of the Committee on purchasing the Herd Book was then called for, which was submitted as follows:

Chicago, Ill., Nov. 21, 1882.

Mr. President: We, the committee appointed by your honorable body to confer with Mr. T. L. Miller regarding the purchase of the American Hereford Record, published and owned by him, submit the following report:

The price asked by Mr. Miller is six thousand dollars, which constitutes the good will of said Record, including about three hundred of the first volumes, and five hundred and twenty of the second, which are nearly ready for distribution, Mr. Miller valuing same at three dollars each. His desire is that the book shall be controlled by the Society and not by any one individual or by a stock company, the latter of which we, as a committee, heartily endorse. We further state that Mr. Miller represented to us that this book has cost him twelve thousand dollars over and above receipts. In conclusion we simply wish to say that we place this matter in the hands of this Society for their careful consideration and final action, advising that it take a liberal view of this matter, and pay Mr. Miller what the unanimous voice of this Society shall deem a fair and impartial compensation for his untiring efforts in the Hereford cause.

Respectfully submitted by the committee.

J. M. Studebaker,
R. W. Sample,
W. S. VanNatta.

After considerable discussion, it was moved by Mr. Adams Earl, of Lafayette, Ind., that the report be placed on file as received, but that the committee be not discharged, and that the Chair add four gentlemen to the committee, and that they confer further with Mr. T. L. Miller at once and report to this meeting. The President named Mr. Adams Earl, Mr. C. Gudgell, Mr. W. Hamilton, Mr. E. Phelps (¶ 147) as the new members of this committee. Mr. H. C. Burleigh (¶ 148) was subsequently added. The committee then retired with Mr. T. L. Miller, and after considerable delay returned and made the following report:

TYPICAL FEMALE HEAD, GRACE 58553.
(From a photograph.)

Mr. President: We, the committee appointed to report on the subject of purchasing the Herd Book of Mr. T. L. Miller, have the following report to make:

Mr. Miller has made the following propositions to the Association:

First. Proposes to sell the Herd Book and 800 volumes for $6,000, payable in one year, and to take the obligation of the Society in payment, with individuals' names as security.

Second. Agrees to continue the publication of the Herd Book, the Society shall make the rules for entry, and the entry fees to remain as at present, $1 for entry and 25 cents for each tracing.

F. D. COBURN, Topeka, Kan.

Third. Will sell it to the Society at any time when the Society sees fit to purchase it at a fair price.

Your committee would recommend to the Association the acceptance of the second proposition, and that an auditing committee be appointed, to consist of three, who shall draft a set of rules governing the future entries to the Herd Book and who shall supply Mr. Miller with a copy of the rules.

W. S. VanNatta,
Chairman.
R. Sample,
J. M. Studebaker,
Adams Earl,
W. Hamilton,
E. Phelps,
H. C. Burleigh,
C. Gudgell.

The report was accepted and committee discharged, it being resolved to leave the Herd Book in the hands of its publisher and proprietor, and that a committee consisting of C. B. Stuart, G. S. Burleigh and W. S. VanNatta be made a permanent auditing committee to pass on pedigrees as to their eligibility for admission to the Herd Book, and also to report a set of rules as soon as possible to this Association which shall, upon approval by this Society, govern the admission of cattle to the American Hereford Record.

It was suggested by the President that the Hereford cattle breeders get together a number of show herds of Hereford cattle in the year 1883, to the number of, say, 100 head, and make a circuit of the fairs. An expression of opinion was called for on the subject.

Mr. Joseph Franks, of Cheyenne, Wyo., said: The Western ranchmen were all very favorably inclined towards the Hereford cattle; that they would be pleased to see such an exhibit; that his firm had individually bought over 600 Hereford bulls.

Mr. W. E. Campbell, of Caldwell, Kansas, said the cross of a Hereford bull on a Texas cow gave the best satisfaction. Every one now on the range wants to get Hereford bulls. The risk of Texas fever was more with him than it was further North. He used, beside his Hereford bulls, two hundred Shorthorn bulls in his herd, of which he had had a good many die from consumption and lack of constitution, while among his Hereford bulls he had only lost two, one from Texas fever and one from an accident. He takes all his bulls up in the winter and feeds them, and finds he can feed two Herefords on the same feed it takes to keep up one Shorthorn. He would be pleased to see a large exhibit of Hereford cattle next year, and thought it would do much good.

Major W. A. Towers, of Panhandle, Texas, thought Herefords were as liable to take Texas fever as other cattle. He has been very much pleased with his Herefords, has taken sixty head to his ranch direct from England. His grade Herefords go through the winter much better than any of his other cattle. Is using about an equal number of Hereford and Shorthorn bulls. Thinks it would be a good idea to take a large show of Hereford cattle through the fairs of Kansas and as far west as Denver, Colo.

The President then introduced the subject of grade Hereford steers to show at the Fat Stock Show. There has been such a demand for grade Hereford bulls to go west and south that they have sold at from $60 to $75, and it has made the steers so scarce that scarcely any can be got, and proposed that the breeders present pledge a certain number of calves to be altered for the purpose of getting a supply of Hereford steers. The following responses were made to this appeal:

T. C. Ponting (¶ 149), 100; C. M. Culbertson, 30; J. P. Holmes, 10; Earl & Stuart, 10 grades and 2 thoroughbreds; Thomas Clark, 1; Hiram Norris & Sons, 12; William Powell, 2; Lyon Bros., 2; G. S. Burleigh, 3; Charles K. Parmelee, 20; Burnham & Sons, 5; J. R. Price, 30; A. A. Crane & Son (¶ 150-151), 12; Burleigh & Bodwell, 1 grade and 4 thoroughbreds; Dr. O. Bush (¶ 152), 6; Thomas Foster (¶ 153), 20; Wallace Libbey (¶ 154) 10; A. H. Hood, 3; A. H. Bullis (¶ 155), 4; A. D. Raub, 8; C. Gudgell, 2 thoroughbreds; W. Hamilton, 5; E. Phelps, 5; Fowler & VanNatta, 5 grades and 2 thoroughbreds; Seabury & Sample, 15; making a total of 319 grades and 10 thoroughbreds. Besides these, the Western men put down for

Swan Bros. & Frank, 600; W. E. Campbell, 10; Scott & Hank, 20; total, 630.

The testimonial to Mr. William Henry Sotham was then introduced, and the following amounts, making a total of $535, were presented to him: T. L. Miller Company, $100; C. M. Culbertson, $100; J. M. Studebaker, $100; T. C. Ponting, $25; William Powell, $5; Edwin Phelps, $25; William Hamilton, $50; Thomas Foster, $50; A. H. Hood, $5; W. S. VanNatta, $10; Earl & Stuart, $20; O. Bush, $5; A. D. Raub, $5; G. S. Burleigh, $5; John Gosling, $5; (¶ 156) Thomas Clark, $10; Price & Jenks, $5; Charles Gudgell, $10.

The subject of the members of this Association making up a purse of $2,000 to be offered in premiums for Hereford cattle to be exhibited at the next Fat Stock Show was introduced by T. L. Miller. It was clearly shown how such an amount as this offered in prizes to Hereford breeders, in addition to the amounts offered by the management of the show, would bring out and encourage the feeding of a very large number of Hereford steers. The result of this appeal was the circulation of a subscription paper, of which the following is a copy:

We agree to pay the following sums, provided two thousand dollars is raised, towards prizes to be given for Hereford cattle at the Fat Stock Show in 1883, the classification to be made by a committee to consist of C. B. Stuart, T. E. Miller and C. K. Parmelee, and such classification to be published as early as possible.

There was about $1,500 subscribed to the fund, and the committee was instructed to use every endeavor to make the sum up to the required amount. The meeting then adjourned to Wednesday, November 22, at the same place.

The meeting of November 22d was called to order by the President, C. M. Culbertson. The report of the former Treasurer, Mr. Adams Earl, was called for, who reported $260 on hand. Mr. C. B. Stuart, who was elected Treasurer at the meeting on the 21st, resigned his office for the purpose of having the Secretary also hold the office of Treasurer. T. E. Miller, of Beecher, Ill., was elected Secretary and Treasurer for the ensuing year.

The resolution was adopted that all cattle for entry in the Herd Book, which clearly came inside the rules made for admission of cattle, should be printed by its editor without submission to the Auditing Committee.

It was moved and carried that the Auditing Committee be authorized to draw the necessary money from the treasury to pay C. M. Culbertson, Jr., for his services in examining the American Hereford Record.

The President suggested that the prizes at the various agricultural shows should be given to young cattle, and thus discourage the keeping in the show herds of older cows and large cattle. After considerable discussion and amendments the following resolution was adopted:

Resolved, That it is the sense of this meeting that the agricultural societies throughout the country should make their highest herd prize for young cattle under two years and over one year old; their second prize to a herd under three years and over two years old.

Owing to a misunderstanding in some of the Chicago papers, as to the testimonial given to Mr. Sotham, the following was adopted as the sense of the Association: "The fund that was presented to Mr. William Henry Sotham was for the purpose of rewarding him for his services for forty years in pushing the cause of Hereford cattle, and is in no sense a charity, but is for the purpose of showing the high sense of appreciation of this Society for the services rendered by Mr. Sotham in introducing Hereford cattle."

The attention of the Society was called to the ages of the Shorthorn steers being shown at the present Fat Stock Show, in which it was stated that Mr. D. M. Moninger, of Galvin, Iowa, entered the steer called "Champion of Iowa" as 715 days old, or 1 year 11 months and 21 days old. He shows six teeth of full size, and by the mouth is over three years old. The horns of this steer have been filed and dressed to give him the appearance of being younger than his mouth would make him. There are several other steers entered by Mr. Moninger as two-year-olds whose mouths would indicate a greater age than that for which they were entered.

JUDGE T. C. JONES,
Delaware, Ont.

After some discussion it was moved that the President, Mr. C. M. Culbertson, be requested to see Mr. J. P. Reynolds and take what steps were necessary when the matter of the ages of

Shorthorn steers came up before the State Board.

A vote of thanks was tendered Mr. J. Irving Pearce, proprietor of the Sherman House, for the use of the rooms occupied by this Society.

The Society then adjourned subject to the call of the President and Executive Committee.

(Signed) T. E. MILLER,
Secretary.

Pursuant to a call from the Executive Committee, there was a meeting of this Association at 10 o'clock A. M. on February 28, 1883, at the Sherman House, Chicago. The business to be brought before the meeting, as stated in the call, was as follows:

WM. WARFIELD,
Lexington, Ky.

First. To consider and act upon rules to govern future entries in the "American Hereford Record."

Second. To consider propositions, looking toward the Association owning the "American Hereford Record."

Third. The organization of a life membership society of American Hereford breeders.

Mr. B. Hershey and Mr. Adams Earl were appointed by the Chair to confer with T. L. Miller for the purchase of the "American Hereford Record" and report to the Association.

After some discussion as to the method of making entries in the American Hereford Record and the undesirable form in which the English Herd Book was published, an extract was read from a letter from Mr. John Hill, of Felhampton Court, England, one of the editing committee of the English Hereford Herd Book. The Hon. Thomas Duckham is also one of the editing committee. The extract is as follows:

"I should like to have your opinion on the resolution I proposed at the last meeting of the English Hereford Herd Book Society, which was, that every cow should have four crosses, and every bull four crosses (which is now the rule). Mr. Duckham seconded me, but we were beaten, of course, being strongly opposed by men who have short pedigree animals and who are making up herds from doubtful beginnings. I go even further than this, and I think it would, in a year or two (due notice being given), do better to have no animal entered unless it can trace to a cow already entered in the Herd Book. This would keep the breed pure, and prevent any alien from getting in. If you and other breeders in America approve of my scheme, it would, of course, strengthen my hands much to have their opinions in writing."

Upon hearing the letter read, the following resolutions were introduced and carried unanimously:

Resolved, That we, as Hereford cattle breeders of America, recognizing the great importance of keeping the breed in its purest state, most heartily endorse all that Mr. Hill and Mr. Duckham suggest in the above communication, and that we hope that the English Society may take some such action looking to some such end.

Resolved, That these resolutions be made a part of our records, and a copy of the same be sent to Messrs. Hill and Duckham.

The committee appointed to report of the purchase of the Herd Book submitted the following report from the owner:

Chicago, February 28, 1883.

I propose to sell the American Hereford Record to the American Hereford Cattle Breeders' Association, if they desire to purchase the same, for the sum of five thousand dollars ($5,000), to be paid for within one year of this date, with 6 per cent. interest. In making this proposition it is distinctly understood it shall be and remain in the American Hereford Cattle Breeders' Association. (Signed) T. L. MILLER.

The proposition was accepted by the Association and the money was raised by assessing each member of the Association $2.17 for each thoroughbred Hereford owned by them on July 1, 1883.

The Chair appointed the following members for incorporating the American Hereford Cattle Breeders' Association: W. S VanNatta, C. B. Stuart, C. M. Culbertson, G. S. Burleigh and T. L. Miller. Mr. J. B. Sollitt then introduced the following resolution, which was seconded and carried unanimously: Resolved, That the thanks of this Association are due and are hereby tendered to T. L. Miller for the able and effective manner in which he has got up and published the American Hereford Record; also for his able and vigorous defense of the Hereford interest through the last eleven years.

The rules governing entries to the American Hereford Record were then brought up by the committee, and adopted. Mr. C. B. Stuart moved that Mr. Miller be appointed a committee of one, as he is contemplating a trip to

England, to confer with the English Hereford Herd Book Association, and act in conjunction with them, taking such steps as may be necessary to place their pedigrees on such a footing as will qualify them for entry in the American Hereford Record. The price of entries in the Herd Book to persons not members of this Association was fixed at $2, and the price of the Herd Book at $5.

The Executive Committee was authorized to employ the Secretary of this Association to take care of the property of this Association and do the work of this Herd Book and pay him for the work done.

A resolution of thanks was then introduced as follows: Resolved, That the thanks of this Association be tendered to Mr. T. L. Miller for the Hereford beef furnished for the dinner this day; and also to Mr. J. Irving Pearce for the use of the club room.

Adjourned to meet again during the Chicago Fat Stock Show in November next.

(Signed) T. E. MILLER, *Secretary*.

In pursuance of the authority given us by the American Hereford Cattle Breeders' Association, to confer with the English Hereford Herd Book Society, in regard to the rules that should govern the admission of cattle to the Herd Book, we attended the Herefordshire Agricultural Show that was held at Abergavenny, England, in June, 1883. During this show the Hereford Herd Book Society held its annual meeting. It was held in the committee room on the show grounds. Mr. J. H. Arkwright presided. After the report of the Secretary, Mr. S. W. Urwick, other business was transacted and the Society had its attention called to the rules governing the admission of cattle to the Herd Book.

Mr. Miller (America) said he had been requested by the American Hereford Cattle Breeders' Association to seek an interview with the Council of the Hereford Herd Book Society, to see if some arrangement could be made so that the pedigree entries of cattle in the Herd Books published by both societies might be made to correspond. He would be very glad to meet the members of the Council to consider the question.

Mr. Chairman: "Have we your book?"

Mr. Miller: "Yes, I sent a copy to each member of your editing committee, I believe. There were two volumes." The Chairman said he "had not received a copy." He "should think that the proposed conference would be very desirable."

Mr. Miller said he "did not know exactly what the rules of the Society were, but in the Association to which he belonged four crosses for dams and five for bulls were laid down, while in England it was only three crosses for dams and four for bulls. He wished to confer with the Council to see if a uniform arrangement could not be made. The members of the Association desired to know, not only that an animal was thoroughbred, but how, and why it was thoroughbred. He would like to obtain the views of the Hereford Council upon these matters, so that he could report to the Association."

The Chairman said that "when the Society began, a mistaken interpretation was put upon one of the rules. When the mistake was found out, they went back to the strict interpretation of the rule, so that the Society had not reduced the standard."

THOMAS SMITH,
Crete, Ill.

Mr. Miller thought that "it would be eminently advisable for the two societies to be in accord in this matter."

The Chairman: "We tried it once, but we were beaten. However, I will propose that we have a special meeting." Mr. Robinson proposed that it should be a general meeting of the members. The Chairman: "I think the Council should consider it first and then it should be referred to a general meeting of the members." Mr. A. Rogers seconded the resolution, and it was carried.

The Society had their annual dinner, after getting through with their business, under a monster tent. The dinner was very largely attended. After this they proceeded with the toasts and speeches, which, as usual on such occasions, were large in number and long in duration.

CHAPTER XXI.

Cattle Frauds—Comparative Treatment in England and America

(¶ 157) Taking the general proposition that the Shorthorns won the sweepstakes in both 1879 and 1880, it would carry much weight for the breed if none of the facts in connection with it were known, but when the age and breeding of the winning animal is made a matter of record it will modify very much the credit to be given for this award.

The sweepstakes steer at Chicago Fat Stock Show for 1879 was shown as a three-year-old thoroughbred Shorthorn and called "Nichols." It appeared later from a statement from Mr. J. W. Prescott, of North Middletown, Ky., in the "National Live Stock Journal," that he was the get of a bull without pedigree, or known descent, costing $40 at eighteen months old and out of a grade cow. The steer was sold October 3, '78, weighing 1,464 pounds, and said to be, at that time, three years old. He was shown in November, '79, as being dropped March 15, '76, in the three-year-old class, weighing 2,060 pounds, as a thoroughbred Shorthorn.

This steer won in three-year-old class of thoroughbred Shorthorns, in the sweepstakes ring for three-year-olds, the best of any breed, and for the best beast in the show of any age or breed, and the "National Live Stock Journal" cup.

In three-year-old classes he was not entitled to show, and the question would turn upon his merit, considering his age, as to whether he was entitled to the awards as the best beast in the show. As a three-year-old, he would be entitled to the first place considering his maturity and ripeness, but being a four-year-old he was not entitled to the first place in either class.

This Kentucky steer Nichols, shown at the Fat Stock Show in 1879 as a thoroughbred Shorthorn, three-year-old, taking the first prize in his class, taking the sweepstakes prize for the best three-year-old of any breed, and taking the champion prize as the best beast in the show, was shown in 1880 as a grade Shorthorn, four years old (¶ 158).

So far as we are informed, these awards stand without a challenge or question by the Illinois State Board of Agriculture with the following testimony on record:

J. W. Prescott sold this steer to Mr. Daniel Nichols, October 3, 1878, weight, 1,464 lbs., and as about two years and a half old. The steer was bred by Mrs. Mary E. Grimes, now Mrs. Prescott. When Mr. Prescott told his wife what age he had given, she then told him that the steer was a three-year-old. (¶ 159)

On the show ground in 1880, Doctor Paaren examined his mouth, with other parties, and pronounced the steer, from such examination, to be nearer six years old than five years old. When he was exhibited as a three-year-old thoroughbred he was called by the exhibitor a "Young Mary," but when pressed to show his pedigree, the breeders' certificate simply stated that he was a thoroughbred, got by a Shorthorn bull out of a "Seventeen cow." This breeder's certificate was probably (¶ 160) from Mr. Nichols, who never bred him at all, and the statement that he was a "Young Mary" steer was made by Mr. Graves, the exhibitor of the steer in 1879.

This is substantially the record of the steer as revealed by apparently reliable testimony. On this record the steer had no right on the show ground in 1879 or 1880, as we will endeavor to show.

His first winning in 1879 was as a three-year-old thoroughbred; he was a four-year-old and a grade.

His second winning was in the sweepstakes, he being four years old, his competitor three years old. He had no right in this class.

His third winning was in the grand sweepstakes for best animal in the show. In this class any age or breed had a right to show. Had this steer been a three-year-old of his weight and finish the award might have been properly made, but being a four-year-old, weighing 2,060 pounds, the probabilities are that if shown at his proper age he would not have won. But having shown under a misrepresentation, as to breeding and age, he had no right to his winnings of 1879. This being true, the first premium given to him as the best three-year-old thoroughbred Shorthorn should have been taken from him and given to John B. Sherman for "Eddie Morris," or to F. W. Hunt, for "Thad Stevens," and the grand sweepstakes should have been taken from him and placed on some other steer, if the Board could determine. It should have been given to one of the first premium animals in the classes; if this could not have been determined, it should have been held for some special meritorious animal.

The above is a fair statement in reference to this steer for his record in 1879. On this record we would claim that he should have had no standing in the exhibition of 1880, and it was on this record that C. M. Culbertson and T. L. Miller made their protest against his being shown at the show in 1880.

In reference to the entry and the action of the Board in admitting him to compete for the champion prize in 1880, the Board placed themselves in a position where their action was a proper subject for criticism.

The steer was entered in 1880 as a four-year-old, while Mrs. Prescott, who bred the steer, says he was a five-year-old, and Dr. Paaren, supported by others (¶ 161), says he was nearer six years old. The testimony of J. W. Prescott, until it shall be disproved, was sufficient warranty for rejecting the animal from entering for the champion prize in 1880. Especially is this true when Dr. Paaren was selected by the Board to determine the ages of the animals on exhibition.

It is clear that an animal may be, for a two-year-old, in his weight, character and quality, the best animal in this class or in the show, but if the animal has another year on him and has the same weight, character and quality he would not be the best animal in the class or in the show. The fact is self-evident; for at the Smithfield Show in London in 1879 the champion prize went to a two-year-old, not because he was the best steer or best animal in the show without regard to age, but considering his age he was the best animal in the show.

The protest against this steer secured another verdict, and that was that the grade Hereford steer, two-year-old, belonging to T. L. Miller was the best steer in the show if the Nichols steer was ineligible.

There is no reasonable doubt but that the Nichols steer was over five years old, weighing 2,465. The two-year-old Hereford was three years younger and weighed 1,845 pounds, and his carcass was worth as much (or more) on the block as the Nichols steer. The grade Hereford steer here mentioned, Conqueror, was the best of the six two-year-old Hereford steers exhibited by us. There was not an uneven spot in him; he was evenly covered by thick, firm flesh.

Why did the Illinois State Board accept this steer Nichols as a four-year-old and permit him to be exhibited as such? Did they require the cancellation of the awards on the animals to which, of a right, they belonged? Why did they permit the steer Nichols—when the exhibitor did not know how old he was, and Mrs. Prescott had said he was dropped in 1875—permit him to show as a four-year-old? Why did the Board go on and publish the cut of this Nichols steer, and the reports of the committee, when they had the evidence before them that he had been shown under fraudulent representation?

It would have been difficult for the Board to explain their actions in this matter in a way that would have satisfied the public or competing exhibitors. The facts in reference to his breeding and age which disproved every statement of the exhibitor were before the Board before their report was published, and their attention was called to these facts; still the report was made and no reference made to the misrepresentation. This thoroughbred of 1879 was permitted to exhibit as a four-year-old grade in 1880, when satisfactory evidence was before them that he was a five-year-old.

To show how the English courts of law looked at a similar matter (¶ 162) of false pretenses, or of selling an animal with a false pedigree, it will be well to read carefully the trial of "Allsop vs. Hopkins," in England, for damages resulting from the sale of a Shorthorn bull, with fraudulent pedigree. Mr. Stavely Hill, the attorney for the plaintiff, in stating his case to the court and jury, gave the following definition of the origin of the Shorthorn breed and of the English Shorthorn Herd Book. We reproduce this report because it gives a clear, concise statement of the origin and character of the Shorthorn breed of cattle.

¶210A

J. B. GREEN,
Marlow, Herefordshire.

"Mr. Stavely Hill, in opening the case to the jury, after referring to the important issues raised by the pleading, proceeded to give an interesting account of the origin of the Shorthorn breed, and of the Herd Book kept by the society. He pointed out that towards the close of the last century some experienced farmers had noticed that the breed of large cattle in the south of Durham and in the north of Yorkshire had capabilities of great improvement, so as to render them more valuable for dairy purposes, and at the same time that they would have greater capabilities for putting on flesh. With this view Messrs. Colling of Ketton and Mr. Bates of Kirklevington crossed their breed; the one, it is said, with a polled Galloway and the other with a Kylo, or West Highlander (¶ 163)—and after several judicious crosses succeeded in producing the present best established and illustrious families of the Shorthorn tribe. It would be clear that this breed must be considered an artificial breed, and there would be a tendency to breed back to the older breed, which would take every opportunity of reasserting itself; and, undoubtedly, if left to a state of nature those points of the animal which might be called its natural characteristics would reappear, while the Shorthorn points would in proportion be lost. In these circumstances it was necessary to keep a constant watch with a view to eliminate any reappearance of the older breed displaying itself, either by a dark nose or a straight horn or other characteristics, and to keep for breeding purposes only those animals which showed the permanent Shorthorn features, and for this purpose the Herd Book has been established as a register of those animals which might be relied on by the breeder to maintain the character of the tribe. The bull in question in this action, professed to be a bull of five crosses, and to be thus eligible for entry in the Herd Book; and the learned counsel illustrated the difference there would be between the produce of such an animal and an animal such as that he should prove the bull really to be, viz., one got by a pedigree bull upon a dairy cow, by pointing out that in the latter there would be much doubtful blood as would amount—to illustrate by means of a chess board—to one-half of the board, while in the former the doubtful blood would be equivalent only to one square, or one-sixty-fourth of the board."

G. H. GREEN,
Of the firm of J. B. & G. H. Green,
Marlow, Herefordshire.

Those who have used the Shorthorns to cross upon other breeds will recognize the truth of the attorney's statement when he says:

"It will be clear that this breed of cattle must be considered an artificial breed, and there will be a tendency to breed back to the older breed, which would take every opportunity of reasserting itself; and, undoubtedly, if left to a state of nature, those points of the animal which might be called its natural characteristics would reappear, while the Shorthorn points would in proportion be lost."

In the action referred to, Mr. Allsop obtained judgment for $3,750, which was not only the direct damages, but the constructive damages for the produce of a bull in the herd of Mr. Allsop, being the difference between the value of the calves by this bull, and what the value would have been had he been what he was represented to be.

The bull in question had been shown at the Birmingham Animal Show in March, 1875, *as a thoroughbred;* when these facts, brought out in the trial, came to the knowledge of the Birmingham society, that society *commenced an action* against Mr. Hopkins, which was tried at Warwick Assizes, February 18, before Lord Chief Justice Cockburn; of which we present the following report:

"His lordship having summed up, the jury retired to consider their verdict at half-past four, and shortly after, returned into Court with a verdict of guilty, and his lordship, in sentencing the prisoner, says: The jury have found you guilty of obtaining £20 ($100), given by the Agricultural Society for the exhibition of the best animal in the class. I think it is a pity to see a young man of your respectability and position, standing convicted of fraud, yet I do not think the jury could have arrived at any other conclusion. You possessed, no doubt, a very fine animal, but then you had no pedigree, and by the conditions of the exhibition you were not entitled to take a prize. You manufactured a pedigree—that no reasonable man can doubt. A man who does that to gain a prize, gains that prize by false pretenses, and he robs not only the society, but the man who ought to have taken the prize, just the same as if it were taken out of his pocket. If you were not sensible that this was a dishonest transaction, then I am sorry for you, for you did not properly appreciate the turpitude of the act. That it was robbing another competitor is the real nature of the case, and if such frauds were allowed to go on unpunished there would be an end to open and honest competition. The jury have recommended you to mercy, and I take that into consideration. A sentence of three months at hard labor is the lightest sentence I can pass upon you."

It is well to note here that the Birmingham Society did not sit tamely down and let the exhibitor of this bull come back to their show the next year and carry off the Sweepstake prize; but commenced an action against Mr. Hopkins, and put a stop to his further depredations.

The "Century" for February, 1881, discusses the different methods now in vogue, by which parties take that which does not belong to them. They say: "He who takes by stealth what belongs to another, is a thief. He who takes by violence is a robber, and a robber is properly supposed to disappear with other predatory animals before the progress of civilization. But this is a superficial judgment. The force that unlawfully dispossesses men of their property, passes through many transformations, but no force is more persistent. Men are plundered nowadays, plundered in America, more than in England in the days of Robin Hood. There are men among us, beside whose robberies, those of the brigands of Italy and Greece, and the Bedouins of the desert are merely pleasantries. Of all the triumphs of invention, none are more wonderful than that by which the hard earned gains of millions are forcibly conveyed to the treasuries of the robber princes. No business is more highly organized, more stealthily pursued, more successfully managed than the business of robbery. Yet under all this elaboration of method, it is robbery; and nothing worse or better. The peculiarity of the modern method of robbery, is the employment by robbers of the state, as their enforced agent and accomplice."

The "Century" here opens and develops methods that are of interest to us in the discussion of the live stock interests of America.

We have called attention to the fact, that about sixty years ago, the Shorthorn influence, or, using the term of the "Century," "Shorthorn robbers," used the English Society, termed the Society for the Diffusion of Useful Knowledge, as their enforced agent and accomplice, and as an endorser for their schemes. Ten years later they used the State of New York as their enforced agent and accomplice—or as an endorser. Later they used the State of Illinois. In November, 1880, standing under the authority of the State of Illinois, as President of the State Agricultural Society, the President permitted these Shorthorn robbers to fraudulently enter and exhibit their animals, and under these fraudulent exhibits, take money that did not belong to them, and statistics were published under the authority of the State of Illinois, based upon fraudulent data. The "Century" closes the article, from which we have quoted, by saying: "What have the people to say about these practices? They do not appear as yet to have anything to say. The robber princes are held in high esteem, they go about to the colleges, some of them, and doctors of law and doctors of divinity grovel at their feet. If any Mordecai has refused to bow down before them, his name has not been reported. Men whose riches have been increased by spoiling their neighbors, are held up as shining examples for the imitation of youth. So long as teachers silently endure such iniquities, it is not to be expected that the people will cry out against them. But the day is sure to come when plain men will clearly see that no man can get with clean hands in an ordinary lifetime a hundred of million of dollars. That such an enormous pile so suddenly collected must be loot, not profit. That will be a day of reckoning indeed, for the robbers and for the judges, and for legislators and public teachers who have been their accomplices. In the meantime these facts are to be borne in mind, for we have among us a class of men, who in their rapacity, are bound on enriching themselves by forcibly seizing the property of their neighbors; and they have learned how to use, for this purpose, the organized force of the state. Some means must be found of putting a stop to them. Unless this be done speedily the respect for law, on which social order rests, will not long survive." The writer whom we here quote, had in mind, when writing, that class of operations known as star route frauds, but it is equally applicable to the cattle ring, that organize their interest under the authority of the state and under this authority organize their society and shows, select judges in their interest, and publish the awards rendered by such judges, as an evidence of merit in their cattle, or standing between fraudulent entries and a proper investigation in regard to them. There is a growing demand throughout the country that these and similar frauds shall be exposed.

JOHN PRICE,
Of Court House, Herefordshire.

CHAPTER XXII.

THE "BREEDERS' JOURNAL"

The amount of opposition that the Hereford breed of cattle encountered, during their upward march to popularity, in America, can be but faintly imagined, indeed, it does not seem possible at this day that the agricultural and stock journals were so completely dominated by the Shorthorn men prior to 1880.

We were, from the commencement of our experience in breeding Hereford cattle, liberal advertisers in the live stock journals, and while this, to some extent, opened their columns to us, to advocate the cause of the Herefords, we were continually handicapped by their desire to do or say nothing to offend their Shorthorn customers.

The following quotation, taken from the "National Live Stock Journal," shows the standard by which the editor of that journal measured merit. It evidently refers to the Herefords as a meritorious, but not widely popular breed of cattle, and takes the ground, by implication, that the Shorthorns are the best breed, mainly, because they are so popular. The article referred to, says: "In conversation with an intelligent and enthusiastic breeder of a meritorious, but not 'widely popular' breed of stock (and here we give formal notice, that we frankly decline inviting criticism, by being more special), we frankly told him that as far as money making was concerned, we should advise him to select a more popular breed. If he desired to do the most he could for his neighbors and the agriculture of his country, his course would naturally be determined by his belief, that his favorite breed was the best for farmers generally. The breed in question has been in this country for at least half a century, but is little more known now than it was twenty years ago. Our friend admitted the facts, but marveled at the obtuseness of the farmers. If any breed has been and continues to be, popular and fashionable over a wide region of country, it is useless to deny it the possession of substantial merits. The Shorthorns have, for instance, had their popularity greatly helped in many ways, but had it not been for substantial merit, this breed could not, by any combination of influences, have been made to gain and retain so widespread a popularity.

"It is quite within the possibilities, however, that some other breed, which has not been so skillfully handled, is a better one for general use than the Shorthorns. It now strikes us we should let their work speak for them, rather than to indulge in reasoning to show what they ought to do. If they were mainly fitted for beef production, we should try to make good exhibits at fat stock shows."

This standard of merit would not do for the Hereford. In our contest for recognition, we were compelled to call things by their right names, and if gross injustice was done to so characterize it.

The "Breeders' Gazette" was established in 1881, by Mr. J. H. Sanders, a year after we established the "Breeders' Journal." In their first number, Mr. Sanders gave the keynote, as to the stand the "Gazette" would take, in reference to breeds of cattle. It is as follows: "While the Herefords, Devons, Jerseys, Holsteins, Polled Angus, Ayrshires and other breeds of cattle will each receive due attention and fair treatment in our columns, yet it cannot be denied that there are ten men interested in Shorthorns to where there is one engaged in breeding any other variety. And recognizing this very general distribution of Shorthorn blood throughout the country, we shall endeavor to make "The Breeders' Gazette" indispensable to the breeders of that strain of cattle. We shall keep a vigilant eye upon the business in the States, Canada and England, and shall always have the latest possible Shorthorn news. The Book (Shorthorn) question will receive due attention, and the views of many of our leading breeders will be published through our columns. Valuable historical sketches of the various tribes to be found in the herds of America will appear from time to time, and a series of articles on the celebrated bulls of the past is now in course of preparation. While thus incurring great expense in order to make a Shorthorn department unequaled by any of our contemporaries, we feel sure that we shall command warm enough support to justify the outlay. At least, we shall strive to deserve it."

If the Hereford cattle were to succeed, there must be a way provided, so that they could break down the opposition to them, by letting it be known broadcast over the United States, their true merits; and also the facts as to manipulated shows, and have an advocate devoted to their interests. Not because they were Herefords, but because they were the most economical feeders and best beef cattle that could be obtained, either on grass or in the stall.

To meet this want, we commenced the publication of "The Breeders' Journal," the first number being issued March, 1880.

Our opening keynote was as follows: "The Breeders' Journal" will be published in the interest of the meat production of America, the world as a market. Economy of production and value of product being the test of merit; and will rest its claims to public favor on the merit of these issues, and the ability with which they shall be maintained."

We claim that we took much broader ground than did our competitors in this line.

The "Breeders' Journal" was a success as long as we published it, being until the close of 1887; and was of the greatest assistance in placing the Herefords before the public.

Mr. Sanders was editor of the "National Live Stock Journal" before he established the "Breeders' Gazette," and his sympathies and the sympathies of his son who succeeded him, have always been with the Shorthorn breeders. Flinging the standard which we quoted defiantly to the breeze, was to our mind manly, but after flaunting the Shorthorn flag in this emphatic way, there was little excuse for the "Gazette's" contemptuous flings at the "Breeders' Journal" as being a "trade circular."

If the "Breeders' Journal" was a Hereford trade paper, the "Breeders' Gazette" jealously guarded its reputation as being the Shorthorn breeders' own and only real, authorized, revised version of the Shorthorn trade journal.

An editor, like any other man, may—probably must—have his preferences, and for these we should not criticise them. It is only when they deny their partiality, and forgetfully cry aloud their fairness and their equal treatment of all breeds that they merit criticism, and then we have not spared them.

When an equal amount of money, time and labor is spent by an agricultural publication to gather, edit and publish data for the Hereford breed, as it expends or has spent on the Shorthorn breed, then we shall believe in the impartiality of their editors, and not before. And in this connection we can say to the impartial student that a study of Hereford history and achievements, will be found a subject surpassing in interest and antiquity that of any other line of investigation open to the explorer of live stock lore.

T. LEWIS,
Of "The Woodhouse,"
Herefordshire.

CHAPTER XXIII.

The Shorthorn Men Endeavor to Get Down Easy

Those who remember the grade Hereford steer "Conqueror" will say that he was a most wonderful animal. We propose to give some of the matter that was written about him and our other show steers, as we think that the Hereford steers exhibited in the years 1879, 1880, 1881, and later were largely conducive to the pronounced success of the Hereford breed of cattle.

The show of Herefords made at the Fat Stock Show, Chicago (1880), was made by C. M. Culbertson, G. S. Burleigh and myself. Mr. C. had one pair of steers that he bought, but the rest were his own breeding. We made our show of our own breeding. The cow "Maid of Orleans" was only three years old and weighed 1,750 pounds. Our six two-year-old steers, two thoroughbreds and four grades, were short two-year-olds and weighed from 1,600 to 1,845 pounds each, off the cars. These animals were all the get of the old imported bull "Success." The "Maid of Orleans" was put up to feed in May, and made, up to the time of the show, 500 pounds. The two heaviest two-year-olds made in eleven months, 800 and 850 pounds.

Mr. Burleigh's were his own breeding and feeding. Mr. Gillette and Mr. Moninger had each entered a yearling for slaughter, but refused to kill. The steer "Conqueror" was dropped on or about the first of August, 1878, being at the time that he was exhibited in 1880, 27 months old, weighing 1,845 pounds. The entire lot was choice, one of which was slaughtered at the show and another went to Detroit, where he was slaughtered by Messrs. Smith & Co. The other four returned to Beecher to be held for another year.

The six steers and "Maid of Orleans" were sketched by Mr. Dewey, and were lithographed in colors with the old bull "Success," making a group of eight head, being the old bull and several of his get. (See color plate.)

The steer "Conqueror" was not fed until January, 1880, with more than ordinary keep, and at fifteen months old, did not weigh over 1,000 pounds. He had never been kept up, but had run in the yard and pasture until that time. He was taken from the cow and raised on the pail and fed corn and oats, ground, until beets would do to give, when he had some of these sliced with his meal, having for the first six months some oil cake.

The "Mark Lane Express" has said that a Hereford calf cannot be taken from the cow and raised on the pail, without detriment to the character of the bullock. Four of the above-mentioned six bullocks were raised on the pail and two on the cows. We recognize the "Mark Lane Express" as one of the ablest and fairest of the live stock journals, but think some of their claims are based on customs, rather than on facts. We should have been very glad if we could have placed "Conqueror" on the Smithfield show ground, not with the entire confidence that we should win there, but that we might have compared our best with the best of England.

Mr. Burleigh's yearlings showed early maturity and ripeness to perfection; one a thoroughbred, the other a grade. So with our grade, which took the sweepstakes for best yearling in the show and the "Farmer's Review" gold medal.

This exhibit of Herefords should have gone far to settle the question of early maturity, and one other, that is, light hind quarters, compared with fore quarters. Still another charge of the Shorthorn men was confuted, to-wit: that the Herefords were small. Of the three-year-olds that were slaughtered, the Herefords were over three hundred pounds heavier than the three-year-old Shorthorn; and of the two-year-olds, the Hereford was one hundred pounds heavier than the Shorthorn, and in both classes the Herefords had the advantage of greater live weight.

Another thing the Shorthorn exhibitors and reporters made a great ado about was the forcing process to which the Herefords had been subjected; they claimed that "It was not so much a contest for Herefords against Shorthorns, as it was skillful feeding and forcing, against the methods of the common farmer." We find this expression in the "National Live Stock Journal," and it was a common expression among the Shorthorn men during the show. On what base does this claim stand? Simply that the Herefords were the best cattle, and they took this way of letting themselves down easily. We are not disposed to find much fault with the "National Live Stock Journal" for using these terms. Their report was generally fair, as regards this show, but they might have gone further and said that Cobb & Phillips had a pair of yearling Shorthorns that they had fed on gruel and milk from birth, and still a Hereford grade of Miller's that had roughed it all his life until two months before the show, beat them on the sweepstakes. Let us see who these simple farmers were, who came to the show with their cattle, having used only the methods of the common farmer.

There was Mr. Gillette, a "common farmer," not much of a breeder, not skilled (only having had about forty years' experience, with only a small herd to choose from—of about 1,000 bullocks), one should not expect much of him (?) Then there was Wm. Sandusky of Catlin, Ill.; he was only a "common farmer," with "common farmer methods." Sure, he had been a Shorthorn breeder ever since he was old enough to do anything, breeding thoroughbreds, and an exhibitor of cattle, for we don't know how long—perhaps not more than twenty years. That is not much and he should not have learned much in that time (?). Then there is D. M. Moninger, of Albia, Ia., another "common farmer," with a "common farmer's methods." He claimed to have been breeding and exhibiting Shorthorns for a quarter of a century. But that isn't much; a man apparently couldn't learn much in that time; he exhibited a show of fat bullocks at the Iowa State Fair, and the beauty of the show the press extolled throughout the breadth of the land; he selected only one out of the entire Iowa show to bring to Chicago (?). There was John B. Sherman, General Manager of the Union Stock Yards, Chicago, another of these "common farmers," who fed according to the "methods of the common farmer," and nothing more. Scrub cattle only, they ought to have said, still, Mr. Sherman had been picking the choicest animals that he could find; had a palace of a stable—large, roomy boxes, every comfort and convenience, and employed an expert herdsman from Scotland to feed them; just "common methods"—that is all (?). We had forgotten: Mr. Sherman did buy three yearling steers from Brown Bros., Sangamon, Ill.—"common farmers," with "methods of the common farmer," whose cattle were numbered by hundreds and acres by thousands. Bah! This claim at this day is so utterly silly as to be unworthy of notice, except to show what lengths a subsidized organ could go in its excuses for the cause it advocated.

Then there was Wm. Scott. How ridiculous to term him a common farmer with common farmers' methods. He bred Shorthorns and exhibited, and sent a cow to be slaughtered, and there is every reason to suppose that he had graduated above any common farmer, but judging him by his cows and taking his organ at its word, he must be nothing more. But then there were J. H. Potts & Son; should nothing be expected from them from their almost unrivaled experience in feeding and showing, boasting of $40,000 in winnings with Shorthorns? Should we have allowed them more time to learn that by and by something might be expected from them? Messrs. Cobb & Phillips, Kankakee, Ill., were they just new beginners, without experience, used only to "common methods"? The veteran experience of Amos F. Moore, of Polo, Ill., we were to suppose must count for naught, as well as that of Mr. Graves, of Kentucky; for, taking the Shorthorn excuse-makers, we should not expect much from a man living in that state where they have only been using the improved breeds of cattle a little less than a century.

J. WILLIAMS,
Llansannor Court, Glamorganshire, Wales.

We might continue this sarcasm, and speak of Mr. Dun, of Ohio, Mr. Bassett, of Illinois, Mr. Higmon, of Illinois, Mr. Green, of Indiana, Mr. Weidman, of Illinois, Mr. Taylor, Mr. Winn, Messrs. Willard & Son, Mr. Bidwell, Mr. Ross, of Avon, Ill. Here is a list of breeders that may challenge the world for experience and, we presume, in knowledge. All, with two exceptions, breeders of thoroughbred Shorthorns for years. And of the two exceptions, J. B. Sherman had an experience in cattle matters second to no other, and a reputation worldwide. The other exception, Gen. Ross, a breeder of thoroughbred Devons and of large experience.

Now, no animals in the show had received more careful and intelligent feeding than the three yearling steers bred by Brown Bros., and fed by John B. Sherman. The steers had made good gains, but they failed to put on that heavy, ripe flesh that the Herefords had and there was no use accounting for the better

PHILIP TURNER,
Of "The Leen," Herefordshire.

quality of the Herefords in any other way than that they made better use of their feed, put on flesh of a better quality, and in the better parts than do the Shorthorns. And when the two breeds come together, and have fair judging, the Shorthorn must go to the wall, while the credit must go to the breed and not to the feeder. We do not mean to say that good care and good feeding does not count. It does count. We have had experienced and careful herdsmen, our cattle have been well handled. But is this not true of the Shorthorns also?

Mr. Culbertson won, we think, whenever he had an opportunity to show against the Shorthorns, and Mr. Burleigh, of Iowa, took the champion prize in Nebraska this same year for the best beast in the show, and in this Fat Stock Show, the Hereford honors were divided among the three Hereford exhibitors, Culbertson, Burleigh and Miller.

Failing to make an impression on the "common farmer" claim, the Shorthorn scribes broke out in another place.

To the Editor of the "Journal": I noticed in your last number an illustration of Mr. T. L. Miller's steer "Conqueror," a very grand bullock indeed; but you denominate him a grade Hereford, as if produced by crossing a Hereford bull upon the common scrub cow of the country—the same as that class of cattle produced which we call grade Shorthorns, by breeding our thoroughbred Shorthorn sires to the common scrub cows of the country. Is not the term grade Hereford a misnomer? And should he not be properly called a cross-bred, as I understand either his dam or grandam was a Shorthorn cow? This being true, why not call him a cross-bred Shorthorn-Hereford, as the term "grade Hereford" misleads; the great bulk of cattlemen everywhere understanding the term "grade" to imply the union of thoroughbred and scrub stocks, and not the union of two thoroughbred stocks, even of different tribes.

The steer himself, except in color, bore many strong Shorthorn characteristics, especially in flank and quarter—a comparison of these points with the pure bred Herefords on exhibition readily discovers these facts.

Again, his rapid gain per day over his pure-bred Hereford cousins, even the prize winners, as set forth in your very instructive tables in the last number of the "Journal," lends additional illustration and force to the fact, that from somewhere besides his Hereford blood he is indebted for his capacity to lay on flesh rapidly, and thus come up to the point of early maturity.

I for one, am perfectly willing that Mr. Miller should have all the glory possible—and he deserves a great deal for the *skillful feeding* of his stock exhibited—but I do seriously object to his borrowing any of our Shorthorn blood to build up his Herefords with, and then parading it through the press of the country as an entirely Hereford variety. "Render unto Cæsar the things that are Cæsar's."—T. C. Anderson.

We find the above in the "National Live Stock Journal" for January, 1881. We met the same thing in the Kansas City "Indicator" and "Farmers' Magazine," of Louisville, Ky. This statement was made by prominent Shorthorn breeders at the Fat Stock Show of 1880. The leading exhibitor at that show stated to quite a crowd that had gathered round him, that Mr. Miller had selected Shorthorn cows from which to breed these bullocks. This same exhibitor said that the cow "Maid of Orleans" had Shorthorn blood, and when asked on what authority he made this statement, his reply was that "somebody had said so."

A letter from the herdsman of Messrs. Lee

& Reynolds, of New Mexico, was published during the show week, stating that the Herefords were giving the best satisfaction. A few weeks previous to this, the "National Live Stock Journal" said that on a careful inquiry among a large number of cattlemen from the plains they found the Herefords were giving the best satisfaction; and still Shorthorn breeders reported and Shorthorn journals circulated that the Herefords were being rejected.

But to return to Mr. Anderson. His statement is made here as though there was no question as to the truth of what he is saying. Mr. Miller never purchased a Shorthorn cow and never bred a Hereford bull to a Shorthorn cow, and there was no cow on his place that had the appearance of a Shorthorn.

The dam of "Conqueror" was a grade cow, bought of Mr. E. Parsons, of Pittsfield, Ohio, seven years previous to this time. While we never knew or inquired how this cow was bred, her appearance would indicate some Devon blood.

We wrote to Mr. George W. Probert, of Pittsfield, who negotiated the purchase of the cow, to call on Mr. Parsons and ascertain, if he could, how the cow was bred. This letter was written on March 11, 1881; he replied on the 16th of March, as follows:

"I saw Mr. Parsons to-day. He says the cow with the white face and upturned horn (dam of 'Conqueror') was sired by 'Fairboy' (4574) 475, Mr. Parsons' old Hereford bull, bred by T. Aston, formerly of Elyria, Ohio. Her dam was sired by a Devon bull, said to be near full blood; her grandam was sired by a Hereford bull and out of a common cow of unknown breeding."

The ages and breeding of our other steers were attacked. Of the four grade two-year-old steers that we showed, "Putnam," "Rob Roy," "Bachelor" and "Conqueror," in 1880: "Putnam" was shown in 1879, weighing 1,152 pounds; "Conqueror" was estimated at that time to have weighed about 1,000 pounds. He had up to this time run in the yard, with fair, generous keep, but not crowded; and was, as our recollection serves us, a younger steer than "Putnam," and "Putnam" had been registered as July 12, 1878.

The steer "Kansas" was bought in Kansas, no account taken of his breeding, as he was not intended as a show steer. When we returned from England in October we found that Mr. Watson had put him up to feed. This was done the week before the State Fair, as he informed us. The only facts we had as to age was his mouth, and that showed him to be one year old; his two-year-old teeth came in December. While a bullock may show by his mouth an older age than his actual age, we believe that in no instance does the bullock carry a younger mouth than the actual age would warrant.

The following letter, copied from the "Correspondence" columns of the Kansas City "Commercial Indicator," was written in answer to other claims made by T. C. Anderson, from the breeder of the steer "Kansas," and states that he was calved in December, 1878, which agreed with the time his two-year-old teeth came:

Irving, Marshall Co., Kan.
Jan. 9, 1882.

Your correspondent, Mr. T. C. Anderson, still endeavors to impress upon the mind the idea that it is given him to expose fraud, and especially the imaginary fraud of T. L. Miller. Now, as I understand the matter, the charge against Mr. Miller is that he does not give the breeding and age of the steer "Kansas." In order that Mr. Anderson may be gratified and the public correctly informed, will say that we bred and raised the steer "Kansas," and sold him to Mr. T. L. Miller. We will therefore give his breeding and age. The sire of "Kansas" is a thoroughbred Hereford bull; his dam was a red and white spotted cow. We should judge from color, horns, build and weight, that she was a cross between the native and Texas; she was about six years old at this time. The steer (Kansas) was calved in December, 1878, and judging from the quality of the cow, we did not consider a calf from her desirable to keep for a grade bull, and we therefore castrated him at an early age. He thrived well with fine promise. The cow being farrow, I shipped her to Kansas City in the summer of 1880, with a car load of cattle, and she was sold as a cull for $1.75 per 100 lbs., and weighed 850 lbs. Judging from the quality of the dam and the outcome of the steer, it is fair to conclude that the merit he possesses was derived from his Hereford blood.

JOHN MORRIS,
Of Lulham, Herefordshire.

It appears to exercise the Shorthorn mind that a grade Hereford steer from Kansas, with only ordinary care up to a few months previous to the Fat Stock Show of 1880, should take sweepstakes over the pampered thoroughbreds and grade Shorthorns of the blue grass regions for the best steer one year old and under two. At this time "Kansas" was 696 days old and weighed 1,580 lbs., his average gain per day

WILLIAM TUDGE,
Adforton, Herefordshire.

being 2.27 lbs. (See "National Live Stock Journal," December, 1880.)

Trusting this will be sufficient explanation to set at rest this cry of fraud against the steer "Kansas," one of the five premium grade Hereford steers, we will leave the charge against the steer "Will" for other parties to explain why he did not make more than 724 days in age from November 10, 1879, to November 7, 1881. WALTER M. MORGAN & SON.

We were awarded and still have in our possession the gold medal offered by the "Farmer's Review," of Chicago, Ill., for the best yearling in the Fat Stock Show at Chicago, 1880. We won it with the Hereford yearling steer "Kansas." The following remarks were called forth from Mr. W. W. Corbett, editor of the "Farmer's Review," by that occasion: "Mr. Miller: I do not know how long the dream of becoming a farmer and stock breeder had haunted your mind before you entered upon the career; I do not know how long you had made a study of the characteristics of the different breeds of cattle, and of principles upon which successful breeding and feeding depend, but I do know, that it is not very many years ago that I saw you almost daily in the streets of Chicago, elbowing your way through the crowds with the same determined look upon your face that you wear to-day, but the silver that now tints your hair was then hidden away in your pocket, if you had it at all. You were then prosecuting an intricate business in the midst of hundreds of the shrewdest, most active and energetic men, who have amassed great fortunes, and done so much to build this wonderful city, and give to it its commercial reputation. You won a most commendable business reputation, and I believe a comfortable fortune. The next I knew of you, you had purchased a magnificent farm at Beecher (¶163) and were preparing to stock it. You did not know as much about Herefords then as you do to-day. But there was among your advertisers that stalwart old Hereford hero, our old friend Sotham, who, with pen and tongue, had been fighting an unsuccessful battle for many years against rival interests. Let us give the scarred old veteran the credit of directing your enthusiastic efforts into this channel. Then you took off your coat and went to work. Your business tact and shrewdness you took into a new field. You took all your pertinacity and combativeness. At the time of your first purchase, the white face and symmetrical body of the Hereford was a curiosity at a Western Fair. You boldly entered the arena and threw down the gauntlet to a noble race of animals, owned and backed by the money and the brains of as shrewd and able a body of men as grace any pursuit in any country. I mean the Shorthorn breeders of the United States. And now your favorite breed is known to all men, and your animals march proudly from every noted fair ground bedecked with the blue and the red.

"I know that your determination and combativeness have led you into many a wordy conflict. A great many people think that 'you do protest too much,' and I have no doubt your name will go down to Shorthorn posterity as the Great American Protester. But after all, I do not believe that there is a competitor of yours here to-night but will willingly pay tribute to the grand contest you have made and the victories you have won.

"You, with your able coadjutors, Powell, Morgan and Culbertson and many others who have more recently enlisted in the service, can retire from it in your later days with the proud conviction that your efforts will result in adding untold millions to the future stock interests of the farmers of this country, by introducing prominently the pure blood of the Herefords, to mingle with the hardy races of the plains. Indirectly you are accomplishing another thing. You are stimulating the breeders of other races of cattle to strain every nerve to add still more perfect points to their animals, as witness the noble specimens that have been so unsuccessful against you at this most important exhibition.

"Mr. Miller, as the representative of the progressive agricultural press, which has so heartily joined hands with the able, earnest and devoted members of the state Board of Agriculture in furthering the live stock interests of the country, I have the pleasure, through them, to present to you the prize gold medal of the 'Farmer's Review,' so justly won by that meritorious young animal that represents in his own body the success of a long line of careful breeding in the old world, your own skill as a feeder, and the adaptability of our climate and productions to give to every domestic animal the touch of perfection.

"It is the hope of the donors that this beautiful medal will add something to the satisfaction that must result from your competition at this show, and that those who come after you will feel a blush of pride as they apply to you the motto herein inscribed: 'The cultured mind guides the skillful hand to success.'"

We give the following extract from the "American Stockman's" report of the 1880 Fat Stock Show held at Chicago, to show how the Herefords there exhibited appeared to the general public:

Now for a fresh family, the Herefords. Must we assign to them the premier place in the hall? Yes, we must; there is no getting away from it; year by year they creep in, and year by year they become more deserving of the premiership. Let us begin with "Alex," Mr. T. L. Miller's three-year-old (and Mr. T. L. Miller is the father of the Herefords in America). Look at that charming countenance, and that great, full, prominent eye, and there you will observe what so delights the judge of the tip-top grazier. The little fellow when he moves actually groans under the load of prime marbled beef. What a bosom. How he twists and straddles those little timbers of his to carry about that great carcass of 1,920 pounds of porterhouse steaks. Gentlemen, there is no offal—you cannot find it, although you look for it. Examine those crops and chine. Aren't they wonderful? Touch that back rib and loin. When had you your hand on the like? Not for a long time, as Saturday's judgment on the carcass will prove. Perhaps he is a little narrow on the quarters, but the rump steaks are there in abundance, and ripe at that; his twist is deep, and he carries his beef to his hamstrings, like a Berkshire pig. Well done, "Alex."

Mr. Miller is surely far seeing, for he called his last two-year-old steer "Conqueror." Come to the front, "Conqueror," although you are yet in the meridian of your fame, and let me describe you to those who have only heard of you but not seen you. Look what an attractive gait he has, how beautifully he walks, how he improves in everyone's estimation the longer he is looked at; everyone can see that his very victory is a foregone conclusion, that he is the best animal in the hall, and that his meat is worth more than the champion Shorthorn winner by at least two cents per pound. "Conqueror" is thick and deep down to the ground. He is smooth in the extreme; he is lengthy and yet compact; he is fleshy and yet fat, and beef from head to heel. What more is wanted? The remainder of Mr. Miller's herd—numbering nine steers and one heifer in all, were an admirable display, one rarely to be met with; as several good judges remarked, the best lot of finished cattle ever seen, shown by one man in America; all bred and fed at Highland Stock Farm, property of T. L. Miller, Beecher, Ill. (¶ 188)

T. J. CARWARDINE.
Late of Stocktonbury, Herefordshire.

We will not indulge in any rude remarks about the judges in the defeat of Mr. T. L. Miller's heifer, the "Matchless Maid of Orleans." No, it is not becoming; they were old men, and left home without their spectacles. It is invariably the case, the cross-bred sections form one of the best features of the show and you will see by the next exhibition that several of the "Beechers" are bound to take a prominent position in 1881. Perhaps they, instead of the remarkable Kentucky steers, will get away with the much coveted "National Live Stock Journal" silver cup. An eager crowd viewed the movements of the judges with much interest in the award for the championship this year; it will be viewed with no less anxiety next year. The battle is now fairly begun between the Herefords and Shorthorns; each have their favorites, and justly so. Stand by your colors until you see who is the winner, but take care the black Polled Angus do not take part in the battle of beef and delay the decision. I am impartial. There is plenty of room in America for all breeds.

In grade Herefords, Mr. C. M. Culbertson produced a grand animal, ripe and well finished, champion of his class as the best three-year-old of any breed, beating Mr. Miller's three-year-old "Alex." So decided the judges on Thursday the 18th inst., but little "Alex" told them he was riper, requested to be dressed and hung up by the heels and the public would see that the Highland Stock Farm steer was a better and more finished carcass than the Hereford Park one. Alas, it turned out true.

Mr. Editor, my time is limited, otherwise from memory I should like to enter into details on the merits of many a worthy animal, such as General Ross' lovely Devons, and John B. Sherman's mammoth steers that appeared at your Chicago Fat Stock Exhibition. But I cannot conclude without taking particular notice of Mr. G. S. Burleigh's three one-year-old steers, one thoroughbred and two grades, raised and fed in Iowa by their breeder. They are perfectly marvelous as to maturity yet full of robustness and a good future. Mr. Burleigh is about the last man I want to tackle in one-year-old competition, and that is a good deal to confess.

LORD WILTON (4740) 4057
Bred by William Tudge; sold at auction for 3,800 guineas, equal to $20,000 (1884). (From a painting by Gauci.)

CHAPTER XXIV.

THE "BREEDERS' JOURNAL" ON THE CHICAGO FAT STOCK SHOW OF 1880

We find in the "Breeders' Journal" the following editorial:

(¶ 189) The Fat Stock Show (1880), held in Chicago, was in many respects one of the best exhibits of stock ever held in this country. This show is held under the management of the Illinois State Board of Agriculture. The Board is composed of nineteen members, one from each of the Congressional Districts of the state, and, with few exceptions, they are Shorthorn breeders.

The time is not far in the past when the Shorthorn breeders had no one to dispute the claims of their breed. The Society's plans were made, so far as the cattle department was concerned, with the express reference to the adjustment of the claims among the different families of Shorthorns and of their different owners. It is yet difficult for them to get away from their preconceived notions and preferences. Not only is it difficult, but with many there is a determined purpose not to depart from them, probably from the fact that they conceive their breed to be the best, and that they must stand guard over them and protect them from the claims that the Herefords are making upon the position that they have held. That this is so is not surprising. In the past there was none to dispute their claims, and many think now that those who dispute them should be treated as heretics were in the olden times.

It has been the practice in times past at the winter meetings to appoint judges for the different classes. These judges seldom acted, and their places were filled from among the visitors on the grounds, sometimes a good selection, but often the contrary.

At the inauguration of the Fat Stock Show the selection has been made by each member of the Board nominating one or more of the butchers in his district, and from these the Board select their judges. These butchers have a country experience; as a rule they do not slaughter or cut first-class beef or mutton. It is such as is quoted as "butcher's stock" in the Chicago markets. This stock ranges in price from two cents to three cents at the present time for live weight, while good shipping steers will range from $3.50 to $5 per hundred, and choice shipping steers from $5.25 to $6.50 per hundred. This latter class comes near the quality that is shown at the Fat Stock Show, and is always in demand at the long price. It is not such a quality as their trade demands. It is not such as they are acquainted with or want. Preferring the "butcher's stock," they are not prepared to give their judgment as to the best shipping grades. These judges are selected largely from districts in which Shorthorns have held for years the dominant position, and the parties selecting them are Shorthorn breeders; they are likely to be friends of the members selecting them; the members' friends are the exhibitors of Shorthorns, and without charging dishonesty, it is fair to presume that their associations lead them to give the preference to Shorthorns. This, then, is the feature of the present system of judging. The judges are selected by Shorthorn breeders from the districts where the Shorthorns are dominant. They butcher only Shorthorns and their crosses, and their experience is in butchering what is technically termed "butcher's stock." They are not acquainted with shipping steers and their value. Two members of the Board reside in Chicago, but for some reason there were no Chicago butchers selected. Among these Chicago butchers are those who have a knowledge of the wants of our best markets, and those of England as well, and are therefore better able to judge as to the value of a first-class animal. These men have slaughtered and cut animals from every breed, while the judges selected have never cut a Hereford.

There is one feature, however, in the experience of the city butchers of large practice; they know the comparative value of other breeds with the Shorthorns.

It may seem a singular fact to outsiders, per-

THE RODD, PRESTEIGNE, RADNORSHIRE.
Home of the Rogers, now occupied by Aaron Rogers.

haps it may seem a singular fact to some of the exhibitors, perhaps it may seem strange to some of the members of the Board that in a city of 500,000 inhabitants no butcher could be found that was suitable for a judge at this Fat Stock Show. Or, if none of these practiced butchers here were suitable, that there might not have been found at the Union Stock Yards, among some of the men who are buying and shipping thousands upon thousands of these higher quality of bullocks, one or two men that might have been suitable to act as judges on these better qualities of cattle. Especially does this seem singular, when one of the members (J. P. Reynolds) is a resident of Chicago, has a large experience, and is eminently intelligent on stock matters; and another (H. D. Emery) who has been a resident of Chicago for more than a quarter of a century, and the editor of the leading agricultural journal of the West during all this time. These gentlemen have been members of the State Board for many years and are probably among the oldest members. When these facts are stated, there will be many who will be surprised that there should not have been found two or three competent men in the great metropolis of the state, the greatest live stock market in the world, and that none but country butchers could be found that would meet the requirements of the Board.

We have endeavored in this notice to be moderate in the expression of our views. This much in review of the judges of the Fat Stock Show.

We are now prepared to notice the exhibitors and their entries:

J. D. Gillette, Elkhart, Ind. .	Exhibited	47	head.
T. L. Miller, Beecher, Ill. . .	"	10	"
C. M. Culbertson, Chicago.	"	2	"
Wm. Sandusky, Catlin, Ill.	"	3	"
D. M. Moninger, Albion, Ia.	"	1	"
J. B. Sherman, Chicago, Ill.	"	6	"
Wm. Scott, Wyoming, Ill. .	"	2	"
Hiram A. Bassett, Jefferson, Ill	"	1	"
J. H. Potts & Son, Jacksonville, Ill.	"	1	"
J. S. Higmon, Rochester, Ill.	"	2	"
John Weedman, Farmer City, Ill.	"	1	"
Cobb & Phillips, Kankakee, Ill	"	2	"
A. W. Taylor, Lake Forest, Ill	"	1	"

J. G. Willard & Son, Harristown, Ill.	Exhibited	9 head.
Thos. Bidwell, Gurnee, Ill.	"	2 "
G. S. Burleigh, Mechanicsville, Ia.	"	3 "
L. F. Ross, Avon, Ill.	"	5 "
J. H. Graves, Kentucky	"	2 "
Total		100 head

The quality was from fair to choice. The interest centered largely between the Herefords and the Shorthorns, and while the Herefords were in the minority, it may be fairly claimed that in quality and ripeness they excelled the Shorthorns. Of this we shall speak more fully before closing our report. The stock to be exhibited were mostly in their stalls by the 13th. The exhibition opened on the 15th. Monday morning the Shorthorn men reported that Swan Bros., of Cheyenne, Wyo., were sick of Herefords and were going to dispose of them and use Shorthorns. This was probably done to counteract the effect of the growing popularity of the Herefords. C. M. Culbertson telegraphed Messrs. Swan to know whether this Shorthorn report was true. The following is the telegraphic correspondence:

Chicago, Nov. 15, 1880.

To A. H. Swan, or Swan Bros., Cheyenne, Wyo.:

Shorthorn men are circulating a report that you have had enough of Herefords and will hereafter buy Shorthorns. I think this is done to influence public opinion at fat cattle shows. A full and prompt answer is important. Answer at my expense. C. M. Culbertson.

Cheyenne, Wyo., Nov. 15.

To C. M. Culbertson, Room 3, Board of Trade Building:

We are using 150 Hereford bulls. Intend to increase to 300. We are also breeding Herefords in Iowa, having a thoroughbred herd there. We are not disgusted with Herefords; on the contrary, will increase our number fast as possible. Swan Bros.

The first two days were given mainly to showing in classes, which called forth but little excitement or comment. Some of the Shorthorn exhibitors had some little question as to where the ribbons should be put.

On Wednesday the show opened with grades and crosses, and the interest centered about this ring.

First premium for three-year-old steer was awarded to J. H. Graves, of Kentucky, for grade Shorthorn.

Second premium to C. M. Culbertson, for grade Hereford.

First premium for two-year-old to A. F. Moore, for grade Shorthorn.

Second to J. H. Potts & Son, for grade Shorthorn.

First to D. M. Moninger, of Iowa, for yearling, for grade Shorthorn.

Second to T. L. Miller, for grade Hereford.

The judging in this class was clearly unjust, especially as to the two-year-olds. T. L. Miller's two-year-olds were very nearly perfect. The lightest one was dressed, an account of which will be given hereafter.

The judges in this class, after passing the grade Herefords, were discharged, as we understand, for incompetency. There is a semiofficial denial of this, but it is immaterial as to what the reasons were. It would not have been done but for a reason, and the one stated above is the most charitable.

A protest was made by Mr. Miller, and argued before the Board. There was no difference of opinion as to the injustice of the award; this, the Board admitted, but they could not

AARON ROGERS,
The Rodd, Radnorshire.

see a way to remedy the error, whereupon the protest was withdrawn and the Board relieved of further responsibility.

To test this award, one steer of J. H. Potts was bought by Mr. Miller for the purpose of dressing, and thus test the judgment of the judges. The privilege of dressing the steer at the Exposition was asked of the President, but he refused to permit it. The steer was then sold to W. Smith & Co., of Detroit, Mich., where the dressed and live weight will be given.

Following this exhibit came the sweepstakes for the best three-year-old steer, the best two-year-old steer, the best yearling steer, and the best cow of any breed. C. M. Culbertson took first premium on three-year-old; T. L. Miller, first premium on two-year-old; T. L. Miller, first premium on one-year-old. Thus it will be seen that the Herefords in these classes not only competed with the grades and crosses of both Hereford and Shorthorns, but also the thoroughbreds of both breeds were added,

ANXIETY (5188) 2238.
Bred by T. J. Carwardine. (From a painting by Dewey.) The man in the picture is Geo. F. Morgan, who selected Anxiety for importation to America.

making the Shorthorn show much stronger than before. The result was a decisive victory for the Herefords, the judges deciding in their favor.

In the cow class the award was to the Shorthorn cow, over "Maid of Orleans," a Hereford cow, belonging to T. L. Miller. This was pronounced the most unjust award of the week. But Mr. M. has arranged this class so that the judgment of the committee may be tested. Both animals have gone to Detroit, where they will be dressed by W. Smith & Co., and an account kept of the weights. Thus, in two-year-old steer class and cow class, when against Herefords, the test on the block will be taken. Many of the Shorthorn breeders charged Mr. M. of protesting and finding fault when the awards were against him; but in this case he will be able to show whether he was right in finding fault or not. The next show was for the champion prize for best bullock or cow in the show. It was admitted that the contest was between the Kentucky steer, a grade Shorthorn this year, and four years old or over, and Mr. Miller's "Conqueror," two years old.

Messrs. Culbertson and Miller protested against this steer showing, as he had been shown under false age and pedigree last year, but the Board decided that he should be shown. The award was made to the Kentucky steer. As the protest was made, the Board ordered another award in case the protest should be sustained, and the next would have been T. L. Miller's "Conqueror." Messrs. C. and M., recognizing that it would be a good deal of work in proving his age and breeding, withdrew their protest, but it did not rest here. There was a good minority, if not a majority of the exhibitors, that believed the Kentucky steer, after his record of last year, should not have been admitted. He was not less than five years old. Dr. Paaren, after examining his mouth, pronounced him nearer six years old than five. He weighed 2,465 pounds, while the two-year-old Hereford weighed 1,845 pounds, and had made 2.27 pounds per day from birth and took the silver cup for the largest gain per day of any two-year-old in the show.

We give in another place an account of this Kentucky steer, as shown last year. The slaughtering test was the only remaining test to be made in the cattle department. We give a table showing the particulars of this, but we wish to call attention to the following facts:

That of the three-year-old steers that were slaughtered, there was one grade Hereford, one thoroughbred Hereford, one grade Shorthorn. The grade Hereford weighed at the time of slaughtering 1,812½ pounds; the thoroughbred Hereford 1,850 pounds, and the grade Shorthorn 1,512½, the Hereford steers weighing 300 pounds more than the Shorthorn.

It will be noticed that there was less difference between the fore and hind quarter in the Hereford than the Shorthorn, and the heads of the Herefords were lighter than the Shorthorns.

The two-year-old Hereford was 146 pounds heavier and had shrunk from the time of leaving home, 45½ pounds, while the Shorthorn had shrunk 99 pounds.

The yearling Hereford steer had a clear field. Mr. Gillette and Mr. Moninger had each entered yearling Shorthorns for slaughter, but both declined to kill and could not be induced to bring the steers out.

The awards were to T. L. Miller, for best three-year-old; J. D. Gillette for best two-year-old, and G. S. Burleigh for best one-year-old.

DRESSED MEAT OF THE THREE-YEAR-OLDS.

	LBS.
C. M. Culbertson's steer weighed	1,256
T. L. Miller's steer weighed	1,250½
J. D. Gillette's steer weighed	1,037½

DRESSED MEAT OF THE TWO-YEAR-OLDS.

	LBS.
J. D. Gillette's steer weighed..........	947½
T. L. Miller's steer weighed..........	1,050½
G. S. Burleigh's steer weighed (one year old)	816½

The Shorthorn men will get but little comfort out of this show. But they think they will next year, for they have agreed to pick up the best steers they can find in the United States, and put them in the hands of Messrs. J. H. Potts & Son, for feeding, and to beat the Herefords at the next Fat Stock Show. We have endeavored to make a report of the Cattle Show as fairly as possible. We are, however, making the report from a Hereford standpoint, and invite the closest criticism as to the facts. We shall expect to renew this report after we get returns from Messrs. Smith & Co., of Detroit, of the dressing of the Potts steer, the Scott cow, and the "Maid of Orleans." We shall also make investigation in reference to the Graves steer, of Kentucky. These three awards reversed, the Shorthorns would have been entitled to just what the Herefords did not show for.

It was admitted that as fine a lot of cattle was never shown on any show ground by one exhibitor as the six two-year-olds shown by T. L. Miller.

We present the tabulated statement of the slaughter test:

THE SLAUGHTER TEST OF THE CHICAGO FAT STOCK SHOW, 1880.

NAME OF ANIMAL.	Weight at home......	Living weight at show	Dressing weight......	Per cent net to gross.	Left forequarter......	Right forequarter....	Left hindquarter......	Right hindquarter....	Head	Hide..................	Feet..................	Entrails...............	Paunch................	Liver, heart, tongue, pluck, beef, cheeks.	Total offal............	Blood and shrinkage.
"Mossy Coat," Hereford.	1860	1812½	1256	69.20	331	320	300	305	32	104	19	108	161½	61½	481	75
"Chub," Shorthorn...	1600	1512½	1037½	68.59	285	285	230	237½	27½	87	16½	79	146½	47½	404	71½
"Alex," Hereford	1910	1850	1250½	67.59	327½	318	295	310	32½	111	18½	120	184	64	580	69½
"Putnam," Hereford.	1652	1607½	1050	65.31	272½	272½	256½	249	28	90½	16	112	185	55	486	65
"Blank," Shorthorn..	1560	1461	974½	66.70	260	253½	227½	233½	28	90	20	74½	174½	47	434	53
"Monroe," Hereford.	1265	1217	816½	67.09	206	203½	203½	203½	23	84	15	70½	114	46	352½	48
"Grand Chunk," Sh. H.	....	1435	917	63.90	.....	.....	.....	.....	21½	70	14	70	256	61	492½	26

To sum up the Waterloo defeat that we gave the Shorthorns at this show—the Herefords in sweepstakes took first for three-year-old bullock; first for two-year-old bullock, and first for one-year-old bullock.

The awards which we earned, and did not receive, were in grades and crosses; first on three-year-old bullock; first and second on two-year-old bullock, and sweepstakes for best cow in the show.

The slaughtering of "Mossy Coat," three-year-old grade Hereford, belonging to C. M. Culbertson, and the sweepstakes to same (bullock) establishes his claims. The sweepstakes to T. L. Miller's two-year-old "Conqueror," contingent best animal in the show, establishes his claim in grades and crosses. Sweepstakes to T. L. Miller's yearling establishes his claim.

One of the winning grade Shorthorns was dressed at Detroit, with T. L. Miller's two-year-old, and the Shorthorn cow taking the sweepstakes was slaughtered with T. L. Miller's Hereford cow, thus bringing these awards to the test of the block. The honors thus won gave the Herefords a prestige for 1881 that they never had before. The report of the dressing of the Hereford cow "Maid of Orleans," will be found in the following letter from Wm. Smith & Son, of Detroit, who were one of the oldest and most reliable firms of butchers in the country. Their exhibit of Christmas beef was the largest and best made in the United States.

As will be seen by the figures given in this letter, the Hereford dressed the wonderful amount of meat to live carcass of 70.48 per cent. She was butchered in the market and hung up entire. So it was impossible to weigh the carcass until it was cut down and quartered.

Detroit, Mich., Jan. 29, 1881.

Dear Sir: We cut the heifer down yesterday, after hanging 36 days. She weighed as follows:

One hind-quarter	273	pounds.
One hind-quarter	276	"
One fore-quarter	285	"
One fore-quarter	289	"
	1,123	pounds.

All the butchers in our market agree with

us that the carcass must have shrunk at least 40 pounds. This would make 1,163 pounds, the most wonderful exhibit we have ever seen. She weighed alive the morning she was killed, 1,650 pounds. She cuts up well, but not so well as the steers, she being very much fatter than they. Her lean meat is the tenderest and most juicy we ever saw.

After a careful examination into the merits of the different animals slaughtered by us for Christmas, we consider the Herefords far ahead of any of the other breeds for quality and profit to the butcher and consumer.

We are very much elated with the success of our show this year, and our intentions are to increase it next year. We will try to buy all your fat cattle next fall. Please let us know when you are ready to sell.

Very truly yours,

WM. SMITH & SON.

A. P. TURNER,
Of "The Leen," Herefordshire.

CHAPTER XXV.

A Conspiracy That Failed

We give this as it appeared in the "Breeders' Journal," only striking out what we consider non-essential at this date:

At the Chicago Fat Stock Show of 1880, the stock was to be gathered by the 11th of November, to be measured and weighed, that the catalogues might be prepared by Monday, the regular opening day of the show. This four days of preparation was enough to show the Shorthorn conspirators that something must be done, or the Herefords were to sweep the board; and on Monday morning rumors were rife that the grade Herefords were from Shorthorn cows; that the grade and thoroughbred Herefords had been subject to a forcing process, while the Shorthorns had come to the show from the hands of common farmers and common farm methods.

Tracing these reports, those who were circulating them said, "they heard so," and this is all that could be made; none were ready to stand father to the report. Another report that the plainsmen were sick of Herefords, and that the Swan Bros., of Cheyenne, were sick of them and were to get rid of their Herefords and replace them with Shorthorns. This report was met early Monday morning following the other reports. The circulators had no other authority, than that they had heard so. A telegram from Swan Bros. branded that a lie. A letter from Lee & Reynolds' ranch showed that the Herefords were doing better than any other cattle they had.

All the usual appliances were in use by these conspirators during the week, that they know so well how to practice.

Some time after this show one of the conspirators (T. C. Anderson, of Kentucky) comes out into the light and he brings two others out—one of these, a Shorthorn breeder and horseman of Wisconsin, another, a banker of Kentucky. We will leave out the banker and give some attention to Anderson and the horseman, and their victim, Mr. Wm. Watson.

Mr. Anderson starts out with the assertion that the dam or grandam of Mr. Miller's bullocks were Shorthorn cows. This I pronounced false, and challenged him to prove. He says, "It is sufficient to say his authority is based upon an assertion made and repeated to Robt. B. Ogilvie, of Madison, Wis., James M. Bigstaff, of Mt. Sterling, Ky., and to myself, by one whom we believe to know the facts." Another time he says "The statement was made by one who was often at my place." I have been desirous of having this man define his position and give his authority. I have had letters from friends advising that Anderson had given the name of his informer, and that it is Wm. Watson, until lately, my herdsman. In the process of dissection, I will give my first attention to Mr. Watson.

Mr. Wm. Watson is the son of Mr. Hugh Watson, a prominent breeder of Angus cattle in Scotland, a man who stood high as a breeder and as a gentleman. His son, William, inherited these qualities. He was placed at the best schools and with the best breeders of England and Scotland to fit him as a breeder. When educated he was placed on a farm and in a position where he could use the inherited and acquired advantages, and was achieving a substantial success. He married and had two children.

From Scotland he went to Australia, assuming large responsibilities. Here he remained a few years and we next find him in Oregon, in the employ of a Mr. Reid, a wealthy citizen of that country, with the management of a large stock of cattle and sheep. From here he went to California, where the American Mrs. Watson met him. Soon after this, we find him in Kansas, in charge of a large stock of Shorthorns and Berkshires.

These positions were important ones and Mr. Watson was entirely competent to fill them. There was only one drawback—the curse of liquor had followed him through all his wanderings, through all his misfortunes, wrecking the hopes of his father, his family and his business. These, I understand, only in outline. My first knowledge of Mr. Watson was the receipt of the following letter:

T. L. Miller.—Dear Sir: I always read with interest your articles regarding the Herefords, a breed I had a great deal to do with in Australia. They would live and grow fat where Shorthorns would starve, and at times we had journeys of 2,000 miles to drive to Melbourne market. The percentage of death among Shorthorns used to be enormous on a long journey, while the Herefords used to go right through, almost as fresh as when they started.

B. ROGERS, "THE GROVE," HEREFORDSHIRE.
Breeder of "The Grove 3d," sold for $7,000.

I take the "North British Agriculturist." I clip from the number of June 12th an account of the Bath and West of England Show, where you will see the Herefords came off gloriously victorious, winning the challenge cups over the Shorthorns in bulls and cows. They have beat the crack Shorthorns of the day.

WM. WATSON.

After this he met me at the Fat Stock Show in 1879, and made application for the position of herdsman. I engaged him. Soon reports came to me of his habits. I brought these reports to him; he denied them. I said to him, "Whatever may have been your past, can I depend upon you for the future?" He assured me that I could. I tried him. His old habits proved too strong for him. I talked to him and he promised reformation. One who has a desire, might feel proud to lift such a man from the vice of intemperance to manhood.

On my second visit to England, taking my wife with me, I put Mrs. Watson into my house as housekeeper, and him to take the waiting on the table, as carver, etc., hoping thus to keep a restraint upon him, coming three times a day to the table, but it lasted hardly ten days, before he got two of my most trusted young men into a quarrel in a saloon, and this was the last of restraint.

On my return from England in October, I said to Mr. Watson that hereafter I must run my business without whiskey, and he assented. Our supervisors had refused licenses, and our saloons were closed. I earnestly hoped by this help to keep him from whiskey; but he would contribute, and have the hands contribute, and buy by the gallon, and have it on the place; while assenting to my plans, and professing to work with me, he was really against me. In this unfortunate condition the Shorthorn conspirators found him a willing tool for their work.

In the fall of 1878 one of the conspirators was an exhibitor of Shorthorns at Minneapolis. Col. W. S. King had the management of this show and had taken special pains to get competent and reputable judges; among them were Geo. Murray, of Racine, Wis., and also another man in whom Col. King had great confidence, who was a cattle buyer of Minneapolis, and another, a Shorthorn breeder of Iowa. The committee were entirely unknown to me, but their manner of judging in classes showed an intelligent knowledge of their business. I was willing to accept them for the sweepstakes judges. The Shorthorn conspirators objected, and Mr. Geo. Murray, one of the most reputable of Shorthorn breeders in the country was ruled off. A Mr. ———, then of Milwaukee, later of Minneapolis, was named to fill his place. When the question was put to him whether he had any interest in Shorthorns, he replied "yes," he was a Shorthorn breeder. Can you judge impartially as between Shorthorns and Herefords? No, did not think he could. He was excused. Another man who was not interested in Shorthorns was named; he could judge fairly as between Shorthorns and Herefords; had no interest or prejudice that would interfere with impartial judging. He was a horseman, and retained. The judges gave the first to the Shorthorns, the second to my Herefords. The first prize herd had no right to the premium, whether as compared with my herd or other Shorthorn herds.

At the Shorthorn conspirator's special request, a Young Herd premium was offered, and this special premium he had made special arrangements to take off; but it was too glaring and he failed to do so, and my young herd carried off the honors. The man who was tendered as a judge in the place of Mr. Murray, was a personal friend of the conspirator before and after judging.

The Shorthorn exhibitors claimed that my winnings were due to greater skill in fitting, and if they could secure Mr. Morgan to the Shorthorn interest, Miller and the Herefords would be nowhere, and the Shorthorns could regain their former position. They placed before Mr. Morgan the beauties of Bow Park; the respectable position that could be his at that renowned institution. But Mr. Morgan could not be bought or seduced.

But this partial failure at Minneapolis did not discourage the conspirators; they arose to the surface again in 1880, at the Fat Stock Show at Chicago, and took advantage of the misfortune of my unfortunate herdsman. To make the fullest use of their tool they must attempt to prove their assertions as to my Herefords by him. "Miller and his Herefords," they said, "owe their success over the Shorthorns to the skill with which Watson has fed them," and now their desire was to get Watson away, and have Miller where Delilah left Samson. They pictured to him the beauties of the blue grass pastures of Kentucky; the high standing of the Shorthorn breeders in Kentucky—what a respectable standing he could have among them and in that society. "Break with Miller and the Herefords and your future is secure," and unfortunately he sold himself for less than thirty pieces of silver, and no doubt regretfully wished to hang himself many times before he was done with them.

I have charged that the Shorthorn men gained the position they hold by other means than those of merit. I present the course of the conspirators at Minneapolis in 1878, as an evidence.

I present the facts of the American Fat Stock Show at Chicago in 1880, and these jobs must attach to the exhibitors of Shorthorns at those shows. These were but illustrations of the plan in which they picked up the best bullocks and show cattle in the country and put them in the hands of the best feeders they could find, and by manipulation of the judges come prepared to beat the Herefords.

Judas took his pay in hand; my unfortunate assistant, in a time of weakness, took his in promises, and as soon as through with him they spurned him, as did the purchasers of Judas. Benedict Arnold turned traitor for revenge, and lived, cursed by those he betrayed and those who bought him. Happily, "Uncle Willie" (¶ 195), as he is now affectionately called by his many friends in all breeds, saw the error of his way, his inherent excellent qualities again predominated, and returning to his "first love," he attached himself again to his father's breed—his first and true love, the Aberdeen-Angus cattle, and has achieved a lasting reputation and position, worthy of his splendid abilities, that outshine all error, and that make even this forced reference to earlier follies repugnant to the writer. All's well, however, that ends well.

It has never been our aim to stand on the defensive. Like Mr. Sotham, we ever proposed to make this Hereford controversy, to the best

MRS. SARAH EDWARDS,
Of Wintercott, Herefordshire.

of our ability, an aggressive one. The Shorthorn advocates for a time adopted the policy of letting us alone, hoping that by the let-alone policy, the Hereford movement would die of itself. But in pushing their claims, they would say to stock journals, "If you permit this Hereford question to be discussed you must look to them for support. We won't support any journal that is not wholly in the interest of our breed." Their patronage was large, and it had the influence they desired.

But the time came when they could not ignore the Hereford claims. They must meet the issue, and in their usual manner they did. Mr. Anderson, of Kentucky, opened the ball. On Nov. 1, 1880, he says: "The effusions of

T. EDWARDS,
Wintercott, Herefordshire.

T. L. Miller (the Hereford advocate) appearing from time to time in the agricultural press of the country, have puzzled me not a little to discover why a gentleman of his sense and evident research, casting aside as if for naught the experience of the British farmer, and of the older states of our nation, could prefer the Herefords (with their heavy necks, heads, forequarters and light hind-quarters), to the Shorthorn, with his well nigh universally admitted superiority for any purpose whatever, for which the cattle kind is intended."

The above is Mr. Anderson's opening of the fight; and we wish the reader to note especially the date of this letter. It was written on the first day of November, 1880; and we wish to call attention to another fact: that the Fat Stock Show opened on the 11th to receive stock, and on the 15th to receive visitors. These dates you will fix clearly in your minds, and we wish Shorthorn breeders as well as Hereford breeders to take note of this—and not only Hereford and Shorthorn breeders, but farmers of England and this nation to note the fact, for we propose to make a case that had its origin at Side View Farm (One-side View, may we say), a case that will stamp Mr. Anderson as a conspirator.

After they had got Mr. Watson away, and magnified his fancied grievances against me, the conspirators wrote severally to the "Kansas City Indicator" several communications, to which I replied as follows in the same paper, in a communication addressed to the editor:

"I have your journal with Anderson's letters of 21st and 28th, and Watson's letter of 28th. Passing Watson's letter with the statement that his remarks, as a whole and in detail, are false:

"Mr. Anderson, after all his discharge of billingsgate, closes his letter of the 28th with the statement that the real question at issue is, that Miller's show cattle had Shorthorn dams and Shorthorn grandams, and with the denial of there being any Shorthorn conspiracy.

"I accept these issues. I gave in the April number of the 'Breeders' Live Stock Journal,' and 'National Live Stock Journal,' the breeding of 'Conqueror's' dam; she was got by Parson's Hereford bull 'Fairboy'; grandam got by Devon bull, nearly full blood; great-grandam got by Hereford bull; great-great-grandam, a red cow of unknown breeding. Will Mr. Anderson find the proportion of Shorthorn blood? (¶ 196)

"The age of 'Conqueror' was stated substantially correct. He was one of four grades, two-year-olds, one of which, 'Putnam,' was showed in 1879 as a yearling, and his age was stated by Mr. Morgan as one year old, July 12, 1879. He weighed at the 1879 show 1,152 pounds; and 'Conqueror' was estimated at that time to weigh 1,000 pounds, and was the younger steer as my recollection serves me, and his weight would indicate this.

"'Kansas' was exhibited as a yearling on the record of his mouth; he had a yearling mouth, and did not change until December. This is all I know of him. The other two steers were from common cows. Mr. W. E. Campbell, of Caldwell, Kansas, and Mr. A. B. Matthews, of Kansas City, have both seen the cows and can testify as to how much of Shorthorn character they carry. Mr. Fielding W. Smith, of Woodlandville, Mo., has also seen the cows, and I shall be very glad to show them to anybody who will take the trouble to call on me. So much for Shorthorn dams and grandams.

"I shall most cheerfully submit four of my two-year-old steers to the examination of an impartial committee in comparison with other two-year-old steers of Shorthorn breed exhibited at Fat Stock Show of 1880, to determine the age from the mouth of the several animals.

"As to Mr. Anderson's denial of any conspiracy; the circumstances at Minneapolis, Chicago, etc., interviews with Watson, the statements they got from Watson, and the continued correspondence Anderson holds with Watson, are against him. When Mr. Watson left my employ, he immediately writes Anderson that he can use his name for authority; Mr. Anderson writes Watson to come to Side View; Mr. Watson goes. The circumstances certainly favor the conclusion that there was a conspiracy, and Mr. Anderson is the victim of an unfortunate set of circumstances if the charge is not true.

"The time has come when the Shorthorn advocates must meet the issues on their merits, and not by suborning witnesses or forming rings by which to secure partial and partisan judges. Browbeating and bulldozing will not answer.

Yours truly,

"T. L. Miller."

We had taken up the past history of the Hereford and Shorthorn progress and followed it from 1742 when Mr. Tomkins commenced the improvement of Herefords, and Mr. Charles Colling the improvement of Shorthorns, in about the year 1775, or a little earlier, to 1834, when Mr. Henry Berry and his associate breeders of Shorthorns, took the machinery of the Society for the Diffusion of Useful Knowledge, using Prof. Youatt as the author to write up the Shorthorns and write down the Herefords. Anyone that will follow the case cannot come to any other conclusion than that Shorthorn ascendency was secured and maintained by a conspiracy of the Shorthorn breeders to advance their interest.

We had proposed at first to continue and bring down this history another decade from 1834 to 1851, but a personal and persistent fight was forced upon us and we were obliged to take it up and leave our purpose of bringing up the past and take the present—not that we proposed to relinquish our plans, but simply to lay them aside temporarily. It was our purpose to show the course that Shorthorn men have pursued to give their breed a prominence, that they never were entitled to; and no amount of abuse, or any number of conspirators or conspiracies could deter us from this object or change our purpose.

It was said in some leading journal about this time, that if President Garfield should disturb the Star Route conspirators or contractors, there would be a war waged on him and lies told about him that would surprise the public. We were prepared for something similar upon ourselves. We may have had men about us that could be bought; we may have had such men sent to us that we might take them into our employ, and they from that vantage point declare the lies that were used, but we proposed to weed out this class and employ men wholly in our interest, for we did not propose to employ men working wholly in the Shorthorn interest. At this time Mr. Wm. H. Sotham was writing and publishing in the "Drovers' Journal" his experience with Shorthorn men and their plans, and the means they used against him and the Herefords, and we called the attention of cattle breeders, stockmen and farmers to this series of articles, which we have preserved in this book, and to the fact that the "Drovers' Journal," beside these articles, was well worth the attention of every man interested in live stock matters for the general information on these subjects.

With these statements, we gave our attention to the issues forced upon us by the advocates of the Shorthorn interest; we quote again from the Anderson letter of November 1, 1880:

DE COTE (3060) 2243.
Bred by T. Edwards. (From a painting by Gauci.)

"So far as his (Miller's) assertions are concerned, about the control of agricultural societies and the use of the press of both this country and England and being engaged in dishonorable practices to keep the Hereford down and the Shorthorn up. * * * There is an old saying that right will finally assert itself and it is a very queer state of affairs if for all these long years, both in England and America, wrong ideas have been prevailing."

Again he says: "What authority has Mr. Miller for saying that the Herefords will graze and feed at less cost than the Shorthorns?"

Without quoting further, we inserted Mr. Anderson's letter in full in the "Breeders' Journal," that we might not be charged with garbling, and took up the fight as a business. When we charged a conspiracy, and the purchase of witnesses, Mr. Anderson felt it necessary to come out with a personal statement to his customers, explaining how he came to be in Chicago at all; it was merely accidental. Well! We will give Mr. Anderson the benefit of this explanation, which is as follows:

HELIANTHUS (4841) 1549.
Bred by the Earl of Southesk, Scotland. Weight at 4 years, 3,000 pounds. (From a water-color etching.)

ANDERSON'S APOLOGY TO HIS CUSTOMERS.

Side View, Montgomery Co., Ky.,
March 28, 1881.

To the Cattle Breeders of the West:

For some months a controversy has been pending concerning the respective merits of the Shorthorns and Herefords, which in an unguarded moment I permitted my opponent to divert from the point at issue and change into a matter personal.

As I have made and proved great charges concerning the breeding and ages of the Hereford cattle belonging to one of the most prominent if unscrupulous (as it has been demonstrated) breeders of that tribe in America, if not in England; and as many of you are my personal acquaintances, it might not be amiss to point out the way; how, by accident, I discovered the frauds which my opponent was practicing upon the Illinois State Board of Agriculture and upon the credulity of the cattle breeders of America.

Without any design whatever on the part of James M. Bigstaff (a fellow Shorthorn breeder and President of the Exchange Bank of Mt. Sterling, Ky.) and myself to attend the Fat Stock Show, held in Chicago, we left Mt. Sterling on Nov. 17, 1880, to attend Col. Robt. Holloway's sale of Shorthorns at Alexis, Ill., on the 18th, and to be in Youngstown, Ohio, on the 20th, on private business matters. We attended Col. Holloway's sale on the 18th, and after the sale was over we were prevailed upon by friends to attend the Fat Stock Show on the 19th, as by leaving Alexis at night we could be in Chicago by morning, and laying over there until the afternoon still be in Youngstown by the 20th. Thus seeing we could fulfill business engagements, we stopped over at Chicago a portion of the day on the 19th. By the merest accident, soon after our arrival we met Robt. B. Ogilvie, of Madison, Wis., a thorough gentleman, with all that implies, and an old friend, and we three attended the Fat Stock Show together.

After examination of the live stock, I remarked the patent dissimilarity between the quarters, flanks and necks of the said T. L. Miller's cattle, as noticeable between the grade and pure-bred Herefords, and expressed a desire to know something of the breeding of the cattle. Mr. Ogilvie asked me if I knew Mr. Miller; I replied I did not, when he remarked, I'll introduce you to him. On looking around for Mr. Miller he did not see him, but remarked, I see Mr. Watson, a very intelligent gentleman, Miller's superintendent, and perhaps he can tell you of their breeding. He then introduced me to Mr. Watson, when I immediately inquired if he could tell me the breeding of certain cattle, pointing them out. He said he could, and immediately commenced with "Conqueror," he being nearest at the time, telling me of his two crosses by Hereford bulls, and that his grandam was a Shorthorn cow, a second of his premium steers being by a Hereford bull and out of a Shorthorn cow, and so on through the lot, describing each steer's breeding and what he had fed him. I asked only such questions in the natural way as any breeder would be likely to ask, concerning more than ordinary cattle he was examining or had under consideration.

This was the first time I ever saw Mr. Watson and I never saw him again until this month of March, 1881. Bigstaff and I hurried off on our trip, that afternoon, and I never saw or spoke to a Shorthorn exhibitor on that day, except young Potts, and with him only passed the civilities of acquaintanceship, nor did I think anything more specially concerning Miller's Herefords until I saw his puffs and blowings in the press, claiming all in all for the Herefords and ranting concerning the Fat Stock Show, when I reminded him of the Shorthorn relationship of his cattle, and that the Shorthorns deserved part of the praise. (¶ 197) On his denying my assertion of the Shorthorn relationship of his cattle, I wrote for the first time in February to Mr. Watson, reminding him of the breeding of Miller's cattle as he had told me at the Fat Stock Show, and requesting him to put in black and white his statement concerning the breeding as made to me at the show in presence of Ogilvie and Bigstaff, as Miller had denied the same publicly. I sent his letter under cover to a prominent man in Chicago, for him to mail to Watson, believing if Mr. Miller should get Watson's mail and see a letter post-marked Kentucky, he would not deliver it, as a man who will deny the breeding of his cattle is not above tampering with private letters. Mr. Watson answered this letter in due course of mail, reaffirming the Shorthorn breeding of Miller's cattle and added about their being exhibited under false age.

This is the unvarnished statement of the whole affair and concerning which T. L. Miller cries bribery, conspiracy, etc. No man on earth ever heard or did I ever offer Mr. Watson one cent or any other amount of money, directly or indirectly, position, emolument, or anything else, for the consideration that he would tell me this or that concerning T. L. Miller's or any other man's cattle. Any other statement and from any other source whatever, that conflicts with the above in regard to the way I got my information concerning the Shorthorn relationship of the cattle which T. L. Miller, "Beecher," Will County, Ill., exhibited at Chicago, last November, is a falsehood out of whole cloth and the retailer of it a slanderer *per se*. Yours with respect,

THOS. CORWIN ANDERSON.

It will be noticed that his aim in this article is to show that he had no special thought of Miller or the Herefords, and that it was the merest accident that he should be at the show. Comparing this apology with his letter of Nov. 1, 1880, as follows:

Side View, Montgomery Co., Ky.,
November 1, 1880.

Editor Kansas City "Indicator":

The effusions of T. L. Miller (the Hereford advocate), appearing from time to time in the agricultural press of the country, have puzzled me not a little to discover a reason why a gentleman of his sense and evident research, casting aside as for naught the experience of the British farmers and the farmers of the older states of our own nation, could prefer the Herefords (with their heavy necks, heads, fore-quarters, and light hind-quarters) to the Shorthorn, with his well-nigh universally admitted superiority, for any purpose whatever for which the cattle kind is intended.

His letter of the 19th in your issue of October 28, just received, makes it perfectly clear, however, when he says, "There is not a prominent Shorthorn herd in America that can furnish milk enough to raise their own calves"; that it is a piece of ignorance on his part; or that his expression is father to his own wish, in these respects. To illustrate, Mt. Sterling, Ky., and within a radius of ten miles, probably contains a larger number of Shorthorns than any other equal extent of territory in America,

GEO. LEIGH,
Aurora, Ill.

yet so far from Mr. Miller's assertion being true, it is just to the contrary; for there is not only not a nurse cow in any one of the herds, but the cows raise their own calves, and in many cases the cows have to be stripped after the calves have done nursing, from the fact that the calves cannot take all the milk. Can Mr. Miller say so much for his much puffed

Hereford herd, or can he name a Shorthorn herd in Kentucky that does not raise its own calves with its own mothers? We think not, never having heard of such a herd in this state. I was well acquainted also with the Shorthorn herds of Jackson County, Mo., until two seasons past, and there, as here, the cows raise their own calves. I also visited a very prominent herd in Mr. Miller's own state (Illinois), this past season, and found the Winslow Bros.' herd not only raising their own calves, but also giving much milk in excess, which was being made up into quantities of butter and cheese for market. I have also just returned from a visit to several of the prominent Shorthorn herds in Ohio, viz., those of Hills, Jones, Andrews, and others, and found the cows raising their calves and no assistants. Therefore, I repeat, without fear of successful contradiction, that Mr. Miller knows nothing of the prominent Shorthorn herds of the United States, when he says they don't give sufficient milk to raise their own calves, and his assertion is father to his own wish, that such should be the case. Further, will Mr. Miller deny that the prominent dairies of England are composed of high-grade and pure-bred Shorthorn cows? If he does, the London "Agricultural Gazette" and "Journal" do not so report, and they are the standard stock papers of that country, so I am informed. I have understood this to be true also of the large dairies of New York.

C. W. COOK,
Odebolt, Iowa.

Again, when he says, if the claim was good for anything, it would not add anything to the value of the breed to go to the plains of Texas, he is in great error, and his assertion proves him anything but a practical cattleman, for on the plains and in Texas it is doubly necessary that an abundant flow of milk be kept up. There the calves depending entirely upon their mother's milk for a steady growth, if they are deprived of it, will be stunted, and all practical cattlemen know the result of cattle being stunted in the first eight months of their existence. They never outgrow it, are that much longer in preparation for market and worth that much less when they go to market.

So far as his assertions are concerned about "the control of the agricultural societies and the use of the press," if they mean anything, they mean that the agricultural societies and press of both this country and England, and, in short, everywhere (the objectionable features and uses to him being the same everywhere), are engaged in dishonorable practices to keep the Hereford down and the Shorthorn up. There is an old saying that right will finally assert itself, yet from the time when interest was first felt in cattle kind, down to this date the Shorthorn has always, in the estimation of the very great majority of beef producers and butchers, been considered the superior of the Herefords; and it is a very queer state of affairs if, for all these long years, both in England and America, wrong ideas have been prevailing, that self-interest should not have so adjusted itself in this respect as to be in accord with the opinions of Mr. Miller.

Again, what authority has Mr. Miller for saying that the Hereford will graze and feed at less cost than the Shorthorn, and when fed be worth more money?

If this be true, why has not the English farmer more generally adopted the Hereford than the Shorthorn? The Hereford first arrived in America about the year 1816 or 1817 (a very few years behind the Shorthorn), and why have not the farmers in the older states where the Hereford was first taken adopted him instead of the Shorthorn? Certainly, Mr. Miller will not allege that the agricultural press and societies have kept the intelligent Yankee farmer, as well as the British farmer, in the dark all these long years. Could it be possible that our prejudices, fed from these sources, have been cheating our pockets for near a century in America, and from the time the memory of man runneth not, in Queen Victoria's realms also? From the best light I can get the Hereford is very near, if not quite as old a tribe of distinct bred cattle as the Shorthorn in England—there they have been bred, grown and marketed side by side for over a century, both originally confined to small districts, but now the Shorthorn habitation is co-extensive with British agriculture and the Herefords more restricted than ever, growing less yearly on account of the steady encroachments of the former on his native territory. The English farmer is given the credit, I believe, of being the best farmer in the world, his yearly rental often exceeding in price the cost for fee simple of our improved western farms. Why is it, if Mr. Miller's assertion be true, about the Hereford feeding and grazing for less than the Shorthorn, and when fed worth more money, that the English farmer himself, generally, don't grow the Hereford instead of the Shorthorn? It can't be now, after growing them for a century, they could not have increased the Hereford numbers so as to supply themselves if they agreed with Mr. Miller. Nor can it be the high price of the Hereford per head that the English farmer, in the ordinary circumstances, cannot purchase them, for no one knows better than Mr. Miller that there never has been a time, nor is it so now, that the best class of thoroughbred Hereford cows in England cannot be bought for as little money per head as the lowest classes of thoroughbred Shorthorn cows. Could it be possible that the English farmer does not know his own interest? He is certainly a great booby if Mr. Miller's assertion be true. I said before that the Hereford put in his first appearance in America in about the years 1816 and 1817, and his arrival was but a few years behind the Shorthorn in this country. If they are such excellent cattle, what has become of all the descendants of all these early importations? Strange the farmers of the older states, where the Herefords were first imported, did not discover their usefulness as best beef and cheap consumers and perpetuate the tribe by breeding them. The Shorthorn cow imported "Young Mary" alone, that arrived in America seventeen years after the first importation of Herefords, has more known descendants than the descendants of all the early imported Herefords put together. The truth is, the Herefords were weighed in the balance and found wanting by the farmers of the older states; they sold them off to the butcher for what they would bring as a bad investment, and this is the reason that there are not more descendants of the early importations of Herefords in the older states where they were first imported. Just as after a thorough trial by the farmers in the new states they will be disposed of. The idea of such an animal as the Hereford, heavily developed (as a tribe) in the head, neck and fore-shoulders (the waste and least expensive portions of a beef) improving to any great extent our American scrubs and Texans (that are also heavily developed in the same parts) for any purpose whatever except work-oxen and freighters (these two classes need big heads and necks), seems extremely ridiculous, and no one knows better than Mr. Miller that the average farmer in England, as in our older states, has discarded them, and that the new states will also after a thorough trial. Wherefore the bold assertions derogatory of the Shorthorn—a beast unequaled and without a rival by the united testimony of intelligent farmers in every clime, for any purpose whatever for which the race is adapted and intended?

Very truly, THOS. C. ANDERSON.

We here give Mr. Anderson the benefit of his opening and closing arguments. We have no right to expect, when we follow the trail of this class of men, that they will stop at any means to obtain their ends; and we shall bring the best record of Shorthorn men to show that this has been their practice. It is possible that Mr. Anderson was not familiar with our former herdsman's habits when he made the compact with him; it may be that he did not know him when he took him into his family. It may be that Mr. Anderson sent him away from Side View because he proved less important as a witness than estimated; it may be that taking Watson into his family had a bad look to it and gave support to the charge of a conspiracy. It is immaterial why Mr. Watson left Side View, Kentucky. He left and knew not where he was going. Mr. Anderson probably regretted that he took this job in hand, and he had to make other apologies in his attempt to make the public believe that he intended an honest fight.

The following appeared in the "National Live Stock Journal," June, 1881:

Editor "National Live Stock Journal":

Mr. A. Matthews calls attention to my steers "Will" and "General," as they appear in the reports of the Fat Stock Show. He says the "General" was entered as dropped Nov. 28, 1877, and on Nov. 10, 1879, he would have only lacked eighteen days of being two years old; and instead of being 612 days old he was 712; and as he weighed 1,397 pounds, instead of gaining 2.28 pounds per day, he gained 1.96.

I find by reference to the Society's report, page 37, Hereford steers one and under two years, his age is given as 712 days, gain 1.96. I find on page 96 of the same report, Lot 6, Sweepstakes ring, he is reported as 712 days old, gain 1.96. Mr. Matthews ought, with his large practice in detail statistics, to have become proficient in figures, and he should be very careful not to put forward false or careless statements.

Again he says: Mr. Miller entered two steers at the show, Nov. 15, 1880, "Alexander" and "Will." Now if "Will" of 1880 was the same "Will" as that of 1879, he was entered at the show of 1880, Nov. 15, "Will," age 1,018 days, weight 1,650 pounds, gain per day 1.62 pounds, and if "Will" was 1,018 days old on Nov. 15, 1880, he would have been, 370 days before that, Nov. 10, 1879, 648 days old, instead of 500; and as he weighed at the show of 1879, 1,114 pounds, his gain per day would be 1.71 pounds, instead of 2.23.

This steer was entered in 1879 as dropped June 28, 1878, and his age as 500 days is correctly stated, and his gain per day 2.23, is correct. On reference to the copies of my entries as made in 1880, his age is stated as June 28, 1878, and his age at the show of 1880 should have been 870 days old, gain per day 1.90, instead of 1.62, as credited to him. I have written to Secretary S. D. Fisher to refer to my entries, and advise me whether the error was his in computing days, or mine in giving age. He was not a winning steer, and the discrepancy had not been noticed by me.

¶229

CANE IS PRESENTED TO THE MAKER OF PROTESTED ENTRIES.

Again: "Mr. Miller won half his premiums with grade Herefords, and it will be time enough to discuss the question as to which cross will produce the best animal when Mr. Miller shows grade Herefords with no Shorthorn blood." (¶ 199)

In your April number I gave the breeding of "Conqueror," showing seven-eighths Hereford and one-eighth Devon and native. Will that be satisfactory to Mr. Matthews? But then what matters it, if by putting the Hereford on the Shorthorn we can make the top steer? Why not accept the issue, and take the Herefords and improve the Shorthorns? They must take the Hereford or Scot. They tried the Scot in the time of Colling, Bates and Booth; and the Kentuckians used the Herefords in 1817 to 1830. The Collings, Bates and Booth Scotch cross has not availed to fix character and quality, but the breed has gone back to the original loose, coarse animal. Collings found that with the old Teeswater cattle there was no certainty as to the kind of produce he would get from them, and only by using Scots could he succeed.

It is a well established fact that for sixty years the "Seventeens" have held their position as the best of all the Shorthorn families, and they owe it to the Henry Clay Herefords; and if Shorthorn breeders will get over their bull-headedness, and recognize the fact that the Herefords can improve any breed they are put upon, and accept and adopt them to improve their long-legged, wheezing and consumptive animals, they will be on the road to success.

I wish it to be distinctly understood that I do not claim greater weight for the Herefords over the Shorthorns, but I do claim equal weight. I claim better quality of beef, and I claim greater economy in cost of making beef on a Hereford carcass than on a Shorthorn; and the difference in favor of the Hereford is fully 25 per cent.

I am trying to get some bullocks from the Texas cow by a Hereford bull, and I want these to fit and show against thoroughbred and grade Shorthorns, and propose to win with them; and I am seriously considering whether I will show again until I get them.

The Shorthorn breeders have overrun the country with their breed, until almost every scrub in the country is tainted with the blood. And if a good winning steer got by a Hereford bull be brought forward, they will swear that he owes his quality to the moiety of Shorthorn blood; and if they have not reputation to go before the community on their own statement, they will hire someone to come forward and swear for them that the dam or grandam is a Shorthorn cow.

To show the standing of the Shorthorns of to-day, I have before me the report of Galesburg sales as follows: First day, 42 head, average $85; second day, 50 head, average $86; third day, 11 head, average $127; 103 head averaged $90. Thornton's circular shows a steady shrinkage of the average price of Shorthorns in England.

Your correspondents are becoming quite interested in this Shorthorn-Hereford controversy. There was a time when they thought if left alone it would die of itself. Mr. Matthews is a wordy man and when he has been fighting in the family I have not taken much interest in him except to enjoy the family quarrel, from the old standpoint that "when rogues fall out, honest men get their dues."

He has been a long time in picking up my letter of 1878. He says, "This is pretty tall blowing, that he (Miller) intends the report shall go to the mountains, spread over the plains, over the ocean, and come back again with greater force and volume."

This was one of the prophecies that I had

forgotten, at least it was out of my mind. You will recognize, Mr. Editor, from your present standpoint, that it was something more than blowing. What must have seemed such at that time sounds very much, as we read it now, as though the Hereford enthusiast of 1878 had a truer conception of the future than others were willing to credit him with, and I am rather surprised that Mr. Matthews should bring this prophecy to light, for it has literally been fulfilled.

I will not undertake to burden your columns with following Mr. Matthews through all his figures to prove that the Herefords are good for nothing. It is sufficient that showing against the Shorthorn cattle with Shorthorn judges, they have won a position within less than ten years that the people of the mountains, the plains, Australia and England recognize. The press quote their popularity in Australia. "Bells' Weekly Messenger," of London, recognizes their importance, and gives a special place to reporting their progress. You are obliged to say, that from a very general inquiry, the cattle men of the plains prefer the Herefords, and I can assure you that this preference has come back with a force and volume that is satisfactory; and to you, who know the influences that have had to be met, it must be somewhat of a surprise that the Herefords have met this success.

I will notice Mr. Matthews in his quotations of my advice to Shorthorn breeders of Kentucky. He says: "Mr. Miller advises the Kentuckians to cross their Shorthorns with the Herefords, saying they will make more money. He must have forgotten that the Kentuckians tried the Herefords over sixty years, and seem to want no more of them."

Does Mr. Matthews know that the best cattle of Kentucky to-day are descendants of those Herefords of sixty years ago? And that to-day, according to Kentucky Anderson, the mountains of Kentucky and Tennessee are full of these in-and-in bred white faces, that the Shorthorns, after sixty years, cannot breed out?

Mr. Matthews follows my advertisement, where I quote my premiums taken in November last at the Fat Stock Show. If the Hereford, by crossing on other cattle, whether Shorthorns or natives, makes a better steer or better bullock, then, are they a better beef breed? The thoroughbred is not now, and never will be used for the butcher's block. They are of value as they are capable of improving the Shorthorns and common or native cattle of the country, and this, the Hereford will always do, and, so far as I am concerned, it will be my aim to show this; and this I well understood when I advised the Shorthorn breeders to put the Hereford bull on their thoroughbred Shorthorn cows, and I wish the Shorthorn breeders to give us the credit of doing it. They have had the country for sixty years, and if we can take the produce of their bulls and make better bullocks let them recognize the fact, and instead of still forcing their breed upon the stockmen of the country accept the issue and verdict.

Mr. Matthews then quotes the "Mark Lane Express." I recognize this journal as one of the fairest and ablest published in England or America—in fact, I consider it stands nearly alone in its impartial statement of facts. At the same time I may differ with them on their facts. They claim that the Shorthorn as a combined meat and milk producing machine, has no equal and no superiors. Now, while I have great respect for the expressed convictions of the "Mark Lane Express," I consider their claim of superiority as a combined machine, while second rate as beef and second rate as milk, is not tenable. It can only be so from the fact that they have some other Shorthorn than what we know in this country; and even then I cannot see how two second-class things can make a first class, and the time will come when they will abandon this position.

The old Yorkshire cow is a good milker, and if taken out of the dairy at an early age—say at six or seven years old—she will go into beef at a reasonable cost, and have performed good service as a milker. This I can readily believe, but then the old Yorkshire cow is a very different beast from the Shorthorn, crossed up, and bred in-and-in with Duke bulls. But even this cow, if kept in the dairy until ten years old, would be fed at a loss with very rare exceptions. I have put questions to the "Mark Lane Express" touching the merits of cattle and I am under many obligations to them for the courtesy, and, I may say, patience, with which they have met my queries. When younger than now one of the things I gave myself a good deal of trouble about, was to get a carriage that would make me a convenient combined machine for business and family use. I never succeeded. I could get along with the thing when I had one child—even when I had two I could manage—but beyond that I could not combine. The farmer wants an animal that will produce beef and milk at the least cost and at the greatest value. I don't expect the "Mark Lane Express" to say anything against its convictions, or a Shorthorn breeder to say anything against his interest.

Very truly yours, T. L. Miller.

CHAPTER XXVI.

TEN EVENTFUL YEARS; A CONSTANT REQUEST FOR TESTS—1871 TO 1881.

(¶ 199) A history of Hereford cattle would not be complete without giving some account of that with which they had to contend in the way agricultural journals met their claims of excellence, as a beef breed.

It will also be of interest, we think, to read what efforts were made by Hereford breeders to have a Fat Stock Show founded, where a public exhibition and competition between the breeds, would demonstrate in an authoritative way what the Herefords could do in America.

The following statement of the case, as made by us, in 1881, will give some facts of interest, and also records of value:

It is now nearly ten years since I engaged in the breeding of Hereford cattle, and from the time I commenced until within the last eighteen months, the Shorthorn men have predicted the collapse of the Hereford boom, nor have these predictions come from Shorthorn men only; but also from my friends; and those that wished me well, have feared the verification of their prophecies. I ask the patience of my friends when I review somewhat the course that has been pursued during these ten years; and of the Hereford breeders I ask forbearance if it should seem that I give to myself a too prominent place in this movement. I here make the statement that the general success of the movement and the success of my associate breeders, has a place in my plans and aims, hardly second to the success of my individual interest; and for this reason there is room for all that may engage in this business during the present generation of breeders.

In following the movements of the Herefords for these ten years, I shall be as brief as possible. The breeders then engaged in the raising of Hereford cattle resided mainly in Loraine Co., Ohio, and vicinity; and of these Mr. John Humphries, of Elyria, deserves special mention. Mr. H. was an Englishman, and, in connection with Mr. Thos. Aston (¶ 200), another Englishman, brought over from England two cows and two bulls. The importation was a joint one, and the animals were named as follows: "John Bull" (3885) 464; "Curly" (801) 14; the cows were, "Victoria" 186, and "Duchess" 15, of the American Hereford Record. "Victoria" was bred by Mr. Turner, The Noke, Leominster. "John Bull" was bred by Mr. E. Price, of Pembridge, and "Curly" by Mr. T. Roberts, of Ivingtonbury. Other bulls were imported by Messrs. Humphries and Aston, and still others by Mr. F. W. Stone, of Canada.

It would be an interesting study to take the American Hereford Record and follow the produce of these two cows. It would be a surprise, even to those who are familiar with breeding, to find what a large proportion of Herefords in the country, previous to late importations, trace to these two cows.

The importation referred to was made in 1851—thirty years ago. Mr. Aston had sold out and returned to England before I entered on the breeding of Herefords. Mr. Humphries was still breeding, though in feeble health, and continued a breeder until about two years ago, when he died, having at the time of his death a very choice herd, though not a large one.

This stock had been bred by Messrs. Humphries and Aston, and others to whom they made sales, but no record, beyond the private one of the breeders, had been kept. My purchases were made from both branches. I should state that in dividing the stock, the bull "John Bull" 464, and the cow "Victoria" went to Mr. Humphries; "Curly" and "Duchess" to Mr. Aston.

I had the stock that I bought recorded in the English Hereford Herd Book, and this led to the recording by other breeders. When I commenced the breeding of Herefords, the Hon. John Merryman was a breeder in Maryland, the foundation of his herd being from Mr. Sotham's importations. Mr. F. W. Stone, of Canada, had the largest herd then in America or Canada. There were individuals from the

Sotham importation scattered over the country. Mr. Phelps, of Pontiac, Mich., had a few from this importation, and the Crapo estate, at Flint, Mich., had one of the largest herds in the states. Mr. Shaw, of Chautauqua, New York, had a herd. In Maine there were several breeders, among them Mr. H. C. Burleigh. As my stock came to a marketable age, I took them to Colorado and Texas for a market. The Colorado sales resulted in giving the breed character and prestige, and whenever the bulls were used they gave satisfaction. I think it was the fall of 1875 that two car loads of grade Hereford steers were sent from Denver, Col., to Buffalo, N. Y., direct from the range, and sold for 7 cents, bringing nearly $80 per head. These were three-year-old steers.

C. S. COOK.
Of the firm of C. W. Cook & Sons, Odebolt, Ia.

It has been found that the Herefords came through the winters on the plains in better condition than steers of any other breed. This, I think it is safe to say, is becoming nearly or quite the universal opinion wherever they are known. The expense of taking bulls to western or southwestern centers was heavy, but there seemed no other way to reach a trial of their capacity for the range. It proved successful and I have no doubt it was the shortest road to success. The growing popularity of the breed provoked opposition mainly and almost exclusively from the Shorthorn men; and in those days I had an idea that it was the province of live stock and agricultural journals to bring forth the facts in regard to the merits of different breeds of cattle. I was then so modest and unassuming that I questioned the propriety of a breeder setting forth the merits of his breed. It would of necessity be termed partisan and one-sided. Under this feeling, in a business letter to Geo. W. Rust, the then editor and proprietor of the "National Live Stock Journal," I said: "I recognize the leading position of the Shorthorn interests in this country; but for some time it has seemed to me that as a journalist you ought to find something to say in the interest of other breeds. As a breeder of Herefords, and believing them to be the best breed for this country, I am somewhat surprised that they do not command the attention of the stock journals and agricultural societies."

I then quoted facts in reference to their standing. Mr. Rust commented on this as follows:

"It is not for us to urge the introduction of either Shorthorn or Herefords, or Devons or Galloways, to the exclusion of each other in the beef producing districts. If there has been more matter published in relation to Shorthorns than other breeds have been favored with, it has been because the gentlemen handling this description of cattle and their grades are relating their experience more generally, and because they have succeeded in awakening a public interest in these cattle, which is continually bringing out all sorts of inquiry about them from all parts of the country. Of course this makes considerable of the Shorthorns in the papers. But how about Herefords? We know, of course, that they are a highly meritorious race of cattle; that their owners regard them as even rivals of the Shorthorns; but when a person asks a question about some matter connected with Shorthorns it is not proper for us to suggest that perhaps Herefords will answer his purpose as well or better. If we would direct his attention to them by some statement as to their merits, where are we to obtain the facts? Are we to say that during fifty-two years in the early history of the Smithfield Cattle Club the Herefords won 185 premiums against 82 won by the Shorthorns? The statement is a good one for Herefords, if true, but will scarcely bear repetition on all occasions, and at best is only one item in the long statements which American farmers require to convince them of the superiority of any race of cattle over their commonest scrubs. Who are raising Hereford steers? What do they weigh at different ages? How early do they mature for the market? Where are they sold? What did they bring? What did scrubs do on the same keep? and what did they weigh, and what did they sell for on the same day? and what was the net difference in profit resulting from the use of Herefords?

"Then, again, if some person, by some sort of accident, acquires a suspicion that the Herefords are a superior race of cattle, and has a disposition to buy them, where can he get animals of strictly reliable blood, and where is the public record by which he can satisfy himself of the authenticity of the pedigree offered him? If he buys Jerseys, he can find two herd books published in this country. If he buys Shorthorns he can find three. If he buys Devons he can find one."

In the "National Live Stock Journal" for July, 1876, pp. 303-4, the following letter appears:

"To the Editor of the 'Journal': In the last number of the 'Journal' you ask: If we would direct (an inquirer's) attention to them (the Herefords) by some statement as to their merits, where are we to obtain the facts? * * * What do they weigh at different ages? How early do they mature for the market?

"It is proper to suppose that a journal of the standing of the 'National Live Stock Journal' would have all standard published reports of the breeding and fattening of cattle and other stock; and if Mr. Jones comes into the office and inquires as to the merits of a particular breed of cattle, or writes to the office for such information, you could refer to reports, and give such information, if such information existed.

"In this country the Herefords have not such a record published and compiled as would be conclusive and full as to their merits; but still there is a fair record in this country. But, to go to England, there is a record, classified, compiled and published, in the proceedings of the Smithfield Club, which is perhaps the most complete and trustworthy that exists so far as it goes. And in quoting from this report I will assume, first, that the merits of a beef-producing breed must be judged by what it can produce as a beef steer, and the record it can show, for its oxen and steers, is the true test by which the trial shall be made.

"I might go outside of the record of the Smithfield Club, to the record of the other English societies, where the Herefords have been shown as breeding stock—and at some other time perhaps I may; but at this time I will confine myself to the Smithfield report of the cattle mainly:

"The first show of this Society was held in 1799. The records of that year were incomplete, but from other sources I learn that Mr. Westcar took the first premium with a Hereford ox. In 1800 the Herefords took 4 premiums; 1801-2 the record is incomplete; 1803 the Herefords took 4 premiums; 1804 the record is incomplete; 1805 the Herefords took 2 premiums; 1806 the record is incomplete; 1807-8-9 the Herefords took 7 premiums. Up to this date it does not appear that the Shorthorn oxen or steers took a premium. Without going into detail for each year, the record up to 1851 shows that: The Hereford oxen and steers took 185 premiums; the Shorthorn oxen and steers took 82 premiums; and to 1857 the Hereford oxen and steers took 9 gold medals; Shorthorn oxen and steers 10, and from 1857 to 1867, Shorthorn oxen and steers took 15 gold medals; Hereford oxen and steers 11.

"In 1869 Herefords and Shorthorns, as to weights, compared as follows:

	Herefords.	Shorthorns.
Under 2 yrs. 6 mos. weighed	1,781 lbs.	1,648 lbs.
Under 3 yrs. 3 mos. weighed	1,936 lbs.	1,976 lbs.
Over 3 yrs. 3 mos. weighed	2,228 lbs.	2,200 lbs.

showing the Herefords in two classes heavier than Shorthorns.

"In the year 1875 the two breeds compared as to weights as follows:

Under 2 yrs. 6 mos.	1,485 lbs.	1,565 lbs.
Under 3 yrs. 3 mos.	2,036 lbs.	1,982 lbs.
Over 3 yrs. 3 mos.	2,169 lbs.	2,281 lbs.
Extra stock	2,524 lbs.	2,290 lbs.

"The three heaviest cattle shown in 1875 were:

1st, Hereford, weighing.......... 2,624 lbs.
2d, Shorthorn, weighing.......... 2,444 lbs.
3d, Hereford, weighing........... 2,420 lbs.

"I have confined myself to figures showing the comparison between the two breeds of Herefords and Shorthorns. It is with considerable labor that I have brought together the information that is scattered over the country in reference to Hereford cattle. As a breeder I have no right to present my claims to the public without something in the way of facts to back my claims. Because I can see so much in the Herefords I have no right to complain because others don't see as I do. I am surprised at times, but I am so sure of the time coming I can afford to wait.

"As early as April, 1873, I offered to help make a record for the Herefords and Shorthorns, as compared with each other, as beef cattle, through the 'Journal.' I have in several ways repeated the proposition from time to time since. I now repeat the offer made in the 'Journal' of April, 1873, over the signature of Miller & Powell. It is true that there is no Hereford Herd Book published in this country, but there is one published in England, and that there may be a copy accessible to inquiries at the live stock headquarters in this country I have ordered from the publisher a set sent to the 'National Live Stock Journal,' of Chicago, Ill.

"A prominent breeder of Shorthorns, while admitting to me that the Herefords were better grazers than the Shorthorns, says: 'You had better take the Shorthorns.' 'Why?' 'Because it is easier to go with the current than against it.' I have heard quite lately of this same breeder paying quite fancy prices for Shorthorns.

"There is one point I should be glad to call your attention to, and that is the make-up of committees at our fairs. In England the Hereford men claim that they have been unfairly beaten on sweepstakes and champion prizes, and for years have sought to remedy this, especially at the Smithfield. Mr. T. Duckham, a prominent Hereford breeder, and publisher of the English Hereford Herd Book, succeeded in securing the passage of the following resolution, to-wit:

A. E. COOK,
Of the firm of C. W. Cook & Sons, Odebolt, Ia.

"'Resolved, that the judges of each breed select an animal that, in their judgment, was suitable for competition for the champion prize.

"'Resolved, that each committee should depute one of their own members to act as judge on the champion prize or regard.'

"These resolutions were passed at the meeting in February, 1876, and will be in operation at the next December meeting. If the same or similar resolutions should be passed by the managers of our State Societies, it would be a good stride towards an equitable award on sweepstakes premium.—T. L. M."

Editor's Remarks: "Of course our correspondent will understand that our question as to what Hereford cattle weigh at different ages was not made because of our inability to refer to statements upon this point in the agricultural text books of the day. It was made for no other purpose than to indicate the absence of statements covering individual experience with these cattle from day to day, month to month, and year to year. It is these statements, and the average results indicated by them, that carry conviction to the public mind, rather than results achieved by different breeders in fitting animals for show. The general public places the highest estimate upon what can be accomplished where cattle are handled in a practical manner for a practical purpose—*i. e.,* when fed for beef by practical men with the sole object in view of producing the best beef in the largest quantity, with the greatest economy. But sensible people require something more than the assurance that, in feeding animals for show where they are forced by all the arts and appliances the owner can command, this one or that one was made to object a greater degree of obesity than another. The statements of farmers handling these cattle, supplying daily almost fresh details concerning their merits, is what is wanted. We desire to indicate that the gentlemen handling Herefords were, in this respect, neglecting the true interests of their cattle.

"As to the proposed test of the comparative merits of Shorthorns and Herefords as feeding cattle, we fear our correspondent will never be gratified by receiving an acceptance of his proposition. No person has such an interest in Shorthorns, considered with reference to the aggregate interest, as will justify him in engaging in this matter. Besides, the reputation of Shorthorns is so well established that there would be but little for them to gain in such a contest. Our correspondent, however, has two courses open to him, either of which seems to us much more consistent and in every way preferable to a 'challenge' for a public and formal test. First, he can go on quietly and feed some Hereford steers for market, and when they are sold let the public know what they weighed, what they sold for, etc., and just how they were handled. If handled as ordinary prudent farmers handle their steers, reasonable weights will go much further towards increasing the reputation of the cattle than a few hundred pounds of extra weight secured by possible forcing or pampering. It would be desirable to have a statement of this sort, accompanied by a statement of what good native steers did alongside of them, for the contest is not so much between Herefords and Shorthorns as between them and the common, inferior stock of the country. If the general farmer can be made to understand that there is sufficient difference between Herefords and natives to justify him in the expense of securing Hereford crosses, the future of the Herefords will be bright enough, no matter what that of the Shorthorns may be. The second course open to our correspondent, if he is really desirous of a test between Herefords and Shorthorns, is to buy some Shorthorns and feed them alongside of the Herefords, and then publish the results. If there should be any Shorthorn breeder dissatisfied with the result, he could buy some Herefords and repeat the experiment."

Since the above was in type the following has been received from the same gentleman:

"I hand you below the weight of Hereford fat cattle shown at the Smithfield shows 1871, 1872, 1873:

1871.	Under 2 yrs. 6 mos...........	1,631
1871.	Under 3 yrs. 3 mos..........	1,856
1871.	Over 3 yrs. 3 mos............	2,307
1872.	Under 2 yrs. 6 mos............	1,634
1872.	Under 3 yrs. 3 mos............	1,858
1872.	Over 3 yrs. 3 mos............	2,137

"I have not the weight classified as to ages for 1873, but the average weight of all ages was 1,934 lbs. I also hand you the average weight of the fat cattle shown at Bingley Hall, Birmingham, in 1873, and these weights include all ages, as at Smithfield, the same year, 1,950 lbs.

"The heaviest steer shown at Smithfield in 1871 weighed 2,360 lbs.; in 1872, 2,626 lbs.; in 1873, 2,538 lbs. The heaviest steer showed at Birmingham in 1873 weighed 2,536 lbs."

And again, from the "National Live Stock Journal" for August, 1876, pp. 349-50:

"To the Editor of the 'Journal': Referring to your remarks on figures given you on Herefords in the July number, I had been led to suppose that the results at the Smithfield show were the most authoritative of any in existence, and I had supposed that you had been living in hopes of seeing just such a show and list in this country.

"Now, it is true that in this country the Herefords have not made such a record as will be conclusive as to their merits, at least so far as the number is concerned, but I hand you herewith a memorandum which will, perhaps, show what they could do. But is not the English record pertinent and to the point? As to comparing Herefords with Shorthorns and natives, you have established the fact to the satisfaction of the majority of thinking, practical men, that they are better than the natives, that the Shorthorn bull on the native cow is an improvement, that the cross produces a better beef steer, which by grazing and feeding will make a better quality and greater quantity at a less cost. I assume this to be true and therefore do not care to travel that ground over again. But taking that point as settled, I now assume and claim that the Herefords make a better cross, in the hands of a practical man for a practical purpose, with the sole object in view of producing the best beef in the largest quantity with the greatest economy.

"I believe it is admitted by the best informed cattle breeders that the Herefords are more hardy and better grazers than the Shorthorns. In referring to the Shorthorns I have no wish to depreciate their merits. I admit them. They are a fine race of cattle, and the breeders of these cattle are entitled to much credit for what they have done. There is perhaps no interest in which the farmer has received so large a benefit as has come from the efforts of the Shorthorn breeder; but is not this step one that leads to something higher and better?

"Now you are aware, perhaps better than anyone else, that in England—a very small country compared with this—they have several breeds of cattle which are pre-eminently the best for the district in which they are bred, and if this be true of England may not the supposition be raised that it may be true of this country? For seventy-five years the Herefords and Shorthorns have been the leading beef breeds in England. There are several large shows held where the breeding stock has been shown, but at Smithfield the test has been on the fat animal, and the Hereford as a fat ox or steer has occupied a leading position, and I assume that it is between these two breeds in this country that the choice must be made.

"Rev. J. R. Smythies, a prominent Hereford breeder in England, made the following offer through the 'Mark Lane Express,' in 1849. He offered to show four Hereford steers, whose ages should not exceed two years and three months, and four whose age should not exceed one year and three months, at the next Smithfield show, in December, against eight Shorthorns and eight Devons of similar ages, for a sweepstakes of one hundred sovereigns for each lot, with this stipulation, that each lot shall have been bred by one man, and that they shall have lain at grass at least four months that summer, without having had anything but what they got there. He says:

"'But this is not all. I am willing to test their hardiness as a breeding stock, as well as their feeding properties. In order to do this, I propose to turn my two-year-old heifer, which gained the first prize at Norwich, into a pasture with the two-year-old Shorthorn and two-year-old Devon heifer which obtained the first prize in their respective classes, and let them remain there until the next meeting of the Royal Agricultural Society, at Exeter, next July, giving them nothing but what they can get, except a little hay from the 5th of November till the 5th of May, the heifers being shown for sweepstakes of 100 sovereigns each.'

"In February, 1849, Mr. Smythies made, in

the 'Mark Lane Express,' the following offer: 'To place four Hereford calves on the first of May next in the hands of any respectable grazier in the Midland Counties, against four Shorthorn and four Devon calves, no calf to be more than four months old on that day. The twelve calves to be turned to grass together, and to have nothing but grass until the 20th of October following, and then to be put in stalls and fed as the grazier thinks proper—but the food to be weighed to each lot—until the following May, when they shall be number belonging to any two breeders of Shorthorns or Devons, in any part of Great Britain, for 100 sovereigns.

"These offers were made but never accepted. As to my offer made through the 'Journal,' you fear that I will never be gratified by its acceptance:

" 'No one person has such an interest in Shorthorns considered with reference to the aggregate interest as will justify him in engaging in this matter. Besides, the reputation of Shorthorns is so well established that there

JOHN PRICE (COURT HOUSE, PEMBRIDGE, ENG.).
The Elkington cup and other valuable prizes won by this veteran breeder.

again turned to grass, and have nothing but what they can get there until the 1st of October; then to be again taken into stalls, and the food given them weighed as before. The whole to be shown as extra stock at the Smithfield show, and after the show to be slaughtered, the four beasts which pay the best to be the winners.' Mr. S. also offered to show 100 Hereford beasts, the property of Sir Francis Lawley, on the first of January, 1849, and the same number, which were the property of Mr. Aston, Lynch Court, on the same day, against an equal will be little for them to gain in such a contest.'

"This is perhaps true. But is there in this country such a record for the Shorthorns as will show that they are entitled to the position they claim? The different branches of the family claim the merit is in them only. I go on quietly, as you suggest, and feed some Hereford steers, but such tests have very little value.

"The test which I propose will have a value, not only to the breeders themselves, but to the stock interests of the world, and it is a test that may be made at a profit to the breeders. On the ground proposed it must be an even test, but I am not writing for the acceptance of the offer. Herefords are being used in Colorado and Texas. In due time the steers will come forward. Your closing proposition is for me to 'buy Shorthorns and feed them by the side of my Herefords, as a test, and if there should be any Shorthorn breeder dissatisfied with the result, he could buy Herefords and repeat the experiment.'

"It would not be conclusive. The Smithfield record complete comes the nearest to being authoritative of any that has come under my notice. It is complete in giving the live weight of animals, but fails in this, that it does not give the cost at which these weights were obtained.

"Judge Downing, of Denver, Colorado, informed me that he sold six Hereford grade steers in June that were four years old in the spring, weighing 1,800 pounds each, and twelve others and three heifers weighing a fraction under 1,500 pounds each, none of them having been fed at all, having made their weights on grass alone, except they may have been fed some hay sometimes in storms.

"Mr. Church, who lives near Denver, has turned off thirty or forty grade Hereford steers for several years past, at three years old, averaging about 1,250 pounds each, that have never been fed anything but what they themselves have taken from the range, and one lot of these steers was sold in Buffalo at 7 cents a pound.

"Judge P. P. Wilcox, now of Denver, says that his cattle ran with a herd in which there was a grade Hereford bull, and from him he had several white-faced calves, and that these white-faced calves were as good at two years old as his others at three.

"Another prominent stockman in southern Colorado says: 'The Hereford cross on my native cattle has been very satisfactory. They stand the winter well, take on flesh rapidly, and are really the best cattle for these ranches that I have ever had anything to do with.'

"Mr. J. Humphries, of Elyria, Ohio, one of the oldest and most careful breeders of Herefords in this country, turned off at different times to the butchers cows and steers, and generally at Christmas, the cattle having run to grass through the season and fed grain for only two or three months as follows:

Seven cows, average weight........	1,271 lbs.
Two-year-old steers, average weight.	1,425 lbs.
Three-year-old steers, average weight	1,914 lbs.

"Such figures are good as far as they go, but they are not conclusive or satisfactory. The Hereford breeders in England have repeatedly sought to make a test, where the feed and care should be the same, but have failed to obtain. Now in England the Shorthorns do not stand so much in the lead of all other breeds as to numbers as they do in this country. At the meeting of the Bath and West of England Society, held in June, at Hereford, there were shown of Herefords, 160; of Shorthorns, 55; all other breeds, 88. Without going into detail as to the show you will permit me to speak of 'Tredegar' (5077) 2478, a Hereford bull that has quite a noted record in the show ring. He was shown here in the class of bull, cow and offspring. It is said of him he was first shown at Newport when a calf, and took first in his class; also at Bath and West of England, at Bristol, first; next at the Bedford meeting of the Royal, he won similar honors, and first at Dudley, of the Worcestershire meeting; first and special at Ledbury; first at Croydon of Bath and West of England's Society, and the champion prize as the best bull of any breed; beating that well known prize winner in the Shorthorn classes, 'Sir Ingram;' then at the Taunton meeting of the Royal, when he repeated his previous performance, and at Hereford, in the show following, took the special prize offered for best Hereford bull; at the Worcestershire Society he also won first prize for the best bull in the yard.

"In the cows it says of 'Rosalind,' unquestionably the most beautiful cow in the yard: 'She was five years old, having bred three heifer calves, having from a calf stood first and second in her class, and won some seven special and champion prizes in the time.'

"It does seem to me that the breeders of two such breeds of cattle would find a pleasure in testing their merits on a scale that would attract the attention of the world, not only of stockmen, but of all men. And the breeders of America should feel that they were the men, and that America was the ground on which to make the test, and this Centennial year, the year to inaugurate and perfect the plan which should test the merits of these cattle. I will not undertake to mark out a plan, but will you not give your influence for such a trial? One hundred thousand dollars capital would be sufficient—perhaps $50,000 would answer—perhaps less. Such a test would be worth more than all the awards that will be made within the next ninety days—or all the awards that have been made for the last twenty years by all the agricultural societies in the United States.

It is well to do as our fathers have done, if they have done right. It is easier to go with the current, but it is sometimes wiser to go against it. Won't you try it on this question?

"There is now open to the world, and brought into the world, a stock country the like of which was never known before. It changes or will change the whole system of breeding and the question must and will be solved as to the breed of cattle best suited for it.

"Beecher, Ill. T. L. MILLER."

Editor's remarks: "We have no interest in any breed of cattle, and the claim of superiority over the Shorthorn advanced by our correspondent on behalf of the Hereford, is a matter which we shall leave Shorthorn breeders to discuss or refute for themselves. We place but a slight estimate upon the results of the Smithfield show, for the reason that the animals are fed for the prize without any reference to cost—they are stuffed and pampered, and brought into the very highest state of obesity, regardless of expense. The result in a contest of this kind depends quite as much upon the judgment of the feeder as upon the merit of the animal. What is wanted is information about cattle handled in a practical sort of way, with a practical end in view; such, for instance, as information about what general farmers can do and are doing with their stock. We should be pleased to see such a show as that of the Smithfield Club inaugurated in this country. Not because we believe it would settle any difference of opinion among rival breeders as to whether the stock of one was superior to that of the other, but because it would demonstrate to the general farmers of the country that there is a substantial benefit to be derived from the use of improved blood. With the big steers on exhibition weighed by judges, and all the details of their keep published, the general farmers, realizing the impossibility of approaching such results with their native stock, would be encouraged to secure better blood. Of course there are other respects in which it would be of advantage, but the above is the principal benefit to be derived from it.

"We know nothing of the challenges to which our correspondent refers—why they were made or why they were not accepted. We beg to say, however, that in general challenges are so worded as to seem fair on their face, while coupled with 'conditions' which no one can accept. We do not know that the challenges of Mr. S. were coupled with such conditions as they were only briefly stated; but the offer to feed 200 beasts might not have found a man in England able to comply with the conditions.

"And, after all, a trial or two establishes nothing. It is only from a large number of trials, embracing all manner of conditions, that a reliable conclusion can be formed. These experiments would, of course, be interesting, although, considered by themselves, not very important. We should be pleased to see the experiment made, as we are always pleased to have the details of carefully considered feeding operations, no matter what kind of animals are the subjects.

"We think there is a public record in this country showing the Shorthorn to be entitled to the high estimation in which it is held. It is the record made up by the general farmers of the country who have made a practical test of the merits of the breed for practical purposes. So far the test has simply been between the Shorthorn and the scrub. There is no public estimate as to the relative value of Shorthorns and Herefords, because the public have not sufficient experience with the Herefords to form such an estimate.

"Our own opinion is that it is not so much a question as to which breed is superior to the other in itself, as to which breed will make the most valuable cross upon our inferior native stock, and for the present there should be a market for all the bulls of both breeds, without engendering any rivalry between them.

"The details given by our correspondent, of Hereford beeves, etc., are full of interest. We shall cheerfully publish more of the same sort as received. If facts enough of this sort are furnished, the general farmers of the country will form an opinion that will not be very far out of the way. We presume the grades to which our correspondent refers in speaking of Colorado were from the Texan stock on the dam's side; or at any rate from dams not fully up to the standard of our native cattle. And it should also be remembered that Colorado pastures are somewhat lighter than our own."

Again, in the "National Live Stock Journal" for September, 1876:

"Editor of the 'Journal': 'And, after all, a trial or two establishes nothing. It is only from a large number of trials, embracing all manner of conditions, that a reliable opinion can be formed. These experiments would, of course, be interesting, although, considered by themselves, not very important.'

"Thus the 'Journal' says, on page 350, in reference to the offers made by Mr. Smythies to test the Herefords against the Shorthorns and Devons. What were these offers:

"First, to take four Hereford steers whose ages should not exceed 2 years 3 months, and

"Second, four Hereford steers whose ages should not exceed 1 year 3 months, against eight Shorthorns and eight Devons of similar ages.

"Third, to take his two-year-old Hereford heifer, which won first at the Royal in her class, and turn her into pasture with a two-year-old Shorthorn and a two-year-old Devon, first winners in their class at the same meeting.

"Fourth, to place four Hereford calves with four Shorthorn and four Devon calves, no calf to be more than four months old.

"Fifth, to show 200 Herefords—100 from each of two herds named—against an equal number of Shorthorns or Devons belonging to any two breeders in any part of Great Britain.

STEVEN ROBINSON,
Lynhales, Herefordshire.

"Now, what would these tests have shown if the offers had been accepted? The first and second would have shown what breed would have made the largest gains in a given time under the same circumstances. The third would have shown the hardihood of the different breeds, by taking animals that had been forced to large weights by high feed and turning them from it to roughing it. The fifth would have shown what the different breeds had done under ordinary breeding care. Would not, then, these tests have shown the merits of the different breeds under consideration, so far as the wants of England were concerned—that is, would it not have been made a prima facie case?

"It is of no special interest to the public what my opinion may or may not be, unless I can show a reason for the faith that is in me. And a test is not conclusive unless made with the conditions alike. Mr. Jones, a good, careful breeder and feeder, will raise a half dozen good grade Hereford steers, and carry to large weights. Mr. Smith, with less care, will raise a half dozen Shorthorn steers and make light weights. Now, these tests made in the usual manner by practical men, don't prove anything. Is there, then, any better way to test the merits of the different breeds of cattle, than to place each other precisely under the same circumstances? To the large majority of steer breeders it does not matter what Mr. Pickrell can make thoroughbred Shorthorn steers weigh, or what Mr. Miller can make a thoroughbred Hereford steer weigh, and while it is important that Mr. Pickrell and Mr. Miller may know what their respective breeds can do, still it is very much more important for the public to know what the grades are doing and can do.

"With this in view we made the following proposition through your 'Journal' in April, 1873, to-wit:

" 'It is our belief that the Herefords are preeminently that stock that must be used for improving the large herds of Colorado, Kansas and Texas. As beef producers in this country, at least, the choice will lay between the Shorthorns and Herefords. In the hands of some men it will be the one, and in those of others the other; and that this question may have a fair solution we hereby propose to any Shorthorn breeder to select 100 or 200 cows in this state, one-half to be served by a Hereford and the other half by a Shorthorn bull and the progeny to be cared for alike and shown at our State Fair at two, three and four years old. We propose further to select in Colorado, Kansas or Texas, from 500 to 2,000 cows, and serve one-half with Hereford and the other half with Shorthorn bulls, and their progeny to be kept alike, and a portion to be brought to our State Fair at two, three and four years old each year. The choice as to how the stock shall be kept we will give the Shorthorn breeders. All to be kept alike; and the experiment may be for one or five years.

"'We believe that beyond the interest the experiment will have to the Hereford and Shorthorn breeders, it will result in a very great good to the entire stock interest of the country. Signed, MILLER & POWELL,

"'Beecher, Ill.'"

"Now it is important, in a test that shall be of value, to make the basis alike, that is, to start with the same grade of cows—and that the circumstances of their keeping should be alike. The proposition of Miller & Powell above stated would secure this. If Mr. Jones starts with a high-grade cow and Mr. Smith with a scrub, the test is not equal; now if no one breeder has an interest sufficient to test this, then let our own State Society offer premiums for fat steers without regard to breed or grade, that will induce the farmers of the state to compete for such premiums. If any of the breeds that are before the public are worth anything to the public, it consists in their value to produce a beef steer. Let our societies make a class for fat cattle, say: Under 2 yrs. 6 mos.; over 2 yrs. 6 mos., and under 3 yrs. 3 mos.; over 3 yrs. 3 mos., and under 4 yrs.; over 4 yrs.; for fat cows that have bred at least four calves, and offer a first, second and third on each. There will be fifteen prizes, and the farmers to compete for—say as follows:

Steer 2 yrs. 6 mos.	$100, $60, $40
Steer 3 yrs. 3 mos.	$100, $60, $40
Steer 4 yrs.	$100, $60, $40
Steer over 4 yrs.	$100, $60, $40
Fat cow	$100, $60, $40

"There is only $1,000, and it is my belief that it will create more interest than all that is now done. I would suggest these premiums as only preliminary to something better, and let the committee to pass on such awards be butchers of experience.

"It is undoubtedly the large and main feature of the fairs—the cattle interest. It is desirable to have pure breeding, and of such quality as to carry character to their produce, and it is desirable to follow this produce and see what it is like. In 1874 the Illinois State Society paid in premiums to:

Beef breeds, thoroughbreds	$1,860
Fat cattle	30
Dairy breeds	745
Dairy products	Nothing
Horses	3,500

"I have never been in the management of a State Fair, and have no doubt that there are reasons that govern the action of the management that I cannot see, but from my standpoint there would seem to be occasion for a change. Especially does this seem to be true in the product of the beef and dairy breeds of cattle.

"T. L. MILLER."

The foregoing is substantially the case as made in 1876. The Shorthorn interest had been accepted and the cattle interest of the country adjusted to it; and it was not surprising that the introduction of another breed that promised to interfere with and supersede that interest should meet with the most determined and bitter opposition. This had been manifest from 1839, when Mr. Sotham introduced the Herefords into this country at Albany, N. Y. With the press and agricultural societies in their control, it was not strange that they should be intolerant, and that a personal fight should be made upon anyone who should press the claims of a rival breed. With the agricultural societies under their management, there was no difficulty in arranging their judges and obtaining the awards. This year (1876) the Illinois State Board passed on the comparative merits of the Herefords and Shorthorns, five prominent Shorthorn breeders being judges.

When the agricultural journals permitted the discussion as to the merits of the Herefords as compared with Shorthorns, it was not surprising that prominent Shorthorn breeders should say, "If you propose to make a Hereford journal of your paper, let the Hereford men support it." It was not surprising that a leading breeder and writer should say: "The question of merit as to which is the best beef breed, has already been decided, and should not be opened or discussed." It was not surprising that under such a pressure, journalists that had been led to believe the claims of Shorthorn breeders valid, should be influenced, and that while they did not refuse articles that were intended to unsettle old beliefs, found it difficult to find room for them in their papers. That a journal in the interest of the Hereford movement should seem a necessity, was not strange. The necessity existing, the "Breeders' Live Stock Journal" was established and put upon its merits, and it met the necessity—not under any false statements, but declaring in unmistakable terms its belief in the merits of the Hereford breed of cattle, and its determination to press those claims, and show why they were the best. The establishment of this journal was not without the expenditure of a large amount of labor and money. As the movement that placed the Herefords on the plains, and put them at work where they could show their produce, had proved a large success and hastened the acceptance of the breed, so the establishment of the "Breeders' Live Stock Journal" cemented and systematized the work.

These movements were my individual and personal enterprises, that brought returns with them to a certain extent; but they contributed largely to the success of all breeders. They were a necessity of the hour, and the results have shown the wisdom of the course pursued. The publishing of the American Hereford Record and the compilation of the work was undertaken and carried through by individual enterprise, without any other aid than the payment of a fee for entries that was thought sufficient.

In thus reviewing the course of the Herefords during an eventful decade we cannot but feel that the means used to advance their interest was fairly well selected, and the results reasonably successful. That the movement was a success there is no doubt. This is evidenced by the character of the men called together at the Hereford Convention; and another evidence is found in the manner in which the Herefords withstood the severe weather of the range winters, which put the seal of permanency upon the movement. Each step that was taken was not in the beaten track of others, but on the contrary, was independent of the movements of other breeders.

The placing of the cattle on the plains was done in the confident belief that they would stand the test to which they would be subjected, and subsequent events have fully proved this. The establishing of a journal for the advocacy of a single interest, and without a general guarantee fund, is another experiment that is usually deemed hazardous, and one that usually demands a guarantee from the interest which it is to support, but the "Journal" became a success without such guarantee. These ends were obtained without other aid than the merit inherent in the movement.

Another step was taken outside of the ordinary methods pursued by breeders, and that was the establishment of an exhibition at the heart of this great cattle interest. A commodious building was erected in which to place specimens of the breed, and there submit them to the judgment of the practical cattlemen of the world. The seeming necessity of such a movement lay in the fact that the Shorthorn men had for half a century been organizing state, district and county agricultural societies, and under these inaugurating fairs or live stock shows, and under the management of these Shorthorn judges we must exhibit our stock, or, if we were not satisfied, stay away.

There was then a necessity for some means by which we could place our stock before the world of cattle breeders. We could have organized an opposition agricultural society, but that would have required large expense, and would have been subject to the same drawbacks as political reforms, to-wit, all the dead-beats and old fair hacks would have jumped on, and we should have been little better off, and another reform would soon have been needed. We established a simple machine, and it was intended it should be made the best exhibit possible, and be at the same time self-sustaining. We believe it the best move that could have been made. We thus brought our own cattle before practical men. We had hoped—had, indeed, expected—that the managers of the Union Stock Yards would have gladly extended to the Herefords accommodations for such an exhibit; but they thought it would be establishing a bad precedent, and declined to do so. We were thus obliged to take the most convenient location we could get, using for a time such accommodation as could be obtained, till the wisdom of the movement was fully established.

This could hardly be termed a new movement, as John B. Sherman had had such an exhibit of Shorthorns for several years, selecting the best specimens of that breed to place in his show stable. Thus the two breeds were brought together in comparison—and this not for one week, but for fifty-two weeks in the year. The move was undoubtedly a good one, and gave the world of cattlemen an opportunity to examine the Herefords and compare their merits with specimens of Shorthorns that they were familiar with and can find in the yards. At the close of the year 1881 we summed up the situation as follows:

"Reviewing the past, what is the outlook for the future? The standing of the Herefords is fully established, and the inquiry for them was never so urgent as now.

"The breeders of cross-bred and grade bulls have sold almost their last animal, and if there are any left they are in out-of-the-way places. The number of thoroughbreds in the hands of breeders is not large, and there would have been none but for the fact of a late opening of spring and consequent late demand for grades. In conclusion, the past has been reasonably satisfactory, and the outlook for the future is very encouraging."

CHAPTER XXVII.

Quality in Beef; Sound Constitution Indispensable to it

It is a fact in history that never, in the years that preceded these Fat Stock Shows in America, had the attention of the public and the interest of the people been so aroused in the matter of good beef, and good and profitable beef cattle, as at this time.

Many, from not having examined into the subject, thought good beef was a matter of chance and not the result of earnest and intelligent effort. The Chicago "Tribune" had the following to say, in regard to the interest awakened and the quality of beef:

"That the Shorthorn and Hereford forces are being marshalled, thoroughly equipped, and better disciplined and prepared than ever before for the great conflict of the breeds, none will deny, who have the least interest in this commendable strife between the two rival breeds of beef cattle. The general public in both continents will never duly appreciate the great debt they owe the Chicago Fat Stock Show, for the improvement from year to year in the quality of beef that will be exported from this market as the result of the emulation in this normal school for the feeder and breeder.

"And as other countries must compete in the European markets with this class of goods, they must keep pace with our feeders or retire from the trade.

"The Shorthorn breeders have not, until very recently, manifested much interest in this contest, which has already created a boom in the Hereford market, that has at least seriously alarmed the thousands of breeders of this favorite class of cattle, who have deceived themselves with the impression that they had nothing to fear from the handful of whitefaced breeders, or their cattle.

"The number, influence, wealth, unparalleled energy and enthusiasm Hereford advocates have developed so suddenly startled the stock breeders of the country, who have been compelled to recognize the excellence of the comparatively new candidate for public favor. The previous indifference of the Shorthorn men as to the possibility of their defeat is in striking contrast with their evident anxiety and the earnest efforts being made at this date to spare no pains or expense to regain their former prestige.

"The grade Hereford steer Conqueror, the Hereford cow Jennie (¶ 201), and other remarkably fine whitefaced bullocks exhibited during the past four years at the Fat Stock Show, have opened the eyes of the most prejudiced friends of Shorthorn cattle as to the necessity of preparing a much better class of cattle for future shows, or gracefully confess a most humiliating defeat."

The butchering and comparing of the different animals at the Fat Stock Shows brought intelligent reports from able men and made it a matter of record; and brought to the people's notice that fat could be marbled between the lean; and that it was not necessary to eat blue, tough meat, if only the right quality of steers were fed.

It was our great aim and effort to show that Herefords possess the requisite qualities for a beef animal, in perfection.

Prof. Tanner of England, in a lecture delivered before the Marshbrook Improvement Society, on the economical production of meat, states the case as follows:

"We find that the old breeds of cattle, the unimproved breeds of cattle, differed very materially from those of recent date. In other words, we have modified by careful breeding the powers of the animals we raise upon our farms for doing this work. The difference to be observed in the local breeds, which were existing in different parts of the country forty or fifty years ago, were largely traceable to the local influences of soil and climate. Since then we have adopted improved and more valuable breeds, and they have succeeded just in proportion as they have been introduced into districts that suit their peculiarities of character. For instance, the produce of the Highlands of Scotland were of a hardy character, and as we come afterwards to other exposed lands, we have the Welsh and the Hereford breeds of stock; then, going into more temperate districts we have the Shorthorns, and the more tender but nimble Devons, and so on, to the Channel Islands stock. But the point is this: that all this produce obtain their points of character by being brought up under certain conditions to which they had adapted themselves, but when the Shorthorns were sent to Scotland they were unsuited to the districts and they perished."

Prof. Tanner goes on to say: "Now those habits, those points of character, which were existing in the different local breeds, were modified to meet the requirements of more advanced farming. For instance, the old system of allowing the cattle to remain until they were four, five, or six years old, before they were completed for the butcher, gradually gave place to a much more rapid habit of growth, a growth which was more prompt and speedy throughout its entire range, and the result was that the animal was finished and ready for the butcher at a much earlier date. The tendency of late years has, therefore, been in the direction of so altering the character of the animals we breed that they become quiet and docile, quite disposed to take their food without restlessness, and also to make good use of the food which is supplied to them. I know perfectly well that you can appreciate the weakness of the character, both of the Shorthorn stock and also many of our delicate breeds of sheep, especially the Leicester. And how has this difference arisen? Largely because the alteration made, in the case of the Shorthorn cattle, in early years had been carried to an extent which has probably—but I will not say probably—which really, has resulted in their obtaining a weaker constitutional character. Where breeders of Hereford cattle have excelled is this, they have kept their cattle under more natural conditions, and as a consequence by keeping their cattle under those conditions, they have submitted to what the Shorthorn breeder would say was a sacrifice, during this time, because they have not made their stock into such rapid fat-formers as the Shorthorn cattle are.

SIR JAMES RANKIN, M. P.,
Bryngwin, Herefordshire.

"If you breed cattle so that they have small, feeble, or imperfect lungs, you must not expose such cattle to rough, cold weather with the same impunity that you might if they had larger lungs and were fully capable of maintaining the warmth of the body. If you give them small lungs they become unable to maintain that warmth under conditions of considerable difficulty, and having reduced the power of the lungs, if you do expose it to inclement weather, it is said at once of the animal that it is too delicate, and why? Because you have reduced the lungs so greatly that the animal cannot keep up its warmth, and becomes 'delicate,' and that is only another phrase for saying that it is predisposed to disease. Permit the animal to become delicate, therefore, and it becomes predisposed to disease. And just in proportion as you limit the exercise that cattle take, by keeping them in their confined stalls, you diminish the size of the lungs, and you make them incapable of resisting the consequence of exposure. Now our Shorthorn cattle are notoriously more delicate than your Herefords, and why are Herefords more hardy? Because they have been bred and brought up under conditions which allowed them greater freedom, which permitted them to take more exercise, and so they are more hardy.

"This is one point; and there is another point which has helped to make the Hereford breed what it is—there has been greater care shown for the production of milk. The fatal error which was accepted by the majority of Shorthorn men was this: We are breeding to produce meat, and milk is of secondary importance; we want pedigree animals which will be of very high value; you can get milk from very cheap cows. Now, there has not been a more fatal opinion in connection with agriculture than this. The production of milk has been looked upon as of secondary moment, and just in proportion as the production of milk has been neglected, so will you find the breeding powers of the animal become weaker and more uncertain. It is, therefore, because of the greater care that has been shown in the management of Hereford cattle, in reference to the production of milk, and also the more natural way in which calves have been allowed to be suckled by their cows and to run with them that you have maintained the breeding powers of these animals, in greater excellence and in greater power. The stronger constitution and the better breeding character of the Hereford stock is largely due to the fact that while you have carried improvements forward, you have not carried them to such an extent as to endanger the stability of the breed for the hardy character which it originally possessed, and for many years past, it has been my pleasure to refer to the Hereford breed as representing the cattle of which there was a sufficient power for the production of meat rapidly, without it being accompanied by weakness through over feeding and neglecting the milk, and, depend upon it, the great future of this breed will be determined by this moderation that you have shown in the past. There is nothing peculiar to the Hereford breed, any more than the Shorthorns, to prevent you going too far in the alteration you make in the structure of the body, and in its general healthy character. It is quite possible for you, and you may have seen instances of it, to produce a Hereford quite as tender and as delicate as the most frail of the Shorthorn breed. But the interests of breeders in the past has been shown by carrying the improvements to a reasonable extent—making them good flesh formers, good feeders, but not overdoing it by bringing your animals into too delicate a condition."

Prof. Tanner concluded by saying: "I know that some of the breeders of Shorthorn cattle will think that I am doing them an injustice, in reference to the quality of Shorthorn beef, but it is my duty to speak the truth, to the best of my ability and knowledge, whether it pleases or not; and I feel convinced, that the excessive limitation which is placed upon the Shorthorns, tends greatly in the direction of producing beef which is decidedly of an inferior character. We know, however, that it is a point of excellence with Hereford breeders, that you are rather liberal in regard to exercise —you allow your stock to take rather a liberal amount of exercise, and this is really a loss of food producing power; but you cannot have it both ways. The great end is to obtain that happy medium that gives you the rich flesh, accompanied by a fair proportion of fat, and not to fatten to such an extent that the muscular growth shall attain to a fatty degeneration."

We present Dr. Sprague's report on marbling qualities of beef from the dressed bullocks at the American Fat Stock Show and from others obtained at the markets. We would call special attention to the clause which we have placed in italics:

"So, after thoroughly weighing and sifting every feature of the structure, of fiber and fat, in the two breeds as represented by specimens, we conclude that superiority must be determined by EARLY MATURITY, POUNDS OF GAIN UPON A GIVEN AMOUNT OF FEED, RELATIVE WEIGHT OF OFFAL AND OF BEST TO POOREST PARTS."

Or as we would state it, economy of production and value of product. There is no doubt that the Doctor reached the rule that must be the true test. But a further and more careful investigation reveals the fact that by the touch and handling, the quality can be determined with tolerable accuracy. The butcher of long experience by the eye and handling, will determine whether the animal will die well or not; he will determine with almost unerring certainty on the points. This fact is well established, but there is as much difference in butchers as there is in any other calling. We will find butchers that by the outside of the animal will tell almost to a certainty what is under the hide, and this class of butchers should be called for judges. Another point: When a society shall reserve the right to take competing bullocks that are brought before them for slaughter, then will they reach results that will be valuable. Dr. Sprague says, "The butcher is the only one that could establish the presence of these desirable qualities" (marbling). We advocated this and urged upon the attention of the Illinois State Board of Agriculture the necessity of making this test. An offer was made in the following language in 1879 in a communication to the Board:

"The object of your Board is to determine, among other things, the comparative value of different breeds of beef animals. Among other ways, expert judges of beef animals are called upon to pass upon the merits of such animals as are brought before them. This will determine the merits excepting the errors of judgment and prejudice in favor of different breeds. To correct such errors and prejudices if any exist, I will submit my cattle, that have come in competition with other breeds, to the direction of your Board to be slaughtered."

We agreed with Dr. Sprague, that the butcher's test was needed. The Board could offer such premiums as would make it an object and inducement for breeders and feeders to bring their cattle to such a test.

The Duke of Bedford did this over seventy years ago, and the correctness of that judgment has been recognized, from that time to the present, by the graziers and butchers of England.

At the request of the State Board of Agriculture of Illinois, Dr. G. Sprague of Chicago prepared an exhaustive paper giving the results of a series of careful microscopical examinations of samples of meat, cut from several of the carcasses butchered at the Fat Stock Show held at Chicago in 1880. It will be remembered that the breeders of Shorthorn and Hereford cattle entered several head of select steers of the two breeds in a contest for the slaughter prize, and also that during the deliberations of the awarding committee the feeling ran very high as to the relative superiority of the meat. The microscopical tests were made at that time, but not so thoroughly as the State Board desired. The able paper of Dr. Sprague has been condensed as follows, and the result of that gentleman's examinations at-

LOVING CUP.
Presented to Geo. F. Morgan, inscribed as follows: Presented to Geo. F. Morgan, Esq., by a few breeders in England, in recognition of his labors to establish the Herefords in America, July 25th, 1883.

tracted great attention from the producers, meat cutters, and consumers of the country.

Dr. Sprague introduces the paper by anatomical comparisons, which, although technically of vast importance, do not relate to the merits of the meat examined, and he then states that:

"The specimens examined under the microscope, and also subjected to tests by cooking, were taken from the loin at the point of division of the fore, and from the hind quarters of the two and three-year-old Shorthorn grades, and from corresponding parts of the Hereford grades of like ages, these being animals slaughtered and dressed in competition at the recent Fat Stock Show in this city. Specimens from same part of carcass were procured of a butcher shop on Adams street, where only high-grade steers are cut up, and of low grade from two South Division shops, and also from the wholesale department of West Jackson street. A careful examination in every state from moisture to extreme dryness, running through a period of ten days, shows no discernible difference in color, structure of fiber, or the equable distribution of the fat vesicles and fat among the muscular fibers of the show steers. So after thoroughly weighing and sifting every feature of the structure of fiber and fat in the two breeds, as represented in the high grades presented as specimens, we conclude that superiority must be determined by early maturity —pounds of gain upon a given amount of feed; relative weight of offal, and of best to poorest parts, and by the test of public taste, based upon the exterior form and finish, as from outward appearance some prefer one breed and others the other. The specimens of high-grade steer beef procured on Adams street showed nearly all the qualities of high marbling of the exhibition steers, and there was no difference in coloring, fineness of fiber, distribution of fat vesicles, nor in flavor or odor when cooked, though there was some difference in the degree to which the fibers were filled with nutritive substance.

THE SCRUB MEAT.

"But a comparison of the scrub meat, or that part of it known as scalawags, with the high grade, prompts us to ever in the future turn our backs upon and refuse to take upon our plate a cut from the class referred to. The lean fibers from this class of carcasses are like rubber, and are held together by a glutinous substance; and not by cellular tissues and fat vesicles, filled with fat and easily separated, as in the case of the high-bred steers. The specimen pieces of the scrub meat—and this held good with the six examined, commenced to curl as soon as the drying process commenced, and in three days the pieces were curled up at the edges and thoroughly dry; whereas the cuts from the high grades remained moist and juicy, and lay flat upon the paper, though on hand three or four days before the others were procured. When dry the cuts of the scrub meat had no flexibility, would break before they would bend, and when cooked were without fat, juiciness or flavor, while the specimens from the high grades maintained all these qualities after having been kept for a full week exposed to the air."

The Doctor made a practical test by cooking samples of all the meats referred to, and his deductions from the tests are simply a corroboration of the opinion that the HIGH-GRADE MEATS ARE THE ONLY ONES FIT FOR THE TABLE.

Upon the question of the excessive fatness, about which so much has been said and written, the examiner, after minutely describing where this excess of fat is to be found, and the immediate cause thereof, says:

"It would be very proper, and a step in advance, to offer a prize on the beast, or rather to the man who should breed and rear the beast, capable of showing the highest specimen of marbling with the least accumulation of fat, merely as such, outside of and measurably away from the muscles. Men can express their likes and dislikes, it costs nothing to do this, but mere expressions of opinion cut no figure in so strictly a practical matter as growing meat for the million. The true inwardness of the subject is, as a rule, very imperfectly understood, though the ways of getting at a better understanding are being gradually found out. The idea entertained by some, generally vegetarians, that fat accumulated in the system is the result of fatty degeneration of muscles, is erroneous in every particular. Fat in the human system and in all animals is as natural an element as are muscles and skin. That some portion of the muscular system may (as is occasionally the case with the heart, which is (hollow) muscle) be the seat of deposit of an unnatural amount of adipose matter, causing the muscles to become pale and the heart's motion feeble, is no proof that all fat presents evidence of degeneration. The blood in perfect health contains fatty elements which in its rounds are given off and taken in by the fat vesicles, these latter holding it in store. The blood also carries the material out of which finger nails, hoofs and hairs are made, leaving this at the proper place to replenish waste; and it would be just as proper to charge the hair and the finger and toe nails with being a degeneration from muscular substance as to charge this to the usual accumulation of fat in the system of man or beast.

"Cattle breeders have always been divided in opinion as to the superiority of meat grown upon the frames of bony steers which are pasture-fed. Some breeders claim that meat thus made must of necessity be of a better quality on account of the continued outdoor exercise afforded and the natural food which the animal takes, while just as thorough and practical men insist upon the correctness of their views, that

A STALL-FED STEER

would produce as good, sound, hardy muscle and as choice meat as could be shown upon an animal fed in the open air and upon natural food." After stating that no one questions that exercise hardens the muscles, the Doctor says:

"The notion entertained by some that an animal cannot be mainly kept in a stall and be made to produce healthy, sound meat is a nonsensical assumption not sustained by physiology, nor is it by experience. That such an animal is liable to fatty degeneration is equally nonsensical. The evidences of health in the stall-fed ox are just as apparent, under reasonably favorable circumstances, as are the evidences of soundness in an apple grown in a fixed position on a limb."

Many of the most prominent breeders of the country pin their faith upon the idea that they can by handling ascertain and determine whether an animal will show marbled meat and other signs of good blood and thorough breeding. Because a steer presents a soft glossy hide, and the proper fullness in parts that indicate good blood and equally good care, these gentlemen have been known to go so far as to wager large sums of money that animals showing these outside signs of excellence would, upon being butchered, show highly marbled meat. In fact, they have professed to be able to locate this choice beef before death. Dr. Sprague held that this was not only impossible but preposterous, and that

THE BUTCHER'S TEST WAS THE ONLY ONE

that could establish the presence of these desirable qualities. Upon this subject he says:

"We acknowledged to having looked upon these meat tests with no little interest, in that we hoped to prove that the outward signs of a fine handler would point quite unerringly to the quality of the meat fibre and the marbling of it. So we selected a steer having a soft, mossy coat and mellow skin, with the other indications of a fine handler in a marked degree, upon which to apply the tests so long recognized and acted upon by breeders. This steer, upon being cut up, gave no evidence of any higher marbling of flesh than either of the others, no matter what their touch, under the hand. Failing to find corroborative evidence in this regard, we applied the microscope with care, in the hope of discovering a muscular fibre susceptible of minuter division by far than the fibre of other steers in the show that came under the butcher's hand. In this we may have been in a measure successful, as the minute filaments of the muscular fibre of the fine handling steer appeared under a lens of high power to be a little finer than others possessed, though the difference was not pronounced, was, in fact, slight, not manifest enough to base a distinction upon. This will be conceded when it is considered that the slight difference, if any existed, was only discovered through a very high magnifying power applied to a single filament, being the 100th division at least of a single minute muscular fibre. One object of this examination has been to discover, if practicable, outward

¶ 235

GOOD BOY (7668) 76240.
Bred by Earl of Coventry.

signs that would point with tolerable certainty to the inner structure during the life of the animal, that we might be able to predict, while the animal was upon its feet, what its fibre would be upon the block, and the extent to which this would be marbled.

"Hence, it is quite fitting, at this stage of the discussion, to refer to the fact that the deer, as well as many other animals, have exceedingly soft coats and pliable skins, yet no marbling of the flesh. These facts have their bearing, and point distinctly to the need of more lessons and closer study. We should think it doing violence to intimate that a steer of coarse outward texture and heavy bone, would show flesh of equal quality with one possessed of a moderate bone and general marks of fine texture upon the surface. But, as for any difference observable in the exterior structure of the steers shown at the late exhibition appearing to indicate a finer flesh fibre, or better degree of marbling, we think the evidence brought out through the tests upon the interior structure point not very distinctly to either beast.

"It occurs to us to hint, in conclusion, that the Board of Agriculture should require experiments upon low as well as high-grade meats, and that these tests should include the careful cooking and testing of the cooked samples. This test is applied to all other kinds of food entered for prizes. Our bread, butter, cheese and preserve making do not now need the safeguards of prizes and tests nearly so much as do our meats, as the differences in the latter are so imperfectly understood, even among the better class of consumers."

We give herewith the outlines fairly representing these two breeds—Hereford and Shorthorn—at the heart-girth, (¶ 201a) supposing them to be cut in two parts, at that point. It will be noticed that drawing a line horizontally across the centers of these two cuts, the greatest weight of the Hereford is above, and of the Shorthorn below such line.

We represent also a cut of a Hereford cow, and have drawn the lines as they appear, and marked and numbered the different cuts. (¶ 201b.)

The line running from the neck to the buttock would correspond with supposed line across the circular outline, and above this we may term the top.

In this cut may be found the prevailing or usual form of the Hereford. They are always better topped, or in other words they are the best in the best parts.

The parts numbered 1, 2, 3 and 9 are always better in weight and quality in the Herefords than in the Shorthorn; 10 and 11 are equally good as to weight, and better in quality. On this the advocate of the Shorthorn would take issue, for if in any point they have a show of quality with the Hereford, it is at this point; 12 (the thigh or round) with the Hereford is heavier on the inside and lighter on the outside, while the Shorthorn is lighter on the inside and heavier on the outside, giving the Hereford the largest amount of good beef here.

The parts 4 and 5 are better in the Hereford than in the Shorthorn, and in the Hereford better than 2 in the Shorthorn; 3 is where the fore roasts are cut from, and where the Hereford makes his weight of fore quarter 40 per cent thicker meat, and of the best quality. The top of the cow in the circle outline—broad and full in the crops—is the rule with the Hereford and the exception with the Shorthorn. And good meat is carried nearer to the horns and hocks on the Hereford than on the Shorthorn. The value of the Hereford on any market where known is above that of the Shorthorn because he is better in the best places.

CHAPTER XXVIII.

Teeth as Indicative of the Age of Cattle. Some Fat Stock Show Comparisons

We think it will be granted without controversy, that if a three-year-old is shown as a two-year-old, and a four-year-old as a three-year-old, and a five or six-year-old in the four-year-old class, there is ground for a very decided protest from the owners of the cattle that are exhibited at their right ages. In this connection, at the risk of being tedious, we shall introduce considerable testimony as to the reliability of the teeth as indicators of age in cattle, and as to irregularity and error in the ages of Shorthorn steers, against which Herefords were unfairly compelled to show. We will also reproduce cuts of teeth to better bring out this method of computing ages. This we first brought out in the "Breeders' Journal." In those days the "battle of the breeds" was a reality, "a condition and not a theory," as some "milk and water" adherents of the cattle trade would have us now believe.

Steers were exhibited at the Fat Stock Shows, that, allowing all due latitude for reasonable variation, carried around with them in their mouths conclusive evidence as to their fraudulent ages. It only needed to be utilized and made a part of the showing to put them in the class to which they belonged.

To show that the age of animals has been made a study by educated and perfectly reliable men, we were at much pains to procure from the highest source obtainable the latest and best information of the age of animals as shown by their teeth.

While in England we found that Prof. Brown, who is at the head of the Veterinary Department of the British Government, had published under the auspices of the Royal Agricultural Society cuts showing the ages of meat-producing animals. These cuts agree substantially with the cuts published herewith. From these cuts exhibitors, judges and visitors at cattle shows may be enabled to judge of the correctness of the ages of animals which are exhibited.

The steers exhibited by the T. L. Miller Co. in November, 1882, at two years and nine months, showed only four permanent incisors; at three years and nine months only six; and so far as we have examined we have found the same results.

Mr. John Price, of Court House, Pembridge, England, had two steers, one three years and six weeks, which had four full incisors, and the adjoining calf teeth were still firm in their sockets. Another steer at two years and five months had four permanent incisors, and two outside ones not yet fully grown. It will be noticed that Prof. Brown and his associates acted as inspectors for English exhibitions with power to disqualify animals from exhibition.

"Early maturity," says Prof. Brown, "is the *sine qua non* of breeders and exhibitors of farm stock, and it is one of the objects of agricultural societies to encourage them in their efforts to produce breeds which reach a state of perfect development at a comparatively youthful period.

"In order that judges of stock may be in a position to decide between animals of undoubted merit, the age is in all cases to be taken into account; and it is presumed that of the competing animals, which are in other respects equal, the youngest will receive the palm.

"Whether or not the practice is invariably in conformity with this theory may be open to question, but there is no doubt that the principle is correct, and in order that it may be carried into effect it is necessary to inform the judges of the exact age of each animal in every class.

"No argument is required to show that the breeder's certificate would afford the most reliable evidence of an animal's age if scrupulous exactitude and honesty were inherent human qualities; it is equally obvious that in their absence the evidence of a certificate is more calculated to mislead than to assist the judgment.

"It will probably be conceded, that persons who are concerned with the breeding and feeding of animals are not always free from risk of error, and it certainly can be proved by the logic of facts that all are not scrupulously exact in such matters as the statement of an animal's age. It therefore becomes necessary to supplement or correct the exhibitor's certificate, by such evidence as the animal itself affords; and by common consent, the periodical changes to which the teeth are subject, are accepted as indications of age, only second in value to positive proof of the date of the animal's birth.

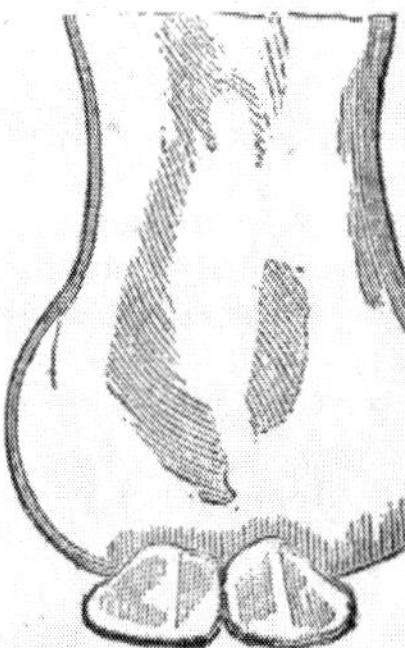

¶204

FIG. 1. CALF'S TEETH AT BIRTH.

"How far the cultivation of breeds, by artificial selection and high feeding, has influenced the development of the teeth, in common with other organs connected with nutritive functions, can only be inferred from the great difference which exists between the accounts of the older veterinary writers on dentition of animals and the facts which are familiar to the few experts of this generation who have taken the trouble to investigate the subject for themselves.

"In this country the most popular writer on veterinary science was Mr. Youatt, whose works on the horse, ox, sheep and pig were published by the Society for the Diffusion of Useful Knowledge. Youatt's remarks on the teeth are copied almost verbatim from M. Girard's work on dentition; and it may be without hesitation asserted, that if Girard's description of the teeth of the ox, sheep and pig were correct at the time when they were written, an improved system of breeding and feeding must have occasioned a very remarkable change in the rate of development of the teeth. In the year 1850 I commenced some investigations on the animals which were bred on the Royal Agricultural College Farm, for the common purposes of supplying milk and meat, and not especially for purposes of exhibition. The cattle on the farm were chiefly Shorthorns, the sheep were Cotswolds, and the pigs Berkshires, and none of the animals were at that time referred to as pedigree stock, nor were they fed on the forcing system. The observations of the first few months of 1850 sufficed to establish the fact that the teeth of cattle, sheep and swine were developed at much earlier periods than those which were stated in Youatt's remarks, copied from Girard.

"From the animals on the college farm the inquiry was extended to Mr. Stratton's valuable stock near Swindon, and to Mr. Kearsey's ram flock at Rodmarten. Among the more highly cultivated animals on these farms the process of dentition was not found to be more forward than among the stock on the college farm; and it is not generally more forward at the present time, after an intervening period of over thirty years of high feeding and careful breeding, in the improved races of cattle, sheep and swine than it was then.

"The contention of exhibitors is, that exceptional development is so frequent during dentition as to disturb any calculations which are based on a rule. In reply it may be stated that the most competent observers do not accept that view. On the contrary they are aware that the alleged exceptions do not often bear a critical investigation. Numerous inquiries have from time to time been made in compliance with the urgent demands of owners of disqualified animals, and those who have been most sedulous in searching for the truth are aware how vague and incomplete the evidence in support of the owner's certificate has been in most cases. Not uncommonly the entry has been proved to be incorrect, and in the few cases where the decision of the expert has been reversed, it has been done, only, on the plea that the exhibitor should have the benefit of the doubt.

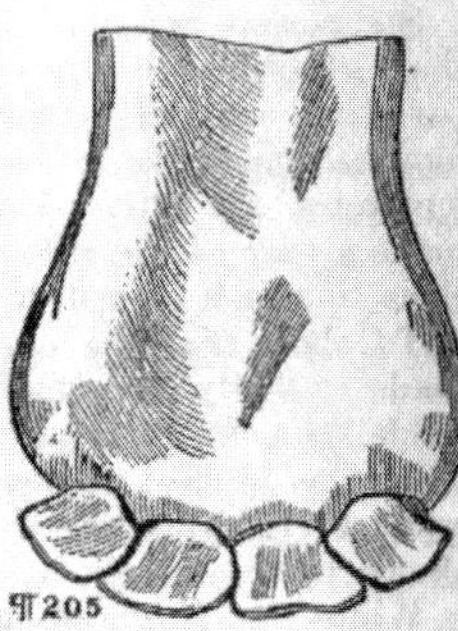

¶205

FIG. 2. CALF'S TEETH AT SECOND WEEK.

"Whenever a considerable number of disqualifications occur at the principal agricultural shows, the aggrieved exhibitors avail themselves of the aid of the press to vindicate their systems of recording the ages of their animals, and to show how impossible it is for any mistake to happen. In no case, however, within my own recollection, have any useful facts been brought to light as the result of these periodical effusions.

"On the occasion of the exhibition of the Smithfield Club in 1881 several pens of pigs were disqualified, and some of the exhibitors expressed their views on the subject of dentition, as indicative of the age, in the agricultural press. Mr. James Howard, M. P., wrote a letter to the 'Agricultural Gazette,' from which the following paragraph is quoted:

"'According to my own observation the dentition varied much according to families, and to a considerable extent in the same family, but in the majority of cases dentition in the same litter is tolerably uniform when three months old; at six months the boars, and particularly the more robust ones, show a marked forwardness; at nine months the dentition is often very varied; at twelve months some of the pigs of the same litter will show a state of dentition from two to three months in advance of others, and also in advance of the condition laid down in Prof. Simond's treatise.'

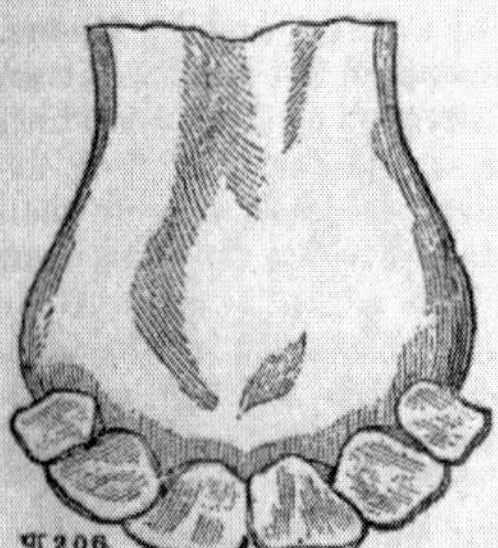

¶206

FIG. 3. CALF'S TEETH AT THIRD WEEK.

"This definite statement of the results of his own observations, by an eminent breeder and exhibitor, was far too important to pass over without notice. It will be observed that Mr. Howard's remarks had reference to the most critical ages, viz., six months, nine months and twelve months, and I, perhaps not unreasonably, indulged a hope that I might obtain an exact account of the particular variations which had been noted at the ages named. I therefore wrote to Mr. Howard for the information. In reply Mr. Howard enclosed a memorandum from his farm manager, expressing at the same time his regret that he did not take notes, adding that the farm manager is a most reliable and truthful man.

"The memorandum from the farm manager is as follows:

"'Britannia Farm, Bedford, Mar. 1, 1882.

"'The subject of dentition as indicating the ages of pigs was first brought under my notice when a pen of your pigs were disqualified at Birmingham show some years ago, and to prove for my own private satisfaction whether it could be relied upon, I from time to time examined litters when in my possession. I should think my examinations extended for about three years, and I know proved conclusively to my mind that no reliance can be placed on the dentition of a litter as a true test of the age. I kept no account or notes of my examinations at that time, not thinking it would come to such prominence as at the present time. The results were as given by you in a former letter, from which the enclosed paragraph is taken.'

"The enclosed paragraph referred to was the paragraph (quoted above) from Mr. Howard's letter to the 'Agricultural Gazette.'

"Another inquiry, which was instituted about the same time as the above, ended in an equally unsatisfactory manner.

"Mr. Sanders Spencer informed me that he had in his possession a pig of one year old which had the lateral permanent incisors. As such an abnormal state of dentition had not been seen before an opportunity was sought to inspect this dental phenomenon. Unfortunately the application was made too late. Mr. Spencer wrote in response:

"'Holywell Manor, St. Ives, Hunts,
"'February 6, 1882.

"'The yelt I mentioned to you on Thursday failed to breed, so she was sold to our village butcher, to whom I will today apply and try to obtain from him or the purchaser the lower jaw of the pig and will forward it to you if not damaged, as they usually are by the butcher when chopping off the snout. I believe I did not mention to you the circumstances that one of the yelt's central incisors (temporary) was broken off, so that the abnormal dentition may have been caused by accident.'

"The promised specimen was not obtained, but in reply to another letter of inquiry Mr. Spencer wrote as follows:

"'After finding the yelt had cut her permanent lateral incisors, I examined her molars, but found nothing unusual in their development. Nos. 1 and 2 were very much worn, and seemed quite ready to move for the permanent ones. No. 5 looked very white, but was apparently fully grown.'"

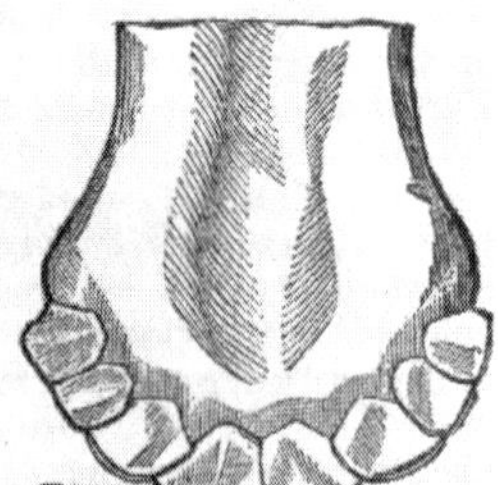

¶207

FIG. 4. CALF'S TEETH AT 1 MONTH.

"It appears from the above remarks that the state of dentition in respect of the molar teeth was exactly what it is expected to be at the age of one year, while the incisors indicated the animal to be six months older. That such a specimen was lost must remain a subject of regret.

"It is not intended in anything which has been stated to question the fact of the occurrence of irregularities in the dentition of the animals of the farm. On the contrary, my subsequent remarks will contain references to numerous and remarkable exceptions to the rule of development of the teeth of different animals. It is, however, contended that the expert is perfectly familiar with the exceptions and knows exactly how to make allowance for them in forming his opinion of an animal's age. Any one of common intelligence can become an expert in judging the age by the teeth if he chooses to devote some years to the patient and critical study of the subject, but until he has thus qualified himself he ought not to assume the right to criticise opinions based on evidence which he cannot appreciate. The events of the last exhibition of the society at Reading, without referring to similar cases which have been sufficiently numerous in past years, suffice to prove that what ought to be the unquestionable evidence of a breeder's certificate cannot always be accepted without hesitation. At the last show eight pens of pigs were absolutely disqualified, and exhibitors were cautioned in respect of the pigs in seven other pens. Pigs which were shown under the condition that they should not exceed two months had the dentition of three months and six months respectively. Pigs shown under six months had the one-year-old teeth well up, and other in the same class had the dentition of eight and nine months. One man was seen to move a pig, after the inspection was finished, from an old class into a younger one. The attempt to secure for the animal a better position than it was entitled to really led to its disqualification, a consequence which, however unpleasant for the exhibitor, is not the most serious outcome of the attempted fraud. Such proceedings as those which have been referred to naturally tend to excite universal suspicion, which may often fall unjustly and without sufficient reason.

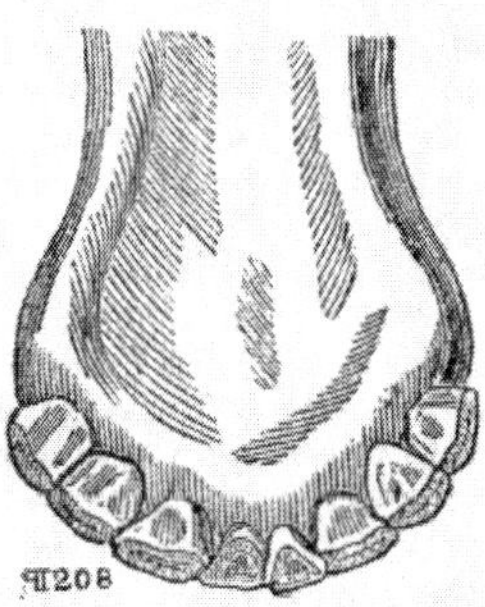

FIG. 5. CALF'S TEETH AT 8 MONTHS.

"In one sense it may be considered satisfactory that from the first disqualifications have been recorded chiefly in respect of the pigs of a few exhibitors, who year after year, in spite of warning, persisted in trying how far they might presume on the forbearance of the inspectors, urging in reply to all remonstrances that it would not answer for one or two to 'show straight' unless they all agreed to do so.

"The great majority of exhibitors during the last twenty years have not given any grounds of complaint, and in many instances, notably among the pigs exhibited by Her Majesty, by the Royal Agricultural College, and by Messrs. Howard, the development of the teeth has been rather below than in advance of the rules which have been laid down as the result of long experience. Altogether the evidence of more than a quarter of a century justifies the statement that the evidence of age which is afforded by the teeth, without being absolutely irrefragable, is the most reliable, under all the circumstances, which can be obtained."

THE TEETH OF THE OX.

"In the front of the mouth of the ox there are eight incisors or cutting teeth in the lower jaw only, an elastic pad of fibrous tissue, covered with mucous membrane, takes the place of teeth in front of the upper jaw. The incisors may be distinguished as centrals, or first pair; middles, or second pair; laterals, or third pair, and corners, or fourth pair; the same term being equally applicable to the temporary and permanent organs.

"Temporary incisor teeth are easily distinguished from permanent, chiefly by their size. The fangs of the temporary incisor teeth are much shorter than those of the permanent incisors, but this fact is not to be recognized until the teeth are removed from the jaw. No question is likely to arise in the mind of the examiner as to the distinction between temporary and permanent organs; in fact, the common term broad teeth, as applied to the latter, sufficiently indicates their prominent feature.

"Molar teeth are named first, second or third, according to their position. In the temporary set there are three molars on each side of the upper and lower jaws, and in the adult these teeth are changed for permanent organs, while three additional teeth, the fourth, fifth and sixth in position, all of which are permanent teeth from the first, are added, making the full set of permanent molar teeth six in each side of the upper and lower jaws.

"In the mouth of the calf at birth the temporary teeth, molars and incisors are all so far advanced that they may be seen in outline under the gum, and commonly the cutting edges of the incisors and a few points of the molars are uncovered.

"The advance of the teeth and the receding of the gums proceed very rapidly after birth, and at the age of one month the temporary teeth, viz., eight incisors in the lower jaw and three molars in each side in the top and bottom jaws, are fully developed.

"No accurate opinion of the age of a calf can be formed from the observation of the state of dentition between the ages of one and six months, when the fourth molar is cut; but during this period the jaws expand, the incisor teeth gradually become less crowded, and the space between the third molar and the angle of the jaw increases as the fourth molar, which is the first permanent tooth, advances to occupy its space.

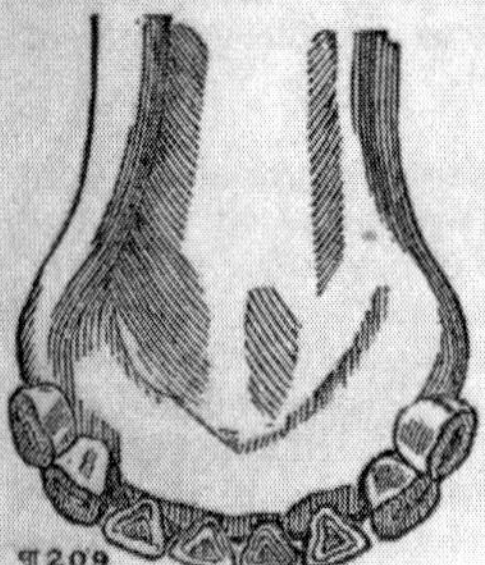

¶209

FIG. 6. CALF'S TEETH AT 11 MONTHS.

"At the age of six months the fourth molar is well developed, but it is in close contact with the angle of the jaw, and the posterior surface is not quite free from the covering of the gum.

"Between six and twelve months old there are no important dental changes, the incisor teeth become worn, and as the jaws increase in size there is more space left between them, but it is not possible to assert from the state of the incisor teeth whether an animal is under or over the age of one year. (Fig. 7.)

"Shortly after one year the fifth molar begins to make its appearance, and at fifteen months it is well up. The appearance which the fifth molar presents at this age is very much like that of the fourth molar at the age of six months. The new tooth is in close contact with the angle of the jaw, and the gum covers the extreme posterior part of its surface.

"No change occurs in the incisors, except that which is caused by the wear of the teeth and the growth of the jaw, until the age of one year and eight or nine months, at which time the two central teeth are loose, and the first broad teeth sometimes begin to project through the gum. In very forward animals the central permanent incisors are cut at the age of a year and seven months, but they are never level with the other incisors before one year and ten months, and their perfect development is indicative of the age of two years.

"The illustration (Fig. 10) was taken from a Shorthorn heifer at the age of one year and ten months, and may be accepted as indicative of the general appearance of the incisors at that age.

"While the first pair of permanent incisors are advancing to take the place of the temporary teeth, the sixth and last permanent molars push their way through the gum, and at the age of two years are in position. Any error of opinion as to the age which might arise from the premature cutting of the central permanent incisors may be corrected by reference to the state of the molars.

"The sixth molar takes the position described, and the first and second permanent molars take the places of the temporary teeth; this change, in my experience, does not usually occur until the animal is a month or two over two years of age.

"From two years and three months to two years and six months the second pair of broad teeth—the middle permanent incisors—occupy the place of the corresponding temporary teeth in all the cultivated breeds. Instances of late dentition present themselves from time to time in which the middle permanent incisors are not cut until the animal is approaching three years old. There is consequently a possible variation of six months in the time of the appearance of these teeth.

"In the illustration (Fig. 12) the ordinary

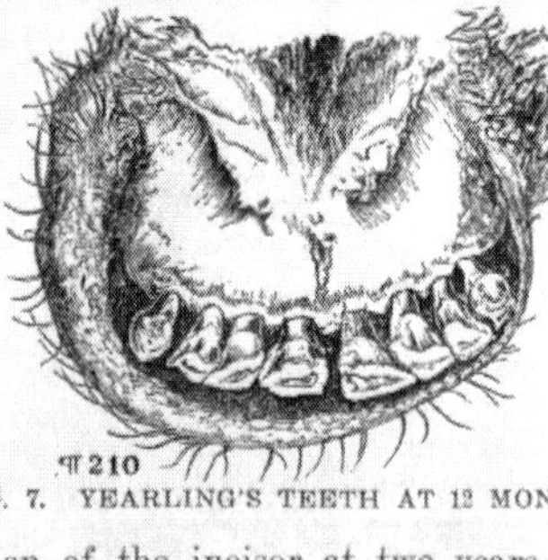

¶210

FIG. 7. YEARLING'S TEETH AT 12 MONTHS.

condition of the incisor at two years and six months is shown.

"It must be understood in reference to the appearance of the second pair of broad teeth that an expert, looking at a mouth which corresponds to the drawings, will conclude that the animal is two years and a half old, but if he is required to certify that the age is under or above that period he must proceed to inspect the molar teeth, and take into account the animal's pedigree, its sex, and its general condition of development.

"If the animal in question is a bull of one of the cultivated breeds, and has been forced to a state of early maturity, it may be expected that the second pair of permanent incisors will have been cut at two years and four months; and if either of the anterior temporary molars remain in their places, the conclusion that the animal is under two years and a half will be strengthened.

"Shortly after the first and second molars are cut the third makes its appearance; occasionally it appears before the others, and the animal at the age of three years will have three anterior molars nearly level with the other teeth, but showing no signs of wear.

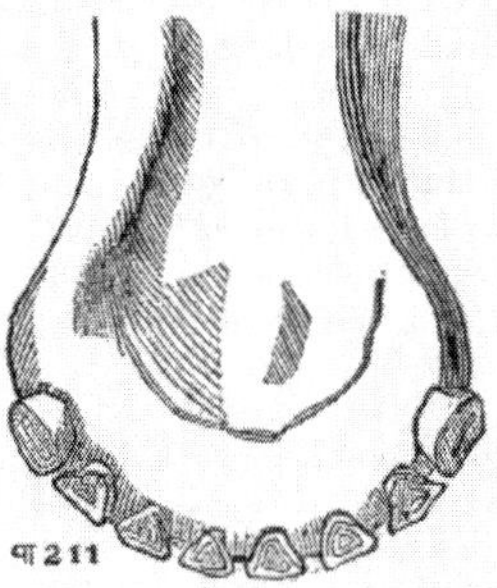

FIG. 8. YEARLING'S TEETH AT 15 MONTHS.

"The eruption of the third pair of permanent incisors may occur at any time between two years and six months and three years of age. In cultivated breeds they are present as a rule before the animal is three years old, and occasionally they will be found well developed after two years and a half. In fact the anterior molars afford more reliable evidence of the age between two and a half and three years old than is furnished by the incisors.

"The fourth pair of broad teeth, the corner permanent incisors, are more subject to variation in the time of cutting than the third pair. In well bred cattle they take the place of the temporary teeth soon after the completion of the third year, but in bulls they are not uncommonly present at two years and ten months old, while in some instances they are not cut until three years and nine months old. Very little reliance, indeed, can be placed on the evidence of the corner incisors, and the examiner is compelled to refer to the molar teeth for the purpose of correcting his opinion.

"With the eruption of the corner permanent incisors, the fourth pair of broad teeth, the permanent dentition of the ox is completed, and after this period the changes in the form of the teeth which are due to attrition will assist the examiner in forming an opinion of the age, but no exact estimate can be based on such evidence. The five-year-old ox will show a considerable amount of worn surface in the central, middle and lateral incisors, and the cutting edge of the corner teeth will be marked by a line of wear, but no one would attempt to determine whether or not an animal were under or above the age of five, and as the years increase the difficulty of judging the age by the appearance of the teeth is not diminished. The teeth become narrower and more widely separated from each other year after year, but the changes are not sufficiently well marked to enable the examiner to accept them as reliable evidence of age."

The Illinois State Board of Agriculture were given a chance to put themselves on record as to whether they wished to have their records of any value as to weights, gain per day and early maturity of the cattle exhibited at the Chicago Fat Stock Show. It must be remembered that the money of the people of Illinois was used by this board. We appeared before the State Board of Agriculture at their annual meeting held at Springfield, January, 1883. We then presented them with the following written statement of facts:

"To the Honorable State Board of Agriculture of the State of Illinois: At the late Fat Stock Show held at Chicago I called the attention of President Scott to the fact that cattle were on exhibition at fraudulent ages, and requested of him that he would bring the information to the board to the end that they might take such action in the matter as would protect the board, the exhibitor and the public from the effect of such fraudulent entries. Whether he brought the matter to the notice of the board or not I am not informed, and, so far as I know, no action was taken on the information.

"The interview that I refer to was on the first day of the show. I again called his attention to the fact that D. M. Moninger of Iowa had on exhibition a steer called the 'Champion of Iowa' at 715 days old, or one year 11 months and 15 days old; that the steer had six full teeth, while he was not entitled, if his age was correctly stated, to have more than two full teeth. That J. H. Potts & Son had a steer entered as 715 days old that had but two full teeth, and that the T. L. Miller Company had a steer entered as 714 days old which only had two full teeth. These two steers had teeth that corresponded with the age that they were entered at, while the teeth of Mr. Moninger's would indicate that he was full three, if not a four-year-old steer.

"I had several of the steers examined, but I

wish now to call your attention to the fact that not only was the steer 'Champion of Iowa' entered at a fraudulent age, but that I have ascertained that the steers 'Tom Brown' and 'Grinnell,' so called, and numbered in the catalogue published by your society 111 and 113, were butchered by Messrs. Eastland & Duddleston, of Chicago, and that the teeth of these two steers, which are now in my possession, show 'Tom Brown' to have been at least five years old, the steer 'Grinnell' to have been at least four years old, by the usual standard authorities, instead of two and three years old, as shown in the catalogue, and these ages may be established by comparison with living animals of today. I also call your attention to the entries of Messrs. H. & I. Groff, of Elmira, Can., for animals Nos. 15, 17, 18, as 902 days old, 1,265 and 1,305 days; the first entered as two-year-old, the others as three-year-olds.

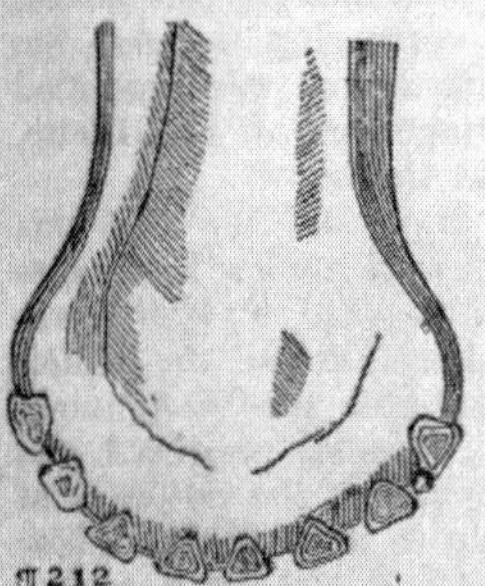
¶212

FIG. 9. YEARLING'S TEETH AT 18 MONTHS.

"These animals were slaughtered by Mr. John Ford, and from him I purchased the heads, from which I took the teeth which I presented to you today. These teeth, by the usual standard authorities, would show them to be from five to six years old, and by comparing them with living animals of today these conclusions would be sustained.

"Believing that your board, with the exhibits you are making, should go forth to the world with truthful results, and that it is your desire to protect the exhibitors who comply honestly with your rules, and that you will not seek to avoid the responsibilities, by the technical form in which the information may reach you, I have appeared before you with the witnesses, to-wit, the teeth of the bullocks. And I charge these exhibitors, to-wit: Mr. D. M. Moninger and Messrs. H. & I. Groff, with the entering of the cattle under false and fraudulent ages, for the purpose of deceiving the committee, the public and to obtain a reward by misrepresentation. T. L. Miller."

The report of the committee to whom the matter was referred placed themselves on record as follows:

"To the State Board of Agriculture:

"Your committee, to whom the communication of Mr. T. L. Miller in regard to the matter of determining the age of animals exhibited at Fat Stock Show of 1882, and complaining of the action of the president of this board in a particular case involving that question was referred, beg leave to say that, primarily, Mr. Miller himself failed to do what, by his own statement was clearly his duty, in not bringing his complaint to the attention of the board in a manner recognized by the established rules of which he was undoubtedly cognizant. Mr. Miller could not have been ignorant of his own plain duty as an exhibitor, and having failed to protest in any proper form against the exhibition of any animal or animals he mentions, the board has not, in his mere verbal complaints, any ground to institute an investigation, or to do any act which could be made to imply a doubt as to the correctness of the statement of any other exhibitor. In other words, your committee believe that the rules relating to this subject are clear and practical, and that Mr. Miller perfectly understood what his rights and duties were in the premises; that the president did exactly what the board had a right to expect its executive officer to do, and certainly gave not the slightest occasion for the complaints made by Mr. Miller, and therefore recommend that no further action of this board in the premises is demanded.

J. Irving Pearce,
John Virgin,
John P. Reynolds,
Committee."

It can be seen from this that they cared nothing for the facts in the case, but shielded themselves behind a technicality, that a written protest was not entered the day the verbal protests were made. We then wrote the following on the subject, which will more fully explain how much those in interest knew of the subject

¶213

FIG. 10. YEARLING'S TEETH AT 22 MONTHS.

before and during the continuance of the Fat Stock Show:

"Editor 'Breeders' Journal':

"My relation to this interest is well known, but in this article I propose to confine myself clearly and closely to the Fat Stock Show of 1882. There has been, through the live stock journals, in the interest of the Shorthorns, a great blowing of trumpets and a heralding as to what that interest was to do at the coming show. On the 13th of November exhibitors commenced coming in with their stock, and between that and the 16th items of interest in relation to the entries were generally known. When I came into the exhibition building on Thursday morning, the 16th, I was informed that large numbers of the Shorthorns were entered under their actual ages. Ascertaining in a few instances by examination that these reports were true, I called upon President Scott and found him presiding at a meeting of the State Board of Agriculture. He met me in the ante-room (¶ 202) and I advised him that, from the best information that I could get, there were a great many cattle in the show older than the entries would indicate. I asked him to bring the matter before the board and have them take the initiatory steps to ascertain the truth of these reports. He said to me that the proper way to get it before the board was by protest. I replied that it was not proper to put that responsibility upon the exhibitors, and offered to place the animals I had on exhibition under any test that the board might see fit to make, and put myself and herd under any examination that they might direct. Mr. Scott, however, called my attention to this section: 'In case of protest, notice must be given to the superintendent of the department before or during the examination of the animal or article protested, and a written statement setting forth the reasons for protesting, verified by affidavit, must be filed with the secretary on the day that the notice is given.'

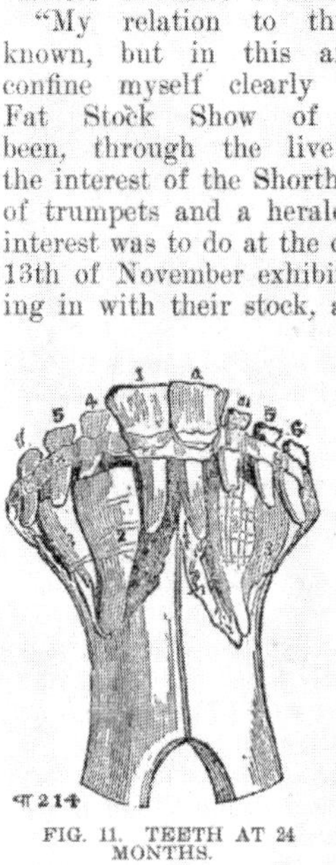

FIG. 11. TEETH AT 24 MONTHS.

"After leaving Mr. Scott I met Mr. J. H. Sanders of the 'Breeders' Gazette,' and called his attention to the fact that entries of animals were made under their actual ages, and said to him that I had just come from Mr. Scott, to whom I had given notice of the fact, and asked him (Sanders) to use his influence to see that examinations were made to see if such fraud existed, and if so to see that it was exposed, to which proposition he gave his assent, and said that he would do all that he could in the matter. Nothing was done by the board. Mr. Sanders at an early day became the medium through which a cane was presented to Messrs. D. M. Moninger and J. D. Gillette, Shorthorn exhibitors at this show.

"I had determined two years ago that this matter of entering animals under their ages should be exposed. A year ago my health was such that I was not able to attend the exhibition, but of the animals that were slaughtered that year the mouths were all preserved and properly labeled and given to the secretary of the board, by him taken to Springfield. These showed that the Shorthorns were of a greater age than the Herefords. The secretary says that he sent them to his house to be boiled for the purpose of taking the flesh from the bones and thus putting them in a shape for preservation. This process destroyed them and their identity. I have only to say in reference to it that this being the true version of their destruction it showed a want of judgment. I had in the course of the last year given notice that I should make special efforts to expose any fraud that should be made by entering cattle under their actual ages. Wishing, however, to take no advantage of the board, or of the exhibitors, I thus gave President Scott notice of the fact that fraudulent entries were made, and called the attention of Mr. Sanders, the editor of the leading Shorthorn organ, to the fact, and there rested the case until the last day but one of the show, when I called President Scott's attention again to the fact that Mr. D. M. Moninger, of Galvin, Iowa, had a steer entered as a yearling, called the 'Champion of Iowa,' and numbered in their catalogue as 112, which was a long three-year-old or over, and cited to him the fact, that for the age, he should have but two permanent teeth, while in fact he had six; and that, besides this error, there were others that were much older than they were entered for. He again called my attention to the fact that it was a duty of the exhibitor dissatisfied to come before the board with a protest, as specified in Section 5, (¶ 203) heretofore quoted. I advised him that I did not come before the board as a protestor, and called his attention to Section 12, which reads as follows:

"'Awarding committees are instructed that

if they have good reason to believe that any exhibitor, by false entry or otherwise, attempts to deceive the committee or the public and obtain an award by misrepresentation, they shall report the fact at once to the superintendent of the department, who shall report the same to the board, who may expel such exhibitor for fraud for at least two years.' He declined to take any action except I should come to him as prescribed in Section 5.

"I then went to the Hon. John P. Reynolds,

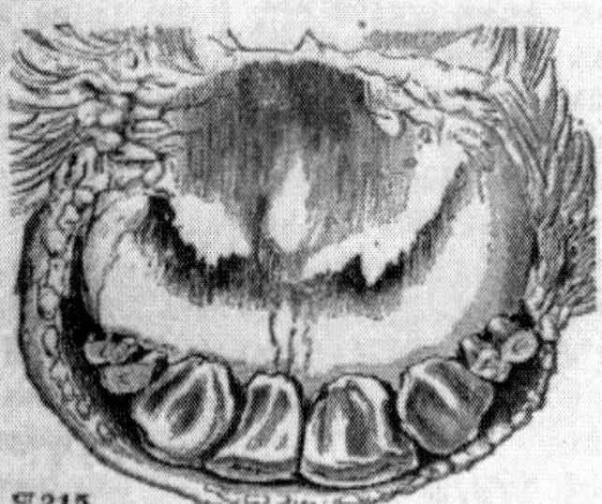

FIG 12. TEETH AT 2 YEARS AND 6 MONTHS.

a member of the board, and called his attention to these facts, and to the fact that I had seen the president and he declined to take any action, and that I should bring the matter before the Hereford Society, which was to meet that evening, and ask of them to take some definite action in reference to these frauds. He afterwards saw Mr. C. M. Culbertson and referred to the conversation I had had with him, and asked of Mr. Culbertson that the Hereford Society should take no action, and that he would see that an investigation was made into these facts. In the evening of this day at a meeting of the Hereford Society I brought up the fact of this steer in particular, and stated that there were many other entries in the show that were fraudulent, and asked what action might be taken on the merits of these entries. Mr. Culbertson then reported the request of Mr. Reynolds that no action should be taken, and that he would endeavor to have an investigation made in reference to the charges as to the fraudulent entries of the Moninger steers. At the meeting of the board on the following day the matter was brought up. As I understood, President Scott opposed the action being taken for the reason that it had not come before the board in the form and manner prescribed by the rules, to-wit: by protest of a competing exhibitor. This view appears to have prevailed, and no action was taken to ascertain ages of the steers on exhibition at this show. The probabilities are that but for this promise of Mr. Reynolds the Hereford Society would have taken some definite action to determine the ages of the Moninger steers and others. The Hereford Society owed it to themselves to secure a thorough investigation as to the ages of the cattle on exhibition. It was generally understood among the exhibitors that the Shorthorns were entered under their actual ages. Mr. Moninger's cattle in particular were examined in comparison with other cattle in the show, especially with the Herefords.

"This was known as well by some members of the board as by the exhibitors at the show. In my interview with Mr. Scott I took the ground that it was the board's business to protect the exhibitors, and that they had no right to compel the exhibitors to protect themselves. Following these efforts to get justice from the board, I found that two of Moninger's steers had been butchered and three cattle exhibited by Messrs. Groff, of Canada, had also been slaughtered by Messrs. Eastland & Duddleston. These were 111 and 113 of the exhibition catalogue; 111 known as 'Grinnell' and 113 as 'Tom Brown.' The others were 15, 17 and 18 of the exhibition catalogue, and were slaughtered by Mr. Ford. These teeth I took before the State Board at its meeting on the second Tuesday in January; first in an interview with Messrs. Scott and Reynolds at Mr. Scott's rooms in the Leland House, Springfield, and afterwards the full board at the agricultural rooms in the State House. The indications as to the age of these animals was discussed as shown by these teeth, and afterwards put in form, formally charging fraud upon the exhibitors, as per the statements herewith, and with the report of the committee, Hon. J. P. Reynolds, J. Irving Pearce and John Virgin.

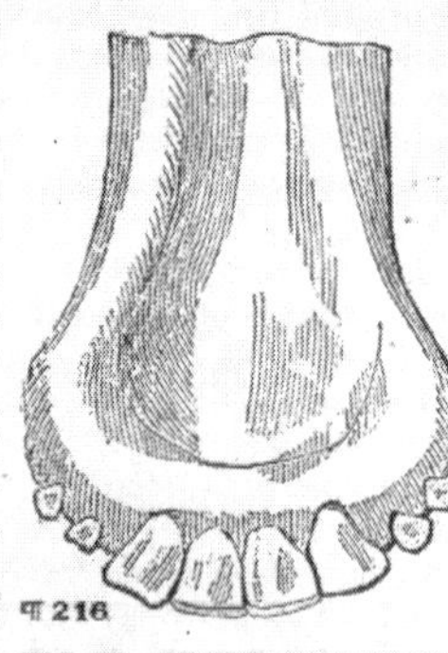

FIG. 13. TEETH AT 3 YEARS.

"This is substantially the case as it progressed. I was desirous that the board should know these facts, and hoped that when it was brought before them they would recognize the necessity of taking action on the information. With the same hope in view, the matter was brought to them on the last day but one of the show; and still with the same hope, action was deferred by the Hereford Society, and with the same hope I brought the proof as exhibited by the teeth of these five animals to Messrs. Scott and Reynolds individually and before going before the board. I had no doubt that when the matter was fully presented to the board they would take the facts as the basis upon which to start an investigation to determine whether the frauds had been committed or not. I confess to some surprise that the committee named should have made a report that would carry the impression that if I wished to bring this matter to a test it should come in shape of a protest as the cattle were brought into the ring for exhibition. It is true that I had the right to make a protest as suggested by the committee, but it was not at all obligatory upon me to do it. It is true that the board have a rule that requires of the judges if they think an exhibitor by false entry or otherwise has attempted to commit a fraud, shall report it to the superintendent of the cattle department, and he shall report it to the board, and they may expel such exhibitor from exhibiting for two years, but this would not be supposed to confine the board in taking cognizance of fraud to these two processes.

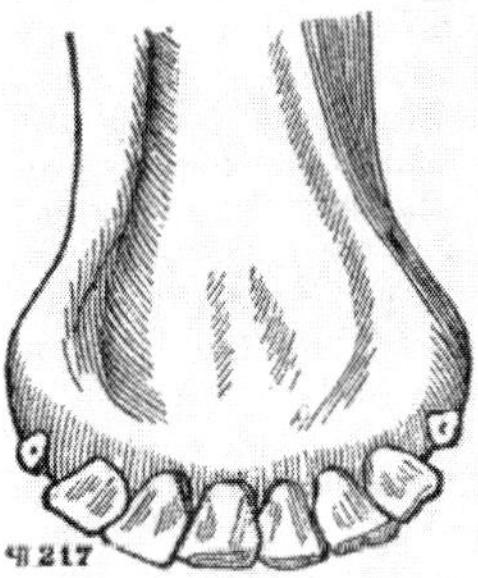
¶ 217

FIG. 14. TEETH AT 4 YEARS.

"I prepared engravings of the teeth, which are recognized by long-established authorities as indicating very closely the age of cattle, at different periods of their existence, and with these engravings the views of Prof. Youatt's History of British Cattle, which was published in 1834. He says at page 279, in reference to the rings on the horns indicating the age: 'These rings, proving the first growth of the horns at the base, have been considered as forming a criterion by which to determine the age of the ox. At three years old the first one is usually observed. At four years old, two are seen, and so on, one being added each succeeding year. Thus is deduced the rule that if two were added to the number of rings the age of the animal would be given. These rings, however, are perfectly distinct only in the cows. They do not appear in the ox until he is five years old, and often in the bull they are either not seen until five or they cannot be traced at all. These rings are not all distinct even in the cow. The three-year or first may be so, but then comes a succession or irregularities of surface that can scarcely be said to be rings, and which it is impossible to count. Another circumstance which must also be taken into account, is that, the heifer goes to the bull when she is two years old or a little before; after that time there is an immediate change in the horn and the first ring appears, so under the rule a three-year-old would carry the mark of a four-year-old. To this may be added that after the beast is six or seven years old these rings are so irregular in their appearance, and so little to be depended upon, that the age indicated by the two horns is not always the same. I have repeatedly seen a difference of one year, and in some instances I could not make the horns agree by two years at least. Further, regarding this process of nature, it is far too irregular for any certain dependence to be placed upon it; it is a mere general rule with far too many exceptions. There is also a certain instrument called the rasp, the use of which has been said to have made many an arm ache a little before a large cattle fair. What human being can tell whether the ring farthest from the head has or has not been removed, or whether the second may not have followed the first? If the rasp is fine and gently used, and a little dirt with or without soap is rubbed over the part, there is nothing to tell tales, except the rather too great smoothness of the horns thereabouts, and this is said to be obviated by giving the whole of the horn a smooth and polished appearance. We have never liked these small, smooth, polished horns. That art had been at work no one could deny, and we were uncharitable enough to suspect that the removal was oftener employed in the removal of a defect than in the heightening of a beauty. Dealers are not so bad as horse merchants, but strange stories have been told of them. We are less scrupulous in describing this deception because we shall presently speak of a method of judging cattle where no roguery can lead us astray.' This is Prof. Youatt's view of the unreliability of judging the age of cattle by the horns. 'As

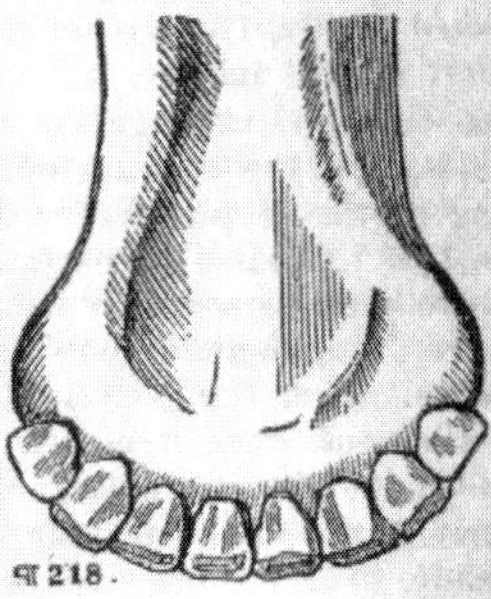
¶ 218.

FIG. 15. TEETH AT 5 YEARS.

to the teeth,' he says, 'far surer marks are presented in the teeth, and where there can be little deception from the hands of dealers (or exhibitors), for their interest would generally lead them to give a more youthful appearance than nature has allowed.' (I have preferred to commence with the ages of the calf as indicated by the teeth, although these have not an especial bearing upon the question which induced the presentation of these evidences.)

"Youatt says: 'The mouth of the new-born

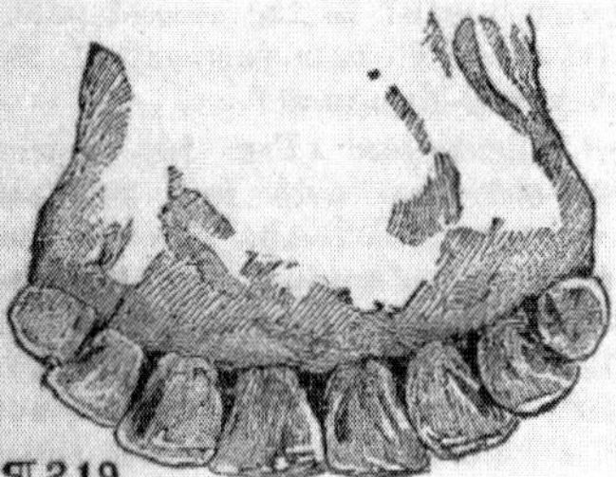

¶219

FIG. 16. TEETH OF SHORTHORN "CANADIAN CHAMPION."
(Exhibited as 3 years 6 months 15 days.)

calf presents an uncertain appearance, depending on the mother having exceeded or fallen short of the period of utero-gestation. Sometimes there will be no vestige of teeth, but generally either two central incisors will be protruding through the gums, or they will have arisen and attained considerable bulk. (Fig. 1.) About the middle or close of the second week a tooth will be added on either side, making four incisors, as seen in Fig. 2. At the expiration of the third week the animal has six temporary incisors or front teeth, as seen in Fig. 3. At a month the full number of incisors will have appeared. (Fig. 4.) These are the temporary or milk teeth. The enamel will be seen covering the whole crown of the tooth, but not entering into its composition, as in the horse; and it will be observed that the edge is exceedingly sharp. The only indication of increasing age will be the wearing down of these sharp edges, and the appearance of the bony substance of the tooth beneath.

"'The two corner teeth will scarcely be up before the center teeth will be a little worn. At two months the edge of the four central teeth will be evidently worn; yet, as the wearing is not across the top of the tooth, but a very little out of the line of its inner surface, the edge will remain nearly or quite as sharp as before. At three months the six central teeth and at four months the whole set will be worn, and the central ones most of all, but after the second or third month the edge of the teeth will begin to wear down, and there will be more of a flat surface, with a broad line in the center.

"'About this time a change will begin, but very slowly to be seen. The central teeth will not only be worn down on their edges, but the whole of the tooth will appear diminished. A kind of absorption will have commenced. There will be a little but increasing space between the teeth. The face of the tooth will likewise be altered, the inner edge will be worn down more than the outer, and the mark will change from the appearance of a broad line to a triangular shape. The commencement of this alteration of form and diminution of size may be traced to about the fourth month. Our cut gives a representation of the two central incisors at eight months. (See Fig. 5.) The central teeth are now not above half the size of the next pair, and they are evidently lessened. At eleven months the process of diminution will have extended to the four central teeth in the manner represented in the cut. (See Fig. 6.) The vacuities between them will now be evident enough. Cut (Fig. 8) gives the mouth of a steer fifteen months old. Cut (Fig. 9) gives the curious and diminutive appearance of all the incisors in a bullock eighteen months old. It would appear difficult for him to obtain sufficient food to support himself in good condition. It is somewhat so, and it may be in a great measure owing to these changes in the teeth and the difficulty of grazing that young beasts are subject to many disorders from seven to eight months upwards, and are so often out of condition. They contrive, however,

¶220

FIG. 17. MOUTH OF "CANADIAN CHAMPION."

to make up for this temporary disadvantage by diligence in feeding, and, to allude for a moment to another animal, we have known many a not only broken-mouthed but toothless ewe to thrive as well as any of the flock, for she was grazing all the day and ruminating all night.

"'At this time, eighteen months old, the corner teeth will not be more than half their natural size; the center ones will be yet more diminished, and, as the cut very plainly represents, the vacuities between them will be almost equal to the width of the teeth. The faces of the teeth also, such faces as remain, will be lengthened; the triangular mark will diminish, principally in the central teeth, while a line, more or less deeply shaded, will begin to appear around the original mark.

"'All this, while the second set of teeth, the permanent ones, have been growing in their sockets and approaching towards the gums, but not, as is said to be generally the case with other animals and with the human being in particular, pressing upon the roots of the milk teeth, and causing them to be absorbed, until at length, losing all hold in the socket, they fall out. The process of absorption commences here in the whole milk tooth, and as much in the crown or body of it as at its root.

"'The process of general diminution seems now for a while retarded; it is confined to the

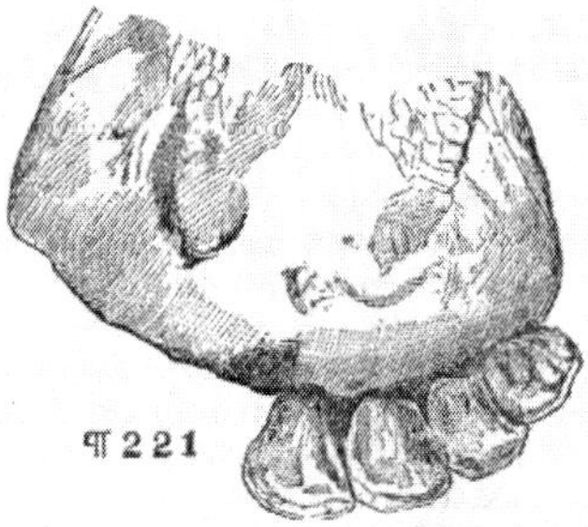

FIG. 18. TEETH OF "KING OF THE WEST."
(Exhibited as 3 years 7 months 15 days.)

central teeth, and they gradually waste away until they are no larger in the body than crow quills. About the expiration of the second year, or a little before, the milk teeth are pushed out, or give way, and the two central permanent teeth appear.' I have here given Prof. Youatt on the growth and absorption of the teeth generally termed calf teeth up to the time that the animal reaches two years old. Prof. Youatt says: 'It will be seen here that about the expiration of the second year the milk teeth are pushed out and the central permanent teeth appear.' We wish our readers to note the process of teeth formation for the first two years of calf life. I have now reached a point where I begin to touch upon the teeth indications as to the age of cattle entitled to enter for exhibition at the Fat Stock Shows.

"The illustration (Fig. 11) gives the mouth of a two-year-old. 'The two permanent central incisors are coming up, and the other six milk teeth remain. The bone front of the lower jaw is taken away in order that the alveoli, or cells for the teeth, may be exposed. The second pair of incisors have almost attained their proper form. The third pair are getting ready, but the jaw is not yet sufficiently widened for the development of the fourth pair. The process of absorption will still be suspended with regard to the two outside pairs of teeth, but will be rapid with regard to the second pair, and a little before the commencement of the third year they will disappear.'

"The illustration (Fig. 13) represents a four-year-old beast with four permanent incisors and four milk teeth. Now the remaining milk teeth will diminish very fast, but they show no disposition to give way, and at four years old there will be six permanent incisors, and often apparently no milk teeth, but if the mouth is examined the tooth that should have disappeared and the tooth that is to remain until the next year are huddled close together and concealed behind the new permanent tooth. They often are a source of annoyance to the animal, and the tooth whose turn it was to go must be drawn.

"It is proper here to state that a four-year-old mouth is as represented (in Fig. 14). It contains six permanent incisors and two milk teeth. Now this latter engraving (Fig. 14) corresponds with the mouth of Mr. Moninger's steer called 'Champion of Iowa,' entered as a yearling, or 715 days old. At the commencement of the fifth year the eight permanent incisors will be up, but the corner ones will be small.

"Another illustration (Fig. 15) gives a five-year-old mouth, or perhaps one a month or two over five years; so that the beast cannot be called full-mouthed, namely all the incisors, until it is six years old. 'It will be seen, however, in this mouth of five years, that the two central pairs are beginning to be worn down at the edges, and that in a flat direction or somewhat inclining inside.'

"I will not follow Mr. Youatt further, as I reach the full limit in which cattle should be exhibited for premiums at this show, and the further age is exhibited by absorption and teeth growing longer and having the appearance of being longer with spaces between; that is, there commences from this time on a gradual absorption and wearing away of the teeth.

"Referring to the December, 1882, 'Breeders'

Journal,' it will be noticed that H. & I. Groff entered an animal, No. 15, of the exhibition catalogue, as dropped May the 27th, 1880, and another, No. 17, May 29, 1879, and No. 18 as April the 19th, 1879. This makes No. 15 two years six months and two days old. I present a cut (Fig. 20), which shows the mouth of this animal, 'Young Aberdeen,' stall 15, and I believe him—Prof. Youatt being the authority—to be five years old. I also present a cut (Fig. 21) showing the outside of 'Young Aberdeen's' teeth, and that absorption has already begun. I also present cut (Fig. 16), which is taken from the mouth of the steer 'Canada Champion,' who stood in stall No. 17 in the exhibition catalogue, and a cut (Fig. 17), which shows the outside of the same mouth and from the same steer 'Canada Champion.'

"Another cut (Fig. 18) was from the mouth of the steer entered in the exhibition catalogue as No. 18. One half of the mouth was destroyed in slaughtering, but the other half answered equally well as though the whole were present. The cut (Fig. 19) shows the outside of the same mouth. Nos. 17 and 18, by referring to the exhibition catalogue, were entered as follows: No. 17, as dropped May 29, 1879, and as 1,265 days old, or three years six months and five days old; No. 18, dropped April 19, 1879, 1,305 days old, or three years seven months and fifteen days old. My next cut (Fig. 22) is the mouth of the steer entered in the exhibition catalogue as 111, and as dropped February, 1880, and as 1,011 days old, or two years nine months and 21 days. Another cut (Fig. 23) shows the front view of the teeth of this animal, both cuts showing the full mouth of eight teeth. Cut (Fig. 24) is from the steer entered as No. 113, dropped August 28, 1879, and 1,174 days old, or three years three months and four days old. Fig. 25 shows the outside of the same mouth. Figures 22 and 23 were from the

FIG. 19. OUTSIDE VIEW OF MOUTH OF "KING OF THE WEST."

mouth of the steer that Mr. Moninger called 'Grinnell.' The Figures 24 and 25 are from the mouth of the steer Mr. Moninger called 'Tom Brown.' From the outside view of 'Tom Brown's' mouth it will be perceived that the teeth begin to show space between them, giving them a long appearance, and showing how absorption has commenced. Now I wish to call attention to the exhibit of T. L. Miller's, as shown on page 715 in the December number of this journal. The Messrs. Groff's 'Young Aberdeen,' No. 15 in the catalogue, two years and six months old, has a full mouth. Nos. 78, 79, 80 and 81 (Miller's exhibit) are two years eight months old, and two years seven months old. Each of these animals (Herefords) had

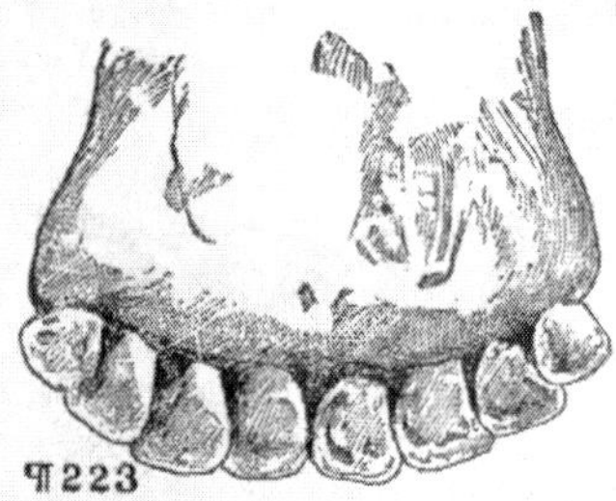

FIG. 20. TEETH OF SHORTHORN "YOUNG ABERDEEN."
(Exhibited as 2 years 6 months 2 days.)

only four permanent teeth. They are entered as practically the same age as Nos. 15 and 111 (Shorthorns), which had full mouths of eight teeth.

"Referring again to Nos. 83 and 84 (Herefords), T. L. Miller's exhibit, one three years eight months, and the other three years nine months old, reaching well up towards four years old, had but six permanent teeth each, while 17, 18 and 113 (Shorthorns) had a full mouth of eight teeth, and a clear indication that absorption had already commenced. I wish to call attention still further to the fact that the following numbers in the exhibition catalogue, 85, 86, 87, 88, 89 and 90 (Herefords), all reached up well towards four years old, no one of which had but six permanent teeth. Mr. Moninger's 'Champion of Iowa' (Shorthorn) had a mouth corresponding to Fig. 14, while No. 74 (Hereford) of the exhibition catalogue, dropped Dec. 1, 1880, 714 days old, or one year eleven months and twenty-four days; and entry 22, dropped Nov. 30, 1880, 715 days old, or one year eleven months and twenty-five days old; neither of these had but two permanent teeth. From the indications of teeth development in many animals I have already examined, where the birth of the animals was recorded at the time of the birth, I find very little variation from the authority of Prof. Youatt, laid down fifty years ago. Several of the animals that were exhibited by

Mr. Moninger and others in the Shorthorn interest were examined, and from the teeth indications, show the entries to be as fraudulent as those presented by the exhibition of these cuts. I ask the careful attention of painstaking and experienced cattlemen, who keep a correct record of the births of their cattle, and I ask of these men, if they will give me the facts that will prove or disprove the correctness of Prof. Youatt's authorities. I shall strike off an edition of these cuts, and a form for entries, showing the ages from two to four years old,

¶224

FIG. 21. OUTSIDE VIEW OF MOUTH OF "YOUNG ABERDEEN."

as recorded at the time of birth, and as indicated by the teeth. In figuring the years, months and days of the ages of the above steers, I have taken the age in days and divided it by thirty, to give the number of months, and that by twelve to give the number of years. I propose to follow these issues until the exhibitors at the Fat Stock Show shall each and all of them come with honest ages, honestly entered.

"I propose to follow these until the State Board shall not rest on technicalities and compel exhibitors to protect themselves; I expect to follow it until the acting president, for the time being, dare act upon information, and investigate fraud by whomsoever perpetrated. I expect to follow these issues until it will be disreputable for any exhibitor to carry away funds not honestly obtained. I expect to follow these until it shall be disreputable to participate in cane presentations to exhibitors who come before this show in a fraudulent manner. From this on I shall not be an exhibitor at this show until these things are accomplished, but I do recognize that the show is one of the most important ever inaugurated in this or any other country, and I shall do all that is within my power to make it respectable, reputable and authoritative. I asked some years ago an English Hereford breeder, why they did not make a more reputable show at Smithfield, in London, England. His reply was that the show was so strongly under Shorthorn influence that there was no chance to show successfully. I hope that this influence is broken at this show. At any rate, I moved at the Hereford Society meeting in November, that $2,000 be raised to be offered in premiums to the Hereford exhibitors at this show. I wish to say that I believe this sum will be raised; Hereford breeders owe it to the exhibitors to see that they are properly compensated for their labor and expense in fitting for this show. The successful exhibitor reaps a benefit by the honors he may obtain at this show, still it is true, nevertheless, that breeders, as a whole, receive a benefit. No one knows better than myself the expense and care that there is in preparing and going through with a successful exhibit. While I am not an exhibitor, I shall expect to contribute freely for this purpose; when I am an exhibitor, if the time shall come again, I shall expect then my share of contribution and hope that I may be the recipient of at least a portion of the fund. It is hardly to be expected that Hereford exhibitors can show yearlings against Shorthorn three-year-olds, or Hereford two-year-olds against Shorthorn four or five-year-olds, or Hereford three-year-olds against Shorthorn five or six-year-olds; in the past Hereford exhibitors had this to do, at least they had a large difference against them. With the State Board composed of Shorthorn breeders and very largely under Shorthorn influence the exhibitor who proposes to expose the workings of these influences can hardly be expected to receive impartial treatment. I

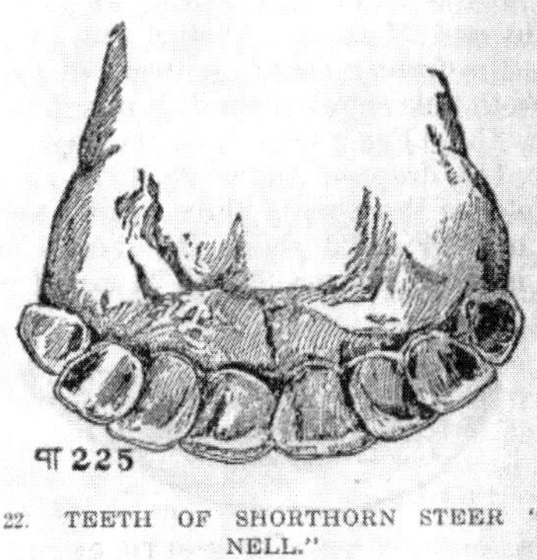

¶225

FIG. 22. TEETH OF SHORTHORN STEER "GRINNELL."
(Exhibited as 2 years 9 months 21 days.)

recognized this long ago, and feel that I can meet these issues more effectively by not being an exhibitor than by being one. I am not disposed to make any promise at present as to what I may or may not do. I do expect, however, to do all I can to place the Herefords in the position which they are entitled to, the leading beef producers of the world.

"I am yours truly,

"T. L. MILLER."

Directly after the Fat Stock Show (1882), we found the following statements in substance in several of the Chicago papers:

"The Illinois State Veterinarian, Dr. N. H. Paaren, was called upon to decide the question raised during the Fat Stock Show by Mr. T. L. Miller, of Beecher, as to the age of some of

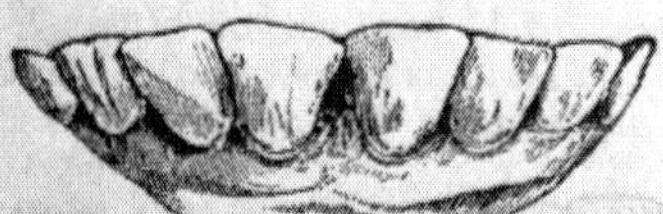

¶ 226

FIG. 23. OUTSIDE VIEW OF MOUTH OF "GRINNELL."

the steers exhibited by Mr. D. M. Moninger, of Iowa. Dr. Paaren decided, after a careful examination, that there was in the condition of the mouths of the animals nothing to prove that they were older than Mr. Moninger stated. He considered that a system of feeding which would make a steer less than two years old weigh nearly a ton would hasten the development of the mouth in a corresponding degree. In this view he is supported by a number of eminent veterinarians."

Hardly believing that Dr. Paaren could have examined the steers, and given as decided an opinion as this article would indicate, I wrote him the following letter:

Beecher, Ill., Dec. 2, 1882.

Dr. N. H. Paaren,

My Dear Sir: I notice by the Chicago "Tribune" that you examined some of the steers of D. M. Moninger as to their ages, and that you found nothing to prove they were older than Mr. Moninger entered them for. Does the article referred to in the "Tribune" properly represent your action and your views? If you will, be so kind as to advise me of the fact, and at whose solicitation you examined them, and when, and which of the steers you examined. The ages of these steers will have the fullest canvass, and I am having cuts prepared representing the teeth of cattle from birth to ten years old, as given by the best authorities, and I propose to verify this testimony by actual comparison with living animals from a large range of herds in this country and England. You are probably aware that Mr. Moninger entered a bullock 715 days old, and that Mr. Potts entered another 715 days old. Mr. Moninger's steer had six teeth fully developed, while Mr. Pott's steer had two. You will recognize that Mr. Pott's steer has a mouth corresponding with the age he gave him, while Mr. M.'s had a mouth that Youatt, on page 322, gives to a four-year-old, and that Youatt and Martin give to a three-years past, and that Allen, in his American Cattle, gives to a four-year-old. Very truly yours,

T. L. MILLER.

I received from him in reply the following:

Chicago, Ill., Dec. 8, 1882.

T. L. Miller, Esq., Beecher, Ill.,

Dear Sir: Your favor of December 2d, addressed to the care of the Prairie Farmer Company, just received. I only call at the above office once a week which accounts for delay in receiving your letter.

In regard to your question about the age of certain bullocks, at whose solicitation I examined them, and when and which of the steers I examined, I have only this to say: Having accidentally learned that a controversy existed in regard to the age of certain animals exhibited at the recent Fat Stock Show in Chicago, my curiosity led me to have some of Mr. Moninger's steers, said to be subjects of controversy, pointed out to me, and with the assistance of a bystander I examined their mouths as to their age. In two of the

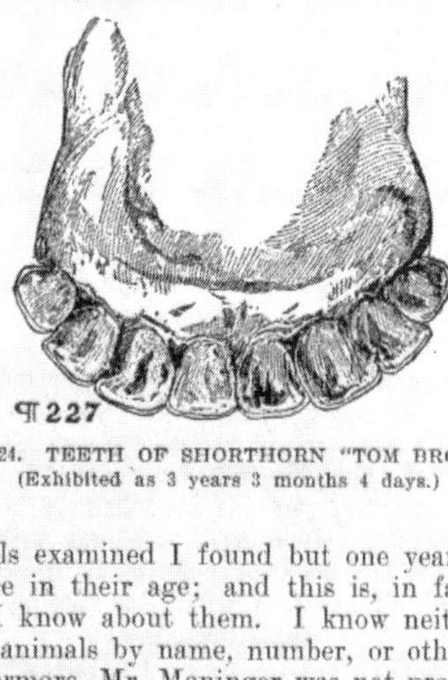

¶227

FIG. 24. TEETH OF SHORTHORN "TOM BROWN."
(Exhibited as 3 years 3 months 4 days.)

animals examined I found but one year's difference in their age; and this is, in fact, all that I know about them. I know neither of these animals by name, number, or otherwise; furthermore, Mr. Moninger was not present at the time; my examination was not made at that gentleman's instigation; in fact, I am not acquainted with him and never spoke to him to my knowledge, and he knew nothing about my intention to examine any of his stock. I repeat it—the examination of said bullocks was entirely a private matter with me, not intended for publication or use in any way, and it is much against my wish to be drawn into the controversy in a matter with which I had nothing whatever to do, officially or otherwise. That I may not be misunderstood in this matter, I will say in conclusion, that I should have had no objection to examine the said animals in the presence of both you and Mr. Moninger; in fact, would have preferred that to my present position.

Yours respectfully,

N. H. Paaren, M. D.

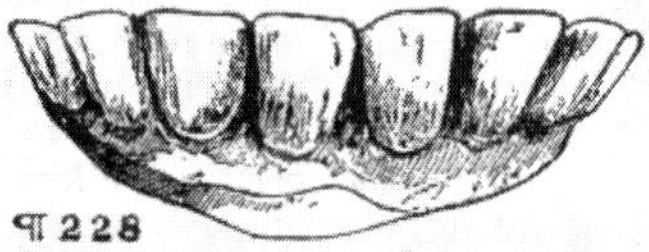

¶228

FIG. 25. OUTSIDE VIEW OF MOUTH OF "TOM BROWN."

CHAPTER XXIX.

¶205A Hereford Items in the Early Eighties

The decade ending 1890 saw great activity in the Hereford interest, and if full extracts could be made from the agricultural and stock journals of that time, they would make good contemporaneous history.

We cannot hope to give a full account of all the movements of Herefords in those days, but will give some extracts from the press of that time.

HEREFORDS WANTED IN AMERICA.

At the Shorthorn Convention, held at Lafayette, Ind., 1880, Judge Jones reported that Shorthorns were preferred over Jerseys, Herefords and other breeds in England. Did the Judge get this information from Shorthorn breeders in England? Did he learn, while there, that there were some 500 head of Herefords bought for exportation to America, and that among the purchasers were Mr. Cochrane, of Canada; Mr. Earl, Mr. Raub, Mr. Fowler and Mr. Sample, all of Lafayette, Ind., men of large wealth, of large experience in the cattle business as packers, graziers and feeders? Did he learn that C. M. Culbertson, of Chicago, a packer of thirty or forty years' experience, and a large grazier and feeder, and that Mr. Ben Hershey, a man of large experience and great wealth, and one who owns a large herd of Jerseys in Iowa, were large purchasers? That T. L. Miller, of some reputation and experience as a Hereford breeder, was a large purchaser? That Messrs. Simpson & Gudgell, of Missouri, were there purchasing Herefords? All of these from this country. Did he hear of any Shorthorns being bought for America? The Judge may say that these were Americans, and he was saying what Englishmen preferred. We will quote what the "Mark Lane Express" says: "There are few graziers in England who would buy Shorthorn bullocks if they could get Herefords."—*The Breeders' Journal.*

ECONOMY OF KEEP.

The Kentucky "Live Stock Record" said that five Hereford bullocks, three years old or over, could be fed where four Shorthorn bullocks could.

HEREFORDS WANTED ON THE PLAINS.

The "National Live Stock Journal" said: "Breeders of Shorthorn cattle may not safely shut their eyes against the fact that the Herefords have made tremendous strides in public favor within the past five years, and that such of our ranchmen, on the western plains, as have tried them, almost unanimously give them the preference over the Shorthorns, because, as they express it, they are 'better wrestlers,' that is, they are better adapted to the conditions under which cattle are compelled to exist on these plains than are the Shorthorns. This, we feel called upon to say, is the almost universal verdict of the ranchmen that we have met, in the past two years, and we have met very many of them."

HEREFORDS AT LONDON.

The "Mark Lane Express" of England, in noticing the letter of a Shorthorn breeder, written from England, and appearing in the "Breeder's Gazette," saying that London salesmen could not sell Hereford bullocks in that market, says of the writer that it is no wonder that he did not give his name, and that the only difficulty the London salesmen experience with regard to Hereford bullocks, is in getting enough of them; and that, as grass-fed beef, there is nothing that comes into London that can touch the Hereford. We should not be at all surprised if the writer of the article referred to was a resident in or near Chicago, instead of being a steward and judge at Smithfield. If he had held the two positions, then the charge that has been made of Shorthorn influence in the management of that show has this to support the charge.—*Breeder's Journal.*

PEDIGREE CRAZE.

The "National Live Stock Journal" struck a blow at the Shorthorn pedigree craze and practically endorsed Mr. Sotham's teachings at this time by saying:

"Time was when the supposed to be expert in pedigrees would walk about the sale ring, catalogue in hand, and when a magnificent specimen of the Shorthorn race, compact, robust, smooth, blocky and vigorous was led in, would shake his head ominously and say to his neighbor, in a patronizing way: 'Oh, yes, she is a grand cow certainly; but look here—' and away back, some six or perhaps ten generations removed, he would put his finger on the name of Mrs. Motte, and say: 'Seventeen.' That

RARE SOVEREIGN (10499) 8118.
Bred by Earl of Coventry.

would settle it; and the grand cow, a prize winner herself, and descended from a race of prize winners, would be knocked down at $150. And then, when a little long-legged, effeminate looking heifer was led into the ring, wheezing and coughing, half-dead with tuberculosis, the expert's eye would brighten and he would sing out: 'Here's royalty for you! Look at that pedigree—perfectly straight, without an outcross,' and the devotees of fashion would vie with each other in their eagerness to show that they knew what they were about, until the worthless beast was carried well up into the thousands of dollars. They bought this Shorthorn because she was straight Bates, and rejected that because she had this or that outcross five or ten generations back, although of the two the latter was much the better animal. A good pedigree is one which commences with a good animal standing before you, and runs back through an ancestry consisting of good animals only; and the better the individual and the longer the pedigree made of such individuals only, the better the pedigree."

HEREFORD BREEDERS SHOW UNTIRING ENERGY.

(¶ 206A)

Mr. Coburn, at that time an active Shorthorn partisan and supporter, wrote as follows:

To the Editor of the "Breeders' Gazette":

The breeders of Shorthorns have not, at any one of the four Fat Stock Shows held at Chicago, been represented at all as the numbers and well known merits of their cattle would justify. The Shorthorns have been frequently beaten by the Herefords—of which there are comparatively few—not, as we think, because the Herefords are so much better in any respect, but, first, because they are valuable cattle, and secondly, but not least, because they are in the hands of a few men who believe in them, and who are energetic, rich and intensely aggressive. They have money—some of them money made in other than the cattle business—and they know how to use it along in obtaining the choicest specimens of their favorite breed, or to combine it with wind, oil cake and printer's ink, in the exact proportions that will best accomplish their ends—viz., capture the prizes from the Shorthorns and persuade people to buy the Herefords. They have had a goodly share of success, and they have earned it, not by having a breed of cattle pre-eminently the best, but by what the world calls their enterprise; something that has much to commend it, especially when something besides wind is to be in it. We like the "untiring energy" feature of their methods, and like their cattle; but what we started out to say was that the Shorthorn breeders must be seized of that same spirit and go to work, not necessarily to make anything new, but make the most of the material already in their hands, or at future Fat Stock Shows they will be left so far out in the cold that nothing less than a search-warrant will find them in time for the next round-up. Unless they rouse themselves to a sense of the true condition of these affairs; to the fact that, while they have been resting on their old-time honors and laurels, the other fellows have made ready to retire with their baggage, cups, cake and all the fresher laurels and honors, they are, indeed, to be commiserated.

The zeal and spirit possessed and displayed by the Hereford men in the last few years, backed up by plenty of money and a few good animals, will win whether the Herefords are generally the best cattle or not. The owners of Shorthorns owe it to themselves and to the great interest they represent to make a proper effort to maintain, at least, the fair fame and name of their life-long favorites. It can be easily done, and there will be no better time to make a beginning in that direction than now. The competitors are already in active motion with a view to coming out first best. What can the Shorthorn men say in response?

There ought to be three hundred model

Shorthorns, representing the herds of twenty different states, at the Chicago Show of 1882, and five hundred in 1883.

F. D. Coburn,
Topeka, Kan.

AMBROSE STEVENS SAYS "OLD WAY'S THE BEST."

The "Kentucky Live Stock Record," in an article on Cattle in America, says: "Now, what are our resources for the great improvement of cattle? Mainly it rests with the Shorthorns. Of the improved breeds fitted to add to the beef production, he is now the chief one. He exists in so much larger numbers than does any other breed, that he must of necessity be the chief source of reliance in improvement, even if it be claimed that another race was better. But no other race has shown equal value in improving inferior cattle. Seventy years since, the Shorthorn was only known in a very limited region of his own country—England. Now he dominates in it everywhere, every other race as the source of improvement of our inferior cattle. He has been in America, in forty to fifty years, the chief source of improvement of our inferior cattle. He has in this country at present such a lead as an improver that no child to-day born will see his position diminished. . . . .

"But there is a movement to introduce more largely the Hereford as a source of improvement. There will, this year of 1880, be not less than five hundred of them imported to America. We are glad to see this. We shall have the Hereford put alongside the Shorthorn in some measure, here, to show their relative value. No one who has seen Herefords will deny their great merit as beefmakers. But the question is, will they make head against the Shorthorn any better in America than they have in England? In the native home of the two races, the Shorthorn has almost already supplanted the Hereford as the source of improvement in market cattle. Can the Hereford make head in America against the Shorthorn, while he has failed to do so in his own country? We think not." . . . .

Commenting on this, the "Breeder's Journal" said:

The editor of the "L. S. Record's" Cattle Department has been conversant with the means that have been used to give the Shorthorns popularity in England and America, and to cry down and write down every other breed from

MISSOURI FARM SCENE.
"Noontime."

1834 to this time. He knows that Rev. H. Berry wrote what purported to be Youatt's History of English Cattle. He knows that Rev. Henry Berry was editor of the "Farmer's Magazine" of London, and a prominent Shorthorn breeder at the same time. He knows that at this time (1834) the impartial historian

STOCK BULLS ON A KANSAS FARM.
Property of the late C. S. Cross, Emporia, Kan.

could have found material in the record of the Herefords that would have placed them, in England, so far ahead of the Shorthorns that they (the Shorthorns) would not have a respectable second-rate position. He knows that the Shorthorn breeders were the active organizers of the Royal Agricultural Society of England in 1839, and it was organized for the purpose of forwarding the Shorthorn interest. And he knows that it has been held and worked by the Shorthorn breeders of England from then until now. He knows the New York fair was organized in 1841 for the same purpose, and has been worked in that interest from then until now, and that for a large portion of that time he was an active worker in that interest in that society. He knows that from 1834, whenever the Hereford and Shorthorn should come into fair contest, whether in England or America, that the Hereford would take the first or top place. Very few men in America are so familiar with the process by which the Shorthorns have taken position in this country and England as Ambrose Stevens, cattle editor of the "Kentucky Live Stock Record"; and still, knowing these facts, he claims for them preeminence. There is not a Shorthorn breeder in Kentucky that dare bring their cattle to a test with the Herefords. [And the "Journal" might have added that Mr. Stevens was the editor of the revised and abridged American edition of Youatt.—T. L. M.]

HEREFORDS SPREADING.

The American "Cultivator" (1883) said:

"Though the Hereford breed of cattle has not as yet been extensively introduced into this section of the country, its excellencies are commanding the situation at many other points, notably in England, Australia, South America and in our western country. It is a matter of record that not only in the London market have they been quoted at from one to two cents a pound above the Shorthorns, but the record of the Smithfield show is, that the Hereford steer has a record over the Shorthorn, and the same record shows that the Hereford steer has made as good weights as the Shorthorn at any given age. And now the Bath and West of England Society has awarded the two champion prizes for the best male and female in the show, to the Herefords. Coupling this with the fact that during the same record he has always brought better prices, and another established fact that he has always been a more economical feeder and grazier, is it not strange that the press and agricultural societies have not been more ready to encourage them?

"A recent sale of one hundred Hereford bulls in England for shipment to the grazing regions of Buenos Ayres shows the estimation in which this famous stock is there held. The Herefords have made more rapid progress in public favor at the west in the last five years, than ever was made by any other breed of cattle in America in the same time. In Colorado and Wyoming there are several herds of from 30,000 to 70,000 head, that are using all the Hereford bulls they can get, and already at the Union Stock Yards at Chicago and at St. Louis and Kansas City Stock Yards, these steers are commanding the top prices, while five years ago they were not known in these yards. In five years more they will be quoted at all the markets, as they have been in the London market in England for the last one hundred years or thereabouts.

"The Hereford cattle are tough, hardy and thrive on a diet both in quality and quantity that would be unprofitable in the Shorthorns. The cattle are very large sized, make excellent beef, are fair milkers, especially when crossed with other kinds, and are withal quite handsome, being red bodied with white markings and white face, the latter being an invariable mark of the kind."

BUTCHERS REGULATE THE AMOUNT OF MEAT CONSUMED.

The "Pittsburgh Stockman" is a good paper

for cattlemen. The following is from its columns (1880):

"Butchers regulate, to a greater extent than most people imagine, the amount of meat consumed in the city markets. Let the supplies of stock be light or liberal, they are in some way disposed of, without customers seeming to be stinted or overstocked. Given the same number of families to supply, the retailer will work off one-half more meat on them at one time than another. The facts of the matter are about thus: The butcher goes to market and finds cattle scarce and high, and is satisfied that he cannot come out even on his purchases. But his trade must be supplied, and he calculates his needs, selects a few steers, and prepares to make a very little beef go a long ways; his customer calls for two or three pounds of meat, and the piece is intentionally cut a half pound lighter than ordered, and the buyer 'makes it do.' Again he goes to market and finds it overstocked. Cattle are plenty and cheap, and he sees a good thing, buys heavily, and goes to work to dispose of the product. A piece weighing two or three pounds is again wanted at the block, and by cutting a little thicker three or four pounds are handed over to the buyer, who again 'makes it do.' Three out of four customers will say nothing about the extra pound or two, and a skillful salesman will thus work off hundreds of pounds extra in a single day. The same tricks are employed in small stock. We have known a single butcher in this city, who usually kills sixty to seventy-five sheep in a week, to use nearly double that number in a week of glutted markets. The heathen Chinee is not alone in his peculiarity 'for ways that are dark, and for tricks that are vain.'"

JUDGE JONES, THE DOUBTER (¶ 207A).

We took the following from the "Kansas Farmer" (1881), written by the Shorthorn breeder and advocate, Judge T. C. Jones, of Ohio: "Speaking of the demands for Herefords in this country, Judge T. C. Jones, writing to the 'London Live Stock Journal,' says that 'it is to be observed, in the first place, that it is of recent origin, that it cannot be predicted what the future of the business will be, or whether the Herefords will be so well adapted to the hardships and privations of our wild ranges as now anticipated. The Hereford breed, like other improved wild breeds of British cattle, has been much improved of late years, and will not endure the hardships it was compelled to undergo in some quarters fifty or a hundred years ago; and it is not likely that it will prove more profitable than crosses of the Shorthorn breed, under a system which allows a large percentage of the unimproved cattle of the plains to perish from hunger and the severity of the cold every winter. It is, moreover, believed by a majority of intelligent observers, that the range method of producing beef—barbarous alike in its influence upon man and beast—will prove an ephemeral business. The grass in these wild and unenclosed districts is scanty and really nutritious but a few months in the year, so that vast ranges

BARNS AND FARMING LANDS.
Weavergrace Farm of T. F. B. Sotham, Chillicothe, Mo.

are required—it is said from three to five acres—to graze a single sheep. The most of the varieties being annuals, the re-seeding necessary is each year, of course, diminished by heavy stocking; we, therefore, have reports of increasing scarcity of grass in all the older districts. It is, in fact, the general opinion of men best informed in regard to the system of beef production, that it has already reached its maximum.'" To which our "Breeder's Journal" added: "Judge Jones will do well to get as well posted on Herefords as on Shorthorns, and he will then know that the Herefords are doing well on the plains; and that, while Shorthorns are dying, Herefords are thriving; and by referring to our September number he will see that grade Herefords from the plains are selling at $5.25 per hundred pounds, while grade Shorthorns from the same range are selling at $3.60."

JOHN V. FARWELL,
Chicago.

WILLIAM WARFIELD'S WARNINGS (¶ 208A).

The "National Live Stock Journal" published an article from the pen of Mr. Warfield, of Lexington, Ky., the great Shorthorn breeder and advocate. Mr. Warfield, in answering his own inquiry, "Is the future of the Shorthorn interest to be a great one?" opens his article as follows:

"That is a question which has been presenting itself to my mind for solution over and over again during the last few months. Clearly its answer depends on many things. But all these many cluster about and depend on just one, namely, the conception which Shorthorn breeders have of what Shorthorn breeding is—what immediate purpose they have in prosecuting the work—to what ulterior end they make it tend. In other words, it is exceedingly plain that the future of the Shorthorn interest is now in a peculiarly serious sense, in the hands of Shorthorn breeders. It will become just what they make it become. And it is not so much energy that is now needed in them, as intelligence, wisdom, and above all, honesty—honesty to themselves and the interest intrusted to their charge.

"Breeders may determine that there is nothing but money, and money for them, to be gotten out of Shorthorns. If so, the future of the interest is not a bright one. Men will begin to look at everything from the point of view of gambling; will, if possible, rouse fancies and fashions, for this color or this strain, which they can supply, and having roused them, will pander to them and lose no endeavor to keep them at fever heat, and the end will be—the end of Shorthorns. Breeders in this case will be simply beasts of prey, preying on the community. Their interests ought not to succeed; and by the stern logic of history, which always works itself out, their prosperity will be feverish and very short-lived. If this is what the Shorthorn business is to be made, I for one, want none of it, and will be among the first to cry out upon it, and to help choke it out, like a noisome reptile.

"Breeders may, however, and doubtless will (I have great confidence that they will) take another view of the matter. They may remember that beef is now the real staff of life. They may remember that America is to supply and is already beginning to supply the world with beef. They may remember that the Shorthorn race has, by constant proving, been shown to be the best beef-producing race of cattle, and that it has in it still further, perhaps indefinite, capabilities of improving in the same direction. And, remembering all this, they may determine to prosecute the business, because they see in it a livelihood, and, mayhap, riches; but they will determine to prosecute it for this purpose and after this fashion, viz., in order that they may furnish a race of stock that will make the most and best beef at the least cost, and so that they may improve the stock of the country by their breeding. If breeders look at the matter in this way, they are benefactors of their country and of the world. Their efforts ought to succeed, and by the fine logic of history their prosperity will be steady and long-lived. In a business point of view, this case differed from the former one exactly thus: That was the true Shorthorn policy, this is the true long-sighted policy. While it is morally true, therefore, that there is an indefinite deal more of nobility in this cause than in that, it is equally economically true that there is an infinite deal more of money in it also. It offers no sudden wealth, but it offers steady progress in both kinds of success.

"This, then, is what I mean by saying that

it depends on the conception of breeding taken by breeders, as to what the future of the Shorthorn interest will be. Let me refer before closing to what seem to me some dangers now standing in the road of progress in the right direction, or threatening to take up position there. And, first of all, I would name the inordinate rage of fancy pedigree—often to the exclusion of the better pedigrees. If we were to breed to stock some nobleman's park, with slim, deer-like fawns, or to tickle his fancy with beautifully printed lists of names in royal succession, such a rage would be legitimate and correct; but if we are to breed for the beef-producing interests of America, where no nobleman has a park, it cannot fail to be ruinous. Not approved paper, even though it reach the commendation of all experts (a thing hardly possible to realize), but the butcher's block is the true test of a pedigree; and that pedigree is necessarily best which produces the most and best beef, not that which has attached to it the longest list of 'fancy' names. Oh, that all could come to see this.

"Next, the dangers from foolish fashion threatens us. The rage for color is now reaching a point which would be laughable were it not serious. Beyond all question roan is the Shorthorn's peculium, beyond all question it is its covering of beauty. Yet, imperious fashion demands a dark red, which is correlated almost hopelessly with harsh hair and harsh handling. As a consequence hundreds of the best bulls born are every year sacrificed on account of color alone. The moral is evident. Gentlemen, if you breed for color you will get—color; but what we want is beef. To get it we must breed for it and it alone, and cast off all silly burdens, carrying which, will delay or hinder the attainment of our one great end.

"I do not mean the rise of Hereford or Aberdeenshire interest as a danger to the Shorthorns. We are alike laboring for the same end, and I welcome aid in beef-producing from all other breeds. The introduction of these new breeds will help—not clog—our progress, and their interest and ours need never clash. They cannot clash unless one or the other of us lose sight of our great common end, and seek to pander to fancy or fashion. I am for beef, and I am for that breed which will produce the most and best of it. That breed I believe to be the Shorthorns, and on, I think, thoroughly tested grounds; but I welcome all others to trial on our broad pastures. Shorthorn breeders have not them to fear—they

SCENE ON THE FARWELL RANCH, TEXAS.
Cows of the T. L. Miller herd in early spring, after a hard winter.

need fear only themselves. Let each ask himself the question, 'For what am I breeding?' His answer to that will declare whether he be an enemy or a friend to the interest.

"William Warfield."

IMPORTATION OF GUDGELL & SIMPSON.

On the 15th of April, 1882, Gudgell & Simpson, of Independence, Mo., shipped 71 head of thoroughbred Hereford cattle from

CHAMPION OX AT SMITHFIELD, 1884.
Cross-bred Hereford sire, Shorthorn dam. Age 3 years 6 months. Weight 2617 lbs. Bred by Chas. Doe, Shropshire. Exhibited by R. Wortley, Norfolk.

England to quarantine them at Quebec. They came by the way of the steamer Texas. Mr. Vaughn was in charge of these cattle.—*Breeders' Journal.*

MR. B. HERSHEY'S IMPORTATION.

"Wallace's Monthly" has the following to say of an importation of Hereford cattle: "Our friend, Mr. B. Hershey, of Muscatine, Ia., has been across the water and has made a large importation of Hereford cattle, numbering some sixty odd heifers, and eight or ten young bulls. They were selected under his own eye, and the best of judges who saw them on their arrival at Quebec, pronounced them a very choice lot. The special object in going so largely into this breed is their supposed better adaption of the vicissitudes of ranch life on the plains. It is claimed that the comparison between this breed and the Shorthorns, on the plains, has been fully made and satisfactorily settled, and that the Herefords are altogether better. This corresponds with our own judgment of the two breeds. There can be no doubt, that as a tribe, they possess more vigorous constitutions and are better able to take care of themselves than the Shorthorns, but an intimation of this kind to a Shorthorn man is like intimating to Charley Foster that some of the Lexingtons have gone blind, and he hurls back at you such choice epithets as 'blockhead,' 'ignoramus,' 'fool,' and every other pet name short of 'thief.' Both breeds have been tried on the plains, and after trial, the verdict is all against Shorthorns, and all in favor of Herefords."

T. L. MILLER'S IMPORTATION.

The second importation of Hereford cattle by T. L. Miller consisted of 114 head. We went to England in May, 1880, and after visiting the principal Hereford cattle breeders, purchased during the month of June. We returned to the United States on July 1st, and visited Washington and had an interview with Secretary Sherman and the President, with a view to have the rule modified, that required ninety days quarantine from the date of landing of the cattle in America, but failed to reach any modification of existing orders. Afterward the time of quarantine was changed to ninety days from date of shipment of cattle.

We then returned to England, shipped the cattle on the 19th of September, via steamer Gallian, Capt. Moen, Master, to Baltimore. They were taken in clean box cars from the dock to the eight hundred acre farm belonging to the B. & O. R. R., seven miles out. Here they remained three months, under the care of Mr. Thos. Smith (¶ 209A), who went from our farm to England to take charge of the shipment.

At the expiration of the quarantine, Mr. T. E. Miller went to Baltimore to bring the cattle home. The cattle were loaded, at a private side track on the farm—from a stock yard built for the occasion, into clean new box cars. The B. & O. road sent the train through to Chicago as a special, on fast passenger train time, running a half hour behind their fast "Oyster Express," and ahead of their through passenger train. The train started at 2:30 P. M. Saturday, September 8th, for Illinois and reached the farm at Beecher, where all were comfortably stabled by Monday evening, January 10th.

Thus these cattle were moved by a special train from Leominster, England, to a steamer at Bristol, and put directly from the cars upon the steamer, and upon arrival at Baltimore kept entirely away from any other stock, and the same plan was pursued until the arrival at Beecher. On the whole route of 5,000 miles they never were in public shipping yard or highway, never an animal sick, and arrived at their destination in as good condition as though they had been quietly on the farm.

Captain Moen was to be especially commended for his care in commanding his steamer, through a very rough voyage that came during the equinox. In the heaviest weather he put his steamer about, and ran back for twenty-four hours to ease the cattle from the effect of the very heavy sea.—*Breeders' Journal.*

EARL & STUART IMPORTATION.

The *Breeders' Journal* said: "A large importation of Hereford cattle was made by Thos. Clark for Earl & Stuart of Lafayette, Indiana, and himself. They were shipped from Liverpool to Portland, Maine, on March 9, 1882.

In the lot are fourteen animals from the herd of Messrs. J. B. & G. H. Green, (¶¶ 210A-211A) Marlow Lodge, Leintwardine, England. Four from the herd of John Price, (¶ 212A) of Court House, Pembridge, England. Two from the herd of T. Lewis, (¶ 213A) the Woodhouse, Hooden, Herefordshire, England. Four females bred from the old Wm. Tudge (¶ 216A) herd of Adforton, were purchased from Mr. John Williams (¶ 213B) of Llansannor Court, Cowbridge, Glamorganshire. Five cows and calves from the herd of Mr. Phillip Turner, (¶ 214A) The Leen, Pembridge, England. Eight heifers from the herd of Mr. J. Morris, (¶ 215A) Lulham, Madley. The largest draft from one herd was twenty-six animals from Mr. T. J. Carwardine (¶ 217A), Stocktonbury, Leominster, Eng. To mention all the animals would not be possible in the space allotted to this subject. Venus (V 12, P 152) 10033 by Lord Wilton (¶ 218A) was considered the plum of the whole collection.

MR. ADAMS EARL'S IMPORTATION.

The "Breeders' Journal" said, in 1881: Mr. Adams Earl of Lafayette, Ind., made an importation of a choice herd of Herefords this year, and had at the head of his herd the two-year-old bull Grateful 2d, got by Mr. Aaron Roger's (¶ 219) old bull Grateful, which has stood at the head of all breeds in England for some time past. Mr. Earl's herd has come over in good condition, and are very choice animals.

MR. H. C. BURLEIGH'S IMPORTATION.

The "Breeders' Journal" said in 1880: Mr. H. C. Burleigh, of Fairfield, Maine, has selected for himself and Mr. J. R. Bodwell, of Maine, fifty-three head of Hereford cattle, from the herds of Queen Victoria, T. Duckham, T. J. Carwardine, (¶ 220A) P. Turner, W. S. Powell, A. Rogers, W. Price, A. P. Turner, (¶ 221A) and B. Rogers (¶ 222A). The above-named parties are among the best breeders in England. Mr. Burleigh has been a breeder of Herefords for many years. He has been a successful breeder and exhibitor. This purchase will place his herd among the first on this side of the water. Besides the Herefords, Mr. Burleigh has brought over a few Shropshire Down sheep. Mr. Burleigh spent several weeks in the West looking for Herefords before he went to England.

CHAMPION STEER, ROAN BOY.
Bred by C. M. Culbertson, Chicago Fat Stock Show, 1883.
Sire a Hereford bull, dam a grade Shorthorn cow.

CHAPTER XXX.

Hereford Items in the Early Eighties—Concluded

HEREFORD COW LEONORA.

Taken from the "Breeders' Journal," 1880: The Hereford cow Leonora, belonging to Mrs. S. Edwards, (¶ 223A) of Hereford, England, was injured on her return from the Smithfield Show, so that it became necessary to kill her. We are not aware of the reasons that induced Mrs. Edwards to fit and show this cow, and several others that she has fitted and shown in the last five years. Of one thing it is quite certain, that she has given herself a large reputation in the cattle world, but three or four more good cows have been lost to her herd.

This has been the practice of the Shorthorn breeders in England and America, of fitting their best heifers for the show ring, and keeping them there as long as they could walk. A few years since, one such was kept in the show ring so long she could not walk—having a carriage built in which she could ride from the cars to the show ground—and such an animal was exhibited as a breeding cow, and for breeding purposes, and awarded first honors. Such awards, on such animals, have been published as an evidence of merit for the breed, and it is not two years since a prominent advocate of the Shorthorn interest pointed to these awards and challenged the Hereford breeders to show their cows in competition with them.

There was a necessity, apparently, why Mrs. Edwards should bring the pick of her herd to the show ring, and that the Hereford cow Jennie should come upon the show ground at Chicago and vindicate and sustain the character of the Herefords for superiority as well in the cow as in the steer class. This has been done on both sides of the Atlantic, and the two winning cows pronounced the best of any breed that has been exhibited. The cow Leonora in England and the cow Jennie in America have vindicated the character of the Hereford breed.

We give herewith the winnings of Leonora. In Mr. Miller's recent purchases in England he bought a half brother and a half sister of this cow. Leonora, bred by Mrs. Sarah Edwards, widow of the late T. Edwards (¶ 224A), of Leominster, calved August 11th, 1875, was the winner of the following prizes: Second prize at Birmingham, 1876; first prize at Liverpool, 1877; first prize at Bristol, 1878; first prize and champion prize for best Hereford female (¶ 225A) at London, 1879; meetings of the Royal Agricultural Society of England: first prize as one of a pair at Hereford, 1876; first prize at Bath, 1877; first prize and champion prize for the best female in the yard at Oxford, 1878; first prize at Exeter, 1879; meetings of the Bath and West of England Agricultural Society: first prize and champion prize for the best female in the yard at Hereford, 1877; first prize and champion prize for the best Hereford female, also champion prize for the best female in the yard at Kingston, 1878; first prize and champion prize for best female, also champion prize for the best animal in the yard at Hereford, 1879; meetings of the Herefordshire Agricultural Society: first prize as one of a pair at Owestry, 1878; first prize and champion prize as one of a pair, also champion prize for the best Hereford female in the yard at Ludlow, 1878; first prize at Shrewsbury, 1879; meetings of the Shropshire and West Midland Agricultural Society: first prize at Kidderminster, 1877; first prize and champion prize for the best cow or heifer in the yard, also champion prize for the best animal in the yard at Malvern, 1879; meetings of the Worcestershire Agricultural Society: first prize at Dursley, 1877; first prize at Cheltenham, 1879; meetings of the Gloucestershire Agricultural Society: first prize and champion prize for the best female in the yard at Newport, (Lord Tredegar's show) 1877, 1878; first prize as one of four at Leominster and Ludlow, 1876; first prize as one of a pair, and champion prize as the best animal in the yards at Leominster and Ludlow, 1877; first prize and champion prize for the best animal in the yard at Ludlow, 1879; first prize and champion prize for the best Hereford at Birmingham (fat show), 1879; first prize and champion prize for the best Hereford at the Smithfield club (fat show), London, 1879. Total of her winnings, $3,250.

FIVE HUNDRED DOLLARS BUTCHER'S PRIZE. (AN OPEN LETTER.)

No. 3 Board of Trade Bldg.,
Chicago, Ill., Sept. 25.

"To Whom it May Concern: Some of the breeders of Hereford cattle have subscribed $500, which they have directed me to deposit with the treasurer of the Chicago Fat Stock Show, to be called 'The Butcher's Prize.' The donors have also instructed me to issue an open invitation to the parties engaged in raising the fund for the best Shorthorn steer at the Fat Stock Show, and to any and all persons interested in breeding or owning Shorthorn cattle, to deposit an equal sum of $500 with said treasurer of the Fat Stock Show, the whole sum, $1,000, to be awarded to the best butchers' carcass among the breeds entered for the 'Butcher's Prize' at the next Fat Stock Show in Chicago, November 16 to 23, 1882. The contest for said Butcher's Prize shall be governed by the following conditions: (¶ 226A) First, all animals entered must be pure bred, or grades not less than half blood. Second, all animals contesting must be slaughtered at the Fat Stock Show at such time as the managers of the show shall set apart. Third, the slaughtering of each animal entered and the award of the Butcher's Prize shall be made by a practical butcher or by a committee of practical butchers to be selected by the following referees: John B. Sherman, Superintendent of the Union Stock Yards; Wm. H. Monroe, a regular buyer for eastern and foreign markets; and John Adams, a regular commission salesman of Chicago. The names of the judges selected shall not be made public until after the 2d day of November, 1882. Any two of the three referees above named may act in the absence of third, and shall make the selection of the judges as above, and in case two of said referees are absent then the third shall select two reputable Stock Yards commission merchants to act as referees, and they shall then select the judge. Fourth, the invitation is also extended to animals of the polled breed, Devons, and all other breeds of beef cattle, pure breeds or grade not less than half bloods, upon their owners or representatives depositing a like sum of $500 for each or either breed with said treasurer of the Fat Stock Show, the whole to be governed by the conditions of the invitation.

"Fifth, the said sum or sums of five hundred dollars must be placed in the hands of said treasurer of the Fat Stock Show on or before the first day of November, 1882, and said deposit must be accompanied by a written notice of the acceptance of this invitation, which notice must designate the name of the breed in whose behalf the invitation is accepted, and a copy of said written notice must be at the same time mailed to the undersigned. Sixth, all matters of dispute, whether involving the construction of this invitation, or any question growing out of this invitation, shall be referred to said referees, whose decision shall be final. Seventh, when the judge or judges appointed by said referees shall have arrived at a conclusion, they shall make their decision known to said referees, who shall announce in writing over their own signatures the winner or winners of said Butcher's Prize, and, upon presentation of said certificate, the treasurer is authorized to pay over to said winner or winners said sum or sums of money so deposited as aforesaid, after deducting any necessary expenses of said contest. Eighth, all entries for the Butcher's Prize must be made to said treasurer of said show in writing on or before the 1st day of November, 1882, together with the name of owner, name and breed of entry, with the age in years, months and days as nearly as possible, and a full statement of the breeding of each animal. Should this invitation not be accepted in accordance with requirement, then said sum of $500 shall be held by said treasurer to be awarded to the best Hereford carcass, pure bred or grade not less than half blood. The contest and award shall be made at the time and under the conditions above set out,

BENTON'S CHAMPION.
Bred by Fowler & Van Natta. A champion at the Chicago Fat Stock Show, 1883.

GRADE HEREFORD CALVES, HALF, THREE-QUARTERS, SEVEN-EIGHTHS AND FIFTEEN-SIXTEENTHS BLOOD ON STRAIGHT TEXAS FOUNDATION.
(Purchased by T. F. B. Sotham for Eastern feeders.)

except that the entries shall be made in writing to the said treasurer on or before the 16th day of November, 1882. Should there be no entries for said Butcher's Prize, then the said treasurer shall hold said money subject to the order of the undersigned. C. M. CULBERTSON,

"President American Hereford Association."

"Mr. C. M. Culbertson, above named, has this day deposited with the undersigned, treasurer of the Chicago Fat Stock Show, the sum of $500, to be called the 'Butcher's Prize,' as stated in the above invitation, a copy of which invitation has also been placed in my hands.

"JOHN W. BUNN.

"Dated at Peoria, Sept. 25th, 1882."

RISE AND PROGRESS OF THE HEREFORDS.

Taken from the Chicago "Daily Tribune," December 30, 1882: "There is, in the rapidity with which the Herefords have spread over this country in the last five or six years, something which is, if not marvelous, at least as remarkable as it must be gratifying to the champions of that magnificent breed of cattle. How much of this has been due to the persistent and pugnacious use of the press cannot, of course, be definitely ascertained, but without great merit to warrant the claims put forth by friends of the breed it is quite certain it could never have gained the place it now occupies as the favorite on many a ranch in the West, and as a highly prized and profitable breed on the farms of the fertile Middle and Western States.

"In their determination to accept no second place on the list of beef-making breeds, and the fierceness of their onslaughts upon all rival breeds, the Hereford breeders give abundant evidence as to the faith that is in them. It cannot be denied that they have made a very creditable show of a right to the first place. Their claim, that the success they met when they carried off three of the four butchers' prizes awarded at the last Fat Stock Show, fairly indicates the general superiority of the Herefords, cannot very well be gainsaid, until some proof to the contrary shall have been furnished. The fact that many Herefords were slaughtered at that show, while but few Shorthorns were killed, will not be accepted by those people who know that the total number of Hereford cattle available for such a test is comparatively small, while the Shorthorns number many thousands. Aberdeen-Angus, Galloways, Norfolk and Suffolk polls, and other recognized breeds are not expected to appear in competition as yet, since there are few cattle of those breeds which can be spared for slaughter.

"If the ensuing year does not bring out Shorthorns which can bear off the honors from the Herefords, in this the final test of the value of a beef bullock, many who are now undecided will conclude that, how smooth and beautiful soever the Shorthorn may be, he lacks the ability to make, when highly finished, the best and most profitable beef. That the Hereford can scarcely be surpassed in ability to withstand hardship is well known; that he will make a fair return for any amount of care, however slight, has been abundantly shown by experience. That the breeders of Herefords are content to rest their case upon the single argument of the superiority of their favorites as beef producers, as do the friends of the Scotch cattle, and do not think it necessary to add the claim that they are good milk and butter producers, will be by some, especially, perhaps by the cattlemen of the plains, considered additional proof of the confidence of the Hereford breeders in the strength of their cause.

"Whatever the outcome of the contest may

HEREFORD.

SHORTHORN.

Competing steers and comparative types, Smithfield winners, 1879.

be between the two breeds, one result may be counted upon as certain; the scrub will be, ere long, driven almost completely out of sight and out of mind. Although the Herefords and the Shorthorns have locked horns with no such purpose definitely in view, the result will be the same as though they had combined their forces to push the scrub from the field."

HEREFORDS NET TO GROSS.

Taken from the "Breeders' Journal," April, 1882: The five steers winning the sweepstakes

JOHN SCHARBAUER, MIDLAND, TEXAS.
and his favorite bull, valued at $5,000, Sir James 65916, by Corrector.

prize offered at the Fat Stock Show last fall by Marshall Field for best five steers in the show, and won by the T. L. Miller Company, of Beecher, Ill., were sold to Wm. Smith & Son, of Detroit, and slaughtered by them for Christmas market. They made the following record:

		Dressed Meat to Gross Weight.	Beef and Valuable Offal.
"Conqueror,"	dressed................	69.13%	83.84%
"Will,"	dressed................	68.13%	83.66%
"Kansas,"	dressed................	68.09%	86.44%
"Bachelor,"	dressed................	66.72%	83.49%
"Washington,"	dressed................	66.05%	81.68%

SOME GOOD HEREFORD GAINS.

"The monthly gain of some of T. L. Miller's Herefords, that are in training for the Chicago Fat Stock Christmas Show of 1880, makes quite a favorable showing. Much is being said about one of John B. Sherman's two-year-old Shorthorn steers gaining 120 lbs. last month. Mr. Watson says he can beat that, as last month Maid of Orleans, two years old, packed on 140 lbs.; Conqueror, 22 months old, 122 lbs.; Bachelor, same age, 110 lbs., and Tom Smith, 23 months, 106 lbs. How's that for an average of Herefords?"—*Breeders' Journal.*

MARSHALL FIELD PRIZE.

Taken from the Chicago "Evening Journal": "There was more than usual interest manifested in the Fat Stock Show, although the storm had the effect of keeping many away that would have wished to attend. The great feature was the competition for the $250 prize offered by Marshall Field for the best five head of cattle of any age or breed. Six herds were entered by the following gentlemen: T. W. Hunt, Ashton, Ill.; John B. Sherman, Chicago, Ill.; J. D. Gillette, Elkhart, Ill.; T. L. Miller, Beecher, Ill.; Morrow & Muir, and H. C. Nelson. Mr. Miller's Herefords carried the day, the prize being awarded to him. There was no little jubilation over the result among the believers in Hereford stock, which they say has risen in value 100 per cent with this victory. Outside of this contest there was no particular feature of interest in the show to-day, except what has been seen during the week."

IMPORTATION OF HEREFORDS IN 1883.

"While many Herefords were imported in 1883 to our different quarantine stations in the United States, namely Baltimore, New York, Boston and Portland, there were still more brought to Quebec, Canada. By a ruling of the United States Treasury Department, cattle for importation into the United States were allowed to be held the ninety days required, on Canadian soil. The following importations were sent to Quebec:

"Geo. Leigh (¶ 227A) imported that year in steamer Texas fifty-six head of Herefords, sailing April 21st from Liverpool. Gudgell & Simpson brought over twenty Herefords in the same steamer. M. H. Cochrane shipped 105 head on the steamer Quebec, sailing May 19. C. W. Cook (¶ 228A) of Odebolt, Iowa, (firm of C. W. Cook & Sons) (¶¶ 229A and 230A), imported 300 head of Hereford cattle on the steamer Quebec, sailing from Liverpool June 30th. Mr. Cook's cattle were selected from numerous herds by Mr. Conant, of Illinois, who had entire charge of the shipment, both on the ocean and in quarantine. Thos. Brittian imported thirty-seven and H. C. Burleigh imported 186 head of Herefords on the steamer Texas, leaving Liverpool on August 1 for Quebec. T. L. Miller imported 108 head of Herefords on the steamship Mississippi, sailing from Liverpool August 18th. Geo. Leigh imported 135 Herefords on the steamer Texas, sailing from Liverpool September 12th. C. M. Culbertson imported 115 Herefords on the steamer Texas, sailing September 12th, from Liverpool."—*Breeders' Journal.*

TESTIMONIAL HERD.

Looking back as we write, the author cannot

but reflect that there are some parts of the road which we have traveled, in trying to advance the beef interest of America, to which we can look back with undisguised pleasure. A man is to be commended for his efforts to make "two blades of grass grow where one grew before," but many times he does not have his efforts appreciated. We cannot pass the incident of the presentation of a testimonial from the English Hereford cattle breeders, without expressing our gratification that our efforts to give the Hereford cattle their proper position was appreciated by our English friends. The following letter from the Hon. Thos. Duckham to the "Hereford Times" was no doubt the beginning of the movement that culminated in 1883. After the opening of the letter, in speaking of the growing popularity of the Herefords, he says:

"I must say that the meed of praise is due to T. L. Miller, of Beecher, Will County, Illinois, U. S. A., whose untiring perseverance in the cause merits every feeling of gratitude which the breeders of Herefords can award him.

"Not only has Mr. Miller patronized the Hereford by establishing the largest and, I think I may add, the best herd of Herefords in the United States, but he has so placed them before the stock owners of that great country, at the various exhibitions and by his sound judgment, untiring perseverance, great talent as a writer and great influence as a gentleman of position; he has claimed for them (and to a certain extent successfully) a front rank in the great state exhibitions of the United States. It was Mr. Miller who protested against sending his Herefords to the International Exhibition at Philadelphia in 1876 unless an English judge was selected to aid in awarding the premiums, which protest resulted in my being sent out by the British Commission. It was Mr. Miller who first introduced the Herefords into the far West to improve the vast herds in the Western states. It was Mr. Miller who resolved to obtain the necessary information and publish a Herd Book for Hereford cattle in America. It was Mr. Miller who, knowing the value of the press in the promotion of any great object, established a paper in the interest of Herefords.

"Knowing all this, from many years of correspondence with that gentleman, I feel that I should be wanting in a proper discharge of duty to him if I did not place those facts before your readers. I may add further, that

HEREFORDS IN ARIZONA.

Mr. Miller was here in June last, when he purchased about 100 of our far-famed, red-with-white-faces, and although far past the meridian of life, he hastened back to use his best endeavors to obtain a removal of the restrictions of a ninety days' quarantine, imposed upon that side, upon animals imported from this country. I fear his usual success has not attended his efforts in that direction, and I hear he is now returning, if not returned to England, to arrange for shipping his purchases.

EX-GOVERNOR S. B. PACKARD,
Marshalltown, Iowa.

"There are several breeders of Herefords in the United States of much longer standing than Mr. Miller, but it was left to him to bring them to the front in the manner in which they are now brought. Mr. Miller only commenced as a Hereford breeder in 1871. In 1872 his then partner, Mr. Powell, a Herefordshire man, came to England and purchased a few Herefords, among them the handsome young cow Dolly Varden, bred by Mr. Morris, Town House, Madley, and her two offspring. Mr. Powell soon entered upon another business, and Mr. Miller was most fortunate in securing the assistance of one so thoroughly conversant with the management of a herd as Mr. George F. Morgan.

"The excellence of Mr. Miller's herd which he sent to the International Exhibition, and the manner in which Mr. Morgan brought them out and placed them before the judges, were themes of admiration to all who beheld them. After I discharged my duties at that exhibition, I visited Mr. Miller and numerous other Hereford breeders in order that I might make their personal acquaintance and see how the various herds acclimated, but at no place did I see them so admirably managed as those under Mr. Morgan's care on Mr. Miller's farm.

"I would here respectfully suggest that the Hereford breeders pay some compliment to Mr. Miller when he returns to this country, in acknowledgment of the very valuable assistance he has rendered to them in creating a demand for their pedigree animals, which, although of great and unprecedented proportion, is now only in its infancy.

T. DUCKHAM.

Baysham Court, Ross, England.
Aug. 17, 1880.

The occasion of the writer's visit to England in 1883 was taken advantage of by the breeders of Hereford cattle in that country to present him with a valuable testimonial to show their appreciation of his labors in securing to the Herefords in America their present position as the best breed of beef cattle.

The "Hereford Times," England, August 4, 1883, has an editorial in reference to the matter, from which we quote: "Within a very short period three very important gatherings of breeders of Hereford cattle have taken place in Hereford, each being in the nature of a celebration, and each of such a character as to warrant the hope that a new era of prosperity for the renowned white-faces is being entered upon. The first was the banquet of Mr. Price, of the Court House, Pembridge, upon the occasion of his splendid victory at Birmingham, when he carried off the Elkington Challenge Cup. (¶ 231) The second was the farewell dinner given a week or two ago by Mr. Burleigh to the breeders from whom he had been making his extensive purchases for America. The third, and most important of all, has taken place in the present week, when the union between the English and American breeders of Hereford cattle has been cemented by the banquet and presentations to Mr. T. L. Miller, of Beecher, Ill., and his righthand man, Mr. George F. Morgan, who have done more than any others to establish the Herefords in universal favor, and raise them to that position of pre-eminence which their excellent and varied qualities entitle them to hold.

"Advantage was taken of Mr. Miller's presence in England—a visit made purely in the interest of the breed, and to enable Mr. Miller the better to enhance their popularity in America—to entertain him at a banquet, and the gathering took place on Wednesday last, Sir Joseph Bailey, M. P., being the president, Mr. Duckham, M. P., and Mr. S. Robinson (¶ 232) the vice-presidents, and the numerous company also including the Lord Lieutenant of the County (Lord Bateman), Earl Coventry (whom we are glad to see is lending the Hereford breed an increasing measure of his valuable influence), and the member for Leominster (Major Rankin) (¶ 233) and generally the leading breeders and farmers of the district.

"The presentation to Mr. Miller consisted of an illuminated address and a purse of 200 sovereigns; that to Mr. Morgan, of a very handsome silver cup. (¶ 234) * * * In the limits of an article like this, it is impossible to enter into a detail of the steps to be taken to raise the Herefords in universal estimation; and indeed it would be uncomplimentary to assume that the breeders need to be especially informed on this point. They know what Mr. Miller intends to do with the £200 which they have presented to him, and this is an example of what we mean. He intends to form a herd to be known as the 'testimonial herd,' and managed in a thoroughly systematic way, with the aid of which he will demonstrate, among our agricultural cousins over the Atlantic, 'the capabilities of the Hereford breed.' They know, too, what Mr. Miller has done in the past; how, in the face of strong prejudice and the great power of Shorthorn influence in America he has made the merits of the Herefords felt, and by 'fearlessly presenting' those merits has achieved successes for the 'white-faces' which, ten years ago, would have been thought impossible.

"We would say to the English breeders of Herefords, 'support, by all means in your power, the hands of men like Mr. Miller.' "

The account of the presentation and complimentary banquet to the writer was given in the same paper, reporting the numerous toasts and responses. Sir Joseph Bailey, M. P., the chairman, proposed the toast of the evening, "The health of our guest, Mr. Miller," in which he made the presentation and closed by saying: "Mr. Miller is already somewhat advanced in years, but when he grows to be an old man, it will be a proud thing for him to say that it had been his lot to reinvigorate and improve the cattle and herds of America."

Lord Bateman proposed the toast, "The health of the breeders of Herefords in England and America." In the course of his remarks he advised farmers to follow his example and dispose of their half-bred herds in favor of stock of the pure Hereford breed.

HEREFORD BEEF IN NEW ENGLAND.

Printed by the "Breeders' Journal" from a New England paper:—"Volumes may be printed and read upon this subject, but the most convincing argument is that which appeals to the pocket-book, and offers 8½ cents per pound, live weight, for choice Hereford, and but 4 cents for the scrub that has eaten his head off in his maintenance. Hundreds of practical instances may be seen on any cattle day in our great markets.

"There is a good profit in rearing good beeves, and an absolute loss in keeping coarse, inferior animals. As an illustration, quite a remarkable steer was brought into Watertown market two weeks ago by Wells and Richardson, raised from the calf by Joseph Nye, of Fairfield, Me. This animal, a grade Hereford, though two months less than three years old, weighed, when landed in the stock yards, 2,030 pounds. The purchaser was R. H. Sturtevant, of the well-known Quincy market firm of H. Bird & Co., dealers in beef. There is no better expert in beef in this city than Mr. Sturtevant, who pronounced this steer as near perfection in shape and quality as was ever landed for sale in this market. Mr. Hathaway, one of the largest dealers in cattle in this market, was of the same opinion, and offered 8½ cents per pound, live weight, for the steer, or upwards of $172, to ship to England; but H. Bird & Co. wanted him at still a higher price to cut up for their best city trade. Mr. Sturtevant informs us that this fine steer dressed 1,604 pounds, counting meat, hide and tallow, a most remarkable result, showing a shrinkage of only 21 per cent. The loins cut from this steer weighed between 100 and 106

ROMEO (6646) 6420,
At 18 months. Bred by T. J. Carwardine.

pounds each, with suet in. The beef was of excellent quality, well marbled, and showing that as good cattle can be produced in New England as anywhere in the world. The loin which weighed 106 pounds carried 28 pounds of suet. Farmers should bear in mind, however, that it will not pay to rear or feed coarse, rough, leggy cattle, in this part of the country at least.

"Few cattle are received in this market that show so small a shrinkage as 21 per cent. Some of the best quality of Western steers, such as

MR. JOHN PRICE'S ELKINGTON CHALLENGE CUP WINNER, AT 3 YEARS.

are now selling at 7½ to 8 cents per pound, live weight, will shrink only 25 per cent. These cattle the butchers call 'rattling good' steers. A shrinkage of 26 to 27 per cent. shows well-fed and profitable steers. Some good cattle shrink 30 to 35 per cent., and common New England cattle about a third. Coarse cattle, old cows and the like shrink 40 per cent. or more."

NEW ENGLAND'S LIVE STOCK INTEREST.

Mr. E. E. Parkhurst discussed in the "American Cultivator" the question whether New England can compete in beef raising with the Western cattle range, and says: "Our markets are daily supplied with beef from the West, produced under great difficulties and transported to our markets over long lines of railroads and often suffering from hunger, thirst, and cold, while New England beef, when it is ripe, can be put into the market as fresh as when it leaves the stables, making it much more desirable as food for the consuming classes. With good and well-bred stock and intelligence and care in feeding, we need not fear competition from any one or any section of our country. The farmer who has lost money by raising steers and feeding high to get them ready for the butcher at 24 to 30 months has yet to be heard from. The farmer who cannot make his farm grow better by such farming had better go West, or change his occupation."

HEREFORDS IN ENGLAND.

The Earl of Coventry, who has taken a decided interest in breeding cattle, has given his views on Herefords in an English paper, and says: "That for early maturity and aptitude to fatten the Herefords are very remarkable." This, combined with "wonderful constitution and general hardiness," ought to convince every man who wants to realize the most money on his grass that he cannot afford to be without Hereford cattle to convert his grass and grain to money. The Earl of Coventry, Croome Court, Worcester, England, is very wealthy, but breeds Herefords for practical purposes. He has some of the best cattle in England, and has been very successful with his stock in the show ring. The cut of Rare Sovereign (¶ 236), a Hereford bull bred by him, was published in the "Breeders' Journal." His bull Good Boy (¶ 235), is a most wonderful, thick-fleshed and massive animal. These facts make the Earl's opinion very valuable. He says:

"As you ask me for my opinion of Hereford cattle, I can only say that after an experience of twelve years, I am still as great an admirer of them as I have ever been. It is true that in consequence of the prohibitory duty in America the demand for this breed has not been so great as in former years, but I believe the check is only a temporary one, and that we shall soon witness a return to fair and steady, if not sensational, prices again. On my land I find that for their early maturity and aptitude to fatten, Herefords are very remarkable, whilst their wonderful constitution and general hardiness would point to them as being specially adapted for crossing with other breeds in our colonial dependencies."

It was on the Earl of Coventry's estate that John Price, of Ryall, did most of his work.—*Breeders' Journal.*

JUDGES AND JUDGING.

While I have undertaken to show the character of the judging, and the manner in which our American agricultural and live stock fairs are run, I have had very little to say about the judging in England. "The Agriculturalist," a live stock journal published in London, has the following to say on this subject. While it does not touch the two breeds, Herefords and Shorthorns, it is a point to show what is thought of the English shows:

"This vexed question, of long standing, was again brought prominently to the front during the Bath and West of England Show at Brighton. On that occasion it referred more particularly to the decisions in the classes of Guernsey cattle. In fact, every one who professes to know what a Guernsey bull or cow is, was simply shocked at the incompetent judges who were appointed to adjudicate in regard to this breed.

"English breeders, as well as breeders from the islands, have protested in strong terms against the decisions to which they came. There are certain points in all breeds of animals which practical men look upon as of importance. These points ought not only to be understood by judges, but they ought to be recognized by them, and thereby acknowledged in the form of the awards they make. But the breeders in this case say the judges were so oblivious of the required points of Guernsey cattle that they awarded the prizes to inferior animals, and left unnoticed, or merely commended, those which were of the orthodox form.

"This comes of the way the judges of agricultural societies are appointed. In this country the authorities of societies have apparently merged into bands or cliques, each member of which seems to say, 'If you will give me a helping hand I will do as much for you some day.' There seem to be no bold critics or censor of this baneful practice or organized cliquism. In America a year or two ago the same personal favoritism prevailed. But they treat these matters better in America than we do in the old country. The system was vigorously attacked in the American agricultural and live stock journals. One principle laid down by a leading critic was, that 'any one—no matter who he was—who made an application to a society to be appointed as judge should not be taken any notice of.'

"Yet the judging of our live stock is mainly entrusted to men who push themselves forward, or apply to the chairman or members of the council for the post. As we have intimated, the whole subject has emerged into a system of cliquism or something worse. Breeders of different kinds of stock have their crotchets, and they take care to get themselves elected as members of the council of the leading agricultural societies. They then exercise their influence in electing judges who will favor their own crotchets. Thus some of the worst faults that the modern herds and flocks of animals have displayed during the present generation have been encouraged, and are now being perpetuated.

"We cannot dwell upon these great faults today, but we will do so on an early occasion. One symptom, however, we may mention, which is, the outrageous prices to which interested noble lords, rich tradesmen and their agents have puffed up certain tribes of animals, the main end and object of which is, or rather ought to be, to produce animal food at the market price per pound over the scales. This is why we are having the ground cut from under our feet by European, American, Antipodean and other breeders and feeders. Both animal food—be it beef, mutton or pork—and such animal's produce as cheese and butter, are now being sent from foreign countries of superior quality in many respects, and in quantities that make home efforts by tenant farmers, who are victims of the false system in question, anything but successful to either themselves or home consumers. If our great agricultural societies were left more open for the services of practical men there would be no difficulty in appointing judges who would give encouragement to the breeders of more paying animals—those that have a greater tendency to produce flesh than fat."

GOLDEN TREASURE (V. 15, p. 126).
Bred by Earl of Coventry.

CHAPTER XXXI.

Roots as Cattle Food

(¶ 237) The value of roots as food for cattle is too well known to need any argument to support the assertion. We have in our own experience found them invaluable as an aid to make ripe bullocks and keep our herd of Hereford cattle in the finest condition. We will state briefly our method of raising sugar beets, as this variety of roots was what we finally adopted as giving the best results. (¶ 238)

As early as possible in the spring we ploughed the ground deeply, and then harrowed thoroughly, going over the field enough times to pulverize the soil thoroughly. Then we spread over the ground, as evenly as possible, well-rotted manure. Next the ground was thrown into ridges, three feet from center to center apart, taking great care to have them straight. This was done by an ordinary plow, and by going twice in the furrow, throwing the earth both ways. This left the ridges with all the manure under the rows.

We found it advantageous to do this as early as possible, so that the field could stand in this condition for ten days or two weeks. There were two reasons for this: the manure had a chance to assimilate some with the earth and so push the young plants more rapidly when planted, and also the weeds had a chance to start.

Before planting we started a single horse through the rows, dragging a plank about ten feet long edgewise to level off three ridges at a time, leaving a flat top on which to sow the seed, and it also destroyed all young weeds on the ridges. Following this with a seed drill the ground was in fine condition and the seed over the fertilizer.

When the plants were half an inch high, cultivation commenced with a single five-toothed cultivator, then were side scraped with steel hoes.

The secret of success being preparing the ground thoroughly first, crop kept free from weeds and soil kept loose.

At the second hoeing the plants were thinned to one in a place and the vacant places filled with the plants which were pulled out.

In a root field of the T. L. Miller Company, the field contained ten and three-quarter acres, the yield was 275 tons of sugar beets. (¶ 239)

Labor: Plowing, dragging, seeding, cultivating and hoeing was $15 per acre, $161.25; harvesting, 54 days pulling, topping, hauling and putting in cellar, 35 days man and cart, $115.70. Total cash for labor, $276.95. (¶ 239) It will be seen by the above that they cost a small fraction over $1 per ton. The ground was well manured; of this we have not the data for cost, but for the use of land and the manure, fifty cents a ton would be ample estimate.

EXPENDITURE ON 18 ACRES OF BEET LAND.

The following statement shows our cost of producing 310 tons of roots, mainly of the sugar beets. The average is $1.37 per ton, without charging for the use of the land; or to charge $3 per acre for the use of the land, it would be $1.50 per ton, giving 300 tons of good feed on eighteen acres of land:

Two ploughings at $2 per acre........	$72.00
Harrowing at 50 cents per acre........	9.00
Rolling at 25 cents per acre..........	4.50
Rising drills, 13½ days at $3 per day...	40.50
Drilling, one man and horse, 2 days at $2 per day....................	4.00
Artificial manure, 5 tons, charge half..	60.00
Seed, as per bill....................	50.00
Ten boys hoeing ten days, at 50 cents per day	50.00
One man and horse scuffling 4 days, at $2 per day....................	8.00
TOPPING BEETS.	
Six boys 14¾ days at 50 cts. per day....	$44.25
Three women 5 days, at 75 cts. per day.	11.25
HAULING BEETS.	
Two men 15 days at $1 each per day...	$30.00
Two boys 15 days at 50c each per day..	15.00
Four horses 15 days, at $1 each per day	60.00
Total cost	$458.50

Total yield of roots, 310 tons.

CHAPTER XXXII.

Challenge to Test all Beef Breeds on a Broad Scale

It was our aim to have a conclusive test made that would go far to prove authoritatively the position that we took, viz., that the Hereford was superior to all other beef breeds in "economy of production and value of product." We made the following challenge in 1881, to test the merits of the Herefords and Shorthorns by selecting 200 cows in the State of Illinois and breed one-half to Hereford bulls and the other half to Shorthorn bulls, and bring the produce before the Illinois State Agricultural Society, and there determine the merits of the two breeds for beef production.

We also offered to select 2,000 cows in Colorado and breed one-half to Hereford and the other half to Shorthorn bulls, and bring the produce to a test that should determine their merits. Neither of these propositions has been accepted, and neither will be. We then made another proposition to exhibit before the Chicago Fair Association in 1881: "25 dry cows that have bred and suckled their calves since Nov., 1880; 25 calves with them; 25 yearling heifers." Of the cows we expected to show in the contest we said:

"None have been grain fed, except when milking before grass grew this spring; and none having been grain fed since the 15th of May, and while dry in the winter were on hay and stalks only. The heifers were on grain in the winter, but run in the yards, and since the 15th of May have been on grass only.

"I will show the above named stock against animals of like age and condition, selected from any Shorthorn herd in Illinois, Kentucky, Missouri or Iowa, for a premium of $250 on each class, the cattle to be owned by the exhibitor on the 1st day of August, 1881.

"The manner in which the cattle shall have been kept shall be certified to under oath by the owner and one of the men who has had charge of the same.

"The premium money, if not provided by the Society, to be provided, one-half by the Shorthorn interest and the other half by the Hereford interest.

"I will in addition show my old bull Success and twenty-five cows or heifers of his get against any Shorthorn bull and twenty-five cows or heifers of his get; all of the cows and heifers to have been on grass since May 15th, and without grain, for $250 or $500.

"I will show Conqueror, Bachelor, Will, Washington and Kansas, bullocks shown at the Fat Stock Show last fall, against a like number of similar ages shown at that show, for $250 for best bullock, and $250 for best five bullocks."

The following from the "Breeders' Journal," prepared by the author, shows a further proposition on the subject: "The discussion as to the merits of the different beef breeds has been active. At the commencement it would have seemed as though the Shorthorns were firmly fixed, as the leading breed for the production of beef, and this held for several years after these discussions commenced. A leading writer, T. C. Jones, in their interest said, as to the merits of breeds, that 'it had already been decided that the Shorthorns held the first position, and it was not a question to be again opened.'

"This same writer at another time did compare the breeds, and when we took this up and invited a full and frank investigation as to merits, his reply was that 'life was too short to enter upon a partisan discussion,' and declined the invitation.

"In discussing the merits of the different breeds in England, with three prominent Scot and Shorthorn men, we said this, that the Hereford would go from pasture to the top of the London market, while neither the Scot nor the Shorthorn could go there except through four to six months in the stall. This our opponents admitted. It is here that the merits of the Hereford stand pre-eminent; as a grazing beast, and from grass, carrying his product to the top of the market.

"While in England we visited the grazing districts. There are such districts in England where no breeding is done, but bullocks are bought at from one to two years old, mainly the latter age, and put in a straw yard and on cake during the winter, and on grass in the spring, and from these pastures to the London market in July, August and September. In talking with one of the leading graziers in Buckinghamshire, he said that whenever he took 'a rear of Herefords they would all go to market at once'; that is, they were even in quality and character, while with Shorthorns there had to be from two to three drafts to get them off. There is then but one reason for the character the Herefords have taken, and this one is merit.

cattle that dare make the test of merit from birth to the butcher's block with the Hereford.

"This test having been refused by old and experienced breeders, we tendered some months since the proposition to Messrs. J. V. Farwell (¶ 240-241) and T. W. Harvey, who have been for the last two or three years establishing respectively, herds of Polled-Angus and Shorthorn cattle. We could have taken a farm on lease within forty miles of Chicago, well adapted to experimental work in the breeding of cattle. We proposed to these gentlemen to take one hundred cows and to breed one-third of them to Shorthorn, one-third to Polled, and one-third to

COMPANY GATHERED AT CHADNOR COURT SALE, 1883.
(The X mark indicates Mr. George Pitt.)

Hereford bulls, for a term of ten years. We laid before them the estimated cost of such an experiment; $15,000 would have been the investment needed, $5,000 for each. We proposed to take one-third for the Herefords if each of these gentlemen would take a third each for the other breeds. We laid before them the fact that such an experiment was needed, and that it would be a great benefit to the world if such experiment could be made, that would test, during a term of years, the character and merits of the three breeds of cattle. They, however, de-

"America is pre-eminently a grazing country, and that breed which can make meat of first-class quality, from grass, must always be the top animal, and if the same beast is a first-class feeder, it will need only a trial to make him the top beast of the world. We have been so fully impressed with this from the start, that we have offered to place him in competition with other breeds, letting them choose the manner in which they should be bred and kept, and we would take them, in any way. This has been declined, and we say that there is not another breed of clined to enter upon such experiment. We think that when friends of these breeds come before the public with their claims for superiority, it would be their choice to place their cattle under such tests as would determine their merits, and it has been our aim from the start to bring about such an experiment, if possible.

"In relation to the discussion of the merits of the several breeds of cattle, especially the Shorthorns and Herefords, it has been said that 'there is room for all breeds.' The 'Live Stock Journal' says, 'this is probably true, there is room not only for thoroughbred cattle and the best breeds, but there is room for scrubs.' But this does not meet the question as it should be met. A man may use a wooden plow, if he has no other means of cultivating his ground; he may stir it with a stick rather than not move it at all. But he would not do this if he had a good pair of horses and a good improved plow to work it with. It is well to use the old spinning-wheel if there is no other means to make your clothing to cover your nakedness. But the old spinning-wheel would not be used if the improved machinery could be had, and steam or water to propel it, and what is true of these and other industries is also true of the live stock interest of the world. There is no room for the poor when the better can be had."

The "Chicago Tribune" copied and made the following comments on the foregoing. "The above is sound in theory. The difficulty is to decide the question of which is the best breed, taking into consideration the circumstances of the intending buyer. Using the machinery as an illustration, it would be more just to compare two spinning wheels, rather than the spinning wheel and the jack; two plows of modern make, rather than the stick of prehistoric ages with the finished steel implement of to-day; two engines of similar form and power rather than unskilled brute force tugging to pull a load, with the highest product of the engine builder's skill swiftly hurrying a train carrying thousands of tons over hills and vales. But much of the improvement made in breeds of animals is due to the keen competition between breeders struggling to gain a first place in popular favor. The fight has made partisans of friends. The partisans labored with, and aroused an interest in the minds of those, who, but for this, would have remained indifferent, and induced them to use better blood; to the great benefit of themselves and the world in general. Whatever the

SCENE AT CHADNOR COURT SALE,
Dilwyn, Herefordshire, 1883.

result may be to the breeders engaged in it, the battle has done immense good to others, and it is to be hoped that the time will never come when all will agree that any one breed or family of cattle is better than all others."

We made an earnest effort to arrange a test of merit with Mr. James Gaines, of Ridge Farm, Illinois, on a sufficiently large scale to be of value, and considerable correspondence was published at that time on the subject. Referring to the correspondence between Messrs. Gaines and Miller: Gaines & Sons proposed to take two car loads of steers from their herd for ten years; they afterwards say that they fed 1,400 steers. The selection of thirty steers from this number would not give any authoritative test of the breeds. It is a well established fact that the Shorthorns are very irregular and uneven as to size and quality. Thirty bullocks out of the large number grazed and fed by Messrs. Gaines & Sons would be likely to be very good steers, but their proposition as to the selection would not give any information as to the cost of making; neither would it give any information as to the general character of the breed; and again, there is no such lot of Herefords in the country to select from, and therefore it would be an unequal test.

The T. L. Miller Company's proposition was to take a given number of cows of uniform character, and give them a given amount of land, keeping a correct account of all expenses; this would, if continued for ten years, have been a very correct and authoritative test. Such a test was proposed by T. L. Miller and published in the "National Live Stock Journal" some twelve years previous to this correspondence.

That proposition had been renewed from time to time, but found no one to accept it, and it was again renewed at this time. We proposed to take 100 acres of land on their farm at Beecher, and select 40 breeding cows to be served by a Hereford bull; these cows and their calves to be kept upon the product of such land, and the produce to be marketed in December of each year, after they were two years old, at the Union Stock Yards in Chicago, if any Shorthorn breeder would take a like quantity of land, and a like number of cows, to be kept and marketed on the same conditions. We also invited any breeder of Scotch cattle to join in such tests on same conditions, each party to come under their

COMPANY ATTENDING "THE LEEN" SALE, 1883.
The celebrated Grove 3d (5051) 2490, at 9 years, sold for $4,250, appearing at the left. His breeder, B. Rogers, standing at his side, umbrella in hand.

individual bond, to make and keep a correct statement of expenses each year, and from the sales of each lot, $500 to be taken and invested in some good securities, until the expiration of the contract. The herd that showed the largest returns at the end of the ten years should be entitled to this fund. We suggested that the original cows should be replaced by heifers bred from this lot, the draft cows to be fed and marketed with the produce. Should this proposition be accepted, and it should seem best to the parties interested, to select land within fifty miles of Chicago, on the line of some leading railroad, and divide this land into parcels and apportion to each breed an equal quantity, we would have accepted such an arrangement.

It was not the intention of the T. L. Miller Co. to limit this to simply three herds, but the Shorthorn and Polled cattle men were allowed, if they chose, to make five herds from each breed, and breed and feed against five or ten herds (or any number not exceeding that).

Our object being to get the most authoritative test that it was possible to have. We contemplated that such arrangements should be adopted as would secure an honest administration for each herd. We believed that it was due to the public, that such a test should be made, and were anxious to submit the merits of the claims that were made for the Hereford breed of cattle, to the most severe and searching tests that could be made.

THE GROVE 3D (5051) 2490.
Bred by B. Rogers.

CHAPTER XXXIII.

Champions Contrasted—American vs. English; Ancient vs. Modern

We have devoted considerable space to our American Fat Stock Shows because the brunt of the battle between the beef breeds of America was waged at these shows.

The account of an English Hereford steer and an English grade Hereford steer, will make interesting reading, demonstrating, as it does, the beef quality of the thoroughbred Hereford and the great merit of the Hereford cross.

We take from the "Mark Lane Express" of Dec. 15, 1884, abstracts in reference to the Smithfield Show of that year, referring to the champion ox, a cross-bred (Hereford sire, Shorthorn dam), age 3 years, six months and two weeks, weight 2,617 pounds. Winner of the Norwich and Smithfield Club Champion Cups; bred by Mr. Chas. Doe, Burwarton, Bridgenorth, Shropshire; fed and exhibited by Mr. R. Wortley, Suffield, Aylsham, Norfolk. (¶ 242) Our readers will recognize the form and character of this animal as purely Hereford. In other words, the Hereford bull has given his character and quality to the steer.

The value of any pure breed consists in their ability to improve other breeds and to improve the common cattle of this and all other countries. Our steer Conqueror at Chicago in 1881, C. M. Culbertson's Roan Boy (¶ 243), in 1883, Mr. Wortley's cross-bred at Smithfield in 1884, and Fowler and Van Natta's Benton's Champion (¶ 244) at Chicago, in 1884, are notable examples of the value of the Hereford cross. The fact that the Shorthorn men claim the honor for Shorthorn blood is not worth talking about.

Let us make bullocks, by crossing the Hereford bull upon all other breeds, until our live stock shall give the top place to the WHITE-FACED BULLOCK, whether the cross is on the Shorthorn, the Devon, the Scot, the Texan (¶ 245), the Spanish, or any other breed. And let us in this establish the fact that the Hereford cross will improve every other breed for producing bullocks, the test being, "economy of production and value of product."

The "Mark Lane Express" said: "There is a very pronounced undercurrent of opinion—led and fostered by the example which has been set on American soil—to the effect that in the immediate future the prize schedule of our leading Fat Stock Shows should be so framed as to encourage the production of the earliest maturity, at the least possible cost; and to this end there must inevitably be classes made, and adequate prize money offered for competitions based on early maturity, in connection with a minimum cost of production, to be determined by a slaughter test. In the United States they have instituted classes for 'early maturity,' for 'cost of production,' and for the 'slaughter test'; but it would seem that a competition might easily be established in connection with the Smithfield Club Show, and particularly so whenever it is available for entries from the other principal shows—in the shape of one collective class for early maturity, as demonstrated on the block, to be determined by a set of judges composed solely of butchers, with separate honors dependent on 'economy of production,' as attested by breeders or feeders. The butcher's function should be to determine the best carcass; the committee should demonstrate by a simple statement on paper which had cost the least, relatively to age, when the butcher's verdict has been given. Something in this direction, and embracing these salient points, will certainly have to be done if the great London Fat Stock Show is to keep pace with or within measurable distance of—similar educational institutions on the other side of the Atlantic.

"But there have recently been two very striking instances in connection with show-yard animals which tend to show more clearly that maturity does not necessarily consist of mere 'fitness to kill,' although we submit that in both cases the question of cost could alone determine where the feeding process should have ceased, in order to afford the most perfect lesson in respect of 'economy of production.' The two animals referred to are Mr. John Price's Hereford ox, with which he won the first Elkington challenge cup at Birmingham, and Mr. R. Wortley's cross-bred Hereford-Shorthorn ox, which won the Smithfield Club champion plate last week. The statement showing the statistical record of Mr. John Price's ox was given in our report of the Birmingham Show, but it will be necessary to repeat it in substance here for comparison with a similar statement in respect to Mr. Wortley's ox, the records of the two years referring to the two successive exhibitions of the animals in each case:

MR. J. PRICE'S HEREFORD.

Date.	Exhibited at	Age in days.	Weight in pounds.	Average daily gain.
1881	Birmingham..	970	1,918	1.97
1882	Birmingham..	1,335	2,342	1.75

MR. R. WORTLEY'S CROSS-BRED.

Date.	Exhibited at	Age in days.	Weight in pounds.	Average daily gain.
1883	Birmingham..	924	1,889	2.04
1884	London......	1,289	2,617	2.03

"Now, these two animals afford very remarkable evidence in support of the growing opinion, to which we have called attention above, that mere fitness to kill, at an early age, does not constitute 'maturity' in the true, or possibly even in the economical sense of the word. These two animals show a daily rate of increase between their 2½ and 3½ years form, which is highly suggestive as to theory as well as very startling in point of fact. The cross-bred has actually maintained his 2½-year-old rate of increase for one whole year longer—for the difference of one point in the decimals is due to the value of the remainder—which we take to afford a wholly unprecedented experience in our reported show-yard animals. Clearly the 3½-year-old form of these two remarkable prize winners is a great improvement on their 2½-year-old form—an improvement, however, which cannot be equitably demonstrated by the figures. For example, the Hereford steer as a steer was a more perfect animal than was the cross-bred, so much so that we contended at the time that he was as good as a beast need be made; whereas, the cross-bred was merely a most promising frame-work—imperfectly covered—that could not possibly have stood a chance with the Hereford had they been competing side by side at the same age. Our personal recollections of the two animals—refreshed by reference to memoranda made at the respective dates—enables us to say this much. Yet the best Hereford ox yet shown in a fat stock show comes in rather a bad second to this marvelous cross-bred. However,

SALE RING AT STOCKTONBURY, LEOMINSTER, HEREFORDSHIRE, ENG., 1884.

AN ENGLISH SALE RING, AUGUST 28, 1884. "LORD WILTON" (4740) 4057, AT 9 YEARS OLD, SELLING FOR 3,800 GUINEAS ($20,000).

the Hereford men have no cause to grumble; it was their blood 'on top.'

"Our friend, Mr. T. L. Miller, of Beecher, Ill., U. S. A., who is the most persistent and consistent advocate of Hereford blood, pure or adulterated in every possible way, provided it is 'on top,' will have a most unique example to quote in favor of his long time argument to the effect that the Hereford bull on the Shorthorn cow would, could and should work wonders. As a matter of fact it has produced a champion prize winner at Chicago in 1883, and a champion prize winner in London in 1884; the latter being such a winner as no show has ever before produced. This is saying a good deal, but we think it can be sustained. Look at 2.03 pounds per day from birth, for the heaviest beast in the show; only 148 pounds lighter than the elephantine 'Welsh Jumbo,' which, at Birmingham, stood, tail, if not head, above the six feet high partition boards! Has there ever been weight for age to equal this? Again, the most remarkable feature about this animal is the gain of 6 cwt., 2 qrs. (728 pounds) in the interval of one year and one week between the Birmingham Show in 1883, and the London Show in 1884. It was not at all difficult to find fault with this animal. He had weak points, from the point of the shoulder to the outside flank and thigh, together with a comparatively indifferent cut of rump steak; but all this was simply nothing when considering the top, together with his immense substance through the heart, and wherever substance is valuable—with but slight drawbacks. Whatever his faults may have been, his weight was almost all in the best places, and such a weight in such proportions we may not see again. Such results two years following in America and in England should go far towards bringing about the life-aim of T. L. Miller, which—so far as we understand it—is to put the Hereford bull 'on top' of all the cow stock of the United States, from pure-bred Shorthorns to the native scrubs and the long-horned, half Spanish cattle of far-away Texas. Indeed, his enthusiasm does not stop here; he would put the white-face trademark on all the cattle of the world."

This cross-bred ox took, in addition to the champion plate of £100 ($500), the £50 ($250) cup as best ox or steer, the £30 ($150) breed cup as the best cross-bred, and the first prize of £25 ($125) in his class; in all, £210, or $1,050. He also took the champion prize at Norwich.

To show the difference in gain per day, and contrast Mr. Wortley's Hereford steer with the Shorthorn steer shown in Chicago, we quote the following from the English "Live Stock Journal," which was published under the title of "A Tale of Two Champions": "It may be of interest to compare the returns made for a year's feeding by the two animals which may be deemed the champions of the season here and in the United States.

"The American champion, 'Clarence Kirklevington,' an almost pure Bates Shorthorn, had this advantage over Mr. Wortley's half-bred ox; that the trial continued until his carcass was examined in the butcher's shop. 'Clarence Kirklevington,' at his death, had a live weight of 2,400 pounds, at an age of 1,372 days. The Islington champion had a live weight of 2,589 pounds at 1,292 days—a marked superiority for the Britisher.

"'Clarence Kirklevington' had for his last year's feeding increased 355 pounds; Mr. Wortley's ox in the same period, waxed 811 pounds, or more than double, and most wonderful of all, the Islington champion weighed at Norwich, on Nov. 20, 1884, 22 cwt., 3 qrs., 13 lbs. (or 2,561 pounds), and had on Dec. 6, in spite of traveling, etc., put on 31 pounds more, showing a steady increase still going on."

COMPANY ATTENDING STOCKTONBURY SALE, AUGUST, 1894.
Lord Wilton (4740) 4057 and two of his sons.

Referring to the above comment on "Kirklevington" carcass and the credit given to the Bates breeding, it is due history to say that it was one of the worst carcasses in the show; and the award was the great and crowning blunder of a series of blunders committed by the judges of the Fat Stock Show of 1884.

In this connection we would contrast, or rather compare the champion ox at the Smithfield Club's first show in 1799. At seven years old, winner of the first prize at the first meeting of the Smithfield Club in 1799. Bred by Mr.

Tully, Huntington, Hereford; fed by Mr. Westcar, Aylesbury, Bucks. (See illustration.)

The English Journal, "Agriculture, the Tenant Farmer's Journal," says of the ox as follows: "In 1799, the Smithfield Club was founded. From the Herefords this winner came, and a noble ox he was, as we find an authentic record that in height he stood 6 feet 7 inches, his girth having been 10 feet 4 inches while his weight was 247 stone, or nearly 31 cwt. [3,450 pounds, American weight—T. L. M.] This reads far above the figures that serve to note proportions at the present time, but as to this it is well to mention that this ox had attained the age of seven years, consequently his full growth had been obtained. There is a striking difference between the Hereford ox of 1799 and 1884, although both belong to the same pure breed, but an interval of eighty-five years separated them. During that somewhat long period much was done to improve the type and characteristics.

DOWNTON CASTLE, HEREFORDSHIRE.
(Photo of 1902.)

The ox of 1799 had worked three or four years as was the custom in those days, whereas the present show specimens for the greater part are spared all efforts in seeking their food, and they never work. It is not our desire to underrate the oxen of olden time. As working animals they might have been immensely superior to such as we now possess, and no doubt they were. But for show purposes work is decidedly against the ox, inasmuch as it bares the shoulders, and thins the neck, both of which are serious defects in the opinion of the best judges of stock. If we make this allowance for our champion, and a little also for the hollow-back labor frequently occasioned, we find him a great, fine ox, and can picture him to have been almost a team of himself, although according to our modern standard of symmetry he might be somewhat below it. Be this as it may, we can, however, point to the fact that Herefords have been vastly improved upon those of last century, and that their reputation was never so wide and pronounced as at the present time. As we propose giving several illustrations in this journal, giving the early type of Herefords, a few remarks on the accepted history of the breed may be of interest to our readers.

"The origin of Herefords is obscure, but it dates back many centuries, and in all probability to the reign of King John. Speed has recorded the existence at that period of 'white cattle' in the adjoining county of Brecknock, and there is no authentic account or tradition, even, that the black cattle of the Principality were other than we now find them; it is pretty safe to conclude that no admixture of white breed has ever taken place. From time immemorial Welsh cattle have been black, and if occasionally on the borders a Hereford cow produced a chance calf with black markings, that calf never became a breeding animal, but was drafted at the first opportunity.

"But should these white cattle have been introduced into Herefordshire—and their existence at the period mentioned seems beyond reasonable doubt—the origin of the present Herefords is comparatively easy to account for. Now, of what was a Hereford primarily composed? The answer of the physiologist, viewing the type of to-day, would be that the parents were red and white. Nothing is more certain than that the markings of the Hereford have been in a way created by most careful and continuous breeding. The white face with red body were not nature's colors, they were not spontaneous, but came as the result of mating the white breed referred to with the red cattle that possibly had long occupied the valley of the Wye, Lugg, Frome, Arrow and Teme.

"During the Saxon era the county of Hereford was strongly guarded by the English against the inroads of the Welsh chieftains, and this, no doubt, led to the introduction of red cattle from other parts of the kingdom. In these early times herds of cattle usually followed in the wake of armies, and it is extremely probable that the aboriginal breed of Sussex and Devon, which stocked the southern counties, was dispatched thither in considerable numbers where many of them bred, and thus stocked the Hereford district. But whatever might have been brought about after this manner there is no uncertainty as to the dark red and mottled-faced cattle of Herefordshire having been somewhat plentiful in the last century. For several hundred years the horned stock of the country was comparatively isolated from its kindred breed by the inferior cattle of the large dairying districts in the counties of Gloucester, Wilts and Somerset. Owing to this circumstance, and being surrounded by sorts in no respect of equal merit, the breeders in Herefordshire were fain to be content with what they possessed, and thus, by degrees, purity of blood and fixity of type were established in their cattle. Much uncertainty has existed as to the type of Hereford that prevailed in the past century. In respect of this there should not be any. For ages there has been no record or tradition of any blending of blood, and in the absence of such evidence, we may fairly conclude that no fresh blood has been sought since the amalgamation of the white cattle of Brecknock with the red stock of Herefordshire in the reign of King John, which produced in the process of time, the famous race now known everywhere as 'Herefords.'

"So recent as a hundred years ago there was no recognized color for the breed; they were nearly every color; *black, black tints and brindle alone* EXCEPTED. Nobody set much store on color, the test of value having been in the size and working powers of the oxen. It was then in the matter of color precisely what it is now with cart horses; for oxen then were bred quite as much for draught purposes as for beef, and but few thought of fattening three-year-old steers. To have done so would have been looked upon not only as a waste of working power, but also as a great loss in the production of beef. The yeomen of the country were exceedingly proud of their noble teams of oxen, and one vied with another in breeding and working the best in his neighborhood. Herds of repute were then many, but a fine choice herd was then termed a 'good stock' and to have a good stock was considered a sufficient guarantee of purity of breed. The grand old ox mentioned is a good representative specimen of the Tully's variety of Herefords."

Then again we find in the "Hants and Berks Gazette" (1884) as follows:

"How far is early maturity likely in future to influence decisions at fat cattle shows, and to what extent ought it to influence them, are points not unlikely to become of pressing importance. A decision in the agricultural hall last week bears on this point and is particularly noteworthy, as it brought two of the sets of judges into direct collision.

"The Shorthorn judges gave the breed cup to the Queen's remarkably wealthy heifer, thus placing her above all the first prize steers, as well as Mr. R. Stratton's first prize cow, but

HAMPTON COURT, LEOMINSTER, HEREFORDSHIRE.
The seat of J. H. Arkwright.

when another bench had to decide for the champion plate the judges did not take the usual course of deciding between the best steer or ox and the best heifer or cow, but had up the whole of the first prize animals for further inspection. This resulted, it is true, in the champion plate being awarded to Mr. Wortley's beast, already pronounced to be the best male, but as reserve to the champion plate, they selected Mr. Hugh Gorringe's Shorthorn steer under two years, which had won the first prize, consequently the latter was pronounced to be the second best beast in the show, although he had been previously beaten for the Shorthorn breed cup by the Queen's heifer. The circumstance is peculiarly relevant to the point of inquiry I have just made, inasmuch as prime quality united to early maturity was what chiefly distinguished Mr. H. Gorringe's steer—characteristics which another set of judges had not placed completely in ascendency over the riper early maturity or a few months older combined to such greater wealth possessed by the Queen's heifer.

"In speaking recently at a meeting of the Marshbrook Agricultural Society, Mr. J. Hill, of Felhampton, spoke of the rare excellence of the first cross between a Hereford bull and a Shorthorn cow, mentioning that the champion ox at Chicago last year was of this variety as well as the crack animals in England this year, which has filled highest positions at the Norwich and Smithfield Shows. No doubt Mr. Hill's opinion is worthy of being paid high deference to in assuming that such a cross is the very best for grazing purposes that can possibly be made, but it should be remembered that there are very few crosses in the past in general farming, where both sire and dam have been absolute pure-bred, owing to the dam being usually considered far too valuable to breed to any bull but one of her own breed. The progeny of pure-bred Devon cows mated with Hereford sires would probably be superior for grazing purposes even to the Hereford-Shorthorn cross, unless the Shorthorn cow was very much superior, as much better quality would be derived. Further, it may be worthy of note that Mr. Miller, when in England about a year since, remarked what valuable grown beasts might be bred by many in the western shires, who have Devon herds, if they were only to cross them by Hereford bulls.

"Another observation of Mr. Hill's in the speech alluded to is worthy of note, as it was a reply to the strictures of Mr. Coleman, in his report on prize farms in the Royal Agricultural Society's Journal, on the method of calf-rearing as adopted by Hereford breeders. He said the Royal judges were unable to perceive how it could pay to keep a cow all the year just to rear one calf, but he thought if the superior early maturity of the calf was considered, full compensation would be obtained for the sacrifice of milk by obtaining a much better beast. This, however, may be a matter of opinion. The point of a better animal being produced by continuing the system than by hand rearing, was scarcely disputed by Mr. Coleman and the Royal judges. All they ventured to doubt was whether too much is not sacrificed for the early maturity and better beast, especially as it comprehends the probable impairment of the cow's milking properties in the future. The question is probably one scarcely admitting of satisfactory decision without absolute experiments."

CHAPTER XXXIV.

HEREFORD MOVEMENTS BEGIN TO COMMAND THE OPPOSITION'S RESPECT

The investigation of the opposition to Herefords, and the means used by Shorthorn breeders from the time of Youatt until the present, would shadow and dim the operations of the Star Route conspirators. These operations, it will be noticed, have been directed against the Herefords and those interested in them with the utmost venom and vindictiveness. When Mr. Youatt, at the dictation of the Shorthorn breeders, put forth all the facts he could reach to advance that interest, and suppressed the facts that would advance and maintain the claims of the Herefords, from that time until now, these tactics have been practiced. In 1834 Youatt wrote in the interest of the Shorthorns; in 1839 the Royal was established and run in the Shorthorn interest; in 1851, control of the Smithfield Club was obtained and run in this interest; in 1841, the New York Society was established and officered and run for the benefit, and in the interest of the Shorthorns, by Shorthorn breeders.

All these schemes have been continuously practiced. Stating these facts as a matter of history, we will take up and discuss the operations of the Illinois State Agricultural Society and its management. We wish to be fairly understood in the position we take. When we charge the Shorthorn breeders of this country and England, we speak of them as writers would speak of political parties when they are charged with corrupt practices. It does not follow that the rank and file are included in the charges of corruption; but it is of those who manipulate these corrupt and fraudulent plans. Their supporters are led and influenced into supporting these leaders and their plans. So it is with the large bulk of Shorthorn breeders. They accept Youatt as authority for the early standing of the Shorthorns. They inquire for some work that will give them information that may be considered a reliable guide, and they will be directed to Allen's work on American cattle, or to Ambrose Stevens, Youatt and Martin. Would we be saying too much, that they have done as much as they could to bring the Shorthorns to the notice of the reader, and as little as they could to bring the merits of the Herefords before the public? There is much that either might have found if they had wished. As early as 1840, H. S. Randall quoted Youatt against the Herefords in his controversy with Mr. Sotham. He ought then to have known that Youatt was the editor for the Shorthorn work.

The first show held under the direction of the Illinois State Board, we think, was in 1853; and we think no one will question the fact that it was largely under the control of the Shorthorn breeders and that the claims of that breed were pushed; but of the earlier years of this society we have not much to say. The premiums, so far as we know, were uniform for Shorthorns, Herefords and Devons.

This was true in 1872, when the Herefords made their show under Miller and Powell, and for 1873 and 1874; but in 1875 the Shorthorn premiums were raised about 70 per cent and the Herefords reduced about 33 per cent; and during four years a majority of the judges for herd and sweepstakes premiums were Shorthorn breeders, and they ignored the Herefords as completely as though they had not been on the ground. While the Hereford premiums were reduced in 1875, the Jerseys were permitted to stand as before, and Shorthorn advocates would, under these circumstances, proclaim awards rendered by Shorthorn judges as evidence of merit. And this is not peculiar to Shorthorn breeders of Illinois, but it is the rule, whether in Illinois, New York, Ohio, Kentucky or England.

In 1876 the Illinois State Board put the Shorthorn premiums back to what they were previous to 1875, but did not change the Herefords. In 1877, the Board still further reduced the Shorthorn premiums and advanced the Herefords, making them even, and placed the milk breeds on the same basis, in which condition they have remained until now.

But this equalization was opposed by the Shorthorn breeders outside, and by their friends in the Board; and while all breeds were permitted to compete for herd and sweepstakes premiums, they were obliged to do so under Shorthorn judges; and Shorthorn breeders and advocates said to the Hereford breeders, "When you can win before our agricultural shows, then we want them." We have mentioned in detail, elsewhere, the show for 1878.

In 1879 the judges were selected previous to the fair, and no protest was permitted. This year, and the year 1877, the judging was fairly conducted—both 1877 and 1879 followed two specially partisan years, the first under W. J. Neely and Col. Judy, and the second under Col. Judy, both Shorthorn breeders. We are not advised fully, or are not authorized to speak for the Board, when we say that the partisan action in 1876 and 1878 brought the Board to the determination to do something to check the manipulations of the Shorthorn men. Still, we believe that it is true.

The inauguration of the Fat Stock Show was an innovation upon the Shorthorn plans. They had been showing fat and barren cows; they had selected the top animals from different herds, and often these show herds represented a fair fortune; and while the country was full of these show herds and winning as breeding stock, they could not bring a respectable showing of cows that would win before the butcher.

At the first Fat Stock Show in 1878, the Hereford cow "Jennie" took the sweepstakes premium for best cow in the show. The Hereford cow won the sweepstakes fairly in 1879, though there might have been a chance for a difference of judgment; but in 1880, of all the partisan work that had been done, this was the most glaring; and as an evidence that we do not express simply an opinion from a Hereford standpoint, we will say that of all the Shorthorn writers that undertook to get comfort out of that show, not one, as far as we know, made a claim for winning the sweepstakes for the best cow in the show.

The judging in 1880, when the different breeds came into competition, was first on grades, and was wholly and entirely indefensible. The judges were discharged for cause, but

COURT HOUSE, PEMBRIDGE, HEREFORDSHIRE.
Residence of J. Price. Mr. Harry Yeld and Mr. John Price in the foreground.

still the verdict was permitted to stand. A protest was made against the awards on grounds of gross injustice. This was admitted. "But," said the Board, "we have no way to set aside an award except on account of fraud." One of the Board very correctly said, "A court, whenever a jury brings in a verdict contrary to evidence, sets it aside, and the jury is discharged. We," said the member, "have discharged the judges, and it follows that the awards should be set aside."

The next day the Herefords, in competition with thoroughbreds and grades, took the sweepstakes for each age—on two and three-year-old steers. We have already alluded to the award on the cow. For the best beast in the show the Shorthorns took this under protest, and the Herefords took the reserve vote for this also.

LORD WILTON (4740) 4057 AND TWO OF HIS DAUGHTERS.

We would now call attention to the dressing of the bullocks. The Shorthorn breeders claimed greater size, heavier hind-quarters as compared with the fore-quarters, with smaller heads, feet, etc. Those who tried to detract from the Hereford victory did not discuss the details of this test. There were entered for dressing, one Shorthorn, 3 years old; two Herefords, 3 years old; one Shorthorn, 2 years old; one Hereford, 2 years old; two Shorthorns, 1 year old; one Hereford, 1 year old. Of the yearling Shorthorns, Mr. Gillette entered one and Mr. Moninger the other, but both refused to dress. Mr. Gillette entered and dressed both the Shorthorns.

The Shorthorn breeders, it will be noticed, did not propose to bring their breed to this test, except Mr. Moninger, and he refused to fulfill. The test by dressing is the one that comes the nearest to a perfect measurement.

While the Shorthorn breeders claimed greater size, the three-year-old bullocks offered by them were 20 per cent lighter than the Herefords; and the two-year-old was 10 per cent lighter than the Hereford two-year-old. The hind-quarters of the Hereford premium steer were 40 pounds lighter than the fore-quarters; while the hind-quarters of the Shorthorn were $108\frac{1}{2}$ pounds lighter, while of the two-year-olds the Herefords showed a difference of 40 pounds, and the Shorthorns $52\frac{1}{2}$ pounds.

Another point the Shorthorn advocates made was as to heavy heads. By an examination of the figures it will be found that even here the percentage is with the Herefords. The Hereford exhibitors met every point in contest and showed greater merit for their breed.

These tests, however, do not touch the real question at issue between the two breeds, to-wit, the cost of production and value of product. As a result of these showings, however, the accessions to the number of Hereford breeders at that time were large and consisting of men of large means and great experience in cattle raising and feeding.

The arguments used to show the unparalleled excellence of the Shorthorns are often of the flimsiest nature. In the Seventh Annual Meeting of the Iowa Improved Stock Breeders' Association, Mr. Pliny Nichols presented the claims of the Shorthorns and stated his position. First, to show the merits of the Shorthorn as a combined machine for milk and beef; second, his merits for beef alone, and third, his value for the improvement of natives and other breeds. On the first he claimed that 75 per cent of cattle for beef and the dairy in England are of Shorthorn blood, quite oblivious of the fact, that on the same principle, he might claim that as 75 per cent of the cattle for beef and dairy were scrub stock, therefore it was best. We quote Mr. C. F. Clarkson, speaker at the same convention, who preceded Mr. Nichols, as follows: "Our modern breeders claim that they have improved this stock (the Shorthorn) in grace and form, while they sacrificed the milking qualities, until it requires the aid of some other breed to raise their calves."

SAMUEL GOODE,
of Ivingtonbury, Herefordshire, for many years in Australia.

With this quotation we will pass the claims to a combined machine and notice Mr. Nichols on the Shorthorn for beef alone. He said, "For early maturity, size, hardiness, prepotency, cost of production and value of product, the Shorthorn excels all others." If Mr. Nichols had closed this schedule of merits by saying he does *not* excel all others, he would have made a statement that he could have maintained.

The contest that year (1879) for the champion prize at Smithfield for the best beast in the show, was between a Hereford and Shorthorn under two years and six months old. They were of the same weight and ages. On this point there was no difference, but the award was given to the Shorthorn, and there were those who questioned the correctness of the award. Knowing that a Hereford always sells for more money per pound in London, we questioned the justice of the decision and had cuts made of the two steers. (¶ 246) The "Mark Lane Express" gave it as their opinion that the award was right, "although," they say, "the Hereford was the firmer fleshed animal," and a little further questioning elicited the fact that the two bullocks going to the butcher, the Hereford would have sold for the most money. If then the champion steer of this show was a two-year-old, and the choice was between two, a Hereford and a Shorthorn of the same age and weight, and the Hereford was of a quality to command the larger price in the market, he should have had the champion place. This must have been known to Mr. Nichols when he came before the Iowa Society. He might, with the same propriety, point to the awards of the Iowa Agricultural Society as an evidence of merit for his breed.

Mr. Nichols then took the Chicago Fat Stock Show and averaged it for three years, and said, "Of course exceptional cases can be given where the Herefords have attained good weights and creditable gains per day from birth, especially when they have resorted to the Shorthorn blood to make the cross on."

Now, Mr. Nichols and other Shorthorn breeders ought to have known that the claim was not valid. The truth is that the Shorthorn on the native cattle of the country, has not produced a uniform bullock. There is no reliability to be placed upon the character of the bullock bred from native or scrub cows by the use of the Shorthorn bull, and we must give the same uncertainty to the heifers as to the bullocks, and these heifers we must use—heifers that no Shorthorn man would admit bore any resemblance to the Shorthorn type—and when we put the Hereford bull upon these heifers and produce winning steers, they turn around and claim the merit for the Shorthorn. But it is not true, and the claim that was made by Anderson and endorsed by Nichols as to the bullock at the Fat Stock Show of 1880 was fraudulent.

Said Mr. Nichols, "The Hereford breeders have by such extraordinary means captured scarcely one-third the premiums as against Shorthorns." The managers of the Fat Stock Show offered a premium, first and second, for the best grade three-year-old steer; for the best grade two-year-old steer; for the best grade one-year-old steer; and sweepstakes for the best of any breed three years old; for the best of any breed two years old, and for the best of any breed one year old; for the best cow of any breed three years old, or over; for the best beast in the show. Here are eleven premiums in the grades, six first and second. The Herefords took two seconds. The judges, in this case, the president discharged for cause; and when the Herefords came into competition again with these and the thoroughbreds the next day under a committee selected with greater care, the Herefords took every premium, and when we take the fact that the judges in the first case were dismissed for cause, it was not in good taste for such Shorthorn men to claim honor for the awards. On the sweepstakes cow a special committee was appointed, and the Shorthorn cow taking the award was nothing more than what would be termed a butcher's beast at the yards, and for the best beast of the show, the Shorthorn steer taking the award had been fully discussed in the press; and the admission of the animal was not generally deemed creditable to the Board or the Shorthorn men. Out of eleven premiums the Herefords took five, and were entitled to eleven. For dressed bullocks they took two out of three. Out of fourteen premiums they took seven and were entitled to thirteen. The Shorthorns took on grade steers four premiums on three steers under a committee of judges that were dismissed for cause and under another committee the Herefords reversed these awards.

Shorthorns may have improved the common cattle of the country, perhaps, as Mr. Nichols said, $10 a head, or an addition of $400,000,000 to the cattle of the country. This was pretty extensive figuring, and we promised to take the Hereford and put him upon this Shorthorn produce, and add $20 a head to the value, and leave Mr. Nichols to estimate the result. This is the point we made and continue to reiterate, that beef can be made on Hereford cattle at 25 per cent less cost than on Shorthorn cattle.

Mr. A. S. Matthews, in his efforts to bolster up the Shorthorns and depreciate the Herefords, gave an account in the "National Live Stock Journal" of the Shorthorns at Smithfield.

In 1879 the champion prize for the best beast in the show was between a Hereford and a Shorthorn under two years and six months old, to which we have referred. The Hereford was the equal in weight and gain per day, and was worth more on the market, and still he was left without the honor. For what reason? Because of Shorthorn influence in the management and in the judges. The facts given as to the judges in the show of the Illinois State Board, is as applicable to the Smithfield, and from this reason, we presume, the Hereford men have kept from the show.

Mr. Matthews then devoted considerable space to show the greater weight of the Shorthorns. If he had taken the history of Mr. Charles Colling with the early Shorthorns, he would have found that a good portion of his life as a breeder was devoted to experiments to reduce the size and coarseness by crossing with the Scotch breeds.

Youatt says: "He (Colling) was sensible also of the difficulty of breeding, with anything like certainty, large good animals. He found the Teeswater like all other extravagantly large animals, frequently of loose make and disposition." The Scotch cross helped him out of his difficulty, and the same is necessary now. Writers in those days gave the Scot credit, but not so the writers of my time.

Quotations were made by Mr. Matthews from the "Mark Lane Express" to support his Shorthorn claims of important victory at Smithfield, 1879, and we gave him the benefit of that journal's remarks on the Bath and West of England Show in 1878, where they said: "We consider the Shorthorn cow class to be a disgrace to the breed, and therefore to the breeder. We are quite unable to discover the 'grandeur' and the 'superb character' of these old crocks, which

DALE TREDEGAR (5856) 14682, AT 10 MONTHS.
Bred by H. J. Bailey, exported to Australia.

some of the Shorthorn fanciers appear to have the faculty of discerning, and do not hesitate to record them as being just a rough lot of cows; if their blood is of the bluest, their carcasses are of the ugliest, and not worth anything beyond contractor's price when they come at last to the shambles. We cannot help thinking that to an unprejudiced mind there must be evidence of a something quite outside of agriculture, and quite useless to the rent-paying farmer, in this Shorthorn 'fancy.' We are sensible of the improvement that has been effected already in the rank and file of our cattle throughout the country by the use of this Shorthorn blood; we do not wish to detract one iota from its legitimate merits, but simply to point out wherein it becomes sometimes a matter for the ridicule of non-believers. We see prizes awarded systematically to animals which are not calculated directly to improve the production either of meat or milk in their descendants, and we are told that there is some marvelous power or virtue stored up in their veins, and that although their bodies—the casket—are unsightly,

ROYAL HEAD (4490) 15765 AT 2 YEARS 8 MONTHS. Bred by J. Williams, Herefordshire. Exported to Australia.

their blood—the jewel—is pure, potent, and almost priceless. Well, we simply do not believe it, as they put it. We are perfectly well aware that any 'terribly in-bred' weed, a wretch to look at, but having an unexceptional pedigree, will, if matched with mongrel-bred stock, produce a result, which is far, very far, in advance of the mongrel-bred dams; but so would any absolutely pure-bred animals. Therefore, we think that farmers, those who have the production of beef or milk in view, have a right to expect something which is calculated to effect their object in a direct manner; no breeder of bullocks would give herd room, much less a high price, for the bluest blood bull which did not carry a frame the character of which it was desirable to transmit. When a lot of highly-bred but not correctly fashionable young bulls are to be bought for about 30 pounds sterling ($150) apiece, and here and there one which has been bred correctly to fashion fetches 3,000 pounds sterling ($15,000) and would not be worth one shilling more to the food producer, then we think we are justified in saying that rent-paying farmers have already drawn the line between business and fancy by refusing to give more than a business price for a fancy article. We should be glad to see every young Shorthorn bull now in the breeders' hands sold to tenant farmers, who would use them for meat or milk as their requirements might decide, and are firm believers in the general usefulness and superior adaptability of the breed to any other; but we feel it a duty to point out the ridiculous position the Shorthorn breeders seem content to occupy at our great shows by the mixed quality and low status of many of the animals exhibited, and which gives opportunity for pointing the finger of scorn, and leads to such questions as, 'When is the bottom of this Shorthorn humbug likely to drop out?' We hear a good deal about the 'alloy,' and if correctly informed, it means an infusion of Scots blood of some kind, and to our mind the very thing these Shorthorns are now needing is another infusion—a strong one of some alloy which will give them the thickness of flesh, the wealth of hair, and the butcher's form they so seldom possess, and then we should be prepared to expect great things of the renovated blood."

The same paper, in speaking of the Herefords at this show, says:

"They are not so numerous as the more fashionable breed, but the quality throughout is excellent. In the aged bull class there are five animals of which the Hereford men need not be ashamed.

"The heifers in milk or in calf numbered only three, but two of them were such animals as it was worth while coming to Oxford on purpose to see. Mrs. Sarah Edwards' 'Leonora' is one of the most perfect animals that has been shown for years; if she had been a Duchess Shorthorn, a poem would have been composed in her honor, and translated into several languages by this time. But no Shorthorn that we have ever seen was cast in such a mold as this Hereford heifer."

On the awarding of the champion prizes at this show we quoted for Mr. Matthews' benefit, from the same authority, the following account:

"The whole conclave of judges came into the ring to decide which was the best male horned animal in the yard, and here the Shorthorn men were in a hopeless minority—cornered in a manner that does not often happen to them in a show yard. In vain they contrasted the strong points of the Shorthorn with the weak ones of the Hereford. It was all to no purpose, the rest could not get away from the Hereford, whose wonderful rib and forehand were too much for the Shorthorn, and a show of hands showed an overwhelming majority for the Hereford. There is no doubt that had the Shorthorn judges been of sufficient numerical strength, they would not only have prevented this Hereford triumph, but also that which followed, when the best female animal was to be decided upon, for it went sadly against the grain to

(FAC-SIMILE OF MR. MILLER'S CELEBRATED CHROMO.)
(¶355) "SUCCESS" (5031) 2, IMPORTED IN 1873 BY T. L. MILLER.
This Lithograph in colors, freely distributed, was one of Mr. Miller's effective methods of drawing attention to Herefords.

award even such a heifer as Mrs. Edwards' 'Leonora' the championship over the Shorthorn 'Diana'; but it had to be done, and we think there were few outsiders who were not thoroughly satisfied. These championship awards were an unquestionable streak of lean for the 'fancy,' and we may depend upon their not allowing such a thing to occur again if they can possibly help it. This, after the Hereford victories at Paris, is about as much as they will be able to bear with patience. We should be glad to see sweepstakes judging of this kind occur more frequently."

The honors will be shifted from the Herefords to the Shorthorns, and from the Shorthorns to the Herefords, and again the Polled-Angus may win over both, so long as the judging is upon the animal without regard to cost; but when their merits are measured by the "cost of producing and value of product," there will be but one result, and that will always be in favor of Herefords.

Mr. Matthews understood very well that the Shorthorns, as a breed, had been bred to "Duke" bulls until in the language of the "National Live Stock Journal," they were a "coughing, wheezing, consumptive lot"; and in the language of Thornton's circular, "weedy offspring, delicate in constitution, and ill-adapted for crossing purposes."

Mr. Matthews' statement as to conditions in England were, however, taken up by the "Mark Lane Express" itself, which denied emphatically Mr. M.'s assertion and said decidedly that the Shorthorn breed had not yet driven the Hereford breed off one acre of ground. In its issue for July 4, 1881, it said: "There is a lively controversy respecting the comparative merits of Shorthorns and Herefords going on in the columns of the 'National Live Stock Journal' (Chicago), between Mr. A. S. Matthews, of Wytheville, Va., and Mr. T. L. Miller, of Beecher, Ill., and they both quote the 'Mark Lane Express' as supporting them in their respective arguments. In this controversy we have no wish to enter; we shall enjoy seeing them fight it out. Neither of these gentlemen requires any assistance from us, but they each of them call for a remark or two from us in respect to some of their statements.

"To begin with Mr. Matthews, as his letter comes first in the 'Journal' and is replied to in the same number by Mr. Miller: In referring to our report of the last Islington Fat Stock Show, Mr. Matthews quotes what was there said of Mr. Grissel's mixed-bred steer, 'by a Shorthorn bull out of a non-pedigree cow,' namely, that he was precisely the sort of animal with which the British farmer must hope to win in that struggle with foreign competition, an animal which can be grown by a system of mixed husbandry, which has the production of meat and milk on arable land for its basis.' Now, if this is precisely the sort of animal that the breeders and graziers on all arable farms of mixed husbandry in the United States must grow if they expect to send beef, live or dead, to England at a profit, and these arable farms of mixed husbandry on which cattle can be bred, grazed and fattened profitably, include all the land east of the Rocky Mountains, and north of what may probably be termed the 'Cotton Belt,' in reply to this we have no hesitation in saying that where conditions of mixed husbandry exist in the United States, equivalent to those which constitute the essential features of mixed husbandry in England, the Shorthorn breed of cattle should, according to our ideas of the subject, be equally suitable and serviceable there as here. But we were under the

JNO. G. IMBODEN, DECATUR, ILL.
Celebrated expert judge.

impression that east of the Rocky Mountains and north of the Cotton Belt there existed in the United States vast plains of grazing lands on which large herds of cattle are bred, reared and fattened, and it has been in reference to these supposed localities that we have ventured to express the opinion, that Herefords would do better than Shorthorns. If 'all the land east of the Rocky Mountains and north of the Cotton Belt,' which will include part of Wyoming, part of Colorado, nearly the whole of Montana, the whole of Dakota, Nebraska, and part of Kansas, not to mention such states as Minnesota, Iowa, part of Missouri, is devoted to mixed husbandry of the English type, then we have hitherto been greatly in error, for it was the district west of Chicago and east of the Rocky

PHILIP D. ARMOUR, CHICAGO.
The great packer and philanthropist was always a patron and lover of the Herefords.

Mountains in which we imagined the Herefords could profitably be bred by allowing the calves to suck their dams, and reared for beef alone. There is a mistake somewhere. However, Mr. Matthews says nothing about the districts within or south of the Cotton Belt, and as we imagine these include Arizona (¶ 248), New Mexico, Texas and Arkansas, in some of which localities, notably Texas (¶ 247), cattle are reared for beef alone, there must be at least a fair chance for the Herefords on the other side of the Atlantic.

"We have one more remark to make respecting Mr. Matthew's statements. He says: 'After the Shorthorns have pretty much driven the Herefords out of England—for they don't seem to have increased in numbers at all; in fact, several other breeds seem to be gaining fast on the Herefords, especially the Polled Norfolk, Sussex and cross-breds—it will be something strange if the Herefords are to supplant the Shorthorns in America. What if Mr. Cochrane of Canada did buy about forty Herefords in England last year? What of it? I hardly suppose that forty Herefords would cost as much as Colonel Cannon gave for one of Mr. Cochrane's Duchess heifers at Dexter Park last June, namely, $8,000.' Now, the Shorthorn breed has not yet driven the Hereford breed off one acre of ground. The Hereford breed of cattle is certainly anything but under a cloud just now. The breed does not extend its area, that we are aware of, except possibly, in Cornwall, but it holds its own in its native districts. For purity of blood, uniformity of character and excellence of beef points the Hereford excels the Shorthorn by a long way, and every score of Hereford bullocks sold into the Midland grazing districts displaces so many Shorthorns; if forties could be had where only scores are obtainable, the displacement would be in the same increased ratio. Mr. Matthews has not an iota of evidence to support his assumption that the Shorthorns are driving the Herefords out of England—except across the Atlantic to take their place. So far as displacement is concerned, the Shorthorns have crowded out mongrels to a great extent, but much remains for them to do in this respect; and they are pushing the Longhorns—a fine old race of cattle—to sure and certain extinction in course of time, because they contend on equal terms. The Devons and the Norfolk Polled cattle are also being crowded by the Shorthorns. But nowhere have they displaced the Herefords."

To add a little more Shorthorn testimony as to the condition of the breed in these years of which we are writing, and when these controversies were being waged, we give a report, made by Mr. George Y. Johnson, secretary of the Kansas State Fair, to a Shorthorn Convention held at Topeka:

"Gentlemen of the Kansas State Shorthorn Association:

"Your last year's essayist, Hon. F. D. Coburn, came before you, very properly, as the advocate or partisan of no particular breed, because he is not a breeder of cattle. I come before you under different circumstances—as a man, who, having measured his ability by his pocket book, his ability to rear by his acreage, his ability to grow by what his land will produce; measuring with some degree of accuracy, based upon average intelligence, the relative profit of a limited means devoted to breeding and raising cattle.

"After careful study and investigation, with the aid of such information as I was able to procure, I made my first venture in grade Shorthorns and a thoroughbred bull, and have carried the breeding up from minimum grades until I find the last product to be an animal possessing 127 parts Shorthorn and one part native. I here became convinced that a remark made by our friend, Colonel Robert Holloway, of Illinois, to me years ago, is forcibly true, viz., 'A fault in form can be remedied by judicious breeding, but a stain on a pedigree can never be wiped out.' Hence, I purchased thoroughbred females of registered, faultless pedigrees, and have now no interest financially in any breeding cattle except Shorthorns and their grades. This much to show you that I come before you as a Shorthorn partisan, and that whatever I may say upon this subject is said from the standpoint of a Shorthorn advocate, yet trusting that this partiality may not blind my eye like those of a fond parent, so that no imperfections are discovered or recognized in my pets.

THE BREEDS AT KANSAS CITY AND CHICAGO IN 1883.

"I find, by a careful study of classes at the Kansas City Fat Stock Show of 1883, that, excluding car-load lots, there were 121 animals entered individually; of these the following classification is made: fifty-five grade Shorthorns, twenty-two thoroughbred Shorthorns, eleven grade Herefords, ten thoroughbred Herefords, five thoroughbred Angus, two thoroughbred Galloways, thirteen grade Galloways, two grade Holsteins, and one thoroughbred Holstein.

"Taking thoroughbreds and grades and classing them altogether, we find the following average results:

KANSAS CITY.

BREEDS.	Av. age in days.	Av. weight.	Av. daily gain from birth.
One year and under two.			
27 Shorthorns	630	1245 lbs.	1.97
11 Herefords	598	1276 lbs.	2.13
Two years and under three.			
12 Shorthorns	892	1549 lbs.	1.73
3 Herefords	992	1736 lbs.	1.75
3 Angus and Galloways....	1,004	1435 lbs.	1.42
Three years and under four.			
12 Shorthorns	1,252	1963 lbs.	1.56
2 Herefords	1,386	2165 lbs.	1.56
2 Angus	1,313	1871 lbs.	1.42

CHICAGO.

BREEDS.	Av. age in days.	Av. weight, lbs.	Av. daily gain from birth.
One year and under two.			
48 Shorthorns	633	1244	1.96
48 Herefords	534	1179	2.01
Two years and under three.			
30 Shorthorns	903	1723	1.91
22 Herefords	942	1543	1.64
1 Angus	999	1815	1.31
Three years and under four.			
26 Shorthorns	1,370	2041	1.49
7 Herefords	1,361	2052	1.50
1 Angus	1,380	2355	1.70

"From this it will be seen that the results at Kansas City and Chicago are practically the same, and hence in treating specifically we include the other also. In the sweepstakes rings, the committee, composed of old, experienced butchers, awarded as follows:

"In the ring for two years old and under three there were at Kansas City fifteen entries, viz.: Ten Shorthorns, one Holstein, two Herefords, and two Galloways. The premium was awarded to 'Benton's Champion,' a grade Hereford, or more properly a Hereford and Shorthorn cross. There were nineteen entries in the ring three years and under four, as follows: Fifteen Shorthorns, two Herefords, one Angus and one Galloway. The premium was awarded to Starlight, a grade Shorthorn—seven-eighths Shorthorn, one-eighth native. From all the foregoing we find the scales pretty evenly balanced between the Shorthorns and Herefords, the black cattle coming in third.

JAS. A. FUNKHOUSER,
Plattsburg, Mo.

The preponderance really is in favor of the Herefords.

A POOR EXCUSE.

"It is a poor excuse or explanation to say that most of the premium animals claiming to be Herefords grades are really crosses with the Shorthorns, as the following query suggests itself to every unprejudiced man: If the Hereford is an inferior animal, why should the cross upon the Shorthorn (a superior animal) produce an animal superior to the superior? It may be some solace that at the Chicago show Starlight, a grade Shorthorn, took the sweepstakes prize as best carcass in his class of three and under four years old, and was also awarded the grand sweepstakes prize as best carcass in the show; but even this is somewhat tamed and dimmed by the fact that the vote of the judges was a tie on first ballot between Starlight and a black steer, Waterside Jock, and another, and

PURE BRED STEER FRED (BY FORTUNE 2080).
Bred by J. S. Hawes, Colony, Kan. Champion dressed carcass, Kansas City Fat Stock Show, 1885.

that the umpire called in, cast his vote for Starlight—leaving two votes for Starlight, one for the black steer and one for a Hereford. Hence, whatever may be our individual opinions, the fact remains that it was a plurality, and not a majority vote, awarding him the grand sweepstakes.

WHAT'S THE MATTER WITH OUR LORDLY SHORTHORNS?

"From all this, are we not brought face to face with the question for causes? Why is it that the lordly Shorthorn, having held undisputed sway and acknowledged superior, peerless, is having to bow before the bald faces and black, hornless Polls? We must admit that there is a fault somewhere, and that it is about time to cease much of this mutual admiration business, and leave off this tooting of Shorthorns with immeasurable pedigrees, and look to the causes of these defeats. I feel a delicacy in saying the following, as it should have been proclaimed in every Shorthorn meeting for the past ten years by much older breeders than myself.

"We seem to be crying aloud to make a noise, to attract attention from something; a kind of 'Great is Diana of the Ephesians' uproar; and while we are at this foolishness, the Hereford, Angus, and Galloway men are intelligently and persistently pushing the claims of their pets, until they are within our very entrenchments and calling for unconditional surrender. Have we not rested under the shadow of a great fame, and relied upon the ancestry of our breed to carry us through everything, long enough? Have not Shorthorn breeders carried on a senseless war upon strains of blood, out-crosses, line-breeding, belittling all others except their own, and bred many herds into degeneracy? Have they not sold for bulls to perpetuate faults, animals that would not make good steers? Have they not bought and sold with too much regard for line pedigrees, and not enough care for the pedigree upon the animal's back?

JEALOUSY VS. ENTERPRISE.

"Is it not time we were breeding with less care for the strain and more for the merits of the animals? Are there not too many jealousies which cause a lack of unity of action? Look at the energy and enterprise of our competitors in ransacking the United Kingdom for a 'world-beater,' and when he arrived too late to ship by the ordinary way, sent him from the seaboard to the interior—a 2,400-pound steer at a cost of $250, that at Kansas City the Angus breed might have a representative in Black Prince. See the action of the Hereford men in putting the knife to an agreed number of their choicest bull calves, and that they might prepare to make common cause of the contest, contributing funds liberally to bear the burden in common. Is it any wonder that such animals as these should meet and vanquish so large a number of Shorthorns, castrated because their owners could not sell them for bulls, bred without intelligence, and by accident developed into fair animals? How does such a line of policy compare with the agreement and its document which, I am told, was made with a prominent firm of Shorthorn breeders and feeders, viz.: To furnish them a specified number of Shorthorns to be reared and prepared for the show ring and block, and then attempting, after admitting the parties to be superior feeders, to dictate how their respective animals should be treated, until said feeders refused, in disgust, to take the animals at all?

"WHAT ARE YOU GOING TO DO ABOUT IT?"

"I am informed that the National Association of Holstein Breeders have agreed to furnish and prepare a specified number of choice animals (thoroughbreds) for competition, both in the rings and on the block, at the fat stock shows, and the claim of the breed as an especial beef animal is new. Similar action was also taken by the Hereford Association.

"Now, in the language of Boss Tweed, 'What are you going to do about it?' Will we leave these important matters to chance in the future as we have in the few years just passed, or will we meet brains with brains, intelligence with intelligence, money with money, and enterprise with enterprise, in the production of superior, well-bred, carefully selected, and thoroughly fed representative types of our breed? Or shall we retire from the Show Ring, disgraced, shamed, humiliated, and hooted out, as the old-fashioned fogy advocates of a by-gone breed of antiquated cattle, and be with them laid away to mould, worn-out relics, outgrown specimens, to make room for a new order of things for the progressive American, with his progressive animals, enlarged and improved?

"Returning to my text, I am a Shorthorn breeder. Despite their neglect and abuse by their owners and advocates, I believe their capabilities of development exceed all others, and that they are best suited by nature for my purpose and locality. I never expect to become a Shorthorn King, and to any of my Kansas friends entertaining such an opinion of their own future, I will say, whatever I can do to aid them to accomplish that end, I will cheerfully, yea, gladly, do; but as an humble breeder of plain Shorthorns I am interested in this issue, and while I am not able, financially, to bear the brunt, nor have I the feeding experience to qualify me to enter into this competition, yet, some one must enter in for us all. The burden should be proportionately borne in these things for the common good. The gauntlet has been thrown down and has been taken up. The warfare is waged. The combatants are falling, first on one side and then on the other. Victory wavers. The reinforcements are in view, coming up to the assistance of our opponents. Shall we let our representatives go down, or shall we reinforce them?

"Such a policy as outlined, pursued by the advocates of the different breeds, has caused the present state of things. As a consequence, grade Herefords and grade Polled bulls are readily sold to-day at from $100 to $200 per head; and grade Shorthorns are slow of sale at $40 to $60 per head; and thoroughbred Herefords and Polled bulls sell at auction at from $350 to $1,400, and Shorthorns at $100 to $500. Therefore it becomes a question that interests every breeder of Shorthorns away down deep in the vacancy of his pockets."

PRINCESS B. 1777.
Bred by G. S. Burleigh, Vassalboro, Me.

CHAPTER XXXV.

Hereford Cattle in England in 1884, as Reported by Our Consuls

The United States government requests information from time to time from its consuls, stationed all over the world. The information gathered by these men is of value, as giving the situation at the time of which they report. We have taken some of these reports, given by consuls in England, in response to a call for information on this subject, because of their value and incidentally to show that facts in regard to the worth of Hereford cattle are easily accessible to the investigator.

CATTLE BREEDS OF THE UNITED KINGDOM.

"The great importance of the information called for to a vast number of people and of interests in the United States led me to seek out an authority of undoubted experience and ability in England, to furnish the desired data in behalf of American agriculturists and others. I adopted this plan for the reason that, in order to make it specially useful, the report should be full and reliable in every respect.

"As a matter of course I could not be able to equal an adept in this particular line of investigation, for the reason that the subject is one covering such a wide field, and one beset with so many difficulties, that only one having an extensive acquaintance with English breeders and breeds of cattle could do the subject justice.

"I was most fortunate in securing the services of Mr. James Long of Hetchin, England, a well-known authority on agricultural subjects, both in England and on the continent, who has prepared the accompanying clear, strong, and exhaustive report.

"It will be found that great care and attention have been given to this report, and that its impartiality and fairness are beyond question. Where so many interested dealers in and breeders of cattle have to be consulted, it is important that the facts about such breeds be stated by one who is perfectly free from bias in any respect. This has been done in this report, and I submit the same with full confidence that Mr. Long's acquaintance with our agricultural interests through this valuable mass of information, will lead to a desire on the part of our agriculturists to follow up the results of his future investigations, as they may hereafter be given to the public.

"Albert D. Shaw, Consul.

"United States Consulate, Manchester,

"Feb. 19, 1884."

The annexed particulars, by Mr. Long, referring to the only pure races of cattle known in the United Kingdom which are essentially British, will be found in almost every case very complete and answer every question put in the circular. The exceptions are the Shetland, the Galloway, and the Sussex, about which it is most difficult to obtain technical information. Some twenty Sussex breeders have been addressed, but their answers are not entirely satisfactory, though the information given will be found reliable in every way. The Shetland is an almost entirely unknown race, and the Galloway, to which I desired to give a fuller place, I hope to supplement; the editor of the Herd Book, who is collecting information, promising to send it to me shortly. In all, the fifteen British breeds are treated and the information given is based upon that furnished by nearly a hundred of the leading breeders in the country, and which has been arranged by the writer, who has added much which an extended experience has enabled him to rely upon. It will be noticed that almost every breeder speaks of his own race as the best; this is natural enthusiasm, and I have in some cases been compelled to slightly tone the rather exaggerated praise bestowed upon one breed in opposition to another. Particulars are added with reference to breeding, feeding, soiling, shipment, and scientific dairy instruments, and centrifugal cream separators, which will be found very complete, the two last named subjects being especially familiar to the writer, who had "investigated them in each European dairy country." Drawings of wood-cuts were annexed as well of these machines and instruments and of the chief races of cattle. The references to the Herefords are as follows:

HISTORY OF THE HEREFORDS.

"There can be no two opinions on the question of what Hereford cattle are; they are undoubtedly a distinct and pure breed of great antiquity. Their early history is like that of many others, rather shrouded in mystery, but it is generally allowed that there has been a breed of cattle, red and mostly with white face and markings, for at least two hundred years in the County of Hereford, and the neighboring counties. When crossed with other breeds the potency of the Hereford blood (pure for centuries) is distinctly proved, as it is an exception for any calves to come any other color than the red with white faces. This has come from Hereford bulls on Black Welsh cattle, Ayrshire and Shorthorns; again, if a Shorthorn bull is put to a Hereford cow, the produce follows the dam in color, and cases have been seen, where the produce of the Hereford bull with the black cattle come black, but still they have had the correct Hereford marking as regards the white face and legs.

VALUABLE QUALITIES OF THE HEREFORDS.

"Their milking properties have been so long neglected in the interest of beef, that they are usually not deep milkers, but give very rich milk. In all cases a cow should be milked regularly and stripped quite clean. No doubt this has much to do in forming good milking tribes of cattle, by encouraging the milk-giving organs as far as possible. Where calves are allowed to suck in the open field this cannot be attained, and is one great cause of the Herefords not giving so much milk as they would under other circumstances. As beef makers they are quite at the top of the market, as market quotations record best Scot and Herefords as being usually quoted together. The calves are usually allowed to run with their dams during the summer, and this gives them a good start, but it is too often lost sight of that they should be kept growing on, when weaned, instead of stunted during the winter and following months.

"The Hereford fairs have long been noted for bringing together the best collection of bullocks in England, and are attended by dealers and graziers from far and wide, as they are highly valued in our great grazing districts. Breeders of Herefords claim for their favorites that they are among the most hardy of all breeds of cattle, can be fed on less feed, and thrive on coarse, rough food, and thus are particularly adapted for countries where it is impossible to take special care of the cattle through bad seasons and winter months.

"Herefords, except in a few instances, have been bred entirely for beef. One great object for breeders is to have their animals as wide on their chine as possible, so as to carry good, full crops when fat, and no cow will milk deeply unless it is made like a wedge—the lighter the neck and forepart, the better. If attention were paid to the Hereford as it has been to the Shorthorn, they could be trained to milk well and deeply, and the richness of the milk is not gainsaid; but whether they would excel the Shorthorns or become equal to the best of them it is difficult to say, nor do I think it worth while for breeders to try. At all events so thinks another Hereford man. They stand first and foremost as a beef-producing race, and perhaps it is as well they should for the present take their stand on that; but if any breeders fancy taking up the milk line, they will probably in a great measure succeed.

KIRKLAND B. ARMOUR,
Kansas City, Mo. President American Hereford Cattle Breeders' Association, 1897-1898.

HEREFORDS FOR CROSSING.

"A celebrated breeder of Herefords in England recently addressed the following queries to a gentleman who had tried the cross of a Hereford bull on Shorthorn dairy cows for several years:

"Question 1. Of calves got by a Shorthorn bull or by a Hereford bull, which fatten the quickest and which are the most valuable if sold fat to butcher?

"Answer. I consider those got by a Hereford bull.

"Question 2. Of heifers got by a Shorthorn or Hereford bull, which do you consider the best for milk, having regard both to quantity and quality both as regards cheese and butter?

HERO (5964) 4352.
Bred by J. Price, Pembridge, Eng.

"Answer. Heifers got by a Hereford bull are, I consider, equal to the pure bred Shorthorn for the production of milk, both as regards quantity and quality.

"Question 3. As to produce got by a Shorthorn or Hereford bull, do you find any difference as to their gain of flesh or ability to thrive both at grass and in yards? And if so, state fully your views thereon.

"Answer. My experience tells me that produce got by a Hereford bull out of a Shorthorn cow feeds quicker both on grass and when put up to feed.

"Question 4. Do you find any difference of size in the produce; and, if so, which are the largest animals—the Shorthorns or those the result of the cross with the Hereford bulls?

"Answer. Produce obtained by the cross, as mentioned in No. 3 (viz., by a Hereford bull) is the larger of the two.

"Question 5. Do you think there is any difference as to hardiness, or as to liability to disease between the Shorthorns and the animals resulting from the cross with the Hereford bull; and if so, to which do you give the preference?

"Answer. Undoubtedly the produce obtained by using the Hereford bull is the hardier and has my preference.

"Question 6. Does the offspring of the cross with a Hereford bull generally follow the marking of the sire or of the Shorthorn dam?

"Answer. I find that the offspring obtained by the cross with the Hereford bull follows the sire in color in five cases out of seven.

"Besides which answers the gentleman added as follows:

"'Having some three years ago bought some Hereford cattle from you, I think you might like to know that they have done remarkably well, though I find it takes a long time to make a name as a Hereford breeder. At the same time I bought the Herefords from you, I purchased ten Yorkshire dairy cows—Shorthorns—from Mr. Gothorp, near Bedale, Yorkshire, and after these cows had calved I determined to try a cross of the two breeds, which I did by using the Hereford bull I bought of you on the Shorthorn cows. The result was beyond my expectation. I reared the calves on skim-milk, etc.; they had a little cake till they were six months old, when they took their luck. At eighteen months old I gave them four pounds each per day when grazing (this would be in September). On the 12th October I put them up to feed, giving them eight pounds of cotton cake and linseed cake mixed, and six pounds of meal with pulp each per day. The week before Christmas I sold two of them, averaging £21 10s ($107) each, and also some Shorthorn bullocks, which I had also bought from Mr. Gothorp. These latter were three months older, and only realized £19 15s (or $99) per head, though similarly fed. In the second week of January, I sold some more of the cross-bred bullocks (they were then twenty-three months old) and they averaged £24 5s 6d (or $121) per head and the remaining Shorthorn bullocks averaged £22 17s (or $114) per head, being, as the others, three months older. I certainly am of opinion that the bullock obtained by this cross is better than the pure-bred Shorthorn for the quick production of beef. I have also some heifers of this cross about to calve, and they carry plenty of flesh, and promise to make equally as good milkers as their dams. I consider the result of the cross satisfactory, especially on this poor, cold clay soil, the grass of which (as you know) will not feed a mouse.'"

"As to their milking qualities, says a tenant farmer, 'no doubt breeders have neglected them almost entirely, as it is the usual custom to rear the calves on the cows, and beef, not dairy produce, is, as a rule, the end aimed at.' This is, however, true in a great degree of other breeds when the best tribes are kept for breeding purposes, and it is a question whether a Hereford does not give as much milk, and perhaps even of a richer quality, than the crack tribes of other breeds, except those bred especially for milking purposes. There are few Hereford dairies kept, but from my own experience, I believe, by selection, that a grand milking herd could soon be established. No one will, who has tried the experiment, agree to the statements that the Herefords do not cross well with other breeds. The Americans have found it out, and now assert that they can sell their Hereford grade steers for more money than those of other crosses. It may be true that they have not been very extensively tried, but the experiments that have been tried will soon 'get wind'; in fact, they have already, and the demand is entirely increasing in consequence.

"One great proof of the Hereford being a pure and distinct race, is that although crossed with whatever breed may be desired, the true Hereford marking is sure to show itself, and if an animal has only a quarter strain of blood, the Hereford marking is still there. The great object in America now is to improve the cattle as beef producers, and to put the good roasting pieces on the narrow-chined, bad-backed cattle of the plains; this they believe, and rightly, too, that the cross with the Hereford will do. Another great point in favor of the cross is the power of the Hereford to endure knocking about and rough usage better than more delicate cattle, and this is of the very greatest importance when considering the vast distance the cattle have to travel through America, and by sea, before they reach this country as beef. A recent purchaser of a large herd of Herefords in this country writes that they had a very rough passage out, and the hatches had to be all battened down, but had no losses, and all arrived in capital condition, none the worse for their knocking about. He could only account for this from the fact that this breed of cattle could stand such usage better than others, or he should have had serious loss. As to their not feeding so well in stalls or attaining such great weights there is proof from many trials, and from Smithfield statistics, that they are little behind, if not equal to any other known breed of cattle.

"Says another authority:

"'The Shorthorn has no quality superior to that which the Hereford possesses; if it has let it be fairly shown. Take each point in order. Both breeds have been well tried, both as grazers and feeders. It is acknowledged that the Hereford is the best grazer, and it is asserted in this country and America that four Herefords can be fed on the same feed as three Shorthorns. There is evidence to show that the milking quality of the Hereford is as good as the high-class Shorthorn, and their milk is much richer. The London market bears testimony to the superiority of the Hereford meat by always quoting it in advance of the Shorthorn. Their early maturity and weight and age has been tested again and again, and there is little difference in either breed. The merit of the Hereford for crossing purposes has been disputed, and now it is an indisputable fact that they are fast gaining ground in the good opinion of graziers. A great many bulls are now sold to dairy farmers to cross with their Shorthorn cows, as they say they can get their calves ready so much more quickly for the butcher, and if kept on for bullocks they beat the ordinary run of Shorthorns in aptitude to fatten and in quality. To mention a few instances: Eight Hereford grade steers were put up to feed, and sixty Shorthorn grades were picked out of a six hundred lot, and then the best of the eight and the best of the sixty were killed as a beef test. A large cattle breeder used nothing but Shorthorn bulls to three hundred cows, and could only make some £3 to £4 ($15 or $20) of his grade yearling bulls. The same man now by using Hereford bulls to the same cows, has sold his yearling bulls at £15 ($75) each. Again, another farmer who used to make £6 ($30) each of his grade Shorthorn heifers, makes £16

ANXIETY 3D (ALIAS SIR GARNET) (6181) 4466.
Bred by T. J. Carwardine.

($80) each of his grade Herefords at the same age. The fact that these men are no breeding enthusiasts, but practical American beef-producing farmers, goes a long way to show the turn things are taking in that country.

WEIGHT AND VALUE OF HEREFORD CATTLE.

"'At the last Smithfield show, Hereford steers in the class under two years weighed: first prize, 13¾ cwt. [or in American weight 1,540 lbs., Eng. cwt. being 112 lbs.—T. L. M.] at twenty-two months; second prize, 14 cwt.

(1,568 lbs.) twenty-three and one-half months; and third prize nearly as high. The weights were tolerably even in all classes. In steers under three, first prize was 17 cwt. (1,904 lbs.) at two years seven and one-half months old; second prize, $16\frac{1}{4}$ cwt. (1,820 lbs.) at two years and eight months. In the class under four, first prize weighed $17\frac{3}{4}$ cwt. (1,988 lbs.) at three years eight months; second prize, $18\frac{1}{2}$ cwt. (2,072 lbs.) at three years four months. In heifers, first prize weighed $14\frac{3}{4}$ cwt. (1,652 lbs.) at three and one-half years; second prize

ROYAL 16TH (6655) 6459.
Bred by J. B. & G. H. Green.

weighed 17 cwt. (1,904 lbs.) at three years two months. The winning cow was $20\frac{3}{4}$ cwt. (2,324 lbs.) at eleven and one-third years.

"'No particulars of value can be obtained as to the performances of pure Hereford in milk, butter or cheese. It is not used for draft of any kind, and it is chiefly bred in the west of England, Herefordshire, and Worcestershire, although many successful breeders are scattered throughout the country upon all soils. Herefords are driven to all the great midland fairs for farmers, who purchase them largely for fattening. The chief grasses grown are clovers, vetches and the best perennials. That the Herefords will do well on heavy as well as light land is now admitted. We can point to cases within our own knowledge, where at the Christmas markets Herefords brought in to fatten, have beaten everything else in realizing top prices, although in a county where they are comparatively little known.'

MILKING QUALITIES OF THE HEREFORDS.

"The milking qualities of the Herefords have no doubt been seriously neglected in the past, and are similarly treated by breeders generally at present; but there is no reason for doubting that as milkers the existing herds show a very considerable improvement. As a rule the Hereford cows, when contrasted with extremely large bulls and oxen, are somewhat small, but of course in no way small as we apply this term to Kerries, Ayrshires, or Channel Islands cattle. The cause of the undevelopment of good milking qualities in all Hereford herds is not far to seek. The soil of the locality which saw the breed originate, is admittedly not suited to dairy cattle, consequently there is not that attention given to the improvement of the herds as milkers as would be the case were they in a district better suited to further their dairy properties. In its original habitat the custom which prevails is to regard the steers as the source of pecuniary profit, and whereas in most other parts it is the general practice to give the females the preference in rearing, it is much more usual for both male and female Hereford calves to be similarly treated, the preference being given to the males. This practice largely obtaining, is obviously calculated to prove detrimental to milking properties. The outcome of all this is that, as a rule, the Hereford is wanting in dairy qualifications. But, on the other hand, the exception does not strengthen the rule, even if it proves it, for where pure-bred Hereford stock is kept purposely for dairy requirements, where the good milkers are kept, and the bad and indifferent are weeded out, it is soon very obvious to the most prejudiced that high milking qualities are resident in the Hereford."

HEREFORDSHIRE AND HEREFORD CATTLE.

REPORT PREPARED FOR CONSUL DOCKERY, OF LEEDS, BY MR. JOHN KERSLEY FOWLER, PREBENDAL FARM, AYLESBURY, JUDGE AT PARIS EXPOSITION, 1878.

DESCRIPTION OF HEREFORDSHIRE.

"In writing an account of this very valuable and beautiful tribe of cattle it is necessary to give a description of the county which gives its name to the breed and also of the soil and climate, as well as the general characteristics of the district, as this particular breed of cattle is especially adapted to certain localities in England, and although I will not venture to affirm that they will not thrive under other climatic and geological circumstances than their own county, from my own personal experience, I find that they are more adapted for those districts which partake more or less of the character of Herefordshire.

"This county is situated in the west midland district of England, adjoining the Welsh counties, and is bounded on the north by Shropshire, on the east by Worcestershire and Gloucestershire, on the south by Monmouthshire, and on the west by Radnorshire and Breconshire. It will, therefore, be seen that it has no sea coast, but the river Wye running through the county gives it communication with the sea, through the Bristol Channel. It is well supplied with railway communication, the Midland giving it a direct route to the north, and the Great Western to the south and west, and also to the metropolis. The city of Hereford itself is, also, connected with the Northwestern line, via Malvern and Worcester, thus giving the county every means of supplying the various grazing districts of England with numbers of excellent store cattle, as also for the dispatch of fat animals to the markets of the great metropolis and the teeming populations of the many thriving towns in the North.

"The soil of the county is varied, the larger portion is a red clay, as also strong loam. Around the town of Ross, where some of the choicest specimens of the breed are found, the soil is a loamy gravel or light loam. The old red sandstone forms also a considerable portion of the county, and some of the hills are limestone. The valleys are particularly adapted for the feeding of cattle, as they are moist and rich, and the soil is of a mixed character, from the continuous washing away of the hills, and the debris finding its way to the lower grounds and forming a rich alluvial deposit well suited for the production of the finest grasses. The hillsides and higher portions of the county are eminently suited for the breeding and rearing of cattle, and the comparative mildness of the climate is favorable for the health and early development of the calves.

"The acreage of the county is 532,890 acres, divided into or about the following proportions:

"Orchards, 27,000; woodlands, 37,000; and the remainder for agricultural operations. According to the last Government returns there were under—

	Acres.
Corn crops	95,299
Green crops	32,813
Clover and rotations grasses.	34,108
Permanent pasture	265,661
Bare fallow	11,247
Hops	6,416

"It will therefore be seen that the permanent pasture far exceeds all the other portions of the land put together. The population in 1881 was 118,147. Very few of the people are employed in manufacture, but many find employment in the autumn in hop and fruit gathering.

THE GREAT CATTLE FAIR IN HEREFORD.

"The city of Hereford is situated somewhere near the center of the county, and is in latitude 52° 4′ north and longitude 2° 54′ west. The climate is on the whole temperate. The city is small and has been the seat of a bishopric from the earliest times, for more than twelve hundred years. The cathedral is very beautiful, but does not rank among the largest of the English fanes. It has portions of Norman work in it, and since its restoration has been made one of the handsomest interiors in the Kingdom. The city proper is rather poor, but some of the streets and the market place are large and spacious, and at fair time their appearance is very wonderful. Every portion of the streets, even up to the cathedral yard itself, is crowded with the 'white-faced beauties' of the county; while Shropshire, Monmouthshire, Breconshire, and even Gloucestershire send their contingents. It is indeed a remarkable sight, being different to anything of its class in England, as the thousands of cattle brought together are all of one type, deep brownish reds with white faces, and some other portions of the body and tips of tail white. There is no interspersing of Shorthorns or other breeds; an occasional Devon is seen, but that seems to be an accident, and the shouting of

PRINCE EDWARD (6616) 7001.
Bred by T. J. Carwardine.

drovers, the bellowing of the cattle, and the general hum of conversation whilst the deals are made, form a singular and very amusing sight. The great fair takes place in the third week in October, and as many as from 8,000 to 9,000 head of cattle have been brought for sale during that time. Some years ago, dealers like Carwardine, Pardington, Jones, Knight, and Price were accustomed to bring some hundreds, and generally sold them to the graziers of the midlands or to other dealers who brought them up to the great markets at Banbury, Aylesbury, and Northampton, where there was always a ready sale. The trade now seems quite changed, and but few good animals ever reach the midland markets, as the graziers themselves go down by rail in a few hours and buy largely of the breeders or dealers, who get together on their own premises lots of from 30 to 100 for their selection, and it is only rarely that men can be suited at the old markets.

HISTORY OF THE HEREFORD BREED.

"I am greatly indebted to the writings of the late Mr. Dixon, a well-known agricultural writer, for much of the information contained

DOWNTON ROSE (V. 10, p. 172) 4486.
Bred by T. Fenn, Downton Castle, Eng.

in this paper, as well as to my good friend Mr. Duckham, member of Parliament for Herefordshire, who was the editor of the Hereford Herd Book, and who has done as much, or perhaps more than any other man, to bring this noble race of cattle prominently before the public at the present time, who has given me so much valuable information, and I cannot do wrong in quoting from these most reliable authorities for many statements which I shall make in this paper. I will also give you my own personal experience as a grazier, and judge at the Royal and other agricultural shows, where I had many opportunities of getting well acquainted with this breed.

"Old Fuller, who was a quaint writer of more than two hundred years ago, says of Herefordshire, 'that it doth share as deep as any county in the alphabet of our English commodities, though exceeding in 'W' for wood, wheat, wool, and water,' and, that 'its wheat was worthy to jostle in pureness with that of Heston, in Middlesex, which furnished manchets for the kings of England, and its Wye salmon were in season all the year long.' And before his day 'painful Master Camden' described the county as 'not willingly content to be accounted second-shire for matters of fruitfulness.' Yet both writers are silent as to cattle, and Drayton sang of 'Fair Suffolk's maids and milk,' of the hogs of Hampshire, and the calves of Essex, and how

Rich Buckingham doth bear
The name of 'bread and beef;'

yet he says nothing of these attributes of Herefordshire.

"Many writers were of opinion that the Herefords were descended from cattle from Devon and Normandy, which were of a deep reddish brown color, and that the white faces were an accident from a singular sport of the breeding of a white-faced bull by a noted breeder of the last century, Mr. Tully, of Huntington, near Hereford. The story I have heard related as follows: That the cow-man came to him, on his coming out of church one Sunday, and told him that his favorite cow, who was daily expecting to calve, had produced a bull calf with a white face, and this had never been known before. Report says the master ordered it at once to be killed, as he dared not let it be known that he had such a stain of blood in his well-known herd; but the man begged him to go and see it, as it was the finest calf he had ever seen. Mr. Tully when he had seen it, agreed with his man that it was a wonder, and that he would, out of curiosity, rear it. He did so, and he proved to be a very remarkably fine animal, and he used him on all his best cows, and his progeny became celebrated for their white faces.

"Many old chroniclers say that the county was noted for its breed of white cattle on the banks of the Wye as far back as the tenth century, but they had red ears, and it is recorded that Lord Scudamore in, or about the year 1660, introduced some red cows, with white faces, from Flanders, and this may have been the reason that the noted Tully bull, after a lapse of more than a hundred years, might have cropped up, as a sport, from the well-known deep red cattle of the country.

"It must not be considered that the white face is the only type of the purity of this breed, as the mottled face is considered by many breeders as of greater value than the pure white, and I can myself testify that some of the finest cattle I ever grazed, and some of the best I ever saw, have been mottled faced, in fact those of the last named type have shown the greatest aptitude to fatten, on the grass, of any, and many graziers have told me the same.

"Mr. Eyton, of Eyton Hall, Salop, was the founder of the Hereford Herd Book in 1845, and when he commenced it, he found it necessary to divide the Herefords into four distinct classes, viz., the mottle faced, the dark grey, the light grey or white, and the red with white face, yet, after the lapse of only thirty-eight years, people question the purity of the breed, if they have not the characteristics of the well-known white face and markings.

"Mr. Duckham says, 'the present uniformity of the color is due to the influence of the bull,' and this is a remarkable corroboration of my views, expressed in a paper on 'Breeding, facts and principles,' which I read at a meeting of the Central Farmers' Club, some few years since, when I propounded the dictum (which, by-the-bye, was not new), 'that the male exercised the external characteristics, and the internal organization followed the female,' in nearly every class of animal.

"Long before the commencement of the Herd Book, the Herefords had made 'a reputation and a name,' by being continually successful at the Smithfield Club annual fat stock show, from its establishment in 1799, by Mr. Westcar, of Creslow, near Aylesbury, Bucks, and who for twenty years in succession won the premium prize with Hereford oxen, against all kinds of cattle. I had not an opportunity of knowing Mr. Westcar, as he died before my day, but I had been for many years on intimate terms with his relative and successor, Mr. R. Rowland, who gave me many interesting stories of Mr. Westcar, who was, undoubtedly, the first man to bring the Herefords to the front against all the world. I remember Mr. Rowland telling me, whilst standing in the midst of the far-famed Creslow Great Ground, and on a spot marked by a clump of trees, where Mr. Westcar's lifeless body was found, he having fallen dead from his horse, how the Duke of Bedford, in the latter part of the last century, was down with Mr. Westcar to Hereford in his carriage and four post-horses, taking two days for the journey, and stopping one night on the road at the well-known country inn, the Staple Hall, at Witney, and accompanied by Lord Berners, in another carriage and four, with some ladies and other members of their families, to attend the great fair at Hereford, and where the Duke desired Mr. Westcar to order dinner for a hundred persons at the principal hotel, and to invite all the more celebrated breeders and dealers to meet him. He described the annoyance of some of the dealers at the noblemen being brought down to see these grand bullocks, which they had only seen in the Creslow pastures, as it had the effect of raising the price of the cattle in the fair at least £1 ($5) per head. After dinner his grace and Lord Berners announced their desire, to have from ten to twenty of the best cows that could be found, and two bulls, to bring into Bedfordshire, there to establish a herd on their estates. Lord Berners, who was a breeder of Longhorns, gave up the breed and took to Herefords. This visit of the Duke of Bedford, with the continued success of the breed in the show yard at Smithfield, by Mr. Westcar, brought them prominently into notice, and fairly established their merits.

"Sir Brandreth Gibbs, the honorary secretary of the Smithfield Club, in his history of the club, states that at their first show Mr. Westcar's prize ox measured 8 feet 11 inches long, 6 feet 7 inches high, 10 feet 4 inches girth, and that he was sold for 100 guineas. This animal was bred by Mr. Tully, of Huntington, and weighed 247 stone (1,976 lbs.) dead weight, 8 pounds to the stone [making, according to the English rule, the live weight 3,458 lbs.—T. L. M.] Enormous as the dimensions of this ox were, they were far exceeded by another Hereford, fed by Mr. Grace, of Putlowes, near Aylesbury, which was 7 feet high, 12 feet 4 inches girth and weighed 260 stone (2,080 lbs.), dead weight (or 3,640 lbs. live weight). Mr. Duckham mentions that about the years 1812 or 1813, Mr. Potter sold for Mr. Westcar at the Metropolitan Christmas market, fifty Hereford oxen that averaged 50 guineas ($250) each, making 2,500 guineas ($12,500); and he mentions that Mr. Smythies, of Marlow, Salop, obtained the following extract from Mr. Westcar's book for the sale of twenty Hereford oxen at different periods from 1799 to 1811, and which I can corroborate, as the same was shown me by Mr. Rowland, when visiting him at Creslow. The list was confined to those which sold for £100 ($500) and upwards:

PEERESS (V. 12, p. 152) 10902.
Bred by T. J. Carwardine.

Date.	Oxen Sold.	Value.
Dec. 16, 1799,	2 oxen to Mr. Chapman	£200=$1000
Dec. 4, 1800,	1 ox to Mr. Chapman	147= 735
Dec. 13, 1800,	1 ox to Mr. Harrington	100= 500
Nov. 26, 1801,	6 oxen to Messrs. Giblett & Co.	630= 3150
Nov. 26, 1802,	1 ox to Messrs. Giblett & Co.	100= 500
Nov. 31, 1802,	1 ox to Mr. Chapman	126= 630
Dec. 4, 1802,	2 oxen to Mr. Horwood	200= 1000
Dec. 4, 1802,	1 ox to Mr. Chapman	100= 500
Dec. 19, 1803,	1 ox to Mr. Reynolds	105= 525
Dec. 19, 1803,	1 ox to Mr. Giblett	105= 525
Dec. 5, 1804,	1 ox to Mr. Giblett	105= 525
Dec. 4, 1805,	1 ox to Mr. Giblett	100= 500
Nov. 26, 1811,	1 ox to Mr. Chandler	105= 525

"The whole 20 sold for £2,123 ($10,615) or an average of £100 6s ($501.50) each. I have also seen at Mr. Ledbrook's, who succeeded Mr. Grace at Putlowe's a few years since, when the

VENUS (V. 12, p. 152) 10133.
Bred by T. J. Carwardine.

price of meat was lower than in the beginning of the century, 50 oxen tied up for Christmas at the end of November, for which he had bid £2,500 ($12,500); the price was rather under 5s ($1.25) per stone (or about 9 cents per pound live weight), but this would have made them average over 200 stone (2,800 pounds each, alive) per head. The class of animal I have been describing is now no more. They were five-year-old worked beasts, and even older which had been for two or three years harnessed in the yoke, and had therefore attained great size. Working in the plow is now comparatively rare, and early maturity is the aim of all the best farmers in England, and the Hereford breeders are not likely to be left behind. It is a rare thing now-a-days to purchase a Hereford steer at a fair over three years old. When I began farming, thirty years ago, I bought a lot of beautiful three-year-old Hereford steers in October at £13 10s ($67) each, in poor condition. I gave them the run of the straw yard, and three pounds of oil cake per day, and turned them out to grass in May, and sold them in August and September at from £23 to £24 ($115 to $120) each, giving me some excellent manure and a good profit on the animals. The price of this class of beast rapidly rose, and now they can scarcely be bought under £21 to £22 ($105 to $110) each, and as they only make about £26 to £27 ($130 to $135) each when off the grass they do not pay enough. I once went to a Hereford fair at Easter and bought 10 of the finest old worked beasts I ever saw at £29 10s ($147) each. They were large fine framed animals, and when they arrived at Aylesbury, Baron Mayer de Rothschild saw them and begged I would let him have them, and I consented on condition that he gave me a round of one of them for my Christmas dinner the same year. He took them to Montmore, and some made £46 to £47 ($230 to $235) each at Christmas and others went off the grass in October at £38 to £40 ($190 to $200) each, but such aged beasts are not found now. Amongst the most noted graziers of these cattle was the late Mr. Senior, of Broughton pastures, near Aylesbury. This gentleman was a very successful exhibitor of Herefords after Mr. Westcar's death, but of late years he grazed Sussex beasts, as he could not get the worked animals from Herefordshire. Mr. Duckham and other writers on Herefordshire cattle say that the county is not by any means a good grazing district, but eminently adapted for breeding and rearing cattle, and that no class of animal thrives so well, when changed onto the fine pastures of Buckinghamshire, Leicestershire, and Northamptonshire.

"As Mr. Westcar's name and his residence at Creslow has been so often quoted by all writers upon the Herefords, I must be pardoned for giving a slight sketch of this famous grazing district. 'The great ground,' as it is called at Creslow, is, as before stated, about 330 acres and is very undulating, and bounded on two sides by a brook, a tributary of the Thames, and on the other two sides by a large double ox fence, with large elm trees affording shade to the numerous head of cattle grazing there. I have seen nearly 250 head of horned stock and 500 sheep and lambs, with 20 mares and foals, grazing in this one field, and all getting fat. It is jocosely said that the cattle are turned into the field in May and by the time they have walked around the inclosure they come out fit for the butcher. The old mansion has formerly been a monastery, and the estate belongs to the Lord de Clifford, in whose family it has been for some centuries, and it is stated that Rosamond de Clifford, 'Fair Rosamond,' was born there. Nothing can exceed the rich pastoral beauty of this district. From the upper ground the eye wanders over the far-famed vale of Aylesbury, the old town, the 'Aegelsbireg' of the Saxons, standing in the midst of the rich pastures of Whitechurch; Quarrenden with its ruined chapel of the fifteenth century; and Fleet Marston, in which parish is Putlowes, formerly mentioned as the rival of Creslow as a feeding pasture, and a rare tract of grass land stretching away for more than 15 miles along the valley of the Thames.

"Sir Brandreth Gibbs, in his History of the Smithfield Club, mentions an incident of some interest in 1825. There was a sweepstakes between three Herefords belonging to the Duke of Bedford and three Durhams belonging to the right honorable Charles Arbuthnot, which was won by the Herefords.

"Mr. Duckham says that from the establishment of the Smithfield Club in 1799 to 1851 all the different breeds and cross breeds were shown together, but since that time they have been exhibited in distinct classes, and, as far as can be learned, during the time they were shown together, the Hereford oxen and steers won 185 prizes; the Shorthorns 82; the Devons 44; the Scotch 43, the Sussex 9; the Longhorns 4, and the cross breeds 8; thus showing that the whole of the prizes won by all the other breeds and crosses in the Kingdom were 190, or only 5 in excess of the number registered by the Herefords alone.

"Mr. Discau says that during fifty-three years to 1851 the Shorthorns by their females made up considerably to the total of the Herefords, as they numbered 174 prizes to the Herefords 207.

"It is interesting to know how the Herefords have retained their former renown, by their comparatively youthful prowess at the present day. We find that Mr. Heath showed his grey beast at Birmingham, winning first honors, with a girth of 9 feet 7 inches; and his Hereford cow at three years and ten months measured 9 feet in girth. Mr. Shirley's gold medal steer at two years and seven months girthed 8 feet 7 inches. And he averred that up to seventeen months old he had had only an ordinary calf and stock treatment. It will therefore be seen that the breed is not only not deteriorating but is likely to maintain its position against all competitors.

THE HEREFORDS AS DAIRY CATTLE.

"Having said so much of the feeding qualities of these animals, I must now allude to their milking qualities. Generally they are not considered such good 'fill-pails' as their rivals the Shorthorns or Ayrshires, nor such butter producers as the Channel Islands breeds, yet their butter making qualities are of a high order. I quote from Mr. Duckham, who says Mr. Read of Elkstone finds the Herefords retain their general aptitude to fatten, and that in the team they are excellent, and that they have been used for dairy purposes for nearly fifty years on the farm, and that he raises his calves by hand after a few days old.

"Mr. James, of Mappowder, Court Blandford, Dorset, says that Hereford dairies are becoming very common in that county; that they let nearly 100 cows to dairy people, and that if he buys one of any other breed to fill up the number they always grumble. His system is to let the cows at so much per year, finding them in land and making the hay; the calves being reared by hand with skim milk and linseed until three months old, and they are then turned out to pasture.

"Mr. Oliver, of Penhallow, Cornwall, says: 'I rear my calves on skim milk. It is generally said Hereford cows are bad milkers. That is contrary to my experience. My cow Patience, bred by Mr. Cooke, of Moreton House, had given 14 pounds of butter in a week, and Blossom, bred by Mr. Longmere, Buckton, Salop, gave 22 quarts of milk yielding 2½ pounds of butter per day.'

"From Ireland and Scotland, reports show that excellent results have been attained. It is fair to say that my own experience is contrary to the opinion that they are better for the dairy than Shorthorns, as when I was judge of cattle at Hereford some few years since, there was a milking competition, and we had all the competitors in the class very carefully milked, and

VICTORIA 1053 AT 9 YEARS.
Bred by T. L. Miller.

both the first prizes were obtained by Shorthorns of high-class pedigree, beating all competitors, even including Ayrshires and Jerseys.

THE HEREFORD IN FOREIGN COUNTRIES.

"The Herefords have proved themselves well adapted for foreign and colonial countries. Mr. F. W. Stone, of Guelph, Ontario, says:

"'I am an extensive breeder of Shorthorns,

which breed I think very highly of; but I have also purchased some Herefords from Lord Bateman's and Lord Berwick's herds and am highly pleased with them. The climate is very variable, varying in twenty-four hours from 30 to 40 degrees, but the Herefords stand the changes equal to any breed.'

"Mr. Edwards, Knockalva, Jamaica, says that for many years they had no change of blood till 1853, when Sir Oliver (1732) and Malcolm (1646) were imported, and that they did the greatest service in the island; that this breed are good workers, hardy and of great aptitude to fatten. Mr. Merryman of Maryland, and Mr. John Johnston, of New York, testified

SIR GARNET (6180) 2489.
Bred by B. Rogers.

to the breed standing the variations of the climate remarkably well. Mr. W. Dangau, from Hunter's River, Australia, in addition to their feeding powers and hardiness of constitution, found they were excellent in traveling long distances and that they would do from 250 to 300 miles better than any others. I have therefore shown that the Herefords are admirable for foreign countries. Amongst the most noted strains of blood I find Leopold (1) and Wellington (4), which bull was sold in 1816 for £283 ($1,415), from whom the mottled faces are mostly descended, and Victory (33), which was a dark grey, and Cotmore (376), which was a white-faced bull, and Brockswood (485), which was a light grey, were all specially noted in the first number of Mr. Eyton's Herd Book.

"Mr. Dixon remarks that there were not many points of difference between the dark greys and the mottled faces, the latter of which were known as Ben Tomkins' sort; and that Rev. Mr. Smythies, of The Lynch, was one of the best and most spirited breeders of his day, and offered to show a hundred Herefords against the same number of Shorthorns from any herd in England. All these remarks show that much pains and infinite care has been taken in perfecting this noble breed, and for the best lines of blood the Herd Book must be consulted.

"The breeders put their heifers to the bull at from eighteen months old to two years, and the calves generally run by the side of their dams for several months. The cows are put to the bull at a certain time, so that they may generally come due to calve in the early spring, and to meet the grass; although some others like the cows to calve about October or November, housing the calves and keeping them on with a little milk and cake, so as to be strong by the summer. Some breeders think that by letting the calves suck the mothers it prevents the cows coming into season for the bull as early as if they were weaned at once, but from inquiries I have made I find but little difference in it. This is contrary to my own and other breeders' practice, as I have found the cow lies barren, especially Shorthorns, for some months after calving, if the calf lies night and day with the dam. Several Herefordshire breeders are in the habit of giving their calves, at a very early age, good, old beans, which should be given whole, and in a few days they begin to crack them after rolling them about in their mouths, and secreting that frothy saliva which seems to be so conducive to a calf's well doing. I have tried the plan and can speak highly of the practice. No food can be better, as beans are peculiarly fitted for forming bone and muscle.

"On the whole, I believe the Hereford breed as a flesh forming animal is second to no breed in the world. The meat itself is equal, when well fed, to the best Scotch, and every authority proves they do well when imported into other climes. In England it is found that the best grass lands are most calculated for their flesh development, and when tied up, liberally fed, and well cared for, they can hold their own in the show yard against any breed in the country. As dairy cattle the Shorthorns beat them, but, taking all things into consideration, England may well be proud of her white-faced Herefords.

"John Hersley Fowler.

"Prebendal Farm, near Aylesbury,
"Jan. 7, 1884."

Consul S. B. Packard, of Liverpool (¶249), reported:

"HEREFORD CATTLE. This breed takes its name from the county where they were first bred, but they are to be found also in the adjoining counties. They are also grazed on most of the great grazing farms of the midland counties, and there are also breeding herds in Scotland and Ireland. The Queen's celebrated herd is kept near Windsor, Berkshire. This breed adapts itself easily to the severe climate of the North, as well as the milder climate of the South. In America, some are to be found in ranches 6,000 feet above the sea level, and no better proof can be given of the hardiness of the Herefords. Of this breed the Earl of Coventry says: 'I have observed Hereford cattle for twenty years, but I only commenced forming a herd nine years ago. During that period I have tried them alongside pedigree Shorthorns and other breeds of cattle, and I am so satisfied of the superiority of the Hereford breed for feeding purposes, that I have disposed of other sorts. They are a hardy breed, doing well out of doors all the year around. Their quality of meat is very superior; they have less rough meat about them than the Shorthorns, hence, first-class butchers prefer them to other sorts.' (Oct. 21, 1883.) They are a perfectly pure race of cattle and have been brought to their present excellence by the judicious selection of both male and female animals, and not by the introduction of crosses of other breeds. This strictly pure blood gives them the great value they have for improving other breeds.

"COLOR. The distinguishing color is red with a white face, chest and belly, white flank, and white tip to the tail; white on the legs, white mane and often white line along the back. The red with white face is invariable, and the white predominates, more or less, on different animals. There are also grey Herefords, but these are now confined to one or two herds.

"HERD BOOK. The date of the first Herd Book is 1845.

"INCREASE. The demand for exportation principally for the United States and Canada has increased the stock of the district, owing to more farmers breeding." Consul Packard had a tabulated form that he copied in his reports, with each British breed, showing the products, average weights and other important statistics very useful, in this form, for comparing the breeds, from which we take the following statistics for Herefords:

AVERAGE WEIGHT AT MATURITY.

Cow	12 to 14 Eng. cwt.=Am. wt. 1344 to 1568 lbs.
Bull	16 to 20 Eng. cwt.=Am. wt. 1824 to 2240 lbs.
Ox	20 to 22 Eng. cwt.=Am. wt. 2240 to 2464 lbs.

Age at maturity: 3 years.

How long bred pure: From a very remote period.

Annual average pounds of milk: Nine thousand five hundred pounds.

Milk to pounds of butter: 30 lbs. to 1 lb. butter.

A good cow has been shown to yield 14 pounds of butter per week at grass.

One gave 55 pounds of milk, yielding $2\frac{1}{2}$ pounds of butter per day.

Meat Product: 1,770 lbs.

Labor: Little or none.

Method of housing: Open yards during winter, with a run out by day; summer, out in rough pasture.

Feeding: Hay, straw, and roots in winter; rough pasture in summer.

Breeding: Heifers have calves at two-and-a half years, and continue to breed until they are old.

Grasses: Clover, rye grass, meadow, fox-tail, and English natural grasses.

Live weights of fatted cattle of this breed:

Herefords.	Oxen (over 3¼ and not over 4 years old).	Steers (over 2½ and not over 3¼ years old).	Heifers (not over 4 years old).
No. 1*............(pounds)	2,394	1,724	1,621
No. 2............(pounds)	2,135	1,862	1,764
No. 3............(pounds)	3,024	1,884	1,855
No. 4............(pounds)	2,500	1,778	1,832

*No. 1 ox of the above table is the property of Mr. J. Price, and was the winner of the Elkington challenge cup, which has never been done except by this Hereford.

"PRICE. At the recent total dispersion by auction of two old established herds the average price was just $375, including cows, bulls and suckling calves. At one sale, the leading bull

TREGREHAN (6232) 6203.
Bred by Maj. Carlyon.

sold for $4,139; at the other sale 12 two-year-old heifers averaged $652 each; the highest priced cow was $1,329; there were 117 animals in one sale and 91 in the other.

"The soil of Herefordshire is various, from clay to light sandy soil, much of which is of inferior quality. The substratum is principally limestone, clay and gravel. The temperature at the altitude of 100 to 300 feet above sea-level is in summer 60°; in winter 39°; the mean during the year, 49°."

REPORT OF CONSUL LATHROP, OF BRISTOL, ENGLAND.

"I have the honor to enclose a report on Hereford cattle in answer to Department circular of the 18th of July, 1883. This consulate has in its immediate vicinity three breeds of cattle, viz., Devons, a fine tribe of Shorthorns, and Herefords.

"I have selected the latter breed as the subject of my report, to the exclusion of the other two, for the following reasons: (1) On account of the wide celebrity already enjoyed by these two breeds, making a report unnecessary; and (2) on account of the fact that the Hereford seems to be, of all breeds in the United Kingdom, the one most suited to the needs of the stock of the United States. While much of what I have written is undoubtedly familiar to our breeders, yet I trust that this report may contribute somewhat towards diffusing widely a knowledge of the great merits of this sterling breed.

CHERRY 24TH, 2410.
Bred by J. B. & G. H. Green, Herefordshire.

"Hereford cattle in the herd are a peculiarly impressive sight. Their grand development, their firm agility and light activity, their intelligent faces and placid expression, and possibly more than anything else, their wonderful similarity to each other, all combine to make a spectacle pleasing to even the most indifferent observer. He cannot fail to note how closely they conform to a common type, and that type a striking one. Its main feature is suggested when I say that they are oftener spoken of as 'white-faces,' or 'red-with-white-faces,' than as Herefords. But Hereford cattle have not always thus assimilated so closely to a common type. Up to well within the present century there were four distinct varieties of the breed, differing widely from one another in appearance, but they have succumbed so completely to the 'red-with-white-face' that a Hereford not thus marked is as rare as a white crow.

"The origin of the breed is doubtful. The best authorities consider it aboriginal; others claim its importation from Normandy or Flanders; others, again, think the climate and conditions of Hereford County have made what they have, out of an animal that originally inhabited the shire of Devon. Be its origin what it may, its environment in Hereford County and surrounding counties has resulted in one of the finest beef producing breeds of cattle in the world, nor is the breed to be despised for the dairy under conditions more favorable than are to be obtained in its home county.

"The authentic history of the breed begins about the year 1800. In the year before this, occurred the first cattle show of the celebrated Smithfield Club, and a Hereford ox was the winner of the first prize; a more general acknowledgment of merit, then, than now, because at that time, and indeed up to the year 1851, all breeds were shown in competition with each other. This ox was 6 feet 7 inches high, 10 feet 4 inches girth, and dressed 1,976 pounds meat; his success was maintained by the breed so well that up to 1851 the Herefords are credited in Smithfield Club records with one hundred and eighty-five prizes for their oxen and steers against one hundred and ninety for all other breeds together, including Shorthorns, Devons, and Scotch. The records for prizes won by Hereford cows and heifers is, however, by no means so good, being twenty-two for them against one hundred and eleven for all other breeds. Mr. Duckham in his interesting and valuable little work on the breed, comments thus upon this disparity between the success of the males and females. He says:

" 'This is certainly great falling off compared with the oxen and steers and goes far to prove the correctness of my remark respecting the study of nature's laws in the cultivation of the soil, and of the adaptation of stock to it. The soil of the county of Hereford being neither applicable for dairy or feeding purposes, those who have cultivated it for ages, made it their duty to breed steers and oxen which should by their superior quality and aptitude to fatten, command the attention of the distant grazier.'

"Herefordshire has 550,000 acres. About 100,000 acres are utilized neither for pasture nor agriculture; the balance is divided equally, almost, between these two pursuits. The substratum is a light red sandstone, and the soil generally is a deep red heavy loam, sometimes with some clay in it. The surface of the county is hilly, and averages about 250 feet above sea level. There are some small but beautiful and fertile valleys. The culture of tree fruits, notably apples, and of hops is largely pursued. Damp fogs prevail at some seasons and help to keep the grass beautifully green all the year round.

"Mr. Southall has kindly furnished me with the following particulars of temperature, rainfall, etc., the results of his own observations at his house in Ross, the southern part of Hereford county:

Temperature.	1882.	1883.
Absolute maximum	84°1′	77°0′
Absolute minimum	19°6′	18°8′
Average maximum	57°1′	56°9′
Average minimum	42°3′	41°6′
Mean	49°7′	49°25′

* * *

"The temperature reached the extreme height only on three or four days in the year of 1884, and in 1883, on one day only. The rainfall amounted in 1883 to 31.52 inches, being 1.34 inches more than the average. There were in this year 197 days on which rain fell.

"The ideal Hereford is thus described by Mr. Duckham: 'The face, throat, chest, or lower part of the body and legs, together with the crest, or mane, and the tip of the tail, a beautifully clear white; a small red spot on the eye and a round red spot on the throat, in the middle of the white, are distinctive marks which have many admirers. The horns are of a yellow or white waxy appearance, frequently darker at the ends; those of the bull should spring out straightly from a broad, flat forehead, whilst those of the cows have a wave and a slight upward tendency. The countenance is at once pleasant, cheerful and open, presenting a placid appearance, denoting a good temper and that quietude of disposition which is so essential to the successful grazing of all ruminating animals; yet, the eye is full and lively, the head small in comparison to the substance of the body. The muzzle white, and moderately fine thin cheek. The chest deep and full, well covered on the outside with mellow flesh; kernel full up from shoulder point to throat; and so beautifully do the shoulder blades blend into the body that it is difficult to tell in a well-fed animal where they are set on. The chine and loins broad; hips long, and moderately broad; legs straight and small. The rump forming a straight line with the back, and at a right angle with the thigh, which should be full of flesh down to the hocks, without exuberance; twist good, well filled up with flesh even with the thigh. The ribs should spring well and deep, level with shoulder point; the flank full, and the whole carcass well and evenly covered with a rich mellow flesh, distinguishable by yielding with a pleasing elasticity to the touch. The hide thick, yet mellow, and well covered with soft, glossy hair having a tendency to curl.'

"A glance at the cuts presented here will show us immediately how closely the animals whose portraits have been selected to accompany this article answer to this description. The bull Romeo (¶ 250) is perfect. He was bred by Mr. Carwardine, of Leominster, in Herefordshire, and was sold in 1882 to Messrs. Earl & Stuart, of Lafayette, Ind., where he now is.

"The ox pictured here (¶ 251) was bred by Mr. J. Price of Pembridge in Herefordshire. He won the Elkington Challenge Cup at Birmingham in 1881, and again in 1882. This challenge has never before been won twice by the same animal, and, in recognition of his great feat, the portrait of this ox is to have

EDWARD PRICE,
Father of John Price, Court House, Pembridge.

the place of honor, the title page, of volume 14 of the Hereford Herd Book, just about to be issued. The general rule is to admit to the Herd Book only cuts of such animals as take first prize at a Royal Agricultural show. The thirteenth volume, I mention here, contains the names of 199 breeders, of whom 11 are either in the United States or Canada. The fourteenth volume, which is to be issued in February next, contains, I am informed, a much larger number of breeders' names. I hardly think it necessary, but still I venture to suggest that no American owner or breeder of Herefords eligible for entry should omit to register them. The Herd Book is under the control of S. W. Urwick, Esq., of Leominster, and all breeders of these cattle are under obligation to him for the accuracy and completeness of the work. I take pleasure in acknowledging here

ANXIETY (5188) 2238.
Bred by T. J. Carwardine, Herefordshire. Founder of the Anxiety family. (From a drawing by Dewey.)

the obligation I also am under to Mr. Urwick for assistance rendered and information extended in connection with this report.

"The two cows portrayed here are both royal prize winners at late shows. Golden Treasure (¶ 252) has a little too much white for a perfect Hereford, but in other respects she is all that a pure-bred Hereford should be.

"Herefords were formerly used considerably in the yoke, where 'they combined the activity of the Devon with the strength of the Shorthorn.' There, as well as in grazing, their placid, quiet temper rendered them doubly valuable. In those old days when they were put to the yoke, when the demand for meat was not so pressing as now, nor money requiring so rapid a turn-over, they were often kept until six or seven years old; and their flesh developed 'that beautiful marbled appearance caused by the admixture of fat and lean which is so much prized by epicures.' But the Hereford is now considered ready for the market at from twenty to thirty months old. Grass with a little oil cake is all they need, and their ability in grazing and facility for fattening make the steers much sought after to graze in the midland counties for the London market. They are in their prime at three, but will grow up to four, and their live weight at maturity is from 1,800 to 2,500 pounds. The calves are dropped generally from April to July. Yearling heifers are seldom put to the bull. The calves run on their dams for 6 or 7 months and are rarely weaned on oil-cake. The young steers are fed upon grass, and get turnips and cut straw and sometimes a little oil cake in winter.

"I subjoin to this report a table showing the live weights of all the cattle of all breeds exhibited at the eighty-sixth annual show of the Smithfield Club in December, 1883, prepared by me from the official catalogue. It is presented more as a matter of interest than for any deductions that might be drawn from it. 'The youngest and oldest Hereford classes as a general thing at these shows,' Mr. Duckham writes me, 'are the heaviest of any exhibited.' The superiority in weight of the younger classes proves, of course, their earlier development; the superiority of the oldest indicates that eventually they attain a greater size than other breeds. But I have already said that Herefords are not commonly allowed, for various reasons, to obtain the age which in the past made them so remarkable for their size and weight.

"Another reason for the lack of 'tall figures' in these days for cattle weights is the partial abandonment of the time-honored practice of feeding up stock until it becomes so fat as to be literally useless for any other purpose than to take a prize. Mr. McDonald, in his report to the Royal Agricultural Society upon the stock exhibited at the society's meeting at York in July, 1883, says on this point: 'Overfeeding has been disappearing somewhat in recent years. There is still too much of it, however.' He says elsewhere: 'Preparation for modern show yards is a severe ordeal and only good constituted animals can endure it. It leads to many breeding mishaps and failures—but when one finds the sires and dams of so many of the prize winners themselves in prize lists, as was the case at York, one is forced to the conclusion that successful showing and breeding go hand in hand to a considerable extent and to a larger degree than is commonly imagined; and one is led to believe that high feeding is not so detrimental in skillful hands to successful breeding as is generally imagined.'

"'Herefords, and only Herefords, are found in Herefordshire, Shropshire, Monmouthshire, Radnorshire, Breconshire, and also in Worcestershire, and Montgomeryshire. Large numbers are also found in Cornwall and Ireland, and there are herds of them in many other counties. They are seldom crossed with the Shorthorn, though they are said to blend well when it is done; the same statement holds good with the Ayrshires. Hereford on Devon has been tried, resulting in a progeny inferior in some respects. Hereford on Alderney is said to produce satisfactory results, improving the cow of the first cross as a feeder and not injuring her milk in quantity or quality. A cross with the West Highland Kyloe was a failure, but with Galloway Polls was a great success.' These statements of the results of Hereford crosses are taken from a prize essay from the Royal Agricultural Society made by the late H. H. Dixon, a notable authority on such matters when alive.

"Evidence establishes beyond question that the Hereford, when removed to almost any climate, does not degenerate as a beef producer. The females, too, of the breed are found most satisfactory for the dairy, under different conditions than can be found in their home countries. Both of these conclusions are contrary to an opinion I have heard many express to the effect that Herefords deteriorate away from home. But I have observed that while such an opinion seems very general, it is maintained by those without special knowledge of the breed, and I think it an inherited prejudice, which a little investigation would disprove to the satisfaction of the holder. 'Old prejudices die hard' is true and trite. In Bedfordshire and Dorset herds have been maintained for many years, fifty in some cases, and these herds are fully up to the standard of the homebred ones; in every case, that is, in which due care has been taken to get an occasional infusion of fresh blood. In the wet and changeable climate of Cornwall the breed is established largely and maintains its reputation, though Devons and Shorthorns are said to deteriorate there. In the counties near London, Surrey, Cambridge, and Kent, Herefords have done well; also, in Wales and Scotland. They withstand the severe climate of the latter country without any seeming difficulty, and will live where many Shorthorns cannot. In Ireland they are much esteemed and their number is constantly increasing. They maintain in all these places their characteristics of early development and rapid and even fattening.

"The breed seems to stand the heat with the same indifference it does the cold. In Jamaica the progeny of some imported Hereford bulls have proved the most valuable and useful stock in the island; and the heat of Australia has not affected in the slightest degree the characteristics of the large number of Herefords there. Of their success in the United States I shall speak further on.

"I have thus far considered the Herefords mainly as a butcher's breed. I will now speak of their qualifications for the dairy. In the shire of Dorset, one of the crack dairy districts of England, producing a butter much sought after, there are many Hereford dairy herds. The owner of the largest of these herds wrote twenty years since as follows: 'Our herd of Herefords has been established nearly thirty years, and so far from their being degenerated with us they are much improved and Hereford dairies are becoming very common in this county. In proof that they are good for milk with us, we let nearly 100 cows to dairy people, and if I buy one of any other breed to fill up the dairy, they always grumble, and would rather have one of our own bred heifers. Our system is, we let our cows at so much per year, finding them in land and making the hay; the calves being reared by hand with skim milk and linseed until three months old, when we take them and allow a quarter's rent of the cow for the calf at that age; they are then turned into the pasture.'

"The proprietor of this herd and writer of

EARL OF SHADELAND 22D, 27147.
Bred by Adams Earl. Called "the Record Breaker," never being defeated.

this letter was Mr. James, of Blandford, Dorset. His son writes me under date January 19th, 1884, that the same system is still pursued and that the Herefords are as great a success as ever. He says: 'My late father and myself have kept and bred Herefords for dairy cows for forty-eight years and have always used the best blood we could get. I have won a number of prizes for "dairy cows" and "dairy cows and offspring" against Devons and Shorthorns.' Mr. James further says: 'In a cold, wet, sour place there is nothing like the Herefords; their good coats are a protection in the winter. The butter that is made is a splendid color and taste. There are lots of Hereford dairies in this county. In the year 1881 I sold two bulls to cross Shorthorns, and I know parties who have crossed Herefords with Shorthorns and have come back again to the Herefords.'

"Mr. White writes me from Wiltshire as follows: 'I keep a dairy of ninety pure-bred Hereford cows, which breed has been kept on this

RUDOLPH (6660) 13478.
Bred by P. Turner, Herefordshire.

farm for the last sixty years, and I have at all times endeavored to obtain the milkiest strain I could, and I think I have now a herd of cows more adapted to dairy purposes than any other Hereford herd in this country. I have made fair trials between the Hereford and Shorthorn as to profit, and I give my decided preference to the former.'

"The testimony from these two herds is the most valuable that could possibly be obtained as to the dairy capabilities of the breed, as they are the largest, and have been longest established of any in existence. I have received letters similar in tenor to the above from various smaller breeders in different counties, and I have not been able to discover an instance where breeders or dairymen have changed back to other breeds after starting in with Herefords. I had hoped to send with this report some figures giving actual milk and butter products, etc., but I must leave them for a supplementary report, as I have already detained this one over a month, waiting for the promised statistics. But it is not so much as milkers that the United States are interested in the breed, but as beef producers; and that in this capacity they are indeed largely interested will be proved when I say that the extraordinary demand for pedigree Herefords from the United States in the last few years has so increased the price of these cattle that the Hereford breeders are looked upon with envy by other breeders throughout the kingdom as having 'struck a bonanza.' That Herefords will repay a large expenditure is undeniable. Their tremendous development of flesh, their activity as feeders, their insensibility to changes of climate, their hardiness, their quiet and placid tempers, are just precisely the qualities needed for the improvement of our western, Texas, and Spanish cattle. Their bulls, too, have a marvelous faculty of impressing their qualities on their get, and there is many a half breed Hereford which is absolutely indistinguishable in appearance and quality from a pure bred one, so completely is the influence of the dam eliminated. Another point. The Hereford is especially strong just where our western cattle are weak, viz., in the development of the flesh on the back. The back of a well-ripened Hereford steer has been compared to a table, and the back of a Texas steer to a wedge. Volumes could not say more.

"There were two remarkable sales of Herefords during the past year; one, the dispersal of Mr. Pitt's (¶ 253) herd at Chadnor Court, (¶ 254) and the other the dispersal of Mr. Turner's herd at The Leen. (¶ 255) Mr. Pitt established his herd in 1842, from four celebrated cows of the day. I present a little statement of the amounts received by Mr. Pitt at this sale.

Number of Animals.	Value.	Average.
32 cows with two calves	$13,450.00	$420.00
25 calves	4,961.63	198.47
12 2-year-old heifers	7,818.05	651.50
3 2-year-old heifers	1,113.67	371.22
12 yearlings	3,995.87	333.00
7 bulls	2,800.00	400.00

"Ninety-one animals averaged about $375 each. The average of $651.50 for 12 two-year-old heifers has never before been equalled in England in any breed.

"Mr. Turner's herd has been established for about eighty years, his grandfather being the founder. Since 1854 Mr. Turner has won with individuals of his herd 111 first prizes, 60 second prizes, 11 third, beside 52 special prizes. He received an average of $369 apiece for his animals, his cows and calves averaging a little higher. His chief stock bull, The Grove 3d (5051) 2490 (¶ 256) brought over 830 gs. ($4,150).

"These two herds were exceptionally fine and had a wide celebrity. Many of the animals went to the United States, making, with all others forwarded, a total of 1,800 pedigree Herefords sent to the United States from February 1, 1883, to February 1, 1884. This includes one lot of 300 sent to Baltimore in

January of this present year. The question naturally comes up now, whether this demand for the Herefords is a fancy or a fashion, likely to die out and let down prices. It is worth considering. I have said that the principal demand for the Herefords in the United States was as beef makers, but I did not intend at all to intimate that their merit as milkers was overlooked. On the contrary, many breeders in the United States are enthusiastic over them as a dairy breed and quite a number of wealthy men are forming herds. Some go so far as to claim that the Hereford is the coming breed, which is going to carry all before it, and that the Shorthorn will eventually fall before the Middlehorn, just as the Longhorn went down in the past. The high esteem in which the breed (¶¶ 256A, 256B, 256C, 256D, 256E) is held in the United States and the growing appreciation of its merits which exists in England preclude, it seems to me, the possibility of a fall in prices in the near future, or, in fact, for many years. It is even possible that for a time prices may go higher than now. A man writes me from Hereford County thus: I am now looking out for a lot of pedigree Hereford cattle for America; they are more difficult to get, as the demand has been great and prices are much higher.

"How to Export Herefords.—By far the larger proportion of the Herefords sent to the United States go via Liverpool, though several large herds have been sent by way of Bristol. I am of the opinion that in many cases better facilities could be obtained via Bristol than are obtained via Liverpool. One reason I have for this opinion is the fact that several of the steamers plying in the lines from Bristol to New York are unusually high between decks, and extremely well lighted and ventilated—an important matter. Another advantage is that cattle can be brought from Hereford in the cars directly alongside of the ship's deck.

"The Great Western Railway Company quote the following to me as about their average rates for transporting cattle from the town of Hereford to Bristol or to Avonmouth decks (a port of Bristol):

Half wagon load, consisting of 4 cattle	$ 5.46
Small wagon load, consisting of 7 fat cattle	8.20
Medium wagon load, consisting of 8 fat cattle	9.23
Large wagon load, unlimited (holding about 10)	11.00

"The Great Western Steamship Company, plying between Bristol and New York, inform me that their rates average from $25 to $30 per full grown animal. The ship provides water and stalls and their bills of lading contain this clause: 'Ship not accountable for mortality or accident from any cause whatever.' A herd of 109 Hereford cattle was carried on this line some time ago at the following rates: Cows and heifers, $24.33; calves, $12.16 each; suckling calves, $4.86 each. A herd of fine Jersey cattle carried on this line subsequently, when freights were higher, paid an average of $30.50 each for full grown animals. In order to take proper care of valuable cattle in ocean transit, there should be one man for each twelve cattle. Competent men for this purpose can be hired in England for about $1.25 per day and all expenses paid, including a pass back to the port of departure. When a steamer gives rate for carrying cattle, the pass over and back for a certain number of cattle tenders is included. The foreman in charge of the tenders would of course get more than $1.25 per day, but in most cases he is the American agent or buyer, or is connected in some permanent capacity with the farm or the business of the purchaser. Such cattle of course are insured to their full value almost invariably, and are admitted, being breeding animals, into the United States free of duty. Certificates from a veterinary surgeon and from the Consul at the point of departure invariably accompany such consignments.

"Herefords in the United States.—Any account of Herefords in the United States would be incomplete without mention in con-

PRINCESS (V. 13, p. 152).
Bred by T. Marston, Herefordshire.

nection therewith of the name of Mr. T. L. Miller, of Beecher, Ill. He was the first, or one of the first, to perceive what a boon to the stock of the United States the almost unknown Hereford would be, and for years he has persistently and enthusiastically advocated him in his 'Breeders' Live Stock Journal.' In 1873 he imported from Hereford a two-year-old heifer, Dolly Varden, with a calf at foot. She has brought a live calf every year since, one of the first being the bull 'Success,' hitherto acknowledged as the best Hereford bull in the United States, and still alive and active. Dolly Varden and Success have been repeatedly exhibited and never beaten, whilst the get of Success has in several instances brought $1,000 per head.

"The 'Hereford Times,' of Oct 18, 1883, says: 'To this purchase of Dolly Varden and her calf, combined with the indomitable energy and perseverance of Mr. Miller, the brisk demand, present high favor and repute in which Hereford cattle are held is attributable.' Mr. Miller's neighbors in Illinois are following his example in importing Herefords. Messrs. G. Leigh & Co., of Beecher, Ill., have bought eight animals within the past few weeks from the herd at Felhampton Court; Mr. Culbertson, of Chicago, two; and Mr. J. V. Farwell, also of Chicago, 16, all from the same herd.

"ACKNOWLEDGMENTS.—In conclusion I wish to acknowledge the extreme courtesy with which my requests for information have been responded to by the breeders of Herefords. It is not always easy for a consul to obtain information. His requests sometimes are met with discourtesy, sometimes with indifference; but in this case only five letters out of about a hundred and thirty-five dispatched by me remained unanswered. In every case in which I had a personal interview with Hereford men, except one, every facility was placed at my disposal for a thorough investigation. I have to acknowledge, especially, the kindness in connection with this report of Thomas Duckham, Esq., M. P.; of S. W. Urwick, Esq., secretary of the Hereford Herd Book; of J. Bowen Jones, Esq., of Shropshire; of Lord Moreton, M. P.; of E. G. Clarke, Esq., of Bristol; of N. J. Hine, Esq., assistant secretary of the Smithfield Club."

From the table prepared by Consul Lathrop, showing the respective weights of the cattle exhibited at the eighty-sixth annual show of the Smithfield Club, December, 1883, we condense the following table of weights for the five heaviest steers under two years old of each breed, as showing the early development and heavier weights of the Hereford yearlings over all other breeds, their winnings evidencing their earlier maturity:

Herefords.	Shorthorns.	Sussex.	Devon.	Cross-breds.
1588	1544	1540	1186	1536
1558	1498	1537	1162	1456
1468	1426	1425	1140	1455
1424	1423	1424	1115	1392
1364	1369	1370	1028	1365

While mere weight decides nothing, without consideration of the cost, this table, taken with other statistics, shows that there is not and has never been any preponderance of weight—age considered—of the Shorthorns, over the Herefords and other large breeds of British cattle.

THE EQUINOX 2758.
Bred by J. Merryman, Maryland.

CHAPTER XXXVI.

Herefords as Dairy Cattle

Those who have had experience with Hereford cattle know that some families are fine milkers, and that the milk of all the Hereford cows is rich in butter. Our experience with grade Herefords as milkers has been highly satisfactory; these grades proving to be good milkers and the quality of the milk rich. One herd of thoroughbred Herefords that we bought of D. K. Shaw, of New York, were all of them good dairy cows.

We have also given elsewhere much testimony on dairy Herefords, especially in the preceding chapter. Mr. Sotham, in his controversy with the Shorthorn scribes, mentioned that a Hereford cow was champion dairy cow at the first show of the Royal Agricultural Society at Oxford, 1839; in a letter to the Albany "Cultivator" for January, 1844, headed "Herefords—their Dairy Properties, etc.," Mr. Sotham says:

Messrs. Gaylord & Tucker: According to promise I send you the result of my dairy; but it will not be any criterion to judge from, although the actual weight is taken from the book of Mrs. Sheldrick, who has kept a correct account of every pound made.

I had no dairy to keep my milk in the month of February, when six of my cows were milking, and not one pound was made from them during that month. Three of them calved in January.

There were nine three-year-olds, two four-year-olds and one seven, milked during the month of March. I sold Cherry, a three-year-old heifer, on the 2d of April. From that time until the first of October I milked eleven, at which time I sold my milk to the milkman. The following is a true statement:

From March 1st to Oct. 1st, butter........	1,456½	lbs.
35 cream cheeses, 3 lbs. each, equal.......	105	lbs. butter
113 quarts cream, sold, equal to...........	113	lbs. butter
	1,674½	

My opinion is that the same cows next year will make nearly double the quantity; for my first cow, Lucy, 4 years old, calved Nov. 28, and made 8 lbs. 2 oz. last week; the only cow I have now in milk. I quote from the "Mark Lane Express" the following remarks from a speech made by the celebrated breeder, W. Fisher Hobbs, Esq., at the annual meeting of the East Sussex Agricultural Society, Oct. 11, 1843:

"He could not, however, conclude without making a few observations as a successful candidate. It had frequently been his good fortune to appear before them in that character, but he never felt so much pleasure from the circumstances as on the present occasion; for it must be admitted on all hands that except in a few instances the competition was very good indeed, and in some cases very severe struggles.

"He was the more happy as a great victory had been gained for a breed of cattle for which he was a strong advocate. He did not think, as judges were generally prejudiced in favor of Shorthorns, that Herefords could have been so successful as they had that day been.

"When he first became a farmer he was determined to have a good breed of cattle. He first tried Shorthorns, because he thought they were the best; and at a sale in Suffolk he purchased several, better than which could not be obtained. He also purchased some Herefords, and kept them together for twelve months and the result was most decidedly in favor of the Herefords. He was, therefore, compelled, contrary to his own wishes, to give up the Shorthorns and take to Herefords; and he had from that time continued to do so, being satisfied that with his soil and climate they paid the best. (Hear, hear.) He trusted the farmers whom he was addressing would do as he had done, and judge for themselves what description of stock was best suited to their farms; and when they were satisfied that they had a breed which would prove most profitable to them, he would advise them to keep to them; and if they came here to exhibit them and were occasionally unsuccessful, he would advise them to go home with a determination of meeting with more success on a future occasion." (Cheers.)

Be it remembered that Mr. Hobbs gained the first premium as the owner of the *best cow in milk, of any breed, with a Hereford. What will the pen depressers of milking Herefords say to this?* Will they not have to put on their "studying caps" to find some endorsers for "*Youatt?*"

Again, I refer you to a sale of Herefords in the "Mark Lane Express" Oct. 30, page 12, the property of Mr. John Hewer: "An in-calf cow, Lady, by Chance, was knocked down for

QUEEN OF THE LILLIES 4367.
Bred by T. E. Miller, Beecher, Ill.

100 guineas ($500); two-year-old Victoria at 87 guineas ($435); yearling heifer, 40 guineas ($200); an aged bull, Dangerous (the sire of some of my heifers), 100 guineas ($500); Lofty, and bull calf, 51 guineas ($255), and several others at similar prices."

What does this say for the Herefords? Will it not "prove" that some *writers know nothing about them?* and will it not teach *some of them* to search for facts before they *"abuse the Herefords?"*

I refer you again to a sale at Algarkirke of an excellent herd of Shorthorn cattle belonging to Mr. Rogerson, same paper for Oct. 23, page 12: "Nonesuch, two years and six months old, was knocked down at £27 10s. ($136.68), a very fine animal; others fetched from £20 to £25 ($100 to $125). Altogether the sale has been highly satisfactory, and reflects equal credit on the breeder and the auctioneer. *Young Spectator* was on the ground and was much admired."

Such comparisons as these are the true standards for other breeders and it is gratifying to find such authority to substantiate the facts I have previously stated. I have an own sister to Lady, several half sisters to Lofty, and two half sisters to Victoria; and they came from the bleak hills of Gloucestershire, from the herds of William, John and Joseph Hewer. I do not think an "*extraordinary*" milker is a profitable animal for the farmer. If I can keep my cows up to from 6 to 8 lbs. each, for nine months, I will be perfectly satisfied; nor do I think that three herds in twenty will do more, unless selected entirely for that purpose. We all know that a great quantity of milk requires to be forced with extra feed; and when such a cow is dried for the grazier or feeder, I want him to keep an accurate account of food consumed, and the price sold to the butcher.

One more statement and I have done. I do not think there are many breeders in this country who are willing to give a remunerating price for a good animal. The butchers have far more spirit than the breeders, and my object in future shall be to serve them. If breeders want this stock, let them go to England and fetch them; they will then know their true value. At present, I intend to make steers of all. Nor do I intend to show another animal for a premium in this country, so long as a combined prejudice exists.

I conclude by saying that the Hon. Erastus Corning has rendered me every assistance with his purse, and deserves as much credit for this noble spirit as the best of the noblemen in England. They know agriculture is the only support of the country. Directing members of this country know equally well the benefit they render it by such praiseworthy transactions, and such are the true supporters of a prosperous and enterprising country. It is such men only that can place America in the exalted station she ought to occupy.

WM. H. SOTHAM.

Hereford Hall, near Albany, Nov. 30, 1843.

A notable instance of Herefords used as dairy stock in England has been mentioned in the Consul's report, that of the experience of the Messrs. James of Dorsetshire.

Several years since Mr. Duckham, in writing us of the Herefords, gave an account of their use by the said Mr. James, of Mappowder, near Blandford, Dorsetshire, England; referring to their use at Mappowder to show their adaptability to different sections of the country. When in England we had occasion to visit his herds and some of the neighboring herds. We learned from Mr. James that his father established this herd in 1837, and that, at his father's death in 1857 or 1858, he came into possession of the herd and has bred them from that time until the present, a continuous experiment of nearly one-half of a century—from 1837 to 1884. We were desirous of knowing as nearly as possible the character of this work. We found that he used bulls from such breeders as Mr. Turner, of The Noke; Mr. Jeffries, of The Grove; Mr. Stephens, of Sheep House; Mr. Stedman, of Bedstone Hall; Mr. E. Price, of Court House; Mr. Thomas Rea, of Westonbury; Mr. Tudge, of Adforton; Mr. Phillip Turner, of The Leen; Mr. Thomas Rogers, of Coxall, and Mr. Myddleton, of Llynaven.

These breeders, from whom Mr. James selected his sires, will be recognized as among the leading ones in England; Mr. James does not milk or work his dairies himself. He keeps eighty cows in milk. These are rented to two dairymen, 40 cows to each. The cows commence calving in March, and the calves put upon the pail until the 15th of May; to which time the milk is used for making butter, and from that time for making cheese and butter. The bull calves go to the butcher, and the heifer calves are carried on to keep the herd good. The rental is so much per head for each cow, and if from any cause the cows of another breed are brought in, the renters are not as well satisfied as when they have the entire complement made up of Herefords. Mr. J. has not kept a record of their breeding, but the animals show as good a character as any Hereford herd we saw in England. This is especially shown in the three-year-old heifers, the two-year-olds, and the yearlings; and had a record been kept the character of the cattle would have placed Mr. James' herd among the best. There were nearly or quite 20 three-year-old heifers that had lost their calves; and Mr. James, instead of holding them for further breeding, was grazing them for the butcher.

It was early in August when we were there, and we had seen no better three-year-olds. They would have compared favorably with any others, as would the yearlings; all were raised on the pail to the 15th of May each year, and then on grass from that on. The cows of the herd are drafted out for the butcher at from 6 to 8 years old, unless an exceptionally extra cow for the dairy is kept further on. The rule is, however, to draft for the butcher at from 6 to 8 years. They are then grazed and fed, and bring the top price for butcher's beasts, either from local butchers or at Smithfield.

Mr. James is satisfied from this half century's experience that he can make more money from thoroughbred Herefords than from any other breed, or mixture of breeds. The spread of this blood in the vicinity of Mappowder for 15 or 20 miles is constantly on the increase, and the different dairies seen, passing through the country, show a large proportion of white-faced cattle. Since the herd has come into the hands of the present Mr. James, he has taken some pains to exhibit cattle largely at the county and district fairs, and has kept a record of the winnings, a copy of which we took. He has not only taken prizes at local shows, but at the Bath and West of England, the Royal, Smithfield, and the Birmingham Shows; and whether showing in competition for dairy honors or in competition for the butcher, he has been a very successful exhibitor. In the dairy competition his competitors have been largely of Shorthorn and Devon blood, and these tests have been made strictly in view of the dairy class of animals. The use of Herefords at Mappowder for half a century as dairy cattle, or, if you please, as general purpose cows, combined milk and beef, has given a record that scarcely any other breed can show for the two purposes of beef and milk combined.

Anyone at all conversant with the production of beef, milk, butter and cheese who will come and examine this herd and inform themselves as to its best doings, will recognize that Messrs. Berry and Youatt were not properly informed when they made the statement that the Herefords were not a desirable milk breed, and those who have adopted the theory promulgated

SILVIA (V. 17, p. 288) 8649.
Bred by P. Turner, Herefordshire.

through "Youatt's History of British Cattle" would find occasion to change their views after becoming well posted as to the results of these 48 years' use of the Herefords in the dairy.

We here make the statement, based upon this experiment, that the Hereford used as a dairy cow may be made a success; and that the bullock produce from such a dairy, and the draft cows, will pay more money than any other breed. And we call attention especially to the following premiums awarded to the animals from this herd:

PRIZES WON BY MR. J. W. JAMES, ON DAIRY-BRED HEREFORDS.

1859 First prize at Sturminster for Yearling Heifer.
1859 First prize at Bath and West of England Show for Yearling Heifer.
1859 First prize at Bath and West of England Show for Extra Stock.
1860 First prize at Dorchester for Yearling Heifer.
1860 First prize at Sturminster for Yearling Bull.
1860 Second prize at Sturminster for Fat Cow.
1860 Second prize at Sturminster for Fat Heifer.
1860 Second prize at Sherbourne for Fat Cow.
1860 First prize at Yeovil for Fat Cow.
1860 First prize at Birmingham Fat Stock Show for Fat Heifer.
1860 First prize at Smithfield Club Fat Stock Show for Fat Heifer.
1861 First prize at Sturminster for Bull.
1861 First prize at Sturminster for Bull.
1862 First prize at Sturminster for Steer.
1862 Second prize at Sherbourne for Steer.
1862 First prize at Yeovil for Steer.
1862 First prize at Dorchester for Steer.
1863 First prize at Bath and West of England Show for Yearling Steer.
1863 Third prize at Bath and West of England Show for Yearling Steer.
1863 Second prize at Bath and West of England Show for Two-year-old Heifer.
1863 Second prize at Sturminster for Cow and Offspring.
1863 Second prize at Sturminster for Cow.
1863 First prize at Sturminster for One-year-old Heifer.
1863 First prize at Wincanton for Two Dairy Cows.
1863 First prize at Sherbourne for Two-year-old Heifers.
1863 First prize at Sherbourne for One-year-old Heifer.
1863 First prize at Yeovil for Two-year-old Heifers.
1863 First prize at Yeovil for One-year-old Heifers.
1864 First prize at Sturminster for Cow and Offspring.
1864 First prize at Sturminster for Two-year-old Heifers.
1864 First prize at Sherbourne for Two-year-old Heifers.
1864 Second prize at Sherbourne for Dairy Cow.
1864 First prize at Dorchester for Fat Heifer.
1864 Second prize at Dorchester for Fat Heifer.
1865 First prize at Bath and West of England for Two-year-old Heifers.
1867 First prize at Sturminster for Fat Cow.
1867 First prize at Sturminster for Cow and Offspring.
1867 First prize at Sturminster for Pair Two-year-old Heifers.
1867 First prize at Yeovil for Pair Two-year-old Heifers.
1868 First prize at Bath and West of England for Dairy Cow.
1868 First prize at Bath and West of England for Two-year-old Heifer.
1868 First prize at Bath and West of England for One-year-old Heifer.
1868 First prize at Dorchester for Dairy Cow.
1868 First prize at Dorchester for Two-year-old Heifer.
1868 Second prize at Sturminster for Cow and Offspring.
1868 Second prize at Sturminster for Fat Cow.
1868 First prize at Sturminster for pair Yearling Heifers.
1868 Second prize at Sherbourne for Fat Cow.
1868 Second prize at Sherbourne for Dairy Cow.
1868 First prize at Sherbourne for Pair Yearling Heifers.
1868 First prize at Yeovil for Three Dairy Cows.
1868 First prize at Bath and West of England for Fat Cow.
1868 First prize at Bath and West of England for Fat Heifer.
1868 First prize at Bath and West of England for Best Cow.
1868 Second prize at Bath and West of England for Fat Cow.
1869 Second prize at Bath and West of England for Yearling Heifer.
1869 First prize at Stalbridge for Stock Bull.
1869 First prize at Stalbridge for Yearling Heifers.
1869 First prize at Stalbridge for Fat Heifer.
1869 First prize at Sturminster for Fat Heifer.
1869 Second prize at Sturminster for Bull.

HORACE (3877) 2492, AT 15 YEARS.
Bred by J. Davies. (From a photograph from life by Bustin.) Horace has been unsurpassed as an improver of the breed.

1869 First prize at Sturminster for Pair Two-year-old Heifers.
1869 Second prize at Sturminster for Pair Yearling Heifers.
1869 Second prize at Sherbourne for Bull.
1869 First prize at Sherbourne for Fat Heifer.
1869 First prize at Sherbourne for Pair Yearling Heifers.
1869 First prize at Sherbourne for Pair Yearling Heifers.
1869 First prize at Yeovil for Bull.
1869 First prize at Yeovil for Pair Two-year-old Heifers.
1869 Second prize at Yeovil for Yearling Heifers.
1869 Second prize at Yeovil for Fat Heifer.
1869 First prize at Bath and West of England for Fat Cow.
1869 First (extra) at Bath and West of England for Fat Heifer.
1869 First prize at Bath and West of England for Fat Cow.
1869 First prize at Dorchester for Fat Cow.
1870 Second prize at Birmingham for Fat Heifer.
1870 Second prize at Smithfield for Fat Heifer.
1870 First prize at Plymouth for Fat Cow.
1870 First (special) at Plymouth for Best Cow.
1870 First prize at Southampton for Fat Heifer.
1870 First prize at Sturbridge for Fat Cow.
1870 First prize at Sturbridge for Fat Heifer.
1870 First prize at Sturminster for Fat Cow.
1870 Second prize at Sturminster for Fat Heifer.
1870 Second prize at Sturminster for Pair Yearling Heifers.
1870 First prize at Sherbourne for Fat Cow.
1870 Second prize at Yeovil for Fat Cow.
1870 First prize at Yeovil for Pair Two-year-old Heifers.
1871 First prize at Bath and West of England for Bull.
1871 First prize at Dorchester for Bull.
1871 First prize at Dorchester for Three Heifers.
1871 First prize at Stalbridge for Bull.
1871 First prize at Stalbridge for Cow and Calf (Dolly V).
1871 First prize at Stalbridge for Pair of Heifers.
1871 First prize at Sturminster for Bull.
1871 First prize at Sturminster for Pair Two-year-old Heifers.
1871 First prize at Sturminster for Dairy Cow and Offspring.
1871 First prize at Sturminster for Pair Yearling Heifers.
1871 First prize at Sherbourne for Bull.
1871 First prize at Sherbourne for Pair Two-year-olds.
1871 First prize at Sherbourne for Pair Yearling Heifers.
1872 First prize at Bath and West of England Show for Two-year-old Heifers.
1872 Second prize at Bath and West of England for Little Bull.
1872 First prize at Stalbridge for Two-year-old Heifer.
1872 First prize at Sturminster for Two-year-old Heifer.
1872 First prize at Sturminster for Pair Yearling Heifers.
1872 Second prize at Sherbourne for Pair Two-year-old Heifers.
1872 First prize at Sherbourne for Pair One-year-old Heifers.
1873 First prize at Bath and West of England for Two-year-old Heifers.
1873 Second prize at Bath and West of England for Calf.
1873 First prize at Sturminster for Bull.
1873 First prize at Sturminster for Two-year-old Heifers.
1873 First prize at Sherbourne for Two-year-old Heifers.
1873 First prize at Smithfield for Fat Heifer.
1873 Second prize at Bath and West of England for One-year-old Heifer.
1874 First prize at Sturminster for Bull.
1874 First prize at Sherbourne for Bull.
1874 First prize at Yeovil for Bull.
1875 First prize at Bath and West of England for Two-year-old Heifers.
1875 First prize at Royal Agricultural Society's Show for Two-year-old Heifers.
1875 First prize at Dorchester for Bull.
1876 First prize at Sturminster for Fat Heifer.
1877 Second prize at Sturminster for Bull.
1877 Second prize at Sturminster for Bull.
1877 Second prize at Shelbourne for Fat Heifer.
1877 Third prize at Smithfield for Fat Heifer.
1878 First prize at Smithfield for Fat Heifer.
1879 First prize at Sturminster for Dairy Cows.
1879 First prize at Sturminster for Pair Two-year-old Heifers.
1879 First prize at Sherbourne for Pair Two-year-old Heifers.
1880 Second prize at Blandford for Dairy Cow.
1880 Second prize at Blandford for Two-year-old Heifers.
1880 Third prize at Blandford for Two-year-old Heifers.
1880 Third prize at Blandford for Yearling Heifer.
1880 Second prize at Sherbourne for Dairy Cow.
1880 First prize at Sherbourne for Pair Two-year-old Heifers.
1881 First prize at Sturminster for Dairy Cow and Offspring.
1881 First prize at Sturminster for Dairy Cow.
1881 First prize at Sturminster for Two-year-old Heifers.
1881 First prize at Sturminster for Pair One-year-old Heifers.
1882 First prize at Dorchester for Bull.
1882 First prize at Dorchester for Dairy Cow.
1882 First prize at Dorchester for Two-year-old Heifers.
1882 First prize at Sherbourne for Dairy Cow.
1882 Second prize at Sherbourne for Cow.
1882 First prize at Sherbourne for Pair Two-year-old Heifers.
1882 Second prize at Sherbourne for Bull.
1882 First prize at Yeovil for Pair Two-year-old Heifers.
1882 Second prize at Yeovil for Bull.
1882 Second prize at Yeovil for Yearling Bull.
1882 Second prize at Yeovil for Cow and Offspring.
1882 First prize at Sturminster for Yearling Bull.
1882 First prize at Sturminster for Pair Two-year-old Heifers.
1882 Second prize at Sturminster for Cow and Offspring.
1882 Second prize at Sturminster for Fat Cow.
1883 First prize at Sherbourne for Three Dairy Cows.
1883 First prize at Sherbourne for Bull over 2 years.
1883 Second prize at Sherbourne for Bull under 2 years.

BULLS THAT HAVE BEEN IN SERVICE AT MAPPOWDER FROM 1837 TO 1884.

1836 Bot, son of Old Sovereign, bred by Mr. Turner, The Noke.
Goldfinder, bred by Mr. Turner, of Westhide.
1839 Young Cotmore, bred by E. Jeffries, of The Grove.
Young Sovereign, bred by Mr. Turner, of The Noke; dam Countess Cow.
1846 Wallace, bred by Mr. Turner, of The Noke.
1877 Bertram, bred by Mr. Turner, of The Noke.
Chance, bred by Mr. Stephens, of Sheephouse; dam from a cow purchased at Mr. Jeffries' sale.
Valentine, bred by Mr. Stedman, Bedstone Hall.
1857 Statesman (1744), bred by Mr. E. Price, Court House.

DICTATOR 1989.
Bred by T. L. Miller, Beecher, Ill.

1859 Happy Lad (2561), bred by Mr. E. Price, Court House.
Sampson (2208), bred by T. Rea, Westonbury.
Dundee (3080), bred by Mr. Tudge, of Adforton.
1869 Leopold (3912), bred by J. W. James.
Prince Chance, bred by T. Rogers, of Coxall.
1875 Taurus, bred by Mr. Tudge, of Adforton; sire Gamester (4594).
1876 Baron Munchausen (5207), bred by T. Myddleton, Llynaven, Clun, Salop.
Chance, bred by J. W. James.
Wonder (3602), bred by T. Rogers.
Baron the 5th (5735), bred by J. W. James.
1881 Northern Light, bred by Mr. Lester, Cefn Ila, Usk, Monmouth.
1879 Lord George, bred by J. W. James. Not entered in Herd Book.

At Sherbourne, England, at the Agricultural Society's Show held in 1883, Mr. James, Mappowder Court, Blandford, was very successful in showing Herefords against Shorthorns as dairy cattle, taking first and second prizes for Herefords, with a first for three dairy cows, of which one took the first prize in the cow class at Sherbourne the preceding year, and the other a first prize as a heifer.

From the foregoing account it will be seen that Mr. James and his father before him had a dairy of eighty Herefords for a half century, that were successful beyond ordinary dairy herds, for the purpose for which they were used.

The "Mark Lane Express" of May 24th, 1880, says: "A correspondent writing from the Vale of Blackmore in Dorsetshire states that he finds Hereford cattle the better in that district than any other breed, both for dairy purposes and for beef, and keep themselves in better order, with the same amount of food and under the same conditions of life. They are not such large milkers as some other breeds, but their milk is richer than that of the Shorthorns."

In a paper on "Pedigree Cattle in Dairy Herds," by Mr. Joseph Darby, in Vol. 15, third series of the Journal of the Bath and West of England Agricultural Society, the following testimony is given to the values of the Hereford cross-bred cow as a milker.

"The Hereford has never been considered much of a dairy animal, yet strange to state, when allied to the Shorthorn, the effect causes latent lacteal fertility of the breed to spring at once into full development. Probably this fact has not been stated so authoritatively before, but those who doubt may be referred to Mr. E. C. Tisdale, Wolland Park Dairy, Kensington, and it will be found that two or three of his very best cows are of this variety. In fact, the one styled therein 'Old Hereford' yielded twenty-three quarts per day in the first month, and twenty-two quarts in the second and third, going on in profit for eleven months, and averaging for the whole of that period fifteen quarts per day. There is not one in the entire list that affords anything like such a record as this, but another Hereford-Shorthorn, styled 'Comalis,' averaged 14.61 quarts per day for 19 months. Facts like these cannot be brought too prominently before the public eye."

It does not appear how these cows were bred; whether by Hereford bulls out of Shorthorn cows or by Shorthorn bulls out of Hereford cows. But the facts, as stated, show the value of the Hereford blood in a cross with the Shorthorn for the dairy, because none of Mr. Tisdale's pure-bred Shorthorns were such good milkers as these crosses, whatever they may have been—Hereford-Shorthorns, or Shorthorn-Herefords. It is quite certain that the cross of the Hereford bull on the Shorthorn cow will produce a better grazing animal than the pure-bred Shorthorn, and the facts given above show that cross-bred Herefords have beaten pure-bred Shorthorns—to put it only in that way—in one of the most noted herds of dairy cattle which England can show. This shows that what we have constantly advocated is absolutely correct, namely, that where there is sufficient keep, the Hereford bull on the top will produce the most useful cattle to meet general requirements all over the world.

WABASH, PURE-BRED STEER.
Champion 2-year-old at Chicago, 1882. Weight 1940 lbs.

CHAPTER XXXVII.

Herefords in Australia, New Zealand, South America, West Indies, and Jamaica

The Hereford cattle have been successful in all the countries to which they have been taken. Their history in other countries than ours proves their value as beef animals and their hardiness and strong constitution, their worth as grazing beasts.

We quote the following from the "Mark Lane Express," 1880:

"Reports are continually coming to hand relative to the success of the Hereford breed of cattle in various parts of the United States, and they do equally well in the Australian colonies. In an excellent report on the live stock of the colony of New South Wales, Mr. Alexander Bruce wrote as follows in 1876: 'The Herefords have proven themselves in Australia to be an excellent race of cattle, with distinctive type, characteristics and form, thoroughly established, and capable of being transmitted to any other race with which they may be interbred. They are principally located in the north of Sydney, in the eastern and northeastern portions of the colony. The effect of the introduction of Hereford blood into our herds has so far been satisfactory, and if a constant supply of pure bulls of this breed is kept up, in the herds in which it has been tried, still further good results may be expected, as the Herefords are hardier, more active and carry heavier and thicker coats than the Shorthorns. They are better suited than the Shorthorns for inferior pasturage and outlying runs, and they stand the road and winter better. A good many owners are now breeding from Herefords, and it is to be expected that they will eventually displace the Shorthorns on the lighter soils and colder country, and on the more distant runs.'

"This is precisely the same sort of testimony we see almost daily in the United States press; and from occasional notices in Queensland and New South Wales papers we find that the Herefords are found to be better 'wrestlers' as well as better beef makers in Australia as well as in the United States."

We give the following extract from a letter of a friend in Australia: "I am glad to see that you still stick up for the 'white faces.' I have taken the 'Hereford Times' about 45 years, and I see that you are a 'thorn' in the sides of the Shorthorn breeders. I am glad to find that you are succeeding so well in America with the Herefords. It seems to me from the reports that you are pushing the Shorthorns off the ground. I wish I could say the same here, but I am sorry to say that that cursed thing called *fashion* takes the lead in everything. The Shorthorn is the most aristocratic, as most of the noblemen of England breed them. The rich colonists here follow their example. Now that the ports are thrown open, some of these rich men have gone to England and purchased many of the best Shorthorns they could find. I believe it is nothing but pride, to show the English noblemen the depth of their purses. One gentleman in this colony has bought about 20 head, which are now in quarantine. He breeds Herefords as well as Shorthorns. He bought his Herefords from me and told me he could keep three Herefords where he could keep two Shorthorns, but he must 'follow the fashion of the nobles.'

"I do not think I have written you since I was at Sydney last April twelvemonth. I was there judging the Herefords. There were some first-class Herefords there. I intended to have shown a pair of heifers at the Sydney exhibition, but the breeders were determined to hold their show in April. My heifers were too young to show for a prize. I offered £5 5s. for the best pair of heifers of any breed under 24 months old. I am very sorry that I could not go to see the show. I gave the prize on purpose to put the Herefords against the Shorthorns, and am happy to tell you that the Herefords were victorious. There was but one pair of Herefords against all the Shorthorns. The white-faces were so good that there was no comparison. The Shorthorns had not a chance with them.

The Herefords were shown by Mr. Frank Reynolds, of Tocal. He is the leading breeder of Herefords and Devons in New South Wales. He will be able to give you every item of information about them, as I could not. I sold a bull and three heifers to go to West Australia; the same person had one bull from me two years since; he turned out so well that he came all the way—1,500 miles by sea—for more. They are the first Herefords sent to West Australia; they all arrived safe, and are in capital hands; they will be well done by. I allowed him to take some of the best I had, so that they should have every chance to make a good start.

ARCHIBALD (6290) 11129.
Bred by A. Rogers, Herefordshire. Weight 3,000 lbs. (From a painting.)

"I must now tell you how my little herd is getting on. I have the vanity to think that I have as nice a herd as I ever saw on one farm. It consists of about 200 head, all so near alike that it is difficult to know one from the other. I can and do make double off my beef to what any breeder in the colony does, as I make as much off my two-year-and-a-half old steers as any of them make off their five and six-year olds. G. PRICE."

We find the following in the "Queenslander," Australia. It will be seen that Australia has good Herefords, but the feature of greatest interest is that Lady Claire 4116, the dam of Lord Wilton, at 16 years of age, should make a trip from New Zealand to Queensland. There is no doubt but what some of the best of the Hereford breed of cattle have gone from England to Australia, and there is no doubt but that they will take an active part in the present effort to improve cattle in that country.

"Messrs. McConnel & Wood's recently imported Herefords from New Zealand deserve more than a passing notice, inasmuch as they include some of the highest class of Herefords ever landed on these shores. The draft contains a cow now become celebrated throughout all parts of the civilized world where the Hereford breed is known. This cow is Lady Claire (Vol. 9, p. 336), the mother of the celebrated Lord Wilton (4740) 4057. (¶ 257) This bull is recognized throughout England and America as the best living representative of the Hereford breed of cattle, being alike unapproachable as a show animal and as a stock-getter, no bull of any breed having begotten so many prize-takers. Lady Claire was purchased by Mr. Stuckey, of New Zealand, before the immense value of Lord Wilton as a sire had been fully appreciated. Otherwise she never would have been allowed to leave England. She is by Marmion, bred by Mr. Tudge, her dam being Lady Adforton, by Pilot, by The Grove. She is in calf to Chippendale (6865), a bull of equally valuable strains. Chippendale—now in New Zealand—is by Lord Wilton, dam Judy (Vol. 11, p. 157) 14419. He also was a most successful show-yard bull, and his value as sire is known by the fact that after Mr. Stuckey had purchased him he was offered £500 ($2,500) on his bargain by an American buyer. His stock have realized long prices in England. Chippendale is the first of Lord Wilton's stock that have come to the southern hemisphere.

"Another of the importations is Duke of Chippendale, a 17-months-old son of Chippendale, from Amethyst (Vol. 9, p. 216), a celebrated cow of P. Turner, The Leen. This bull contains eight strains of the celebrated Sir Benjamin. Mr. Stout, another young bull selected by Mr. Wood in New Zealand, is also by Chippendale, and of the Amethyst tribe. The twins Castor and Pollux, nine months old, are by George 4th, by Coomassie, their dam Charity 2d, being by Horace 3d (5386), by the celebrated Horace (3877) 2492. We learn that Mr. Wood purchased four more Chippendale calves not yet old enough to wean. This addition to the Durundur herd is a most wonderful one, and had our ports been open for the importation of English stock the enterprising owners could not have selected in England an equal number possessing more valuable or more fashionable strains of blood."

HEREFORDS THE BEST CATTLE.

A correspondent of the "Queenslander," Mr. Reginald Wyntham, writing from Leeksfield, says:

"I have sold scores of Hereford bulls myself for use in Shorthorn herds, and hope to sell many more. A buyer came to me only a few days ago wanting Hereford bulls for his Shorthorn cows, his Shorthorn bulls having all died during the late drouth, and I ventured to say that there were few cattle breeders who have not seen and admired the first cross between the pure Hereford bull and Shorthorn cow. As Australia is the finest cattle country in the world, particularly the Queensland portion of it, and as the day must come, sooner or later, when they will be worth double what they now sell at, we shall do well to consider the merits of all the different breeds, that the more suitable breeds may be kept on the different classes of country. In a drouth it is a fact—and a well known fact, too—that Shorthorns will die long before Herefords and Devons, or, I believe, any other breed. New Zealand, I should say, is better adapted for the Shorthorn, where an acre of land will provide him with sufficient grass and water, and where he will not have to travel a great distance to market; but in Queensland, on very many runs, the Herefords or Devons should be kept."

HEREFORD CATTLE IN AUSTRALIA.

Referring to the assertion of a writer, that Hereford cattle grew wild when turned on ranges in New Zealand, Mr. A. J. McConnel, of Queensland, in a contribution to the "Live Stock Journal," London, says: "With regard to the Hereford cattle of New Zealand, I wish to say that all I have seen were extremely quiet and docile, and could not, by any stretch of imagination, be said to have any inclination to wildness or even unsteadiness, and I have heard enough of their ways to convince me that the ferine Hereford cattle, of which Proteus writes, must be the result of some terrible neglect. With respect to the Hereford cattle of Australia I must ask you to receive a novel writer's statement with great reserve. It is really not worth alluding to, for the paragraph in which the author writes of the unsteadiness of the Hereford cattle was merely a repetition of the usual talk on cattle stations in years past, when this particular trait of wildness was insisted upon by men who had never seen a well-bred Hereford herd. I have the author's authority to say that his only experience, in a large way, of Hereford cattle was with a neglected herd where some white-faced bulls had been used. These by no means could be called pure, and that after a short period of careful management this particular herd was quieter than the herds on any neighboring properties. I should like to add a few words about my own experience. On the property with which I am connected Hereford cattle have been bred since 1865 in mountainous and broken country, and care was taken at the start to get the best purebred bulls obtainable in New South Wales, where Herefords had been bred from imported stock since 1825. Before we commenced with Herefords we had a very quiet and highly bred Shorthorn herd. We made the change because we were of the opinion that the Herefords were better grazing cattle and more profitable altogether, and the event has proved that we were right. But the point I wish to make clear is, that the very quiet and tractable herd of 7,000 Shorthorn cattle has been succeeded by a herd more tractable still of Hereford cattle. The same care and the same treatment was shown to the cattle in both instances. There can be no doubt that a Hereford is more agile in his movements than any breed except the Devon. That I concede readily, and I hold it to be one of his excellencies, for it enables a grazing Hereford in Australia to range about for food in times of scarcity, and do well, when a softer breed gives up the struggle for existence. I suspect the truth to be that when we hear of an inclination to wildness on the part of any cattle of any breed, if we knew the circumstances, they would show that neglect of owners and a want of knowledge in working numbers of cattle together on the runs is at the bottom of

FISHERMAN (5913) 76239.
Bred by T. Rogers, Herefordshire.

this inclination to wildness. I affirm that the Hereford is not more prone to ferine habits in Australia than any other breed. I have seen neglected Shorthorn cattle as wild as it is possible for cattle to be—in fact, they were useless, for they could not be got when wanted."

A CHALLENGE.

"To the Editor of the 'Queenslander':

"Much has been written and published in your columns during the past two years of the merits and demerits of the different breeds of cattle. The majority of the writers are gentlemen engaged in breeding Hereford cattle for stud purposes, some of whom have had very little experience with the cattle they were (to use a colonial phrase) 'cracking up,' as they had, to my knowledge, been engaged in their production only a few years. The object with all Hereford advocates, clearly, has been to prove that their cattle will subsist and make

TARQUIN (12717).
Bred by P. Turner, Herefordshire, first prize yearling bull R. A. S. E., 1887.

condition on poorer pastures than Durhams; that they were of sounder constitution, and hence better able to withstand the effects of a long drouth than Shorthorn cattle. Well, it seems to me, Mr. Editor, that all the correspondents possible will never lead the public, for whose benefit you permit the ventilation of the subject in your columns, to a just conclusion, for gentlemen who have invested a considerable capital in Hereford cattle, with a view to making a profit by the disposal of their male calves as young bulls to their neighbors, are quite sure to do vigorous battle against other breeds, but I am of the opinion that sound, practical tests afford the best proofs of all.

"Let the Herefords and Durhams be brought into competition, under exactly similar conditions, and the results made public, and the question of which is the best breed for our climate and pasture will quickly be settled. We have had one such test, the result of a challenge I issued through your paper, and I will now, with your permission, issue another challenge, as we have now unfortunately a most favorable opportunity of proving which breed has come through the drouth the best. I have a lot of pure-bred Durham cattle here in a paddock that is stocked at the rate of a beast to six acres. About 150 of these are breeding cows, and I would show any number of them, not being less than 15 per cent. of the whole, for condition against a like proportion of Herefords, not being less than ten head, to be selected from any one herd in Leechhard district that has been sustained by the natural pastures for at least two years. The only conditions I impose are the cows competing must be four years of age or over and breeders, the test to take place within two months.

"W. K. PEBERDY.

"Gellenbah, 22d Jan'y, 1881."

THE CHALLENGE ACCEPTED.

"*To the Editor of the 'Queenslander'*:

"I have to-day seen W. K. Peberdy's challenge of the 22d of January. If I read it aright it is merely a question as to which breed of cattle show the best condition after the late drought. I will therefore take to Rockhampton twenty Hereford cows that are four years, with their twenty suckling calves by their sides, and show them against W. K. Peberdy's twenty Shorthorns with their twenty calves. They shall be there before the 22d of March, on a date to be named by W. K. Peberdy, the best conditioned cows to win the prize, which I propose to be a ten-guinea cup, such as I gave at the last show, and the loser also to give ten guineas to the hospital.

"I am glad to see Mr. Peberdy so very philanthropic as to breed Shorthorns for other purposes than for sale and to make money. Unfortunately I and Archer Rignall Windom and Frank Reynolds and Loder are compelled to breed for the dirty dollars. Of course the cattle shown are to be pure breds, out of the two stud herds. I name Albert Wright of Nullalbin as judge on our side, and I recommend one of these three gentlemen as umpire, namely, Vincent Dowling, John Button, or James Ross.

"I am sir, etc.,

"BEARDMORE OF TOOLOOMBAR.

"22d of Feb'y, 1881."

THE HEREFORD-SHORTHORN CONTROVERSY THAT FOLLOWED.

The acceptance of this challenge, it was supposed, would result in a trial that would be valuable to the Hereford interest in that country.

Under date of 13th of March, Mr. Peberdy replies to this and objects to showing calves with the cows, and that the distance to Rockhampton was some 80 miles further for him to travel than for Mr. Beardmore, and then proposes to issue another challenge to Mr. Beardmore, privately, through his agent at Rockhampton. The second challenge by Mr. Peberdy is as follows:

"To definitely settle which is the best breed of cattle for the central districts of Queensland, Herefords or Durhams, I now issue the following challenge to Mr. O. C. I. Beardmore, he having written viciously against Durham cattle and warmly in praise of Herefords.

"The conditions of the challenge are as follows:

"1st. The cattle exhibited by Mr. Beardmore shall be pure Herefords, and the cattle exhibited by me shall be pure Durhams.

"2d. The definition of the word 'pure,' as applied to the different breeds or pens of cattle exhibited, shall be taken to mean that the cattle exhibited as Herefords shall have no admixture of Durham or Devon blood in their veins; that the cattle exhibited as Durhams shall have no admixture of Hereford or Devon blood in their veins. Should either pen contain one or more beast that, in the opinion of the judges or umpire, is not 'pure,' then the pen shall be condemned, and the opposing pen be awarded the prize.

"3d. That judgment shall be given on different points of merit and awarded by points as follows:

"Condition of cows, ten points.

"Best cattle for pastoral purposes generally, ten points.

"Aggregate weight of pen, five points.

"Greatest yield of milk, five points.

"4th. That the judges be requested to make their award on the merits of the pens placed before them by the exhibitors, independent of any prejudice they may have against either breed.

"5th. The cows exhibited shall be in full milk, but it shall be optional with exhibitor whether he pens the cows' calves.

"6th. Should the judges desire it, the whole of the cows shall be milked in their presence, after being fourteen hours apart from their calves, to decide their qualities as milkers.

"7th. Should the points be awarded in equal number to each breed, then the umpire and judges shall examine the cattle again as one body, and shall, having taken into consideration the points of the contest, decide which is the best pen of cattle on the whole, and award accordingly by majority.

"Should Mr. Beardmore be disposed to accept the above challenge as it is given, I will meet him in Rockhampton with ten cows of mine to compete with ten cows of his for a piece of plate to be selected by the victor, of the value of $100, at any time he may please between now and the date of the P. A. and H. show in the above-named place, or during the week that show is held. It is, perhaps, better that Mr. Beardmore should understand that I will not travel my cows that are rearing calves 150 miles to meet him, but if he accepts the challenge, I will truck them down. I omitted to add that the cattle exhibited by either party shall have been his property for, say, six months prior to the issue of this challenge. I am quite willing to accept Mr. A. Wright as judge on Mr. Beardmore's side, and will also accept either Mr. Vincent Dowling or Mr. James Ross as umpire; and will appoint either Mr. Cunningham, of Rannes Station; Mr. Wood, of Calliangal, or Mr. Lamond, of Stanwell, as judge on my side. I shall not consider myself bound by this challenge unless I receive notice of its acceptance fourteen days before the date fixed for the contest. I may add, I make condition a leading point to suit Mr. Beardmore; I also make weight a minor point, so that it may tell against myself, and destroy any objection he might otherwise have to the condition.

"(Signed) W. K. PEBERDY."

To which Mr. Beardmore makes the following reply:

"To the Editor of the 'Queenslander':

"Sir: I have to-day received a copy of a second challenge from Peberdy on the subject of Herefords vs. Durham cattle. In your issue of the 9th of February Peberdy issued a challenge which I accepted without alteration and left him to name the day. To that I adhere, and will do so two months longer; after that, I shall have to wean my calves and not be in a position to produce twenty mothers with their progeny, and I now again call Mr. Peberdy to stick to his challenge as then issued or hand over two guineas to the hospital and then retire into obscurity.

"With regard to his second challenge, my Herefords are not milkers, and therefore I could not compete for milkers. My Herefords are not broken in, and could not be put in a yard singly to be handled, and I have yet to learn how cows are to be in full milk and have no calves. Peberdy's words are: 'The whole of the cows to be milkers.' In his fifth clause calves are to be optional, in his sixth it is pretty evident there must be calves in the contest. Again, Mr. Peberdy states that he can't drive his cattle 150 miles, but must truck them. Now, all Hereford men have contended that Herefords are better travelers—this is one of the very points I took up—and Peberdy's desire to avoid traveling his cattle down, shows the weakness of his breed of Durhams. As he says he has 150 miles to travel to go to Rockhampton while I have only 90 miles to go, I will travel my Herefords another 60 miles to make up the 150, and so give him no cause of complaint. Again, what does he mean by making 'condition' a leading point to suit Mr. Beardmore. It is 'condition' and 'condition' only that he 'challenged on' in his letter of the 9th of February. I look on Mr. Peberdy's challenge as a piece of 'blow' and nothing more, and that he has no intention of bringing things to a point. I am, sir, etc.,

"BEARDMORE OF TOOLOOMBAR."

(There is no date to Mr. Beardmore's letter.—Ed. Q.)

ELTON 1ST (9875) 11245.
Bred by Earl & Stuart. Celebrated son of Sir Richard 2d (4984) 970A.

We have several other letters pertaining to this controversy. We give but one of these, and that signed by John Fulford, of Lyndhurst, and dated the 21st of April, and which appeared in the "Queenslander" of May 10th:

"To the Editor of the 'Queenslander':

"Sir: I note in Mr. W. K. Peberdy's challenge letter of 22d January a sneer at the limited experience of some of the Queensland breeders of Herefords, who, he says, 'have been engaged in their production only a few years.' I do not know to whom he refers when he makes that statement, but he evidently loses sight of the fact that the majority of the Hereford breeders were breeders of Shorthorns before they commenced to breed Herefords, and therefore can lay claim to a larger experience than men who have bred Shorthorns only, and should be in a better position to know which breed is most suitable to this country. All cattlemen know that no breed of cattle were more abused than Herefords were for many years in Australia and the fact of their coming into favor at all—considering how little capital has been invested in them, and how few have been imported, compared to the fortunes spent in importing Shorthorn bulls and cows from England, and animals bred in the Colonies—speaks volumes to the mind of the unprejudiced person. However, it has not taken our American cousins many years to find out the superiority of the Hereford breed over Shorthorns under conditions almost identical with our own Queensland—namely, where the cattle have existed on grass alone, as they do in Texas, Colorado and other states, and where they are subject to drought, cold and wet, and have to travel long distances to market, as most of the Australian cattle have to do. No one can say that Hereford cattle have as yet had a fair trial in Queensland, as putting a few white-faced bulls into a herd for a year or so, and then reverting to Shorthorn bulls is not giving them a trial at all, and with the exception of a few herds that is all the trial they have had. However, I am told that Messrs. Archer, of Gracemere, who, I think, were the oldest breeders of pure Shorthorns in North Queensland, are using nothing but Hereford bulls in their general herd, and many others are following their example, I believe, in the neighborhood of Rockhampton; and I know of several stations further north that are doing the same. Mr. Peberdy says that correspondence on the subject should come from proprietors of fattening paddocks, drovers and butchers, and not from the breeders; but I think if we had to wait for their views (in print) on the subject we should not be enlightened very soon. But I can give one instance of the estimation Hereford cattle are held in, by one practical man, who, I believe, fattens largely, by referring to an advertisement in the 'Queenslander,' signed J. M'Connel (of Durundur, Brisbane River), a short time since, wanting to purchase a thousand Hereford store bullocks. He, too, is an old breeder of Shorthorns, yet he evidently prefers Herefords for fattening on his country. Mr. Peberdy evidently thinks his challenge to the few Hereford breeders of the Leechard district is, as he terms it, 'a true invasion of the Hereford stronghold,' while to me it appears as a very cheap means of letting the public know that he has a certain number of pure Shorthorn cows, bred in South Wales, for which he paid large sums of money; and it might be some of those identical cows, that probably had the advantage of English grasses and hand-feeding (and in consequence may be well-grown, heavy cows), that he wishes to show against the Herefords in his district, and his country may be (and probably is) very superior to the country the Herefords are pastured on. I have no desire to write against Shorthorn cattle or their breeders, as they are a breed of cattle I hold in high estimation, when in their proper place, and that, I take to be, is where English grasses abound in summer, and where plenty of fodder is available in winter, and the few favored parts of this great colony of Queensland where there are no frosts, and the best descriptions of natural grasses and fattening shrubs and herbage; and, as the despised Hereford has not had a fair trial under the latter conditions, I claim that no one can say that the breed is not quite equal, or superior, to the Shorthorns for any part of Queensland. I will now assert what I can prove to be the fact—that we lost far less cattle on this run during the late drought than any of our neighbors within a radius of 100 miles, who have herds of, say, 5,000 head or over on one run, and I may state that we have the largest herd in this district (on one run)—namely, about 20,000 head—in fact, the usual 5 per cent allowed annually for losses for the past twelve months.

"I am, sir, etc., JOHN FULFORD.

"Lyndhurst, 21st April."

(Mr. Fulford begged hard to be allowed to say one word more on this subject, and our resolution gave way. It must not be supposed, however, that we have reopened our columns for a continuance of the controversy.—Ed. Q.)

We also give the "Queenslander's" report of a Charters Towers Show. It will be noticed that this Hereford question has become fully as interesting in Australia as in America, and the discussion is likely in the end to bring that breed which has the greater merit to the front.

"Unlike the National Shows, the cattle are here displayed in all their native simplicity, all purely grass fed, and exhibited in yards. Although few beyond Mr. Parr's laid claim to purity of blood, they were, with very few exceptions, when looked at from a general standpoint, a magnificent collection. The bulls were few in number, and far from first-class, a passable animal, originally from the Mount Noorat herd, Victoria, being awarded principal honors; and Mr. Hann showed a promising Sir Roderick calf. Owners, no doubt, were unwilling to risk bringing valuable bulls from a distance. The bullocks, cows and heifers, however, more than made up for the deficiency in the bull classes.

"The principal exhibitors were Messrs. Murray-Prior, of Bulli Creek; Fulford (Messrs. Barnes and Smith), Lyndhurst; Hann, Vanneck; J. and W. D. Clark (Monahan), Lolworth; Glissan, Dottswood; B. C. Parr, Newstead; and Allingham, Hill Grove. It was well known throughout the North, thanks to the extensive circulation of the 'Queenslander,' that this show was to be the scene of a grand tilt between the 'Ballys' (Herefords), and the red, white, and roans (Shorthorns), and the interest of this great contest never flagged from the time the cattle entered the show grounds until their carcasses were displayed in the butcher's shop. It may be safely asserted, that as much money was depending on the butcher's scales, in this contest, as is usually risked upon an ordinary cup race.

"In the Shorthorn division, Mr. Parr's blood told in the heifer class—13 entries—Mr. Vanneck securing second. Mr. Parr also carried off the 100-guinea prize with a pen of three highly bred heifers, which were, however, in very low condition in consequence of the drought. Messrs. Clark, of Lolworth, secured the prize for the best cow bred north of Rockhampton, and the championship of the yard, with a grand roan cow of beautiful shape, and showing lots of quality.

"In Herefords, Messrs. Barnes and Smith (Mr. Fulford) had matters entirely their own way. Their pen of three fat bullocks, entered to do battle with the Shorthorns, and a pen of six bullocks, were as near perfection as can well be imagined of purely grass-fed cattle.

A. C. REED,
Evanston, Ill.

"Up to this point all had been plain sailing with the judges, each particular breed having been judged within itself. On entering on the miscellaneous division, however, cattle of 'any breed' came into competition, and the work of the judges, therefore, became much more difficult. For the best pen of three heifers in this division, there was an exceedingly good class, the judges commending the whole class. Three Herefords here secured a victory, Mr. Fulford's pen having been placed first, Mr. Parr's second, and Mr. Monahan's (Lolworth) third. Then came the great contest of the meeting, the best pen of three bullocks, number 0 brand—that is, four-year-olds—and although Messrs. Clark,

Vanneck, and Glissan entered, it was clearly seen from the first that only Mr. Murray-Prior's Shorthorns and Mr. Fulford's Herefords were in the race. Mr. Prior's three were beautifully matched, pure white, massive, and well topped up. Mr. Fulford's were well selected. These were judged on Thursday morning, and the whites awarded the coveted prize. They were slaughtered on Friday morning in the presence of a large concourse of breeders and the shop (Harvey's) in which they were hung up was for a time besieged with persons eager to obtain results.

"The weights were as follows: Prior's, 1,003 lbs., 982 lbs., and 1,020 lbs., a total of 3,005 lbs., giving an average of 1,001.66 lbs.; Fulford's, 1,024 lbs., 996 lbs., and 933 lbs., a total of 2,953 lbs., giving an average of 984.33 lbs. Mr. Prior, therefore, won by an aggregate of 52 lbs., and an average of 17.33 lbs. The plucky owners, however, were not content to accept this as a final decision between the two rival herds, but arranged a meeting on the same show grounds next year with No. 0 cattle, which will be then 5-year-olds. It is but justice to Mr. Fulford to state that his cattle had been driven 240 miles to the show, being 100 miles in excess of the distance traveled by Mr. Prior's, and that they were got together on the 1st, 2d, and 3d of May. For the champion prize, for the best fat bullock on the grounds, any age or breed, Mr. Fulford was first, with a splendid cross-bred animal, which on being slaughtered weighed 1,079 lbs.

E. E. ESSON,
Peotone, Ill.

"The above weights [all dead weights, T. L. M.] were considered really good, considering the severity of the season, and they point to the Kennedy as being one of the best cattle districts in Australia. There were only two exhibitors of Devons—Messrs. Hann and Farrenden. These exhibits, however, showed evident signs of Shorthorn origin, Mr. Hann's showing most of the Devon characteristics."

The battle of the breeds has come in Australia, as well as in America. The above account of the Charter's Towers Show shows the Herefords first on the best pen of three heifers, and the champion prize for best fat bullock in the show, the Shorthorns taking first in the best three bullocks, showing an aggregate of 52 pounds greater weight in the three carcasses. The Shorthorns were driven 140 miles, and the Herefords 240 miles to the show.

In response to a letter from the author, we received the following from an Australian correspondent:

"Editor 'Breeders' Journal': I am in receipt of your favor dated Dec. 10th, and also the six copies of your journal, which you so kindly forwarded; you may rest assured that I will do my best to get the 'Journal' a large circulation in Australia. I never lose an opportunity, when cattlemen are about, of showing your paper, and numbers have promised to write for it, having given them your address.

"I am extremely glad to see the forward movement of the Herefords in America. Now that they have got a start, and such a start, nothing can stop their onward progress. Their good qualities alone would send them to the front as beef producers. Our herd has been established over forty years, and every year makes me love them more, and, I am happy to say, every year is a blow to prejudice and an advance to the Herefords. I think in the course of time they will occupy the first rank in Australia. It is a great drawback to Australia, the total prohibition of stock from England, and I am sorry to say that the cattle trade in Australia has been very bad for some years, prices ruling very low, and now many parts of the country are suffering from a severe drought, which will test the Shorthorn hardiness.

"What price do you charge for your best stud bulls, and could arrangements be made to ship at San Francisco, and what would be the probable cost of so doing? They have admitted sheep from America and I do not see how they can prevent cattle from coming. However, I will inquire. I will be glad to send you some information about cattle when I can find time, and opportunity occurs, and I enclose you a copy of a letter received from a large and wealthy squatter, who, being a practical man, makes his opinion doubly valuable. I would like to get from you, if possible, all the numbers of your journal from the start, if you can lay your hands on a copy of each, and let me know the cost and the best way of remitting the money, so I can send it.

"Would be glad if you had an agent here in this country. Continue my subscription. I start this week to New Zealand to recruit my health, and on my return in a couple of months I will communicate with you again, and it is possible that next year I may be able to run across and see you. Yours very truly,

"F. Reynolds.

"Tocal, Patterson, New South Wales, Australia."

The copy of the letter that Mr. Reynolds refers to is as follows:

"F. Reynolds—My Dear Sir: Press of work and absence from home prevented my replying to yours of the 6th inst. I have been working Hereford cattle since 1859, and consider them better suited to the requirements of squatters on inland stations (ranches), especially those situated long distances from market. If in good paddocks of artificial grass, close to market, or stall fed, I consider Herefords quite equal to Shorthorns. They may not, perhaps, be quite so heavy, but they fatten more rapidly and at less cost, and die better than the Durhams—namely, they cut up less waste, and the meat is certainly superior. There can be no doubt of the superiority of the Herefords on stations remote from market. They travel better and do not cut up. Being more active, they can do the long dry stages more easily. They are naturally more hardy, and after a long journey die better. Their meat is less livery, and they don't lose their inside fat like the Durhams. My old drover will tell you that if he had a Hereford in his mob of fats, this beast always looked best at the end of a long journey.

"On large stations where at any time the water is scarce, and stock compelled to travel long distances from the feed to the water and back again, the difference between Herefords and Durhams is very marked, the Herefords traveling in good condition long after the Shorthorns are poor. Again, after a long drought, the Herefords pull themselves together more quickly and get their condition in less time. I have heard men argue that Herefords are wilder than the Durhams and in nine cases out of ten I have found they knew nothing personally of the matter, never having had a herd of Herefords, but they had been told. My experience teaches me that if Herefords are properly looked after, they are actually much quieter in the lots and quite as steady on the runs as the Durhams. I have worked both, and I also know from my own knowledge that a neglected band of Durhams is worse to deal with than Herefords under the same conditions. I am, of course, speaking from actual experience.

"Not long since, with my partners, I purchased three stations in Queensland, all Durham cattle, about 16,000. My first active move was to remove the Durham bulls and substitute Herefords. This will give you, better than a volume of writing, my opinion of the respective breeds after 23 years' work among both; and after a careful study, I have come to the conclusion that Herefords are the best and most profitable all-round cattle, less subject to disease, and better constitutions, quite as steady on the range and steady in the yard. They cross well with the Durhams and are better mothers. A Hereford cow will never leave her calf as long as a Durham. If you wish for a proof of the merits of the breeds, just put 100 of each breed into paddocks equally badly grassed—that is to say, in a dry time—and I will guarantee you will find the Durham fall away much more rapidly than the others; in fact, the Herefords will be in fair condition when the Durhams are actually dying. Both lots, of course, are to be in equally good condition when put in the paddocks.

"In conclusion, I am quite certain if breeders would put aside prejudice and give Herefords a fair trial they would find all I have said true enough, and be content to drop the Durhams in the future.

"You are at liberty to make any use you like of this letter. Yours very truly,

"Vincent Dowling.

"Lue, Rylestone,

"N. S. W., Australia, 1882."

A later letter shows the Hereford movement growing.

Editor "Breeders' Journal": The Sydney Show has just closed. The Herefords were not in large force, but were well represented by drafts from the herds of Mr. Frank Reynolds, who had about a dozen very good ones. Your old friend, Mr. Chas. Price, was a judge at this show on Herefords, and bought a splendid bull calf bred by Mr. Frank Reynolds. The Herefords are growing in popularity, and those who have been breeding for some years are greatly encouraged.

J. H. McELDOWNEY,
Chicago Heights, Ill.

The "Breeders' Journal" is fully appreciated by them, and they are very glad to hear through it of the great success the Herefords are making in America and England. They wish to establish an Australian Hereford Record, and will be glad if you will give them some facts and a plan for doing so.

Mr. Frank Reynolds and Mr. Chas. Price are both having a better demand for their young bulls than ever before. All that try Herefords are fully satisfied that they are the best cattle for our runs, and it looks as though "our time was coming." I shall endeavor to see that you have a largely increased subscription list from Australia. Yours respectfully,

QUEENSLAND.

Adelaide, Australia, Sept. 19, 1883.

HEREFORDS CONSIDERED THE BEST IN AUSTRALIA.

We find in the "South Australian Register" (1883), published at Adelaide, an account of a discussion as to beef breeds before the Royal Agricultural and Horticultural Society of South Australia. In the discussion we find the following from Mr. Chas. Price, of Hindmarsh Island, who had been a breeder of Hereford cattle for nearly or quite half a century, and had a herd of about 200 thoroughbred Herefords, bred for breeding purposes.

J. GORDON GIBB,
Lawrence, Kan.

Mr. P. said he had "had a good deal of experience in the breeding and rearing of Herefords and Shorthorns and had given some careful attention to the relative qualities of the two breeds. The Polled Angus breed he had had nothing to do with. They were all agreed that the best beast was that which would produce the most and best beef on the smallest quantity of feed, and he held that the Hereford possessed these qualifications. It will live and thrive where the Shorthorn will not do at all. He has brought his Herefords out at two and one-half years and that could not be done with the Shorthorns. He had looked over his sales for the past ten years and found that his steers had averaged $62.50, at two and one-half years. He found that Shorthorns would not average that or anything approaching to it. He was positive he could carry three Herefords where he could not carry two Shorthorns. He had one Hereford cow that had produced fourteen calves before she was fifteen years old, and would have her fifteenth before she was sixteen years old, and was now fit for the butcher. He started with four Hereford heifers and bought some Shorthorns. He got a Hereford bull. A friend from Melbourne, named Saddler, who had been buying cattle, came to look at his cattle and told him that it was madness to try to breed such cross-breds with a Hereford bull from those cows, and that he had better get a Shorthorn bull. From his experience he found that his Shorthorns lost flesh, that they only bred once in eighteen months, whereas the Herefords bred once in twelve. Last year when the rains were six weeks to two months late, the Shorthorns would have died where the Herefords looked well. He had no change for his cattle, and there was not one living on that dry feed that was not fit for the butcher. He considered the Hereford the best cross and worth all the cattle in the world for that. It might be put to a scrubber, and even then a good deep breed would be got. With the Shorthorns, long-legged, useless brutes were got. He had been at Sydney Shows for years, and the Hereford bullocks had taken first prize in every year he had been there. He maintained that there was no difference in the weight, and to prove this quoted statistics from the Smithfield Show as follows:

"In the year 1869 the Herefords under two years and six months weighed 1,731 pounds; Shorthorns the same age, 1,648; Herefords under three years and three months, 1,936 pounds; Shorthorns, 1,976 pounds; Herefords over three years and three months, 2,228; Shorthorns, 2,200. In 1875 Herefords under two years and six months, 1,485 pounds; Shorthorns, 1,565 pounds. Under three years and three months, Herefords, 2,036; Shorthorns, 1,982 pounds. Over three years and three months, Herefords, 2,161; Shorthorns, 2,281 pounds. Extra stock, Herefords, 2,524 pounds; Shorthorns and Norfolk cross, 2,290 pounds. Three of the heaviest cattle shown in 1875 weighed, first, Hereford, 2,624 pounds; second, Shorthorn, 2,444 pounds; third, Hereford, 2,420 pounds."

SUPERIORITY OF HEREFORDS AS RANGE CATTLE.

In publishing the "Breeders' Journal," we received a large number of congratulations and commendations from parties in the different parts of the States and Territories, on the course that we have taken in advocating the Hereford interest in this country, and not only from America, but from Australia and England as well. Among these, none were so highly appreciated as one from the veteran Hereford advocate, Mr. William H. Sotham, in which he expressed great satisfaction at our course in exposing the frauds at the Fat Stock Shows. Probably next we esteemed the endorsement of a veteran breeder, nearly 80 years of age, in Australia, who has been a breeder of Herefords in that country during an ordinary lifetime, expressing the warmest sympathy and commendation for the course that we have taken in advocating the merits of the breed that he has given a life work to. In speaking of his experiences in that country, we find that it is not dissimilar to ours in this country; while in some parts of Australia the Herefords early became popular, and are becoming the dominant breed, still they have met the Shorthorn opposition, that has retarded their movements, and it must be admitted that the Hereford men have seemed to lack that nerve and push that was necessary to overcome the opposition; but Hereford breeders in Australia as well as England felt the influence of the Hereford movement in America, and began to move with more strength and system for the permanent establishment of their breed of cattle in that country. We valued highly also a letter from that veteran breeder, Mr. Duckham, at Baysham Court, in Ross, Herefordshire, England. Another from Mr. Hill, Felhampton Court, Salop, England, both speaking warmly in commendation of our efforts. Mr. Hill had quite a trade, and has had for some years, with South America.

About this time we received a letter from Mr. Fulford, of the firm of Fulford & Williams, Springhill Station, Queensland; in referring to a purchase of Herefords from Mr. J. Barling, of Casino, New South Wales, he says: "The cattle arrived at Springfield a short time since, after a journey of nearly 1,500 miles, with only the loss of three head, and were in excellent condition on arriving, notwithstanding that for the first 500 miles grass and water were very scarce, and that quite half the cows were suckling calves. As an instance of the superiority of the Hereford over the Shorthorn as a traveler, I will mention that just as our cattle were passing near Rockhampton a draft of pure Shorthorn heifers started from there for the Flinders River, and traveled almost the same stages for about 600 miles, when they parted company. The Shorthorn heifers were miserably poor, while the Herefords that had traveled from Richmond River, New South Wales, to Rockhampton, then 600 miles in company with the Shorthorns, and then about 300 miles to the end of the journey, arrived in excellent condition, half of them having calves at foot, while the Shorthorns were dry heifers. As there have lately been some interesting letters in the 'Queenslander' in reference to Herefords vs. Shorthorns, I have taken the liberty of sending you these particulars, thinking they may be of interest to some of your readers.

"I may add that we have a herd of about 17,000 Hereford cattle in Lindhurst, not having used anything but Hereford bulls for ten years past, owned by Messrs. Barnes & Smith Bros., of which I have been manager for nearly twelve years, and I think that I may say that for quality and docility the herd will compare favorably with any herd in North Queensland, worked under similar conditions."

Thus was the Hereford movement receiving evidences of prosperity in all parts of the world. A reference to the map of Australia and South America will show somewhat the relation that this movement has to the meat production of the world. Australia as well as South America and our western range from Panama to Manitoba, are great grazing districts, and these ranges must supply the beef of the world. The testimony is but one way by all those who have used the Herefords on these ranges.

HINDMARSH ISLAND AND MR. CHAS. PRICE'S HEREFORD HERD.

The "South Australian Register" (1884) gave an account of the farm of Mr. Charles

BEAU REAL 11055.
Bred by Gudgell & Simpson, Independence, Mo. Favorite stock bull of Shockey & Gibb.

Price, of Hindmarsh Island, and its surroundings. We give this full statement that our readers may see the conditions under which one of the largest Australian herds of thoroughbreds has been bred and reared. Mr. Price was then a man something over eighty years old, and we think was on the island for something over forty years. He was one of the most active

Hereford breeders in the world, and one who took a very active interest in all that pertained to the Hereford breed of cattle, especially in its struggle for merited recognition.

He had the usual difficulties on the show grounds and elsewhere in carrying forward his operations; still, he never questioned the ultimate outcome, and bred (making a fair paying business) for bullock breeding mainly. He, however, built up his herd, retaining the females until he had something over 200 head.

ANXIETY 4TH 9904 (6283).
Bred by T. J. Carwardine, sire of Beau Real.

We give this full statement from the "Australian Register" that it may be seen that it is not the most favorable country in which to develop a breed of cattle, but it will be found that under just such conditions as these, one of the best herds of Herefords in the world has been reared, and Mr. Price lived to reap very large returns in Australia for his breeding stock. With these remarks we give the statement as appearing in the journal referred to:

"Every South Australian who takes the slightest interest in live stock of any description must have heard of Mr. Charles Price and his herd of Herefords, but for several years past very few have had the pleasure of seeing any of the beautiful animals, because they are rather difficult to get at upon their island home.

"Being one of the privileged few whose business it is to see all that is to be seen and is worth seeing, I availed myself of an invitation from Mr. Charles Price and visited his farm at Hindmarsh Island on the 17th ult. Hindmarsh Island is about eight miles long, and averages two miles in width. The River Murray, after passing through the middle of Lake Alexandrina, divides, one arm going down by Goolwa to within about half a mile of the sea, and then runs eastward parallel with the beach, from which it is divided only by a sandhill, for several miles, until it reaches the Murray mouth, where it is joined by the other branch, and the space of land enclosed is known as Hindmarsh Island. There is another large island to the eastward, known as Mundoo Island, and a number of small islands are to be seen dotting the surface of the river in various directions. The water is sometimes perfectly fresh, often brackish, and occasionally perfectly salt, so that the farmers have now and then been put to great straits for want of fresh water, and once a large number of cattle died through scarcity of it. At the same time there were wells existing at the western end of the island, at the base of the sandhills, in which there was and still remains an inexhaustible supply of beautifully fresh, sweet water. These wells are merely oblong pits sunk at the foot of the large sandhills, having planks around the sides to keep the sand from breaking in, and the water is the result of soakage. The soil, as is usual where sandhills exist, is clay at a foot or two below the level of the land upon which the sandhills rest, and the water is chiefly the result of absorption from the hot, moist atmosphere in summer time. When the clay is gone through, in most cases, a brackish or salt-water stratum is encountered. Upon Mr. Price's farm there are fresh pools and salt pools within a few feet of each other; but the salt water is always upon the lower level.

"From the ferry, which is close to the western end of Hindmarsh Island, a metaled road proceeds southward for about a mile, and then turns eastward through the length of the island. The whole of the land from the road to the river on the right-hand side of the road for two and a half miles belongs to Mr. Price, and he rents a large block on the left proceeding from the ferry. Further along on the west side there is a large block which has lately been taken into the possession of Mr. West-Erskine. This land has been terribly mismanaged, and through being broken up too much the fibrous roots of the grasses and other herbage which bonded it together have been destroyed, so that hundreds of acres may be seen on a windy day traveling from place to place, or flying about in the air. If the wind is from the south the river receives the greater portion, but if it comes from the north then it becomes the property of Mr. Price, covering up the herbage to the depth of several feet. In some places for perhaps fifty acres the sand has blown away for several feet in depth, and pieces of limestone stand up like the 'boys' left by contractors for railway cuttings to show the depth to which they have excavated, except that there is no sod left on top, to which the excavators are so partial that they will sometimes even add two or three spits rather than not show enough. The sand which is carried by the north wind into Mr. Price's land to-day may be removed into the river by the south wind a week or two after, and thus a deal of harm is done. The metaled road is also covered in many places with several feet of sand. The Council, Mr. West-Erskine, and Mr. Price are all doing what they can to prevent this undesirable state of things, but it is probable that it will not be stopped until the shifting sands have been covered with boughs and planted with grasses.

"There are very few trees now upon the island, but the settlers are inclined to encourage their growth. The Shea oak seems to thrive there, and in one place where a large tree has been cut down there are a hundred or so of seedlings, so that if the settlers chose to obtain seeds and sow them there would be good hopes of covering the place in a few years. Mr. Price introduced the tobacco tree (really a species of tobacco), which grows to a height of twelve to eighteen feet in favorable localities. This took kindly to the sand and soon every one was anxious to use it along his boundaries as a hedge and breakwind; but the people did not cut it back enough, and the trees grew up with long, straggling branches, which broke down with the winds, and now the plant is being neglected on all hands. I have seen the tobacco tree made into a very effective wall or breakwind, about ten feet high, by continually stopping all branches that break beyond bounds. The Agave Americana, or 'American Aloe' or the century plant, and the common Opuntia, or prickly pear, have also been used with some success for stopping the shifting sands. Attempts have also been made with couch grass and by sowing with rye grasses and other plants to fix the sands, but too often before they get roots the wind comes and blows seed and soil away together.

"Mr. Charles Price has been an enthusiast in respect to Herefords ever since his boyhood, and by this time ought to be a pretty good judge of what a Hereford should be. He believes that he now possesses 200 beasts which will compare favorably individually with anything in the Australian colonies, and as a herd, with any in the world, and probably he is not far out in his estimate of their character. He was brought

PRIZE WINNING HERD OF 1885, PROPERTY OF THE IOWA HEREFORD CATTLE CO.
Modesty 2d, 24284. Melody 16th, 18549. Stately 2d, 18522. Forelock 17999. Washington (8152) 22615.

up at Bringewood, on the Downton Castle estate, which belonged to the late Mr. T. A. Knight, and was acknowledged to be one of the best judges of that breed in those times. This gentleman always kept fifteen picked Hereford cows for the use of the Castle—for milk supply—and of their calves four of the best heifers were annually selected to keep up the herd, and the remaining calves, whether male or female, were sold to such of the tenants on the estate as desired to have them at the uniform rate of

HESIOD (6481) 11675.
Bred by P. Turner, Herefordshire, founder of the Hesiod line.

$10 each, which was the butcher's price for calves.

"Mr. Price's father was a dairyman and his farm was close beside the Castle, so that his sons, who had taken a great liking for the Hereford breed, knew pretty well to a day when there would be a calf for sale, and being always among the people who had to do with the breeding of the animals, they had already gotten a pretty fair notion of what constituted a perfect Hereford. They knew the pedigree of each of the fifteen cows, and could give a good idea of what might be expected from each cow. When the calves were a fortnight old, Mr. Knight decided whether they were to be kept for his own herd or whether they were to be sold and the Price boys generally came in ahead. It was not very long ere they got together a large herd of pure-bred Herefords of the 'Knight' strain, and they found the business so profitable that they gave up the dairy business and devoted themselves to breeding pure Herefords, to which they have continued to give their almost undivided attention ever since, both at home and here, the Knight strain being the foundation of the herds in both places, and the purity of the strain is being maintained with scrupulous care by the family, notwithstanding the fact that many thousands of miles of sea separate the one herd from the other.

"There is no indecision when I have to describe Mr. Charles Price's character. He is most decidedly gone upon Herefords. His sitting room is decorated with numerous portraits of Hereford cattle, and his library is filled with books and periodicals relative to Herefords. He is filled up with Hereford lore, which leaks out at every opportunity. Among the portraits of cattle is one of a bull named 'Cotmore,' which was the property of Mr. Jeffries, of The Grove, Herefordshire, and won the premium at Oxford in 1839. This animal, Mr. Price informed me, was the grandest Hereford bull ever calved, and if the portrait is a faithful representation it seems to fully bear out the encomiums passed upon it. The next portrait is that of Emperor, the winner of the Royal (England) Agricultural Society's Show at Shrewsbury in 1874, and bred by Mr. Thomas Sheriff, of Coxall. Next to this comes the portrait of 'Conqueror,' attached to which is a short interesting history. 'Conqueror' was bred in the same herd as the celebrated 'Cotmore' above mentioned. He was purchased by Mr. Price when he was farming in Herefordshire, and after he had used him for three years he took the bull to market, where he was offered $160 for him by a butcher, but he wanted ten dollars more, and had to take him home again. He let him for seventeen cows, at $5 each, and then took him to Leominster, in Herefordshire, where he let him from October till Christmas for $100, and the next fall let him from Christmas until May for $100.

"The next year the same person hired him for $250 and when that term was expired Mr. Price sold him for $250. Another portrait connected with the history of Herefords in South Australia is that of Bringewood 2981 E. H. B., which was bred by Mr. J. Price, of Bringewood, and was purchased from him by Mr. Charles Price, of Hindmarsh Island, brought out here, used by him for four years and then sold by auction in Adelaide, to Mr. J. H. Angas, for 117 guineas ($585), being then the first and only pure-bred Hereford bull imported into the colony. There are several other portraits hanging around and one group of Herefords and grade Herefords which the "Breeders' Journal" (America) has thought worthy of illustrating in colors. They are very nice beasts, especially the three pure-breds, but there are better on Hindmarsh Island. Gratitude, or love for a perfect specimen of his class, induced Mr. Price to have a portrait of a bull, which he owned quite a short time ago, named 'Monarch,' which was the sire of fully 200 head of calves, each of which was a splendid, level, square and perfect type of what a Hereford should be.

"The pure strain of Hereford has been maintained in this herd all through, as will be seen by the list of bulls employed, since he first commenced in South Australia. First came Bringewood (2981), the best bull that could be found in England, and the first Hereford bull imported to South Australia. As stated before he was sold to Mr. J. H. Angas after four years' service on the Island. Neptune, his progeny, was used for a short time only on account of being too nearly related and Mr. Price not being favorable to in-and-in breeding where a bull of equal quality can be got from elsewhere. The successor was Mount Aitkin, then Hindmarsh Island, and Merman took his place, these being followed in due time by Chief of the Isle—by Mount Aitkin out of Alpha. After him for one year came Goolwa and Neptune, the last having been sent back to the island by Mr. Richard Holland, of Turret Field, South Australia, previously to sending him away to Sydney for sale. Next came Monarch, whose sterling qualities caused his portrait to be taken, which portrait, by the way, is rather marred by exhibiting him full face to the spectator. After him came Myrtle Bank, and then Benefactor, who will very shortly give place to a young bull named Earl Tredegar. Thus it will be seen that during a period of less than thirty years twelve bulls have reigned over the Hereford herd, and if the herd book is turned up it will be found that each has a famous pedigree or ancestry attached to its name.

"Mr. Price landed in South Australia in 1853, and hearing that 2,000 acres were being surveyed on this island he started to have a look at it along with Mr. Samuel Goode (¶ 258), now of Goolwa. It was a wild place then, and the first night they stopped with the blacks, who were fishing. He was pleased with the land—or sand—and having secured what he wanted has never had occasion to regret his choice. He pities the poor fellows who have to grow wheat for a living, and does not envy those who have to grow wool for a subsistence. He thinks he can produce one Hereford where another man would have his work to do to maintain three sheep, and does not think that the cattle waste as much food or cut up the land as badly as the sheep would do.

The Herefords are in splendid condition, owing to the great quantity of grass, which is even now quite green—a thing that has never been noted in the previous thirty years' history of the place—and there is not one cull in the whole herd. One beast is so like the other in shape, size, color and appearance that it is not possible to tell them apart. Mr. Price himself, who is always among the cattle, can only tell by such minor distinctions as would escape the casual observer—such as a small spot of red near the eye or some other trivial distinction.

"I have given the history of the first Hereford bull that was imported into South Australia, but the story of the first cow was not accompanied with so fortunate results. In 1863 Mr. Charles Price purchased the best cow he could find in England, and paid £60 ($300) for her, which was the highest price that had been paid up to that time for a Hereford. She was named 'Maid of Coxall' by North Star, and was bred by Mr. T. Rogers, of Coxall, Herefordshire. She was in calf to Battenhall before named, and had a heifer calf by her side named Queen of the Ocean. On the voyage, during a storm, Maid of Coxall was washed out of her box and both of her hind legs were broken, so that she had to be killed, but Queen of the Ocean was saved, and thus became the first pure-bred Hereford female in the colony. She was calved in 1863 to Bringewood and has produced a calf every year for fourteen years. Her pedigree is found in the Hereford Herd Book, and also in the New South Wales Herd Book.

"Mr. Price calculated upon raising fifty head every year and for some time past his calculations have been verified. Last year he was a little short of his number through an accident to the bull. This was only a temporary affair and this season his calculations will again most probably be carried out. The following is the history of the Queen of the Ocean's progeny:

R. G. HART,
La Peer, Mich.

"Alpha, a heifer by Bringewood, sold for 120 guineas to Messrs. Kirkham Bros., along with her heifer calf Resalama. This Mr. Price informs me is the only cow the firm ever bought, but they also purchased the bull Hindmarsh Island, then aged about five years, at the same time paying 100 guineas for him. Since then the brothers have been famous for their fat cattle, and they are now a very respectable herd.

"Neptune, a bull by Bringewood, sold when a yearling for 150 guineas to Mr. R. Holland, afterwards used by Mr. C. Price, for a short time, and then sent to Sydney, where he was sold.

"Mermaid, a heifer who was shown several times and always took a prize. When about four years old in returning from a show she fell over a cliff and was killed. On the day of her death Mr. Price received a letter from Mr. Frank Reynolds, Tocal, Patterson, N. S. W., offering 150 guineas ($750) for her.

CLEM STUDEBAKER,
South Bend, Ind. Firm of F. H. Johnson & Co.

"Merman, a bull by Bringewood, sold to Mr. W. J. Maidment as a yearling for 100 guineas ($500).

"Chief of the Isle, a bull, by Mount Aitkin (4800), sold to Mr. Haywood Porter for 150 guineas ($750) as a yearling.

"Goolwa, by Hindmarsh Island (4647), a bull, sold as a yearling at Sydney for 82 guineas ($410).

"Queen of the Murray, a heifer by Goolwa (4647), had one calf and then died.

"King of the Murray, by Monarch (206 in N. S. W. Herd Book), sold as a calf for 100 guineas ($500) to Mr. Ramsey, Western Australia, who purchased two bulls and three heifers from Mr. Price.

"The next two bulls were made into steers, in consequence of no purchasers being ready for them.

"Queen of the Murray, by Monarch, now rising two years and in calf to Benefactor.

"Altogether, Queen of the Ocean has produced ten bulls and four heifers, seven of which realized over 800 guineas ($4,000), and what with prize money and other odds and ends in the shape of steers and heifers still unsold, her progeny has proven to be a valuable source of income.

"The next cow worthy of special notice is Effie, by Bringewood, out of Pigeon, imported from Tasmania. Pigeon was purchased by Mr. Price from Mr. John Chambers in 1853, along with two other heifers, for $180. She is a splendid animal, and was a great bargain. Pigeon's first calf was a heifer named Jennie Deans, which was sold with a calf to Mr. J. H. Angas for 200 guineas ($1,000). Effie was the next offspring of Pigeon and was dropped on January 9, 1872, and had her first calf on December 3, 1874 (Dolly Varden, by Merman). Since then she has given a calf every year, and is now in calf again. She is a splendid cow, with every point from head to tail, from back to hoofs, hair, horns, skin, color and everything else to character, but this is equally true in respect to the whole herd. They are so alike in every particular that a stranger can scarcely distinguish the worst from the best.

"Sturt Pea is another beautiful cow, got by Monarch, out of Effie, by Bringewood. She was calved on October 17, 1877. Her first calf was Sweet Melon (another first-class beast, now running on the farm, and one of the grandest heifers that was ever calved). On January 15, 1882, she gave birth to a bull calf, which was sold along with eighteen other bull calves to Messrs. Grant & Stokes for a run up in the far north. Her last calf by Benefactor was also a bull and was dropped on July 30, 1883, and she is now in calf again. Her progeny are all like the rest, without a bad point in any part.

"Effie 2d was calved on August 15, 1879, by Monarch, out of Effie. She is full sister to Sturt Pea, and is now in calf to Benefactor.

"Little Dorrit, by Mount Aitkin, out of Jennie Deans, took the ten-guinea cup at Adelaide some years ago for best Hereford cow. She had her first calf in 1877, and has given a calf regularly every year since. Little Dorrit 2d, by Goolwa (4617), Little Dorrit 3d, and Little Dorrit 4th are all the progeny of this fine cow, and are all good specimens of pure-bred Herefords.

"Both the bulls now in use on the farm were

purchased in Sydney at the time of the late Melbourne Exhibition and are of good pedigree and splendid character. The following particulars will be interesting to breeders:

"Benefactor, calved February 16, 1876, bred by Mr. F. Reynolds, Tocal, Patterson, N. S. W., sold to Mr. F. C. Coyder for $375 on April 25, 1877, and purchased from him by Mr. Charles Price, who has had him three years. There are between 60 and 70 calves on the farm belonging to him, and another 60 or 70 on the road for the present season. He took the first prize at the Maitland Show in 1877. Benefactor, by Sir James (4975), by Royal Hero (4094), dam Fairy, by The Captain (1409), imported g. d. Fancy, by Thurston (1422), imported g. g. d. Favorite, by The Captain (1409), g. g. g. d. Wanton, imported, by Radnor (1366).

"His successor, just commenced duty, was also purchased by Mr. Price, on August 30, 1883, from Mr. F. Reynolds. His pedigree is as follows: Earl Tredegar, calved at Tocal on September 17, 1883; sire Dale Tredegar (5856), imported; dam, Last Day 1st, by Sir James (161 N. S. W. H. B.); g. d. Last Day, by Sir Hercules (160 N. S. W. H. B.); g. g. d. Eattie, by Garibaldi, imported; g. g. g. d. Rebecca, imported (N. S. W. H. B.—Vol. XI, p. 393, E. H. B.) Last Day is dam of Last Day 1st, 2d, 3d and 4th, all noted good animals. Last Day 1st won first prize at the Singleton Show in 1878; was very highly commended at the Maitland Show in 1879; took second prize at Sydney in 1879, and, coupled with Fancy 1st, won the Challenge Cup given by Mr. Charles Price, of Hindmarsh Island, for the best pair of yearling heifers of any breed. She took the second prize at the Singleton Show in 1879 and the first prize at the Maitland Show in 1880. Her daughters, 1st, 2d, 3d and 4th, have taken a number of prizes at various shows.

"To those who are pleased with the sight of really well-bred animals, and others who are desirous of learning all that is known about Herefords, I can confidently recommend a visit to Hindmarsh Island."

AN AUSTRALIAN CATALOGUE.

We once received a catalogue of Hereford cattle bred by Mr. F. S. Reynolds, of Tocal, Patterson, New South Wales. The Hereford bulls were fifteen in number. Many of them are by Dale Tredegar, a bull imported from England. Dale Tredegar (5856) (¶ 259) was calved August 14, 1879, and was bred by Mr. H. J. Bailey, Rosedale, Tenbury, England, was got by Tredegar (5077) 2478; dam Rosa 2d, by King of the Dale (3891). This bull, under the name of Dale Tredegar, has his picture at ten months old in Vol. XI of the English Herd Book. He won first prize at the Royal Agricultural Society's Show at Carlyle in 1880, second at the Bath and West of England Show at Worcester in 1880, first at the Gloucester Agricultural Society's Show in 1880, first at the Hay Agricultural Show in 1880, and third at the S. & W. Midland at Bridgenorth in 1880. It gave us great pleasure to receive a Hereford catalogue from this distant country, indicating as it did the Herefords were making their way to the front in all parts of the world.

AN AUSTRALIAN SHOW.

We have an account of a show held in Australia. We quote from the "Sydney Mail" of April 15, 1884, of the meeting of the Agricultural Society of that city. Of the Cattle Department it says: "In the rosy period between 1874 and 1878, when four figures for a well-bred bull was no uncommon figure, the Agricultural Society was each year favored by very large entries. Alfred Park has had as many as 450 head displayed, and its sale rings used to be crowded. The show afforded proof that in the matter of breeding, low prices have not proven potent enough to lower the stock's quality. In the half hundred or so of the Shorthorns shown, there were several animals which bear comparison with the champions of past years. The score of Herefords show distinct evidences of progress, and the Devons, of which there were about ten, prove unmistakably that the hardy red breed are still favorites. There were over twenty Ayrshire cattle. The judging commenced on Saturday morning.

J. S. HAWES.
Reading, Mass. (Formerly of Colony, Kan., and Vassalboro, Maine.)

We note the awards of the Herefords as follows: Bull, three years old and over, F. S. Reynolds, Kenilworth; for two years old and under three, first and second, F. S. Reynolds; bull one year old and under two, first to Percy Reynolds, second to F. S. Reynolds; bull calf six months old and under twelve, F. S. Reynolds; cow three years old and over, F. S. Reynolds; heifer two years old and under three, Percy Reynolds; heifer one year old and under two, F. S. Reynolds first, W. McLeod, second; heifer calf six months old and under twelve, F. S. Reynolds first, H. McLeod second."

THE "BREEDERS' JOURNAL" ON AUSTRALIA.

Commenting upon the Hereford situation in 1884, we said in the "Breeders' Journal": "Australia is the one competitor with the American continent for the European trade. The inland lines on which cattle must be moved in Australia are as long and as expensive as those in America. The water transportation is an average of twelve weeks or more, while the water transportation from America is less than that many days. The quality of the Australian cattle to-day is better than that of the American cattle from the plains. This gap must be closed; that is, the quality of the plains cattle must be improved, and this can be done no way better than by use of Hereford blood. This for two reasons; there is no other race of cattle so hardy, and which adapt themselves so well to the conditions to which cattle must live under as the Herefords. And whatever may be said of other breeds claiming merit for this purpose on the ground of grazing, there are no other cattle in the world that can compare with the Herefords. This being true the future of Hereford breeding is a sure success.

SIR EVELYN (7263) 9650.
Bred by T. J. Carwardine.

"When these facts are fairly considered and the other facts that are behind these there is a certain and sure profit in the bullock breeding, we can say to all inquirers as to the future outcome of Hereford breeding, make it as large as you can, as successful as you can, and you will always find very large returns."

A LETTER FROM AUSTRALIA.

We received the following from Mr. Price: "T. L. Miller: I must now tell you that I have had an oil painting taken of two of my breeding cows. They were full sisters, got by a Lord Ashford bull. I have sent you a rough copy of them, as I want you to tell me what you think of them. They were taken at the end of the summer, when they had nothing but dry grass for the last five months. They have never tasted a bit of anything in their lives but what they got from pasture, where all my herd run together. You will see that Effie 2d is in better condition than Sturt Pea, as she did not suckle her calf. Sturt Pea suckled a great calf all through the summer. I do not know, but I think I never saw a better Hereford cow than Effie 2d in my lifetime. Most of my cows and heifers are descended from the same strain as these two cows. I think I told you in my last that I had offered ten two-year-old heifers for 25 guineas each. The gentleman took them and four bull calves to commence a Hereford herd. I hope they will manage them well; they are a grand lot. I must now close, hoping this will find you well. I am now far gone in my eightieth year, enjoying first-rate health, taking pleasure in my 'white-faces,' as I have got them to such perfection.

"Yours very truly, CHAS. PRICE.

"Hindmarsh Island, Australia,

"April 27, 1885."

The pictures spoken of were received in good order and they certainly represented grand specimens of beef cows. It hardly seemed possible that they have not been fed grain. They were an inspiration to make it a life-work to breed as good Hereford cattle as has Mr. Price. What can be more satisfactory than for a man to have around him a herd of cattle that he has bred himself, which for massive beef qualities cannot be excelled.

LIGHT ON THE AUSTRALIAN SITUATION.

"Editor 'Breeders' Journal':

"I received a letter from you dated October 15, 1884, saying you have received my name from Mr. Charles Price, Hindmarsh Island, S. A., and a slip of the 'Queenslander,' containing a letter of mine on Herefords from Mr. A. J. McConnel, and I feel quite pleased to think it was made use of by you in your 'Breeders' Journal.' I would have replied to that letter sooner, but wished to be in a position first to give you an account of our doings for half year just past on this station. I may first inform you that we are experiencing our second season of drought, and as the cattle in this country on the large runs (ranges as you call them) have nothing but the natural grasses to subsist upon, of course a short supply of rain means a short supply of grass.

"We have on this station about 22,000 cattle, and they have about 1,000 square miles of country to make use of, the greater extent of which is good grazing country, and I think our management of cattle is very similar to that on your ranges; but I think our cattle do not get so mixed up in this country as they do on your ranges, as in the more settled districts here a great deal of fencing is done; more for sheep than for cattle, to save the cost of shepherds, as sheep in unfenced country without shepherds would be useless, while cattle, if bred in a certain locality, will not roam far, even in unenclosed country.

"I am glad to say that our cattle are standing the effects of the drought exceedingly well, and last year we were able to muster fat cattle at a time of the year (November) when no other station in the district could do so, which I attribute to the fact of our cattle being Herefords and the others in the district Shorthorns. The herd here were originally Shorthorns, the station having been formed in 1863 with Shorthorn cattle from New South Wales. I took the management of the herd in 1870, and the year after we commenced with Herefords, both to breed them pure and putting pure Hereford bulls into the general herd.

"We started our pure herd by purchasing 15 pure heifers from Mr. Frank Reynolds, of Tocal, N. S. Wales, and purchasing a bull named Cato, whose dam, Green Lady, was bred by Mr. Green, of Hereford, England, and imported to this country in calf with Cato, his sire being Julius Caesar (3187). Cato was purchased at two years old in Sydney, and was in use here nearly 13 years, and got us a very fine lot of stock. He was only fed two winters (on oil cake), and I think would have been alive now if we had had suitable feed for him, but having no cultivated land here (except a vegetable garden), and no means of getting fodder from the coast except by horse or bullock teams, at about $125 per ton carriage, we do not as a rule get up any fodder. We have had other stud bulls in use also, bred in N. S. Wales, and added to our stud females by the purchase of about thirty pure cows and heifers from the same colony. We have now about 250 pure Hereford cows and heifers and all the bulls we have in the general herd are pure Herefords, and have been for the last ten years, and the herd now is getting to be all red, with white faces, as I spay the cows when 8 or 9 years old, and also cull and spay all inferior heifers every year.

"We got some rain last December and January that caused a spring in the grass that enabled us to commence work among the cattle, branding and spaying, etc. Since January 28th we have branded close on 4,000 calves, spayed 1,400 females, mustered 360 fat cattle for butcher and 1,100 fats to go to be boiled for their hides and tallow to the coast, and have still about 500 more to muster for latter purpose.

"Our great want in North Queensland is a certain and regular market for fat stock. We have not a large enough population to consume one-half of the stock fattened here, and owing to lack of railways, we have no means of taking them to the principal markets to the south of us, and to take cattle to Sydney overland means about 1,500 miles of a drive, and to Melbourne, a great deal further.

"I consider we are fully 50 years behind you in America in advancement, and we seem to be still crawling on our hands and knees instead of marching. You seem to manage matters more systematically in America than we do, and have better plans of thoroughly testing the capabilities of live stock than we have in this country.

"You will scarcely credit that there are no cattle ever weighed alive at the stock shows in Queensland, and no prizes given for the best carcasses of beef, and, generally speaking, judges at shows go for size more than quality in the fat cattle classes.

LOVELY 2D (V. 15, p. 299), 21977.
Bred by R. W. Hall, Herefordshire. A celebrated Michigan winner.

"I exhibited some fat bullocks about a year ago at a stock show about 200 miles from this, and won every prize I competed for but one. That was for a pen of three bullocks from $3\frac{1}{2}$ to 4 years old. I met five pens of Shorthorn bullocks and prize was given for the heaviest cattle. Cattle were slaughtered and weighed the day after the show. A pen of Shorthorns won that dressed an average of 1,005 pounds each; our three dressed 988 pounds each. But our cattle traveled 100 miles further to show than the Shorthorns, and of course the whole distance was traveled by the cattle on foot. I wanted to bet the owners of Shorthorns that ours were the best butchers' and consumers' cattle and leave it to the decision of the butchers who purchased both pens, but they would not make the bet, but agreed to leave the matter to the judgment of two butchers whom we knew,

CASSIO (6849) 13352.
Bred by P. Turner, Herefordshire. Rated the best bull ever seen in Canada.

without having a bet on it. The butchers, not knowing which was the Hereford carcass and which the Shorthorn when hung up in shop, decided that the Hereford carcass was far and away the best. I may say that we decided to each select one carcass from our three, and have the decision on that. I then offered to enter into a sweepstakes of from $100 to $250 each and show a pen of bullocks this year same ages (3½ to 4 years), the prize to be awarded to the cattle that show the best beef for butchers and consumers.

"The Shorthorn men would not enter into the matter on those terms, so I decided not to be an exhibitor this year. The following were the classes we won in last year: First prize for best fat bullock in the yard, any breed or age; first prize for best pen of three fat bullocks, any breed or age; first prize for best pen of four fat bullocks under 4 years old; first prize for pen of three heifers from 2 to 3 years, any breed; first prize for pen of six fat Hereford bullocks, any age; first prize for best Hereford cow bred in district; first prize for best Hereford bull bred in district.

"Our pen of three Hereford heifers beat a pen of Shorthorns that were awarded a prize of $500 for the best pen of Shorthorn heifers. The latter was a great honor, as the pen of Shorthorn heifers were the best that had ever been shown in North Queensland, and were stud cattle out of a stud herd, while our three heifers were grade heifers (about three-fourths bred), out of our general herd. I could have shown pure heifers from the stud paddocks very much superior to those I won with, but they were too valuable to travel 200 miles to a show and the same distance back again.

"With regard to the great opposition to Herefords from the breeders of Shorthorns in this country. I fear it will last much longer than it has with you in America, simply because the breeders of Shorthorns outnumber the breeders of Herefords in this country by at least five hundred to one, and our leading journal, the 'Queenslander,' stopped all correspondence on the subject last year. We want someone like yourself in this country to start a journal in the interest of Herefords. However, I hope a great deal of good will arise to the breed in this country, when people are more acquainted, generally, with the great success they are having with you in America, particularly if they go on prospering and giving satisfaction on your ranches and ranges, where, I presume, your cattle have to exist on the natural grasses as they do in this country.

"I am quite sure they are gaining ground in Australia, and that Shorthorns are gradually losing favor with cattle breeders, particularly in the poorer pasture districts. Of one thing I am certain, that is, that pure Shorthorns have decreased in value quite 50 per cent. in Australia in the last five years and as no cattle are permitted to be imported from England, it looks bad for the breed.

"One of the largest stud herds in Australia, and also the very best, has lately been sold at auction and realized very poor prices, compared with what cattle from the same herd brought at annual sales four or five years ago.

"I refer to Messrs. Robertson Bros., Colac, herd in Victoria. About eight years ago they bought out another celebrated breeder called Moreton, of Mount Derrimut, and gave him close on $200,000 for about 400 head of pure Shorthorns (his entire herd), and now I see, looking over the sale lately held, that their whole herd of cattle, numbering over 850, nearly all of which were pure bred and 350 of them stud cattle, brought a total of $95,575. In this number were included several out of and got by imported high-class English cattle.

"The highest price brought by a bull was 525

guineas, and highest price for a cow 420 guineas, but the average for 27 stud bulls was a fraction over $410, and for 72 stud cows $440, those being the very pick of the herd.

"There are very few stud Hereford herds in Australia, not more than about ten, I think, and the largest number are in N. S. Wales. However, in the last few years a few of the largest Queensland cattle owners have taken up Herefords, and have been using nothing but them in the general herd, but the majority are still firm believers in the Shorthorn. The Devon is also coming into favor and Shorthorn men fancy that breed crosses better with the Shorthorn than Herefords do. But it is nothing but prejudice that makes them think so, as it has not been thoroughly tested in many parts of the country.

"I see you use Hereford grade bulls very largely in America. I confess that I am surprised at that, as we cattle breeders in Australia are of opinion that it is a great mistake to use a grade bull, particularly a grade Hereford or Devon, and I must confess that I have always seen bad results in this country from so doing, as cattle so bred go to be inferior in the long run. All breeders of any note in this country use nothing but pure bulls. I have forgotten to mention that we have a great deal of the same Hereford blood, both in N. S. Wales and Queensland, as Lord Wilton has in his veins, as the firm I am connected with imported two bulls named Lord Ashton and Lord Brandon, some years ago. The former by Adolphus, dam Lady Ashford, by Carbonel (1525), and the latter by Avon (2393), dam Lady Brandon by Brandon (2972), g. d. Lady Adforton, by Pilot (2156), the last two mentioned cows being the g. d. and g. g. d. of Lord Wilton.

"I fear the length of my letter will tire your patience, so I will stop. I enclose ten shillings to pay subscription to the 'Breeders' Journal' for this year, and if sufficient, for 1886, also, and hoping to hear from you again at no distant date, I am,

"Yours very truly,

"JOHN FULFORD.

"Lyndhurst Station, North Queensland, Australia, June 13, 1885."

STAND THE DROUTH IN AUSTRALIA.

We have the following from Australia, which we commend to Hereford breeders, ranchmen and farmers of America. We would especially call attention of range cattlemen to the facts set forth in these letters and articles. If the range cattle can protect themselves from the storms of winter and the drouths of summer, steady and satisfactory profits can be realized. Another fact as shadowed forth in the articles is that Australian breeders usually, if not always, look for thoroughbred sires to put with their herds.

In Australia, as in America, it took hard times to drive away erroneous prejudice and predilection, that prevented the dispassionate consideration of the value of breeds, as the following correspondence will show:

"Editor 'Breeders' Journal':

"I send you two letters I lately issued in our local press. My challenge, you see, was not taken up. Herefords are steadily gaining ground, and there is a good demand from drovers for our coast country; Shorthorns nowhere.

"This awful drouth is not over yet. Sheep dying in millions; cattle by thousands. It has been on now near three years. The openings are gradually drying up and the big water holes getting lower and lower. Our bullocks at Lales Creek generally average 820 lbs. dead carcass; this year they will hardly average 700 lbs. I fear our interior will yet be a desert, as it must have been when our early explorers, between

H. R. HALL,
Orleton, Herefordshire.

1830 and 1850, went out and returned with the report, 'All desert.'

"Our sugar industry has 'gone to pot.' The German beet sugar and want of reliable black labor has closed many large plantations.

"Yours faithfully,

"BEARDMORE OF TOOLOOMBA."

The following are the letters referred to:

"*Editor 'Daily Northern Argus'*:

"I see by newspaper correspondence that many people would like to see the trial between Herefords and Shorthorns come off, and I think it just a good time to challenge Mr. Peberdy to produce his 20 cows against mine. The season has been the worst ever known up here. On the 29th of November last, the first rainfall for many months, and my Hereford females were very poor, nearly all of them having to feed their calves and support themselves on the bare ground. I have had since then about one month's fair rain, but no floods and no wet season, and strange to say, although the grass appears O. K. it has none of its usual fattening qualities in it. There could scarcely, therefore, be a better time to show the hardiness of these Herefords and their vast recuperative powers.

H. W. TAYLOR,
Showle Court, Herefordshire.

"I now challenge Mr. Peberdy, of Jelimbah, to bring down to the next race at Rockhampton ten Shorthorn cows with their ten sucking calves not under four months old, and ten dry Shorthorn cows (may also have calves if owner prefers, must be breeders), to show against a similar lot of Hereford cows from my herd. The stock book to be produced so that the judges can see they are pure bred. The cattle on both sides to be driven by land, and not railed, to the ground, and as I live nearer to Rockhampton than Mr. Peberdy, I bind myself to drive mine sixty miles extra to make me even with his distance. The judges to decide which are the best breed of cattle for the Central District, and the loser to give the winner a ten guinea cup properly engraved, and five guineas to the hospital.

"I shall also give Mr. Peberdy the right of examining my drovers as to whether I fed my cows on hay during the drouth, or whether I had them slung up in slings in trees to keep them up, and I shall expect the same from him, as I understand he had a very suitable tree near his house occupied with a cow in slings, and had to import hay from Melbourne to keep them alive. The Herefords did not require that. I have only my own herd to pick from; Mr. Peberdy has the pick of N. S. Wales, as he has bought his cows from several herds, and I don't bind him to produce cattle of his own breeding.

"Yours, &c.,

"BEARDMORE OF TOOLOOMBA.

"May 7, 1885."

"*Editor 'Daily Northern Argus'*:

"Mr. Peberdy's reply on the 14th May to mine of the 7th is what I expected, declining the challenge, abusive, spiteful. The gentleman evidently forgets that abuse is not argument, and that this is not the first time he has made an otherwise interesting controversy a medium for personal abuse. I shall write no more in reply, but I shall state a few facts for the benefit of the public.

"Mr. Peberdy might also with advantage remember that running down my cattle does not raise the value of his or further the interest of the Shorthorn breed. Mr. Peberdy also introduces a novel idea that those gentlemen who do not win the cup or stakes contended for shall advertise in the papers that they have not gained it. Will he kindly set the example? He also makes the same mistake made by several Shorthorn men before him, viz., of stating that the Hereford men have got a dose of Durham in their cattle, when they put a few fine cattle in the show yard. Will Mr. Peberdy tell us what quality a Hereford can get from a Shorthorn? Is it color, constitution, traveling capacity, smallness of bone, decrease of offal, depth of brisket? What quality is it? I only know one thing that might be gained (I speak for myself only) viz., a square hind quarter and with it perhaps a few more pounds in weight at the expense of all the foregoing qualities. To conclude, is not this letter-writing of Mr. Peberdy's a cheap way of advertising that he has some 200 cows to dispose of?

"Three mobs of bulls have lately passed Tooloomba from the south, one from the Logan of 70 head, very poor and dying, one of 300 from the Clarence poor, not fit for work, not one sold from the time they left, mostly Shorthorns, and the few half-breed Herefords in them far better in condition than the rest, and lastly a mob of pure-bred Herefords from the Clarence, 200 strong when they left. Of these some ten head were Shorthorns, and the drovers had to sell them for what they could get, because they could not travel with the Herefords, 150 Herefords he sold on the road up to Rockhampton, and as he passed Yaamba, Mr. Shannon, of Salt Bush Park, a Devon breeder, met them, and was so struck with their general condition and beauty after traveling 850 miles, that he bought the lot, 50 head, and a lucky man I consider him to be. Some of those fifty were actually fat and all fit to put in a herd at once. Some 1,200 head of bulls have lately passed Rockhampton, and this lot of 200 Herefords is the only one sold out. No one seeing these three mobs of bulls could help being struck by the difference in their condition on a long journey, and it would appear also that these letters on Shorthorns and Herefords are doing their work, for while not a Shorthorn was sold, the whole of the Herefords were cleared out.

"The last two mobs of bulls came from within eight miles of one another. I would also ask Mr. Peberdy if it is true that Mr. John Living is going in for Herefords? No man in this district has spent more money on or brought up better Shorthorns than Mr. Living, but this last two years' drouth seems to have shown him that Mr. Wright's Nalabia Herefords alongside of his (Living's) Shorthorns have proved the hardier and better cattle.

"Wishing now the gentlemanly (and the printer's devil need not put a ? after it, for it is not required) Mr. Peberdy a pleasant three month's western trip. Yours, etc.,

"BEARDMORE OF TOOLOOMBA.

"June 2, 1885."

THE DURUNDUR HERD.

Some very interesting history of the Herefords in Australia is supplied in the following letter:

"Editor 'Breeders' Journal':

"Sir: I enclose a postoffice order for 5s. in payment of postage of the three volumes of the 'Journal' that you sent me. They arrived a few weeks ago. I called upon Gordon & Goch, and inquired about the pictures of Success. They informed me that the pictures had arrived, and they would distribute them among people who would appreciate them. I received one some time ago with the 'Journal.' I have had it framed. Some little time ago a Hereford Herd Book society for the publication of Australian pedigrees was started in Sydney. I requested the Secretary to send you a copy of the rules, etc. Unfortunately Hereford breeders are few and live far apart in Australia, so we cannot meet as often as we should, and form plans for pushing the breed to the front. At present the continued drouth prevents the sale of cattle on any post. When this drouth ceases there is no doubt there will be a demand for bulls to replace those that have died, and it is probable that Hereford bulls will be used by many who formerly had Shorthorns. The former cattle are proving themselves best adapted for this part of the world, where the seasons are so trying. We also find they mature early. It is rarely that we keep a Hereford bullock to four years old, we sell a great number at three years old. In fact, it is only a question of size; the condition is always there. In years past we owned a very well-bred herd of Shorthorns, the direct descendants of imported stock. They were excellent cattle, of good quality, first-rate color, with big bodies, on short legs; but in a few years the constitution gave way, they became delicate and narrow, in spite of money being spent and care taken to change the bulls frequently, and taking care to get good ones. In 1872 it was decided that Hereford bulls should be introduced and the breed adhered to. The result has been quite satisfactory. We breed and wean a greater percentage, the cattle are quieter, and we can keep more stock on our freehold. It has been interesting to notice the gradual change of the herd as each successive draft of pure sires has been introduced. About the second cross a good number of the females were unshapely, bad colored, and of nondescript appearance. The pure bulls again corrected all that, and now we have reached a stage in which 80 per cent. of the calves are properly marked with characteristic colors and appearance of pure Hereford cattle. They are hardy, vigorous and fat from the start, and in our opinion there is no breed like them for making prime bullocks at 3½ years old, weighing 750 pounds dressed, and this is to be done entirely on grass. All our bullocks are fattened on grass. We buy and fatten a great many store bullocks, and we notice that the few Herefords we get in this way fatten about two months sooner than the Shorthorns. We fatten store bullocks on an average of about nine months. The timber haulers in our neighborhood prefer Hereford to Shorthorn bullocks for their work, and give us five dollars per head more for them than for other breeds. At our shows at Brisbane we were fairly successful with our Hereford cattle. Unfortunately there is not much competition in the Hereford classes. There is, however, a prize given for the best bull in the show, of any breed.

T. S. MINTON,
Shropshire, Eng.

Last year we showed our bull Blood Royal (7424), but as he had to walk fifty-two miles from here to Brisbane and from Brisbane back again, we could not get him very fat. He was beaten by a Victoria bull that came by steamer. He was a very fat animal that could scarcely walk; however, he beat our bull. Last August we showed our bull Prince Leopold under the same conditions as before. This time, however, we were more successful, beating animals of Shorthorn, Devon and Angus breeds from the best herds in Australia. Our bull is nine years old, and weighed seven pounds under a ton. He walked home in five days, and did not seem to feel the journey in the least.

A. E. HUGHES,
Wintercott, Herefordshire.

"I enclose some particulars of our herd. We have a catalogue in the press, and when it is printed I will send you a copy.

"The Cressy Company's herd, from which a good many of our stock are descended, was formed in 1825 by the purchase of a bull and three cows from one of the best herds in Herefordshire. Unfortunately their pedigrees have been lost, but all the cattle bred from them were recorded in the private herd books of the company. Good cattle were bred from these cows, bulls in the early days selling for 80 and 100 guineas.

"The following imported sires were used. The pedigrees of many of them have been lost:

"1. Billy (4353), imported in 1825.

"2. Cressy 1st, imported in 1837. He was bred by Mrs. Jeffries, of The Sheriffs, near Knighton, Herefordshire, and was purchased by Mr. J. D. Toosey when a yearling for 80 guineas.

"3. Trojan (4384), imported in 1833 by Mr. Edward Bryant.

"4. Trojan (5083), imported 1840 by Mr. Thomas Williams, and sold on arrival for 300 guineas to Mr. Toosey.

"5. Hereford, imported with Trojan (5083).

"6. Cronstadt (1198), imported by Mr. Robt. Keate, with the cow Cressida, was bred by E. Williams, Llowes Court. She was by Glasbury (709), dam by Quicksilver (353). She was a fine cow and had good calves.

"7. Zealous (1822), imported by Mr. J. Cox, of Clarendon. This well-bred bull cost 400 guineas in England.

"Mr. Toosey used bulls of his own breeding, and had two fine animals, one called Garibaldi, by Garibaldi (2005), and another called Undergraduate, by the Oxford Lad (4192). Undergraduate was sold for 300 guineas and is now in New South Wales.

"Many of the Hereford cattle in Australia are descendants of this fine herd, which is now dispersed, the late owner, Mr. J. D. Toosey, having died a year or two ago, aged about 80 years.

"It is much to be regretted that our ports are closed against foreign cattle, or we might take advantage of the valuable importations to America that have been going on for some time past. Yours faithfully,

"A. J. McConnel.

"Durundur, Brisbane, Queensland, Nov. 8, 1885."

"P. S.—I omitted to tell you the following interesting conversion of a Shorthorn breeder into a breeder of Herefords.

"A cattle-owner breeds very good Shorthorns, and his neighbor very good Herefords. The two herds mix on the boundary of the respective runs. The owner of the Shorthorns musters his cattle on the boundary to sell the fat bullocks to the butcher, and in so mustering unavoidably collects on the same camp his neighbor's Herefords. To his dismay he finds the butcher at once picks out all his neighbor's bullocks and drafts them first; then the butcher takes the Shorthorns. This has been too much for the breeder of Shorthorns. He has sold his Shorthorns and replaced them with Hereford bulls."

PEDIGREES OF HEREFORDS IN AUSTRALIA.

We find in the "Breeders' Journal" for 1885, taken from the "Queenslander," the following account of the Hereford herds in Australia. There was a plan for establishing an Australian Hereford Herd Book, and we sent to the breeders who had the matter in charge copies of the "American Hereford Record," so far as they were then published. The Association of

Breeders assumed form and they received applications for entry. They, however, found that their breeders had not kept records in all cases, and in some of them, where the records have been kept, owing to the death of parties, the records were lost. There is one fact clear, that many of the best of the breed have been taken from England to that country, and that each herd can show, not the detailed data that would be desirable, but, as a rule, they can show the line of sires. It will be seen that the first importation was in 1825, some twenty years before the Herd Book was established in England.

There is no doubt that herds referred to here are pure bred and the editing committee probably found not much difficulty in determining what were pure bred, and they should give the facts as fully as possible upon which they admit animals to entry, and unless the breeders showed that they raised only thoroughbred sires and dams of the Hereford breed their animals would not find places in the public record.

The admitting of animals of doubtful pedigree is not raising their value, but is lowering the value of all better-bred stock, and a loose policy at the beginning will be felt for all time.

Commenting upon the proposed herd book at the time, we said in the "Breeders' Journal": "There will probably be clear evidence, as it permits of breeding in the majority of the Hereford herds of Australia, although special dates may be wanting, and this evidence should be spread upon the record as fully as possible. We have a right to an interest in the action that the Australians shall adopt at this time, and we urge upon their attention the fact that the policy they now adopt will be felt for all time. It will be felt in the general reputation of their herds, and, what is of greater importance, it will be felt in the character of the herds themselves, or, we might say, in the character of the individuals in their herds. There is now a uniformity in the individual character of the Herefords that does not exist in any other breed. This should be kept in mind, and cross-bred animals will damage this individual merit.

"If, then, the Australian breeders shall have in mind the ultimate good and value of the breed, rather than the supposed value of individuals, in securing a place in their record for animals of doubtful merit and of doubtful breeding, then they will be laying the foundation that will result in great good to the stockmen of their country.

"The Herefords are to occupy the lead in beef production the world over; and we would that we could impress this truth upon the Australian breeders at this time, so that they might fully realize the necessity of building upon such foundation as will stand the test of time, and of the closest scrutiny. With these thoughts we give the article from the 'Queenslander' referred to:

THE FOUNTAIN HEAD OF AUSTRALIAN HEREFORDS.

"There are very few, if any, Australian Hereford herds that do not trace back either directly or indirectly to the Cressy Hereford herd of Tasmania. As this breed of cattle is coming so rapidly to the front, particularly in our coast districts, a brief history of that herd will be of interest to many. Unfortunately the work of compiling a public Herd Book had not been undertaken until the Cressy Herd Book, which was so carefully kept by the late Mr. James Denton Toosey, had been lost, a circumstance which has rendered the work of tracing the pedigree of the herd one of considerable difficulty. Fortunately, however, Mr. A. J. McConnel, of Durundur, a gentleman who is well posted up in Hereford lore, is in possession of a number of letters written by the late Mr. Toosey a short time before his death, and these, which have been kindly placed at our disposal, throw considerable light on the origin and history of the herd. The Cressy Company's herd was founded by the late Mr. Toosey, who

"THE LEEN," PEMBRIDGE, HEREFORDSHIRE.
Home of the Turners.

brought with him from England a bull and three cows. They left England in November, 1825, and were landed in Hobart Town in May, 1826, and were thus the first Herefords that arrived in the Australian colonies. Mr. Toosey had a number of other cattle and horses in his charge by the same ship. The bull's name was Billy, entered in the English Herd Book (4353). The cows were Matchless, Beauty, and one unnamed (the dam of Diana). The pedigrees of the cattle have been lost, but Mr. Toosey states that they were purchased at high figures from one of the first breeders in Herefordshire, and that he bred some of his best stock from them; bulls in the early days bringing from 80 to 100 guineas. Beauty and Matchless were splendid cows, and much admired in the herd. Diana was calved in 1828. She was a very fine cow, and was sold by auction in 1841. Her dam died in 1829. Beauty died

"LYNHALES," HEREFORDSHIRE.
Home of S. Robinson.

early in 1832, having bred five calves. Mr. Toosey used Billy for some years, and then replaced him by his son, Comet, who was born in 1830. Comet's dam was Matchless. He was a splendid bull and good stock-getter. Comet was succeeded in the herd by Cressy 1st, a bull purchased as a yearling by Mr. Toosey for 80 guineas from his breeder, Mrs. Jeffries, of The Sheriffs, near Knighton, Herefordshire. Cressy was a noble bull, well shaped and of excellent quality of flesh. He was used for several years in the herd and then sold by auction. His pedigree, which was a good one, was given to the purchaser, but, unfortunately, no copy has been preserved. Mr. Jeffries was the breeder of the celebrated bull Cotmore (376 E. H. B.), whose live weight at the Royal Show at Oxford in 1839 was declared to be 35 cwt. (3,920 lbs.) The next bull used was Trojan (4384 E. H. B.) He was imported in 1833 by Mr. S. Bryant, and after being in the colony some time, was sold to Mr. Toosey, who described him as a very fine symmetrical bull. After him was added another bull, also called Trojan (5083). He was imported about the year 1840 by Mr. Thomas Williams, of Launceston, and purchased by Mr. Toosey for 300 guineas. He is described as a bull of good quality and great weight. He was a successful stock-getter, and was the sire of Dainty 1st, calved October, 1843; Lady Bird, calved November, 1840; Juniper, calved June, 1846; Trojan was afterwards sold to Mr. Wier, of Victoria, who has his pedigree. Mr. Williams at the same time imported the bull Hereford, who was afterwards used at Cressy with good results.

"After this bull Mr. Toosey used two of his own breeding, Baron and Duke. The former was calved 14th November, 1845, and was by Trojan (5083), dam Blowdy, by Cressy 1st, g. d. Bashful, by Billy (4353); Belinda, by Billy (4353)—Beauty. Duke was calved 1st September, 1848, and was by Baron, dam Darling by Trojan (5083); g. d. Daphne, by Cressy 1st—Damsel, by Billy (4353); Diana by Billy, from imported cow. He was a very good bull and was used in the herd till 1854, when he was sold to Mr. Sloper Cox, of Mudgee, New South Wales, for seventy guineas.

"Mr. Toosey next used Priam, purchased in April, 1850, from his breeder, Mr. David Gibson, of Pleasant Banks, near Launceston. He was by the celebrated Hampton (513), who was sold for 500 guineas on his arrival at Launceston, out of Miss Stockton, a fine cow, imported from England by Mr. Gibson.

"In June, 1854, Mr. Robert Keate purchased for the company the bull Cronstadt (1198), and the cow Cressida, then three years old. These cattle arrived at Cressy in November, 1854. They were bred by Mr. Edward Williams, of Llowes Court, Hay.

"Cronstadt was succeeded by Zealous (1822), a bull purchased by Mr. James Cox, of Clarendon, Tasmania, for 400 guineas, that price being paid in England. He was bred by Mr. George Pitt, Chadnor Court, Dilwyn, and was a remarkably well bred bull, coming from a long line of illustrious ancestors. His sire, Milliam (1321) won first prize at the meeting of the Hay and Wyside Agricultural Society in 1855. His g. d. White Rose, by Young Cotmore (601), won second prize at the Royal meeting at Norwich and was one of the six cows to which the first prize was given in 1855 at Ludlow. Young Cotmore (601) won first prize at the Leominster Agricultural Society's meeting in 1841, and with cow and offspring at a subsequent meeting. His third dam, Rose, won first prize in her class at Hereford in 1845; fourth dam, Blossom 3d, by Young Favorite (460), was a winner of the first prize at Hereford, 1854. She was a dam of the celebrated bull Big Ben (248).

"Young Favorite won first prize in 1837, and

with cow and offspring, first at the meeting of the Gloucester Agricultural Society in 1839. Zealous' pedigree traces back three more generations to the esteemed blood of Mr. Knight, one of the breeders of the best cattle of his time. Two years ago Mr. Pitt's herd was dispersed, when ninety-one animals averaged £77 1s 9d ($385).

"The next bulls used were Cressy 2d (4474), Clarendon, and Garibaldi, all bred at Cressy, the two former being sons of imported Cressida, and the latter out of her daughter Countess, by Cronstadt (1198). Clarendon was by Cronstadt, was calved 1st September, 1860. Cressy 2d was by Zealous and was calved 29th July,

¶313

MAIDSTONE (8875) 79190.

Bred by H. W. Taylor. Invincible in English show rings. Champion at the Paris World's Exposition, 1889. Sold for $7,500 gold. Exported to Argentine, S. A., where he was again champion. Weight 3,200 lbs.

1861. He was a fine animal and after being used many years was sold to Mr. R. Q. Kermode, of Ross, for fifty guineas. Garibaldi was calved 19th November, 1862. He was by Garibaldi (2005), a fine bull imported by Mr. W. Field, of Enfield, Tasmania. Mr. Toosey used Garibaldi for some time with success, and eventually sold him to Mr. James Cox, of Clarendon. The next bull used was Undergraduate, calved 13th June, 1872. He was bred by Mr. Robert McDougall, of Arundel, Victoria, and was got by his celebrated bull, The Oxford Lad (4192), afterwards sold to Mr. George Loder, of Abbey Green, for 600 guineas ($3,000). Undergraduate's dam was the Cressy cow Jessie, by Zealous, she being a descendant of Matchless.

"Mr. Toosey considered Undergraduate a magnificent animal, and sold him to Mr. Vincent Dowling, of Lue, for 300 guineas ($1,500). Soon after Mr. Toosey gave up breeding. Some of his principal sales were the following: Mr. Learmonth, of Groongal, many years ago purchased a number of heifers, and, mating them with English bulls, formed a very fine herd. The cattle there, as seen by a well-known Queensland Hereford breeder a few years ago, were very large and thick. A number of females from this herd were exported to found a herd in New Zealand.

"About forty-five years ago, Mr. Toosey sold a bull to Mr. Hobbler, of the Hunter. This is probably the animal named Hobbler's Trojan, a name met with in some of the Tocal pedigrees.

"About thirty years ago ten or twelve heifers and the bull Young Cronstadt (5050) were sold to Mr. Lyall, of Western Port, Victoria. These heifers were principally by Trojan (5083), and from them, by Jerry (1288), Mr. Lyall bred cattle equal to any in Australia.

"About the year 1868, Mr. J. D. Cox, of Cullenbone, bought seven heifers, most of which were by Cressy 2d. In 1872 he purchased six more, these and four others that he bought in 1874 being by Garibaldi.

"Messrs. McConnel and Wood's fine cow Duchess 5th is from one of these. Mr. Cox sold drafts from his herd to Messrs. J. Gardiner and I. Irving (Tomki). He also sold a cow to Messrs. Archer, of Gracemore, for eighty guineas, and in 1882 Messrs. McConnel and Wood, of Durundur, purchased the whole of his herd descended from the Cressy cattle.

"In 1872 Mr. Reynolds bought some heifers, and Mr. G. Loder a bull called Julius, and later on Mr. Vincent Dowling bought the bull Undergraduate and all the young females by that bull.

"Finally, owing to old age, Mr. Toosey disposed of the whole of the remainder of his herd

HOTSPUR (7726) 21721.
Bred by T. J. Carwardine, Herefordshire. Sold for $3,500.

to Mr. John Taylor, of Winton, near Cleveland, Tasmania, who still continues to breed from them.

"Mr. Toosey, as already stated, kept a record of his breed up to the time of his death, but since then his son states that the Herd Books are not to be found, and must have been destroyed. This is very much to be regretted, as a detailed record of the earlier cattle of the foundation herd of Australia—a herd the blood of which is to be found in every herd in Queensland, and perhaps in every Hereford herd south of the equator—has probably disappeared for all time. The loss is all the more to be deplored that arrangements have almost been completed for compiling a reliable Hereford Herd Book for the whole of the colonies."

THE DURUNDUR EXPERIENCE.

It is the source of great gratification to us when we can secure the opinion of men who have had a large experience in the use of the Hereford in its purity and in crossing on other breeds. There is no question but that our large western range is valuable for the purpose of breeding cattle and producing beef, and the question which they wish to determine is, as to the best breed for the purpose of improvement on that range. The improvement of Texas and range cattle has moved very slowly and is of comparative recent date, but the discussion of these questions in this country has had a very wide influence and is bringing out in Australia the experience and practice that cover a half century's time under similar conditions to those on our own ranges; we find in the "Queenslander" the following article from a correspondent of that journal, which we published in the "Breeders' Journal," leaving out some remarks as to individual animals and giving what might be termed that general experience, which will be of value to our breeders, and to which this correspondent refers. The experiment has been in operation for 36 years, and we refer to the article itself to see with what results:

"I must honestly confess that I have generally viewed the Hereford breed with an unfavorable eye, induced from personal experience of their general wildness when bred or running in mountain districts, though I never denied the excellence of their beef-producing qualities. After some days' experience of the Durundur herd I must admit the error of my views, for a quieter and more kindly dispositioned herd, from aged bulls, through every grade, down to the year's calves, could not be found even in their original homes in the English counties. The cattle I worked amongst in earlier days were a bastard breed, Herefords only in name —*magni nominis umbra*—badly worked and worse managed. At Durundur exactly the opposite is the case. The cattle are worked on a very different system; the consequence is they are as quiet and docile as a mob of milkers, looking upon man as their best friend, rather seeking his company than avoiding it.

"The original source from which all Hereford herds in Australia sprung, either directly or indirectly, is that of the celebrated Cressy herd of Tasmania, started by the direct importation from England of three cows and one bull in 1825. In 1850, Hereford cattle were first introduced into Durundur by the purchase of some pure-bred bulls from Mr. Reynolds, of Tocal, New South Wales. The general herd then consisted of Shorthorns, and the proprietors for some years oscillated between the two breeds, in reality making some costly experiments, aiming more at quality than quantity. In 1873, however, a final decision was come to, and since that time nothing but Hereford bulls have been used, their prepotency being so assured that in a few generations the original strain of Shorthorn blood was entirely eliminated. In 1882 the pure herd was first formed by the purchase of 49 pure Hereford heifers from Mr. Reynolds, of Tocal, having such grand blood in their veins as could be introduced by the imported bulls Lord Ashford, Royal Head (¶ 260), The Captain, Thruxton, with many colonial-bred bulls, all of high breeding and many of them Royal prize winners. Misfortune followed their importation, however, as many of them fell victims to that dire disease, pleuro-pneumonia. Later on, 15 cows were purchased from Mr. J. D. Cox, of Mudgee, New South Wales, whilst in 1883 eight more were obtained from the same gentleman, all descendants of English cows, imported to Tasmania, many of them prize winners themselves or the produce of recipients of show honors.

"Since then the stud cattle have been augmented by the purchases from such well-known and successful breeders as the Messrs. White, of Muswell Brook; G. Rouse, of Mudgee, and G. Loder, of Singleton. These later lots are all undeniably pedigree, descended from such splendid sires as Defiance, Oxford Lad, etc., and all bred from cows, winners of first prizes at the great English shows. These females numbered 123, to which may be added about 200 more, being a choice herd of high-grade cows, bred by thoroughbred bulls from dams that have been carefully selected for several generations.

"The general herd are deserving of much credit, being large framed, of a good uniform color, and very even as regards shape and make. One feature in the raising of Hereford cattle is especially noticeable. They come to perfection for butchering purposes at an earlier age than other breeds. Rarely is a bullock kept after reaching four years of age; in fact, the Hereford beast at three and a half years will fetch as good a price in the open markets as one of the other breeds at five years of age. Consequently the breeder realizes on his stock far earlier than he otherwise would. The introduction of Herefords and the adhering to that breed has been most satisfactory. The Herd Books show a greater percentage of branded and weaned calves, the stock are quieter, and the estate carried a larger quantity. Most interesting must have been the gradual improvement of a herd as each successive draft of pure sires was introduced. Mr. Wood tells me about the second cross. A good number of the females were unshapely, bad colored, and of nondescript appearance. The pure bull again corrected all that, and now the stage has been reached when 80 per cent of the calves are properly marked with the characteristic colors and appearances of the Hereford cattle. They are vigorous, hardy, fat from the start. There is no breed like them for making prime bullocks at three and a half years old, weighing 750 pounds, dressed, and this done entirely on natural grass. Further, there is no breed like them for accommodating themselves to the changes and chances that grazier's cattle are exposed to in the capricious climate of Australia.

"This station buys large numbers of store bullocks yearly, and it is a curious fact that the few Herefords they get in this way fatten about two months sooner than the Shorthorn cattle. From personal experience I know that Herefords travel long distances to market better than Shorthorns, with less lameness, and keeping their condition better; further, that a Hereford beast in hard times will fight for his living where a Shorthorn simply lies down and dies. From inquiry I find the timber getters are always ready to pay $5.00 per head more for Hereford bullocks than for Shorthorns to break them into the yoke, as they find them more active, hardier and more docile.

"The owners of this herd have evidently created an ideal till they have succeeded in producing a highly satisfactory result. The experiments have no doubt been costly, but the primary difficulties appear to me to have now been surmounted, and they are now fairly in

IMPORTED LORD WILTON 5739, AT 3 YEARS.
Bred by C. M. Culbertson. Son of the champions, Lord Wilton and Pretty Face.

the straight run for the goal of success. There is not a doubt that their stock will be constantly sought after by breeders desirous of improving their own herds. The three runs of Durundur, Corondale, and Mount Kilcoy have wonderful capacities for fattening stock—splendid flats and gullies, while the ranges are clothed with herbage to their very summit, and permanent streams of water. It is marvelous to me, how little the residents of Brisbane know about this valuable district.

"When we see a long-established herd of consummate excellence, the result of one man's mind, we naturally ask ourselves the question how that man becomes possessed of the judgment and accompanying qualifications to enable him to do that work, the results of which are before us. Accident could never accomplish such high and uniform results. Some men are amongst stock all their lives and never acquire

REPRESENTATIVE SUSSEX COW.

that knowledge. They can multiply animals already improved for them, but for their lives they could not breed up a family or tribe to any preconceived pattern. This Messrs. McConnel and Wood have done; their good example is well worthy of imitation. The great error many make is expecting, after the purchase of high-bred animals, that great results will come without effort and unsought. The best animal is nothing unless well fed. Their growth and development comes of liberal feeding, and full growth and symmetry from full feeding to fatness.

"The cattle on these runs are a proof of our argument; witness the eagerness with which they are sought after and the quantities supplied. The man who, after buying select animals, retreats to his shell like a turtle under the idea that he is possessed of property which through the innate force of circumstances will retain all the good it has about it at the time of purchase will wake up—Rip Van Winklelike —to find out the error of his ways. This fault Messrs. McConnel and Wood have not been guilty of. Slowly, step by step, with every possible care and forethought, they have sought to improve their stock. Well have they succeeded, and if breeders have any wisdom for their pains they ought to reap a rich harvest in the future; they most certainly have deserved it."

In closing this section of our work we quote again from the special U. S. Consular Reports (page 645) on cattle, heretofore mentioned. Consul Griffin, of Auckland:

NEW ZEALAND CATTLE IN THE UNITED STATES.

"The high class of cattle in this colony, and the low price at which they can be obtained, has very naturally attracted the attention of cattle breeders in the United States. In August, 1883, Mr. A. W. Sisson, of California, dispatched Mr. Rolin P. Saxe, a cattle expert, to New Zealand, to purchase for him a band of pure-blooded Herefords. Mr. Saxe arrived in Auckland in September, 1883, and after visiting several of the cattle districts in the colony, purchased 20 two-year-old heifers in calf and 24 bulls from one to two years from the New Zealand Stock and Pedigree Co., of Auckland. Mr. Saxe was not only surprised at the superb condition of the company's cattle, but at the low prices at which they were sold. They were shipped to San Francisco by the Pacific Mail steamer City of Sydney, in October last, being the first shipment of New Zealand-bred cattle ever made to the United States.

"Mr. Saxe is of the opinion that Hereford cattle can be more easily and economically brought to California from New Zealand than across the continent by railway from Illinois and other states celebrated for this particular breed. In Illinois these cattle sell at from $500 to $5,000 per head, whereas they can be bought in New Zealand at from $100 to $700 per head. * * *"

NEW ZEALAND HEREFORDS.

"The New Zealand Stock and Pedigree Company of Auckland has one of the largest herds of pure-bred Herefords in the world. This breed has long been a favorite one here. They are tough, hardy, and able to pick their food on poor soil, and when two or three years old outweigh any other breed, and are famous for their high-priced meat; that is to say, their loins are well developed, and their yield of succulent porterhouse and sirloin are proportionately heavy. The hindquarters of the pure-bred Hereford are long from the hip backwards. The thighs are large and full and well meated at the hocks. The whole carcass is set square on good, short legs, standing well apart. The flesh is firm, the hide mellow, with soft hair, not too fine, but giving the impression that it can be stretched to any extent.

"The color of this breed is a distinct red with white face, mane, and white breast and legs as far as the knee. As an evidence of how they stand hard feed, it is said that during the long drought of 1878 and 1879, in Australia, about five per cent of the Herefords were lost on a run in Queensland, against 10 per cent of the Shorthorn herd and 20 per cent of the stud Shorthorn. In one large paddock there were 70 Shorthorn and 70 Hereford bulls, one and two years old. The Shorthorns got so poor that they had to be turned out on the run, the paddocks being bare of grass, but the Herefords kept in good, strong condition.

"When Captain Cook first visited New Zealand there were no cattle in the country, but at a subsequent period some were introduced from Australia. In the early settlement of the colony the length of the time occupied in a voyage from England and the many difficulties which had to be overcome by the pioneers prevented any special attention being given to the improvement of the breed of cattle by importation, as that necessarily involved a heavy expenditure of money, not to say anything of the time and patience required to introduce them; but at last the colonists began to improve their herds by the introduction of thoroughbreds from Europe, and I have not the slightest hesitation in saying that nearly all the imported cattle thrive better in New Zealand than in their native homes, and that this superiority is developed to a still higher degree in their offspring."

HEREFORDS FOR THE ISLAND OF JAMAICA.

Hereford cattle have done remarkably well on this island, and purchases of thoroughbreds are still continuing in England for exportation to that island.

On the 10th of November, 1885, a little knot of Hereford breeders gathered at the Southwest India docks to see the shipment, by the ship Catib, of six thoroughbred Herefords to that distant island. Mr. John Malcolm, of Paltalloch at Knockalba, and Retrieve, Penn., in America, was the purchaser. Mr. Malcolm personally superintended the loading, and he, in company with other friends who had come up from Hereford to see them off, were entertained at luncheon by the captain, Mr. Vicary. The lot of Herefords comprised two bulls and four heifers. The bulls were Lemon Boy 2d, bred by Mr. P. J. Hughes, and Benjamin 17th, bred by Messrs. J. B. & G. H. Green, of Marlow, Leintwardine. This bull was a year and a half old, and said to be a good one. The four heifers were all yearlings past, and named, respectively, Miss Hamar 2d, Countess 10th, Miss Silver 7th, Miss Nobleman 8th. They were all bred by Mr. Thomas Myddleton, of Beckjay, Aston-on-Clum, and were an extraordinary good lot. These four heifers were picked out of ten heifers shown by Mr. Myddleton at the Gloucestershire show, they going to the show off the grass and beating a like number shown by John Price and others. These cattle are going to join a large and valuable herd of Herefords that have been bred in Jamaica since 1835.

REPRESENTATIVE DEVON COW.

CHAPTER XXXVIII.

Herefords on the Block

The following items in regard to the beef price of Hereford cattle as they ranged from 1881 to 1886, taken as they are from publications of those dates, give information as to the actual cash value of the Herefords:

HEREFORD BULLOCKS IN LONDON.

Mr. George T. Turner says in a letter to the "National Live Stock Journal": "The Hereford bullock in London is quite a season animal, and comes only as grass-fed beef in the late summer and autumn, then it tops the market, for there is no better grass beef sent to London than the Hereford, except the West Highlanders, which are older and few in number. These generally make prices which are above the ordinary top quotations. The Polled Scotch cattle are stall-fed, and do not come under the same category as the Herefords."

THE LONDON METROPOLITAN CATTLE MARKET.

On Monday, Jan. 16, 1882, the best supplies comprised a fairly conditioned assortment; a few choice Scots brought 6s. per stone, but the more current topping rates for first quality Scots, Devons and Herefords were 5s. 8d. to 5s. 10d.; Shorthorn, 5s. 4d. to 5s. 6d., occasionally 5s. 8d. Some Danish cattle were on offer, and ranged from 4s. 8d. to 5s. 4d., and in a few instances 5s. 6d. At Deptford there were about 1,100 head, comprising Dutch, French and American; the top rates for the two latter were about 5s. 4d. to 5s. 6d.

This, in the language of American reports, would be quoted about as follows:

	Live, per cwt.	Dressed, per lb.
West Highland Scots, extra choice.$	10.75	18¾c
Choice Herefords, Scots and Devons.	10.50	17⅔@18¼c
Shorthorns, extra choice............	10.00	17⅔c
Choice Shorthorns....................	9.50@ 9.75	16½@17c
Danish and French....................	8.25@ 9.50	14½@16½c
Americans	9.50@ 9.75	16½@17c

It is said that the quality of the best American beef is such that when dressed it is often sold as British raised, and that it is impossible to do this with the product of any other country.

HEREFORD COW JENNIE.

The Hereford cow Jennie, that won and took the champion prize as the best cow at the Fat Stock Show in 1878, and won, but did not take it at the same show in 1879, was slaughtered and dressed at Providence, R. I., December 5, 1879, by T. M. Lincoln & Co., of that place. We give below a letter of Messrs. Lincoln & Co. to Mr. Imboden (¶ 261) (one of the best judges developed by the American Fat Stock Show), showing the dressed weight and their opinion as to the character of the meat:

Providence, R. I., Dec. 11, 1879.

Mr. J. G. Imboden.

Dear Sir: In answer to your letter of Nov. 21st, concerning the Hereford heifer, Jennie, will say that she was dressed last Friday. The day before dressing she was exhibited in front of our store, and attracted considerable attention. She dressed a very pretty color, and is very thick, and mixed beautifully. Her percentage of shrinkage was the lightest of anything that has been dressed about here.

Her live weight is given as she weighed the day she was dressed, which you see is 98 pounds less than her Chicago weight. Live weight, 1,622.

DRESSED WEIGHT.

	Lbs.		Lbs.
Meat	1,110	Liver	12
Hide	92	Heart	8½
Tallow	154	Tongue	5
Total	1,356	Total	25½

Shrinkage 16.39 per cent, and meat 68.44. The caul, which weighed 37 pounds, is included in the 154 pounds of tallow. The weight of the meat in quarters, hinds being cut with one rib on them, is 267 and 264 for hinds, and 287 and 292 for the fores.

Yours respectfully,

T. M. Lincoln & Co.

On November 27, 1882, the beef market in London, England, was quoted as follows:

"Cattle, which are not at all plentiful, maintain steady value. Herefords, Scots and Devons were quoted at 18¾ to 19¼ cents per lb. dressed; Shorthorns at 17⅝ to 18¼c."

The "Chamber of Agriculture Journal" (Eng.), of Monday, December 4, 1882, speaking of the cattle market in London, says:

"Herefords sold for 18¾c to 19½c; Shorthorns at 17⅝c to 18⅛c," making a difference of 1⅛c per pound in favor of the Herefords. Danish cattle quoted at 17c; Canadian from 14½c to 15c; Swedish, 14c to 14½c.

HEREFORD STEERS DRESSED.

We republish from an eastern paper the following account of two Hereford steers which were exhibited at the New York State Fair (1882), where they won first and second prizes. They were purchased and slaughtered by John Battersby, butcher, of Albany. The beef of the two weighed in the aggregate 3,370 pounds. The beef in the carcass of the steer George was 63 per cent of live weight, and of Pierre, 69 per cent. With hide and tallow the former turned out 80 per cent, the latter 87 per cent. The lean was finely streaked with fat, and was of a rich texture. The following was the result in detail:

	Pierre. lbs.	George. lbs.
Weight before killing	2,310	2,575
Beef	1,590	1,668
Tallow	147	215
Hide	128	141
Liver	23	19
Tongue	10	10
Heart	7	8
Total	1,905	2,074

This beef was graded at the West Albany market as first-class.—*Breeders' Journal,* 1883.

THE FIRST HEREFORD RANGE STEERS.

Some very fine distillery-fed cattle were recently marketed at Chicago. They were half-blood Herefords, originally from the Wyoming ranch of Swan Brothers; in fact, they were the first fruits of the bulls of that breed which they took out to the ranch some three or four years ago. The lot numbered 75 head, averaged 1,380 pounds, and sold at $6.15 per cwt., which was at least 25 cents per hundred more than any other cattle sold for on the day of their arrival. They sold to Armour & Co. (¶ 262), to dress at Chicago and be forwarded in refrigerator cars to a New York butcher who makes a specialty of handling the best grades of beef. The lot dressed sixty-four pounds per hundred, an extraordinary record, being about eight pounds above the average. The lot attracted much attention, and was inspected by numerous breeders of note, the most of whom came for the express purpose of seeing the first really large lot of Herefords that has been marketed. On the same day and with the same lot was a large shipment of ordinary range cattle that had been slop-fed the same length of time as the half-bloods, which averaged 1,309 pounds and sold at $5.75. A few practical lessons like that, showing in actual figures the value of improved blood, carry more weight than almost any amount of theorizing.—*Breeders' Journal,* 1883.

COMPARATIVE PRICES IN LONDON.

From the "Farmer and Chamber of Agriculture Journal" (England) we get the market reports at Smithfield on the 2d of June, 1884. There were in the Metropolitan Cattle Market on that day about 2,400 beasts, and the best Scots, Herefords, etc., were 5s. 4d. to 5s. 6d. (or 16½c to 17¼c dressed, equal to $9.50 to $10 per cwt. alive) per eight pounds. The best Shorthorns were sold at from 5s. 2d. to 5s. 4d. (or 16c to 16½c per lb. dressed and $9 to $9.50 per cwt. alive); the second quality beasts, which were made largely from Shorthorn blood, were selling at from 4s. to 4s. 10d. (or 12½c to 15c dressed and $7 to $8.50 per cwt. alive).

The above are about the regular comparative quotations that appear in the English journals from week to week. The Scots and Herefords stand within the range of 5s.4d. to 5s. 6d., which means about $1.33 to $1.37½ for a stone of 8 pounds weight. The best Shorthorns stand at $1.29 to $1.33, and then comes in a quality below the best, which makes from $1 to $1.20. Dividing these sums by eight shows the price per pound of meat, making no account of the offal. This second quality is a feature that exists in all markets in reference to the Shorthorn cattle, which brings the average very much below the Scots and Herefords. The Scots and Herefords have that uniform char-

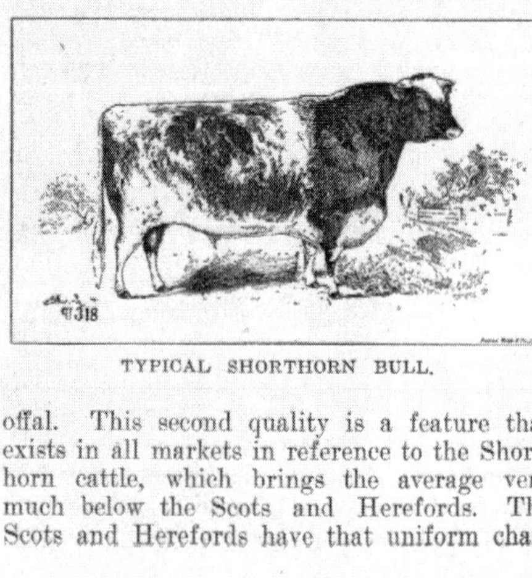

TYPICAL SHORTHORN BULL.

acter that is represented between \$1.33 and \$1.37½.—*Breeders' Journal,* 1884.

ANOTHER SALE OF RANGE STEERS.

The Swans showed at Union Stock Yards, Chicago, Hereford steers raised on the range, and fed at their feeding stables at Omaha distilleries.

A lot of 20 choice two-year-old grade Hereford steers, averaging 1,235 pounds, sold at \$6.75 to Bailey & Co., of Evanston. At the same time a car of natives of the same age, weighing 1,116 pounds, sold to Armour & Co., at \$6.10, and twenty-eight tailings sold at \$5.75. They had sold the week previous sixteen head

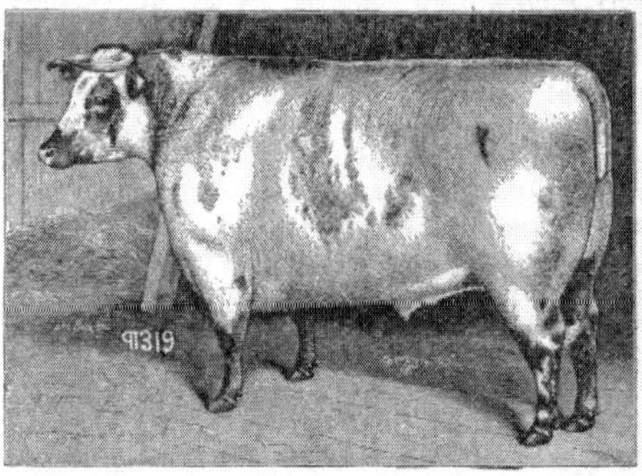

SHORTHORN STEER, SCHOOLER.
Winner in class at Chicago and Kansas City, 1883-4-5.

of grade Hereford steers, averaging 1,443 pounds, to Wolfe & Pfaelzer, at \$6.30. This lot dressed 64½ per cent of beef.—*Breeders' Journal,* 1884.

MR. FUNKHOUSER'S STEERS.

James A. Funkhouser, Plattsburg, Mo., sold in Chicago, on Dec. 10, 1886, seven yearling grade Hereford steers at \$6.50 per hundred. They were shipped to Albany, N. Y., where they were killed, and from Chicago to New York they only shrunk fourteen pounds per head.

Weight of seven steers in Chicago, Dec. 10, 1886, 9,490 pounds; killed by Wiley Bros., Albany, N. Y., Dec. 18, 1886:

	Gross Wt. in Albany.	Weight of beef.	Per cent. net to gross wt.
1 Steer	1280	835	65.23
1 Steer	1310	826	63.05
1 Steer	1350	887	65.70
1 Steer	1350	886	64.14
1 Steer	1390	884	63.60
1 Steer	1410	918	65.19
1 Steer	1300	828	63.69
Total Weight	9390	6044	average of 64.36

Weight of hides, 660 pounds; weight of tallow, 747 pounds; per cent of profitable weight (carcass, hide and tallow) to gross or live weight, average 79.36.

We wish to call attention to the fact that through a journey of 836 miles the seven steers made a shrinkage of only 100 pounds, or 14 pounds each.

It is a peculiarity of the Herefords that they make less shrinkage than any other breed in traveling, whether by rail or foot.

Another feature—the average price for these yearling steers at Chicago was a fraction under \$84.—*Breeders' Journal,* 1887.

SOME MAINE STEERS.

At the New England and Eastern Maine, Maine State and one local fair, North Kennebec, Messrs. Burleigh & Bodwell, Vassalborough, Maine, exhibited 18 Hereford steers on which they were awarded 33 premiums, 24 of which were first prizes. These cattle took every first premium offered for pairs of fat cattle over two years, and every first prize on matched cattle. They were then taken to Boston and arranged in front of the Quincy House, and then arrayed before Faneuil Hall Market, with their beautiful red coats and white faces, their symmetrical forms, combined with the rarest quality, which, added to their great weights, considering age, made this a sight of a lifetime. They were then slaughtered. The dressed weight and shrinkage of these steers were as follows:

Description.	Age. Yrs.	Shrinkage, Per cent.	Net Wt. Lbs.
1 Pair	5	20½	3356
1 Pair	5	20	3253
1 Pair	5	26	2683
1 Pair	4	20½	2829
1 Pair	4	23½	2781
1 Steer	3	19	1403
1 Steer	3	23	1451
1 Pair	3	24	2457
1 Pair	3	22	2341
1 Pair	27 months	24½	1942

—*Breeders' Journal,* 1885.

KANSAS CITY FAT STOCK SHOW, 1885.

Awards were as follows:

For best three-year-old carcass—Texas Jack (Hereford bull on Texas cow); W. E. Campbell, breeder, feeder and exhibitor.

For best two-year-old carcass—Fred (Hereford); J. S. Hawes, breeder, feeder and exhibitor.

For best one-year-old carcass—Kansas (Galloway); M. R. Platt, breeder, feeder and exhibitor.

Sweepstakes for the best carcass in the show—Fred (¶ 264) (Hereford steer); J. S. Hawes, breeder, feeder and exhibitor.

These animals do not show as heavy weights or dress so large a per cent as the dressed carcasses at the Chicago Fat Stock Show, but the question may be fairly raised as to whether, for the consumer, they may not be fully as good.

We call attention to the fact that a cross-bred Hereford and Texan takes the first as a three-year-old, and a thoroughbred Hereford wins as a two-year-old, and also sweepstakes for best carcass; a Galloway taking first as yearling. Thus a two-year-old Hereford takes the sweepstakes at Kansas City and Chicago Fat Stock Shows as the best dressed carcass.

At the Kansas City Show, in the competition between breeds, the Herefords took only one premium, and that went to Mr. Jas. A. Funkhouser for Challenge, under one year old, by Invincible, by Success. But on carcasses they took two out of three, and sweepstakes. That is something like old-time reading.—*Breeders' Journal*, 1885.

DRESSED CARCASSES AT THE KANSAS CITY FAT STOCK SHOW FOR 1885.

Owner.	Breed.	Name.	Live weight at time of slaughter.	Left fore quarter.	Right fore quarter.	Left hind quarter.	Right hind quarter.
W. E. Campbell	Hereford and Texan	Texas Jack	1695	290	289	261½	261
W. Morgan & Son	Hereford	Jock	1760	292½	289	279	272
J. S. Hawes	Hereford	Fred	1725	301	300	261½	255
G. S. Burleigh	Hereford	Star	1300	224	210	196	203
H. Blakesley	Shorthorn	Arabella Beauty	1715	288	273	267	251
W. H. H. Cundiff	Shorthorn	Angelica 13th	1510	238	223	245	231
E. B. Millett	Shorthorn	Queen	1475	234	222	231	228
J. H. Potts	Shorthorn	Hopeful	1535	273	265	245½	240
A. B. Matthews	Galloway	Matthews' Pet	2090	380	369	314	311
M. R. Platt	Galloway	Kansas	1320	222	225	204½	205
W. R. Estill	Aberdeen Angus	Bloom	1180	198	185	195	201

PRIZE BEEF AT DETROIT.

At the late Fat Stock Show (1880) at Chicago the judges decided, we hope, to the best of their ability, but the only true criterion for beef cattle to be judged by is the butcher's block. Recognizing this truth, Mr. T. L. Miller, of this place, wishing to see his Herefords slaughtered in competition with the animals against which he showed, having bought some of the latter, sold three head of Herefords, and the J. H. Potts & Son's two-year-old Shorthorn steer to Wm. Smith & Son, the best butchers in Detroit. Mr. Smith had already bought the first and second premium Shorthorn cows in their classes, to the largest of which cows the able (?) judges also awarded the champion prize for the best cow in the show. These animals with many others were all purchased by Mr. William Smith to make his usual display of Christmas beef. The hanging up of so much valuable meat, and particularly the opportunity which it afforded of seeing Hereford and Shorthorn beef hung up side by side, drew together large crowds. Stock breeders were there from Canada, Ohio, and Illinois, and even men from the distant Western plains.

We think we may safely say that no equal display of high-fed and ripe bullocks, prize sheep and wonderful hogs was ever made by any one man in this country, and those who had seen like exhibits in England said they had never seen the quality of the meat surpassed. In fact, the display took all by storm, and it was universally acknowledged by all who saw it—and there were thousands—that it was impossible to make a better show.

Mr. Smith occupies stall No. 1 and 2 in Mansfield Market, Detroit. His Christmas display consisted of twenty-eight head of prize beef animals, one hundred and twenty head of prize sheep (quite a number of them having won honors at the Chicago Fat Stock Show), and prize Suffolk and Essex hogs.

Mr. T. L. Miller's prize three-year-old Hereford heifer, "Maid of Orleans," hung in a conspicuous place, and was a center of attraction; her massive form and deep meat, combined with rounds and chine that could not be excelled, drew forth the praises of all. The carcasses of the two Shorthorn cows that showed against the "Maid of Orleans" for sweepstakes at Chicago were quartered and hung up, so an exact

SHORTHORN STEER, CLEVELAND.
At 497 days. Weight 1,290 lbs. Most perfect Shorthorn bullock ever produced in America. Practically approximating the Hereford type.

opinion of their value as beef animals could be formed. Mr. Wm. Smith, who, by the way, is one of the best judges of a beef animal, either alive or dressed, in the country, pronounced the meat of the Shorthorn cow that took the sweepstakes not worth within a cent a pound of what the second Shorthorn cow was, and that the Hereford cow "Maid of Orleans" was worth a cent more per pound than the best Shorthorn, making her two cents per pound better than the cow that was awarded the sweepstakes.

Mr. Smith said the meat that he considered the best of the whole exhibit were the Hereford carcasses, and that he could not ask for anything better; the marbling of the meat was perfect, and that the following were the live and dressed weights of three of the steers:

	Live Weight at Detroit.	Dressed Weight at Detroit.	Per cent Dressed to Live Weight.
T. L. Miller's two-year-old Rob Roy..	1436	971	67.6
J. H. Potts & Son's Shorthorn two-year-old..........	1590	1045	65.6
G. S. Burleigh's yearling Hereford..	1300	880	67.6

The above figures show that the Hereford steer Rob Roy dressed two per cent to the hundred more than the Shorthorn steer of J. H. Potts & Son. The quality of the Hereford steer carcass could not be excelled. He dressed a very white, nice color, and the fat was distributed throughout the lean, presenting one of the best samples of marbled meat that we ever saw. Wishing to test the quality of meat by taste as well as sight, we procured some juicy steaks from this carcass, for which Mr. Smith could not hear of pay, and we returned him our hearty thanks. These steaks were cooked at a restaurant, and partaken of by a party of stock breeders, who are well calculated to judge of its qualities, having largely traveled both in this country and Europe. The unanimous verdict was "unsurpassed."—*Breeders' Journal,* 1881.

It was by the constant publication in the "Breeders' Journal" of such facts as these that we were able to get and keep the Herefords before the public. We were among the first agricultural publishers to send large numbers of free sample copies to lists of people we desired to interest, and by this method we not only brought the "Breeders' Journal" to the notice of agriculturists, thereby gaining largely in our paid subscription list, but we brought the Hereford breed and its merits before the farmers and ranchmen of America in a manner never before attempted.

We have been deeply gratified by the fruits born of these efforts, but confess that at the time they were made we were secretly somewhat annoyed that they were not fully appreciated by the very men most benefited.

GRACE, PURE-BRED COW, WEIGHT 1,875 LBS.
Exhibited at Chicago Fat Stock Show. Fed by John Gosling.

CHAPTER XXXIX.

Fair and Fat Stock Show Reports

ORGANIZATION OF THE NEW YORK STATE FAIR AND CATTLE SHOW IN 1841.

The following is the record:

Senate Chamber, Albany, Jan. 11, 3 P. M.

Met pursuant to adjournment, H. D. Grove, Esq., one of the Vice-Presidents, in the chair. The committee appointed to nominate officers made their report, which was read and accepted, and the following gentlemen were unanimously elected officers of the Society for the ensuing year: Joel B. Knott, of Albany, President. Vice-Presidents: 1st District, Jeromus Johnson, of Kings; 2d, Robert Deniston, of Orange; 3d, Caleb N. Bement, of Albany; 4th, Edward C. Delevan, of Saratoga; 5th, Benjamin P. Johnson, of Oneida; 6th, Lewis A. Morrell, of Tomkins; 7th, Willis Gaylord, of Onondaga; 8th, T. C. Peters, of Genesee. Additional Members of the Executive Committee: Alexander Walsh, of Rensselaer; George Vail, of Rensselaer; Henry D. Grove, of Rensselaer; A. L. Linn, of Schenectady; John D. McIntyre, of Albany; Henry S. Randall, of Cortland, Corresponding Secretary; Ezra P. Prentice, of Albany, Treasurer; Luther Tucker, of Albany, Recording Secretary.

On the third Wednesday in March, at a meeting of the Executive Committee, consisting of Messrs Knott, Bement, Vail, Prentice, McIntyre and Tucker, committees were appointed for each county. At a meeting on the third Wednesday of May, present, Bement, Knott, Prentice, Walsh and Tucker, the time and place for holding the first Fair was fixed to be at Syracuse, Sept. 29th and 30th, and the premium list was arranged and published. At a meeting of the Executive Committee of the New York State Agricultural Society, held at Syracuse, Aug. 18, 1841, present: Messrs. Knott, Johnson (of Oneida), Gaylord, Randall and Tucker, the following Viewing Committees were appointed to award the premiums offered by the Society at their cattle Show and Fair to be held at Syracuse on the 29th and 30th days of September:

ON CATTLE.

Class I.—Bulls of any breed, 3 years old and upwards—Henry S. Randall, Cortland; A. B. Allen, Erie; C. N. Bement, Albany; William Garbutt, Monroe; J. D. McIntyre, Albany.

Classes II and III.—Bulls of any breed under 3 years old—Francis Rotch, Otsego; Henry Rhoades, Oneida; J. C. Hathaway, Ontario; Geo. Vail, Rensselaer; George J. Pumpelly, Tioga.

Class IV.—Cows of any breed, 3 years old and upwards—Anthony Van Bergen, Greene; E. P. Prentice, Albany; Thomas Hollis, Otsego; Ira Hitchcock, Oneida; Hiram Bostwick, Chemung.

Classes V and VI.—Heifers, any improved breed under 3 years—Lewis F. Allen, Erie; Silas Gaylord, Onondaga; Thomas Weddle, Monroe; John Gaskin, Otsego; Jonah Davis, Chemung.

Class VII.—Grade cows—Garret Sacket, Seneca; C. S. Button, Wayne; M. Bullock, Albany; Thomas Goodsell, Oneida; William Ottley, Ontario.

Class VIII.—Grade heifers—S. W. Brace, Onondaga; John M. Sherwood, Cayuga; William Alexander, Otsego; D. D. Campbell, Schenectady; Rufus Boies, Cortland.

Class IX.—Cows, native breeds—Myron Adams, Ontario; Thomas S. Meacham, Oswego; Aaron Barnes, Oneida; —— Crane, Herkimer; Tyler Fountain, Westchester.

To Breeders—F. Rotch, Esq., having given the Society $30, for that purpose, premiums will be awarded to breeders as follows: To the breeder of the best thoroughbred bull, $10; to the breeder of the best thoroughbred cow, $10; to the breeder of the best thoroughbred heifer, $10.

FOR WORKING OXEN.

William Gaylord, Esq., having contributed $20 for that purpose, a premium will be given for the best yoke of working oxen, $12; for the second best yoke of working oxen, $8. In awarding this premium, particular reference will be had to the close matching, excellent training and docility of the animals, as well as to their general good appearance. Committee: Abel Baldwin, David Bundy, and Dan Hibbard.

Fat Cattle—Mr. Rust offers a sweepstakes, twenty dollars entry, for the best yoke of fat cattle. Committee: B. P. Johnson, B. D. Noxon, and M. D. Burnett.

A plowing match, under the direction of the Onondaga County Agricultural Society, will take place immediately after the trial of plows, on the second day of the Fair.

The first Agricultural Fair was held at Syracuse, on the 29th of September (29-30), 1841. About 1,200 sat down to dinner at the Syracuse House. An address was delivered by the Hon. Micah Sterling, followed by L. F. Allen and others.

The following are the awards given at this show:

CATTLE.

Class I—Bulls—3 Years Old and Over.

To J. M. Sherwood, Auburn, for his bull Archer, bred by F. Rotch, Butternuts, 1st prize.

I. M. FORBES, Henry, Ill.

To E. P. Prentice, Albany, for his bull Nero, bred by himself, 2d prize.

To C. N. Bement, Albany, for his bull Astoria, bred by himself, 3d prize.

To Silas Gaylord, Skaneateles, for his bull Splendid, 4th prize.

"There were several others animals (in this class) on the ground, possessing, in the estimation of your committee, high grades of excellence, and they only regret that the premiums were not more numerous. Among these your committee particularly noticed the animals of Mr. McIntyre, Mr. Van Bergen, Mr. Fonda, and Mr. Sears."—*Report of the Committee.*

Class II—Bulls—2 Years Old.

To John Johnston, Fayette, Seneca Co., for his bull Royal William, bred by G. V. Sacket, Seneca Falls, 1st prize.

To Thomas A. Clark, Chittenango, for his bull Young Warden, bred by Thomas Hollis, Gilbertsville, 2d prize.

To D. D. Campbell, Schenectady, for his bull Rotterdam, bred by himself, 3d prize.

To Nicholas Garner, Burlington, for his bull ——, bred by himself, 4th prize.

Class III—Bulls—1 Year Old.

To Moses Kinney, Cortlandville, for his bull Daniel Webster, bred by G. V. Sacket, Seneca Falls, 1st prize.

To Enoch Marks, Navarino, for his bull Brutus, 2d prize.

To Benjamin Stoker, Cortland Co., for his bull ——, 3d prize.

To Joseph Baker, Onondaga Co., for his bull ——, 4th prize.

"Your committee beg leave to express their regret that though the exhibition in Classes II and III were very numerous, yet but few of the animals were in what they considered common store order, which rendered the effect of comparison with such as were high fed very difficult."—*Report of Committee.*

Class IV—Cows.

To John M. Sherwood, Auburn, for his cow Stella, bred by F. Rotch, five years old, 1st prize.

To Ezra Prentice, Albany, for his cow Daisy, 3 years old, bred by himself, 2d prize.

To John M. Sherwood, Auburn, for his cow Daisy, 12 years old, 3d prize.

To John M. Sherwood, Auburn, for his cow Pansy, 5 years old, bred by F. Rotch, 4th prize.

To Corning & Sotham, Albany, for their Hereford cow Matchless, imported, an extra prize, equal to the highest premium awarded on cattle.

"Your committee further report that a new and beautiful race of cattle were presented for their examination, the Herefords, imported by a distinguished breeder of cattle, residing in Albany County, which they take pleasure in recommending to the attention of those who desire to improve their stock. Your committee recommend a special premium of twenty dollars for the Hereford cow Matchless, as we consider her a very superior animal, and they would also suggest the propriety of offering and awarding premiums for the best blooded animals of each individual breed—improved Short-horned Durhams, Herefords, and Devons, at their next Agricultural Meeting, in addition to the premiums offered for the best animals of any breed."—*Report of Committee.*

Class V—Two-Year-Old Heifers.

To John M. Sherwood, Auburn, for his heifer Sylvia, bred by F. Rotch, 1st prize.

To E. P. Prentice, Albany, for his heifer Diana, bred by himself, 2d prize.

To Corning & Sotham, Albany, for their Shorthorn and Hereford heifer Eliza, imported, 3d prize.

Class VI—Yearling Heifers.

To Ezra P. Prentice, Albany, for his yearling calf Charlotte, bred by himself, 1st prize.

To John M. Sherwood, Auburn, for his yearling calf Norna, bred by H. S. Randall, Cortlandville, 2d prize.

To John M. Sherwood, Auburn, for his yearling

heifer Dianthe, bred by J. Alexander, Burlington, 2d prize.

To William Fuller, Skaneateles, for his heifer calf ——, bred by himself, 4th prize.

"All the animals on which the above prizes were awarded, with the exception of the Hereford cow and the Shorthorn and Hereford heifer of Messrs. Corning & Sotham, were thoroughbred Shorthorns."—*Report of Committee.*

Class VII—Grade Cows.

To Wm. Ward, Camillus, for his eight-year-old half-blood Holderness cow, 1st prize.

To W. H. Sotham, Perch Lake Farm, for his half-blood Durham cow No. 1, 2d prize.

To W. H. Sotham, Perch Lake Farm, for his half-blood Durham cow No. 2, 3d prize.

To W. H. Sotham, Perch Lake Farm, for his half-blood Devonshire cow, 4th prize.

"The best grade cow which came under our observation belonged to G. V. Sacket, of Seneca Falls, but he, being one of the committee, generously withdrew her from competition."—*Report of Committee.*

Class VIII—Grade Heifers.

To H. S. Randall, Cortlandville, for his roan heifer, bred by himself, 1st prize.

To G. V. Sacket, Seneca Falls, for his red and white heifer, bred by himself, 2d prize.

To G. V. Sacket, Seneca Falls, for his roan heifer, bred by himself, 3d prize.

To H. S. Randall, Cortlandville, for his red and white heifer, bred by himself, 4th prize.

Class IX—Native Cows.

"The Committee on Native Cows would report that very few cows, and those of inferior quality, were to be found in the pens; and they probably not intended for exhibition. They regret that the farmers in this vicinity should have refrained from taking advantage of the very liberal encouragement offered by this Society, by the false impression that cows were going to be brought from a distance which would have eclipsed the cows of this neighborhood. We are unwilling to believe that there are not cows in this village and vicinity that would have honored the exhibition and made a credit to the State. They regret that a matter so important as the improvement of our native cows does not excite more attention. Such cows must, of necessity, be the groundwork of much of the improvement in cattle. If a farmer has a cow possessing some excellent qualities, he is prepared to improve in any desirable point. The general dissemination of high-blood animals renders such crossings easy and cheap, and it is a matter yet at issue whether such crosses will not make the most desirable animals for the common farmer. We want the best native cows for such crosses, and the committee are of the opinion that the Executive Committee of the State Society are holding out liberal encouragement for active competition in the matter of improving our native cattle. In conclusion we would add that we hope that no future committee will be under the necessity of reporting 'no competition,' but let the farmer, the lawyer, the merchant, and mechanic bring forward their best cows, and render it a matter of nice discrimination to decide between them."—*Report of Committee.*

WORKING OXEN.

To Caleb Gasper, Marcellus, 1st prize.

To Samuel Allen, Jr., New Haven, 2d prize.

FAT CATTLE.

To P. N. Rust, Syracuse, for the best yoke of fat oxen, one of which was bred by G. V. Sacket, 1st prize.

BULL CALVES.

To Ezra P. Prentice, Albany, for his thoroughbred improved Durham bull calf Homer, 6 months old, bred by himself, 1st prize.

To Samuel Phelps, Ira, for his grade Devonshire, 3d prize.

TO BREEDERS.

To Francis Rotch, Butternuts, as the breeder of the best bull, prize.

To the same, as the breeder of the best cow, prize.

To the same, as the breeder of the best 2-year-old heifer, prize.

C. FORBES,
Henry, Ill.

"The premium to breeders having been offered by Mr. Rotch, he declined receiving more than a certificate of the award, leaving the money ($30) with the Society to be offered in premiums for the same purpose next year."—*Report of Committee.*

The foregoing we copy from the New York State Agricultural Report, 1841. We call attention to the extracts from the report of committees. This system should be enlarged and revised in present-day shows. Judges should be obliged to give written reasons for awards.

It will be noticed that all thoroughbreds showed in the competitions; that the committee made the recommendation that classes should be made for each breed; that they would not allow the Hereford to compete, but recommended a special premium to the Hereford

ROYAL GROVE (9137) 21500.
Bred by P. Turner, Herefordshire. Sensational bull, exhibited at Chicago Fat Stock Show, 1875.

cow Matchless, exhibited by Messrs. Corning & Sotham; and in looking over the names of the officers of the Society and the names of exhibitors, that they are in many instances the same; among them, Mr. H. S. Randall, who took so prominent a position in opposition to Mr. Sotham and the Herefords.

We are not disposed to charge fraud on these managers, but we have the right to show, as we have done in the Sotham-Randall controversy and the Youatt-Berry History of British Cattle, that these parties were in the interest of Shorthorns, and if Randall, Prentice, Allen, etc., as Shorthorn men, took the control of the New York State Fair, they would, as a matter of course, give the preference to that breed of cattle.

This was manifest in the Rust ox, which took the first premium as best fat ox. While they gave him the premium, they claimed him as a Shorthorn, and gave that breed the credit, whereas he undoubtedly owed what quality he had to the Hereford.

THE ILLINOIS STATE BOARD OF AGRICULTURE RECORD.

From the published records of the Board, we have taken pains to get together the names of the important officers and of the judges on Herds and Sweepstakes in the cattle departments. Leaving nothing to memory, we submit here the names of these judges and their residences so far as the reports of the society give them. It will be remembered that these judges passed upon the merits of cattle of different breeds, competing against each other. In 1872 John P. Reynolds was superintendent, and the Judges on Herds were as follows: John M. Milliken, Ohio; John H. Bacon, Iowa; M. Smith, McLean Co., Ill.; J. Reece, Warren Co., Ill.; W. H. Russell, Marion Co., Ill. Judges on Herds got by one Bull: J. R. Miller, Caseyville, Ill.; John Kelly, Heyworth, Ill.; John H. Potts, Jacksonville, Ill.; J. W. Hopkins, Granville, Ill.; J. F. Coe, Sterling, Ill.

1873: John P. Reynolds, President; Emory Cobb, W. H. Russell, Superintendents. Judges on Herds: G. Barnes, Canton; D. Rankin, Biggsville; J. C. Mosier, Kankakee; H. C. Reed, Princeton; E. F. Smith, Morning Sun, Ia. Judges on Herds from one Bull: M. W. Robinson, Des Moines, Ia.; W. W. Parish, Momence; J. R. Skelton, Skelton; Jos. Kelso, Tazewell Co.; H. J. France, Woodford Co.

1874: John P. Reynolds, President; Emory Cobb and W. H. Russell, Superintendents. Names of Judges not given.

1875: D. B. Gillham, President; W. J. Neely, Superintendent. Herds—Judges: P. A. Coen, Washburn; A. Kershaw, Wayne; W. Noel, Paxton; M. Sumner, Warren; W. Fuller, Clinton.

Herds from one Bull—Judges: Hugh N. Cross, Jerseyville; J. H. Potts, Jacksonville; H. Burruss, Carrollton; H. K. Parr, Seneca; W. Noel, Paxton. Sweepstakes—Judges: J. S. Overholt, Streator; C. L. Hostetter, Mt. Carroll; J. Barnes, Ottawa, Ill.; C. R. Wood, Yorkville; W. H. H. Holdridge, Tonica.

1876: D. B. Gillham, President; W. J. Neely, Superintendent; J. W. Judy, Marshal. Herds from one Bull—Judges: J. D. Van Doven, Fisks Corner, Wis.; J. H. Spear, Tallula; H. E. Williams, Dixon; A. Herford, Perona; J. W. Hundey, Champaign. Herds—Judges: James Mix, Kankakee, Ill.; J. R. Shaver, Ottawa; J. Brown, Galena; J. L. Moore, Polo; H. M. Winslow, Kankakee.

1877 at Freeport: D. B. Gillham, President; Samuel Dysart, Superintendent, J. W. Judy, Marshal. Herds—Judges: S. Riegle, D. C. May, A. Jeffry, N. Hawks, A. J. Wilbeck. Herd from one Bull—Judges: Edward Isett, W. Stocking, Simon Sheaf, J. M. Swaney, F. T. Seward. Judges of Sweepstakes: W. Rathje, John Gosling, Lewis Steward, W. Moffatt, C. D. Hart.

1878: Judges on Herds: D. Gore, Carlinville; Thomas Murray, Polo; Simon Sheaf; W. W. Riggs, Riggston. Judges Herds from one Bull: E. H. Stewart, Marengo; T. C. Sterrett, Warrensburg; H. Tennison; W. Vorhies,

"SIR BARTLE FRERE" (6682) 6419, BRED BY T. J. CARWARDINE, HEREFORDSHIRE,
And five of his sons, yearling grade steers, bred and exhibited by Adams Earl, Lafayette, Ind.

Vorhies; T. W. Shelton. Judges Breeders' Young Herd: L. W. Sheldon, Union; W. Vorhies, Vorhies; T. C. Sterrett, Warrensburg; H. Tennison, White Hall. Judges Sweepstakes: George Reed, Belvidere; David Shaff, Holcomb; J. E. Cronk, Belvidere; Wm. Stocking, Rochelle; Chas. M. Saxby, Freeport.

It will be found, after careful examination, that a large majority of these judges were in the interest of the Shorthorn breed of cattle. It was claimed that it was difficult to get judges that were not interested and partial to special breeds or interests; and, admitting this, we then urged the necessity of slaughtering, to test the award, and to our insistence on disinterested and impartial judging by men not interested in any breed can be traced the incipiency of the Fat Stock Show in America.

We hoped that breed prejudice might disappear when competition was limited to steer and fat cattle destined for slaughter, and that at a Fat Stock Show, "individual merit and prime qualities of meat" might be discussed, weighed and fairly judged. We hailed, therefore, the advent of the American Fat Stock Show with comfort, even though we knew it must be conducted under the same Illinois Board. We were destined to learn, however, that country butchers selected by Shorthorn breeders and their friends were liable to be, as we have felt, too loyal to the appointing power; nevertheless, to the establishment of the Fat Stock Show may be ascribed the commencement of Hereford supremacy.

In 1875 we appealed to the Illinois State Board of Agriculture for equal and even-handed justice at their hands. We were told "you are doing well; you are making headway as fast as you ought to expect; time will cure all your complaints." We replied by giving figures to show that the Herefords had a record that entitled them to an even classification and that we had a right to claim that the State Board of Illinois should not discriminate against them, but promised that, with or without the aid of this Board, the Hereford should go to the front. We said then, "They are reaching it; they will attain it; they will hold it." To prove to the Board that our claim to equal recognition of the Herefords with the Shorthorns was well founded, we prepared and presented to the Board the accompanying table:

GRADE STEER "REGULUS," AT 3 YEARS, WEIGHT 2,345 LBS.
Champion over all breeds, Chicago Fat Stock Show, 1885. Bred and exhibited by Fowler & Van Natta, Fowler, Ind.

Tabular Statement, Showing the Number of Prizes and Amount of Money won by each Breed of Cattle, during the period that all Breeds competed together, from the year 1799 up to 1851; after 1851 the Classification of Breeds took place. Excepting the years 1799, 1801, 1802, 1804, and 1806; the records being incomplete are omitted in all Breeds.

DATE	SHORT-H'NS.				HEREF'DS.				DEVONS.				L'G-HORNS.				SUSSEX.				SCOTCH, etc.				CROSS-BR'D.			
	Oxen or Steers.		Cows or Heifers.		Oxen or Steers.		Cows or Heifers.		Oxen or Steers.		Cows or Heifers.		Oxen or Steers.		Cows or Heifers.		Oxen or Steers.		Cows or Heifers.		Oxen or Steers.		Cows or Heifers.		Oxen or Steers.		Cows or Heifers.	
	No. of Prizes	Guineas	No. of Prizes	Guineas	No. of Prizes	Guineas	No. of Prizes	Guineas	No. of Prizes	Guineas	No. of Prizes	Guineas	No. of Prizes	Guineas	No. of Prizes	Guineas	No. of Prizes	Guineas	No. of Prizes	Guineas	No. of Prizes	Guineas	No. of Prizes	Guineas	No. of Prizes	Guineas	No. of Prizes	Guineas
1799																												
1800					4	52									1	8	1	10	2	20								
1801																												
1802																												
1803			1	15	4	80			2	25	1	10																
1804																												
1805					2	40	1	10	1	10											1	10			1	10		
1806																												
1807					2	30															1	10			1	20		
1808					3	40	1	10	2	40							1	20			1	10						
1809					2	40			2	30			1	20							1	10						
1810	1	20			3	50	1	10	2	30							1	20										
1811	1	20			1	10			1	20							1	20			1	10						
1812					2	40	1	10	1	20							1	20			1	10						
1813	1	20	1	10	2	40											1	29			1	10						
1814	2	40	1	10	2	45							2	45			1	20			1	10						
1815	1	20			4	90	1	10	1	20			1	20			1	20			1	10						
1816	1	20			2	40	1	10	1	20							1	20			2	35			1	25		
1817					1	25															1	25						
1818					3	50															2	35						
1819	1	25			2	40			1	10											2	25						
1820	3	60			4	55	1	10													1	10						
*1821					3	45															1	25						
1822					2	35	1	10	1	15											1	10						
1823	1	20			4	55	1	10													2	30					1	15
1824	2	40			1	15	1	15	1	10																		
1825	3	50	1	10	2	25															1	10					1	15
1826	2	30			3	45	1	15							1	10					1	10						
1827	1	20	1	10	3	45															1	10			1	10		
1828	1	£10	1	£10	4	£65													1	£15	1	£10						
1829	2	40	1	10	4	45	1	£15																				
1830	3	55	1	10	3	30	1	15													1	5						
1831	3	40	2	25	4	50																						
1832	1	20	2	15	4	55															1	10						
1833	2	35	3	25	5	55																						
1834	3	50	2	30	3	30																						
1835	2	30	2	30	5	60																						
1836	1	20	3	35	3	45															1	10						
1837	1	15	4	40	6	95			1	10											1	5						
1838	1	10	2	30	5	85	1	5	1	15	1	5										Sp						
1839	2	35	3	35	8	100									1	20						Cs						
1840	3	70	3	45	5	50									1	5					2	20			2	£15		
1841	5	45	4	50	6	100															4	30						
1842	3	45	4	45	7	100	1	10							1	5					2	20					1	£20
1843	3	60	7	80	6	70			2	25																		
1844	3	60	6	75	8	95	1	5																				
1845	2	25	4	40	5	65	2	25	4	65	1	15																
1846	1	15	7	85*	8	120			1	10											1	10			1	10		
1847	3	65	3	35	2	30	1	5	5	45					1	15					1	10			1	15	2	30
1848	6	70	4	70	4	85	3	25	3	40											1	10						
1849	4	85	6	85	5	50			5	60	1	10									1	10						
1850	6	80	7	95	5	65			3	50											1	10						
1851	1	15	6	75	9	140			3	40											1	10					1	20

* In 1821 there is a Prize of £10 adjudged to a Cow, but no breed recorded.

BREEDS.	Oxen or Steers. No. of Prizes.	£	s	d	Cows or Heifers. No. of Prizes.	£	s	d	Totals. No. of Prizes.	£	s	d
Herefords	185	2758	2	0	22	231	0	0	207	2989	2	0
Short-horns	82	1399	5	0	92	1132	15	0	174	2532	0	0
Devons	44	662	10	0	4	40	10	0	48	663	0	0
Scotch	43	500	15	0	..	....	..	.	43	500	15	0
Sussex or Kent	9	178	10	0	3	36	0	0	12	214	10	0
Long-horns	4	89	5	0	6	63	18	0	10	153	3	0
Cross-breeds	8	108	5	0	6	101	10	0	14	209	15	0

We expressly called the attention of the Board to the following table as proving the standing of the Herefords at Smithfield from the time Youatt's dishonest history appeared till Shorthorn control abolished the competition between breeds.

Winnings before the Smithfield Club from date of Youatt's history to 1851:

Year.	No. of Prizes Hereford Breed.	Guineas.	Dollars.	No. of Prizes Shorth'n Breed.	Guineas.	Dollars.
1836	3	45	225	1	20	100
1837	6	95	475	1	15	75
1838	5	85	425	1	10	50
1839	8	100	500	2	35	175
1840	5	50	250	3	70	350
1841	6	100	500	5	45	225
1842	7	100	500	3	45	225
1843	6	70	350	3	60	300
1844	8	95	475	3	60	300
1845	5	65	325	2	25	125
1846	8	120	600	1	15	75
1847	2	30	150	3	65	325
1848	4	85	425	6	70	350
1849	5	50	250	4	85	425
1850	5	65	325	6	80	400
1851	9	140	700	1	15	75
16 ys.	92	1295	6475	45	715	3575

Every cattle superintendent at the Illinois State Fairs from 1871 to 1874, inclusive, except Mr. Reynolds, was a Shorthorn breeder at the time of holding the position. At Peoria in 1874 there was a vacancy in the Committee of Judges on Herds, all breeds showing, when an outside man was called in. This man was the only one that examined the Hereford herd, and he said afterward that the other members of the Committee would not look at the Herefords. At Ottawa, in 1876, the Hereford exhibitors made a special appeal to the president, D. B. Gillham, to the superintendent, and other members of the Board, that the judges on herds be selected from men that were not Shorthorn breeders—men that should be impartial in their judging. The superintendent was active to get such a committee, and secured the assistance of the marshal of the ring (both Shorthorn breeders) in making the selection. They said to the Hereford exhibitors that they had taken special pains to get an impartial committee and that they had found a referee from the central part of the state. Who? Why, — — ——, the noted Shorthorn breeder, to pass upon the merits of the Shorthorns and Herefords. There were probably no men in the state of Illinois more unsuited to occupy the place of a judge, or to select impartial judges when the Herefords and Shorthorns came into competition. The referee was the head of the Shorthorn interest in Illinois; the marshal was the hand; the superintendent in his fealty to the Shorthorns, was all the Shorthorn men could desire in managing the Cattle Department to secure them in their position.

Now, if the Illinois State Board of Agriculture intended to be impartial, we submit that it was not in good taste to place Shorthorn breeders in absolute charge of the cattle department, and that they ought to have known that Shorthorn breeders alone were not competent to act as judges when Shorthorns competed with other breeds. Men who have been mentioned advocated special legislation in the board in favor of Shorthorns and against all others; while the Illinois State Board so organized their committees that control of the awards—at least a majority of them—was in the hands of Shorthorn breeders.

The Herefords were, therefore, forced to appeal from prejudiced societies, run by self-seekers, to that higher tribunal, the public. In voicing this appeal to the public, we became the target for ridicule. When we voiced the complaint of the Hereford men to the Agricultural Societies, we were called a "chronic kicker," and "anything to beat Miller" became an axiom of the opposition. When we saw certain contemporary Hereford breeders using this "anything to beat Miller" crusade, to further their own immediate ends, we abandoned the show ring and worked the more zealously through our "Breeders' Journal."

Not being able to prevent the publication of the "Breeders' Journal," our opponents attempted to lessen its influence by calling it a "trade circular" and heaping abuse upon its editor, and by subtly currying favor with other Hereford breeders in an attempt to have our work discredited in the Hereford camp. But they were unsuccessful, for although some so-called Hereford breeders allowed their temporary selfish interests to belittle them into a jealousy, for what they conceived to be their own position as Hereford men, the great and overwhelming majority have ever conceded us that recognition that has so gratified us.

J. J. HILL.
St. Paul, Minn. Celebrated railway magnate and lover of fine stock.

No honest, earnest, inspiring advocate of a cause can hope to present its merits without encountering opposition; an intelligent advocate will expect opposition. Honest opposition, from those representing a conflicting interest, is commendable; but underhanded opposition from within one's own camp is treason, despicable treason, that will be found born of jealousy. We would caution future generations of Hereford breeders to avoid the appearance of such contemptible evil. If there spring up Hereford advocates (as there have and will) willing to devote their time, talents and money to spreading the Hereford gospel, *all* that he may do or say may not be pleasing to each individual Hereford man; but take care that the overbalancing good be weighed against the objectionable little, and, above all, sink your jealousy and encourage such a man, for "appreciation is all there is in life."

GRADE HEREFORD STEER "DYSART," AT 3 YEARS. WEIGHT 1,890 LBS.
Exhibited by C. M. Culbertson, winner of the Breeders' Gazette gold challenge shield for best beast in the show bred and fed by exhibitor, Chicago, 1895.

The fair in 1877 was held at Freeport; a new superintendent entered upon his duties here with the determination to give to all exhibitors a fair show, and selected committees with this in view. We were enabled to take on herds open to all breeds, the second premium for the best bull and five cows; and first premium for the best five cattle, male or female, the get of one bull; the second for the best cow or heifer of any age. During this exhibition, in competition for herds and sweepstakes premiums, a prominent Shorthorn breeder of Missouri had an animated conversation with the superintendent, the drift of which I did not hear, but to which the superintendent replied: "Jim, Miller is going to have a fair show." In 1878 the State Fair was again held at Freeport, the regular superintendent being in France. At that show I said to the Vice-President: "I am entitled to have a fair show, and should like to have it without quarrels, and I am willing to submit to any committee of judges that President Gillham and ex-President Reynolds may select." Says he, "That is fair, and I will see that it is carried out."

He went to Gillham and Reynolds, and they made a selection of a committee under which were shown herds, open to all breeds, five cattle, male or female, the get of one bull. I took, with Success and his get, the first premium. This raised a row among the Shorthorn exhibitors. They took their complaints to the marshal of the ring, refusing to submit to the action of this committee, and for the competition for the balance of the herd and sweepstakes premiums, the Shorthorn representative made the selection of the judges and ran them entirely in the interest of the Shorthorns. Failing to get a hearing, I withdrew my cattle from the competition. The result of the Shorthorn men's action at this show I suppose to have been the means of changing the policy of the Board for the Fair of 1879, held at Springfield. They selected with care competent and impartial judges, whose services were paid for, and against whom no protests were to be heard. The Shorthorn exhibitors felt themselves at a disadvantage under this policy, and endeavored to break it up and select judges on the ground, but did not succeed in their scheme. They submitted, the Board refusing to change their plans.

It was not only known to me, but it was known to the Board, and it was known to the exhibitors who have met on the grounds, that, as far as the cattle department at the Illinois State Fair was concerned, the Shorthorn men undertook to control the judges so as to secure the awards themselves. They were in a great measure enabled to do so by their knowledge of men, by their associates, not exhibitors, and by members of the Board in their interest; and against these odds the Herefords had to contend for ten years.

The plan of the Shorthorn men during these years made the competition over the comparative merits of the two breeds a warm and exciting contest; and to avoid this, instead of following out the plan adopted for the judging at Springfield in 1879, by carefully selecting judges before coming upon the ground, they have withdrawn the competition as between breeds, thus shirking responsibilities that properly belong to them to meet.

What breeders and feeders want to know is this very thing, to-wit, the best breed. This test the Shorthorn men do not intend shall be made, unless they can make the judges. To this demand of the Shorthorn breeders, the State Board of Illinois, the State Board of Michigan, the St. Louis Fair Association, the Northern Ohio Fair Association and others have all surrendered, and discontinued the practice of bringing the Herefords and Shorthorns in competition. They demand the right to name the judges, and if not conceded, that the competition between breeds shall not be continued.

1879. The American Fat Stock Show was inaugurated by the Illinois State Board of Agriculture in 1879. We hailed it as the opportunity of the Herefords to exhibit their inherent thrift in a more practical way. The character of the awards at this show was the subject of comment by the press of the country. We have had something to say of them in Chapter XVII.

1880. The Herefords made a large show at the fairs of America in 1880. While most of the societies prohibited the competition of breed against breed, and thus lessened the interest in the shows, still the Hereford made rapid strides in numbers and qualities. At the New England Fair held at Worcester, Mass., J. S. Hawes, of South Vassalborough, Maine, exhibited sixteen head, and Burleigh & Bodwell, of Fairfield Center, Maine, exhibited twenty-six head. Both herds made an excellent show for the breed, Messrs. Burleigh & Bodwell taking the herd premiums. At the New York State Fair held at Albany, the Hereford exhibitors were Messrs. Burleigh & Bodwell, of Maine, and Erastus Corning, of Albany. At

HOTSPUR (7028) 9355.
Bred by J. Price, Herefordshire.

the Pennsylvania State Fair, the Hon. John Merryman, of Maryland, made a very fine exhibit from his herd. At Minneapolis, Minn., Col. W. S. King brought all breeds in competition. The leading Shorthorn herds of the Northwest were there. One of these had a $4,000 Duke bull at its head. The Hereford exhibitors here were C. M. Culbertson, of Newman, Douglas Co., Ill., and Shaw & Bullis, of Minnesota. The exhibit of Herefords was remarkably good, Mr. Culbertson taking the herd premiums over all breeds. At the NEBRASKA STATE FAIR Mr. G. S. Burleigh, of Mechanicsville, Ia.; Messrs. A. A. Crane & Son, of Osco, Ill., and Messrs. Swan Bros., of Cheyenne, Wyo., were the exhibitors. This show of Herefords was remarkably good, both in number and quality. The "Nebraska Farmer" offered a special premium for the best cow, steer or heifer in the show of any breed. This was won by Mr. G. S. Burleigh, of Mechanicsville, Ia., with his Hereford Princess B. (¶ 265). Thus at the only two fairs where breed was permitted to compete against breed, the Herefords

¶330

A TYPICAL JERSEY.

were the winners. At the IOWA STATE FAIR the Herefords were in strong force, but were not permitted to compete with Shorthorns. At the ILLINOIS STATE FAIR, Mr. C. M. Culbertson was an exhibitor, as was Mr. Thos. Clark, of Beecher, and Tom C. Ponting, of Stonington, Illinois. The show was very good. At the ST. LOUIS FAIR, the exhibitors were Culbertson and Clark.

It was out of place, when the stockmen and farmers of the country were making earnest inquiry for the best breed, that the Shorthorns should be drawn off. This list that we have given, though far from complete, shows how wide the range the Hereford had taken, in 1880. From Maine to Cheyenne on the north, and from Maryland to St. Louis on the south. A complete account would have carried us to Kansas City and Central Kansas, and when we further consider the fact that five hundred Herefords were brought from England that year and that none of these reached the show ground except the herd of Messrs. Burleigh & Bodwell, of Maine, the Hereford breeders had a right to feel encouraged by the progress the breed was making.

In 1881, at the Chicago Fat Stock Show, the following Herefords were exhibited:

PURE HEREFORD.

Name of Exhibitor.	Age in days.	Wt. Nov. 4, 1881.	Av. gain per day in lbs. since birth.	Name of Animal.
T. L. Miller Co....	1224	1965	1.60	Will.
T. L. Miller Co....	1242	1930	1.55	Washington.
T. L. Miller Co....	943	1700	1.84	Crystal.
T. E. Miller........	2243	1560	0.69	M'd. of the Mist 1365.
T. E. Miller........	1326	1335	1.00	Lady Elliott 2150.
T. Clark	1777	1410	0.79	May Queen 11th.
C. K. Parmelee....	1082	1215	1.12	Roxana.
C. K. Parmelee....	365	880	2.41	Anxiety 4th.
C. K. Parmelee....	209	500	2.39	Manito.
C. K. Parmelee....	251	520	2.07	Emeline 2d.
C. K. Parmelee....	256	565	2.20	Helena 6th.
C. K. Parmelee....	249	610	2.44	Calico.
C. K. Parmelee....	329	745	2.26	Beatrice.
C. K. Parmelee....	199	500	2.51	Magnet.
C. K. Parmelee....	193	400	2.07	Lady Tuscola.
GRADE HEREFORD.				
T. L. Miller Co....	1190	2145	1.80	Conqueror.
T. L. Miller Co....	1190	2035	1.71	Bachelor.
T. L. Miller Co....	1054	1925	1.82	Kansas.
T. L. Miller Co....	1145	1700	1.48	Tom Brown.
T. L. Miller Co....	1135	1550	1.36	Oxford.
C. M. Culbertson...	1039	1485	1.42	Spot Face.
C. M. Culbertson...	685	1025	1.49	Curley.
C. M. Culbertson...	1056	1675	1.58	Left.
C. M. Culbertson...	1056	1715	1.62	Right.
C. M. Culbertson...	1291	1835	1.42	Broad Horns.
C. M. Culbertson...	2059	1925	0.93	Beefy Back.
G. S. Burleigh.....	267	725	2.71	Rose.
G. S. Burleigh.....	622	1280	2.05	Bailey.
G. S. Burleigh.....	1065	1675	1.59	Ab. Piatt.
G. S. Burleigh.....	1694	2600	1.53	Star.
Ill. Ind. Univ.......	1129	1670	1.47	Burnham.
Ill. Ind. Univ.......	806	1335	1.65	Junior.

The following premiums were awarded:

Lot 2—Hereford, Thoroughbred.

Best steer 3 and under 4 years—First premium, Will; second, Washington; both exhibited by T. L. Miller Co., Beecher, Ill.

Best cow 3 years old and over—First premium, Maid of the Mist, exhibited by T. E. Miller, Beecher, Ill.; second, May Queen II, Thomas Clark, Beecher, Ill.; third, Lady Elliott, T. E. Miller, Beecher, Ill.

Lot 5—Grades and Crosses.

Best steer 3 and under 4 years—First premium, Hereford steer Conqueror, bred and exhibited by T. L. Miller Co., Beecher, Ill.

Best steer, 2 and under 3 years—Third, grade Hereford, Kansas, T. L. Miller, Beecher, Ill.

Best steer, 1 and under 2 years—Third, Bailey, grade Hereford, G. S. Burleigh, Mechanicsville, Ia.

Best cow, 3 years old and over—Second, grade Hereford, Beefy Back, C. M. Culbertson, Chicago, Ill.

Lot 9—Dressed Bullocks.

Best carcass of steer, 3 and under 4 years—First premium, grade Hereford, Broad Horns, C. M. Culbertson, Chicago, Ill.

Best carcass of steer, 2 and under 3 years, grade Hereford—Ab. Piatt, G. S. Burleigh.

Best carcass of steer, 1 and under 2 years—Premium to Bailey, grade Hereford, G. S. Burleigh, Mechanicsville, Iowa.

Lot 10—Dressed Bullocks, Sweepstakes.

Best carcass of steer of any age—Premium $75, to grade Hereford steer Broad Horns, C. M. Culbertson, Chicago, Ill.

Special Premium.

Marshall Field, Chicago, $250, for best five head cattle, any age or breed, awarded to T. L. Miller Co., Beecher, Ill., for five Hereford steers—Conqueror, Will, Washington, Bachelor and Kansas.

A Chicago daily thus comments on this show: "The event of this show was the contest for the prize of $250 offered by Marshall Field, of Chicago, for the best five head of cattle, of any age or breed. It was the largest prize of the show so far as value was concerned. There were seven herds entered by the following exhibitors: Shorthorns: T. W. Hunt, of Ashton, Ill.; John B. Sherman, of Chicago; John D. Gillette, of Elkhart, Ill.; Luther Rawson, of Oak Creek, Wis.; Morrow & Muir, of Clintonville, Ky., and H. C. Nelson, of Canton, Ill.

"The T. L. Miller Co., of Beecher, entered a herd of five Herefords. The ring was completely filled, with the finest animals of the show, and the judges experienced some difficulty in getting about. The gentlemen chosen to decide the merits of this grand herd ring were James Peltz, of Polo, Ill.; Frank Gerpiser, of Springfield, and Edward Leize, of Chicago. The handling qualities of every animal was determined and every point of excellence thoroughly discussed. So far as close searching and complete work was concerned no set of judges ever more fully satisfied the exhibitors or the public of their desire to do even and exact justice. After two hours of conscientious work the unanimous decision was in favor of the Herefords, to whom the prize was awarded. It was, indeed, a glorious and crowning victory, and to say that the fortunate winners were delighted but feebly expressed the state of their feelings. The aggregate of the five winning animals was exactly 10,000 pounds, making the average 2,000 pounds to the ounce—a showing that speaks volumes for the 'white-faced' breed."—*Chicago Tribune.*

William Houseman, writing from England in 1882 to the "National Live Stock Journal," Chicago, said:

"While trade in pedigree stock is, in general, quite at a standstill, and the suspension of business, usual at this time of year, may be marked by the lack of those advertisements of public sales, and of cattle to be disposed of by private bargain (which crowd the columns of the agricultural papers during 'the season') an uncommon commotion has been raised in the Hereford breeding districts by the presence of American Hereford breeders, who are buying up all the best Herefords they can lay hands on for very large shipments in the spring.

"In favor of the Hereford, it must be allowed, as the results of competition with other breeds during this past year, this breed has proved signally successful; at the Royal Dublin Society's show, where not only last year but for two years consecutively, a Hereford bull has won the Chaloner plate, value $775, as the best of any breed; at the Worcester County Show, where a Hereford heifer (Mr. Carwardine's Pretty Face), since exported to America by Mr. Culbertson, won the championship over Mr. Acker's famous Lady Carew 3d, the first prize Shorthorn cow of the Royal Agricultural Society of England, and at some other shows.

1882.—The campaign of 1882 was a memorable one.

The Tippecanoe County Fair at Lafayette, Indiana, for 1882, offered unusually liberal premiums for the beef breeds, consisting of three prizes for aged herds respectively $500, $150, and $50. They also offered a sweepstakes prize for the best herd of thoroughbred beef-breeding cattle, one bull, and four heifers under 2 years old, $100, $50, and $25. Also a sweep-

A WEST HIGHLAND BULL.

stakes for the best bull of any age or breed, the best cow of any age or breed, and the best steer of any age or breed.

The Hereford exhibitors consisted of Messrs. Earl & Stuart, of Lafayette; C. M. Culbertson, of Chicago; O. Bush, of Sheldon, Ill.; C. K. Parmelee, of Wolcott, Ind.; Fowler & Van Natta, of Fowler, Ind.; William Constable, of

Beecher, Ill.; Thos. Clark, of Beecher, Ill.; B. Hershey, of Muscatine, Ia.

Mr. Hershey's herd consisted of five yearlings, four heifers and one bull, which were purchased at the close of the fair by C. K. Parmelee for $3,000.

The Shorthorn exhibitors were J. H. Potts & Son, of Jacksonville, Ill.; Harvey Sodowsky, of Indianola, Ill.; L. Palmer, of Sturgeon, Mo.; Crofton Bros., of Winona; Thos. Wilhoit, of Middletown, Ind.; Wm. Stephenson & Son, of Little Indian, Ill.; Harper & Mason, of Wabash, Ind.

The great interest and excitement of the whole fair was in the sweepstakes ring. Here

¶332

GROUP OF RED POLL CATTLE.

on Friday morning were drawn up in grand array nine aged herds, all two years old and upwards, to compete for the $500 prize. Three of these herds were Herefords and six Shorthorns. Each herd consisted of one bull and four females. The committee after a long and careful examination gave first to J. H. Potts & Son's herd of Shorthorns, while the Hereford herd of Mr. C. M. Culbertson, although considered by many to be the best, had to be content with second. Mr. Culbertson certainly exhibited a grand herd and if his cow, Pretty Face, a prize winner in England before she was imported to this country, and considered an extraordinary plum there, had been the equal in size with the other three females in the herd, we think there would have been no doubt of his carrying off the first prize.

Owing to the time consumed in awarding the above premiums, it was now after one o'clock on the last day of the fair, so the committee that made the above awards was discharged and a fresh committee put to work on the sweepstakes for the best young herd under two years old. Here the Herefords and Shorthorns, as to number of herds, were equal, six competing, three Shorthorn and three Hereford. The superiority of the Herefords was more marked in this ring, and the committee were unanimous in placing the first prize on the Hereford herd of Fowler & Van Natta, the second going to the Shorthorn herd of J. H. Potts & Son, and the third to the Hereford herd of C. K. Parmelee. The young herd of Fowler & Van Natta consisted of entirely American bred cattle. The four heifers were remarkably even, and were universally admired. Two of them were bred by T. L. Miller, and two by T. E. Miller. The bull was a very blocky, heavy-fleshed, good-coated animal, bred by C. M. Culbertson.

In the contest for the best bull upon the ground, fifteen animals appeared in the ring, seven Herefords and eight Shorthorns. The display of Hereford bulls was remarkably fine, while among the Shorthorns there were only two that approached goodness. The prize was carried off by Messrs. Fowler & Van Natta, with their bull Tregrehan, a very smooth, even, short-legged bull. Mr. Constable's bull Hero, which won the sweepstakes in this same ring, the previous year, was shown here again, but failed to win, although he was in far better form than he was the year before.

The following is a list of the premiums awarded:

THOROUGHBRED HEREFORDS.

Bull three years old and over, 1st, Hero, (¶ 267) W. Constable, Beecher, Ill.; 2d, Anxiety 3d, (¶ 268) Thomas Clark, Beecher, Ill.

Bull two years old and under three, 1st, Royal 16th, (¶ 269) Earl & Stuart, Lafayette, Ind.; 2d, Romeo, Earl & Stuart.

Bull one year old and under two, 1st, Prince Edward, (¶ 270) Earl & Stuart; 2d, Anxiety 5th, Fowler & Van Natta, Fowler, Ind.

Bull calf under one year old, 1st, Jumbo, Earl & Stuart; 2d, El Paso Boy, W. Constable.

Cow three years old or over, 1st, Downton Rose, (¶ 271) C. M. Culbertson, Chicago, Ill.; 2d, Lady 3d, Earl & Stuart.

Heifer two years old and under three, 1st, Peerless, (¶ 272) Thos. Clark; 2d, Venus, (¶ 273) Earl & Stuart.

Heifer one year old, and under two, 1st, Miss Broadgauge 2d, Benj. Hershey, Muscatine, Ia.; 2d, Prettymaid, Earl & Stuart.

Heifer calf under one year old, 1st, Rachel, C. M. Culbertson; 2d, Crocus 3d, Fowler & Van Natta.

The above exhibit of Herefords in their class was a beautiful sight, but we think that of all the classes that of the yearling heifers was the finest. Eighteen of these choice young things were brought out in competition, and it was a difficult matter for the judges to select the best. The first prize female, Miss Broadgauge 2d by Horatius, was a very choice animal, with a square, massive top which won her, probably, the premium, although the winning heifer the following week at Crawfordsville was Viola, a heifer bred by T. L. Miller, from his fine old cow Victoria. (¶ 274)

The money that it would have taken to have bought this string of heifers would have been a small fortune of itself, as Messrs. Fowler & Van Natta were said to have refused $5,000 for the five that they had in the ring.

The following week, nearly the same parties showed at Crawfordsville, Ind., with the following results. The breeders of Hereford cattle, thinking it would be better to make a combined show of their cattle, resolved after their exhibition at Lafayette to keep together during the rest of the fair season on the principle that their combined forces could do more to break up the Shorthorn rings than they could if they separated and exhibited in smaller numbers over the entire country. In pursuance of the above resolution ten cars were ordered and the Herefords swept over to Crawfordsville in full force. Such an exhibit of stock had been unknown at this fair, and they were compelled to erect fifty additional stalls for the accommodation of the Herefords. They met here many of the Shorthorns with which they had competed the week before, and in the contest for the sweepstakes prize on young herds under two years old, the Shorthorns did not get a ribbon, the first, second and third premiums being all awarded to Hereford cattle.

The attendance at this fair was very large, and the weather being good was a great success. The prizes to Herefords in their class were as follows:

Thoroughbred Herefords: Bull three years old and over, 1st premium, Hero, Wm. Constable, Beecher, Ill.; 2d, Sir Garnet, (¶ 275) C. M. Culbertson, Chicago, Ill.

Bull two years old and under three, 1st, Tregrehan, (¶ 276) Fowler & Van Natta, Fowler, Ind.; 2d, Royal 16th, Earl & Stuart, Lafayette, Ind.

Bull one year old and under two, 1st, Anxiety 4th, C. K. Parmelee, Wolcott, Ind.; 2d, Prince Edward, Earl & Stuart.

Bull under one year of age, 1st, El Paso Boy, W. Constable; 2d, Jumbo, Earl & Stuart.

Cow three years old and over, 1st, Lady 3d, Earl & Stuart; 2d, Cherry 24th, (¶ 277) C. M. Culbertson.

Heifer two years old and under three, 1st, Venus, Earl & Stuart; 2d, Winnie 4th, Earl & Stuart.

Heifer one year old and under two, 1st, Viola, Fowler & Van Natta; 2d, Prettymaid, Earl & Stuart.

Heifer under one year old, 1st, Crocus 3d, Fowler & Van Natta; 2d ————, C. M. Culbertson.

Sweepstakes for best aged beef herd, 1st, J. H. Potts & Son, Shorthorn; 2d, H. Sodowsky, Shorthorn; 3d, C. H. Culbertson, Hereford.

Sweepstakes for young herd under two years old, 1st, Earl & Stuart, Hereford; 2d, Fowler & Van Natta, Hereford; 3d, C. K. Parmelee, Hereford.

Sweepstakes for best bull, any age or breed, 11 entries, six Herefords and five Shorthorns, 1st prize, Tregrehan (Hereford), Fowler & Van Natta.

Sweepstakes on cows, Harvey Sodowsky, Shorthorn, Indianola, Ill.

It will be seen by the above awards that the premiums were changed somewhat from the Lafayette awards. Sir Garnet is given here the second place, where he was overlooked the preceding week. This was a fine large bull, by The Grove 3d, who is the sire of some remarkably good cattle.

In the two-year-old bull class, Tregrehan is here given first place, while at Lafayette in his class, he was overlooked, while winning sweepstakes for best bull upon the ground. In the yearling, Anxiety 4th is here given first, while at Lafayette he was overlooked entirely.

AN ABERDEEN ANGUS CELEBRITY, BLACK PRINCE.
Bred in Scotland, exhibited at Chicago and Kansas City Fat Stock Shows, 1883-4. This steer did more to bring the Angus breed to notice in America than any other animal of the breed.

In bull calves, the position of the winners is reversed, El Paso Boy here taking the first, and Jumbo the second. For aged cows, Messrs. Earl & Stuart's imported prize winner Lady 3d,

VIOLET. FOWLER 12899. VIOLA 4020. PEERESS. MISS FOWLER. LASSIE 23204.

Grand Sweepstakes Herd over all breeds, Illinois State Fair, 1886. Bred and owned by Fowler & Van Natta, Fowler, Ind.

SIR BARTLE FRERE (6682) 6419.
EARL OF SHADELAND 12TH.
ELTON 1ST (9875) 11245.
GARFIELD (6975) 7015.
THE GROVE 3D (5051) 2490.
Shadeland Stock Bulls, property of Adams Earl, Lafayette, Ind.

takes first here, over Cherry 24th, who did not win anything at Lafayette. Lady 3d had seen her best days as a show animal, breeding as she did a calf every year.

In heifers one year old and under two, the heifer Viola, bred by T. L. Miller, took first here, while at Lafayette she was overlooked.

FAIR AT SPRINGFIELD, ILL., 1882.—Much to the disgust of the Shorthorn exhibitors, the breeders of Hereford cattle, with their grand display of over seventy head of white-faced beauties, loaded their cars at Crawfordsville, Ind., and sped away to the home of the Shorthorns, and pitched their tents at Springfield, Ill. The Shorthorn men wanted to know what right the Hereford men had to come out in such force, and fill the spectators with enthusiasm at the sight of such mountains of beef comprised in such compact bodies, mounted on short, fine-boned legs and covered with mellow hides and good coats of hair. These Shorthorn breeders felt particularly grieved that the Hereford men concluded to exhibit their cattle in mass, and concentrating their forces, thus make a display that the Shorthorn men had no chance of beating. They even went so far as to say it was a bulldozing movement, inaugurated to awe the different fair associations. As far as we heard the different fair managers express an opinion they were anxious to have this grand display come to their grounds, as it attracted many thousands of spectators, who liked to see improvement and did not want to be compelled to always see the same old barren Shorthorn cows from year to year.

The prizes of Herefords in their classes were as follows at this fair:

Bull three years old and over, 1st premium, Hero, Wm. Constable, Beecher, Ill.; 2d, Anxiety 3d, Thos. Clark, Beecher, Ill.

Bull two years old and under three, 1st, Royal 16th, Earl & Stuart, Lafayette, Ind.; 2d, Tregrehan, Fowler & Van Natta, Fowler, Ind.

Bull one year old and under two, 1st, Anxiety 5th, Fowler & Van Natta; 2d, Anxiety 4th, C. K. Parmelee.

Bull calf under one year, 1st, Jumbo, Earl & Stuart; 2d, Exchange, Earl & Stuart.

Cow three years old and over, 1st, Cherry 24th, C. M. Culbertson, Chicago, Ill.; 2d, Lady 3d, Earl & Stuart.

Heifer two years old and under three, 1st, Venus, Earl & Stuart; 2d, Peerless, Thomas Clark.

Heifer one year old and under two, 1st, Prettymaid, Earl & Stuart; 2d, Viola, Fowler & Van Natta.

Heifer calf under one year old, 1st, Crocus 3d, Fowler & Van Natta; 2d, Rachel, C. M. Culbertson.

Sweepstakes, best bull, Hero, Wm. Constable.

Sweepstakes, best cow, Cherry 24th, C. M. Culbertson.

Sweepstakes for best aged herd, Earl & Stuart (Herefords) 1st; C. M. Culbertson (Herefords) 2d.

Sweepstakes for best young herd under two years old, J. H. Pickrell (Shorthorns) 1st; Earl & Stuart (Herefords) 2d.

The show of grade Herefords at this fair was very fine, and in the grade classes the Herefords of Mr. J. R. Price, of Williamsville, Ill., won all he showed for, over all competitors.

Splendid exhibits were made at Missouri, Kansas, and Nebraska fairs in 1882 by F. W. Smith, and Whaley & Young of Missouri, and Fowler Bros., J. S. Hawes, and W. Morgan & Son.

BIRMINGHAM, ENG., FAT STOCK SHOW, 1882.—We give the following account taken from the "Birmingham Post" of the winnings of the Hereford cattle at this English Fat Stock Show:

"Herefords.—This breed is not so strong in point of number as it was last year, the stalls being 24 against 31, but it is again distinguished above all other competitors by contributing the premier animal of the show. Mr. John Price, (¶278) of Pembridge, once more beats everything with his grand ox by Grand Duke—Satin. The bullock has added about four hundred and twenty pounds to its weight by the year's additional feeding, and has developed still further the fine qualities which excited so much admiration at its previous exhibition. At the age of three years it now pulls 21 hundred weight, 1 qr., 14 lb. (2,394 lbs.), and every ounce of its meat is in the

WASHINGTON (8152) 22615.
Bred by A. E. Hughes, Herefordshire.

proper place. As it is bred and fed by the exhibitor, it carries the president's £25 sterling cup, as well as the £50 sterling, as best Hereford, and the Elkington challenge trophy, of the value of 100 guineas. As this last prize has been won by Mr. Price two years successively, it now becomes his own property. It is to be noticed as a remarkable and unexpected circumstance that this repeated victory has been gained by the same animal. The Queen's Shorthorn heifer was the most dangerous rival Mr. Price had to encounter in the final award, and there were loud shouts of applause in the hall when it was announced that the subject had defeated his sovereign—not on account of that feature of the case, for a royal victory would have been just as warmly welcomed, but as expressing the congratulations of spectators on an exhibitor having at length scored the double event required by the conditions of the Elkington Cup. This feat, however, is not the only sign which the show affords of the strength of the Pembridge herd. Mr. Price also takes the second prize for oxen, and both first and second in the younger class of Hereford steers, with bullocks which are all the progeny of Regulus."

The Elkington challenge cup, of the value of $500, won by Mr. John Price with a Hereford bullock, for the best animal in the show, was first placed on offer in 1873, and to be the property of the winner must be won two years in succession. (¶ 279) We copy from the "Mark Lane Express" the following statement: "The cup was won in 1873 by Mr. E. Wortley, in '74 by Mr. Robert Wright, in '75 by Mr. Richard Stratton, in '76 by Mr. Samuel Kidner, in '77 by Mr. Richard Stratton, (¶ 280) in '78 by Lord Lovat, in '79 by Mr. H. D. Adamson, in '80 by Mr. Peter Dunn, in '81 and '82 by Mr. John Price, Court House, Pembridge, Eng."

PURE-BRED STEER, RUDOLPH, JR.
Bred, fed and exhibited by Geo. F. Morgan, winner of grand sweepstakes Chicago Fat Stock Show 1886.

1883.—The Illinois State Fair.—Hereford class awards were as follows:

Bull three years old or over, 1st, Wyoming Hereford Association; 2d, G. S. Burleigh. Bull two years old and under three, 1st, G. S. Burleigh; 2d, T. M. George. Bull one year old and under two, 1st, George Leigh; 2d, Fowler & Van Natta. Bull under one year old, 1st, C.

M. Culbertson; 2d, O. Terrell. Cow 4 years old or over, 1st, C. K. Parmelee; 2d, T. M. George. Cow three years old and under four, 1st, and 2d, Thos. Clark. Heifer two years old and under three, 1st, Thos. Clark; 2d, C. K. Parmelee. Heifer one year old and under two, 1st, Fowler & Van Natta; 2d, Thos. Clark. Heifer under one year old, 1st and 2nd, C. M. Culbertson. In the sweepstakes: Hereford herd to consist of bull two years old or over, and a three-year-old, two-year-old, one-year-old, and under one-year-old female, the prize went to the Wyoming Hereford Association, of Cheyenne, Wyoming. At the head of this herd stood the remarkable bull Rudolph, (¶ 281) a grand animal, which combined great scale, with a smoothness and finish unequalled among his competitors. Best bull of any age, 1st, Wyoming Hereford Ass'n, Rudolph. Best female, any age, 1st, C. K. Parmelee, Princess. (¶282) Sweepstakes: Herd, all breeds competing, 1st, to Fowler & Van Natta's Herefords.

New York State Fair, 1883.—In the Hereford class the herd prize was awarded to the Hon. E. Corning, of Albany, for his bull Comus, and cows Filbert, Edna, Dorcas 5th, Duchess 8th, and Edna 2d. On aged bulls, E. Corning took first with Comus. On two-year-old bulls, J. L. Northrup, of Westfield, took first with Santa Claus. In bull calves, E. Corning took first with Kenwood. For aged cows, E. Corning took first and second with Filbert and Edna. In yearling heifers, E. Corning took first with Edna 2d. J. L. Northrup took first on heifer calves with Charmer 3d. On fat cattle there was a premium for oxen over four years old and under four years old. G. Ayrault, of Poughkeepsie, won first on the former and first and second in the latter; also second for aged fat cows and first for fat heifers. Hon. Erastus Corning, of Albany, took first for fat cow over four years old with his Hereford cow, Topsy Turvy. * * *

In 1883 Mr. W. E. Campbell, an extensive ranchman and cattle raiser of Caldwell, Kans., exhibited at the Kansas State Fair a fine herd of Hereford cattle right off the prairie, where they had grazed the entire season, and had had no other feed of any kind whatever, this fact being established by the affidavit of reliable parties posted on the stalls of the cattle. Notwithstanding these facts, the cattle were in excellent condition, and proved the superiority of the Herefords over all other breeds as grass or range cattle. The Equinox 2758 (¶ 283) standing at the head of this herd had proven an excellent stock getter as well as invincible show bull. He won first prize in the grand sweepstakes ring open to bulls of any age or breed, in a strong field of eighteen show bulls, representing the best Shorthorn, Hereford, Polled Angus and Galloway herds of Iowa, Illinois, Missouri and Kansas. He also carried off the first prize in his class, and his bull calf Jumbo and his heifer calf Lady Maud 4th each won first honors in their respective classes. In the sweepstakes ring for the best bull and five of his calves, open to all breeds, The Equinox again came to the front and won second honors, though his calves were very young and showed to bad advantage on that account. Queen of the Lillies, (¶ 284) out of Jessamine by Winter de Cote, that was first at Bismarck, was assigned a second place here after much hesitation. The winnings of this herd are unprecedented. It was a trump card for the Herefords as grass cattle. The merits of any thoroughbred race are best demonstrated by the quality of their progeny, and to practically demonstrate the superiority and potency of the Herefords over other breeds, Mr. Campbell exhibited the yearling heifer Texas Jane. This heifer was sired by a thoroughbred Hereford, and was out of a little scrub Texas cow. She weighed about 900 pounds, and had all the character and markings of a thoroughbred Hereford. She was universally admired and attracted much attention and comment from

SAMUEL WEAVER,
Decatur, Ill., America's greatest cattle feeder.

the public, to whom she gave the following account of her birth and breeding, through a placard conspicuously posted over her stall:

"I was born on W. E. Campbell's ranch, Aug. 19, 1882, and was at once christened Texas Jane.

"My father was a Hereford thoroughbred,
My mother a wild 'Texas scrub.'
The cross makes me easily fed,
And I am able to rustle for grub.

"Don't stare at the meat on my back,
Or be surprised at my snow-white face;
For it was all the work of papa,
That gave me this Hereford grace.

The Herefords won sweepstakes on bull of any age or breed; first and second sweepstakes on bull and five calves and second in sweepstakes on cow of any age or breed.

THE LAFAYETTE FAIR, 1883.—The best Herefords in America were again pitted against each other at the Lafayette fair. In the show for aged bulls there was Messrs. Fowler & Van Natta's Tregrehan who won the sweepstakes in 1882 as the best bull of any age or breed over a combined show of seven Herefords and eight Shorthorns of the year. Mr. Thomas Clark exhibited Anxiety 3d, a son of old Anxiety; and Mr. C. K. Parmelee exhibited Sir Garnet, a son of The Grove 3d, recently purchased from Mr. Culbertson. The result was, first prize to Mr. Thos. Clark's Anxiety 3d, and the second to Fowler & Van Natta for Tregrehan. There were no two-year-old bulls exhibited. On yearling bulls Messrs. Earl & Stuart took the first prize with Jumbo, and the second with Lord Horace. In bull calves the first prize went to Emperor 2d, owned by Earl & Stuart, and the second to Beckjay Hero, owned by Thos. Clark.

Of cows three years old and upward a splendid exhibit was made. The first prize went to Mr. Thos. Clark for his cow Peerless by Lord Wilton. This cow took the sweepstakes as the best Hereford female of any age at St. Louis, 1882. The second prize went to C. K. Parmelee, of Wolcott, Ind., for his cow Silvia. (¶ 285) In the two-year-old class the exhibit was an extraordinary one, and was most hotly contested, as there is seldom seen such a grand display of in-calf heifers. The first prize was awarded to Princess, a prize heifer, imported and owned by C. K. Parmelee. The

VIEW ON SAMUEL WEAVER'S FARM,
Decatur, Ill.

second prize went to Duchess 12th, also an imported heifer, and owned by Mr. Thomas Clark. In the yearling class, Messrs. Fowler & Van Natta took first prize with Crocus, Messrs. Earl & Stuart taking second with Belle. For heifer calves the first went to a calf by Tregrehan, bred and owned by Messrs. Fowler & Van Natta; the second went to Earl & Stuart for a calf by Lord Wilton.

In the sweepstakes ring for best herd, the Herefords and Shorthorns competed against each other. It will be remembered that at previous year's show, the aged herds consisted of one bull and four cows, and that the judges placed the first prize on a herd of Shorthorns, while the Herefords were forced to be contented with second place. This year, the rule was, herds to consist of one bull two years old or over, one cow three years old or over, one heifer two years, one heifer one year, and one under one year old. This year the fortunes were reversed and the first prize was placed on Fowler & Van Natta's herd of Herefords, at the head of which stood Tregrehan, and among the females were Viola, Crocus and Actress. The second place went to Mr. Wilhoit's Shorthorn herd. The third prize went to Mr. Thomas Clark's Hereford herd, at the head of which stood Anxiety 3d, and the females of which were Peerless, Duchess 12th, Silver and Flossie.

In the young beef herds, under two years old, the Shorthorns did not draw a prize. The first prize was awarded to Messrs. Fowler & Van Natta's young herd of Herefords; the second was awarded to Messrs. Earl & Stuart, who had a choice herd of young things, at the head of which stood the bull Jumbo.

New England Fair, 1883.—The New England Agricultural Society, which held its annual meeting at Manchester, N. H., September 4th to 7th, 1883, had a good display of stock. In the Hereford class there were awarded to J. P. Kimball, of Canterbury, N. H., first and second for two-year-old bulls, second prize for yearling bull, first for bull calf, first and second for aged cows, first for two-year-old heifer, first for yearling heifer, and first for heifer calf. To Seth Quimby, of Bow, N. H., first prize for yearling bull. The sweepstakes gold medal for best bull was given to J. P. Kimball for his bull Major, and a silver medal for his cow Mollie Holmes. Mr. Kimball also took the Hereford herd premium.

Mr. W. P. Small, of Canterbury, N. H., in the grade class showed Herefords and won first and second prize for two-year-old heifer and first for heifer calf, against other breeds. He also took first and second prize for two-year-old steers broken to the yoke, and first prize for yearling steers broken to the yoke. The "Country Gentleman" had the following to say of this fair:

"The Herefords attracted great attention. One point of value operating to make this breed and the Devons popular in New England is the beauty and usefulness of the oxen. The best yokes of working oxen and fat cattle at this show were of these breeds. W. P. Small, of Canterbury, and J. P. Kimball, of the same place, showed over twenty head each, and there was not a poor Hereford in the pens. The New England Hereford breeders call these cattle excellent in the dairy as well as for working and the shambles. Judging from the interest shown at this fair the Hereford is gaining in popular favor in the Eastern States."

At the Iowa State Fair, 1883.—The Hereford breeders were: Wyoming Hereford Association, of Cheyenne; Mr. Ben. Hershey, of Muscatine, Ia., Swan Bros. & Kaufman, of Indianola, Iowa; G. S. Burleigh, of Mechanicsville, Iowa; W. C. McGavock, of Franklin, Mo.

Best aged bull, first prize, Wyoming Hereford Association (Rudolph); second, G. S. Burleigh. Best two-year-old bull, first prize, G. S. Burleigh. Best one-year-old bull, first

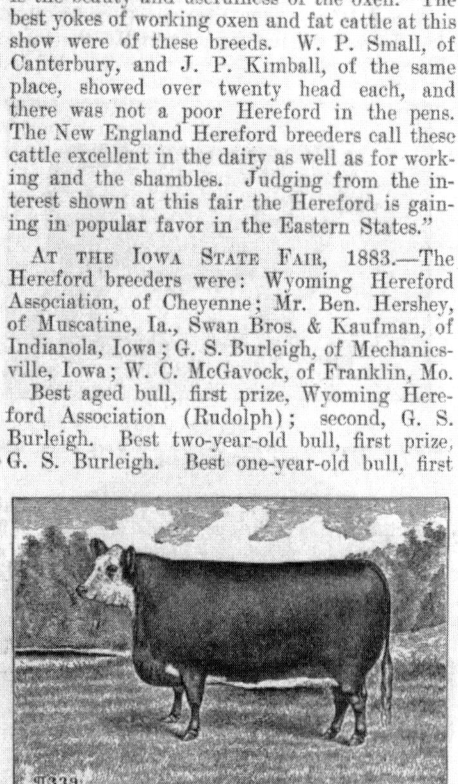

BURLEIGH'S PRIDE.
Cross-bred from Angus sire and Hereford dam by H. C. Burleigh, Vassalboro, Me.

and second prizes, Wyoming Hereford Ass'n. Best bull calf, first prize, Wyoming Hereford Ass'n.

Aged cow, first prize, G. S. Burleigh; second, Wyoming Hereford Ass'n. Best two-year-old female, first prize, Swan Bros. & Kaufman; second, W. C. McGavock. Best one-year-old heifer, first prize, Wyoming Hereford Ass'n; second, G. S. Burleigh. Best heifer calf, first and second prizes, Wyoming Hereford Ass'n.

Sweepstakes: All breeds competing. Third prize, Wyoming Hereford Ass'n.

The Platt Prize Hereford Steer.—The champion Hereford steer at Smithfield, 1883, was bred by Mr. Frederick Platt, of Upper Breinton, Hereford, England. This steer was two years, four months and two weeks and four days old, and was bred from the Hereford stock of Mr. Aaron Rogers on the dam's side. His sire was Horace 2492 (3877). (¶ 286) This bull Horace was the animal that Mr. Platt paid $2,500 for when he was ten years old. He had

PLUSH.
Grand sweepstakes carcass, Chicago Fat Stock Show, 1886. Fed by John Gosling.

been the stock bull in Mr. John Price's herd at Court House, Pembridge, Eng. He was a bull of remarkable substance, and covered with a curly coat of hair that is seldom equaled. He was such a prepotent sire that he conveyed his good qualities to most of his descendants, and many of his get have been prize winners in the show ring.

The steer weighed 1,680 pounds, and won in a ring of ten competitors. There were shown in the same ring a steer bred by Mr. Thos. Lewis, of Woodhouse, sired by Young Sir Frank 2669 (4274); two more steers bred by Mr. Lewis Lloyd, of Monks Orchard, sired by Lord Wilton 4057 (4740), of which so much was said as a stock getter; another, bred by J. R. Hill, by Commander 3209 (4452), the Mrs. Edwards bull. The second prize in the ring, however, went to a second steer exhibited by Mr. Platt, which was also by Horace. The third prize went to a steer bred by the late Mr. Morris, of Weston.

We quote the following from the "Breeders' Journal" report of the Kansas City Fat Stock Show for 1883:

The two important exhibits of live stock in America for 1883, the first at Kansas City, the second at Chicago, have made a long stride toward settling the merits of the breeds.

The Kansas City Show was the first that had been held there, and was a very creditable opening. The Herefords were represented by W. E. Campbell, of Caldwell, Kansas, with a yearling heifer and steer from Texas cows and a Hereford bull and five heifer calves. This exhibit was one of the most interesting and attractive of the show, as indicating the potency of the Herefords and their ability to improve the common cattle of the country. They were of fine quality, and would have passed without many doubts for thoroughbreds.

Fielding W. Smith, of Woodlandville, Mo., had on exhibition five yearling steers, the get of Dictator, (¶ 287) the champion bull at St. Louis in 1881 and 1882. These were out of Shorthorn cows, and were of great substance; one of them was as perfect a beast as there was in the show, and we hope to see him and his master another year, the steer having passed the interim in the hands of a good feeder.

Messrs. Scott & Broaders, of Leavenworth, Kansas, exhibited a half-blood Hereford steer—a very smooth, level animal, not fat enough to win, but an excellent butcher's beast. Messrs. Gudgell & Simpson, Independence, Mo., exhibited a very good pair of pure-bred Hereford steers. Messrs. A. A. Crane & Sons, of Osco, Ill., exhibited a pair of thoroughbreds and one grade Hereford bullock. T. E. Miller, of Beecher, Ill., exhibited a pure-bred, fat Hereford cow. Messrs. Seabury & Sample, of Lafayette, Ind., exhibited a yearling grade Hereford steer of good quality. Messrs. Fowler & Van Natta, of Fowler, Ind., made an exhibit of the grade Hereford steer Benton's Champion. Thomas Clark, of Beecher, Ill., exhibited his yearling grade Hereford steers Nip and Tuck. These attracted a good deal of attention, and were very creditable steers.

The Herefords made a very creditable show, Adams Earl winning on Hereford bullocks three and under four years old, with Wabash (¶ 288) first premium; A. A. Crane & Sons taking second. A. A. Crane & Sons also took first premium on Hereford bullocks two and under three years old. Gudgell & Simpson took first and second on Hereford steers one and under two years. T. E. Miller took first premium on Hereford cow three years old and over; Gudgell & Simpson taking second. Adams Earl took special premium, a farm wagon, for best Hereford animal in the show, on pure-bred steer Wabash. W. E. Campbell took special premium of $25 on Hereford cow.

The Shorthorns were fairly represented, J.

H. Potts & Son taking the champion prize for best steer in the show with the three-year-old Shorthorn steer Starlight. Fowler & Van Natta took champion prize over all breeds for best two-year-old in the show with Benton's Champion. Thos. Clark took champion prize for best yearling beast in the show with the steer Tuck.

With the figures before us, it is well to make the comparison that they enable us to do. The champions by ages for the highest honors—the best beasts in the show—were Starlight, Benton's Champion, and Tuck. Their ages and weights were as follows:

Age in days.	Wt.	G'n p'r day.	Exhibitor.	Breed.	Name of Bullock.
1382	2170	1.57	J. H. Potts & Son.	G. S'horn.	Starlight.
911	1885	2.06	Fowler & VanNatta.	G. H'ford.	B. Cham.
526	1240	2.35	Thos. Clark.	G. H'ford.	Tuck.
1498	2350	1.66	Adams Earl.	T. H'ford.	Wabash.

It will be noticed that Starlight is one year three and one-half months older than Benton's Champion, and only weighed 285 pounds more. The only open question is the one of quality. If Starlight was in quality much better than Benton's Champion, it might be an excuse or reason why the championship should have been given him, but we think that the steer Benton's Champion was the better quality. And if Tuck was slaughtered he would show a better quality of meat than either, though he had not as much weight for age as had Benton's Champion. There is a tendency with judges to give the awards to the larger bullock.

It is perhaps well to compare the merits of Starlight and Wabash. Wabash is 116 days or nearly four months older than Starlight, and weighed 180 pounds the most. Although four months older, he has made a gain per day of 1.66, while Starlight has only made 1.57.

A feature at this show was the sale of Herefords by Mr. Adams Earl, of Lafayette, Ind., making an average of nearly $600—nearly or quite $100 more than the best sale of Angus or Galloways.

The Chicago Fat Stock Show, 1883, opened (to quote the "Breeders' Journal" report), with one of the largest exhibits ever made since its first opening, and with more satisfactory financial results. We have no need to call the attention of our readers to the management of the show held a year ago; to the gross and glaring frauds as to the ages of steers exhibited by Shorthorn exhibitors, and the protection which they had from the president of the board and the Shorthorn element in and out of the board.

We have no need to call the attention of our readeis to the strenuous efforts made by the president and his associates to defeat the resolution requiring all animals that should be exhibited at future shows to undergo a critical examination by a competent veterinary surgeon and two competent and experienced breeders of cattle, to determine their ages by the dentition of teeth, being able to defeat the resolution at the January meeting, but were not able to defeat it at the February meeting.

The show opened on the 14th ult. with the following entries: 95 Herefords or their grades; about 300 Shorthorns or their grades; a few Scots and Holsteins. In the classes where all breeds competed, the premiums awarded to Herefords were as follows:

Grades or crosses: Best steer or spayed heifer three and under four years, 1st, C. M. Culbertson, Roan Boy, grade Hereford. Best steer or spayed heifer two and under three years, 1st, Fowler & Van Natta, Benton's Champion, grade Hereford. Best steer one and under two years, 2d, Thomas Clark, Tuck, grade Hereford. Sweepstakes: Best steer or cow in this class, C. M. Culbertson, Roan Boy, grade Hereford. Sweepstakes: Best steer or spayed heifer three years and under four, judged by feeders, Mr. Culbertson's Roan Boy.

HENRY F. RUSSELL.
Westonbury, Herefordshire.

Grand Sweepstakes: Best steer or heifer any age or breed, awarded to Mr. Culbertson's Roan Boy.

Best lot of eight cattle three years and under four, all breeds competing, 1st, John B. Sherman, Hereford and Shorthorn.

Best twelve cattle one year and under two, all breeds, 1st, Fowler & Van Natta, Herefords; 2d, C. M. Culbertson.

In the competition for cost of production, the Herefords won in the class for steer or spayed heifer two and under three years, first premium on spayed heifer Hattie, exhibited by G. S. Burleigh, Mechanicsville, Ia.

And again in the class for steer or spayed heifer one and under two years, first premium to steer Stonington, exhibited by R. J. Stone, Stonington, Ill.

A large amount of cash in special premiums given at this show from fund collected by the

WESTONBURY, PEMBRIDGE.
Home of Mr. H. F. Russell.

American Hereford Breeders' Association was won by the following exhibitors: Adams Earl, A. A. Crane & Son, C. B. Stuart, C. M. Culbertson, Fowler & Van Natta, B. Hershey, John B. Sherman, Thomas Clark, F. W. Smith, Tom C. Ponting, H. Norris & Son.

The special premium offered by McCormick Harvesting Machine Company, for best five cattle, was won by John B. Sherman, with grade Herefords.

There were a large number of beasts slaughtered. For the best three-year-old carcass we think the Herefords improperly beaten, the first premium being given to the Shorthorns, and also for the best carcass; and we believe the Culbertson steer, Roan Boy, was the best bullock, the most level, and of the best quality. The Potts steer was uneven. If the judges gave the award for the greatest amount of outside fat, then perhaps there was less room for criticism.

THE SHOWS OF 1884.

A good showing of Herefords was made in 1884 at the leading shows. At the ILLINOIS STATE FAIR by Thomas Clark, Geo. Leigh & Co., E. Esson and Jas. Caldwell.

At the ST. LOUIS (MO.) FAIR the Hereford prizes were divided between Messrs. C. M. Culbertson, Earl & Stuart, Wm. Constable, and F. W. Smith.

A determined effort was made this year by the Hereford herds of MICHIGAN, marshaled by the veteran Hereford advocate, Wm. H. Sotham. The leading herds, those of Thomas Foster and Wm. Hamilton, of Flint, were placed in preparation under the management of T. F. B. Sotham, who was given an interest in their winnings. These and other Michigan herds made the rounds of the leading County and District fairs, and at the State Fair made the greatest Hereford show ever seen on its grounds. Mr. R. G. Hart, of Lapeer, and Edwin Phelps, of Pontiac, adding their herds to those above mentioned.

A special fund of $2,000 was raised among the Hereford breeders for Herefords at the AMERICAN FAT STOCK SHOW, and much good work was done by enthusiastic effort.

The Herefords at the ROYAL AGRICULTURAL SHOW at Shrewsbury (1884) scored an unparalleled success; the number of Hereford entries largely exceeded those of any other breed of cattle, and the quality throughout was so marked that they monopolized the attention of visitors. There were 144 entries, but though they have no class for family groups, there were classes for pairs of bulls, classes for pairs of heifers, and classes for four heifers, so that the number of animals exhibited was much larger than that.

The first prize for old bulls fell to Mr. Aaron Roger's Archibald (¶ 289), the second to Lord Coventry's Fisherman (¶ 290).

The first prize for cows in milk or in calf was awarded to Lord Coventry's Golden Treasure.

The first prize for cow and two offspring was awarded to Mr. H. W. Taylor for the cow Rosamond and her produce.

The first prize for four heifers went to the executors of Mr. T. J. Carwardine, and the first for pairs of heifers was awarded to Mr. John Price (¶ 291).

There was never such a show of Herefords made in the world before.

1885.—AT THE FAIRS (¶292.)

The ILLINOIS STATE FAIR was held at Chicago, on the grounds of the Driving Park, near Douglas Park, in the western part of the city. A Hereford exhibit creditable to the breed was made by Messrs. Fowler & Van Natta, of Fowler, Ind.; Iowa Hereford Cattle Co., of Indianola, Ia.; J. C. Bertram, of Aurora, Ill.; C. M.

Culbertson, of Chicago; A. C. Reed (¶ 293), of Goodenow, Ill.; E. E. Esson (¶ 294), of Peotone, Ill.; Jas. H. McEldowney (¶ 295), of Illinois, showing about 60 head all told.

At the Fair at Bismarck, Kansas, the Hereford exhibit was by J. S. Hawes, Shockey & Gibb (¶ 296, ¶ 297), Lucien Scott, G. A. Fowler, all of Kansas (¶ 298).

At the Iowa State Fair, for the sweepstakes for the best herd, there were ten herds shown of the different breeds. Third premium was gained for the Herefords by the Iowa Hereford Cattle Co. (¶ 299). A. A. Crane, of Illinois, and Ford & Drummie, of Iowa, made up a fine exhibit (¶ 300).

The Michigan State Fair witnessed another splendid show of Herefords. This year the exhibit came from the herds of William Hamilton, Thos. Foster, Foster & Pearsall, Edwin Phelps, R. G. Hart (¶ 301), of Michigan, and F. H. Johnson & Co. (¶ 302), of Indiana. There was as usual in Michigan no competition between breeds.

At the Indiana State Fair, the Herefords were represented by the Indiana herds, F. H. Johnson & Co., of South Bend, and the Indiana Blooded Stock Co., of Indianapolis, making a most creditable show.

The Herefords were exhibited at Nebraska State Fair 1885, by J. S. Hawes (¶ 303), Kansas, his herd being headed by Sir Evelyn (¶ 304). The rest of the exhibits were made by Chas. M. Sears, of Aurora, Neb.; E. E. Day, of South Bend, Neb.; A. A. Crane, of Osco, Ill., the latter herd headed by Equinox 2d, a 2-year-old weighing 1,700 pounds. There was no sweepstakes between breeds (¶ 305, ¶ 306).

At the English Royal Show, 1885, there was no competition between breeds. The Herefords were well represented. Exhibitors were: J. Price, Earl of Coventry, H. R. Hall (¶ 307), H. W. Taylor (¶ 308), J. Rankin, M. P.; W. Tudge, C. Knott, T. S. Minton (¶ 309), J. H. Arkwright, A. Hughes (¶ 310), F. J. Gough, S. Miller, A. P. Turner (¶ 311), J. Naylor, J. R. Hill, S. Robinson (¶ 312).

The Show was a very creditable one to the Hereford cattle.

At the Gloucester Agricultural Show, among the Shorthorn exhibitors were some of the best of the Shorthorn breeders. "Bell's Weekly Messenger" says of the Herefords: "They made a magnificent display." The aged bulls were H. W. Taylor's Maidstone (¶ 313), Lord Coventry's Good Boy, H. R. Hall's Hotspur (¶ 314). They stood first, second and third in the order named. Mr. Taylor's Maidstone took the champion prize for the best male animal of any breed in the show, and his Vanity 7th likewise for the best female in the show. "If these honors had not been repeated so often in England this year," says the "Messenger," "we might attempt to enlarge upon these awards." An interesting fact in connection with these sweepstakes awards is that of the three judges, two were Shorthorn breeders.

At the Royal Counties Show, 1885, at Southampton, all breeds in competition, Earl Coventry took the champion prize with his Hereford bull Goodboy, defeating the Shorthorn, Sussex (¶ 316) and Devon (¶ 317) breeds.

Mr. John Price took champion prize for bull and offspring at Brighton.

At the Oxfordshire Fair, 1885, the Earl of Coventry won sweepstakes with Hereford bull Goodboy and Hereford cow Golden Treasure, in competition with all breeds. Mr. Joseph Darby, writing to the "Farmer," says: "Mr. Handley's Self Esteem 2d (¶ 318), which here came into collision with the Earl of Coventry's Good Boy, had for the past two seasons been the crack Shorthorn of the period."

Chicago Fat Stock Show, 1885.—At the American Fat Stock Show, 1885, the Herefords scored a splendid victory. The exhibit was one of the best made at this show, and the Illinois State Board of Agriculture was entitled to great credit for the advancement they had made in the selection of judges. It

PURE-BRED STEER.
Bred by F. Platt, Barnby Manor, Newark-on-Trent, Champion at Smithfield, 1886.

was claimed by the Shorthorn men that they had not suitable preparation for this show. This statement was made by them to account for their defeat, but they proposed to come the next year with a stronger exhibit. It is well to consider whether this claim for want of readiness for this show was valid or not. First in their three-year-old class, Messrs. Morrow &

Renick came with Schooler (¶ 319), Bedford 4th, Renick Sharon, and Weathers, the pick of the Kentucky herds, and high up in the best class, and they ought to be good representatives of the best Shorthorn blood, and we believe were fully up to anything that the Shorthorn men ever exhibited at this show, having individual and uniform merit. Their two-year-old exhibits were very creditable to the breed, and in their one-year-old, they may consider themselves fortunate if they can produce another Cleveland (¶ 320). Under one-year-old they were certainly very weak. In grades and crosses they were represented by such breeders as George Elliott, of Harristown, Ill.; J. H. Potts & Son, Jacksonville, Ill.; B. Waddell, Marion, Iowa; John B. Sherman, Chicago, Ill.; C. S. Barclay, W. Liberty, Ia.; J. R. Peak, Winchester, Ill.; Forbes Bros. (¶ 322 and 323), Henry, Ill.; T. W. Hunt, Ashton, Ill., and John D. Gillette. Our readers will recognize among these some of the leading Shorthorn breeders of the country.

F. PLATT.
Barnby Manor, Nottinghamshire.

Mr. Gillette, especially, came with a world-wide reputation as a bullock breeder, and is supposed to have done more to give reputation to the Shorthorn breed of cattle than any other living man, and while his cattle rank high as grades and crosses, the fact still remains that Mr. Gillette gave his life work to developing the Shorthorn as a butcher's beast, having all this time selected the best sires from among the Shorthorn families.

The Polled Angus breeds were represented by six thoroughbreds and fifteen grades or cross breeds.

The Holsteins had a small representation in thoroughbreds and grades.

The Hereford exhibits were creditable to the breeders (¶ 321, ¶324), both in thoroughbreds and grades. For the winnings we refer to our account in the different classes. The judging for the different breeds in their purity was comparatively a tame affair.

The interest culminated when the three-year-old grades and crosses were brought into competition, and the exhibits were as follows:

Exhibitor and animal.	Weight.	Blood.
C. M. Culbertson, Scott	2140	G. Hd.
C. M. Culbertson, Schuttler	2106	G. Hd.
C. M. Culbertson, Reynolds	2055	G. Hd.
C. M. Culbertson, Dysart	1890	G. Hd.
C. M. Culbertson, Pearce	1945	G. Hd.
J. B. Sherman, Albright	2220	G. Hd.
Fowler & Van Natta, McGregor	2345	G. Hd.
Fowler & Van Natta, Regulus	2345	G. Hd.
Adams Earl, Dick	2160	G. Hd.
Adams Earl, Excelsior	2070	G. Hd.
Lucien Scott, Jersey	1460	Hd & J'rs'y
Ind. Bl'd S. Co., Stonington	1900	G. Hd.
J. B. Sherman, Warfield	2125	S. H. G.
J. B. Sherman, Richland	2185	S. H. G.
J. B. Sherman, Longmore	2220	S. H. G.
John D. Gillette, J. P. Reynolds	2100	S. H. G.
John D. Gillette, Right-About	2050	S. H. G.
John D. Gillette, McCue	2015	S. H. G.
George Elliott, Little Britain	1765	S. H. G.
J. H. Potts & Son, Champion	1940	S. H. G.
J. H. Potts & Son, Surprise	1685	S. H. G.
Swan & Bosler, Jake	2326	S. H. G.
Morrow & Renick, Eales	2320	S. H. G.
John D. Gillette, Barney	2245	S. H. G.
John D. Gillette, Cherry	1750	S. H. G.
B. Waddell, Dan	2080	S. H. G.
Lucien Scott, Boots	2050	H. G.
James J. Hill, Hutcheon	2455	An. G.
James J. Hill, Turiff	2140	An. G.

The first premium went to Fowler & Van Natta, and the second and third to Mr. Culbertson. The exhibit of two-year-old grades and crosses was large and as follows:

Exhibitor and animal.	Weight.	Blood.
C. S. Barclay, White Cedar	1900	S. H. G.
Charles B. Stuart, Texas Boy	1500	G. Hd.
Seabury & Sample, Joe	1705	G. Hd.
Fowler & Van Natta, Dandy Boy	1905	G. Hd.
Fowler & Van Natta, Benton Will	1805	G. Hd.
Fowler & Van Natta, Adams	1930	G. Hd.
Fowler & Van Natta, Peter	1304	G. Hd.
B. Hershey, Prospect	1920	G. Hd.
J. H. Potts & Son, Red Cloud	1810	G. S. H.
J. H. Potts & Son, Victor	1805	G. S. H.
J. H. Potts & Son, Major Taylor	1580	G. S. H.
J. H. Potts & Son, Richmond	1545	G. S. H.
J. H. Potts & Son, Billie S.	1640	G. S. H.
J. H. Potts & Son, Mack	1610	G. S. H.
J. R. Peak & Son, Roan Boy	1685	G. S. H.
J. R. Peak & Son, Wild Bill	1580	G. S. H.
John D. Gillette, Driver	1760	G. S. H.
John D. Gillette, Rocket	1900	G. S. H.
John D. Gillette, Victor	1775	G. S. H.
John D. Gillette, Doc Woods	1865	G. S. H.
John D. Gillette, Col. Mills	1595	G. S. H.
John D. Gillette, Constance	1655	G. S. H.
John D. Gillette, Red Plum	1560	G. S. H.
John D. Gillette, Cuba	1770	G. S. H.
John D. Gillette, Cain	1730	G. S. H.
J. B. Sherman, Gem	1840	G. S. H.
J. B. Sherman, Turk	1620	G. S. H.
Forbes Bros., R. Brodus	1800	G. S. H.
Forbes Bros., Varna	1690	G. S. H.
Lucien Scott, Captain	1920	G. Hol.

In the two-year-old class grades and crosses the Herefords found their weakest place. The first premiums went to J. H. Potts & Son, the second to John B. Sherman, both first and second Shorthorn grades, the third to Fowler & Van Natta's grade Hereford.

The yearlings were as follows:

Exhibitor and animal.	Weight.	Blood.
Adams Earl, Sir Bartle 1st	1470	G. Hd.
Adams Earl, Sir Bartle 2d	1515	G. Hd.
Adams Earl, Sir Bartle 3d	1420	G. Hd.
Adams Earl, Sir Bartle 4th	1295	G. Hd.
Adams Earl, Sir Bartle 5th	1325	G. Hd.
Adams Earl, LaFayette	1220	G. Hd.
Swan & Bosler Co., Roan Dan	1155	G. Hd.
Fowler & Van Natta, Sam	1355	G. Hd.
Fowler & Van Natta, Matthews	1280	G. Hd.
Ind. Blooded Stock Co., Lindall	1555	G. Hd.
Ind. Blooded Stock Co., Leland	1475	G. Hd.
Ind. Blooded Stock Co., Photograph	1290	G. Hd.

Exhibitor and animal.	Weight.	Blood.
Seabury & Sample, Bob	1435	G. Hd.
Seabury & Sample, Mort	1390	G. Hd.
Seabury & Sample, Sam	1185	G. Hd.
Seabury & Sample, Jack	1630	G. Hd.
J. H. Potts & Son, Bob Moore	1325	G. S. H.
J. R. Peak & Son, Tommylin	1140	G. S. H.
J. R. Peak & Son, Henry	1275	G. S. H.
J. R. Peak & Son, Marcus	1155	G. S. J.
J. R. Peak & Son, Richard	1170	G. S. H.
J. R. Peak & Son, Arch	1290	G. S. H.
J. R. Peak & Son, Rover	....	G. S. H.
John D. Gillette, Roy	1470	G. S. H.
John D. Gillette, Champion	1505	G. S. H.
John D. Gillette, Rob	1260	G. S. H.
John D. Gillette, Lightfoot	1280	G. S. H.
B. Waddell, Philip	1065	G. S. H.
T. W. Hunt, Billy	1210	G. S. H.
T. W. Hunt, Reliance	1200	G. S. H.
Lucien Scott, Last Chance	1300	G. Hol.
Wm. R. Estill, Flash	1360	G. An.

The first, second and third premiums went to Adams Earl for Hereford grades, sired by Sir Bartle Frere (¶ 325).

The foregoing tables are interesting as comparing the weights of Herefords with other breeds and prove that by the scales test the average for Hereford has no superior, if indeed they are equaled.

The show of calves numbered twenty-eight entries, one-half of which were by Hereford bulls. The first and second premiums went to J. R. Price, Williamsville, Ill., on Herefords.

For best grades, any age, Fowler & Van Natta's Regulus (¶ 326) was winner. In sweepstakes by ages the show was confined to winning steers in the different classes, and in the three-year-old class the competitors were the pure-bred Shorthorn Schooler, and the pure-bred Hereford, Suspense, and the grade Hereford Regulus. The Fowler & Van Natta steer Regulus took the champion prize as the best three-year-old in the show. It is well to state here that Suspense, the pure-bred Hereford, would have undoubtedly taken this prize but for the fact that he was somewhat overdone.

In the two-year-old class, Leigh & Crane's pure-bred Hereford steer Sampson took the champion honors. In the yearling class the pure-bred Shorthorn steer Cleveland took the champion prize. This beautiful Shorthorn may truly be said to have appropriated the Hereford type.

In the calf class Mr. J. R. Price's grade Hereford heifer was the champion winner. Thus, in four classes, three-year-olds, two-year-olds, yearlings and calves, the Herefords took three of the prizes and the Shorthorns one.

There was considerable complaint among the losing Shorthorn exhibitors that the sweepstakes should be confined to the three first prize steers in the different classes; hence another ring was created by the Board, which was termed the "consolation sweepstakes by ages." It is quite safe to say that no amount of complaint by Hereford exhibitors had ever brought about such a concession.

The four regular winning steers above named, not being admitted to this ring, enabled Schooler, a Shorthorn, to take a first prize as a three-year-old; Mr. J. J. Hill's (¶ 327) Benholm (Polled Angus) to take first as a yearling, and J. H. Potts & Son's Diamond, a grade Shorthorn, to take the first as a calf.

Messrs. Morrow & Renick carried off the prize for the best pen of cattle three years old and under four, with Shorthorns, and the Indiana Blooded Stock Co. the second with Herefords. For the best two-year-old under three, Fowler & Van Natta, first with Herefords; John B. Sherman, second with Shorthorns. For the best pen of yearlings, Adams Earl, with Herefords; second to the Indiana Blooded Stock Co., with Herefords. The best pen of calves, J. A. Funkhouser first, and the Swan Live Stock Co. second, both Herefords.

The Shorthorns won all three prizes for heaviest fat steers with overgrown oxen that can only be produced at a loss. The only merit in the exhibit for heaviest bullock is, that it gratifies the curiosity of a good many city visitors, there being no merit in either of the three bullocks. The Board later abolished this class.

For the greatest gain per day in the three-year-old class, Fowler & Van Natta won both first and second with Herefords. In calves, Mr. Benj. Hershey took second with a grade Hereford.

REES KEENE,
Pencraig, Careleon, Monmouthshire.

For the grand sweepstakes for the best beast in the show, the competition was between Elbert & Fall's Shorthorn steer Cleveland; the Indiana Blooded Stock Co.'s thoroughbred Hereford Suspense; Mr. J. J. Hill's Benholm, Polled Angus; Messrs. Fowler & Van Natta's Regulus, grade Hereford, and a group of cows not eligible elsewhere in the show.

Fowler & Van Natta's grade Hereford Regulus was awarded the honor as the best beast in the show. The "Breeders' Gazette" prize for the best animal in the show, bred and raised by the exhibitor, was awarded to Mr. C. M. Culbertson, on Dysart (¶ 328).

DRESSED CARCASSES AT THE AMERICAN FAT STOCK SHOW, 1885.

Weights of the various parts. Net to live weight, etc.

Name of Animal.	Owner.	Breed.	Live weight.	Four quarters	Right fore-quarter	Left fore-quarter	Right hind-quarter	Left hind-quarter	Blood	Tongue	Head	Feet	Paunch	Liver	Guts	Tallow	Heart	Lungs	Hide	Trimmings	Shrinkage	Per ct. of carcass	Per ct. of valuable offal.
1 Jim	Swan & Bosler	Hereford	1,785	1,199	314	304	286	295	41½	11	38	19	115	14½	48½	106	4	22	97	15	55	67.17	12.20
2 Suspense	Ind. B. S. Association	Hereford	2,220	1,519	398	390	366	364	54	12	45½	24	145	14½	47	131	5	20	125	16	63	68.42	12.30
3 Gottlieb	L. Scott	Holstein	2,145	1,314	345	337	319	313	70	13½	42	26	197	24	49	165	8	23	112	17	83	61.26	13.91
4 Turiff	J. J. Hill	Pol'd Angus & S.H.	2,060	1,404	382	385	317	320	51	10	33	22	149	16	47	124	7	20	108	13	61	68.15	11.84
5 Hutcheon	J. J. Hill	Pol'd Angus & S.H.	2,375	1,614	422	435	376	381	47	13	41	24	165	20	57	143	9	24	113	17	68	67.95	11.70
6 Kinloss	J. J. Hill	Pol'd Angus & S.H.	2,055	1,407	369	371	331	336	56	11	40	21	155	15	39	126	6	20	99	16	44	68.45	11.77
7 Dick	A. Earl	Grade Hereford	2,135	1,498	383	395	355	365	44	11	40	20	118	15	43	139	6	17	111	23	50	70.21	12.50
8 Excelsior	A. Earl	Grade Hereford	2,055	1,417	375	378	329	335	52	11	37	21	141	17	41	112	8	19	94	18	67	68.95	10.95
9 Jesse	L. Scott	Grade Hereford	1,445	929	222	223	244	240	36	8	27	14	106	14	45	104	5½	13	74	15	55	64.28	13.25
10 Regulus	Fowler & Van Natta	Grade Hereford	2,280	1,560	407	401	375	377	51	11	46	22	185	17	57	148	8	20	107	18	30	68.42	12.02
11 Boots	L. Scott	Grade Shorthorn	2,000	1,307	351	346	301	309	62	10½	40	24	150	21½	57	134	8	23	95	13	55	65.03	12.37
12 Lookout	J. D. Gillette	Shorthorn	1,870	1,287	350	349	295	303	51	9	36	23	139	13	44	90	5½	15	97	12	39	69.35	10.77
13 Surprise	J. H. Potts	Shorthorn	1,670	1,086	280	281	262	263	43	8½	32	21	92½	16	47	77	5	17	93	10	122	65.03	9.76
14 Elkington	A. Earl	Hereford	1,650	1,055	273	277	247	258	34	9½	34	17	130	13	53	119	5½	15	94	14	57	63.95	7.75
15 Champion	B. Hershey	Hereford	1,660	1,023	266	274	239	244	30	9	41	19	150	13	35	66	5	14	126	18½	111	61.63	12.41
16 Bendigo	George Lingler	Hereford	1,440	932	247	248	219	218	35	7	41	18	82	13	38	69	6½	12	99	12	86	64.72	12.60
17 Joe	Seabury & Sample	Grade Hereford	1,700	1,128	290	292	269	277	52	11	35	19	132	13	57	103	6	15	93	10	26	66.35	13.05
18 Faith	L. Scott	Grade Hereford	1,415	970	258	259	220	233	36	8	28½	14	100	12	33	95	5	15	64	12	23	68.65	12.15
19 Dandy Boy	Fowler & Van Natta	Grade Hereford	1,800	1,215	316	319	287	293	47	11	38	22	125	16	46	113	6½	16	106	14	25	67.50	13.14
20 Snow Ball	C. S. Barclay	Shorthorn	1,615	1,075	289	291	243	250	44	9	34	18½	136	15	41	103	7	16	84	11½	21	66.56	12.57
21 Roan Turn	W. S. White	Shorthorn	1,935	1,340	344	340	360	326	46	9½	36	21	136	16½	49	105	7½	18	94	11	46	69.25	10.51
22 Roan Boy	J. R. Peak & Son	Grade Shorthorn	1,630	1,113	294	283	268	268	49	8½	32	19½	116	12½	40	93	6	15½	86	12½	27	68.22	12.19
23 Red Plum	J. D. Gillette	Grade Shorthorn	1,470	985	259	264	233	229	31½	7½	29½	17	116	13	29	103	6	16	80	12	24	67.00	13.09
24 Driver	J. D. Gillette	Grade Shorthorn	1,610	1,154	307	294	269	284	31½	11	36	20	120	14½	38	81	5	18	100	9½	34	71.67	12.42
25 Grace	Swan & Bosler	Hereford	1,800	1,253	316	310	311	316	29½	9½	33	15	134	14½	45	99	6	14	89	22½	32	69.61	10.86
26 Reuben B.	Forbes Bros.	Grade Hereford	1,755	1,153	310	311	262	270	58½	7½	37	21	133	16½	47	121	7½	19½	93	6	35	65.70	12.47
27 Varney	Forbes Bros.	Grade Hereford	1,660	1,092	290	290	253	259	49	9	37½	23	123	14	51	102	8	20	92	7	33	65.78	12.71
28 Captain	L. Scott	Grade Holstein	1,880	1,237	323	315	299	300	34	9	39	23½	144	20½	44	129	7	17	111	23	42	65.80	13.61
29 Last Chance	L. Scott	Grade Holstein	1,275	800	205	206	192	197	30	9½	29	20	130	15	45	68	5½	16½	84	8	15	62.74	13.09
30 Ben Holm	J. J. Hill	Polled Angus	1,895	1,353	365	351	313	324	39½	10	34½	20	122	15	46	110	6½	21	80	14½	23	71.39	10.89
31 Rosey	Overton Sea	Sussex	1,325	886	231	234	207	214	27	8½	32	16½	111	14	37	68	5	14	87	16	13	62.62	12.62
32 Cleveland	Elbert & Fall	Shorthorn	1,245	849	210	216	210	213	30½	9	26	15	91	12½	36	74	4½	13	66	9	10	68.19	12.32
33 Bob	J. P. Gillette	Grade Shorthorn	1,225	803	210	212	190	191	24½	9	28½	18½	103	11	35	63	4	12	78	15	21	65.71	9.95
34 Tommy L.	J. R. Peak	Grade Shorthorn	1,100	692	169	170	175	178	25	6½	24½	15	129	11	33	60	4½	14	68	6½	11	62.90	12.63
35 Philip	B. Waddell	Grade Shorthorn	960	595	153	154	145	143	26½	6½	23½	15	103	11	25	20	4	15	78	7	22	61.97	12.24
36 Sir Barth	A. Earl	Grade Hereford	1,455	966	254	248	232	232	33½	9	31½	18	90	15	46	89	5	14½	104	12½	21	66.39	14.22
37 Sam	Fowler & Van Natta	Grade Hereford	1,285	823	207	208	205	202	21	7	29	16½	118	13½	48	80	4½	13½	85	10½	16	64.04	13.74
38 Roan Dan	Swan Brothers	Grade Hereford	1,130	723	184	185	176	176	32	7	27	15	87	11	36	75	4½	12	76	9½	15	63.98	14.38
39 Charlie	Estill & Elliott	Pol'd Angus & S.H.	1,275	834	218	216	198	202	30½	8½	27½	16	119	11½	35	66	5	14½	74	10	34	65.41	12.04
40 Bob Moore	J. H. Potts & Son	Grade Shorthorn	1,270	836	208	209	210	209	29½	7	26	17	116	11½	31	61	5	15½	75	9½	30	65.81	11.65

Some interesting deductions are obtainable from this chart. It offers a splendid opportunity for comparison of various points.

In the dressed carcass contest, for best two-year-old carcass, Seabury & Sample's Joe, a grade Hereford, was winner, which also gained the sweepstakes for the best carcass of any age. For carcass showing the greatest amount of edible beef, Mr. Hill's Turiff won. The carcass exhibit was one of the most interesting in the show. With very few exceptions the carcasses were creditable. The weather was very unfavorable for the best appearance of the meat, and it was decided that for future shows some provision should be made by which meat could be cooled, that exhibitors might have the benefit of the best appearance the different carcasses might make.

KATHLEEN.
Bred by A. P. Turner (sire, The Grove 3d), winner of 15 first and champion prizes; exported to Argentine, S. A.

Not a carcass of all that were killed was properly cooled or set, and this had been true up to this time in the majority of all the slaughtering tests. A table of results is included on opposite page.

The judgment on the whole was fair, impartial and intelligent, and it may be fairly stated that more value may be attached to the awards made at this show than at any former show.

At the KANSAS CITY FAT STOCK SHOW in 1885 (mentioned also in chapter 38), for the best three-year-old dressed carcass, first premium was gained by W. E. Campbell, with grade Hereford.

The Herefords scored again for best two-year-old carcass, J. S. Hawes taking first premium. Mr. Hawes also gained the championship for best dressed carcass in the show with the same two-year-old Hereford.

Although in competition between breeds, on foot, the Hereford took only one premium, that to James A. Funkhouser for Challenge under one year old, for carcasses they took two out of three sweepstakes.

We present a table on the results of the slaughter test:

THE SLAUGHTER TEST AT KANSAS CITY FAT STOCK SHOW, 1885.

OWNER.	BREED.	NAME.	Live weight time slaughter	Weight of blood	Weight of lungs	Weight of tongue	Weight of heart	Weight of liver	Weight of spleen	Weight of guts	Weight of paunch	Weight of feet	Weight of head	Tallow	Hide	Left forequarter	Right forequarter	Left hindquarter	Right hindquarter	Net to live weight	Per ct. of carcass
H. Blakesley	Shorthorn	Arabella Beauty	1,715	51	7½	8½	8½	3	1½	89	106½	16½	29½	151½	78½	288	273	267	261	1,092	63.68
W. H. H. Cundiff	Shorthorn	Angelica 13th	1,510	54	10½	6½	9	13½	2	83½	107½	18	31½	100	82	238	223	245	231	937	62.00
A. B. Matthews	Galloway	Matthews' Pet	2,090	67	10½	7	9	15	2½	94½	103	25	40	178	132	380	369	314	311	1,374	68.79
E. B. Millet	Shorthorn	Queen	1,475	46½	10½	7	7	14	2	77½	125½	16	31	115½	71½	234	222	231	228	915	62.00
W. E. Campbell	Hereford & Texas	Texas Jack	1,695	57	8½	7	6	15	2	66	85½	21	36	134½	112½	290	289	261½	261	1,101	65.00
W. Morgan & Son	Hereford	Jock	1,760	49	9	6½	6	14	1½	77	118	20	35	141½	109	292½	289	279	272	1,132	64.01
J. S. Hawes	Hereford	Fred	1,725	54	8	7	6	6	1½	79	110	20½	34½	109	109	301	300	261½	255	1,127	64.75
J. H. Potts	Shorthorn	Hopeful	1,535	42	8½	6	7	11	1½	58	87½	19	33	124½	81	273	265	245½	240	1,023	67.29
G. S. Burleigh	Hereford	Star	1,300	42	10½	5½	5½	11	1½	70	88½	18½	32½	65½	86	224	210	196	203	833	64.00
M. R. Platt	Galloway	Kansas	1,320	40	7	5½	5	9½	1½	52	81	16½	23½	96	77	222	225	204½	205	956	64.84
W. R. Estill	Aberdeen-Angus	Bloom	1,180	35½	....	4½	5	10	1½	51	73½	13½	18½	91	70	198	185	195	201	779	66.00

THE ENGLISH SHOWS FOR 1885.

The "Mark Lane Express" said of the Bath and West of England Show at Brighton:

"There was one feature in connection with the show at Brighton which did not appear on the surface, but when culminated proved to be of a very important and striking character. This was the large and good class for groups of animals, bull and two of his get, all pure breeds competing. As very many of the animals which formed the several family groups were entered in the ordinary breed sections, no idea of the actual scope of this class could be formed until the several groups had been made up from the rank and file of the cattle and marshaled in the ring; then it formed the finest sight of the kind which any show yard in the country has ever afforded. Those who witnessed the judging of this remarkable competition in the great horse ring will not be likely soon to forget the sight of the twenty-one different lots, comprising groups of Herefords, Devons, Sussex, Shorthorns, and even of the Highland breed. But of this more in its place. It is to be hoped that the Royal and other leading societies will do more to encourage similar competitions, as they are of great interest, and likely to prove practically useful. The Royal does not pit breed against breed in its class for groups, but the Essex society has been doing good work in this way. A prize or prizes, worth winning, and open to all comers —of pure blood—would be sure to make breeders try their best not only for themselves but also for the prestige of the particular breed in which they are interested. This particular competition will be something by which to remember the Brighton meeting, for it was really a show in itself.

C. N. COSGROVE.
Le Sueur, Minn.

"Herefords were in about the usual numbers, all very select, and all exceedingly well shown. In the small classes for older bulls, Mr. H. R. Hall, of Holme Lacy, won with one of Lord Wilton's get, Hotspur (7726), bred by the late T. J. Carwardine; Mr. John Price's Hotspur (7028) (¶ 329), by Regulus (4076), came in second, so that the position of these two animals was the same as at the Essex show. Lord Coventry's Goodboy (7668) was entered in this class, but as he had won the first prize in it last year he was not eligible to compete. Mr. H. W. Taylor's Maidstone (8875), winner of the champion prize at Dublin this year, had really no competition at Brighton, as he did not compete with the two Hotspurs as at Waltham Abbey. In the younger class Mr. J. Price's Pembridge, by Hotspur (7028), was first, as at Waltham Abbey. The class for yearling heifers was commended throughout by the judges, as it deserved to be. Hereford breeders will be not a little indebted to the persistency with which the Earl of Coventry exhibits his stock; the cattle are good, he is always ready to send them far and near, and the successes he has met with are no more than he deserves.

"The competition in the class for bulls of any pure breed with two of their offspring, to which reference has already been made, really demands more space than we can devote to it. The scope of a competition of this sort is really greater than appears at first sight, because the whole of the cattle sections can be drawn upon. As we have said before, the object of it is primarily to show the character of the bull's stock, and also to create a wholesome rivalry between the several breeds. At the recent Essex show there was a precisely similar class, and two of the groups competed again at Brighton, with a different result, as will be presently seen. The competition, which was under the head of 'any breed,' comprised 20 entries of family groups, namely, five families of Jerseys (¶ 330), four of Herefords, four of Sussex, two Devons, two Shorthorns, two Guernseys, and one Highland Scotch (¶ 331). We are not in a position to say positively that all these groups were in their places before the judges, because of the difficulty of making a complete examination whilst they were in the ring, and many of them had elsewhere no *locus standi* as a group. It will suffice to say, however, that they not only 'filled the eye,' but they filled the great horse ring—the only decent ring there was on the ground. As an educational institution it far and away exceeded anything else at Brighton, and we take it that it was the best competition of its kind which has ever been before the British public. The Hereford groups were from the herds of Mr. John Price, Mr. S. Robinson, Lord Coventry, and Mr. T. Duckham, M. P. Mr. Price's group won the chief prizes, which consisted of a first prize in the class, value 30 pounds sterling ($150), and a silver cup offered by the Marquis of Bristol; they were the four-year-old bull Hotspur (7028), the two-year-old heifer Dowager, dam by Horace (3877), and the yearling heifer Venus, dam by Theodore 2d (5707). This group was in competition with Mr. Handley's Shorthorns at the Essex show the other day, namely, the white bull Hovingham (43363), now nearly six years old, sire Sir Arthur Ingram (32490); the noted bull Self Esteem 2d, dam by Sir Robin (40720), and the bull calf Royal Hovingham, dam by Alfred the Great, Self Esteem being probably the best Shorthorn bull of the day.

"At the Essex show the Shorthorn group was placed before the Hereford group, but a protest has been lodged against the award on the ground of an informal appointment of judges. At Brighton this Shorthorn group was simply 'not in it,' and it is worth bearing in mind that the bull Hotspur (7028), which headed the winning group of Herefords, was second in the older bulls' class at Brighton to Mr. H. R. Hall's Hotspur (7726). The second place was awarded to Messrs. E. and A. Stanford, of The Eatons, Ashurst, Steyning, Sussex, for the Sussex bull Goldsmith (391), coming eight years old, the bull Reading (516), over four years old, dam by Clayton (319), and the cow Dorset 8th (2365), over five years old, dam by Dorchester (325). All the cattle comprising this group were excellent, and whilst the win of the Herefords was unquestionably just, the second place given to the Sussex breed was a very popular award."

One of the leading English journals said of this award: "All this goes to show that when the Shorthorns are not made secure by preponderance of influence among the judges, as they have been in the main for years past, they do not win every time when they come in competition with other breeds."

The "Hereford Times" said:

"Breeders of Herefords, and people of the country generally, will be much gratified to hear that the famous 'whitefaces' have again asserted their splendid qualities in a striking manner, the Earl of Coventry having carried off the champion bull prize at the ROYAL COUNTIES SHOW at Southampton, on Tuesday, with his wonderful Goodboy, defeating the Shorthorn, Sussex and Devon breeds. The Royal Counties Society is, after the national society, one of the most important in the country, and this fact makes Lord Coventry's success all the more creditable. That it should not be said we are unduly lauding the breed, or too highly appraising its future, we will let an impartial critic speak. The 'London Times,' in its notice of the Show, says: 'In all the 51 entries of Shorthorns, 48 of the Sussex breed, 26 Devons and six Herefords, there is nothing that is not of a high order. In the interesting competition for the ten-guinea prize, given by Mr. Walter, M. P., for the best bull in these four breeds, the champion is the Earl of Coventry's Hereford bull Goodboy, a surprisingly good animal both in front, along his level and broad back, and his hindquarters, more particularly his remarkable rounds and flank.'"

THE FAIRS OF 1886.

At the Minnesota State Fair, 1886, in the sweepstakes ring, six herds contested for the honors; two herds of Shorthorns, three herds of Whitefaces and one of Devons. The judges in this class were Mr. R. H. Bullis, of Winnebago City; Mr. J. T. Mather, of Illinois, and Mr. Baker, of Hustisford, Wis. The Herefords took all the honors in this contest. J. O. Curry, Aurora, Ill., took sweepstakes for the best herd; Iowa Hereford Cattle Co. second.

At the NEBRASKA STATE FAIR IN 1886, Herefords made a grand show in their classes and in the grand sweepstakes for the best herd of any

GEO. W. HENRY.
Goodenow, Ill.

breed; Shockey & Gibb took first sweepstakes premium for their Hereford herd, consisting of Beau Real, Lady Wilton, Downton Ringdove, by Auctioneer; Pinktie, by Remus, and Sarah Miller, by Success.

The FAIR AT LAWRENCE, Kansas, in 1886, was in all respects a creditable show. The

Herefords were from the herds of Shockey & Gibb, Lucien Scott, Walter Morgan & Son, and George W. West.

BANGOR, MAINE, FAIR, 1886.—This great exhibition of agriculture and domestic industry at Bangor proved to be throughout the greatest success of its kind ever recorded in New England. It was successful at all points, in the list of fine cattle and horses shown, in the extensive collection of farm implements and machinery, in the multitudes it drew from all parts of New England and beyond, and financially. The two agricultural societies whose

FORTUNE (5922) 2080.
Bred by J. S. Hawes, Colony, Kan.

joint efforts, directed by experienced management, secured this triumph and presented the interests of farming in so attractive a light before the public, had the most substantial reasons to be satisfied with their work, which has received such endorsement from the people.

We noticed in the awards made, Burleigh & Bodwell, of Vassalboro, took 15 first and 7 second prizes with first for best herd of Herefords, and first and second for best Hereford bull three years and over.

The working oxen and steers, the town teams, the trained steers, matched oxen, and pulling oxen, with the rest, filled the exhibition park with attractions such as no true farmer could but be pleased with. Such a display of fine stock it was worth the trouble to cross New England to inspect and admire. The equal of it had not been seen in that, if indeed in any other, section of the country before, and it all imparted a solid character to the exhibition which it would not have been possible to supply by an indefinite extent of exhibition of any other kind.

At the IOWA STATE FAIR, 1886, all the departments were well filled, and the cattle exhibits were especially fine, consisting of five herds of Herefords, twelve herds of Shorthorns, seven herds of Holsteins, five herds of Jerseys, one herd of Red Polled (¶ 332), and one herd of Aberdeen Angus (¶ 332A). For the best sweepstake show, all breeds, Swan & Bosler Co. took first premium for grade steer calf, with a Hereford. In Hereford sweepstakes, the Iowa Hereford Cattle Co. took first for best bull of any age and best cow of any age. For grand herd premiums, all breeds competing, the Iowa Hereford Cattle Co. took second. Best herd of three fat cows, any breed, sweepstakes to Swan & Bosler on Herefords.

At the ILLINOIS STATE FAIR, 1886, live stock commission firms at the Union Stock Yards offered grand sweepstakes prizes for the best beef herd; first, $500; second, $300; third, $200. The Hereford herd of Fowler & Van Natta, of Fowler, Ind. (¶ 333), won first prize of $500 over a ring of eleven herds, comprising three Hereford herds, three Black Polled herds and five Shorthorn herds. This same herd took the sweepstakes at Chicago in 1885. It would be difficult for the Black Polls or Shorthorns to present a stronger show than this of 1886.

To describe the fever heat with which both exhibitors and audience awaited the result of this award would be a difficult undertaking. There were Fowler & Van Natta with their Herefords, Adams Earl with his Herefords (¶ 334), Thos. Clark, Herefords; Billy Potts with his Shorthorns; Harvey Sodowsky with Shorthorns; Walker & Son, Shorthorns; Olin & Son, Shorthorns; Anderson & Findlay with Polled Angus; the Brookside Farm Co. with their Galloways, and J. S. Goodwin of Kansas, Polled Angus.

There is hardly a doubt but what each owner was convinced in his own mind that he had the best herd. The onlookers took pride in this show. It was a hardly fought battle, but the winners deserved the honors. When the blue ribbon was tied on the Herefords, the excitement broke out in cheers and throwing up of hats by the appreciative crowd. The second prize of $300 went to Harvey Sodowsky's herd of Shorthorns, and the third prize of $200 to Walker & Son's herd of Shorthorns.

In the ring for young beef herds, any breed, the Herefords were again victorious in this show. There were two Hereford herds and one Black Polled herd comprising the competition. The result was the Hereford herds each received a vote, and the casting vote was placed on Adams Earl's herd of Hereford cattle, making him the winner of the young herd prize.

The sweepstakes for best Hereford bull, any age, was awarded to Fowler & Van Natta's bull Fowler 12899. The sweepstakes for best Hereford female, any age, was awarded to Thos. Clark's Flossie 10915.

INDIANA STATE FAIR, 1886.—The departments were all well filled, and there was an exceptionally good exhibit of cattle in both beef and dairy breeds. Herefords were represented by Adams Earl, Tom Clark, the Iowa Hereford Cattle Co., and the Indiana Blooded Stock Co. Morrow & Renick, of Kentucky, were on hand with their choice herd of Shorthorns, as was also Thos. Wilhoit, of Middletown, Ind., and Jacob Henn, of Illinois. R. C. Auld, of Michigan, was on hand with his Aberdeen-Angus, and Judge Goodwin, of Kansas, also showed a fine herd of "Doddies."

In the beef breeds the contest was for grand sweepstakes. This was a grand ring of cattle. The judging was good except on the young herds, where we were of the opinion Mr. Clark's magnificent young herd was entitled to first honors. In the ring for aged herds the competition was exceptionally strong, there being no less than four herds of Herefords, three herds of Shorthorns, and two herds of Polled Aberdeen-Angus.

After a long and tedious examination of the different herds, Mr. Adams Earl of Lafayette, Ind., was awarded first prize on his grand herd headed by Sir Bartle Frere. This is an honor of which Mr. Earl has a right to be proud, as it is seldom that there is as strong competition as there was in this ring. In the young herds the contest for first place was between Tom Clark, of Beecher, and Adams Earl; although there were very creditable Shorthorns in this ring, it was evident from the commencement that the first ribbon would go to the Whitefaces. Mr. Earl was again successful with as even a herd as we have ever had the pleasure of looking at, Morrow & Renick taking second; how the judges could go by the herd of Tom Clark and give the Renick herd second is "one of those things no fellow can tell," as every individual Hereford was far superior to the Shorthorns. Sweepstakes for the best bull, any age or breed, was given to Washington (¶ 335), owned by the Iowa Hereford Cattle Co. Sweepstakes on cows was captured by the Shorthorns, it going to a roan cow owned by Thomas Wilhoit.

The cattle department at the KANSAS CITY

STRETTON COURT, HEREFORDSHIRE.
Home of the Yeomans family.

FAIR OF 1886 was the feature of the show. The Shorthorns, Angus, and Galloways were in large force, and of good average merit. The Hereford force consisted of 36 head; F. W. Smith, Columbia, Mo., had 11 head; Iowa Hereford Cattle Co., 7; Lucien Scott, Leavenworth, Kansas, 9; Walter Morgan & Son, Irving, Kansas, 8; Mr. Gregory, Missouri, 1.

The judges were: G. W. Henry, Kansas City, a breeder of Angus cattle; Mr. Joe Duncan, Missouri, a Shorthorn breeder, and Mr. Newt. Winn, also a Shorthorn breeder, of Missouri.

The Hereford herd sweepstakes prize was given to the Iowa Hereford Cattle Co. Sweepstakes for best Hereford cow went to W. Morgan & Son. There was a good deal of interest throughout the entire exhibition, though some disappointment manifested among the exhibitors not winning. The exhibit was a good one, and it was not always easy to draw the line. The judging was according to the best judgment of the committees, and may be pronounced satisfactory as a whole.

MR. GEO. W. HENRY'S GREAT STOCK BULL, PRINCE EDWARD
And four of his get.

THE NINTH ANNUAL AMERICAN FAT STOCK SHOW, 1886.

The following breeders showed Herefords: Adams Earl, Lafayette, Ind.—One yearling thoroughbred. Grades: Two three-year-olds, two two-year-olds, one yearling, two calves. Fowler & Van Natta, Fowler, Ind.—Grades: one four-year-old, two three-year-olds, one two-year-old, one yearling, two calves, one three-year-old Hereford Jersey. George Leigh, Aurora, Ill.—One two-year-old thoroughbred, one three-year-old, one two-year-old, one yearling, and one calf; grades. Wyoming Hereford Association, Cheyenne, Wyo.—One two-year-old thoroughbred, one two-year-old Angus Hereford, one two-year-old Hereford Angus. J. R. Price & Son, Williamsville, Ill.—Grades: One two-year-old, two yearlings, six calves. Iowa Hereford Cattle Co., Indianola, Ia.—Two two-year-old thoroughbreds, two yearling grades. C. M. Culbertson, Newman, Ill.—Grades: One four-year-old, one two-year-old. G. W. Henry, Ashkum, Ill.—One thoroughbred calf, two grade calves. Chas. Saunders, Greenfield, Ill.—Five yearling grades. Wallace Libbey & Co., Ottawa, Ill.—Two three-year-old grades. Dutton & Wilkinson, Maple Park, Ill.—Two grade calves. Swan & Bosler, Indianola, Iowa.—One two-year-old grade. Samuel Weaver, Forsyth, Ill.—One thoroughbred calf. Number of exhibitors, 13; number of thoroughbreds, 7; number of grades, 46; total, 53. There were exhibited 71 Shorthorns, 31 pure-breds and 40 grades, by 15 prominent exhibitors. The seven Angus exhibitors included Gudgell & Simpson, of Independence, Mo., who at the same time also owned some Herefords, which we predicted would in time convert them from their error in selecting the Black breed, this prediction being amply fulfilled at this writing. There were nine pure and 10 grade Angus. Four Devon exhibitors showed 13 pure and one grade Devon. There was also one Sussex and four Holsteins. Shorthorns were judged first.

Herefords came next. There were no three-year-olds shown. In the two-year-old class the Wyoming Hereford Association took first on Rudolph, Jr. (¶ 336); George Leigh, second on Slasher; Iowa Hereford Cattle Co., third on Stars and Stripes. In the yearling class, Adams Earl took first, having no competition. In the calf class Samuel Weaver took first on Christmas Gift; G. W. Henry, second on Rossland's Royalty.

Probably no exhibitor ever entered a ring more loyal to true cattle interests or with more faith in the Hereford than Samuel Weaver (¶ 337), of Forsyth, Ill. (¶ 338). Mr. Weaver had attended the 1885 Show and Hereford meeting and went home determined to try his skill on a pure-bred Hereford steer. A month later an extra good calf was dropped on Christmas day. Mr. Weaver made the steer Christmas Gift of it and brought him to Chicago in 1886, a winner, as noted above. Mr. Weaver is an unassuming man of sterling honesty and wealth. His wealth then, as now, consisted of upwards of 4,000 acres of the best land in central Illinois. He had been a cattle feeder, and prepared cattle to successfully top the market at Chicago for forty years; a thoroughly practical and successful man in every way. He looked upon the ephemeral cattle fanciers who had made their money in other lines as men who should be willing to learn from, rather than teach, the veterans. The jealousy of exhibitors disgusted Mr. Weaver and when the President of the Hereford Society, a prominent grain dealer and packer, came up to him and said authoritatively, "You must not expect to beat Mr. Lumberman's calf," it was the "last straw."

Mr. Weaver knew he had the best calf, the judges sustained his judgment in their awards, and, believing that the way of the show yard is hard, and to a man of his means unnecessary, he quit showing in disgust, selling Christmas Gift to an Ohio feeder, who again brought him out a winner in 1887.

We always regretted Mr. Weaver's action in abandoning the show ring, though in our own experience we could sympathize with him, and we have been glad to see that he kept up his interest in the Herefords and that his herd fell into such capable hands as young Mr. Sotham's, and we have had great satisfaction in learning from Mr. Sotham that he owes all his present standing as a business man and breeder to Mr. Weaver, calling his farm and herd at Chillicothe, Mo., "The Weavergrace Breeding Establishment," because of the gracious helping hand extended to him unselfishly by Mr. Weaver, when help was sorely needed. In helping Mr. T. F. B. Sotham, Mr. Weaver unwittingly built himself another monument and at the same time proved a tower of strength to the Hereford cause, in the day of their need.

In closing this resume of the shows of a vital year in Hereford history, we present the different classes where all breeds competed together at Chicago Fat Stock Show, 1886, in full. The tables will be found an interesting study.

WILLIAM A. TADE,
Bonaparte, Ia.

Grades and Crosses. — There was a good show of three-year-olds, but the contest was narrowed down to J. H. Potts & Son's Champion, Morrow & Renick's Cyclone, and George Leigh's roan Hereford steer. The judges disagreed and the referee was called in, who gave the first to Morrow & Renick, second to Potts & Son, and third to George Leigh. The judging was criticised and the correctness of the awards called in question. The following is the list of three-year-old grades and crosses exhibited, together with their age in days, weight, and average gain per day, since birth:

Exhibitor and Name of Animal.	Age in days.	Wt.	Av. gain per day.
J. H. Potts & Son, Champion, Short'n.	1252	2055	1.64
B. Waddell, Jumbo, Shorthorn	1103	1715	1.55
Adams Earl, Hobson's Choice, Her'f'd.	1303	1970	1.51
Adams Earl, Quality, Hereford	1242	1860	1.50
F. C. Stevens, Francis, Holstein	1273	2140	1.69
F. C. Stevens, Hiram, Holstein	1267	2115	1.67
Fowler & Van Natta, Peter, Heref'd.	1237	1620	1.31
Fowler & Van Natta, Benton Will, Hereford	1409	2145	1.52
Fowler & Van Natta, Jerry, Heref'd.	1352	2240	1.68
Morrow & Renick, Gaines, S. H.	1290	1900	1.47
Morrow & Renick, Cyclone, S. H.	1300	1933	1.49
Morrow & Renick, Golden Slipper, S. H.	1115	1795	1.52
Geo. Leigh & Co., Tom, Hereford	1265	2190	1.73
J. Richardson, Robertson, S. H.	1278	2015	1.58
B. F. Waters, Snowball, S. H.	1237	1900	1.53
B. F. Waters, Bob Harrison, S. H.	1329	2020	1.52
B. F. Waters, George, S. H.	1155	1745	1.51
Wallace Libbey & Co., Jerry, Heref'd.	1261	1920	1.52
Wallace Libbey & Co., Tom, Heref'd.	1263	1775	1.40
C. C. Blish & Son, Bob, S. H.	1288	2065	1.60

In the two-year-old class the show was excellent, the Wyoming Hereford Association

F. A. NAVE,
Attica, Ind.

taking first, J. H. Potts & Son second, and B. F. Waters third. The first prize steer, Nigger (¶ 339), was a cross-bred Polled Angus and Hereford, and well entitled to the place. The following animals were entered in the two-year-old class for grades and crosses, their age in days, weight and gain per day:

Exhibitor and Name of Animal.	Age in days.	Wt.	Av. gain per day.
J. H. Potts & Son, Conqueror, S. H.	951	1790	1.88
J. H. Potts & Son, Snowflake, S. H.	968	1780	1.84
B. Waddel, Whiteface, S. H.	761	1485	1.95
Adams Earl, Sir Bartle 2d, Heref'd.	955	1925	2.02
Adams Earl, Sir Bartle 4th, Heref'd.	931	1760	1.90
B. C. Rumsey, Buffalo Tom, S. H.	1081	1950	1.68
C. M. Culbertson, Frost, Hereford	1066	1790	1.68

Exhibitor and Name of Animal.	Age in days.	Wt.	Av. gain per day.
E. T. Doney, Fashion, Devon	927	1885	2.03
Fowler & Van Natta, Matthew, Heref'd	973	1735	1.78
Morrow & Renick, Fred Young, S. H.	1002	1855	1.85
Morrow & Renick, Barnaby, S. H.	992	1605	1.63
Morrow & Renick, Overton, S. H.	1086	1555	1.43
George Leigh & Co., Norris, Heref'd.	932	1550	1.66
B. F. Waters, Nip, S. H.	948	1740	1.84
B. F. Waters, Joe McCarthy, S. H.	943	2045	2.17
B. F. Waters, John Kincaid, S. H.	1094	2025	1.85
B. F. Waters, Washington, S. H.	1043	2010	1.90
Swan & Bosler L. & C. Co., Plush, Hereford	786	1540	1.96
Hugh W. Elliott, Clarence, P. Angus.	971	1730	1.78
Wyoming Hereford Association, Nigger, Hereford	914	1540	1.68
Wyoming Hereford Association, Ranger, Hereford	963	1820	1.89

In the yearling class J. H. Potts & Son took first on Diamond, grade Shorthorn; T. W. Harvey second on Robert Peel, Polled Angus; J. R. Peak & Son third on Frisk, grade Shorthorn. The following animals entered the yearling ring:

Exhibitor and Name of Animal.	Age in days.	Wt.	Av. gain per day.
M. H. Cochrane, Dominionist, Angus.	559	1200	2.15
J. H. Potts & Son, Diamond, S. H.	661	1390	2.10
Adams Earl, Sir Bartle 6th, Heref'd.	694	1565	2.26
F. A. Townley, Pedro, S. H.	367	1085	2.96
T. W. Harvey, Robert Peel, Angus.	696	1465	2.11
T. W. Harvey, Duncan Gray, Angus.	614	1405	2.29
Ned Price, Express, Hereford	713	1470	2.06
Ned Price, Excelsior, Hereford	681	1420	2.09
Fowler & Van Natta, Jacko, Hereford	606	1135	1.87
Morrow & Renick, 14th Amendment, Short Horn	567	1180	2.08
Geo. Leigh & Co., Adams, Heref'd.	558	1355	2.42
J. R. Peak & Son, Ed., S. H.	726	1450	2.00
J. R. Peak & Son, Crown Prince, S. H.	703	1420	2.02
J. R. Peak & Son, Lord Thomas, S. H.	623	1425	2.28
J. R. Peak & Son, Wm. Westley, S. H.	722	1495	2.07
J. R. Peak & Son, Frisk, S. H.	537	1305	2.43
Iowa Hereford Cattle Co., Model, Hfd.	708	1330	1.88
Iowa Hereford Cattle Co., Shield, Hfd.	683	1335	1.96
Chas. Sanders, Jack, Hereford	601	1250	2.08
Chas. Sanders, Buster, Hereford	626	1220	1.95
Chas. Sanders, Curly, Hereford	582	1120	1.92
Chas. Sanders, Clipsy, Hereford	572	1175	2.05
Chas. Sanders, Fred, Hereford	569	1020	1.79
Swan & Bosler Co., Scotch Gray, An.	599	1195	1.99

In the calf class, Fowler & Van Natta took first on Sam Jones, grade Hereford; Samuel Weaver, third on Dandy, grade Angus; J. H. Potts & Son, second on Captain, grade Shorthorn. The calf class consisted of the following animals:

Exhibitor and Name of Animal.	Age in days.	Wt.	Av. gain per day.
Overton Lea, George, Sussex	345	745	2.16
J. H. Potts & Son, Captain, S. H.	345	825	2.40
Adams Earl, Eclipse, Hereford	315	780	2.47
Adams Earl, Exeter, Hereford	264	765	2.96
Ned Price, Bashful, Hereford	350	794	2.27
Ned Price, Bonanza, Hereford	326	845	2.59
Ned Price, Business, Hereford	319	925	2.90
Ned Price, Broncho, Hereford	296	810	2.73
Ned Price, Bruno, Hereford	276	760	2.75
Fowler & Van Natta, Sam Jones, Hd.	338	980	2.90
Fowler & Van Natta, Sam Small, Hd.	338	960	2.84
Morrow & Renick, Bright Light, S. H.	346	690	1.99
George Leigh & Co., Ancutt, Heref'd.	256	715	2.79
Samuel Weaver, Bald Hornet, Heref'd	307	840	2.74
Samuel Weaver, Dandy, Angus	355	945	2.66
G. W. Henry, Rossland's Improvement, Hereford	320	910	2.84
G. W. Henry, Bourbon Prince, Heref'd.	340	775	2.28
Wilkinson & Dutton, Alphonso, Hfd.	280	800	2.86
Wilkinson & Dutton, Fanny Field, Hereford	302	825	2.73

The SWEEPSTAKES for best grade or cross bred was given to the Wyoming Hereford Association's Angus-Hereford two-year-old, Nigger, which was also awarded sweepstakes for best two-year-old, any breed. Fowler & Van Natta took the calf sweepstakes (all breeds) on the grade Hereford calf Sam Jones. THE GRAND SWEEPSTAKES for best beast in the show was awarded the thoroughbred Hereford steer Rudolph Jr., exhibited by the Wyoming Hereford Association.

DRESSED CARCASSES. For best three-year-old, the sweepstakes was awarded to Wallace Libbey & Co., grade Hereford steer Jerry, sired by Monitor 2848, by Illinois 920, by Success 2; his dam was a grade Shorthorn cow.

For best two-year-old, Swan & Bosler's grade Hereford steer Plush (¶ 340) (prepared and exhibited under the management of John Gosling, the veteran and expert) was awarded sweepstakes. This steer was sired by Cheyenne 1912, by Ridgeville Boy 1476, and his dam was a grade Devon cow. The dam of Cheyenne was Peerless 1157, by Success 2. Plush also won the grand sweepstakes for best dressed carcass in the show.

This finished the cattle awards. A crossbred Hereford won the sweepstakes as best two-year-old in the show, and a thoroughbred Hereford won the grand sweepstakes as best beast in the show. On the block, a grade Hereford won sweepstakes in both the three-year-old and two-year-old classes, and a two-year-old grade Hereford won the grand sweepstakes as best carcass in the show. The Hereford breeders had a right to be well satisfied with results of the season and with the show.

It had been a common practice for years for Shorthorn men and others to charge that the forequarters of Herefords are heavier than the hindquarters as compared with Shorthorns. We reprint from the "Breeders' Journal" the following figures as made up by the American Fat Stock Show, 1886. Dressed carcasses, 3 years and under 4:

HEREFORDS.

Exhibitor....	Animal.....	Forequarters	Hindquarters	Difference...	Per cent.....
Adams Earl	Dick	778	720	58	.08
Adams Earl	Excelsior	753	664	89	.13
Lucien Scott	Jessie*	445	484	..	..
Swan & Bosler	Jim	618	581	37	.06
Fowler & Van Natta	Regulus	808	752	56	.07
Ind. Blooded Stock Co.	Suspense	789	730	59	.08
	Total	4,191	3,931	260	.06½

*This animal's hind-quarters weighed 39 pounds more than the forequarters. Per cent gain, .08.

SHORTHORNS.

Exhibitor....	Animal.....	Forequarters..	Hindquarters.	Difference....	Per cent.....
J. H. Potts & Son	Surprise	561	525	36	.07
J. D. Gillette	Lookout	699	598	101	.16
	Total	1,260	1,123	137	.12

POLLED ANGUS.

Exhibitor	Animal	Forequarters	Hindquarters	Difference	Per cent
J. J. Hill	Hutcheon	857	757	100	.13
J. J. Hill	Turriff	767	637	130	.20
J. J. Hill	Kinloss	740	667	73	.10
	Total	2,364	2,061	303	.14½

In dressed carcasses, 2 years and under 3, a similar result was to be seen:

HEREFORDS.

Exhibitor	Animal	Forequarters	Hindquarters	Difference	Per cent
Seabury & Sample	Joe	582	546	36	.06
Adams Earl	Elkington	550	505	45	.08
G. Leigh & Co	Bendigo	495	437	58	.13
Fowler & Van Natta	Dandy Boy	635	580	55	.09
Lucien Scott	Faith	517	453	64	.14
B. Hershey	Champion	540	483	57	.12
	Total	3,319	3,004	315	.10

SHORTHORNS.

Exhibitor	Animal	Forequarters	Hindquarters	Difference	Per cent
C. S. Barclay	Snowball	587	493	94	.18
J. R. Peak & Co	Roan Boy	577	536	41	.08
J. D. Gillette	Driver	601	553	48	.08
W. S. White	Roan Twin	684	656	28	.04
Forbes Bros	Rub'n Bro'dus	621	532	89	.16½
Forbes Bros	Varna	580	512	68	.13
J. D. Gillette	Red Plum	523	462	61	.13
	Total	4,173	3,744	429	.11½

POLLED ANGUS.

Exhibitor	Animal	Forequarters	Hindquarters	Difference	Per cent
J. J. Hill	Benholm	707	637	70	.11

It was publishing such disconcerting facts as these that made the opposition dislike T. L. Miller. It is not pleasant to fight battles; only the sincere belief that we were doing the agricultural interests of America and of the world a real service kept us firm in our resolve to have the Hereford merits fully known. We know that the Hereford needs only to be known to be appreciated, and we have fought his battles, knowing that our efforts would sooner or later be appreciated. The Hereford will never lack for a competent advocate, and we are glad to let our mantle fall upon one so able and well equipped as young Mr. Sotham. He has been doing splendid service and we are glad to see him successful and well supported. We note with great satisfaction that he is a success as a breeder of Herefords as well as he is as an advocate, his great bull Corrector (¶341) having already proven worthy to stand in the company of Success, Horace, Grove 3d, Lord Wilton and other great epoch-marking bulls. The Hereford breeders would make a serious mistake were they not to rally in every possible way to his support.

WEIGHTS OF THE DIFFERENT PARTS OF THE DRESSED CARCASSES AT CHICAGO FAT STOCK SHOW, 1886.

Name of Exhibitor.	Name of Animal.	Breed.	Age in days	Weight at slaughter	Average gain per day	Dressed weight.	Pounds net to gross weight.	Right forequarter	Left forequarter	Right hindquarter	Left hindquarter	Blood	Tongue	Head	Feet	Paunch	Liver	Guts	Tallow	Heart	Lungs	Hide	Trimmings
Three-Year-Olds—																							
J. J. Hill	Wildy	Angus	1372	2165	1.65	1526	70.5	391	410	370	355	50	12	38	21	141	14	32	130	6	19	103	7
T. W. Harvey	Paris Favorite	Angus	1361	2165	1.62	1490	68.8	392	399	357	342	59	12	35	22	141	16	53	144	6	22	105	5
J. R. Peak & Son	Cambridge Geneva	Shorthorn	1348	2050	1.56	1406	68.6	359	363	340	343	55	7	36	21	162	15	34	114	7	16	93	10
Morrow & Renick	Cyclone	Grade Shorthorn	1300	1900	1.49	1303	68.4	342	344	309	308	54	10	36	20	123	14	42	125	6	18	89	22
T. W. Harvey	Nerissa	Angus	2078	2100	1.04	1437	68.4	386	387	327	337	39	10	31	21	140	14	52	150	6	23	89	14
Fowler & Van Natta	Jerry	Grade Hereford	1382	2180	1.66	1473	67.6	402	397	337	337	47	12	45	21	147	14	48	143	7	17	113	15
Adams Earl	Quality	Grade Hereford	1242	1795	1.50	1181	65.8	302	304	285	290	47	16	35	19	158	16	42	106	7	16	94	9
W. Libby & Co.	Tom	Grade Hereford	1263	1760	1.40	1145	65.0	300	299	271	275	54	9	36	19	183	13	47	73	5	15	101	13
W. Libby & Co.	Jerry 2nd	Grade Hereford	1261	1905	1.52	1280	64.5	318	320	296	296	50	9	36	21	205	15	35	83	6	17	118	11
Two-Year-Olds—																							
M. H. Cochrane	Compton	Grade Angus	926	1390	1.50	977	70.3	260	263	225	229	28	9	26	15	81	8	26	78	5	15	74	12
H. W. Elliott	Clarence	Angus-Shorthorn	971	1655	1.78	1143	69.1	302	301	267	273	35	10	30	21	124	12	36	84	5	16	90	11
George Leigh	Norris	Grade Hereford	932	1460	1.66	930	68.0	260	264	234	255	35	9	34	19	124	12	34	46	5	15	90	9
Fowler & Van Natta	Matthew	Grade Hereford	973	1655	1.78	1113	67.2	285	281	267	280	39	10	34	21	125	12	38	55	5	17	102	9
Wyo. Hereford Ass'n.	Rudolph, Jr.	Hereford	883	1530	1.73	1028	67.1	261	266	249	252	42	8	32	18	117	15	38	69	5	13	97	10½
Ia. Hereford Cattle Co.	Richmond	Hereford	981	1725	1.95	1149	66.6	298	296	278	277	40	10	40	22	161	12	43	58	5	17	105	11
Wyo. Hereford Ass'n.	Nigger	Angus-Hereford	914	1575	1.68	1039	65.9	267	270	249	253	39	9	26	16	110	11	34	79	6	15	91	12
Swan & Bossler	Plush	Hereford-Devon	786	1515	1.96	997	65.8	257	269	234	237	39	10	23	18	102	12	44	89	5	17	91	9
J. H. Potts & Son	Conqueror	Grade Shorthorn	951	1725	1.88	1114	64.6	282	282	275	271	47	9	35	20	169	15	45	101	6	19	88	9
C. S. Barclay	Allen	Grade Shorthorn	767	1350	1.78	841	62.3	215	222	201	203	30	8	27	17	165	12	32	50	6	18	82	9
J. W. Morse & Son	Newton	Devon	989	1080	1.18	673	62.3	173	175	161	164	27	8	30	14	105	11	36	49	5	13	68	10
Yearlings—																							
M. H. Cochrane	Dominionist	Angus-Shorthorn-West Highlander	559	1055	2.15	768	72.8	195	196	189	188	23	6	24	15	114	9	30	54	4	11	77	11
William Moffat	Logan	Shorthorn	556	1245	2.33	833	66.9	212	215	212	204	32	8	26	16	110	12	36	38	6	15	74	10
Gudgell & Simpson	Alex	Angus	667	1410	2.13	942	66.8	239	239	218	232	31	8	28	19	125	8	70	51	5	18	89	7
M. H. Cochrane	Mineralist	Angus	641	1350	2.19	902	66.8	231	231	221	219	35	9	27	17	102	12	39	55	5	12	97	10
R. Huston & Son	Grover	Shorthorn	551	1080	1.97	718	66.5	175	181	181	181	24	6	21	15	92	11	28	47	5	15	65	10
T. W. Harvey	Edison	Angus	683	1295	1.95	852	65.9	219	220	193	220	32	7	22	15	113	9	45	73	5	12	85	10
John Gosling	Scotch Grey	Grade Angus	599	1160	1.99	758	65.3	193	189	187	189	25	6	21	14	88	7	37	66	5	15	72	10
T. W. Harvey	Duncan Gray	Grade Angus	614	1355	2.29	877	64.7	219	220	218	220	27	8	25	16	132	11	44	71	5	14	72	11
Overton Lea	Brown	Sussex	649	1215	1.93	761	62.6	205	203	177	176	31	8	35	15	133	11	33	44	5	15	88	9
T. B. Wales	Brookbank Lad	Holstein-Friesian	689	1290	2.02	813	62.2	200	202	204	207	30	9	30	18	131	12	34	54	4	20	81	10

THE ENGLISH SHOWS OF 1886.

The "Hereford Times," in speaking of the SMITHFIELD SHOW OF 1886, said: Another great distinction has this week fallen to the Hereford cattle, a representative of the white-faces having carried off the championship of the Smithfield Club Cattle Show on Monday. The successful animal is the grand ox belonging to F. Platt, of Barnby Manor, Newark, and it made a clean sweep of all the honors within its reach. The show of Herefords was the best seen at Smithfield since 1881. The entries number five more than last year, and among them were some marvelous specimens, especially in the classes for young steers and heifers. In all the youngsters numbered a baker's dozen, the winner turning up in a charming creature with a beautiful head and splendid body, shown by L. Loyd, West Wickham, Beckenham. The animal is just a year and ten months old, and weighs just over 12 cwt. (1,350 lbs.); it had very little difficulty, however, in beating a two months younger beast, the property of H. F. Russell (¶ 342, ¶ 342A), Westonbury, Pembridge, who took third prize at Birmingham. The first prize taker last week, shown by A. P. Turner, Pembridge, and by far the heaviest animal in the class, was now but highly commended. L. Loyd also received third prize. The Earl of Coventry was commended in this class. The steers above two and not exceeding three years were small in number, the prizes going to J. F. Hall, Ripple, Tewkesbury, the third prize taker in Birmingham; J. Andrews, Ivingtonbury, Leominster; and T. Duckham, Baysham Court, Ross. The Queen was highly commended in this class at Birmingham, but this time her representative was without recognition. The steers above three and not exceeding four years were a very fine lot. The principal feature was the first prize taker, the three-and-a-half-year-old ox, weighing 21 cwt. (2,350 lbs.), shown by Mr. Platt. This was, as we have already indicated, a magnificent animal, of perfect shape and superlative excellence in respect to quality. Not a fault could be found with it, and it frequently elicited admiration as it was paraded before the judges. It took first prize in its class, as at Birmingham, but now it went further, and not only received the breed cup for the best Hereford, beating Mr. Turner's famous Kathleen, its Birmingham conqueror, but also won the silver cup, value 50 pounds sterling, offered for the best steer, or ox, and, further, the champion plate of 100 guineas for being the best beast in the show. This was totally different from Birmingham. Mr. Platt's representative only had a "look in" at Birmingham, but now the Bingley Hall champion was just as much out of the running. The win, however, was a popular one, loud cheers greeting the decision of the judges. The champion animal is thus described in the catalogue (¶ 343): "No. 57. Frederick Platt (¶ 344), of Barnby Manor, Newark, Notts, 3 years 6 months 3 days, bred by the late T. Oliver, Hide Field, Weobly, Hereford, sire San Juan (616), dam Beauty. Sire of dam, Mansel (3240)." The full details of its winnings on Monday are as follows: 1st prize in its class, 25 pounds sterling ($125); silver cup as best of its breed, 30 pounds sterling ($150); silver cup as best steer or ox in the classes, 50 pounds sterling ($250); champion plate, as best beast in the show, 105 pounds sterling ($525); and a gold medal. Total, 220 pounds sterling

CLEM GRAVES,
Bunker Hill, Ind.

($1,100). In the older steers the class prizes fell to the same animals as at Birmingham, the second going to R. Keene (¶ 345), Pencraig, Caerleon, Mon.; and the third to W. Groves, Baucott, Wellington, Salop, with two fine animals. The heifers under four years old made a good display, Kathleen (¶ 346) (weighing 2,000 lbs.), a much older beast than Mr. Platt's, but over 3 cwt. lighter, again winning the 1st prize. Many people thought Kathleen would follow up her victory, as at Birmingham, but it was not to be. She is certainly much superior to Vanity VII, a slightly older heifer, belonging to H. W. Taylor, Ledbury, which took second prize. The Earl of Coventry, who was highly commended at Birmingham with Tunic, was now placed third; but the second prize taker at Birmingham, shown by R. Shirley, Baucott, Salop, was not on this occasion noticed. A very good class of cows resulted in J. H. Arkwright, Hampton Court, Leominster, taking first prize, and H. W. Taylor second; the positions being the same as in Birmingham. J. Watson, M. P., Berwick Hall, Shrewsbury, was third—a substantial advance on the Bingley Hall commendation.

At the Bath and West of England Show held at Bristol, 1886, champion prizes were as follows: The best bull in the show, H. W. Taylor, for Hereford bull Maidstone. For the best cow in the show in calf, Hereford, A. E. Hughes; reserve, H. W. Taylor, Hereford.

The Shows of 1887.

Good exhibits were again made of the Herefords at the Minnesota State Fair. The following were the exhibitors: Fowler & Van Natta, of Indiana; the Cosgrove L. S. Co. (¶ 347), and W. G. Sawyer, of Minnesota.

The creditable exhibit at the Nebraska State Fair was contributed from the herds of Shockey & Gibb and J. S. Hawes, of Kansas, C. M. Sears, and E. E. Day, of Nebraska.

Messrs. Fowler & Bassett, G. W. Price, Geo. W. Henry (¶ 348), (¶ 351), Thomas Clark, Tom C. Ponting, of Illinois, and Fowler & Van Natta, of Indiana, furnished the splendid exhibit at the Illinois State Fair. The event of this show was the contest for the grand sweepstakes prize for herd of beef cattle, any breed, which was won again and for the third year in succession by Fowler & Van Natta for the Herefords. Mr. Thos. Clark had in his Hereford cow Peerless 2d the champion female; while Messrs. Fowler & Van Natta's great bull Fowler scored again for the Herefords as champion bull of any age or breed.

At the Kansas State Fair the victorious Hereford herd of Shockey & Gibb headed by

SUCCESS (5031) 2.
Bred by J. Morris, Herefordshire. Favorite stock bull of T. L. Miller.

Beau Real was reinforced by Mr. J. S. Hawes' herd headed by Fortune (¶ 349). Mr. Jackson and Mr. Miedlin also had winning Herefords, Shockey & Gibb's young herd gaining the sweepstakes over all breeds.

The Iowa Hereford Cattle Co., under the management of Henry Yeomans (¶ 350), Shockey & Gibb, of Kansas, Fowler & Van Natta of Indiana, Moffatt & Sons and Wm. A. Tade (¶ 352) of Iowa, put up an excellent exhibit at the Iowa State Fair.

The Tenth Annual American Fat Stock Show, 1887.

In the three-year-old ring, grades and crosses, 22 entries, third prize was awarded to Mr. Ned Price (Hereford). In the same class for yearlings there were thirty-one entries, first sweepstake premium awarded to Fowler & Van Natta, grade Hereford. In the calf ring of twelve entries Mr. Price taking first, second to Fowler & Van Natta, and Mr. Funkhouser the third premium (all Herefords). In the dressed carcasses for the greatest percentage of profitable meat, premium was awarded to Herefords.

And so we close an incomplete but representative resume of Hereford winnings at the period when we closed out our Herefords, retiring to the more favorable climate of Florida.

We had made a good fight, the merit of the Hereford was by this time acknowledged in every part of our country—not only to equal the Shorthorn but to surpass it. The Herefords had been defrauded of their birthright, but we had forced a restoration. About this time the fight had been well nigh whipped out of the opposition, and finding they could no longer control the agricultural societies, the opposition began to use its influence subtilely to do away with the contests between breeds. No Hereford man ever advocated a cessation of the contests between breeds and so long as their opponents could fully control the judging just so long they were anxious to continue these contests. It was not until the Hereford began to get a measure of justice that the Shorthorn men began to carp of the fruitlessness of such contests. Journals that were formerly proud to proclaim themselves the official exponents of Shorthorn lore, were forced by the onward march of the Hereford to take on the semblance of neutrality, but it cannot be gainsaid that they have used their powerful influence to prevent these competitions between breeds that were being more and more conclusively settled in favor of the Herefords.

Some weak-kneed Hereford men, unhappily for the Hereford breed, will get seated in the powerful and controlling positions of the American Hereford Society. They are willing to take every advantage of the work of more intelligent and aggressive Hereford advocates, and at the same time covertly belittle the work of those advocates to the opposition, while suing to meet personal ends, for the opposition's influence and favor. Such treachery is sure to react on politic self-seekers in the future, just as similar double dealing has reacted on their ilk in the past.

What is the objection to the competition between breeds? We reiterate that no true Hereford man desired or desires to abolish these contests, and we repeat that no Shorthorn man demanded their cessation so long as the Shorthorn interest controlled the awarding of the premiums. "A fair field and no favors" is all the Hereford ever asked, and when at last it begins to come within his reach the Shorthorn interest interposes its power and stops breed contests. To give the demands for a cessation of breed contests a semblance of impartiality, they must chiefly come from seemingly neutral sources and so the subtle methods of the past are not abandoned, but having proclaimed itself neutral, the Shorthorn press that wore its collar openly in days past, but wearing it secretly in the heart still, comes forth with the trumped up claim: "Breed contests engender ill feeling, they settle nothing, nor prove anything, one breed wins at one show and at the next the award is reversed. One year one breed wins most and the next year favors another breed. And so it proves nothing for either to win. Let us have peace and let each breed develop within itself."

PRAIRIE FLOWER 1159.
Bred by T. L. Miller.

Pretty argument this, but we know its craftiness and deceit. What does it prove within the

Hereford breed when one year, or at one show, Mr. Nave's (¶ 353) herd beats Mr. Sotham's and at the next the awards are reversed and at still the next Mr. Clem Graves (¶ 354) comes in first best over them both, while at the fourth show honors are equally divided?

What was proven by our herd beating the herd of Mr. Culbertson one week and the tables being turned the next week and our mutual competitor, Mr. Earl, beating both of us a week later, all within the Hereford camp and under honest judges?

This is proved: All parties to the contest

RED CAP 4th, 3507.
A favorite cow of T. L. Miller.

had good cattle; the judgments of their merits varied with different judges. Each must have been good or it could not have been at any time recognized in the contest, and thus so far as the comparative merits of these Hereford herds are concerned the awards have proven nothing else. The whole contest has resulted in throwing the struggle to that only safe tribunal—the public, an infallible tribunal that always sooner or later rewards true merit and genius.

So it has been, is, and ever will be with contests between breeds. The Hereford may beat the Shorthorn at one show, and the Shorthorn may return the compliment at the next and the Angus may then come in and down them both and nothing will be proven conclusively by the awards themselves, but the open competition has afforded an interesting and instructive exhibit, enjoyed by the spectator and profitable to exhibitor and exhibition. The agriculturist, for whom all the exhibitions are supposed to be held, will have a proper and deserved opportunity to compare and form his own conclusions.

Yea, verily, the abolition of breed contests is born of deceit and selfishness. The Hereford and Angus have ever been eager for it. Trace the origin of this abolition crusade to its lair and it will be found within the Shorthorn camp or its followers. They would pose the Shorthorn as the old "cosmopolitan" breed, the "old stand-by," and would have all others considered interlopers, that must not be considered as competitors.

The spirit that dictated Youatt's history and controlled the agricultural societies and press, dies hard—but die it must. But self-asserted neutrals, who declare their impartiality, while working night and day to rehabilitate one breed, and at the same time by every possible way endeavor to make themselves indispensable to that one breed and masters of its lore, to the neglect of other breeds, must not feel hurt or slighted if the advocates of other breeds question their loyalty to them.

We have grown old in the advocacy of Herefords. We do not now, and have not for many years owned a "Whiteface," but we love their merits still and know their peerless value in the upbuilding of the world's agriculture. Let the Hereford breeders and all intelligent cattlemen take an octogenarian's words in all soberness, for they are written in all seriousness and truth.

T. L. MILLER'S HEREFORD WINNINGS.

It is fitting to here record the following list of premiums that were awarded to the T. L. Miller Hereford herd at the Illinois State Fairs from 1875 to 1879 on breeding stock. In the year 1876 they took premiums at this show in their classes only; but going from Illinois State Fair to the Northern Ohio Fair, they took the two herd premiums over one of the strongest Shorthorn shows, David Selsor, of Ohio, being one of the principal exhibitors, and from there to the Centennial at Philadelphia, where they won the first honors in their class.

ILLINOIS STATE FAIR.

1875.

Bull 4 years old or over—1st premium, Sir Charles 543.

Bull 2 years old—1st premium, Prince 861, by Sir Charles 543, out of Beauty 2d; second premium, Success 2.

Bull 1 year old—1st premium Royal Briton 882, half brother to Success; second premium Sir Charles 2d 913.

Bull calf over 6 and under 12 months—second premium, Parsons 857.

Bull calf under 6 months—first premium, Advance 1.

Cow 4 years old and over—First premium, Dolly Varden 5.

Cow 3 years old—first premium, Laura 853; second premium, Katie 1139.

Heifer 2 years old—second premium, Grace 1086.

Heifer 1 year old—1st premium, Charlotte 1102; second premium, Victoria 1053.

Heifer calf—1st premium, Prairie Flower 1159.

1876.

Bull 4 years or over—1st premium, Sir Charles 543.

Bull 3 years—1st premium, Success 2.

Bull 2 years—1st premium Royal Briton 882, half brother to Success.

Bull calf, over 6 months—1st premium, Seward 906, by Success; 2d premium, Uncle Sam 934, half brother to Success.

Bull calf under 6 months—1st premium, Dore; 2d premium, Prince 2d 862, by Success.

Cow 4 years or over—1st premium, Dolly Varden 5, dam of Success; 2d premium, Beauty.

Cow 3 years—1st premium, Grace 1086; 2d premium, Katie 1139.

Heifer 2 years—1st premium, Victoria 1053; 2d premium, Charlotte 1102.

Heifer 1 year—1st premium, Prairie Flower 1159, by Success; 2d premium, Mary Hughes 1149, by Success.

Heifer calf, over 6 months—1st premium, Peerless 1157, by Success; 2d premium Eugenia 1130, by Success.

Heifer calf under 6 months—1st premium Maid of Honor 1145, by Success; 2d premium, Miss Humphries 1152, by Success.

The reports of the Illinois State Board for 1877 and 1876 do not show the names of the winning animals, but simply credit the awards to the owner of the cattle. So far as we remember the animals, we have put them down. It is true of the awards of those two years that they were largely of the Success (¶ 355) stock, and the winnings of the two years must have been at least three-fourths on Success and his get.

1877.

Bulls 4 years old or over—1st premium, T. L. Miller, Success; 2d premium, T. L. Miller.

Bull 1 year old, 3 entries—1st premium, T. L. Miller.

Bull calf under 6 months, 4 entries—1st and 2d premiums, T. L. Miller.

Bull calf under 6 months old—1st and 2d premiums, T. L. Miller.

Cow 4 years old, 7 entries—1st premium, T. L. Miller.

Heifer 2 years, 4 entries—1st and 2d premiums, T. L. Miller.

Heifer 1 year, 4 entries—1st and 2d premiums, T. L. Miller.

Best heifer calf over 6 months, 4 entries—1st and 2d premiums, T. L. Miller.

Best bull, any age, 8 entries—sweepstakes to Success.

Best cow or heifer, any age, 15 entries—sweepstakes to Dolly Varden.

Best bull and five cows or heifers, 1 year old and upwards, and owned by one individual or previously existing firm, 12 entries—2d premium, T. L. Miller, Success and five cows and heifers.

Best 5 calves of any breed, male or female, under 1 year old, and owned by one individual or firm, 8 entries—2d premium, T. L. Miller, Success calves.

JAMES C. WILLSON,
Flint, Mich.

Best 5 cattle, male or female, of any age, without regard to ownership, the get of one bull, the sire to be shown with the herd and considered in making the award; 3 entries—1st premium, T. L. Miller, Success and 5 cows and heifers.

Best cow or heifer of any age or breed, 45 entries—2d premium, Dolly Varden.

1878.

Bull 4 years old, 2 entries—1st premium, T. L. Miller, Success.

Bull 2 years—1st premium, T. L. Miller, a son of Success.

Bull 1 year, 6 entries—2d premium, T. L. Miller.

NANNETTE (V. 11, p. 246) 4511.
Bred by T. Middleton.

Bull calf over 6 months, 4 entries—1st and 2d premiums, T. L. Miller.

Bull calf under 6 months—1st premium, T. L. Miller.

Cow 4 years old—1st and 2d premiums, T. L. Miller.

Cow 3 years—1st and 2d premiums, T. L. Miller.

Heifer 2 years, 3 entries—1st and 2d premiums, T. L. Miller.

Heifer 1 year—2d premium, T. L. Miller.

Heifer calf, over 6 months, 3 entries—1st and 2d premiums, T. L. Miller.

Best bull, any age, 5 entries—T. L. Miller, Success.

Best cow or heifer of any age, 6 entries—T. L. Miller.

Best 5 cattle, male or female, of any age or breed without regard to ownership, the get of one bull, the sire to be shown with the herd and considered in making up the award; 6 entries—1st premium, T. L. Miller, Success, and 5 cows or heifers.

1879.

Bull 3 years old or over, 4 entries—1st premium, Success 2.

Bull calf under 1 year old and over 6 months, 3 entries—1st premium, Dictator 1989.

Cow 4 years old or over, 2 entries—1st premium Prairie Flower 1159 (¶ 356), by Success; 2d prize, Victoria 1053.

Heifer 2 years, 6 entries—1st premium, Highland Queen 1141 (¶ 357) by Success.

Heifer calf over 6 months, 4 entries—1st premium, Miss Filley 1899, by Success.

Best cow or heifer of any age, 8 entries—premium, Highland Queen 1141, by Success.

Bull and 5 cows or heifers 1 year old or over, owned by one individual or previously existing firm; 2 entries—premium, Seventy-six 1093, Victoria 1053, Prairie Flower 1159, Highland Queen 1141, Charity 708, Maid of Orleans 1146.

Five cattle, male or female, of any age or breed, without regard to ownership, the get of one bull, the sire to be shown with the herd and considered in making the awards; 4 entries—2d premium, Success 2, Prairie Flower 1159; Highland Queen 1141, Charity 708, Beatrice 742, Maid of Orleans 1146.

Five cattle of one breed (all breeds competing), male or female, over 1 year old, bred and owned by the exhibitor; 5 entries—2d premium, Victoria 1053, Prairie Flower 1159, Highland Queen 1141, Charity 708, Maid of Orleans 1146.

WINNINGS OF THE T. L. MILLER HERD OF HEREFORDS AT THE AMERICAN FAT STOCK SHOW, CHICAGO.

1878.

Hereford steer 4 years old or over—1st premium, T. L. Miller.

Hereford steer 3 years old and under 4, 3 entries—1st premium, Miller.

Cow 3 years old or over, 3 entries—1st premium, Jennie.

Sweepstakes ring, open to all breeds, for best cow three years old or over in the show, Jennie.

1879.

Hereford steer 4 years old or over, 4 entries—1st premium————

Hereford steer, 3 years, 2 entries—1st premium—

Hereford steer, 2 years—1st premium—

Hereford steer, 1 year—1st premium—; 2d premium, Will, sire Success 2, dam Mollie.

Hereford cow 3 years old or over—1st premium.

GRADES OR CROSSES, OPEN TO ALL BREEDS.

Steer 4 years old or over, 16 entries—1st premium.

Hereford steer, 2 years, 3 entries—1st premium, Will, sire Success, dam Mollie; 2d premium, Washington, sire Success, dam Miss Smith 1083.

Hereford cow 3 years—1st premium, Maid of Orleans 1146, sire Success 2, dam Laura.

Grades or crosses, open to all breeds.—Steer, 1 year, 22 entries—2d premium, Kansas.

Sweepstakes—Steer 2 years, any breed, Conqueror.

Sweepstakes—Yearling steer, any breed, Kansas.

Dressed Carcass.—Steer any breed 3 years, 3 entries—1st premium.

Prize for early maturity, open to all breeds. —Steer 2 and under 3 years, 11 entries—premium, Conqueror.

1881.

Hereford steer 3 years—1st premium, Will, sire Success 2, dam Mollie; 2d premium, Washington, sire Success 2, dam Miss Smith.

Grades or Crosses open to all Breeds.—Steer 3 years, 34 entries—1st premium, Conqueror, sire Hereford bull, dam one-half Hereford.

Grade or cross-bred steer 2 and under 3 years, 33 entries—3d premium, Kansas, sire Hereford bull, dam native cow.

Steer showing the greatest average gain per day since birth, 3 and under 4 years, 10 entries—2d premium, Conqueror, sire Hereford bull, dam one-half grade Hereford.

The Grand Special Premium, offered by Marshall Field, $250, for best 5 head of cattle any age or breed, was given to T. L. Miller for 5 Hereford steers—Conqueror, Will, Washington, Bachelor and Kansas.

1882.

We believed that the condition in which breeding stock must be shown at the public exhibitions in this country was damaging to the stock exhibited and discouraging to the average farmer. We therefore endeavored to secure a class for grass-fed cattle. We have, however, failed to get such a class, managers of fairs alleging that such an exhibit would not interest visitors or be creditable to the management. We determined, however, that we would bring some of our stock from the pastures and exhibit at the Fat Stock Show of 1882. We reproduce from the official statement or catalogue of the show, this exhibit. We headed the exhibit with our breeding stock, giving the age in years and months instead of days.

T. L. MILLER COMPANY'S EXHIBIT, CHICAGO FAT STOCK SHOW, 1882.

Exhibit No.	Sex.	Name.	Weight.	Age.	Breed.	Remarks.
67	Bull	Dauphin 18th	2330	3 yrs. 8 mos.	Hereford.	From the pastures to the show, each cow having raised a calf during the season.
68	Bull	Winter de Cote	2060	3 yrs. 2 mos.	Hereford.	
69	Bull	Success	2030	9 yrs. 9 mos.	Hereford.	
...	Cow	Victoria	abt 1700	9 yrs. 0 mos.	Hereford.	
91	Cow	Highland Queen	abt 1600	5 yrs. 7 mos.	Hereford.	
92	Cow	Charity	abt 1600	5 yrs. 6 mos.	Hereford.	
93	Cow	Beatrice	abt 1500	5 yrs. 0 mos.	Hereford.	
92	Cow	Nightingale	abt 1600	4 yrs. 9 mos.	Hereford.	
78	Steer	King William	1655	2 yrs. 7 mos.	Grade Hereford.	Fed for exhibit.
79	Steer	Wallace	1720	2 yrs. 8 mos.	Grade Hereford.	
80	Steer	Highland Lad	1680	2 yrs. 8 mos.	Grade Hereford.	
81	Steer	Beecher	1815	2 yrs. 7 mos.	Grade Hereford.	
82	Steer	Conqueror II.	1705	2 yrs. 11 mos.	Grade Hereford.	
77	Steer	Bertie	1310	2 yrs. 0 mos.	Hereford.	
77	Steer	Bachelor	1435	2 yrs. 0 mos.	Grade Hereford.	
76	Steer	St. Paul	1232	1 yr. 6 mos.	Grade Hereford.	
76	Steer	Eighty-one	1155	1 yr. 5 mos.	Grade Hereford.	
83	Steer	Pythias	1870	3 yrs. 9 mos.	Grade Hereford.	
84	Steer	Damon	1990	3 yrs. 8 mos.	Grade Hereford.	
85	Steer	Barnum	1980	3 yrs. 6 mos.	Grade Hereford.	Grazed on grass and without other food until Sept., when they were fed cornstalks and light grain feed.
86	Steer	Jumbo	1920	3 yrs. 6 mos.	Grade Hereford.	
87	Steer	Abbey	1710	3 yrs. 6 mos.	Grade Hereford.	
89	Steer	Buck	1450	3 yrs. 9 mos.	Grade Hereford.	
89	Steer	Bright	1600	3 yrs. 9 mos.	Grade Hereford.	
90	Steer	Prince	1560	3 yrs. 10 mos.	Grade Hereford.	
215	Steer	Napoleon	1475	2 yrs. 10 mos.	Grade Hereford.	
216	Steer	Murat	1395	2 yrs. 10 mos.	Grade Hereford.	
217	Steer	Ney	1355	2 yrs. 10 mos.	Grade Hereford.	
218	Steer	Marion	1245	2 yrs. 7 mos.	Grade Hereford.	
219	Steer	Ohio Chief	1470	2 yrs. 6 mos.	Grade Hereford.	
220	Steer	Douglas	1437	2 yrs. 6 mos.	Grade Hereford.	
75	Cow	Princess Alice Maud	1800	10 yrs. 10 mos.	Hereford.	

This table shows an average weight on five 2-year-old past steers of 1,715 pounds, on two even 2-year-olds, 1,372 pounds, and three 1-year-olds, 1,284 pounds. These steers had been stall fed, and would, but for the burning of our barn, have made 100 pounds more weight per head. Our three-year-old grass steers made an average of 1,653 pounds, our 2-year-old grass steers made an average of 1,360 pounds. These two lots may be taken as the standard weights for the ordinary farm management. Such an exhibit is of value and perhaps of more value than our stall fed steers 2-year-old averaging 1,750 pounds. We were criticised by some parties for taking our grass steers to the show, but we had grown used to criticism, and if we had heeded it we should not have been breeders of Herefords, and the Herefords would not have occupied the position they do today. We thought if we could show that with ordinary methods of hay and grass the farmer can put his three-year-old steers on the market at an average weight of 1,650 pounds, and his two-year-olds at nearly 1,400 pounds, he would recognize the Herefords as better beef animals than he has been in the habit of handling from any other breed.

We called attention to the fat cow nearly 11 years old, weight 1,800 pounds. We believed such an exhibit a credit to the breed, breeding for more than the average life of the Shorthorn, and then going to the butcher as first-class beef, weighing 1,800 pounds. It is a credit to the breed that it can fit three-year-old heifers to 1,750 pounds, but greater credit to take an aged breeding cow and make 1,800 pounds of first-class beef.

In this exhibit of our breeding stock we had the three-year-old bull, Dauphin 18th, bred by Mr. J. B. Green, Marlow, England, weighing without fitting 2,330 pounds, and Winter de Cote, a brother to Mrs. Edwards' celebrated Leonora, and then old Success, nearly ten years old, with all the vigor of a two-year-old, and no sign of age upon him. Also four of his heifers, and the nine-year-old cow, Victoria (by old Sir Charles), the dam of Dictator, owned by Mr. F. W. Smith, of Woodlandville, Mo., who won with him the sweepstakes of the best Hereford bull two years in succession, at St. Louis, over the best Hereford bulls of England and America.

The bulls in this exhibit were from ordinary stock keep, and the cows from pasture only. We expected the time was not far distant when stock from the pastures would form an important part of the live stock exhibits of our leading shows, and if this exhibit of ours would hasten that time, we would have accomplished what we intended to do. The advice of the best and most experienced breeders is to never buy fitted stock, and whatever may be the breeder's reputation in the show ring, his home exhibit must support it, or very few will wish to choose from his herd. The five cows exhibited dropped within three years, including the year they were shown, fourteen calves.

Whenever the time shall come that grass fed cattle and cows with calves at foot shall be a feature at our leading fairs, the Herefords will show character to which no other breed can attain.

D. P. WILLIAMS,
Guthrie Center, Ia.

CHAPTER XL.

Letters

FROM:
T. L. Miller, 1881.
W. Danger, 1879.
S. C. Skidmore, 1880.
Geo. Leigh, 1880.
Geo. T. Turner, 1880.
Thos. Duckham, 1880.
H. M. Vaile, 1881.
J. C. Wilson, 1881.
Dakota, 1881.
John Merryman, 1881.
James Funkhouser, 1884.
Hugh Craig, 1884.
S. D. Fisher, 1884.
Theodore Whyte, 1884.
C. M. Culbertson, 1885.

FROM:
D. P. Williams, 1886.
W. S. Ikard, 1886-87.
F. O. Skidmore, 1886.
F. W. Smith, 1882.
W. E. Campbell, 1882.
Henry Lane, 1882.
James W. Cox, 1882.
H. Bowen, 1882.
Kennedy Bros., 1882.
Cephalus Black, 1883.
Amos Bissell, 1883.
H. C. Burleigh, 1886.
P. Mehan, 1886.
A. M. White, 1886.

MR. MILLER'S POSITION.

The following is an open letter from T. L. Miller to the "Farmer's Magazine" of Kentucky (1881):

It is satisfactory when another states your position, if he shall do it correctly; but the editor of the "Farmer's Magazine," in his editorial entitled "What will he do with it?" is not exactly fair in giving me a position, or in stating that position, but on this point we will not quarrel.

The Hereford breeders have not been blockheads. They have attended to the breeding of their stock, and have produced a race of cattle that have held the top price, both with the grazier and butcher. The breeding intelligence and practical brains have not been with the Shorthorns. The agricultural societies and press have been controlled in the interest of the Shorthorns, and that control has not been fairly obtained, because they have been organized and managed, as stated in the preamble, for the purpose of bringing forth the best, while they have been worked to advance the Shorthorn interest.

"The press and the agricultural societies were taken covertly and by storm in the interest of Bates or Booth, and wrong was made to look right and right wrong." This is a fair statement of my position, and I propose to prove it by the record.

Again you say: "Those who have observed Mr. Miller closely, as we have done, must be impressed with the contempt he manifests for all those who think differently from him."

There are writers in the interest of the Shorthorns who are forever pointing to the past history of the Shorthorns as an evidence of merit, as certain persons are pointing to the history of their parents or grandparents, or great-grandparents as an evidence of their ability. For such I have no great respect. I think if you will go back in the Shorthorn history to 1817, that time of the Sanders and Clay importations, and follow the breeding of those cattle through their crosses, you will find that the Hereford blood in that importation and their produce, went a long way towards giving to the Kentucky cattle the character they have had in the past times. You will find to this time those markings that denote Hereford blood; and you will find that with these markings there is a quality that I claim belongs to the Herefords. Now there is no question but that the Herefords were an important element in the cattle interest of Kentucky from 1817 to 1840, and you will find men now living that will recognize this fact. You will find another fact: that the Kentucky reputation for good cattle is based more upon the Seventeens than upon the Bates, Dukes and Duchesses, and when the latter are forgotten this will be your boast. One of your prominent breeders told me in Kansas City several years ago that he knew the Duke bulls were damaging the character of the Shorthorn cattle, but they could not sell any others to a profit. Furthermore, there are none of your best breeders but know that Kentucky cattle are poorer for having used this blood.

W. S. IKARD,
Henrietta, Tex.

What you say of Mr. Goddard will be said of many of your best breeders. They can take the Seventeens and make a breed or family that will top all other families of Shorthorns in the state.

After you have investigated the cattle history of Kentucky from 1817 to 1840 turn your attention to the operations of the breeders in England, at the time they took possession of the machinery of the Society for the Diffusion of Useful Knowledge, and notice with what tact they obtained the endorsement of that society for their breed. Rev. Mr. Berry might have written over his own signature and in his own name the history of the Shorthorns and the history of other breeds of English cattle but that would not have appeared well. He therefore obtained the appointment of Mr. Youatt, a scholarly gentleman of that period, to write their history at his (Rev. Henry Berry's) dictation.

Will there be any question but that the Rev. Henry Berry supervised the writing of the history of other breeds of English cattle that appear in the same volume, "Youatt on British Cattle"? If you should have any doubt on this point after a careful investigation I should be pleased to enter upon this study with you.

After a careful study of Youatt and his history, I would recommend to you a careful examination as to the organization of the Royal Agricultural Society of England in 1839. You will there find that some of the Shorthorn breeders used some of the best and most noble of England's nobility and gentry for the endorsement of that breed, known then by the name of the Shorthorn or Durham breed of cattle; while the breeders and advocates of this breed were the judges and committeemen who managed the awards and you will find that the press published these awards on the report of these judges and of these committeemen.

You will find that your State Societies were under the same sort of management, and you may look into the history of the Herefords and Shorthorns in 1834, at which time Youatt wrote, and you will find that at the public sales of blooded stock and butcher's stock in England, the Hereford sold at longer prices usually than the Shorthorn.

You may then go to the records of the Smithfield Society and you will find on the records of that Society that the Herefords and Shorthorns were competing, and that up to 1834 the Hereford oxen and steers had taken eighty-five premiums amounting to 1,295 pounds sterling ($6,475), while the Shorthorns had taken thirty-two premiums, amounting to 585 pounds sterling ($2,925). This showing was a matter of record in London at the time Prof. Youatt was writing.

In the history of the Herefords Prof. Youatt gives one experiment of the feeding of three Shorthorns and three Herefords, showing about thirty per cent in favor of Herefords. He gives seven pages or more to the report of sales of Shorthorns, and no account of the sales of Herefords, while he might have had access to the records of such sales.

He gives two pages (312-313) to quoting the Duke of Bedford's experiment of feeding different breeds of cattle, but as far as Herefords or Shorthorns are concerned, he says: "Although the Herefords are now established at Woburn . . ." and in a foot note adds: "To the many records of experiments on the comparative fattening qualities of the Hereford and Durham cattle, we will not now refer. One, although not then assigned to its proper author, the present Duke of Bedford, was detailed on page 34 of this work. We will content ourselves with referring to that. The patrons of the Shorthorns, however, have not considered it altogether satisfactory in its details."

This is disposing of the Duke of Bedford's experiments as a historian has no right to do. Will anyone doubt that if these experiments had resulted in favor of the Durham, they would have found a place in this "History of British Cattle?"

Between the years of 1834 and 1840, covering the time when the Royal Agricultural Society of England and New York State Agricultural Society established their fairs, the report of the Smithfield Club shows that the Herefords took thirty-five premiums, amounting to 465 pounds sterling ($2,325); the Shorthorns took fifteen premiums, amounting to 230 pounds sterling ($1,150).

In 1839 Mr. W. H. Sotham brought an importation of Herefords to Albany, N. Y., and nearly every Shorthorn breeder in New York State quoted Youatt against him, and the entire Shorthorn interest of that state fought Mr. Sotham from the start, compelling him to exhibit with discrimination against him and under Shorthorn judges. Mr. Sotham had a fight at that time, with justice and merit with him, but was overpowered by numbers and beaten by such men as H. S. Randall, Lewis F. Allen, Ambrose Stevens, backed by such men as E. P. Prentice, C. M. Bement, Stephen Van Rensselaer, George Vail, J. M. Sherwood, and others, and these used the New York State Agricultural Society to aid in this defeat.

If my success with the Herefords had depended upon the introduction of them among the Shorthorn men, who were aided by the Illinois State Agricultural Society, I would have likely found the same result that Mr. Sotham did. I, however, took my field among the breeders of the plains, and their acceptance of the Herefords secured the success of the enterprise.

The Smithfield continued to show all breeds in competition up to and including 1851, when the Shorthorn men took the management of the society, "covertly and by storm," changed the showing from the old plan of showing all breeds in competition to class showing. From 1840 to 1851, the Herefords had taken sixty-five premiums, amounting to 920 pounds sterling ($4,600), the Shorthorns had taken thirty-seven premiums, amounting to 365 pounds sterling ($1,825). During these years, say from 1800 to 1851, Hereford beef had sold on the London market at a higher price than Shorthorn beef by one or two cents per pound. And during all this time the results of careful trials were published and accessible, showing that Hereford beef could be made at from 15 to 30 per cent cheaper than that of the Shorthorn. These are facts accessible to Shorthorn breeders and editors, and still they make claims that ignore them; and societies that are supported by the State funds give special encouragement to the Shorthorns and discriminate against other breeds Is it singular that I should not honor men that will put forth claims that the entire record will condemn?

Your correspondent, Mr. T. C. Anderson, under date of November 1, 1880, in an article to the "Kansas City Indicator," says: "What authority has Mr. Miller for saying that the Hereford will graze and feed at less cost than the Shorthorn, and when fed are worth more money?" Any breeder having been engaged in that business as long as Mr. Anderson should understand the history of the cattle he breeds, and the breeds with which he competes, better than to put such questions—he ought to be better posted in cattle history. He may be excused, being a breeder, for making the best show he can for his breed, as a lawyer may be for quoting law and facts that will favor his client. But you occupy rather the position of the judge. Your position may be termed a judicial one, and you are supposed to stand as the instructor, and not as the advocate.

You should know the history as the judge should know the law, and when the advocate presents his case you should not give a verdict by silence or endorsement that will work an injury to your readers. I am aware that the position I have taken is revolutionary. I propose it shall be, and had I not the facts to support my position I might be written down a fool. And it becomes you to look well to precedent before you commit yourself to claims that have no basis in equity.

Very truly, T. L. Miller.

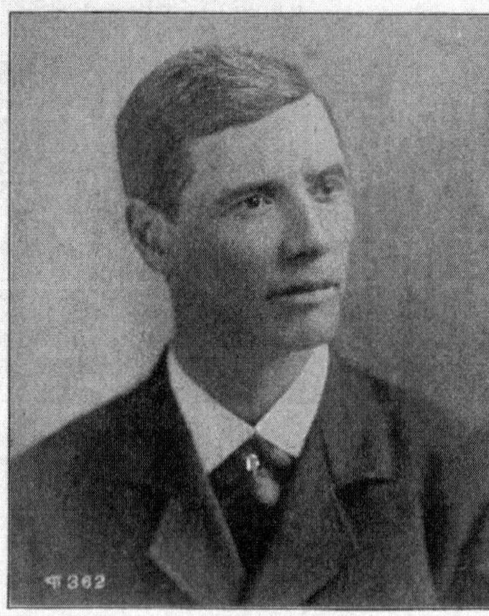

GEO. W. RUST.
Eminent Shorthorn authority.

HEREFORDS FOR CROSSING.

In the "Breeders' Journal," Vol. II, p. 3, Mr. H. M. Vaile, a leading Bates Shorthorn breeder, says:

"While they (the Herefords) are uniform in color and markings, they are not uniform in fattening capacity, and it is the experience of unprejudiced men, that they are a failure in crossing upon other cattle, or comparatively so." To which we replied, "I would credit Mr. V. with an honest intention of stating what he believes to be true. But this statement is wide of the mark. There is no breed that carry their quality so uniformly upon their produce as do the Herefords, and I should be glad to show Mr. V. the produce from widely differing dams."

Mr. Vaile closes by saying, "the Herefords have been in the hands of good breeders; yet, for one hundred years they have not extended their limits, and scarce increased in numbers, until Mr. Miller became their champion. He has given them new life by his great zeal, and I expect this will continue for a time, for they are not without merit. But they never have been, and I cannot believe they will ever be the equal of the Shorthorns." At that time we said: "I would advise Mr. Vaile to spend some time among the cattle men of Kansas City, and talk with them as to the character of the Hereford and Shorthorns. The Herefords will top all other cattle, not because of my zeal, but of the merits of the cattle."

In reply to Mr. Vaile we quote Mr. J. C. Wilson, as follows:

"'Breeders' Live Stock Journal': In the January number of the 'Journal' I see an article headed 'Shorthorns vs. T. L. Miller,' in which Mr. Vaile is quoted as saying: 'While the Herefords are uniform in color and markings, they are not uniform in fattening capacity, and it is the experience of unprejudiced men that they are a failure in crossing upon other cattle, or comparatively so.'

"Now, Mr. Editor, I wish to say that Mr. Vaile cannot be posted in either particular of which he speaks regarding Herefords, when he makes such a statement as the above. It is not the experience of men who breed Herefords, that they fail either in fattening uniformly or

HEREFORD BULLS ON THE TEXAS RANGE.
Property of Mrs. C. Adair (R. Walsh, manager), Paloduro. (From a photograph taken in the spring after wintering without feed or hay.)

in crossing successfully with other cattle. We have been breeding beef cattle for profit for twenty years, and for sixteen years we have bred and fed a large number of grade Hereford cattle, both steers, heifers and cows, and we find them more profitable than either Shorthorn or Devons, with both of which we have experimented.

"In the winter of 1878, in an address before the Farmers' Institute of Michigan, held in Flint, I had the honor, and pleasure also, of giving these experiments in detail; and the address has been published in full in the Agricultural Report of Michigan for that year, to which I would most respectfully refer Mr. Vaile for proof demonstrating the error of the position he assumes with regard to the Herefords, both as regards fattening evenly and crossing successfully with other cattle. An application to Prof. Baird, at the Agricultural College at Lansing, will enable anyone wishing the report to get one.

"Our grade Hereford steers have always surpassed both the others in early maturity, in being more easily and cheaply kept, and in market value when ready for the shambles. I cannot say we were unprejudiced breeders and feeders, for we started out strongly prejudiced in favor of the Shorthorns, but the Herefords, on their merit alone, have worked themselves into our favor so far, that at present we breed and feed none but the white faces. We have now over 200 grades of this breed on our farm, and we are satisfied that they make us more dollars and cents than we could possibly get out of Shorthorns. They will produce more beef in shorter time with less outlay of food and less care, than the Shorthorns. They are a hardier race of animals, stand our winters better, and are less liable to disease than the Shorthorns.

SHORTHORN BULLS ON THE TEXAS RANGE.
Property of the Farwell syndicate. (From a photograph taken in the spring after wintering on hay and cotton seed.)

"JAMES C. WILSON, M. D. (¶ 358)
"Supt. Crapo Farm,
"Flint, Mich.

"P. S.—Our experience with the Herefords is supplemented by that of our neighbors. Mr. Thomas Foster, a celebrated Devon breeder and feeder, has changed to the Herefords, and finds them more profitable. The Hon. Wm. Hamilton, a celebrated Shorthorn man of this country, and also the Hon. S. Howard, another Shorthorn breeder and feeder, have become converted from the error of their ways, and gone into the breeding of Hereford grades and full-bloods. They all confirm what I have just said about the Herefords. W."

AN AUSTRALIAN'S EXPERIENCE.

We take the liberty of quoting a letter (1879), from W. Danger, Esq., of Hunters River, Australia, to the Hon. Thomas Duckham, England:

"Previous to my leaving the Colony for England, I had not any pure-bred Herefords, my importation having been Shorthorns, but Mr. Hobler, who occupied an adjoining station, had Herefords, and his bull occasionally strayed among my cows. The result was, I had a good many white-faced cattle. Thus it frequently happened that five or six of these cross breds were among a draft lot of probably 150 to 200, and when the dealers were taken to them, the white faces were sure to be pointed out as the right sort, and the expression made, 'I wish all of the lot were like those beasts. Prime beasts! Real plums!'

"I had opportunities of seeing them when slaughtered at Sydney, when the carcass butchers told me their quality was excellent; that they equalled the Shorthorns in their rough fat, and that they preferred the white-faced bullocks to those of any other breed.

"I also found that they traveled the long distance from the station to Sydney—say 250 to 300 miles—better than any others; a less number became lame on the journey, which is, I consider, very greatly in their favor.

"For what I have seen in other herds, the Herefords readily became acclimated and fully retain their general character, displaying equal or greater aptitude to fatten than the Shorthorns. And from these combined advantages I lost no time, on my arrival in England, in purchasing and exporting bulls.

"My nephew has now my station, and he has recently purchased a three-year-old bull, bred by Mr. Reynolds, of Tocal, Patterson, Maitland. He was winner of the first prize as a yearling, a two-year-old, and a three-year-old, successively. The Herefords are now more numerous than the Devons, and nearly equal the Shorthorns in number."

EARLY TEXAS EXPERIENCE.

Aransas, Bee Co., Texas.

Dear Sir: Your note is at hand and contents noted. You wish to know how the Herefords we bought of you the winter of 1876 are getting

HEREFORDS ON THE RANGE IN TEXAS.
(From a photograph taken in the spring after wintering on the range without hay or feed.)

MAJARIS CANYON.
Near the Canadian River, Panhandle of Texas.

along. My son and I only kept five head, Hope and Venture, the largest and smallest grade heifer and one grade bull. All five are living and doing well. We like them better than Shorthorn stock. I think you can sell some here in this portion of the country as everybody is pleased with them. They can live and do well where other stock will die. We have some beautiful half-breeds coming on. I would like if you would bring down some Herefords next fall. I want some of your best thoroughbreds. Yours truly,

S. C. SKIDMORE.

MORE EARLY TEXAS EXPERIENCE.

Tascosa, Oldham Co., Texas.
October 17, 1880.

Editor "Texas Live Stock Journal":

I am on the ranch of Messrs. Lee and Reynolds, and thought a general sketch of the improved land and blooded stock owned by them might perhaps help to prove the value of such stock, and that they can be bred and raised with less expense as well in Texas as other states.

These gentlemen have for the last few years been importing thoroughbred bulls from Canada and the Eastern states, and now they have their herd graded up to such a standard of excellence, that it has no equal on this continent. For the past two years they have been unable to supply the demand for young stock. To meet this big and growing demand, they decided last year to make a specialty and breed with the express view of supplying Texas ranchmen with improved and thoroughbred bulls. Being thorough and practical cattle men, well versed in the wants and requirements of the trade, and possessed with unlimited means, they have and will make the venture a success. To show their confidence and earnestness in the business, I would state that for the ranch now occupied by them, they paid something over $40,000, which is, by the way, the best situated and adapted for business of any I have seen. I feel my inability to give a description which will do justice to either ranch or cattle.

SHORTHORNS.—Among their first importations, besides a fine lot of heifers, were twenty head of two and three-year-old bulls, bred by Mr. F. W. Stone, of Canada. A grander and more massive set of fellows it has never been our pleasure to see. Next in order are a lot bred at Bow Park, Canada, the first of which we would mention is the Earl of Airdrie 3d, out of Rose of Racine, by Imported 4th Duke of Clarence, the latter of which is the purest bred Duchess bull in existence and cost in England the snug little sum of $13,000. The Earl of Airdrie is a solid red, unusually large for his age, and of great bone and substance. Next comes Gauntlet, a rich roan-colored yearling, sired by the same famous animal, out of Sanpareil 12th. Gauntlet was shown when a calf, at all the leading fairs, both in Canada and the United States, taking the silver medal at

GRADE YEARLING HEIFERS IN THE ADAIR HERD.
J. A. Brand, Panhandle of Texas.

Ottawa, offered by the Governor-General for the best calf of any kind in the Dominion, and has also taken the first premium wherever shown. There are too many of these pure and excellently bred Shorthorn males in this herd to give anything like a minute description of all of them; we will, therefore, merely mention one other, and pass to the Herefords. That one is Prince Arthur, which is in all probability the purest bred animal on the continent; resembling in form a huge Berkshire pig. Prince Arthur is out of Princess Royal, Imported, by Vanguard, out of the grand Mantaline tribe.

Herefords.—Now, having briefly made mention of a few of the most noted Shorthorn males in use on this ranch, we will also briefly notice the Hereford part of the herd, and in doing so think it sufficient to merely mention that they are the best that money would buy, the most of them being the get of T. L. Miller's Imported Success, Seventy-six and other imported bulls, costing way up in the hundreds per head. There is a great controversy going on between cattle men as to the relative merits of the breeds, and having handled Shorthorns all my life, I am naturally prejudiced against Herefords to some extent, but since experimenting with them on the range for the past year I must admit that the Herefords are unquestionably the best "wrestlers." We are now separating the males and cows, and I only wish the editor of the "Journal" was here to pass his opinion upon the relative merits of each breed. I will state, however, that while the Durhams are in fair flesh the Herefords are rolling fat, and that each have had the same opportunities and facilities; except, it is claimed by the herders, that notwithstanding the better condition of the Herefords that they have done more service during the season than have the Durhams; proving very conclusively that they are the thriftiest and best cattle for Texas. What Texas ranchmen want is a blooded bull that will take his place by the side of the Texan and wrestle through the winter in good flesh, and consequently be in good condition for rendering valuable services early in the spring, which want can be supplied by no other as satisfactorily as by the Herefords. Parties wishing to purchase fine males with a view of improving their herds cannot do better than to correspond or call on the proprietors of this ranch. GEO. LEIGH.

It is only necessary to add that starting in with these excellent and representative stocks of both breeds, the Herefords won the preference of their owners and the Hereford supplanted the Shorthorn on the Reynolds ranches and are still (1898) doing so. T. L. M.

AN ENGLISHMAN'S TESTIMONY.

We give below an extract from a letter of George T. Turner, Knockhalt, England, to the "National Live Stock Journal":

"The Carlisle meeting of the Royal Agricultural Society was once again very unfortunate on the score of weather, but the north country people, who do not mind rain, because they are used to it, came in goodly numbers and 'did' the show as thoroughly as though the sun had been shining all the time. They came on purpose, and the mere rain and mud could not hinder them from seeing all there was to be seen. In spite of good attendance, however, there was a loss to the society of about 950 pounds sterling. There is a grim sort of satisfaction to be derived from the fact of fewer hundreds being lost by the society this year than there were thousands last year. The show was on rather a smaller scale than usual, but it was very select. Nearly all the stock was good, and the Carlisle meeting does not rank second to any former show held in connection with the society in point of interest and general utility.

"Herefords were a long way from home, and were consequently small in numbers, but the classes included some of the best cattle that the county of Hereford can, at the present time, produce. At the head of the list of bulls was Mr. Aaron Rogers' Grateful, which is, without doubt, the best showyard Hereford of the day, now that Mrs. Sarah Edwards has lost the incomparable cow, Leonora. He wears well, too, and is in good show form a couple of seasons after the famous Shorthorn show bull Sir Arthur Ingram has played himself out, both bulls being of the same age. These two animals contested the open championship of the Oxford meeting of the Bath and West in 1878, when the Hereford won easily, and at the same time and place the Hereford cow Leonora won the open championship prize for the best cow or heifer in the yard quite as easily. *These open championships have not been since offered, nor will they be, so long as the Shorthorn men retain the influence they now command, in the councils of the agricultural societies.* Two of the Hereford females which took first prize in their respective classes have, I understand, been purchased for the United States, namely, Mr. T. Middleton's cow, Nannette (¶ 359)—a winner of many prizes—and Mr. Thomas Fenn's heifer, Downton Rose, also a very successful animal in the show yard, and both of them Herefords of the first water. They should prove of great value to their new owners. Hereford cattle are now being sent into the chief show yards of the kingdom in first rate form, and the breed has been brought to a very high state of excellence. The different strains of blood have been fairly intermingled, and families have not been bred in-and-in, as the so-called fashionable Shorthorns have been, consequently they have not lost constitution. I shall be deceived if they do not work themselves into greater favor than the Shorthorns in the great breeding districts of the United States."

MR. DUCKHAM COMMENDS OUR WORK.

To the Editor of the "Hereford Times":

Sir: It was with much surprise that I read in

A MODEL SURFACE TANK IN A TEXAS PASTURE.
(Capital syndicate.)

an article in the "Hereford Times" of the 7th inst., under the heading "Hereford Cattle in America," that "the numerous breeders of pure Herefords may be congratulated on the new market which has been opened to them by the enterprise, judgment and business tact of Mr. Morgan and Mr. Thos. Rogers." Now, sir, I have no desire to detract in the slightest degree from the honors really due to those gentlemen, but quite the contrary, as I know full well the value which is to be attached to the judgment and care of one who can so successfully select and make up animals for the show ring as Mr.

W. E. CAMPBELL, OF WINCHESTER, OKLA.
(Nicknamed "Shorthorn Campbell," and later "Baldface Campbell.")

Morgan, as also, to an auctioneer, who so perseveringly uses his best efforts in the discharge of the duties he undertakes, as Mr. Rogers. At the same time, I must say that the meed of praise so profusely given in that article is due to Mr. T. L. Miller, of Beecher, Ill., U. S. A., whose untiring perseverance in the cause merits every feeling of gratitude which the breeders of Herefords can award him.

Not only has Mr. Miller patronized the Herefords by establishing the largest, and I think I may add, the best herd of Herefords in the United States, but he has so placed them before stock owners of that great country at the various exhibitions, and by his sound judgment, untiring perseverance, great talent as a writer and great influence as a gentleman of position, he has claimed for them (and to a certain extent successfully so) a front rank in the great state exhibitions of the United States. It was Mr. Miller who protested against sending his Herefords to the International Exhibition at Philadelphia in 1876, unless an English judge was selected to aid in awarding the premiums, which protest resulted in my being sent out by the British Commission. It was Mr. Miller who first introduced the Herefords into the far West to improve the vast herds in the western states. It was Mr. Miller who resolved to obtain the necessary information and publish a Herd Book for Hereford cattle in America. It was Mr. Miller who, knowing the value of the press in the promotion of any great object, established an agricultural paper in the interest of Herefords.

Knowing all this from many years of correspondence with that gentleman, I feel I should be wanting in the proper discharge of a duty to him if I did not place those facts before your readers. I may add further, that Mr. Miller was here in June last, when he purchased about 100 of our far-famed red with white faces and, although far past the meridian of life, he hastened back to use his best endeavors to obtain a removal of the restrictions of a ninety days' quarantine, imposed upon that side upon animals imported from this country.

I fear his usual success has not attended his efforts in that direction and I hear he is now returning, if not returned, to England to arrange for shipping his purchases.

There are several breeders of Herefords in the United States of much longer standing than Mr. Miller, but it was left to him to bring them to the front in the manner in which they are now brought. Mr. Miller only commenced as a Hereford breeder in 1871. In 1872 his then partner, Mr. Powell, a Herefordshire man, came to England and purchased a few Herefords, among them the handsome young cow Dolly Varden, bred by Mr. Morris, Town House, Madley, and her two offspring. Mr. Powell soon entered upon another business and Mr. Miller was most fortunate in securing the assistance of one so thoroughly conversant with the management of a herd as Mr. Morgan. The excellence of Mr. Miller's herd which he sent to the International Exhibition, and the manner in which Mr. Morgan brought them out and placed them before the judges, were themes of admiration to all who beheld them. After I discharged my duties at that exhibition, I visited Mr. Miller and numerous other Hereford breeders, in order that I might make their personal acquaintance and see how the various herds acclimatized, but at no place did I see them so admirably managed as those under Mr. Morgan's care.

I would here respectfully suggest that the Hereford breeders pay some compliment to Mr. Miller when he returns to this country, in acknowledgment of the very valuable assistance he has rendered to them in creating a demand for their pedigree animals, which, although of great and unprecedented proportions, is now only in its infancy. T. DUCKHAM.

Baysham Court, Ross, Aug. 17, 1880.

marbled to perfection, after one, or sometimes two winters' feeding in stalls, with ground feed. I have in mind now, one pair of them that I slaughtered, one of which made eighty pounds and the other eighty-one pounds of dressed meat, hide and tallow, to the 100 pounds alive. Those were oxen of six to seven years old, and had been worked before feeding. I notice, however, that now-a-days your prize steers beat it, but for that day I considered it extra fine. So you will readily see that twenty-five years ago

SAMPLE OF A MODEL RANGE HERD.

I preferred the Herefords to anything for fine beef animals. But enough. I am now up here in the richest valley in the United States, which aside from wheat is to be a great stock country in the future. I am just getting a stock farm started to make Hereford beef."

A HARTFORD (CONN.) BUTCHER'S EXPERIENCE (1881).

The following is an extract, from the letter of a gentleman in Dakota, and it will be noticed, that from cutting and marketing, he has had an experience that sustains the position we have taken as to the quality of Hereford beef:

"For many years at Hartford, Connecticut, my old home and birthplace, I handled very many fine cattle, and was for many years in the marketing business, and in that rich city, cut up many prize cattle and always cut first-class beef. But in those days, from 1852 to 1862, I used to get some Hereford cattle, fed by Connecticut valley farmers, and they were always superior beef to the Durhams, or the noted Connecticut red oxen of that day. Fat and lean

A MARYLAND BREEDER'S EXPERIENCE.

John Merryman was one of the oldest breeders of Hereford cattle, and thus stated his experience:

At the New York State Fair, held at Watertown, in 1856, I purchased from Messrs. A. & H. Bowen, a yearling bull, Catalpa, and a heifer, Lilac. My next purchase was from Mr. Sotham, and consisted of thirteen cows and heifers, and two bulls, including Blenheim, 1879.

My next purchase was from the State Board

of Massachusetts, including the imported cow Milton, bred by John Longmore, of Orleton, and her four-year-old cow and heifer Promise. Milton was brought over with the bull Cronkhill, bred by Lord Berwick, and he was sire of Cora and Promise. The same year, I purchased five more head from Mr. Sotham. Curly, the Ashton bull, died while in my possession. I had given Col. Edward Lloyd, Blenheim in exchange for him. He got out of Milton, a bull-calf, Marion, and out of Miss Tully, also an imported cow, a heifer, Hattie. She brought me a number of superior calves by Sir Richard 2d, including Belle McAlpin, dam of Stonie Williams. I have had two bulls from Mr. F. W. Stone, Admiral and Canadian; also imported Sir Richard 2d, who was bred by J. H. Arkwright, Esq. I afterwards used Illinois, son of Success, bred by T. L. Miller, and now have Prince of the Wye, bred by Mr. T. Duckham, M. P. I have been a breeder of Herefords for twenty-five years, and am, perhaps, entitled to the position as the oldest breeder now in the business. Respectfully yours,

JOHN MERRYMAN.

Cockeysville, Md., Aug. 9, 1881.

A SAMPLE CASE FROM KANSAS (1882).

Mr. Editor:

Before consenting to exhibit my Herefords at Wichita Fair, I saw the president—who is an honorable gentleman—and he promised me that no Shorthorn breeder should be put on the sweepstakes awarding committee, but through the intent or carelessness of some subordinate officer, three Shorthorn breeders were chosen as judges of sweepstakes. Two of them were unknown to me; the other was the Hon. Wm. Ross, an old acquaintance, who had been an exhibitor of Shorthorns at previous fairs. From such a personage I expected even-handed justice. My bull had been pronounced the best bull in the show by quite a number of Shorthorn breeders, and was almost the universal choice of the multitude. They were so outspoken in his favor that I thought I had an easy walk-away. Imagine my surprise when I found he was not noticed by the judges any more than a yellow dog would have been outside the ring, and the 1st and 2d ribbons were tied upon Shorthorn bulls.

As I led my bull from the ring a prominent

SAMPLE OF SO-CALLED HEREFORDS TOO OFTEN FOUND ON THE RANGE AND ELSEWHERE.
(Every whiteface is not a Hereford.)

Shorthorn breeder remarked: "Well, the white-faced bull is the best bull on the grounds, if he did not get a ribbon." "That's what I say." "I say so too," and other like remarks from the bystanders, caused a crowd to gather around the deeply-wronged animal and he was viewed over and over again, and in ten minutes he had as many friends as anybody. I felt the injustice so keenly that I could scarcely conduct myself with proper decorum, and when the Honorable Ross came round to see what the row was about, I addressed him thus: "Well, Uncle Billy, I would like you to take a look at my little bull; I don't think your committee has ever seen him. I would have felt highly complimented had you noticed him while in the ring." This was more than his Honor could bear, and he replied: "The Herefords have no right to show against Shorthorns anyhow." To which I replied, "If that is the case, I will leave my cows in their stalls; the public will give the Herefords justice if you Shorthorn judges will not." I then announced that I would protest the awards. The crowd then shouted, "Show your cows; you shall have justice." I consented, and did show them under protest. But the result was the same as in the bull ring and the crowd became so outspoken that the superintendent announced, that, owing to the dissatisfaction among the "outsiders," he would postpone the exhibit of herds competing for sweepstakes honors until the next day and then he would try and secure new judges to pass upon them.

My protest was substantially as follows:

"To the Officers and Directors of the Sedgwick County Agricultural, Mechanical and Stock Association:

"Grounds of Protest: 1st. That the Hon. Wm. Ross is now, and has in the past been a breeder and exhibitor of Shorthorn cattle at our annual fairs, and that he is interested in the success of the Shorthorn breed; and further, that he has shown great prejudice and bitterness against the Hereford breed in certain newspaper articles published over his signature, in the papers of this city, and in our belief, he could not do justice to any animal competing for sweepstakes honors that was not a Shorthorn.

"2. That the other two committeemen are Shorthorn breeders and interested in the success

A MODEL RANCH HEADQUARTERS.
Spring Lake Ranch of W. E. Halsell, Bovina, Tex.

of the Shorthorn breed, and that both of them have acknowledged that they knew nothing of the Hereford breed, one of them stating to myself that he did not know anything about Herefords, whether they were good for milk or anything else, and that the other (Mr. Roberts) made substantially the same statement to Mr. R. L. Houston.

"3. That the Shorthorn bull that said committee awarded the second sweepstakes premium, entered the ring displaying a blue ribbon in token of a first premium in his class, and, although his attendant was requested to remove the same, said display was continued throughout the entire showing for sweepstakes honors, contrary to all rules, regulations and customs governing such exhibitions and may have had an influence upon the aforesaid sweepstakes awarding committee.

"4. That Section 30, of Rules and Regulations, adopted by your society for the government of sweepstakes, was wholly ignored and disregarded by the aforesaid awarding committee, which is evidence of their incompetency to act as judges upon sweepstakes, or of doing willful justice."

After filing the above protest, I went about my business and paid no further attention to it. But the Shorthorn men were better posted than I (this being my first experience, I admit I was not prepared to meet all the tricks), and put their heads together, selected Mr. Ross and Mr. Blakeley (my competitor), to go before the Board and defeat my protest at all hazards. I had left the grounds when the Board met and the Shorthorn men supposed the field was clear. But as I could not be found, my friend, the Hon. John Kelley, who is also the vice-president of the society, took notes, and after the objections were heard, championed the white-face

A "ROUND-UP" ON THE RANGE.

cause. The protest was sustained, and a new committee ordered to be chosen. The Shorthorn exhibitors ruled off quite a number of committeemen, but finally three were selected. Two of them I had never seen before. One was a city butcher; one an Englishman, who had extended experience in England, New Zealand and this country; the other was an old feeder and shipper, and none of the three were interested in any breed of cattle. A number of the Shorthorn breeders pledged themselves not to bring their cattle into the ring again, which caused considerable delay and excitement. But the superintendent enforced the rules and they were compelled to bring out their stock or forfeit their class premiums.

When the bulls were led out a dense crowd gathered around them and it was with difficulty that the police cleared the way. Hon. John Kelley and a number of prominent cattle men then inspected the display and then stepped aside and cast a vote as to where first and second premiums should go. This was done secretly before the judges came into the ring. Scarcely any time was taken upon the first, and the committee was unanimous in favor of my bull. The second went to Mr. Blakeley's walked off without any interruption, followed by a few of their admirers, whose faces looked fully two inches longer than the day previous. I am free to acknowledge that this was one of the proudest days of my life. Not so much on account of this victory, but because I had resented the insult and wrongs done the dumb brute and had secured for him even-handed justice.

Next came the cows. My three-year-old cow, Ella 3d, with a good strong bull calf tugging at her, was awarded first; and last, but not least, my whitefaced herd was declared the best on

FAMILIAR SCENE ON THE RANGE.

Shorthorn, and the bulls that got first and second the previous day got nothing. The action of the committee and the outside judges were precisely the same, and gave general satisfaction. I was completely overwhelmed with congratulations, and the little bull was patted and stroked by many a man that had never seen a Hereford before. "Ain't he a good one." "He is the best bull I ever saw." "Just look at his back." "Yes, and his quarters, too. He comes out behind like a Berkshire pig." "If it was not for his horns he would be perfection, wouldn't he?" and other like exclamations, came from all sides. The defeated Shorthorns

the grounds. The judges were unanimous in all these awards and the bystanders shook their hands and congratulated them upon their judgment.

I supposed the battle over and the victory won beyond all question; but in this I was mistaken. The secretary notified me that the Shorthorn men had protested and ordered him to withhold my diplomas and premiums; that the Hon. Mr. Ross denied writing the newspaper articles referred to in my protest, etc., etc. They then employed a lawyer to manage and attend to their case. I then sought the Hon. John Kelley, but he had gone home and

"DRIFTING." A SCENE ON THE RANGE OF COL. C. C. SLAUGHTER.
North Texas and New Mexico line.

I was left alone to fight it out as best I could. I could see no reason why I could not make all the showing necessary without the aid of a lawyer, therefore, I did not employ one. I sustained every point taken, and the word of the editor and old files of the newspaper proved that Hon. Mr. Ross' memory was treacherous, to say the least. Judge Little finally proposed to withdraw the Shorthorn protest, which was urged upon technicalities only, and to sustain the first judgment, if I would divide the premium money with Mr. Blakeley. To this proposition, I replied: "It is not the money I am contending for—thirty dollars is but a trifle—but there is a principle involved and underlying this matter that I am contending for. I have won the awards by fair and impartial judgment and I am determined to maintain my rights if I have to resort to the Supreme Court to secure them." This was a clincher and after another half day's parley among themselves, they withdrew their protest and the secretary paid me my money. A delay of four days was caused by these proceedings. W. E. Campbell.

Mr. Campbell, now of Kiowa, Kansas, was then at Caldwell; he first introduced Shorthorns, and was known as "Shorthorn Campbell" for many years until the merits of the Hereford became appreciated by him. After the above experience, in true western fashion, to his countless friends and acquaintances he was known as "Bald-face Campbell."

EARLY EXPERIENCE IN VERMONT.

Cornwall, Vt., May 8, 1882.

The year that I was twenty, which was thirty-eight years ago, I visited with two of our best farmers, Corning and Sotham's herd of Hereford cattle at Albany, N. Y. At the same time I saw the Prentice herd of Shorthorns, and have since seen some of the finest herds of Shorthorns in the country, but have never seen a herd of cattle that so completely filled my eye, as Corning and Sotham's Herefords did.

The two farmers that were with me purchased two bulls of Mr. Sotham and I raised a few calves from the bulls out of common grade cows, and from their calves I had steers the fall after they were two years old, on grass alone, 1,350 pounds, live weight; dressed, 800 pounds.

One spring I purchased two steer calves, half-blood, one-year-old, that had been wintered on

(*341) "CORRECTOR" 48976—GREATEST OF MODERN HEREFORD SIRES.
Bred and owned by T. F. B. Sotham, Weavergrace Breeding Establishment, Chillicothe, Mo.

lowland hay. They were not large and not in good condition. I turned them into a pasture with a lot of two-year-olds, with no feed except the pasture; I dressed the following December 616 and 632 pounds, and the finest beef I ever tasted. What the live weight would have been, I do not know.

A few years after these bulls were brought into our town, there was a great Shorthorn fever, and bulls of that breed were introduced and the Herefords neglected, and I never had as fine steers, or saw as fine steers in our town of the grade Shorthorn, as there were of the Herefords, and I have heard several of our best farmers make the same remark.

I had several half-blood cows, some of them very fine, but now and then would have one that would not make good milk cows, but would make nice beef. I am, with most of the farmers in Addison County, engaged in breeding Merino sheep, and at present prices it is for our interest to devote the most of our energies in this direction. But the Champlain Valley in our county is as fine grazing for cattle as any section in the country, and I shall be pleased to see Herefords again introduced.

The chromo of eight premium Herefords you wrote of sending me, I have not received; shall be pleased to receive one.

Yours truly, HENRY LANE.

MR. CORNING'S NEW YORK EXPERIENCE.

The following letter was written to the "Breeders' Journal":

Corning Farm, Albany, N. Y.
May 15, 1882.

Mr. Editor:

In your article on Wm. Henry Sotham, Esq., I noticed an assertion upon which I should like to throw some light and which I expected Mr. Sotham would do in the following May number of the "Journal." Paragraph No. 2, second column, page 194, states: "Recognizing these facts the Shorthorn breeders made special efforts to detach Mr. Corning from the Hereford interest and were finally successful."

To be sure, they were by their efforts successful in proving to Mr. Corning the superiority of the Hereford to the Shorthorn.

Many of Mr. Corning's friends were interested in Shorthorns at that time, and when he took a fancy to the Herefords, they endeavored to detach him from whatever favorable views he had of them, but being a very practical gentleman, he resolved to test the merits of the two breeds himself, by a trial under conditions he knew would be perfectly just and similar to both breeds. At that time Mr. Corning owned a choice herd of Shorthorns, and these with the Herefords were bred from, fed and killed, both

A SAMPLE OF COL. C. C. SLAUGHTER'S (DALLAS, TEX.) LAZY "S" BRAND, NORTH TEXAS.

having the same care and treatment. The animals were weighed regularly, as was also the feed consumed by each.

Mr. C. and his son, the present Hon. Erastus Corning, were so far convinced of the superiority of the Herefords, that the Shorthorns were sold, and the descendants of the same herd of Herefords can now be seen on the farm.

The Herefords in the trial exceeded the Shorthorns in the number of pounds gained and put the gain in the most desirable portions of the animal for market.

In the forty-two years the Herefords have been on the farm, there has never been a Hereford cow but what has given plenty of milk to raise her calf, whereas some of the calves of the Shorthorn cows would have died of starvation but for outside help.

The Corning herd of Herefords are kept in breeding condition only and are exhibited in that state.

As prize winners they rank among the highest in the country, having won the New York State Agricultural Society's Large Gold Medal three times. Imp. Comus has won seven first premiums, Marchioness 2d, Katy, and Victoria 6th have all won first prizes as cows, when competing against some very noted herds.

The demand for Herefords in the last six months has been unprecedented.

While in Colorado some time ago, I asked an extensive cattle man what he thought was the best breed for beef. "Well, there isn't much doubt about that, when a man can get five dollars more a head for white-faced steers, from Kansas and Illinois buyers, than for any other steers, no matter how well-bred we tell them they are." JAMES W. COX, JR.

A NEW YORKER ENDORSES MR. SOTHAM.

Meslina, Orleans Co., N. Y.
June 5, 1882.

Mr. Editor:

The "Breeders' Journal" of April last was put into my hands a few days ago by a friend, and while I have not had time to examine it as much as I wish, I have read the article upon the life and career of my old and valued friend, Wm. H. Sotham, as a breeder of Herefords, and much of which came under my own knowledge. It is a truthful history, and his likeness recalls to my mind many pleasant memories, and I am glad to know that he is yet living and sees his favorite breed taking rank where he always claimed they would if given anything like a fair chance alongside the Shorthorns; many men less resolute than he would have succumbed to the great pressure against him in the early introduction of the Herefords. I have witnessed many an argument between him and the Shorthorn men, besides those wielded by his able pen. Mr. Remington, of whom you speak, is a brother-in-law of mine, and has started a small cattle ranch in Holt Co., Nebraska, and purchased a Hereford bull of Erastus Corning, of Albany. His son, M. C. Remington, Jr., is attending to it. I am still of the opinion there are no better cattle for the shambles or for working oxen than the Herefords, but as beef is now produced so much cheaper in the West, we have to resort to other branches of farming. I purchased some very good animals of Mr. Sotham, one heifer two years old, and calf by her side. Woodbine, illustrated in transactions of the New York State Agricultural Society in 1853, took first prize at Saratoga. I then sold the late John Merryman, of Cockeysville, Md., the first Hereford he ever purchased. I last met Mr. Sotham in Detroit some four years ago, and through his recommendation I purchased some Berkshires of Mr. Smith, with which I have been very successful, carrying off first prize in our state. I should like to be informed of Mr. Sotham's address.

Sincerely yours, H. BOWEN.

SUCCESSFUL IN DAKOTA.

Mr. T. L. MILLER, Beecher, Ill.

Dear Sir: Have intended writing you for a considerable time, but between getting in a new herd of Minnesota cattle, and putting up our hay crop, have had little leisure since the first of May. We are now in position to give the results of our first year's experience with Herefords, and am glad to say it is most satisfactory. We used the yearling bulls we bought of you in June of last year on a limited number of native cows during July and August. Every cow they served had a calf last spring and all are alive and doing splendidly. They all show the Hereford blood distinctly—and some of them are as well marked as their sires. After the season was over, we turned the bulls out on the range and let them run all winter. Did not stable them or feed any hay or grain. They came out this spring—thin in flesh, but in good thriving condition, and filled up very fast when the new grass came. Altogether, they proved themselves a very hardy lot of cattle and well adapted to roughing it on the range. Have used them on our entire herd this season, and if next year's calves prove equal to those we now have, the Herefords will unquestionably have the lead here. All our neighbors conceded to us the best lot of calves to be found in this section.

Some parties who have been using Shorthorn bulls, are negotiating with us for grade Hereford bull calves. Hoping you are enjoying good health and the "Highland herd" is flourishing, we remain, yours truly, KENNEDY BROS.,
Fort Sully, D. T.
Oak Creek Ranch, D. T., Sept. 8, 1882.

GRADE HEREFORD SIRES IN MISSOURI.

Concord, Mo., Jan. 9, 1883.

T. L. MILLER, Beecher, Ill.:

I wish you would be kind enough to tell me what is the trouble, if any, in the breeding of my bull, or rather, I should say, the marking of my grade Hereford calves. I have been told that he (the bull) puts too much red in their faces. Last spring, I had dropped twenty-three grade Hereford calves, sired by Red Cloud 2086 (5528), and out of high-grade Shorthorn cows. They (the cows) are all red, some few with white spots, and deep red roans. The calves are without exception deep reds, with white bellies, white feet, white ends on their tails, and white faces back to their ears, with a broad red spot over the nose and some of them extending up to the eyes, covering one-third of the face. They have short legs, broad hips, straight lines top and bottom, well sprung ribs and deep briskets. In short, everyone was a good one, without a runt or dwarf, and to-day are fat, living on hay, staying out of doors on the prairie. Is the bull at fault? Are the spots all right and as they should be? I have bred to him for this year fifty choice grade cows, and will in a few days have more calves. Now, if the spotted faces are all right, the calves are.

I am a Hereford man from honest convictions and because I believe the Herefords to be *the* cattle. A hedge fence alone separates my herd from a herd of Shorthorns that for pure breeding and fancy strains of blood stands pre-eminent in Central Missouri, but as they emerge from their warm barn and well-filled boxes, showing the kindly attention of an intelligent and good master, and in themselves most splendid specimens of a splendid breed, they excite not the slightest sense of shame, for the jolly little whitefaces, that go scampering away over the frozen snow to the music of the jingling icicles that hang to their soft and mellow hides, and who wouldn't know an ear of corn if they'd meet it in the road. Please answer me the question of their faces, and oblige.

Yours truly, CEPHALUS BLACK.

NOTE.—The man who buys grade Hereford bulls should first convince himself that they are by thoroughbred Hereford bulls, and then that the form of the calf is good; and then, if the calf has a brockled face, it will not hurt him.

MORE NEW YORK EXPERIENCE.

Amos Bissell & Son, of Milford, N. Y., wrote in October, 1883, as follows:

On April 13th last, we placed a pair of grade Hereford steers in a stall together, and told the man who grooms our stallions to see what he could make of them. That day they weighed 2,100 pounds and were so good that two weeks later the butchers offered us $143 for them to kill. Five months and eleven days later we drove them to the village (one mile) and they pulled the scale at 3,180 pounds, but had on a pine yoke which we think will weigh 35 pounds, but is safe at 40 pounds. Thus the cattle gained 1,050 pounds during that time, or an average of 6.44 pounds per day for the pair. Both weighings were honest in every particular and on the same scales. Age, three years last spring. Their daily feed consisted of small potatoes, hay, nine quarts wheat middlings, and two quarts oil meal. August and September, hay, Bassano beets, four quarts of oil meal and twelve quarts of middlings. They are now eating potatoes and some quantity of grain. We hope to grow them a ton during the experiment year.

These steers have no straight breeding in them, but are doubtless three-quarters Hereford, as we have much of that blood mixed through our dairies.

By the way, Mr. Geo. Clarke, our neighbor, had at one time over 300 pure-bred breeding cows of his own importing and breeding—many years before you gentlemen at the West adopted the breed.

MR. FUNKHOUSER A MISSOURI CONVERT.

Mr. Jas. A. Funkhouser, of Plattsburgh, Mo., writes as follows (1884):

In regard to my calves by Dauphin, I will say I have eighteen, ten bulls and eight heifers. The ten bull calves average weight is 711 pounds and their average age is 242 days, or eight months. The eight heifers' average weight is 601 pounds, and their average age is 239 days. They certainly are a choice lot, in fact, the best lot of calves of any breed I ever saw. My Shorthorn friends, even, acknowledge they are an extra lot of calves, but claim they would be good from such cows by any bull. It occurs to me they are acknowledging just what you have been preaching for years, that a Hereford bull bred to a Shorthorn cow makes an improvement. I am well satisfied with the cross, and my only regret is that I did not get the Hereford sooner. I have forty-three cows in calf to my young bull Invincible, and he has proved himself to be a very sure breeder. He was two years old October 18, 1883, and weighs over 1,300 pounds, after a hard season's work. I am very much pleased with him and think he will yet be a credit to his breeder. The heifer I had of you did not do so well last season; this season she has improved amazingly and now I am not ashamed to say she is by Dauphin 18th.

FAVORABLE CALIFORNIA REPORT.

San Francisco, Cal., Sept. 3, 1884.

Mr. Editor:

Last evening I read up what your correspondents have to say about local Herefords, and from experience can say the Hereford has been found the animal par excellence for New Zealand and Australian pastures. After the severe cold weather in South of New Zealand winters he will flesh up in half the time it takes the Shorthorn on rough outdoor feed.

In 1883 Mr. R. P. Saxe was sent from this city by Mr. A. W. Sisson to select from the herd of the New Zealand Stud Company, forty-four head of Hereford bulls and heifers. These arrived without any loss and were sold here at prices from $450 to $750 per head.

Every buyer without exception has expressed his delight with the stock and especially the favorable manner in which they go through the winter, compared with the Durham.

I note in your last number that Mr. A. Galatin, of Sacramento, who purchased from us, is applying for registration of his stock in the American Hereford Record. This may be done with every assurance of the pedigrees furnished by this gentleman being authentic and reliable. They were sent to us direct from the stud and pedigree company.

Last year they sent us twenty-five Herefords and five Shorthorns. The first-named came through by steamer without loss. Of the five Durhams two died on the voyage. We sold one of the Durhams, "Spring," for $400 to Dr. Lillerneranty. He won the blue ribbon with him, at the Golden Gate Fair, September, 1884, and resold him. I note he is again a winner and held at a very high price by the owner.

The Herefords won prizes at Alameda, Stockton, Sacramento and San Jose fairs, and were sold at an average of $290 per head, and shipped in every direction, from San Diego, in the South, to Frazer River in British Columbia, in the North.

From every direction we hear nothing but satisfaction. The company have at Auckland a herd of 800 pure Herefords.

I think we can match, if not show a superior bull to Success, from our herd. The Hereford bull for British Columbia had to foot it inland 250 miles to reach the run.

Yours faithfully, HUGH CRAIG.

AN OFT-REPEATED OCCURRENCE.

Illinois State Board of Agriculture,
Secretary's Office.

T. L. MILLER, ESQ., Beecher, Ill.,

Dear Sir: Yours of the 23d received, and in reply will state that it was surely an oversight, the omitting to print your specials at the Fat Stock Show in the 5,000 edition of the premium list, until the whole edition was printed.

It will appear in nice form in the 5,000 additional editions and also in the 2,000 printed separate from the State Fair list.

It occurred in this way: Your offering was made at the time the Board met in Chicago, and was pasted in the record book. I knew this, of course. All the remaining specials I separated and kept in a file by themselves.

When the printer reached this part of the list and wanted copy, I was very busy with my correspondence and turned the matter over to Mr. Mills to arrange, and when prepared I gave copy to the printer not thinking but all were in. Nor was it discovered, as stated before, that you were left out, until the edition was printed; was very sorry, of course, but among the thousand things to watch, sometimes it will happen that an error or an omission will occur.

Yours truly, S. D. FISHER.

Springfield, Ill., July 25, 1884.

IN THE HIGHLANDS OF COLORADO.

Editor "Live Stock Record":

In your paper of the 30th inst., I notice that Mr. Pryor, the "Shorthorn Stalwart," flings broadcast to all Hereford breeders certain questions which are so evidently propounded, only after deep thought and careful consideration, that it is with great hesitation that I venture to answer them. Mr. Pryor asks:

First. "Do you not all find the Shorthorn cross indispensable when you wish to get good high-grade bulls?"

By no means. For, although the Shorthorn is an unquestionably good foundation, upon which to build up a grade herd, I consider that equally good results as to form, and superior as to color, can be obtained by crossing the Hereford bull with the Devon or red Sussex cows. And I have good reason to think that no animal will catch the butcher's eye much quicker than the grade produce of the Hereford and Polled-Angus, or Galloway. Mr. Pryor asks:

Second. "Are not nine-tenths of all your bulls sold for range purposes from grade Shorthorn cows?"

Taking as an example my sale of bull calves for next spring, a correct notice of which I see in your columns, my answer is, certainly not.

Third. "What proportion of them, if any, are from grade Hereford cows?"

In the above-mentioned bunch the proportion of calves had from grade Herefords cows, ranging from first to fourth cross, is somewhat in excess of two-thirds.

Six years ago I was fortunate enough to buy thirty grade Hereford cows and heifers, presumably first and second cross. I have since bred them to first-class Hereford bulls with most satisfactory results. The increase has been considerably above the average—the death-rate almost nothing. Twenty-six out of the thirty original cows will to-day face the tally; and yet these delicate cattle have been obliged to "sniff the breeze and stem the blizzards" at an altitude of 8,000 feet, without the assistance of either 15 cent cribs of Kansas corn, or succulent alfalfa.

I do not wish to enter into the question of superiority of herds; so long as my own pudding finds favor I am satisfied. But when Mr. Pryor claims, or rather insinuates that the result of a first cross between Herefords and Shorthorns produces a better calf than can be obtained by inbreeding to third or fourth cross Hereford cows, I must respectfully beg to differ with him, always supposing the would-be purchaser to be looking for "Hereford grades."

Yours truly,

THEODORE WHYTE.

Estes Park, Jan. 4th.

Mr. Theodore Whyte was a breeder of Hereford cattle, and Mr. Pryor was a breeder of Shorthorns in Colorado.

MR. CULBERTSON TO ILLINOIS BOARD OF AGRICULTURE.

Gentlemen: Inasmuch as many absurd awards were made at our last Fat Stock Show, which have caused wide comment among cattle men, I beg leave to submit the following expression of my views, feeling assured, from conversation with many there present, that I speak the sentiment of a great majority of breeders, feeders, exhibitors, and cattlemen generally, independent of affinity for any particular breed of cattle. So far as my own exhibit was concerned, I expected nothing, and consequently was not disappointed.

I have just passed through our sixth Annual Fat Stock Show, and I am pleased to say that, from year to year, we have continually improved in the exhibit in the cattle department, our last exhibit being adjudged the best ever seen together on this continent. I wish the same could be said of the judging. This is, and ought to be, the Great American Beef Show, the result and report of which is heralded not only all over this country, but over Europe also, hence, the very great importance of having just awards made. But how and what are the facts in regard to the awards? I do not believe that there is one of you who does not feel deeply mortified and ashamed of most of the more important awards. The errors were so glaring that it would seem no good excuse could be made, or one, at least, that would have any force in it. This show was created for a school, at which all might come to compete and learn. Inducements have been held out to make steers and prepare and exhibit them—the best beef animals to win. We have been promised, from year to year, the best judges obtainable. We have never had them (in my opinion). Illinois, Indiana, Iowa, Missouri, Ohio, Kentucky, Kansas, Minnesota and Canada, have come up here, bringing with them their best cattle for judgment. Has justice been done them? I think you will say *no*. Please pardon me when I say to you, in all sincerity, that if, in the future, you cannot obtain more competent judges, it will not be long before those who have taken great pains to make steers, "using both skill and patience in preparing their cattle," will become completely discouraged and disgusted, and the show, and the purposes for which it was created, will be pronounced a by-word and a mockery.

In my criticisms of the awards I shall touch only on a few of the most important ones, fearing that I may tire your patience. In the first day's judging the Shorthorn steer Schooler took third prize. He was the best two-year-old Shorthorn in the show by all odds—a credit to the breeder and feeder—hard to beat in any ring, and worthy to compete for grand sweepstakes. By this award he was barred from competing for the grand sweepstakes.

In the three-year-old Hereford class but two animals were shown, to-wit, a cow belonging to Fowler & Van Natta, and the steer Hoosier, belonging to Adams Earl. The cow took first prize, and no one was more surprised than Mr. Van Natta himself. The steer Hoosier was perhaps the most skillfully fed and fattened animal in the show, and the most evenly fleshed animal in the show—not a lump as large as an almond could be found on him from his horns to his hocks. He, likewise, was barred from showing for the grand sweepstakes.

In the three-year-old class for grade Herefords, Benton's Champion took third prize. This, of course, barred him from the grand sweepstakes prizes. This Benton's Champion was such an extraordinary good steer that Mr. John D. Gillette publicly said "he was as good, if not the best, beef steer he had ever seen." Mr. Gillette is perhaps the most noted and successful steer breeder and feeder in the West—a man of integrity and rare good judgment.

But the greatest wrong that was done was the awarding the grand sweepstakes carcass prize to the steer Clarence Kirklevington. Of all the carcasses on exhibition—30 in all—this carcass, with but one exception—was the most objectionable of any. After hanging on the hooks for two days and nights, the weather being cool, it was as soft and springy as when first killed—its outer coat of fat handling as soft and slippery almost as a sponge filled with oil. As a proof of what I have said about this, it may be interesting to know what became of this carcass. I find that the butcher, Mr. John Ford, at the special request of the proprietors of the Grand Pacific Hotel, bought the carcass; that the carcass weighed some 1,650 pounds; that he paid 12½ cents per pound for it—costing about $200; that he sent about 600 pounds to the Grand Pacific Hotel; that the hotel sent back to him about 300 pounds; that he got about $90 for 300 pounds from the hotel, and that Mr. Ford estimated that the loss on the carcass would be about $50. If such is the case, he could not have realized quite 5 cents per pound for the balance, 1,350 pounds. One of the proprietors of the hotel said that the meat was the worst they ever had in the hotel. And this, gentlemen, is the grand sweepstakes carcass for the year 1884.

The truth is, gentlemen, that Clarence Kirklevington as a beef animal never deserved a ribbon of any kind. He was a beautiful animal for a picture, groomed and polished to the very highest degree, but as a beef animal he, to say the least, was a delusion. The judges on carcasses, in 1883, gave the grand sweepstakes to a very lean and only about half fattened animal. In 1884 the very reverse was the case.

My opinion is, that no half-fat animal will retain its form in cooking. The marbling or fat strands, or tissues, not being solidified among the lean, will melt out in cooking. The fatty part, it is thought, contains most of the flavor, hence when cooked out leaves the lean minus the flavor. Mr. Kinsley's cook hits the nail on the head when he says: "A beef animal is like an apple—1st, the green stage; 2d, the ripe stage, and 3d, the over-ripe stage—when it begins to deteriorate."

Now who is to blame for all this blundering? I am of the opinion that it is mainly chargeable to the management. You were very unfortunate in choosing your judges, but when you saw their incompetency, which was made very apparent the first hour they acted, I think it was your bounden duty—a duty you owed to all the exhibitors—to have promptly discharged them, and substituted other judges. Mr. Dysart, in 1881 or 1882, when he was the superintendent of the ring, on the first half day of the judging, promptly discharged the judges when it became apparent that they were not competent, and had other judges substituted who gave very general satisfaction.

The only places where you are likely to find expert judges of beef is among men who constantly use none but the choicest of beef, and for carcass judges none so good as the men who do the cooking and serving up.

It was reported that all judges who served in 1883 would be barred from serving in 1884. How does the judging of 1884 average with that of 1883? In the Polled yearling class you allowed a steer to be led into the ring to compete for the prize that had been protested by about all the cattle exhibitors and expert judges on ages, as being one year older than he was entered for. I must say that much surprise and dissatisfaction was manifested on account of this action.

I have from the start taken great interest and pride in our fat cattle shows and have made, from time to time, many suggestions about arranging and making our show and system more perfect. I am still with you, but I must say that if better judging cannot be had, your show will lose the confidence and respect it should have and the people and exhibitors will lose interest in it.

In what I have said I have tried to state facts as they appear to me, and have also tried not to give offense to anyone. If I have said anything that is not true, I shall be glad to make the correction at any time. I have tried to make what I have said to you as non-partisan as it was possible to do, and make my meaning understood.

With best wishes for your board, and all that it implies, and hoping that you may be able in the future to procure judges of note, whose opinions all will respect—judges who, as it were, will be blind as to breeds, who can distinguish the true line of merit, let it be ever so delicately drawn, and will hew to the line, then will you have true educators. Then, and not till then, will you have fully accomplished the purpose for which our fat show was created.

C. M. CULBERTSON.

Chicago, Ill., Jan. 8, 1885.

IOWA HEARD FROM.

Editor "Breeders' Journal": We are butchers of eight years' experience, and during that time, occasionally getting hold of a grade Hereford, found by comparison with all others, they were par excellence as beef cattle, giving a larger percentage of the most valuable cuts to the butcher. We notice also a great difference in heft of hides—no small item to us. The Hereford's hide is 20 to 30 per cent heavier than scrubs or Shorthorns. This, we believe, in part accounts for their great ability to withstand cold. These facts led us to breeding Herefords. About four years ago we commenced quietly saving the best grades we found in our business, and two years ago we crossed them with a Wilton bull, and one year ago we bought four imported cows hard to beat. We have a bull calf from one of them, now ten months old, that is perfect in form, marking and color, and weighs 950 pounds. This is the first thoroughbred dropped to us, and we pride ourselves on our commencement. At our County Fair last fall, with strong competition in Shorthorns, Holsteins, etc., we took sweepstakes on best cow and got one vote of three on a yearling bull, these all being any age or breed. Herefords are now booming in Guthrie County, where but two years ago Shorthorns were all the rage.

Yours respectfully,

D. P. WILLIAMS (¶ 360).

Guthrie Center, Ia., Jan. 28, 1886.

TEXAS TURNING.

Mr. Editor: I write you a few lines about how the Herefords are doing in this country. I bought two heifers and one bull of T. L. Miller and a heifer and bull of Tom C. Ponting in January. They are very fine and doing well. I have six thoroughbred Hereford bulls, all in fine condition, and thirty-five little whitefaces that are looking fine, and more to come. There is more talk about Herefords this season than ever before. At the convention of the North Texas Cattle Raisers' Association at Weatherford the hall was beautifully decorated. There was a beautiful Texas star, just back of the President's stand, with two United States flags supported by a point of the star on each side. In the center of the star—the most appropriate thing that could have been placed there—something that was suggestive of what we are striving for: also of the best breed of cattle, and the breed that is taking the lead and will hold it; the best rustler of all fine blooded cattle and next to the Texas bullock of any breed that has ever been tried in the state—was a beautiful picture of T. L. Miller's noted Success bull. Success to the Convention, success to the business, and success to the Herefords, as the name of the bull indicated. I don't think anything more appropriate could have been put in the place of the picture, as the name of the bull represented all we could ask for.

Yours truly,

W. S. IKARD. (¶361)

Henrietta, Tex., Mar. 20, 1886.

In March, 1887, Mr. Ikard wrote us again as follows:

"Stock have never wintered better so far in Texas. There will not be any loss in this part of the state to speak of, and from what I can learn stock are wintering equally as well all over the state. My Herefords are doing very well indeed. They are fat enough to butcher, and have not been fed any this winter, wintering as well without feed as any common Texas breed of cattle. They are the chief rustlers of any fine breed of cattle that I have ever seen, and they have certainly proved equal to the common Texas bullock as a range animal, rustling for his own living. There was but little raised last season, and no feeding done this winter except to work stock."

MORE FROM TEXAS.

Mr. Editor: Thinking that you would like to hear from this corner of creation I pen these few lines. I have, up to date, lost five head of the twenty-one cattle that I bought of your townsmen and estimable gentlemen, T L. and

T. E. Miller. I have been breeding Herefords now about nine years, having bought five head, three bulls and two heifers, nine years ago in San Antonio, Texas, of Mr. T. E. Miller, and can say, not boastingly, but with truth and candor, that I think Herefords are the cattle for the range. My grades stand it better than even the native Spanish cattle. I am a true friend to the Herefords, and hope to see them universally used as a beef producing animal. The people here are taking to the whitefaced cattle right along. All seem to want them, but the price somewhat scares most of them. They are used to buying other and less desirable cattle much cheaper, and they are loth to pay a good price for a good animal, but it pays better to pay a good price for a good animal than to get a mean one given to you. I have said all I can say in their favor, and none too much, and I hope to see them crowned king, which I don't think will be long now. The black Muley, Durham, Devon, and all others will have to take a back seat; they cannot shine when they come in contact with the Herefords. Long may the "Journal" flourish. I am well pleased with the paper.

Yours truly,

F. O. Skidmore.

Arkansas, Texas, Mar. 30, 1886.

TEXAS GETTING HEREFORDIZED.

I have handled Hereford cattle for two years, and have during that time found that when it comes to rustling they have not only shown the ability to down the Shorthorn and other improved breeds of cattle, but they can successfully hold their own with our native Texas cattle, which have always been admitted to be the rustlers par excellence.

The winter of 1885 we had five Shorthorn bulls at the pen and five Herefords. The Shorthorns were good average specimens of the Shorthorns, the Herefords were four registered ones, and a bull from old Success, but cannot say if he is recorded. The Durhams were fed oats twice per day with all the cotton seed they could eat, the Herefords were fed once a day and sometimes not that, for whenever a bulling cow would come near the pens the Herefords invariably took out after her and would not return for two or three days, but Mr. Durham could not be tempted to leave the sunny side of a corn crib by a herd of bulling cows whilst a norther was blowing, and would always be on deck at meal time with unfailing regularity.

In the following spring the Herefords came out fat, strong and hearty and by March 1st were out on the range busy as bees. The Durhams were also turned out, but they were far from being fat, in fact they were not even in good fix, and later in the season when the warm weather set in, one had no trouble in finding out which breed of cattle, the Durham or Hereford, could stand the heat best, as the former sought the most shady places in the pasture, and stayed there; but the Herefords kept on chasing cows all over the range, and seemed to mind the heat but little, and they just got about 70 per cent of the calves gotten. Last week I brought these same bulls home for the winter, and the Herefords, after an unusually hard summer's work, look first rate, the Success bull Joe being as fat as the steers are on the same range (and cattle are in fine shape here at this time) whilst the Durhams look like a good many bushels of "4 bit" oats will have to be pumped into them before they will be in half the shape the Herefords are now. I would like some admirer of Durham cattle as range cattle to notice the vast difference there is in the grade Hereford and the grade Shorthorn calf raised on the range, and if that don't convince him of the superiority of the Hereford I don't know what will.

The range raised Hereford calf of seven months is a big lusty fellow, with a beautiful red body and white face, whose sole object seems to be how big he can grow and how fat he can keep, and he "gets there," too, whilst the grass raised Durham is a long haired, miserable looking fellow, who grows, too, but mostly in length of leg and poorness, and who doesn't confine himself to the red color of his aristocratic Kentucky sire, but mostly catches on to the plebeian brindle, black or yellow of his Spanish mother, and it is from her that he inherits what few good qualities he has.

The day is coming when Texas will send out many a train load of whitefaced steers, as Texas is a grass country from way back, and all that a Hereford wants is a trial and plenty of grass. Given these, he will be sure to please. This section is fast getting Herefordized, but it will take time, as there are lots of Durham cattle here whose owners are going to "wait and see." Yours, A. M. White.

Woodward's Ranch, Milam Co., Tex., Oct. 29, 1886.

EARLY STRUGGLES IN NEW ENGLAND.

Mr. Editor: In accordance with your expressed wish, I send you a short article upon that—to us—important topic, "Herefords in New England." But first allow me to ask the leniency of your readers for so brief an article on so important a subject on account of long continued ill-health. As perhaps very few of your readers are aware, I will say the second, and by far the most important importation of Hereford cattle into the United States was made by Sanford Howard for the Vaughn Bros., of Hallowell, Maine, in 1830. The first being imported by Henry Clay into Kentucky in 1817. The Vaughn Herefords were a splendid lot of cattle, and were kept on the farm since and now owned by Hon. J. R. Bodwell and occupied by Burleigh and Bodwell for quarantine and sale purposes. These cattle, like the next importations into Massachusetts, Sotham & Corning, Albany, N. Y., and Capt. Pendleton, of Searsport, Me., in 1841, all met terrible opposition by Shorthorn men, of which your humble scribe was one. Never in the history of any state did the worthy "red with white faces" have stronger opposition than in the Dirigo State. But when we look back and see the Shorthorn blood they had to contend with—the best in the known world—such bulls as Young Denton, Comet, Fitz Favorite, and a host of others which stood within three miles of my door in the great grass valley of the Kennebec, you will admit with me the white faced pilgrims had a hard, nay, an unequal battle to fight. It seemed almost like marching a little army of untrained volunteers against a mighty host of veterans strongly entrenched behind fortifications of years in building. But what a mighty change has been wrought in a half century and less. A few men had courage to try the Herefords; among them was Mr. Joseph Underwood, of Fayette, Me., and Wingate Hains, of Hallowell, Me. The former's enterprising sons, G. and G. Underwood, still are breeding a large fine herd, and it is a treat to a Hereford man to sit before the open wood-fire on a winter evening at their hospitable home and listen as they relate their earliest experience (as boys) with Hereford steers. I advise no Western breeder, who visits Maine, to miss a call on the Messrs. Underwood.

It was not until 1854 that any stockman had courage to take a Hereford bull upon the sacred soil of my native town (Fairfield), as Shorthorns had held full and undoubted sway there. But Henry Lawrence, who, by the way, was a great judge of cattle as well as his uncle, conceived the idea of buying the old Pendleton (Hereford) bull Kimroe, then fifteen years old. Quite a number of the best stockmen patronized him largely because they had great faith in Mr. Lawrence's judgment, not from the appearance of the bull, for the poor old fellow had had rough usage since he left his home near Cardiff, South Wales. This poor old bull was used one season and died, but his progeny were not prepossessing in appearance at a tender age, say up to two years old, yet their feeding qualities and net weight was the means of converting many a "shorthorner," of which number the writer of this article was one.

As I said, less than fifty years has made a wonderful change in the cattle interests, and the appearance and breeds in New England. Thirty years ago you would hardly see a Hereford at any show except the old Kennebec; not one at our State Fair; all were Shorthorns. But if you will go with me to the New England, or the Maine State Fair, during the month of September, 1886, I will show you nearly 300 of the finest steers ever seen in any country, and more than seven-eighths of the number will show the unmistakable evidence of the well-bred Hereford. Yes, they will be there, and well did a certain agricultural writer say of them last year: "It was a sight worth seeing, and long to be remembered." There will be whole teams of steers there, and some heifers, that will weigh a ton apiece. What has wrought this change? Necessity is the mother of invention. So, also, competition compels us to look to "cost of production and value of products." Western competition has driven us to give up Shorthorn cattle and take a breed that we can produce five pounds of choice market beef from as cheaply as we could four pounds with Shorthorns, and not only the difference in the amount, but the finished Hereford will command nearly or quite a dollar per hundred more in the market.

Yes, the Battle of the Breeds, so far as the Shorthorn is concerned in the East, is forever settled. Although so strongly entrenched behind seemingly impregnable fortifications, and fighting to the death, their flag is shot away, their munitions of war are expended and what was once a little insignificant squad of red with white faces, have become an overwhelming host, and New England is agriculturally saved. She need not be dependent on any section of our country for beef. But I would not have your readers think that the Herefords have no enemies, or the Shorthorns no friends, even in New England, in proof of which I will cite you to the fact that at the great agricultural show to be held by the New Massachusetts Society at Boston, in October, there is offered a special prize by the State Board of $125 for the best Shorthorn herd, but nothing for a Hereford herd, and this, notwithstanding there is more than four times the value in pure-bred Hereford cattle in New England, that there is in pure

Shorthorns. The fight for supremacy will go on in the West and Southwest, and while the backbone of the Shorthorns is broken there will be a new rival come up, and it will be a rival worthy of your steel. Don't ignore the Sussex, you will find a harder foe than the Shorthorns, or I am very much mistaken upon a three and a half years' acquaintance.

In 1856, although a strong and ardent admirer of Shorthorns, which breed of cattle I had bred, handled and fattened from boyhood, I happened by mere chance to obtain some of Kimroe's stock, *i. e.*, steers of his get, others by sons of him. They fed so rapidly, weighed so heavily and were of such wonderful quality I could hardly believe that my vision was quite right. After being thoroughly convinced of the fact that my sight was all right, then I thought I would decide to go largely into Herefords, when, ever and anon, my long-cherished idea of the perfect animal—the Shorthorn—would cause me for a time to delay my purpose; but not long. I started to breeding Herefords in earnest—after feeding Hereford steers about eight years—in 1866. In 1868 I bought the entire Hillhurst herd of Herefords, of Hon. M. H. Cochrane, the advent of which marked a new era in live stock industry in New England. At the head of this herd was the bull Compton Lad (3764) 1327, which for quality, style and general make-up, was never surpassed. Showing this bull with the herd at all principal fairs from the eastern borders of New England to Pennsylvania, where he won seventy-four prizes, including gold and silver medals, out of seventy-five prizes competed for, and seventeen of these against all breeds, and which was the means of making a host of friends for the Herefords. The march of improvement has been onward. Of the 840 head of stock imported by our firm —Burleigh & Bodwell—too many have been allowed to go West for the interest of New England, yet, the showing here is to be viewed with pride by the friends of the great and cheap producing race of cattle.

The sale of Hereford bulls to the resident stockmen for the year 1886 has been very fair considering it is an off year, and as the years roll on, following each other with rapid flight, so will the panting and puffing engine steed with fire and steam bearing their heavy loads of noble bovine beauties, with the unmistakable signs which the Hereford never fails to impress upon his progeny, to the great manufacturing centers, to the joy and satisfaction of the consumer, and profit and pleasure of the producer.

Yours truly, H. C. Burleigh.

Vassalboro, Maine, September, 1886.

A CHICAGO BUTCHER'S TESTIMONY.

About the first of October (1886) Messrs. Fowler & Van Natta, the well known Hereford breeders of Fowler, Indiana, sent to the Chicago market a car load of yearling half-blood Hereford heifers, that averaged 885 pounds in the yards. They were purchased and slaughtered by Mr. P. Mehan, for his meat market at No. 2911 South Park Ave., Chicago. After disposing of the meat of these heifers, Mr. Mehan wrote the following letter:

Chicago, Nov. 1, 1886.

Messrs. Fowler & Van Natta,
Fowler, Ind.

Gentlemen: In regard to the twenty-five Hereford heifers, bought of you for my market, I wish to say that those heifers killed the best of any cattle I ever slaughtered, and gave the best satisfaction both to myself and customers. They carried their weight on their backs, where it is most valuable, and dressed a larger proportion of high-priced meat than any other cattle I ever used. Yours truly,

P. Mehan.

These twenty-five heifers were raised on grass and went direct from the pastures to the shambles. The above letter, coming as it does from a practical butcher, doing business in the city of Chicago, is valuable testimony for the Herefords. About this time a correspondent of the "Breeders' Gazette" had asserted that the Shorthorns were the best cattle, because they carried a larger proportion of first-class meat than the Herefords. Mr. Mehan says these half-blood Hereford steers raised on grass, "carried their weight on their backs, where it is most valuable, and dressed a larger proportion of high-priced meat than any other cattle I ever used."

Messrs. Fowler & Van Natta with their large herd of thoroughbred Herefords have shown great enterprise in pushing the breed to the front, that cannot be too highly commended. They went to large expense in fitting cattle for show purposes, and have carried off many prizes at the large Fairs and Fat Stock Shows, including the champion herd prize at the Illinois State Fair three years in succession, and several times they exhibited the grand sweepstakes steer at the Chicago Fat Stock Show. The example they have set of making steers of their grade bulls and spaying their grade heifers, and putting them in the market, was followed by many more of our large breeders. If consumers once get accustomed to the Hereford beef they will buy no other, and the only way to get them accustomed to it is for all the breeders to do as Messrs. Fowler & Van Natta did—send their grades to market.

OLD BLOOD IN VERMONT.

We make the following extract from a letter received in 1882, from Mr. Henry Lane, of Cornwall, Vt.: "I saw at Mr. O. S. Bliss', of Georgia, in this state, a very fine oil chromo of Herefords, which pleases me. Thirty-eight years ago I went with two of my townsmen to Albany, N. Y., to see Corning & Sotham's herd of Herefords, and we purchased two bulls that were brought into this town and the stock from them made the finest beef cattle we have ever had."

Commenting on this letter in the "Breeders' Journal" (p. 392, 1882):

We take pleasure in calling attention to this testimony, as showing that the work of Mr. Sotham forty years ago comes to us to aid in our work. There is no way in which Vermont and New England farmers can advance the farming interest of those states so well as by the introduction of Hereford cattle. Two years ago we sold a bull and some heifers to go to Mr. H. O. McKnight, of Locust Hill, Penn. Mr. McKnight, after two years and six months' experience, returned here last week and bought of the T. L. Miller Co., the Hereford cow, Hampton Olive 3d, and a Hereford bull calf from British Lady, the mother of T. L. Miller's heifer Prairie Flower. Mr. McKnight's experience has given him the fullest confidence in the Herefords as a good butter dairy cow and beef breed. The bull purchased by Mr. McKnight in 1879, was by Success, and has proved a duplicate of the old bull. His purchase now is a young bull by Winter de Cote; it will be recollected that Winter de Cote was bred by Mrs. S. Edwards, and is a brother of the celebrated Hereford cow Leonora. The cow Hampton Olive was bred by J. H. Arkwright, of Hampton Court, Leominster, England, and imported by the T. L. Miller Co. in 1880. The price paid for the cow and calf was $900. This gives Mr. McKnight a herd of some twelve females, all thoroughbred and very choice animals. Besides these he has a large number of grades, among them yearling steers weighing about 1,000 pounds. We call attention here to Mr. McKnight's experience, showing, in connection with our extract from our Vermont correspondent, what farmers in an ordinary practice of dairy and stock business can do.

Vermont can breed Hereford grades, making a very good butter cow of the heifers, while the steers will sell to the butcher at from twelve to eighteen months old at $75 or more. Men who, like Mr. Lane, have had an old-time experience with Herefords, will recognize the correctness of our predictions. The grade steers that we have now, 24 months old, would sell in Chicago for $100 per head.

KANSAS IN EARNEST.

Editor "Journal":

I had my white faces at the Wichita, Kansas, fair again this year and "downed the Shorthorns" worse than ever. The Shorthorn breeders made their usual amount of "wind pudding" and abused the white faces as usual, but to no purpose. They wanted to show for "big money" they said; so I proposed to show "The Equinox" against any Shorthorn bull they could bring against him at any time during the week for $1,000 cash, and if they wished they could increase it to $5,000, and I would put up $1,000 forfeit just to accommodate them and put the ball in motion, but these "high-toned" gentlemen preferred eating their own words rather than losing their money and meeting inevitable defeat. The following clipping gives what stock I had there:

"Mr. W. E. Campbell, of Caldwell, Kan., had on exhibition fifteen head of his famous Herefords, among which was noticed the bull Equinox, No. 2758, weighing 1,920 pounds, two years old. Mr. C. values this animal at $1,500, which he has refused. Among his exhibit was the Duke of Rosewood, the Queen of the Lillies by Prince Royal, Ella 3d, Mermaid, Jessamine by Miller's Success, Empress by Emperor, all of which are two years old. The yearling, Third Duchess of Somerset, by Emperor, was also among the collection. Mr. Campbell has over 6,000 head of stock on the range in southern Kansas and the Indian Territory. Mr. Campbell offered to show his bull Equinox for a grand special sweepstakes prize of $1,000 cash, open to exhibitors of both fair associations, but the offer was not accepted."

And the following from the "Caldwell Commercial" of September 14, 1882:

"Our friend, W. E. Campbell, the great Hereford and Shorthorn breeder of Sumner County, better known as Shorthorn Campbell—to distinguish him from the numerous other Campbells, imported nearly or remotely from the land of oaten cakes—visited both fairs with his Herefords, and we are pleased to state, captured the class premiums and carried off the sweepstakes for the best thoroughbred herd, the best bull of any age or breed, and the grand $100 sweepstakes for the best bull of beef breeds, against all competition at both fairs.

"In this triumph Mr. Campbell laid out C. A. Betz's Shorthorn prize bull, which had never been beaten before, and had been ruled out of competition of other associations, because of its supposed overpowering superiority to all animals of that class."

I have just bought two more ranches, cattle ranges, etc., and have not time to write more.

Respectfully yours,

W. E. CAMPBELL.

Caldwell, Kan., Sept. 27, 1882.

P. S.—"The Equinox" was never stabled twenty-four hours all summer, but ran in the pasture with the cows the entire season and was only stabled nights.

E. S. SHOCKEY,
Kansas City, Mo.
Mr. Miller's assistant in the establishment of the "Breeders' Journal," and later of the firm of Shockey & Gibb.

CHAPTER XLI.

The Hereford Versus Shorthorn—Mr. Miller Re-States His Position

As we have had considerable to say about the methods used by Shorthorn men in America to hold their cattle in the position that they claimed for their breed, before the Hereford came into the arena demanding a fair field and no favors, we considerately put what we have further to say on this subject at the end of the History.

In the times of our great activity in the Hereford cause we received a letter from an old ex-editor, who says: "I regret that the fight between the Shorthorns and Herefords has assumed a personal and bitter character, which is so damaging to the breeders taking part in it. The press may gain somewhat from the interest thus excited, but is not the sacrifice of individuals too great and in a great measure unnecessary? If I ever take up the cause of any breed of stock, it will be to advocate its merits, and not to expose the demerits of their opponents. I regret that you have become personally the object of such bitter attack, but at the same time, know of no one who could more successfully ward off the blows."

In reply to the remarks of our ex-editor, we contended that an erroneous situation had been fraudulently established by the Shorthorn interest.

To meet the claims and the demands of the Shorthorn advocates, we went back in history one hundred years, and brought up their record. This record shows that Youatt's History of British Cattle was fraudulent, in that it suppressed facts detrimental to Shorthorns and favorable to Herefords. We brought out the record showing the organization and management of the agricultural fairs to have been in the interest of the Shorthorn breeders. We showed that the advocates of the Shorthorns used Youatt's History of British Cattle and the awards rendered at the Agricultural Societies to support their claims. We charged that the press supported the Shorthorn breed on the evidence thus produced by the Shorthorn managers. We brought forward the record of the Herefords, and this record placed them in the lead of all other breeds as beef producers. The course we have taken has not been based on our own opinion simply, but upon recorded facts, tests and experiments, that are accessible to the Shorthorn advocates as well as to ourselves. These Shorthorn advocates have never met the facts, but have resorted to personal abuse, hoping thereby to mislead the bullock breeders, and by so doing retain their position.

In years past the press accepted Youatt's History of British Cattle as authority; they accepted the awards rendered at the agricultural societies in this country and England as fairly made and rendered; and basing their actions on these facts, gave their support to the breeders of Shorthorn cattle, as did their predecessors and those especially interested in Shorthorns. We claim that the facts we presented were entitled to the attention and examination of said press, and that those failing to give this attention and examination are, or were, parties to the fraud that has been handed down to them from their predecessors.

The press occupies another and more important relation to this question. They are bound to know whether these facts are true; they are bound to investigate; they are bound to give the results of this investigation to their readers, if they would act in good faith. They claim to be disinterested, to have no interest in any breed, and they ask their readers to believe not only what they say is true, but they ask their readers also to believe that they are intelligent and well-informed on these subjects. Failing in this investigation and examination, they perpetuate the fraud that has been handed down to them. We are now prepared to ask, by what terms should the acts of these Shorthorn advocates and these editors, be characterized? We believe that it is fraudulent to present issues that have no foundation in merit or fact; we believe that the editor is bound to know the facts. They assume to be teachers and guides, and taking this position and proving false to the responsibility, they should be held accountable. Our friend says if he "were to take up the cause of any breed he should advocate their merits, and not expose the demerits of other breeds." We should like to know by what standard he would measure the merits of his own, unless he compared them with others.

In entering upon the breeding of beef cattle, we found the Shorthorns the dominant and accepted breed for that purpose, in this country. We appeared before our state agricultural societies and placed the Herefords beside the Shorthorns; we asked our societies to pass upon the merits of the breeds as represented by those upon exhibition. The question, then, was opened, as a matter of course, as to the comparative merits of Herefords and Shorthorns; from this point the discussion was continued. Our State Society discriminated in favor of Shorthorns by giving them larger premiums, and thus, by inference, declaring to the world that they were the better cattle. They submitted the question of merit between the breeds to men selected from or by the Shorthorn breeders to act as judges. The question then as to the merits and demerits of breeds was here opened; we had to accept of it, and when they claimed a half century's acceptance of their cattle as evidence of merit over the Herefords, it *compelled us to go back* of the merits of the individuals as they stood upon the show grounds and inquire as to the methods by which the Shorthorns had obtained their endorsements. We have placed the result of these examinations before our readers and the world. On these facts we appealed to the cattle raisers of America and the verdict has been reversed; the Herefords are now accepted as the best and have taken the place among the cattle breeders of the country to which their merits always entitled them. Such a revelation could hardly be accepted without more or less bitter feeling; it was to be expected that the Shorthorn breeders and advocates, with their associates, certain editors of live stock journals, finding the facts against them, should resort to vindictive and bitter personal attacks to cover the weakness of their position.

Presenting a breed of cattle, as we have, claiming an advantage on the score of economy and value of product over this dominant Shorthorn breed, we were compelled to show wherein this value consisted, and the weakness of the Shorthorn claims. All we asked was a full and fair investigation into the claims and facts that we presented; if they had not merit, then we were not entitled to the verdict; if they had, we expected the Herefords to find acceptance and we expected to have that position as an advocate of these claims that belongs to us.

So influential was the opposition that a few contemporary Hereford breeders, actuated by jealousy, were prevailed upon to cravenly criticize our methods and to belittle our work; while at the same time taking every possible advantage of what we had done almost single-handed. The opposition called us a "kicker," knowing full well, that only a persistent and consistent "kicker" could make the slightest headway against their adverse and perverse machinations.

On account of our activity, this fight became for a time a personal one upon the writer. Having control of the shows our opponents made our retirement from the show ring imperative. Other Hereford breeders might get a semblance of justice, but we could not. Having established an agricultural journal in which our Hereford writings would not be garbled and misrepresented, the opposition press was forced, as it were, to gnash their teeth in their impotent rage and vent their spleen in personal abuse of us.

At this writing (1898) history appears to be repeating itself in the case of the leading Hereford advocate. Let Hereford breeders beware, that they play not into the hands of their opponents by a failure at all times to recognize and publicly acknowledge the value of the work of those who have fought and of those who still honestly, openly and courageously fight the battles of their breed.

With this word of warning brought forth of much experience, and having given our reasons for engaging in the unpleasant task of controverting the position assumed by the Shorthorns, we will reproduce what was said at the time in the "Breeders' Journal," in regard to the court before which the adverse claims were brought.

We will consider our Illinois State Board of Agriculture. Has it been of a partisan character in the past? Has it been run in the Shorthorn interest? We answer these questions again in the affirmative, and will bring more proof to sustain our position from the records of the society.

Up to and including 1870, excepting a few Devons, there were no breeds to dispute the superiority of the Shorthorns. The question as to partial judgments was confined to the breeders of the different families of Shorthorns. We find that Mr. A. B. McConnell, in his inaugural address as president (1865), says that "the society has now entered its thirteenth year, and its advancement has been, and is now, a high source of gratification to its projectors; but since the organization of the society, each successive executive committee have failed in getting full reports from the awarding committees. The simple fact published to the world that a certain animal, article or implement has received the first prize of the society, is shorn of half the value to the public, unless the reasons for the award are given by the awarding committee, fully setting forth its merits and the reasons for the award. If some plan can be devised by which full reports can be received from all the awarding committees, I am satisfied it will add very much to the value and usefulness of our annual exhibitions, and also to the transactions of the society." Dr. William Kile, on assuming the presidential chair (1869), after referring to the great success the society had met with since its organization, says: "We should not forget that every citizen, of himself or by his representative, has the undoubted right to call in question the wisdom of any line of policy we may adopt, and to criticize freely every act we perform. On all questions, therefore, which admit of argument and diverse views, and especially everything which affects seriously any interest committed to our fostering care, let us continue to be circumspect and to act so that every step we shall take shall commend itself to the sound, reflecting judgment of all. As your presiding officer, and as a member who has been identified with the society from its origin, I beg leave to call your attention to the following points: You are aware of the great and delicate duty performed by the awarding committees at our fairs, and have found that a full attendance of those (the judges) originally appointed by the Board is rarely secured. I have thought that perhaps the system practiced in Indiana, of paying committeemen by the day for their services, would be attended with good results, and respectfully ask that you will take the matter under consideration."

We would call attention especially to this recommendation of President Kile. We for many years urged, and in this were joined by all the more conservative and painstaking members and exhibitors, that the procuring of competent judges should be by selecting experts and paying for their services. The objection given was the extra cost of such judges. We shall discuss this further on.

At the fair held in 1871, Mr. Wm. W. Aldrich, of Ohio, appeared on the show ground with a herd of Hereford cattle, and in competition with Shorthorns on the same premium list. But besides class premiums, there were offered for herd and sweepstakes $825, for which the Herefords had a right to compete, but must compete under judges in the Shorthorn interest, and often composed of leading Shorthorn breeders, making their awards a discrimination in favor of Shorthorns.

President D. A. Brown, in his inaugural address that year, says: "The state, gentlemen, has confided the management of its institutions to our peculiar care, and I feel bound by the trust reposed in me to recommend for your consideration the adoption of such rules for the regulation of our exhibitions as will give to them that high moral character their importance demands."

We have quoted from Presidents McConnell, Kile and Brown, showing that they realized the responsibilities in assuming the management of the state's agricultural interest, and that they were under obligations to so manage and direct the operations of the State Agricultural Shows as to bring forward the best, and see that such awards were made as would command the respect and confidence of the world. At this time, 1871, a new act of incorporation was had from the legislature of Illinois, creating a department of agriculture, the objects of which should be the promotion of agriculture, horticulture, manufactures and domestic arts. See "Transactions of the Illinois Department of Agriculture, 1872."

In 1872, Messrs. Miller & Powell were exhibitors of Herefords under the same classification as in 1871. One of the awarding committees in sweepstakes that year was composed of John M. Milliken, of Ohio; John H. Bacon, of Iowa; William M. Smith, of McLean Co., Ill.; J. H. Reese, of Warren Co., and H. Russell, of Marion Co.; a suitable committee to pass upon Shorthorns, a majority of them being among the prominent and leading Shorthorn breeders of the States.

At the reorganization of the society in January, 1873, under the new constitution, a newly elected president, the Hon. John P. Reynolds, in his introductory address, reviewed the history of the old society in its twenty years' experience, calling the attention of the new Board to what had already been done and the responsibilities that rested upon them for the future; and reminded them that the Board is placed in sole charge of the agricultural department of the state, and for the establishment of county agricultural boards, and that whatever the state might do will be entrusted to, or enjoined as a duty upon, this Board. While recognizing the difficulty surrounding the appointment of awarding committees, and prescribing rules to govern their action while serving as such, he respectfully referred the subject to the careful consideration of the Board, when revising their list of premiums in the live stock department.

The published premium list shows that the Board increased the premiums on Shorthorn heifers and cows 27½ per cent, leaving the Hereford classification as before. The awarding committees on herd and sweepstakes premiums were so composed that Herefords had no consideration at their hands. A member of one of these committees said that a majority of the committee would not consider the Herefords.

In 1874, Shorthorns and Herefords, in class premiums, exhibited on an even classification as to premiums.

At the organization of the Board in January, 1875, Mr. D. B. Gillham, president, in his inaugural address, on assuming the duties of the office, says: "It is a noteworthy fact that from the magnitude of the exhibitions of the Shorthorn element in class A, Cattle Department, and the peculiar circumstances surrounding it, the general participation by the people in it, and the interest manifested by all, that our present system of selecting awarding committees fails to render the satisfaction so desirable by the Board, to either their exhibitors or the public. It may be well to effect some change in regard to that department. In calling your attention to the matter, gentlemen, I will leave you to suggest the plan and to act in the premises as your better judgment may direct." The records show that at this fair the Board had advanced the class premiums on Shorthorns about 70 per cent, and reduced the Herefords something like 15 per cent.

For the fair of 1876, the Board reduced the class premiums for Shorthorns to the same amount as in 1874, taking off the increase of 70 per cent made in 1875, still leaving the Herefords with their reduction of 15 per cent, or 50 per cent less than the Shorthorns. Having been subject up to this time to the decisions of Shorthorn judges in the herd and sweepstakes competition, we made a special effort to have this feature of the awarding committees corrected and to this end we made a special application to President Gillham, that the awarding committees for this competition should be selected from impartial and intelligent men. We also made a special request of President Scott to the same end. Our right to such committees was freely conceded. Mr. W. J. Neely, who was the superintendent of the cattle department, was also appealed to, and his first committee was five Shorthorn men, composed of James Mix, H. M. Winslow, of Kankakee; J. R. Shaver, of Ottawa; John Brown, of Galena, and J. L. Moore, of Polo. Protesting informally, Mr. Neely undertook (according to his statement) to get an impartial committee and professed to find a good deal of difficulty in doing so, and was obliged to get the assistance of Col. Judy to aid in selecting and completing the committee, and J. H. Spear, of Tallula, the townsman of Col. Judy, and one of the prominent and winning Shorthorn breeders at previous fairs, was selected to hold the balance of power in this new committee.

In coming to the fair of 1877, the Herefords came on an even classification, the Shorthorn premiums in classes having been reduced about 15 per cent, and the Hereford increased something over 20 per cent. The Hon. Samuel Dysart had been chosen superintendent of the cattle department, and for the first time in the herd and sweepstakes competition the awarding committee were selected with a view to make the awards upon the merits of the cattle exhibited.

In 1878 the two breeds in class premiums stood upon an even basis. When approaching the competition between Herefords and Shorthorns, for herd and sweepstakes premiums (Mr. Dysart then being absent in Europe), we said to Mr. Cobb, of Kankakee, that Herefords were entitled to fair and impartial committees, which he admitted. We proposed to him that President Gillham and ex-President Reynolds be requested to choose the awarding committees for this competition. He said, "That was right, and he would see that such an arrangement was carried out." The committee was selected by these gentlemen, and the Herefords won the first premium for the best five cattle, male or female, in the show, the get of one bull, the sire to be shown with the herd. Officers of the society attached to the Shorthorn cattle interest of the country, then took charge of the selection of the judges, and selected them in the interest of the Shorthorn exhibitors. So clear and unmistakable was the partisan character of this work that the Board changed the manner of selecting judges for the fair of 1879, when the Board was organized, with Col. J. W. Scott, as president.

In his inaugural address President Scott said: "It is needless for me to call attention to the magnitude of the interest involved in an effort to promote the advancement of agriculture, horticulture, manufactures, and the domestic arts in the leading agricultural state of the Union. The state, through the legislature, has been very liberal, and given much latitude to the operations of the Board in creating the same, leaving the free exercise of our better judgment in the promotion of the interest involved, and thereby increasing the responsibilities of this Board. One of the more prominent methods for the promotion of agricultural interest has been, and will be for some years, the exhibition of the annual State Fair. It seems that the great importance of a just and competent award of our premiums would justify a departure from the old custom of selecting committeemen, at least in the more important department of live stock, where the competition is sharp. It is important that some means should be devised to secure competent men to act as judges, whose attendance can be depended upon, and relieve the Board from the necessity of selecting from the visitors men who are frequently not as skillful as the good reputation of the Board should require."

The committee to whom was referred the president's address, reported as follows, on the matter of awarding premiums:

"Fully realizing the importance of more care in awarding premiums, especially in the classes of live stock, we would recommend the passage of the following resolution: 'That the president, together with the superintendents of the several classes, be, and are hereby constituted a committee, and empowered to employ as committeemen such gentlemen of recognized fitness in the several departments as may be deemed necessary, the remuneration in no case to exceed the actual expenses of such gentlemen.' "

Under this resolution, committeemen were selected, who were deemed competent and impartial, to judge the different lots in Class A (cattle department), which committeemen having been selected with care and without outside influences, were to act as judges and no protests were to be allowed. We were advised of this action on coming to the secretary's office on the show grounds at Springfield, before the opening of the show. We gave it our hearty approval and urged upon the superintendent that under no consideration should the arrangement be broken. It was carried out, although the Shorthorn exhibitors undertook to break it up.

For 1880 the class premiums for both breeds were the same, but competition between breeds was suspended.

For 1881 the classification of Herefords and Shorthorns was even; the sweepstakes confined to competition for a herd consisting of bull and four cows or heifers of any age. We have not the data before us for 1882, but premiums in class stand on an even basis.

This, then, in a concise form and up to the year 1882 when we ceased showing, is the record of the society as it stood between breeds. Each of the incoming presidents recognized the necessity of a change in the manner of selecting judges. Each recognized that the system on which they had worked was defective, and so far as the record shows, but one change was made, and that in 1879, and that only held in practice one year. Each and all of the presidents recognized the responsibilities that rested upon this Board, the importance of the issues that were submitted to their management, and that the live stock interest was one of the largest and most important that they had to deal with. They appointed awarding committees at each and every annual meeting, with the knowledge that these committees were not likely to act, and have depended upon judges selected from the crowd of visitors; men who were, in the language of President Scott, "frequently not as skillful as the good reputation of the Board should require." This custom was in vogue for thirty years, and we submit that under such a system the awards had very little value. The Shorthorn interest sought to perpetuate this custom, that under the excitement of the show ring they might select men in their interest.

Leaving the State Fair and the action of the Board in reference thereto, we are prepared to take up the management of the Fat Stock Show. The first of these shows was opened December 2, 1878, the Herefords winning the sweepstakes for the best cow in the show, and no other. The second was held November 10, 1879; third, November 15, 1880; the fourth, November 7, 1881.

The action of the Board in reference to selecting judges for this show was to take them from the different sections of the state, recommended by different members of the Board. Men, as a rule, who had no experience with first-class bullocks, such as would be exhibited at this show, whose decisions did not carry weight or authority where the judges are known. Of the four shows mentioned, we only wish to call attention to a few facts therewith, and these mainly as regards ages at which cattle have been exhibited.

At the show in 1879 the slaughtering test was introduced with results as set forth on page 147, Vol. XVII., Illinois State Report, which we show in a schedule in Chapter 16. Each of the bullocks represented in this schedule was three years old, the grade Hereford at three years six months and fourteen days, with a mouth showing six permanent teeth, the grade Shorthorn represented at the same age, with eight permanent teeth.

At the show in 1880, of steers, three and under four years old, there were slaughtered two Herefords and one Shorthorn. One of the Herefords was three years seven months and thirteen days old, and the other three years eleven months and thirteen days, each with six permanent teeth. The Shorthorn was represented as three years six months and fifteen days, having a full mouth of eight teeth, but somewhat broken, and difficult to tell from the teeth his age.

In the year 1881 the slaughter test revealed in the Shorthorn three-year-olds an older mouth than the age for which they were entered would indicate, while the mouths of the Hereford bullocks corresponded with the ages for which they were entered, as per standard authorities. These mouths were preserved and placed in the hands of the officers of the society for preservation. The facts in reference to these indications were fully before the Board through its officers. An effort was made at the opening of the show of 1880 to have the animals on exhibition examined by a veterinary surgeon, to determine the correctness of the entries that had been made. He entered upon his duties and met with a decided rebuff from one of the leading Shorthorn exhibitors, while the Hereford exhibitors and several of the Shorthorn exhibitors tendered their co-operation in making such examination; they were not, however, completed. This brings us to the show of 1882.

With our experience of the past shows, with reference to entries of animals at fraudulent ages, there came a determination to expose, if possible, such frauds, if any existed at this show, and cattle were generally examined by experienced cattle men, exhibitors and others, and it was found that a large number of cattle, judging from the indications of the mouth, were entered at ages much under what this evidence would indicate.

Among the rules governing the Fat Stock Show was one as follows: Section 5. "In case of protest notice must be given to the superintendent of department before or during the examination of the animal or article protested, and a written statement giving the reasons for protesting, verified by affidavit, must be filed with the secretary on the day notice is given."

Another rule.—Section 12. "Awarding committees are instructed that if they have good reason to believe that any exhibitor, by false entry or otherwise, attempts to deceive the committee or the public, and obtain an award by misrepresentation, they shall report the fact at once to the superintendent of the department, who shall report the same to the Board, who may expel such exhibitor for fraud for at least two years."

Rule.—Section 17. "Decisions of awarding committees shall be final, and no appeal will be considered except in cases of fraud and protest."

These were the only published rules relating to the subject of frauds in entering and exhibition of cattle at this show. Section 12 provided for the disqualification and expulsion of any exhibitor who attempted to deceive the committee or the public, and obtain an award by misrepresentation. It is true that this rule made it obligatory upon the awarding committee to bring this matter to the notice of the superintendent, and that he should bring it to the Board, that they might expel such exhibitor. The life of this rule is that an exhibitor who, by fraud or misrepresentation, attempts to deceive the public, shall be expelled. This would be a fair construction of the rule, and it would be a strained and unnatural construction to say that the Board or its officers could not take cognizance except it come through the awarding committee, and from them to the superintendent, and from him to the Board.

It was a well established fact among exhibitors and members of the Board that entries were made at fraudulent ages, and when we brought this matter directly to the president it was properly before the society, and to say that we had no right to a hearing until we came before them through the prescribed routing of Section 5, of the Rules, had no place in justice or equity. The animals had not yet been before any committee, but they had been entered and a catalogue of the entries prepared. The animals thus catalogued were in the charge and under the control of the society, and they had a right to enter upon any investigation that would tend to advance the best interests of the show. The cattle were entered in classes one, two and three years old, respectively. The exhibitors had undertaken to give the exact age of each animal. This was done among other things, that the society might determine the growth and gain per day, and upon such growth and gain per day an award was to be made. They had schedules which provided for an elaborate detailed statement of the results to be obtained.

The fraud as to age committed the great state of Illinois, through the Agricultural Department, to a fraud and makes them a party to it. The State Board, representing the Agricultural Department of the state of Illinois, were assembled on the opening of the show on the morning of November 16, to make their preparations and arrangements for the week's work that was before them. While thus assembled the facts of these frauds were brought to their notice. We called on President Scott, and advised him of these frauds and asked that he would bring the matter to the notice of the Board, that they might take the initiatory steps to determine the ages of the animals in the cattle department before they should be brought before the committees. When we entered the outer room, President Scott, who was with the Board, left the inner room and met us, the Board still in session, and we tendered our herd of cattle then on exhibition for the opening of such an examination as they might see fit.

Mr. Scott referred us to Rule 5, as his remedy in the premises. We advised Mr. Scott that we came not as a protestor, but to bring a fact as to fraud, which was well known throughout the building in which the show was to be held, and advised him that he might find this report sustained by members of the Board, and exhibitors of Shorthorn as well as Hereford cattle; and urged him to take such measures as would protect the interests of exhibitors, who were there under the rules of the Board.

On leaving this room we met Mr. J. H. Sanders, editor and proprietor of the "Breeders' Gazette," and called his attention to the fact, and asked his aid in having an investigation made in reference to this fraud as to the ages of cattle. Mr. Sanders promised to give all the aid that he could in the premises, and do all he could, if fraud was found, to expose it. By his public act, Mr. Scott, or the State Board, so far as we know, took no action in this matter of fraud, but had the cattle brought before the committees, who made the awards based upon the fraudulent entries.

We wish to state, in reference to Rule 5, that the protest of an animal when in the show ring is substantially a farce. The notice of protest filed, the committee would go on as if no protest was made, the Board would go about their business, close their show, and go to Springfield on the first of January following, where the protestor might appear before them, get a hearing, or, should he find an opportunity to argue his case on the week of the show, there was so much haste, so much business crowding upon the members of the Board, that it was difficult to get at the merits of the question.

We again appeared before the president on the last day but one of the show, and again called his attention to special animals exhibited by Mr. D. M. Moninger, that the Board might take some action that would disqualify Mr. Moninger as an exhibitor, and forfeit the premiums that had been awarded to him. Mr. Scott again referred us to Rule 5, and said that we had our remedy through it. We informed him that we came not as a protestor, but to bring a fact to them of a fraud that was flagrant, and such a one as the Board ought to take cognizance of.

Failing to get satisfaction from Mr. Scott, we then went to the Hon. John P. Reynolds, a member and ex-president of the Board, and called his attention to the fact, and asked that he would give his influence to the exposing of the fraud that had been committed upon the society and the public, and called his attention to Rule 12, as authority for asking that the society should take some action in the premises. We also advised him that as this was the last day of the show, but one, we would be obliged to bring the matter to the notice of the Hereford Society, which would meet that evening, and ask them that they would take action to protect themselves and the public from the consequences of these frauds. Mr. Reynolds went to Mr. C. M. Culbertson, the president of the Hereford Society, and asked him to use his influence so that the Hereford Society should not take any action in the premises, and he would bring the matter before the State Board on the following day and urge an investigation into the charges of fraud that were made.

The matter was brought before the Hereford Society, when Mr. Reynolds' request and promise were made known by Mr. Culbertson. No action was taken other than to request President Culbertson to take such action in connection with the State Board as might seem desirable. The charges were brought before the State Board the last day of the show by Mr. Reynolds, and the investigation was urged by Mr. Dysart and Mr. Reynolds. It was opposed by President Scott and Col. J. W. Judy. No other action was taken, so far as we know, other than to say to Mr. Moninger that charges were made that his steer, Champion of Iowa, was said to be older than the age for which he was entered. On Mr. Moninger stating that the age for which he was entered was correct, the matter was dropped. The steer was entered at one year eleven months and fifteen days, showing a mouth of six full teeth.

There were two other steers exhibited by Mr. Moninger, one as a two-year-old, called Tom

Brown, that were slaughtered in Chicago, and we obtained their mouths. We took these and others that we were able to obtain, went to Springfield and met the new Board at its organization, and asked that they would consider the evidence that we brought before them. We submitted to them standard authorities as to the ages of animals as indicated by the teeth, referring to the information we had given on the week of the show, and submitted the mouths as evidence of the frauds, and asked whether Mr. Scott had not failed in his obligations to the public and to the state in not making investigations in reference to the information that was brought to him as to the fraudulent ages of the cattle exhibited. On motion of Mr. D. B. Gillham, we were requested to put our charges in writing. These charges and the report of the committee to whom they were referred, we presented in Chapter 28, on dentition. The committee saw fit to shield the president behind a technicality of rules of the Board, to-wit: Section 5 and Section 12, which we have given.

It will be remembered that this information was brought to Mr. Scott before the cattle were brought into the ring at all. We could not have acted under Rule 5, and in a manner recognized (as the committee say), by established rules. In fact, we said to President Scott that we did not come as a protestor, but we came for the purpose of placing facts before the society, before there was any complication of awards, and asked that the Board would take the initiatory steps to determine the correctness of the charges, and relieve the exhibitors from becoming prosecutors. The committee reported that we could not have been ignorant of our own plain duty as an exhibitor. Was it fair to say that the exhibitors are under obligations to prosecute, and also that the society or its officers would not recognize fraud except it comes to their notice through their awarding committees or their superintendent?

This is what they substantially said in the committee's report, and what the Board said in adopting this report. As illustrating Mr. Scott, in his peculiar position, we reproduce, in Chapter 28, a cartoon from the "Breeders' Journal," which we had prepared at the time to better illustrate the Board's effort at sustaining him on this untenable and dangerous ground. We repeatedly called attention to the quotations from the inaugural addresses of the different presidents of the Board, in which they recognized the necessity of measures to make their awards authoritative and of value to the public. If it was to be understood that the State Board of Agriculture of Illinois would not take cognizance of fraud, except it come through these two mediums, the power that had been delegated to them by the state should be revoked, and the state legislature should provide some other means to encourage the agricultural and live stock interest.

If no record had been kept of the facts developed at the different shows, grave injustice would have been done the Herefords. It is a matter of fact that all the Hereford two-year-olds showed with half as many teeth as many of the Shorthorn two-year-olds had, at the Chicago Fat Stock Show in 1882. This proves conclusively that the Herefords were much the younger. We give herewith the facts of this matter which we published in the "Breeders' Journal" at the time:

STEERS TO WHICH PRIZES WERE AWARDED, GRADE AND SWEEPSTAKES CLASSES, FAT STOCK SHOW, 1882.

Catalogue No.	Name of Exhibitor.	Age in days..	Age.			Official w'ght.	Gain per day.	No. of permanent teeth..	Breed.	Premiums Awarded.	
			Years....	Months ..	Days.....					Grade class....	Sweepstakes..
113	D. M. Moninger	1174	3	3	4	1945	1.65	8	*Grade Shorthorn	1st—$30	$50
17	H. & I. Groff.	1265	3	6	5	2400	1.90	8	*Grade Shorthorn	3d — 10	..
111	D. M Moninger	1011	2	9	21	1850	1.83	8	*Grade Shorthorn	1st— 30	..
109	D. M. Moninger	1034	2	10	14	1905	1.84	6	Grade Shorthorn	2d — 20	..
22	J. H. Potts & Sons	715	1	11	15	1600	2.23	2	Grade Shorthorn	1st— 30	50
38	Fowler & Van Natta	574	1	7	24	1410	2.45	Calf	....	2d — 20	..
112	D. M. Moninger	715	1	11	15	1665	2.31	6	*Grade Shorthorn	....	..
83	T. L. Miller & Co.	1355	3	9	5	1870	1.31	6	Grade Hereford	None.	
84	T. L. Miller & Co.	1378	3	9	28	1990	1.44	6	Grade Hereford	None.	
78	T. L. Miller & Co.	970	2	8	10	1655	1.71	4	Grade Hereford	None.	
79	T. L. Miller & Co.	978	2	8	18	1720	1.75	4	Grade Hereford	None.	
80	T. L. Miller & Co.	972	2	8	12	1680	1.73	4	Grade Hereford	None.	
81	T. L. Miller & Co.	966	2	8	4	1815	1.88	4	Grade Hereford	None.	
82	T. L. Miller & Co.	866	2	4	26	1705	1.97	4	Grade Hereford	None.	
74	T. L. Miller & Co.	714	1	11	15	1380	1.93	2	Grade Hereford	None.	

*In Chapter XXVIII on Dentition, we present illustrations of the mouths of some of these steers.

"We give a table, showing the catalogue number, name of exhibitor, age in days, and age in years and months, weight, gain per day, number of teeth, and the premiums awarded, and the age as indicated by the teeth. We have also taken the steers as exhibited by the T. L. Miller Co., that were stall-fed, and of the ages three, two, and one year old, with their ages, weight, gain per day, and age as indicated by the teeth. It will be seen by this that Mr. Moninger's No. 113 was entered as three years three months and four days old, weighing 1,945 pounds, to whom was awarded the first premium in the three-year-old grade class, and sweepstakes for the best three-year-old in the show.

"By referring to Nos. 83 and 84 of the T. L. Miller Co. exhibit, they were by the entries six months older than Mr. Moninger's steer, and weighed about the same, but they had only six teeth, while the Moninger had eight. And referring to No. 17, H. & I. Groff, which won third premium in the grade class for three-year-olds, was represented three months younger than Nos. 83 and 84, and had eight teeth.

"Referring to Nos. 111 and 109, D. M. Moninger's, the first represented as two years nine months and twenty-one days old, and weighing 1,950 pounds and having eight teeth, taking first premium for the best two-year-old grade; 109, entered as two years ten months and fourteen days old, weighing 1,905 pounds, had six teeth, taking the second premium for grade two-year-old. For steers one and under two years, the premium was awarded to J. H. Potts & Sons' Red Major No. 22. There were twenty-six entries.

"Referring to 78, 79, 80, 81 and 82, two-year-old grades of T. L. Miller Co.'s exhibit, averaging about two years eight months, and average weight 1,717 pounds, it will be found that none of them had over four teeth. Referring to J. H. Potts & Sons' No. 22 and T. L. Miller Co.'s No. 74, they are substantially the same age and the same mouth, to-wit, two teeth in each. Again, referring to No. 112, D. M. Moninger, one year eleven months and fifteen days old, weight 1,665 pounds, with six teeth. This steer was entered by Moninger in the class of yearlings at the ages and weights as stated. Mr. Potts, competing with him, said to Mr. Moninger, that if that steer No. 112 was exhibited he should protest him. Mr. Moninger did not bring the steer before the awarding committees, which is prima facie evidence of the fraud upon the society in making the entry. The protest of an influential Shorthorn breeder would be too sure to be recognized, while it was tolerably safe to ignore the "kicks" of the Hereford men.

"Mr. Moninger exhibited what purported to be ten two-year-olds, of an average age of two years nine months and twenty-four days, weighing 1,864 pounds, making a gain of 1.84 pounds per day. But it will be borne in mind that seven of these steers had eight permanent teeth and three had six, while the T. L. Miller Co.'s steers averaged two years, seven months and twenty days old, two months younger as per entries, than Mr. Moninger's.

"And we wish to call special attention to the fact that neither of these last-mentioned steers had over four full teeth, a little more than one-half as many as the Moninger steers; or, in other words, Moninger's steers averaged over seven teeth, while Miller's averaged less than four. We submit on that showing whether the Herefords were not entitled to class prizes in this show.

"Again, referring to Section 12 of the Rules of the State Board, governing awarding committees, on page 4 of their fat stock premium list, and page 52 of their animal premium list for 1882, they are instructed, that if they believe an exhibitor, by false entry or otherwise, attempts to deceive the committee or the public, they shall report to the superintendent, and the superintendent to the Board, and they may expel such exhibitor from exhibiting before the society for two years. We intended to put the responsibility of fraudulent entries upon the State Board before the show opened; in fact, we determined this at the show a year ago. The Hereford breeders in the state of Illinois must look forward to the time when they can come before the society at its shows and receive fair and impartial treatment. This is not only to protect their own rights, but that they may bring before the farmers of Illinois the evidence that the claims they make for the Herefords are founded upon a sure and solid basis. The majority of the citizens of Illinois suppose that the awards that are given at their state shows have a value and that they can judge from these somewhat as to the character of the different breeds.

"We have no desire to convict Mr. Scott or any other officer of the Board of using his position to protect fraudulent exhibits, or of using his position in a partisan manner to protect any class of exhibitors. We have no desire to charge or convict the Board of such practices, and that they might be fully protected from such suspicion or such charges, the facts that we show in this table were brought fully to the notice of President Scott at the opening of the show, and, as we suppose, through him to the Board. Again, it was brought before them on the day but one before the close of the show. Again, we followed them to Springfield and sought an interview with Mr. Scott in his room at the hotel, Mr. Reynolds being present, presenting him with the evidence of this fraud and urging upon him that action should be taken that would protect exhibitors in the future.

"One manner suggested was by employing a veterinary surgeon who understood the teeth indications of age in animals, and associate with him two good, practical cattlemen, to examine all the entries that should be presented to them in future shows. He replied to this that such an examination as this would break up the show. Again, we took these evidences before the State Board; we made our verbal statements, and were asked to put them in writing, and the committee appointed by the Board say that Mr. Scott did what the Board ought to expect of their chief officer in refusing to make these examinations. When afterward a motion was made that such examinations should be made in the future, the Board refused to order such examinations.

"The 'Breeders' Gazette,' 'not a trade circular,' says of T. L. Miller's charges before the State Board, in reference to fraudulent entries of cattle: 'The standing of parties accused is such that their motives or veracity cannot be impugned.' Mr. Sanders occupies, in this case, a position somewhat analogous to that of the Tombs lawyer, appearing for his client, and undertaking to get a verdict in his favor by trying to prove the previous good character of his client. As to the report of the committee, Mr. Sanders says: 'They were impartial men, who have no prejudices against breeds or breeders, and the fact that the report was unanimously adopted relieves the Hon. J. W. Scott, the late president, or the retiring members of the Board of Agriculture from any want of a desire to protect the exhibitor from imposition or fraud.'"

In the April (1883) number of the "Breeders' Journal" we said:

"We learn that at the meeting of the State Board in February the question of determining the ages of cattle that were to be exhibited at the Fat Stock Show of 1883 was to be determined by examination. This was the re-opening of what the Shorthorn interest supposed was fixed at the January meeting. They, at that time, determined that such examinations should not be made, but they found in the month that intervened between the two meetings that there was a determination to press that issue. We understand that it was decided to employ a veterinarian and two competent and disinterested cattlemen, not exhibitors at the show, to examine all the cattle that should be placed on exhibition at the next show.

"How much our expose of the frauds perpetrated at the last show had to do with this we are not advised; in fact, it is immaterial. We have had a severe and protracted fight before the State Board to secure for the Herefords an impartial hearing. There have been men in the Board, from the commencement of this controversy to the present time, who have desired to see fair play, and if the inside and outside history of the Board could be written we think that the lobby, or the outside influence, has been the strongest. When Mr. Scott said to us in Springfield in January that such a rule would break up the show, we thought then, and we think now, that he believed that the Shorthorn men would not show under such a rule, and hence the conclusion that the show would be broken up. We do not know that these conclusions are correct, but we believe they are, and we believe that Mr. Scott at the last show, had it not been for this fear of Shorthorn influence, would have undertaken then to have exposed the frauds that were being perpetrated.

"At the February meeting of the Board it was decided that the State Fair for the next two years should be held in Chicago. We believe that this is a wise move, and we understand that the citizens of Chicago are making liberal provisions to make the Fair a success. And we hope and expect that the changes that were made in the management of the Board at its election in September, and the rules that are likely to be adopted by that management, will put the State Agricultural Society of Illinois in the front rank of a reformatory movement that the great cattle interests of the world require. We are not advised as to whether any action has been taken in relation to the selection of judges. This is needed. One of the movements a year ago in this direction was to ask several State societies to send judges to act at the Fat Stock Show of 1883. Among the societies responding to this request were those of Ohio, Michigan, Indiana, Wisconsin and Ohio, each of which have made discriminations against Herefords in favor of Shorthorns for years. It seems strange to a large proportion of fair-minded men, and men well posted in public matters, that these states should run their fairs and shows in the interest of any breed or for partisan purposes. We are not surprised that such men should think we are actuated by partisan interests in making these charges and entering upon this contest. We are not surprised that they should charge it to the fact that we did not get all the premiums we thought we were entitled to.

"So soon as the Hereford breeders came forward to make the exhibits at the State fairs in force we withdrew as an exhibitor. We should have withdrawn from the Fat Stock Show two years before but for the fact that we were strongly urged not to do so. There has been no breeder in the State of Illinois that has done more than we have to make the stock exhibits before the State Board a success. Our exhibits have always been creditable, they have been large. Not one of the breeders of cattle in the State of Illinois represented the State at the Centennial Exhibition in Philadelphia in 1876 but ourselves, and although we have claimed our rights, we believe that to-day a large majority of the Board during the last eleven years will credit us with having prosecuted our claims in a fair and gentlemanly manner, and that a majority of the Board are personally friendly to us. We shall assume that in the future the Board will not need any promptings from us to make its shows, its exhibits and its awards creditable to the State and the world.

"We have made our contest with the State Board of Illinois because Illinois is our home. We had a right to a hearing, and to fair and impartial treatment, that we would not have in any other State or before any other society. We believed if this Shorthorn influence was broken here it would be substantially broken elsewhere, and we see no reason to change our views. The State Board of Indiana has been, if possible, more under Shorthorn influence than the State Board of Illinois. So with the State Board of Ohio, so with that of Michigan and Iowa; and they will do well to profit by the experience of the State Board of Illinois. The Indiana State Board, at its late meeting, for the first time gave other breeds a classification equal to that of the Shorthorns.

"The State Board of Illinois has now a president that has no interest in Shorthorn breeding, a man independent of all Shorthorn influence, as we believe, and we think that the Shorthorn members of the Board and the Shorthorn breeders outside of it did all they could to prevent his election. He was nominated, and the Board has elected the Hon. John P. Reynolds as the superintendent of the cattle department. Mr. Reynolds has been connected with the Board for a great number of years, perhaps from the start. He has been its secretary, its president, and we think that the Board owes to him more of the good there is in its management than to any other man who has been connected with it. We know somewhat of the course pursued by the Shorthorn interest in and out of the Board, and so far as the cattle interest is concerned he has been for a fair, impartial and intelligent management, both as to classification and judging, and we would submit our interest as we have before stated, to his judgment alone. He has undoubtedly exerted a larger influence in the board than we could have done in his place, and with Mr. Landrigan as president and Mr. Reynolds as superintendent of the cattle department we can scarcely doubt but that there will be a fair administration in the management of the cattle department.

"Although the Shorthorn interest will die hard, and will leave no stone unturned to influence action in its favor, the Hereford breeders are strong in the merits and right of their cause, to meet the issues in a fair and candid spirit. They are destined to win the world over, and perhaps they had no right to expect a larger success than they have already attained. While we are not an exhibitor, we trust that the Hereford breeders as a whole will make such a show and such an exhibit at the State fairs and fat stock shows that will show their confidence in the present management. Should Mr. Scott's predictions be realized as to the Shorthorn breeders, let the Hereford men make good the deficiency.

* * * * * *

"Since writing the above we have received from Secretary Fisher a copy of Mr. Moninger's statement, which was read at the February meeting, and is as follows:

" 'To the Honorable Board of Agriculture of the State of Illinois: The cattle belonging to me, referred to by T. L. Miller in the statement made to your Board in the January meeting, were of my own breeding and feeding, ages as recorded in my list of entries at the late fat stock show. A sworn statement of the ages of the steers named, with others shown in the herd, is in the hands of the Iowa State Board of Agriculture.

" 'Tom Brown, winner of the sweepstakes prize for the best three-year-old steer in the show, dropped August 28, 1879, is well known by many cattlemen in Iowa. He was a prize winner at the Iowa State Fairs of 1880, 1881 and 1882. The fairs were held during the first ten days of September each year. His weight when shown as a yearling was 960 pounds; at 2 years old, 1,500 pounds; and at three years old 1,900 pounds. When shown at a few days over twelve months old and 960 pounds weight, is the time, according to T. L. Miller's statement, when he was at least three years old (not far from the age of his sire at that time).

" '*This is a brief statement of the history of the prize winner Tom Brown, that seems to trouble one T. L. Miller greatly, inasmuch as his teeth were wrong, he says, when really it was his back that made him a troublesome steer in the ring to competitors—of the squealing kind.*

"'I will here state that at the Iowa State Fair of 1880, when Tom Brown's show career commenced, Hon. James Wilson, of Traer, was superintendent of cattle. He selected for judges on herds of fat cattle Hon. C. Clarkson, of Des Moines, agricultural editor of the Iowa State Register; Hon. Oliver Mills, of Atlantic, and C. W. Norton, of Durant, Iowa. These gentlemen, as well as many other well known stockmen, such as Col. Scott of Nevada; Robert Miller, of West Liberty; Hon. W. T. Smith, of Oskaloosa, now president of the Iowa State Board of Agriculture, I might refer to, who well remember my exhibit at the Fair, and especially the little yearling steer, since named Tom Brown.

"'Grinnell, No. 111, dropped February 7, 1880, was shown in my herds of one and two-year-olds at the Iowa State Fairs of 1881 and 1882. At the Fair of 1881 his weight was 1,370 pounds, age 19 months; at the Fair of 1882 his weight was 1,800, age 31 months. The weights here given are home weights. This steer was always shown in a herd of five steers of remarkable similarity, and he is not so easily distinguished from his mates. However, the correctness of his age can be proven by a good cattleman who has known and admired him from a sucking calf up to the date of his trial in the ring at the late show in Chicago.

"'The steer Iowa Champion, No. 112, dropped late in November, 1880, was a prize winner in my herd of yearlings shown at the Iowa State Fair of 1882. This is his first appearance in the show ring.

"'The steer Champion, 102, dropped Dec. 7, 1879, is the largest steer that I ever raised for his age. He was first shown as a suckling calf, with his mother, at the Marshall County (Iowa) Fair of 1880; was winner of first prize for best yearling at Iowa State Fair of 1881; was next shown at Chicago Fat Stock Show of 1882 weighing at twelve months old 1,100; at twenty-four months old, 1,700, and at a few days under three years old, 2,200 pounds. This steer, I believe, is not named in the protest. I mention him here on account of his remarkable size and growth. He was, however, beaten by Mr. Gillette's steer Mammoth, 148, weighing 2,220 pounds at seven months younger, and I do not question the age in the least in this ring. However, Mr. T. L. Miller put in his usual protest. As will be remembered, an examination of the teeth of the animals in this ring was made, and the awards announced as made by the judges. This is the ring in which Grinnell referred to, was awarded the first prize.

"'In regard to my exhibit for the late show I will state that they were fed for exhibition from almost the date of birth. I am not one of those who claim to have matured a steer of 1,845 pounds weight at a little over 27 months old and only fed him 12 months preceding the show, as I understand to be the age and care of the steer Conqueror, shown by T. L. Miller at the show of 1880. While I do not say it is impossible in Mr. Miller, yet an exact statement of the prize steer would be of interest to many. As Mr. Miller in his communication states the importance of "going forth to the world with truthful results," I would suggest that your secretary, in editing these results, make marginal notes of explanations where needed, viz.: In the table on page 182, Vol. 18, Report of 1880, in which Mr. Miller reports sale of his show cattle at 12½ cents per pound live weight while Mr. Ross, Mr. Highmore, Mr. Sodowsky and Mr. Scott fail to reach an average sale of 7 cents per pound for their prize cattle, the price for choice exporting cattle for the week of the show not being over 6¼ cents per pound. In Vol. 19, Report of 1881, Mr. T. L. Miller reports sale of his show herd at 12 cents per pound, while Mr. Hunt, Mr. Nelson, and others only reached 8 cents for their best prize cattle.

"'A due regard for Mr. T. L. Miller's earnest desire for "truthful results" might justify some explanation of these tables for the benefit of readers who may not know the superiority of Mr. Miller's kind of cattle and the wherefore of the great difference of price obtained. I would further suggest that these tables, which are valuable if correct, be published in the stock journals. Money is what we are after and if Mr. Miller's cattle will sell in the open market for beef for from 4 to 6 cents per pound more than other breeds of cattle, they are the kind, and he can well afford to retire from—to him, the trouble of the show ring, as he threatens to do. "Truthful results." Truthful Results. Who can better cry fraud than the man who has "been there?" D. M. MONINGER.

"'P. S.—A sworn statement from the man who dressed the steers Tom Brown and Grinnell, is being prepared, and will be sent you. Also a sworn statement from me as to their ages will accompany it. By the way, the steer Grinnell is going to come out the champion of anything on record, dressing over 72 per cent.

"'To S. D. Fisher, Secretary. D. M. M.'

"This was referred and the committee made the following report:

"'To the State Board of Agriculture: Your committee to whom was referred the communication of Mr. D. M. Moninger, of Galvin, Ia., would beg leave to report that they have duly considered the matter; that as no formal protest has been entered against the entries made by him as prescribed by the published rules, no action of the Board should be taken, and the committee recommend that the communication be placed on file with other papers in the case.

J. IRVING PEARCE,
JOHN P. REYNOLDS,
JOHN VIRGIN,
Committee.'

"There is but little to Mr. Moninger's statement beyond what his entries would show. He says a sworn statement of the ages of the steers named, with others shown in the herd, is in the hands of the Secretary of the Iowa State Board of Agriculture. It would have been well if Mr. Moninger could have had a sworn copy of these statements. In reference to the steer Tom Brown, Mr. M. seems to think 'it was his back instead of his teeth that was troublesome to competitors of the squealing kind.' This, we think, is the only reference that Mr. M. makes to the fact that a three-year-old steer carried eight teeth with evidence of absorption already commenced. After some considerable general talk about this steer, the judges and the parties who know something about him in the show ring, we would call attention to the fact that Mr. M. states that Tom Brown received his name some time after those exhibits were made, which may or may not be a convenient way of covering his identity.

"As to Grinnell, a two-year-old steer, he does not undertake to explain how he could grow eight teeth while he was only entitled, according to standard authorities, to four. We wish to call special attention to the remark in Mr. Moninger's statement in reference to this steer, 'that he was one of five steers of remarkable similarity, and he was not so easily distinguished from his mates.' But he suggests that he can prove his age by a man that has known him from a calf. It would have been well for Mr. Moninger to have produced this testimony. General statements are not as easily refuted as certificates giving age by dates, and sworn statements, if not true, subject a man to charge of perjury.

"As to the steer, Iowa Champion, No. 112, he gives no explanation as to how he could have grown six teeth when entitled to only two. There are some general statements made about the steer Champion No. 102.

"On the whole, the statement that Mr. Moninger makes will hardly convince anyone that his entries as made at the Fat Stock Show were truthful. But he promises some further data, and we defer any further notice of this reply until his case is made and placed on record.

"Mr. Moninger undertakes to draw attention from his case by reviving the charges of Tom Corwin Anderson, in 1880. We met these at the time, we believe satisfactorily to everybody who took any interest in the case, and especially to Mr. Anderson. We hope when Mr. Moninger has his affidavits prepared they will be in such shape that they will have some value in and of themselves, and that they may contain some explanation as to how such abnormal teeth developments may be secured.

"The committee in reporting upon this statement, say that as no formal protest has been entered against the entries made by him (Mr. D. M. Moninger), as described by the published rules, no action should be taken, and the committee recommend that the communication be placed on file with other papers in the case. Mr. Miller has never undertaken to make any protest; each and every time that he has come before President Scott or the Board he has stated distinctly that he has come to advise the Board of facts, and not as a protestor."

On page 86 of Vol. 4 of the "Breeders' Journal," appears the following humorous burlesque, written by an unknown Kansas correspondent (thought to be W. E. Campbell), which shows so thorough an appreciation of the truth of the case that it cannot be left out of this history.

TRIAL OF D. M. MONINGER.

"Editor 'Journal':

"You will probably find the following of interest to your readers:

"D. M. Moninger, the great Shorthorn breeder of Iowa, and the exhibitor of the crimson herd of Shorthorns at the leading Western Fairs last fall, was arraigned in Chicago before the Court of Public Opinion at the November Term, Judge Breeders presiding, J. H. Gazette appearing for the defense and T. L. Whiteface for the prosecution.

"The indictment charged the prisoner with frauds perpetrated upon the public, and of obtaining moneys and valuable premiums by false pretenses and misrepresentations to and before the Illinois State Board of Agriculture, on or about the 16th day of November, 1882. The indictment further charges and sayeth that on or about the 16th day of November, 1882, the aforesaid D. M. Moninger did enter, or cause to be entered, a number of Shorthorn steers under false ages and misrepresentations, thereby defrauding honorable exhibitors out of large sums of money and valuable premiums. And the indictment further sayeth that on or about the 16th day of November, 1882, the aforesaid D. M. Moninger did wilfully, knowingly and fraudulently enter the Shorthorn steer Champion of Iowa, as being one and under two years old, when the aforesaid Moninger knew said steer to be three years and eight months old. The indictment further sayeth that the horns of the aforesaid steer, Champion of Iowa, had been filed, scraped, shortened, and otherwise disfigured for the purpose of deceiving the public, in making said steer appear much younger than the said D. M. Moninger knew him to be.

"A motion was made to quash the indictment on account of informality, but was overruled and the defense forced to trial.

"A. Herdsman was the first witness called, and testified that the ages of cattle could readily be determined by an examination of their mouths; that he had examined the mouth of the steer, Champion of Iowa, entered by the defendant as a yearling, and found that said steer had six full grown, second-growth teeth, and that he was about three years and eight months old.

"On cross-examination he said that he had examined the mouth of the Shorthorn steer Red Major, entered and exhibited by Potts & Son as the same age to a day as the defendant's steer; that the Potts steer had but two second-growth teeth, and that the Moninger steer had six second-growth teeth and must have been about two years older than the Potts steer.

"John Ranchman was then examined and testified that he had bought and sold thousands of cattle in Texas and Colorado; that whenever there was any dispute or doubt about the ages of cattle when delivering and classifying them, they were lassoed, thrown down and their mouths examined. If the animal had no second-growth teeth it was classed as a calf or yearling; if two second-growth teeth, as a two-year-old; if six second-growth teeth, as four years old, which is the highest and most valuable class they have, and are commonly called beeves. The two first second-growth teeth appear at about eighteen months, and at twenty-four months they are fully developed, and so on each twelve months, until the animal has what we call a full mouth.

"The counsel for defense then produced some printed matter clipped from the 'Breeders' Gazette' and the 'Chicago Times,' which they asked leave to read and submit to the jury as evidence, to which the prosecution objected, and after some discussion the court ruled that the printed matter could be read, but that the jury were at liberty to use their own judgment as to the value and weight of the same.

"The articles were then read to open court, and represented that the world-renowned Dr. N. H. Veterinary had examined the mouths of the steers in question and found that they were entered in accordance with the ages their mouths or teeth indicated.

"J. H. Gazette then volunteered his testimony; was placed upon the stand and testified that he had known the defendant many years, and had always considered him a very fine Shorthorn breeder. He admitted that defendant's cattle had an unusual number of large teeth for their age, but he thought they had been forced out by the feeding of very hard, flinty corn, which was known to grow on Mr. Moninger's farm; and the crushing of this flinty corn (in his opinion) had forced the second growth of teeth through the gums at a very early age, and thought Dr. N. H. Veterinary based his conclusions upon the same theory, and that the Doctor was very high authority in steerology.

"T. L. Whiteface, believing that Dr. N. H. Veterinary had been grossly misrepresented and used as a scape-goat, to his detriment for the benefit of the defendant, asked that the Doctor be subpœnaed and brought into court, which caused quite a rustling throughout the court room, that did not subside until the Doctor had been sworn and placed upon the witness stand.

"He testified that he had not authorized the statement made in court by Mr. Gazette, much less the articles read by the defense; that he was not interested in the matter, and did not know any of the steers involved, either by name or number, and very much regretted that he had been dragged into so disagreeable a position.

"Court then adjourned until Monday, when it is expected the prosecution will introduce testimony of a startling nature.

"Yours ever, BOVINE.

"Grass Plot, Kansas, Dec. 25, 1882."

That the Hereford winnings made a decided impression on the breeders of Shorthorns is shown by the following from the "Breeders' Journal":

"A writer in the 'National Live Stock Journal' under the heading of 'The Present and Future of Shorthorns', says in reference to showing at the Fat Stock Show that 'in the future we must insist on its being between thoroughbreds of the different rival breeds, so there will be no doubt about how much Shorthorn blood there is in one that is shown as a Hereford grade. It is manifestly unfair to breed the pick of the Hereford bulls imported into this country on some of our best cows, and then exhibit their progeny as grade Herefords. It is no grade, it is a cross-bred. If this practice is continued by the Hereford men, the Shorthorn men should retaliate by selecting several of the best Hereford cows and breed the best Shorthorn bulls upon them, and show the calves of these Hereford cows as grade Shorthorns. And nothing would be settled by this course. It would show only the unfairness which Hereford men have been practicing in the past, and every time the blue ribbon has been awarded on the Hereford grades in the best fat stock shows in this country, doubtless one-half or more of the ribbons should have been labeled, "A Shorthorn cow has proved true to her well-established character, namely, she makes a half-breed better than the average of the sire's race." And we understand that these Polled Angus friends follow their Hereford friends' practice in this regard, and their crosses will be harder to detect, as we are told these Polled Angus bulls are such prepotent fellows that at one sweep they knock both horns and color off the finest Shorthorn cow in the land and bury the unfortunate cow's calf in a dark grave and deny them the privilege of being recorded alongside of their mothers. Retaliation upon these Polled Angus men is practically denied the Shorthorn men, from the fact that we are told that no Shorthorn bull in the world can make these pokers grow out at the top-nots of those black cows, or the orange color to blossom upon their smelling tubes. So it appears at the present time the Shorthorn breeders are put to a disadvantage with these Polled Angus breeders, and the only way left us to beat them is to fight them as we propose to fight the Hereford breeders—thoroughbred against thoroughbred.'—[This is part of an article signed W. H. H. Cundiff.]

"Will our Shorthorn friend state how many thoroughbred Shorthorns have been winning honors at the Fat Stock Show? Can he remember that it was the thoroughbred Hereford cow Jennie that won the sweepstakes for the best cow in the show in 1879? Will he state how much Shorthorn blood the steer Conqueror had? Here we call to his attention that every grade Hereford that has been shown at the Fat Stock Show was a well marked Hereford. Our friend is in trouble, and considers himself at a disadvantage with the Hereford. He must have the white face and the usual markings. If he puts the Shorthorn bull upon the Hereford cow he must bring affidavits upon the show ground if he would claim any merit from the Shorthorn, for the produce will be a Hereford in appearance and character. We cannot at the present time see any way by which he can be successful upon the show ground, except to follow the practice in vogue at the last show, to-wit, showing cattle at from two to three years older than the ages for which they were entered. But we fear even here he will be at a disadvantage, as we understand that the rules of the Board are to be enforced at the next show."

It is hardly necessary to say that at this time the "Breeders' Gazette" was not in any sense of the word a partisan of the Hereford breed of cattle.

We had the following to say at the time, in regard to this matter:

"We have no desire to prolong this controversy, but we wish to state our beliefs. Something about three years ago there was formed a syndicate, a Bates corner in that class of the Shorthorn breed of cattle. We believe that out of that syndicate grew the establishment of the 'Breeders' Gazette,' and that the parties interested in that movement were the main supporters of that journal.

"We believe that out of that syndicate was formed the plan to break the Hereford interest, by charging T. L. Miller with frauds at the Fat Stock Show as to the ages of his cattle. That that syndicate was a signal failure there is no doubt."

From the "Breeders' Journal," April, 1883, we quote: "Mr. J. H. Sanders of the 'Breeders' Gazette,' in noticing Mr. Moninger's statement before the State Board in reference to the entries of his cattle, says: 'We have been favored, as heretofore stated, with a copy of the communication sent by Mr. Moninger, of Iowa, to the Illinois State Board of Agriculture, in reply to the recently published statement of T. L. Miller, of Beecher, Illinois. The action of the State Board in refusing to consider the tardy protest of T. L. Miller, months after the awards had been made on the cattle in question, has generally been endorsed by the public, who have never questioned the integrity or the motives of this body of honorable gentlemen. The untarnished reputation and high standing of the Board, individually and collectively, has not suffered in the slightest degree with the unprejudiced public from Mr. Miller's aspersions. His (Mr. Miller's) motives are generally understood and have been manifest from the beginning of the movement, to advertise certain interests in every possible way, regardless of courtesy or even common decency. Mr. Miller was requested by the officers of the Board at the proper time, during the show, to reduce his verbal complaints to writing and make specific charges against the steers exhibited by the Messrs. Moninger, Groff and others. He was assured that his protest, if so made, should have proper attention, and he well knew that recognized experts would thereupon be called to determine the matter of the ages of the steers in question. But Mr. Miller failed to do this until long after the steers were killed, and now the evidence he presents, the alleged teeth of the steers in question, is no more convincing to the friends of the parties than the positive statement of Mr. Moninger, a breeder, and the other man he names, as to the age of the steers, even if the identity of the teeth has been established, which has not been done.' To which we replied in the 'Journal,' Vol. 4, page 230: 'Mr. Miller appeared before the Board in the morning of the day that the show opened and advised Mr. Scott of the frauds that were being perpetrated upon the Board, exhibitors and the public, and asked that the Board might take the initiatory steps to ascertain the correctness of the entries and protect the exhibitors that were honestly before them under the rules.'

"'Mr. Miller again appeared before Mr. Scott the last day but one of the show, and again called his attention to these facts, and asked that an investigation might be had, and with witnesses, if need be, to prove the fraudulent ages of the steers in question. Failing to get a promise of any action from Mr. Scott, he then went to Mr. John P. Reynolds. Mr. Reynolds promised to give his efforts and influence, as did the Hon. Sam'l Dysart, another member of the Board, which was opposed by Col. Scott and Col. Judy.

"'Mr. Miller secured the teeth of several of these animals that were slaughtered in Chicago, and met the newly elected Board at their meeting in January, and laid before them again these facts and this evidence, and charged there and then that either Mr. Scott had failed in meeting his responsibilities, or that the Board had failed in theirs. These facts were brought to the notice of the Board not in the shape of protest but in the shape of information, which the Board ought to have taken cognizance of the week of the show.

"'Mr. Sanders says "the action of the State Board in refusing to consider the informal protest of Mr. Miller months after the awards had been made on the cattle in question, has generally been endorsed by the public, who have never questioned the integrity or motives of this body of honorable gentlemen." Mr. Sanders knew when he wrote this article that Mr. Miller had so informed the Board on the opening day and the closing day but one of the show, and he knew that Mr. Miller, instead of making a protest, took the first opportunity when the new Board was organized to meet the Board with these facts. Mr. Sanders knew that there was a strong feeling in the Board in favor of these charges having an investigation, and he knows, and knew then, that the investigation of these charges was opposed by the Shorthorn influence, in and out of the Board. He knew that when the effort was made to have the rule passed by which, at future Fat Stock Shows, a proper and authoritative examination should be made as to ages of the cattle on exhibition, it was opposed by the Shorthorn members, including Cols. Scott and Judy, who were perhaps the leaders in the opposition. He knew when he wrote that article that the following resolutions were passed at the February meeting, providing for such an examination at future shows as would prevent the repetition of these frauds:

"'Resolved, That the President be, and he is hereby, authorized to select and secure the attendance of three gentlemen, one of whom shall be an experienced and practical veterinary surgeon, and two of them experienced cattle feeders or breeders of neat cattle, who shall examine every animal entered for competition in Class A at the Fat Stock Show, and report in writing to this Board, previous to the commencement of work by the awarding committees, their judgment as to the ages of the respective animals so examined.

"'Resolved, That the committee on ages shall act independent of the statements of exhibitors as to the ages of their respective entries.'

"When Mr. Sanders says that that action of the Board at its January meeting has generally been endorsed by the public, he knew, when the sentence was penned, that public sentiment had compelled the Board to provide for future examinations. When he says that Mr. Miller's movements are generally understood, and have been manifest from the beginning of the movement, to advertise certain interests in every possible way, regardless of courtesy or even common decency, he knew that he was occupying the contemptible position of a pettifogger.

"It is true that Mr. Moninger entered steers at this show as two years old with eight teeth, as one year old with six, and when Mr. Sanders was asked by a prominent exhibitor at this show to make an examination of the mouths of these steers, he excused himself by saying he could not tell by the teeth the age of a bullock. Still, he claims to be the 'pioneer journalist' in the live stock interest of America. He knew that Mr. Miller went before the Board to advise the Board of these facts. He knew that Mr. Miller advised him (Sanders), and that he had promised to give his influence in aiding Mr. Miller to expose any fraud in the entries of animals that were entered at that show.

"These are facts, and Mr. Miller's record in this matter has been open and above board. Mr. Sanders' position in undertaking to defend Mr. Moninger in these fraudulent entries might be excused in him as an individual, being a personal and intimate friend of Mr. Moninger, but as a journalist there is no excuse. Mr. Sanders might inform himself that two-year-old steers cannot grow eight teeth, or one-year-old six, and he knows that these bullocks had such teeth, and we are not confined to the teeth that we have secured to show this. We can, whenever the time comes and it becomes necessary, bring such an array of witnesses to prove these statements as will make it conclusive before any court in Christendom. Mr. Sanders says that 'Mr. Miller failed to make his protest in a formal manner until long after the steers were killed, and now the evidence he presents (the alleged teeth of the steers in question) is no more convincing to the friends of the parties than the positive statements of Mr. Moninger, a breeder, and the other men he names.' Mr. Moninger's referring to certain names is not presenting them in evidence at all. They have had nothing to say as yet; when they do have, we will discuss the merits of their statements.

"We would call attention to the 'Gazette,' which professes to give a correct copy of the statement made by Mr. Moninger to the State Board, that it has left out a portion, which we have put in italics, commencing with '*this is a brief statement*,' etc., etc."

From the "Breeders' Journal," June, 1883, we quote: "We have heretofore given Mr. Moninger's statement in answer to our charges of fraudulent entries at the late Fat Stock Show. The 'Breeders' Gazette' of the 5th inst. introduces Mr. Moninger as follows: The remarkable percentage of net to live weight made by the grade Shorthorn steers Grinnell and Tom Brown, exhibited at the last American Fat Stock Show in this city by D. M. Moninger, of Galvin, Ia., have already been reported in these columns as tending to set at rest the fraudulent charges as to how the bullocks died, as the English say. We herewith present the affidavits of Messrs. Moninger and Duddleston:

MR. MONINGER'S AFFIDAVIT.

"'The steer Grinnell, winner of the first prize for best two-year-old steer in the Chicago Fat Stock Show of 1882, was dropped February 7th, 1880. The steer Tom Brown, winner of first prize for best three-year-old steer in the same show, was dropped August 28th, 1879.

"'D. M. Moninger.

"'Feb. 24th, 1883.

"'Subscribed in my presence and sworn to before me by the said D. M. Moninger, this 24th day of February, 1883, as witness my hand and seal notarial.

"'Henry Stone, Not. Pub.'

AFFIDAVIT OF GEO. DUDDLESTON.

"'Chicago, March 27th, 1883.

"'This is to certify that I, George Duddleston, a butcher doing business at Nos. 83 and 85 Fifth Ave., Chicago, Ill., purchased the prize steers Grinnell and Tom Brown from D. M. Moninger, of Galvin, Iowa, during the last Fat Stock Show held in this city; that the live weight of the steer Grinnell, allowing fifty pounds for shrinkage (his show weight having been 1,850 pounds), was 1,800, and that said steer Grinnell's dressed weight was 1,310 pounds, a percentage of 32 to 77 of net to live weight. The show weight of the steer Tom Brown was 1,945 pounds, from which 50 pounds is to be deducted for shrinkage, leaving his live weight at time of slaughter 1,895 pounds; the dressed weight of the steer Tom Brown was 1,340 pounds, a percentage of net to live weight of 31 to 70. Having seen the carcasses dressed at each of the Fat Stock Shows, I can truly say that I have never seen so deep a fleshed bullock as the steer Grinnell above mentioned. Furthermore, it is my belief that the ages of the above named steers were substantially as given to the managers of the show by Mr. Moninger.

"'Geo. Duddleston.

"'State of Illinois, Cook County. Attest.

"'Subscribed and sworn to before me this 27th day of March, A. D. 1883.

"'Geo. Braham, Not. Pub.'

"Of which the 'Breeders' Gazette' says: 'The allowance of 50 pounds shrinkage from show weight will be generally regarded as none too large, especially in view of the fact that the champion Hereford steer imported Sir Richard's shrinkage, as we are informed, was 75 pounds. The statement of Mr. Duddleston derives additional weight from the fact that he has a decided preference, as a rule, for Hereford cattle, and his experience as a butcher is such as to render his clearly expressed opinion as to the age of the steers worthy of some credence.'

"Mr. Miller obtained the teeth of the steers from Mr. Duddleston. Mr. Moninger makes affidavit as to the age of the steers, but does not show or explain how a steer at two years old can grow eight teeth; neither does he explain how Tom Brown at three years old could grow eight teeth, while the first was entitled to only four and the latter to six. As to the net to live weight as given by Mr. Duddleston it is foreign to the questions at issue. It will be remembered that the steers had lived something over one month from the time they were weighed; that they ought to have gained 75 pounds instead of shrinking 50. But this is immaterial. Mr. Duddleston's opinion may be valuable, but it would seem to us that Mr. Moninger is bringing to his aid very weak support. We had hoped to see Mr. Duddleston, but have not been able to inquire as to who sought his affidavit and how he could reconcile the fact that these steers carried a full mouth of eight teeth at the age they were represented to be. Not seeing him, we have written him a line, and may be able to give his answer next month. We have received a letter from a responsible party and have had it in our possession for some time, waiting for some testimony from Moninger. The letter is as follows:

"'Feb. 22, 1883.

"'T. L. Miller, Esq.

"'Dear Sir: I attended a farmer's sale of stock, etc., to-day, and while there was engaged in conversation with three gentlemen with whom I am well acquainted. Two of them are breeders of Shorthorn cattle, and the other a large feeder of cattle. The conversation turned upon yourself and your cattle, when one of the breeders above referred to said that a man working in his neighborhood who formerly lived in Iowa, near D. M. Moninger, says you are correct about the age of the steer Tom Brown, shown by Moninger last November at the Fat Stock Show. He said this party stated that he knew the steer well, and that his age is about the same as you consider it to be, and that there are many persons in the vicinity who know the same thing. Yours truly,

"'E. W. P.'

"We are fully satisfied that the proof might be had from several parties as to the ages of the steers exhibited by Mr. Moninger last November. Mr. Moninger undoubtedly relied upon having his steers passed through the exhibition without question as to the correctness of his entries, and we have no doubt that the fact that the steers exhibited greater age than they were entered for, and that the dressing would have exposed this, is the reason why the bullocks were not dressed at the show. It was well known by the exhibitors at this show, and that of 1881, that the Shorthorn bullocks that were dressed in 1881 showed an age by their mouths largely in excess of that for which they were entered. They knew that the Hereford exhibitors had these mouths preserved and placed in the hands of the officers of the State Board as an evidence of the fraud as to age. Our persistency in calling attention to these frauds and bringing forward these evidences (for it will be borne in mind that the 'Gazette' admits that Mr. Duddleston gave us these teeth) has secured action by the State Board which will prevent any such frauds in the future. We are advised by members of the Board that this rule will be rigidly enforced, to-wit: That all cattle offered for exhibition in 1883 will be examined by a competent veterinary surgeon and two practical men to determine the ages without reference to the entries.

"It is now nearly five months from the time these charges were made, before these affidavits were given, and no evidence now beyond what was before the Board and the public on the 16th of November. This is not a matter as to whether a Shorthorn exhibitor or a Hereford exhibitor shall be successful at the Show. It is beyond and above this. It is a question as to whether the State Board of Agriculture of Illinois shall publish to the world, under its authority and sanction, data that are based upon an erroneous and incorrect basis."

Grades and Crosses.—The journals that were advocating the lordly Shorthorn could not bear to have the Herefords win a prize, and when they did win with a steer that was the most distantly related to the Shorthorn they claimed all the glory for their breed.

Strictly speaking, at the date of which we are writing there were very few, if any, cross-bred steers, as we understand the term; that is, a thoroughbred Hereford bull bred to a thoroughbred Shorthorn cow, and there were certainly none bred the other way, or the calf would come red with white face. Our grade Herefords were bred from all kinds of native cows, with a thoroughbred Hereford bull for sire. On this subject we quote from the "Mark Lane Express" as follows:

"The 'Breeders' Gazette' (Chicago), a journal which appears to be *laying itself out as far as possible to further the interests of Shorthorn breeders and fanciers,* admits that the Herefords come nearer 'dividing the honors as beef producers with the Shorthorns than any other breed in America.' This is pretty well from the 'Breeder's Gazette,' especially when it is seldom contended by even the warmest partisans of the Shorthorn breed in this country that Shorthorns excel the Herefords purely as beef producers. However, the 'Gazette' goes on to say: 'They (the Herefords) are an old and well established breed, and have long been known in this country; but it has only been within the past six years that they may fairly be said to have made any material advances in popular favor. Since then, however, the demand for improved bulls, for use in the vast herds on our western plains, has led to the extensive use of the Herefords, and so far as we have heard an expression of opinion, with most excellent results." Certainly the demand during the past three years has been largely in excess of the supply, and prices have risen materially. Within two years past the importation from Great Britain, notwithstanding the hindrances of an expensive quarantine, have very greatly exceeded the sum total of importations to this country previous to that date, and there is no perceptible falling off in the demand for them by our ranchmen. Their great merit as grazing beasts is unquestioned. That they possess great hardiness is clearly evidenced by their heavy shoulders, well sprung ribs and general make-up. As a rule, there is no breed of cattle extant the structure of which more clearly indicates strong vital organs than the Herefords. The breeders of Herefords in this country have manifested much spirit for several years past in pushing the claims of these cattle to the front, and have undoubtedly met with a good share of success at our Fat Stock Shows. It has been frequently alleged that the excellence which has characterized the animals shown as Hereford grades heretofore was due largely to the admixture of Shorthorn blood—that, as a rule, they have not been any more properly grade Herefords than grade Shorthorns, and that they might as well have been shown in the one class as the other. The Hereford breeders have accepted the issue, and have declared their determination to make a bold stand and face the music hereafter with purely bred Herefords at our Fat Stock Shows.'"

Commenting on which the "Mark Lane Express" said:

"No doubt the Hereford men will be able to 'face the music' with butchers as judges; we shall see when the time comes. Meantime, we do not clearly understand the sense in which an animal called a grade Hereford could be with equal correctness described as a grade Shorthorn. We can very well understand that an animal which has any proportion of Shorthorn blood in its veins, whether much or little, would have its merit—should it possess any—ascribed to the admixture of Shorthorn blood, by partisans who are blind to everything but their own hobby; but whether such merit could be so claimed or not, there should be no question as to whether any particular animal was a grade of one breed or that of another. If the sire was of one breed and the dam of the other, the offspring would not be a grade but a cross-bred. We cannot tell what the word grade may be understood to cover, as used in the United States, and it is not used at all in this country; but we should take it to mean a step, or one of a series of steps, in remove from a thoroughbred sire out of a mongrel dam. The question, then, is simply whether these animals at the Chicago show were sired by Hereford or by Shorthorn bulls. If, as we have already stated, they were from pure bred sire and dam of different breeds, then they were crosses and not grades; and if the sire was a 'grade,' the offspring would clearly class as a grade of the breed represented by the sire. As for the merit, let both parties claim it, but there should be no sort of question as to the classification. At our Fat Stock Shows cross-bred, grade and mongrel animals are all classed together as 'crossed or mixed bred,' a very unsatisfactory arrangement to our thinking; nevertheless it covers the whole ground, which the term 'grade' used alone evidently does not."

In closing this chapter we give an extract from the "Breeders' Journal" for June, 1883, written just prior to our departure for England to purchase an addition to our Hereford herd.

"We are glad to have an opportunity to reply to Mr. Sanders and Mr. Moninger's defense. We shall be absent for some months, and should anything come up in reference to this matter, and there is a delay to notice it, it must be attributed to this reason. It would be well for the 'Breeders' Gazette' to explain somewhat in reference to the teeth indications of age. Mr. Sanders claims to be the pioneer live stock journalist of America, and whether the claim is worth anything or not, he should sink his personal friendship and sympathy for Mr. Moninger in his higher duties as a journalist. Mr. Sanders knows, or may know, that no two or three-year-old steer can have eight teeth. The rule requiring the examination of teeth to determine the ages of animals in the future was opposed vigorously by a large portion of the Shorthorn element of the Board, and they were able to defeat such action at their January meeting. It was, however, re-opened at their February meeting and carried.

"In the issue that has been made between the Herefords and Shorthorns during the last ten years, the Shorthorn advocates have never attempted to meet the positions that we have taken, but have undertaken to throw mud and dirt instead. We advised a prominent member of the Board that, while we should do all that we could to advance and forward the success of the shows made under the auspices of the State Board of Agriculture of Illinois, until some of these questions which have given rise to a good deal of controversy were settled and disposed of we should not be an exhibitor. We received for a reply the following, which we feel at liberty under the circumstances to make public. We do withhold, however, the name of the writer. We will say that he is not a Hereford breeder and never has been, and so far as we know, never expects to be. He says:

"'I am anxious to see your herd well represented at our large shows. You have done more to bring a very valuable and meritorious breed of cattle to the attention of breeders and the public than any other gentleman on the continent, and for the persistency in which you have pressed the just claims of a meritorious breed to a successful end, you deserve the praise of the whole agricultural community. Indeed, your zeal and determination in making known the real merit of your favorite breed of cattle has been and will be of untold benefit to the breeders of Shorthorns, for it has aroused them to the necessity of cultivating and developing the better qualities of the Shorthorns, instead of resting their claims on fancy pedigrees. Pedigree is all well enough in its place and for the purpose intended, but real and uniform merit has a value more important.'"

We have done very little in bringing forward notices of commendations to our course in this controversy, but we have much pleasure in many letters we have received. The contest which we have had to meet has been one of trial and annoyance, but our work is done, and inures not to our benefit but to the world, to the breeders of cattle and to the consumers of their product.

Our exposure of these frauds was not to blacken or damage Mr. Moninger, or any other man, but to secure fair, impartial and intelligent judgment in passing upon the claims of the different breeds of cattle; and we wish to state that we never considered Mr. Moninger a sinner above all others, but there came at last a feeling among exhibitors that they must protect themselves, because those who were managing the societies would not protect them. From this on the decisions were, as a rule, fairer, improving as the judgment and intelligence of the managers allow it. This is all that we asked and to this we had a right.

The foregoing is a repetition of part of much similar matter published in the seventies and eighties that at this writing (1898) has already yielded much fruit of vast benefit to American agriculture. We cannot refrain in this connection from quoting a more recent correspondence. There had been a tendency on the part of the Board to lessen its vigilance, and on the 18th of February, 1898, we wrote to Mr. J. Irving Pearce, then president of the Illinois State Board of Agriculture, calling his attention to the necessity of adopting some definite rule to determine the ages of live stock to be exhibited at their annual fairs, and suggested that the teeth indications were the surest method of determining this question, and afterwards sent to him cuts of teeth of horses, cattle, sheep, and hogs, as adopted by the Royal Agricultural Society of England. In replying to my letter of the 18th, Mr. Pearce wrote as follows:

"Chicago, March 4th, 1898.

"Mr. T. L. Miller, DeFuniak Springs, Fla.

"Dear Sir: I have your letter of the 18th ult. before me and note with pleasure the interest you still continue to take in the annual exhibitions of the Illinois State Board of Agriculture. I think it is not probable that there will be another Fat Stock Show held soon in Chicago, the Board acting very wisely in making fat stock divisions to the cattle and sheep departments of the State Fair. The carcass prizes, of course, have to be abandoned, as it would be impossible to save the meat without incurring more expense than the Board can at this time stand. On account of the weather and the likelihood of fat hogs to suffer from the three or four loadings required to get them to the Fair, it was thought impracticable to add a fat stock division to that department.

"Your table giving descriptions of the mouths of cattle of different ages is very interesting to me, and doubtless will be to the Board, when I have an opportunity to have it read at the next meeting. The rule of the Board of Agriculture at present is to appoint a jury of three veterinarians to pass upon the exhibit, as to the ages of cattle, sheep and hogs competing for fat stock prizes. This rule has proven very satisfactory, and has on several occasions 'passed out' an animal for being over age. If you have cuts of the mouths of cattle, sheep and hogs, and would care to loan them, our secretary would, I know, be glad to include them in our Annual Report of 1897, the copy of which is about ready for the printer.

"The progress of the Fair for 1898 is, I believe, beyond your anticipations, and I hope to have the pleasure of seeing you there.

"Very truly yours,

"J. IRVING PEARCE."

Mr. Pearce showed my letter to ex-President John P. Reynolds, and from Mr. Reynolds I received the following, dated March 10, 1898:

"468 LaSalle Ave., Chicago.

"Mr. T. L. Miller,

"My Dear Sir: Our mutual friend, J. Irving Pearce, handed me your letter of 18th ult. to him, touching the show of fat stock at the State Fair, with request that I express to you, as I did to him, my approval of your suggestions as to the careful determination of the ages of the contesting animals by the examination of their teeth.

"Your demonstration years ago fully convinced me that your position on that question is absolutely correct, and that the adoption of your views by the Board is entirely practicable. In no other way does it seem to me possible to insure accuracy as to ages, which fair dealing and justice to exhibitors demand.

"I was under the impression that the specimens of teeth illustrating the subject, which you showed us at Springfield, were left in possession of the Board and placed in the museum. At least there ought to be a complete series there for comparison on occasion of controversy or differences of opinion among the judges. I suggested to Mr. Pearce that possibly you might be induced to meet the Board at some convenient time and give them a talk on the subject. I know of none so well informed and prepared to impart the much-needed information as you are, and hope such opportunity may offer. Very truly yours,

"JOHN P. REYNOLDS."

On the 20th of March I received another letter from President Pearce, as follows:

"Chicago, March 20, 1898.

"T. L. Miller, Esq., De Funiak Springs, Fla.

"Dear Sir: Your good and interesting as well as instructive letter of the 9th came duly to hand. I am glad to know that you keep up your interest in cattle breeding and feeding. You have done more than any one I know to further that interest and do not keep your light under a bushel. I am glad that you or your son are agreeing to furnish me with such information that will enable us to adopt the teeth test. If you can get it to me by the 10th or 11th of April I would be glad. Thanking you for your good letter, and hoping we may see you at our next Fair, I am, Yours truly,

"J. IRVING PEARCE."

CHAPTER XLII.

Mongrel Pedigrees Make Mongrel Breeds; Conclusively Demonstrated on the Range

In December, 1881, Mr. Geo. W. Rust, (¶362) the eminent Shorthorn historian and authority, wrote an article for the "Breeders' Gazette," Vol. I, page 29, on the Shorthorn Herd Book question, and its relation to the Shorthorn breed of cattle—their merits and demerits. In January, 1882, in the "Breeders' Journal," Vol. III, page 43, we quoted from this article that Mr. Rust wrote. Mr. Rust took us to an account for quoting him incorrectly, and that the matter might be fairly before our readers we quoted Mr. Rust as published by the "Breeders' Gazette," on page 29, Vol. I, Dec. 8, 1881, as follows:

"The action of the Shorthorn Association at the late Jacksonville convention will commend itself to the great body of breeders, and the more as it comes to be discussed and understood. The Record of Pedigrees—the integrity with which it is conducted and the degree of confidence placed upon it—exercises so direct an influence upon the prosperity of the breeding interest that there can be no questioning the fact that it should be in the hands and thoroughly under the control of the breeders themselves, acting in some associate capacity.

"The record, as such, should have been instituted in the first instance by them, but through the force of circumstances and the want of proper co-operation, this was not the case. But the interest is now so vast, and has been already so seriously prejudiced, and its extension likely to be so influenced in the future by the unfortunate conditions surrounding the Pedigree Records, that it is an imperative necessity, that the breeders should take these records under their own control and direction, purely as a matter of self protection, if for no other reason.

"There are many ways in which the unsatisfactory conditions of the Pedigree Records have prejudiced the interest of breeders, some of which it may be well to enumerate.

"Through *the lack of system and method in the Herd Book* (Shorthorn) itself, the editor [L. F. Allen, mentioned by Mr. Sotham in his history, Chapter 13 of this volume.—T. L. M.] has been *unable to detect and exclude improper pedigrees from record,* and in a great many cases his *judgment has been warped by his personal interests* to admit pedigrees and classes of *pedigrees which a disinterested person would probably have excluded.* As a result the records have fallen into such shape that none but an expert can tell much about them; and *innumerable pedigrees of a doubtful character have been given the sanction of a record. The general public, unable to discriminate between the true and the false, has, under these circumstances, been imposed upon by unscrupulous persons, who have not hesitated to sell, as genuine Shorthorn, animals that were anything but as represented.* [A pretty incrimination, truly.—T. L. M.] And when a rascal sells an honest man such an animal, the mischief is broader than it first appears; for the honest man has a reputation upon which the produce of *this spurious animal can be sold* to others and scattered far and near into other herds, and as he is unconscious of the fraud that has been put upon him, he innocently passes it along and involves his friends and his neighbors.

"An extensive demand for bulls among farmers and in the grazing regions should bring a rich reward to the breeders who have cultivated and maintained the excellence of Shorthorn cattle, but they find themselves brought into competition with other people who have bulls to sell, which, in the *unfortunate condition of the records,* they are able to substitute for pure Shorthorns, and which they are glad to sell at such prices as would restrict, if not in many cases utterly destroy, the profits of legitimate breeding. Thus, *the public record of pedigrees* [Shorthorn.—T. L. M.], instead of *being a means of public protection, as it should be, has come to be a means of public imposition.*

"But time tries everything; and *the people*

who have been imposed upon with spurious cattle, and those who have been deceived into supposing they were introducing well bred bulls into their grazing herds, fail to realize the anticipations which they entertained. They had heard and read much of the excellence of Shorthorn cattle, and felt they had a right to realize something of it in their own herds. But *time brought only disappointment.* Unfortunately, however, instead of placing the cause of the failure where it belonged, upon the Record, upon the rascals who, through it, had perpetrated fraud upon them, and upon the particular animals which had been imposed upon them—instead of placing the cause of failure where it belonged, they, still relying upon the fidelity of this record, the integrity of the men who sold them the cattle, and the purity of the animals themselves, committed the monstrous error of ascribing their failure to a want of merit and excellence or adaptability in the great race of Shorthorn cattle. And their desire for improvement not being satisfied, there immediately sprung up a demand for some other breed of cattle which does have the power to favorably impress itself upon the stocks with which it is interbred.

"The respective merits of different breeds of cattle need not be brought into this discussion, and would indeed be foreign to it; but the fact cannot well be disputed that the Shorthorns had such a start, both in respect to the estimation of the public regarding their merits and the number of animals and breeders, that the advancement of any other breed of cattle to a position of nominal rivalry should have been very slow and difficult, if not impossible. And it would have been impossible, had the estimates of everybody regarding the character and adaptability of Shorthorns been based upon experience with genuine specimens of the breed. And if the facts could be all ascertained, it would doubtless be found that those whose experience with Shorthorn cattle is said to have been unsatisfactory in an intelligent effort to put them to practical use, have not generally had in their possession well-bred representatives of the breed. They were *imposed upon with impurely bred cattle,* and, disappointed in the results, have acquired impressions concerning the Shorthorns which not only do this whole breed of cattle, but the breeders of it, great ininjustice. The extent of this injustice can scarcely be estimated; but in the grazing regions, individuals here and there are paying for bulls of other breeds two or three times the price for which Shorthorns can be obtained; and among the general farmers everywhere throughout the country the same false impressions, springing from precisely the same source, are operating to discourage the purchase, restrict the demand for and depreciate the price of every Shorthorn calf that is dropped.

"This unfortunate condition of affairs has been growing worse from year to year; and it is high time the breeders, acting through a regularly organized association, should move for a reformation of the record, and the protection of their personal and pecuniary interests, which are being so seriously prejudiced. To hesitate longer about applying a remedy would be little less than criminal.

"There must be a public record of pedigrees. This record must command the confidence of the public. To command this confidence the record must be conducted with honesty, with care, and without bias or prejudice of any sort. *A dishonest management cannot be guarded against as long as the record is the private property of one man, who can admit or reject at pleasure any pedigree offered.* Care in its

T. F. B. SOTHAM.
Chillicothe, Mo.

compilation cannot be secured where the compiler is *responsible to no one but himself for its accuracy.* It cannot be said to be free from bias where the editor who passes upon a pedigree receives a dollar if he approves of it and loses a dollar if he rejects it. It cannot be said to be free from prejudice where the rules for the *admission of pedigrees are made to conform to the views of one man,* and the great body of breeders whose interests are affected have no means of expressing or enforcing their views of the matter.

"Thus it is seen that to secure for the record of pedigrees the character, reliability and quality necessary to such a record, and also necessary for the protection and advancement of the interest of breeders, it must be controlled and published by an association of the breeders themselves—and in no other way can the desired end be obtained; and as the matter is discussed there can be no question about the great body of breeders approving the action had at the Jacksonville meeting of the association, looking for an association record. It can be made purer, more reliable, more accurate, more systematic, more methodical, and, therefore, more intelligible than under any merely private management. At the same time it will cost less money, and be a bond of union between breeders sufficient to create and maintain a strong and active association, the existence and work of which will be a source of pleasure and profit to all legitimate breeders.

"There are some other features of the matter referring to how the record should be conducted, what it should be, and how it can be reduced in bulk and cheapened in price, that are reserved to another occasion."

This is Mr. Rust's entire article, from which we quoted, of which quotations Mr. Rust complained.

We now give our quotation, as found on page 43, Vol. III, of the "Breeders' Live Stock Journal" for January, 1882:

"Geo. W. Rust.—Many who read the above name will remember the man who bore it, as the editor and maker of the 'National Live Stock Journal,' a man who wrote fearlessly and ably. We met an article of his written in the interest of the new movement for a Shorthorn Herd Book. In speaking of those who have endeavored to improve their stock by using Shorthorn bulls, he says: 'They had heard and read much of the excellence of Shorthorn cattle, and felt they had a right to realize something of it in their own herds. But time brought only disappointment. And they committed the monstrous error of ascribing their failure to a want of merit and excellence or adaptability in the great race of Shorthorn cattle; and their desire for improvement not being satisfied, there immediately sprang up a demand among them for another breed of cattle which does have the power to favorably impress itself upon the stock with which it is interbred.'

"Quoting this much from Mr. Rust's article, we will say that it is well stated, and that Mr. Rust has always been considered good authority. But it is due to Mr. Rust to say that he ascribes this failure to the old Herd Books, and not to the Shorthorn race of cattle. But the fact remains, and the farmer or bullock breeder does not care whether this inability of the Shorthorn race of cattle to improve the common or native cattle of the country comes from an original defect in the breed, or the admitting of bogus Shorthorns to record, and therefore giving them a diploma to go forth to deceive the purchaser; or whether it comes from speculative malpractice in breeding. The cause is immaterial—the fact remains the same. They fail to improve the stock upon which they are bred, and the breeders want something else—some breed that will favorably impress itself upon the stock upon which it is bred, if they have to pay two or three times as much as they can buy Shorthorns for.

"Mr. Rust intimates that the manner to reform and improve the Shorthorns is to establish another record—of course, if there is to be any improvement it must be by recording only the best Shorthorns.

"The Shorthorn men are in a bad fix, and they will find it difficult to get out of it by using Shorthorn bulls. Their only sure and quick remedy is by using Hereford bulls upon Shorthorn cows. Many have accepted our advice and are taking this course, and are finding good results.

"A prominent bull dealer who has been selling Shorthorn bulls to Wyoming, advises the bull breeders to breed Hereford bulls to their Shorthorn cows if they wish a better price. And another equally prominent bull dealer and breeder, who has been selling bulls to Wyoming, says Herefords are a failure in Wyoming, but he knows what he states is not true, and many of whom he has bought Shorthorn bulls are buying Hereford bulls, and some of them are getting thoroughbred cows as well.

"We would advise the Shorthorn breeders to take the Hereford remedy and not to try and patch up a breed with consumptive Shorthorn bulls, and establish a cross-breed record. Start new, gentlemen.

"We will quote Mr. Rust again. He says: 'This unfortunate condition of affairs has grown worse from year to year, and it is high time the breeders, acting through a regularly organized association, should move for a reformation of the Record, and the protection of their personal and pecuniary interests, which are being so seriously prejudiced. To hesitate longer about applying the remedy would be little less than criminal.'

"Too late, George. The horse is stolen; no use locking the barn now."

We now quote Mr. Rust, of February 7, 1884, in the "Breeders' Gazette":

HEREFORDS VS. SHORTHORNS—MR. MILLER AND MR. RUST.

"There is no form of misrepresentation more annoying to a man than to be perpetually misquoted, or to have one's expressions garbled and used in a different sense from that in which they were originally employed. And for a year or two past—I don't know how long—I have been annoyed by the persistent efforts of Mr. T. L. Miller, in a paper published by him in the interests of his herd of Hereford cattle, and for the advancement of his private interests as a breeder, to place me in the position of having asserted certain things with reference to Shorthorns and Herefords which I have never said or written, and which Mr. Miller knew I never intended to say and did not believe. During the past fifteen or sixteen years I have had frequent occasion to write something of these cattle, and scanning this matter, Mr. Miller selects isolated sentences here and there, which, by separating from their context and using on a different subject from that of which I was treating, he attempts to make me say, what he would like to have me say. And not only this, but when my language does not entirely suit him, he does not hesitate to change it—to rewrite my sentences—to put parts of different sentences and paragraphs together; and make up such a statement as he pleases, and then to quote the whole as my words; and if they are my words, they must, of course, represent my opinions. In the last issue, now before me, in an article on 'Beef Breeds,' he says:

"'One has but to learn from the Shorthorn journals that their reputation is gone, and they may account for it by bad and dishonest practices, bad breeding or any other reason. The fact remains, and there is no better expression in accounting for it than that of Mr. Geo. W. Rust, in a letter to the "Breeders' Gazette," in one of the earliest issues, in which he says that "farmers had heard much of the Shorthorn breed of cattle and of their ability to improve the common stock of the country, and having tried them and been disappointed, they were seeking some other breed that would accomplish this purpose."'

"Now, I never wrote any such stuff to the 'Gazette' and never expect to, until I lose all my sense of honesty, together with the use of my eyes and my ears. And I do not remember to have written but one article over my proper signature for the earlier numbers of the 'Gazette,' and I find but this one in looking over the files, and the above quotation bears evidence of having been twisted out of that article. It was on the subject of the Shorthorn Herd Book —an entirely different subject from that to which Mr. Miller applies my language, after having perverted it to suit his purpose. It will be found on page 29, of Vol. I; and I trust I may be given space for a brief extract or two.

C. B. SMITH,
Hereford Park, Fayette, Mo.

"I had stated in substance, that the demand for bulls among farmers and graziers should bring a rich reward to those who had cultivated and maintained the excellence of Shorthorn cattle; but breeders found themselves brought into competition with other people with impurely-bred [yet these were recorded as pure—T. L. M.] bulls to sell, which, in the unfortunate condition of the records, they were able to substitute and sell for pure Shorthorns. And then comes the following paragraph:

"'But time tries everything; and *the people who had been imposed upon with spurious cattle,* and those who had been deceived into supposing they were introducing well-bred bulls into their grazing herds, failed to realize the anticipations which they entertained. They had heard and read much of the excellence of Shorthorn cattle, and felt they had a right to realize something of it in their own herds. But *time brought only disappointment.* Unfortunately, however, instead of placing the cause of failure where it belonged, upon the Record, upon the rascals who, through it, had perpetrated the fraud upon them, they, still relying upon the fidelity of the Record, the integrity of the men who sold them the cattle, and the purity of the animals themselves, committed the monstrous error of ascribing their failure to a want of merit and excellence or adaptability in the great race of Shorthorn cattle. And their desire for improvement not being satisfied, there immediately sprang up a desire among them for some other breed of cattle which does have the power to favorably impress itself upon the stocks with which it is interbred.'

"So far as the quotation Mr. Miller assumes to make from me is based upon anything I have said or written, it is based upon those portions of the above paragraph which are printed in italics. It will be observed that aside from the changing of expression, he purposely omits the context and matter coming between the portions which he pretends to take in such a way as to give my meaning precisely the reverse.

"I need not comment upon the unfairness of this sort of misrepresentation, but I wonder whether it is necessary to make the same allowance for other things which the same gentleman states. In his own testimony about the value and merits of Hereford cattle, does he represent the real bottom facts, as accurately as he does what I have written? He publishes from time to time testimonials and letters from people showing the excellence of Hereford cattle, and I wonder if he transcribes what they have written with the same candor and fairness as he displays in rewriting my sentences.

"Is it necessary to resort to these tricks and subterfuges, to descend to this jugglery of words, in order to make and sustain a reputation for Hereford cattle? I hope not. I believe the cattle to be better than one would infer from the tricks and tactics to which Mr. Miller finds it necessary to resort in their behalf. They ought to be good enough to make their way in public, holding up their heads with all other breeds, and standing solely upon their merits, with all the facts known, and everybody's experience and opinion fairly stated and considered. They ought to be good enough for this, but Mr. Miller seems to think they need some additional support, of a kind which he is specially qualified to give.

"I do not suppose my opinion upon the comparative merits of Hereford and Shorthorn cattle is worth anything to anybody, even if I have given the matter some attention, and for that very reason the misrepresentations of Mr. Miller are the more exasperating.

"Geo. W. Rust."

OUR REPLY.

We did not give the article in full from which we quoted in January, 1882, because of its length, and not because we were not willing to put all that Mr. Rust had said before our readers. We submit that the article as a whole is more damaging to the Shorthorns than the quotations that we made, and that we did Mr. Rust full and ample justice in the quotation by saying, "But it is due Mr. Rust to say that he ascribes this failure to the old Herd Books and not to the Shorthorn race of cattle. But the fact remains, and the farmer or bullock breeder does not care whether this inability of the Shorthorn race of cattle to improve the common and native cattle of the country comes from an original defect in the breed or the admitting of bogus Shorthorns to record, and, therefore, giving them a diploma to go forth to deceive the purchaser, or whether it comes from speculative malpractice in breeding. The cause is immaterial—the fact remains the same."

We have always had great respect for Mr. Rust's abilities, fearlessness and integrity, and we may have occasion to still further quote from what he may say or what he has said in times past. Now, if it is true that it is difficult to tell where the bogus Shorthorns are, and where the true ones are, it is immaterial how this difficulty originated. That it exists, Mr. Rust himself admits, and he is still writing in the current numbers of the "Gazette" on the inaccuracies, errors and frauds of the existing Herd Book, and in the article from which we quoted he was urging that the Herd Book should be in the hands and under the control of breeders for the purpose of purifying it of existing errors. Since that time the Shorthorn Society has bought the existing Herd Books with all these widely advertised frauds and errors and adopted them without revision as the standard, and if a new beginner, or an old beginner, wishes to select an animal he must go to that record, and it is folly for Mr. Rust or anyone else to undertake to claim merit and value in the Shorthorn breed and still admit that parties have been selling bogus, impurely bred animals, and that the Herd Books have made a record of these until it is difficult to tell which are pure and which are impure.

We should not take up so much room in this matter except for the high standing that Mr. Rust has occupied in the live stock interest of this country. No man knows so well the frauds that have been committed in the Shorthorn interest as does Mr. Rust. No man understands better the great damage that has been done to the breed through the bad practices and speculative tendencies of those who have been leaders in the Shorthorn movement for years past, and it is true, we think, that no man has done more to expose these frauds than Mr. Geo. W. Rust, unless it is ourselves. And we will here refer to what we said of Mr. Rust in the February number of the "Journal," page 77. In speaking of the Shorthorn Herd Book he says:

"It affords no means of ascertaining anything beyond the names of the various animals in the successive crosses and the names of their breeders, and these mere names repeated in a meaningless way from volume to volume, six, seven, ten or twenty crosses, all detailed with careful perspicuity, long lists of names of bulls and cows with no information as to whether either had anything beyond their paper record to show they were Shorthorns."

Later he says: "I made the statement that while all public records were more or less defective, the system of Shorthorn records, while most important of all because of the number and amount of capital invested in the breed, was worst of all."

We submit the foregoing facts to the public and our readers. The entire article from which we quoted is much more damaging to the Shorthorn interest than where we left it. While it is true that the article was intended to discredit the Herd Book, it still brought out most clearly, and from the most authoritative source, the frauds and speculative practices of those persons who had been engaged in breeding and selling Shorthorns, and the movement which Mr. Rust hoped would remedy these evils, instead of purging itself, has adopted the very record which he condemned.

Mr. Rust when writing this correspondence, was residing in Boulder, Colorado, where he had unsurpassed opportunities to personally inspect the sorry state Shorthorns got into when left to rustle for themselves on the open range. Truly a more forlorn spectacle cannot exist in the eyes of a cattleman than a herd of Shorthorns in winter on the range; too shiftless to range for a living and literally waiting for death.

Our quotations of Mr. Rust seemed to him to make it imperative that he get something in print derogatory to the Herefords to counteract the light we had set him in, and the following is republished from the "Breeders' Journal" for June, 1884:

Geo. W. Rust gives what he terms some "Practical Experiences" as to the Best Plains Cattle, and is somewhat personal in his remarks; but we give the following extracts from the article in question, which appeared in the "Breeders' Gazette" of May 29th, 1884. The article referred to commences by saying:

"I always enjoy a talk with Mr. Carey Culver, whom I call my neighbor, although he resides twenty odd miles away, over on the Big Thompson, who, with his brother-in-law, Mr. Mahoney, a business partner, were among the earliest introducers of improved cattle upon the plains, because he has had an experience antedating and covering more years than anyone I know about, concerning the adaptability of different breeds of cattle for use on the plains, and he never fails to tell me something which interests me.

MURRAY BOOCOCK.
"Castalia," Keswick, Va.

"He was at my house recently and I had a more than usually interesting cattle talk with him, in which he gave me a more connected account of his cattle operations and more of the results of his experience and observation in breeding cattle than he had ever felt inclined to impart. I showed him a recent number of Mr. T. L. Miller's Hereford paper (the 'Breeders' Journal') in which that gentleman states: 'We think Messrs. Culver & Mahoney took Herefords first, but they felt that the price was so high that they could not afford to buy, and being able to buy Shorthorns and Scotch cattle they took them, and this has been true, we know, of some other firms.'

"I expressed my sympathy to Mr. Culver at the poverty which prevented him from buying such cattle as he felt he needed in his business."

We can imagine Mr. Rust meeting Mr.

Culver on the street and condoling with him upon what Mr. Miller had said as to his poverty, and inviting him to his house to show him the "Journal," and warming up on this subject, hoping thereby to elicit something in favor of the Shorthorn cattle. We would say in reference to the foundation for our remarks on this that somewhere in 1874, we should think, we met Mr. Culver at his home, and in discussing the propriety of buying Herefords he said that we held them too high; that he could buy Scotch cattle for less, and he thought they would do him just as well. We will see further on whether Mr. Culver was correct in his conclusions in 1874. It is not Mr. Culver alone, but the ranchmen state distinctly that they cannot afford to buy thoroughbreds (and many of these are perhaps more wealthy than Messrs. Culver and Mahoney), and therefore buy grades, and when they cannot afford to buy grade Herefords they take Shorthorns.

Mr. Rust goes on to state that Messrs. Culver and Mahoney bought, in 1873, Shorthorns, Scots and Herefords; that they used these for about four years in their herd at home and in Wyoming, and at that time sold to Mr. Alex. Swan their herd of range cattle at $24 per head, when the largest price that had been paid before was $16, and that Messrs. Swan were so well pleased with the Hereford cross that they found in this herd that they said to Messrs. C. & M. that if they could get Hereford bulls and put them upon their cows they would buy all the bulls that they could raise, and as the result of this proposition Messrs. C. & M. bought two bulls for themselves and one for a neighbor, Mr. Blore; that Messrs. Swan took the crop of bulls and sold them to other parties (we presume at a profit). Two of the steers that went to Messrs. Swan in the herd sale, it is stated, went to John B. Sherman, of the Chicago Stock Yards, and were shown at the Fat Stock Show. (We think this is a mistake, but as to the quality of the steers there is no question.)

THE HEREFORD CROSS.

Mr. Rust says the cross of the Hereford upon the Shorthorn cow was also productive of the most satisfactory results. These Hereford bulls that Messrs. C. & M. bought to meet the proposition of Swan Brothers to raise bulls for them were bought of T. L. Miller, and Mr. Culver persuaded Mr. Dick Blore to buy one also, and Mr. Blore has been so well satisfied of the advantage of using a Hereford bull in preference to a scrub that he keeps a Hereford bull to this day, and he walks over to Mr. Culver's fence and points out to Mr. C. the grades from Culver's Shorthorn or high-grade Shorthorn cows as specimens of Hereford breeding.

This is fairly quoting Mr. Rust on the Hereford cross on Shorthorn cows as exhibited on the Big Thompson. We will now follow this herd still further in the hands of the Messrs. Swan. Mr. Rust says that these gentlemen (Messrs. C. & M.) bought these different breeds of bulls, Shorthorns, Scots and Herefords, in 1873, and that in about four years they sold to Messrs. Swan for $24 a head, when the highest price at which cattle had sold before had been $16 per head. This would have brought them to 1877. In that year Mr. Swan came to T. L. Miller and bought about fifty Hereford bulls, and we presume and have always understood that his purchase was occasioned by the very good results that he found of the Hereford cross upon the Culver & Mahoney herd, and it appears at the same time he made a proposition to C. & M. for all the Hereford bulls they would raise.

At about the same time he made a contract with Mr. C. M. Culbertson for his crop of grade bull calves, but took a lot of thoroughbreds, and in a year or two after the first purchase from T. L. Miller he bought another lot of fifty, and again in 1881 or '82 a similar lot. These were some of the results that grew out of Messrs. Culver & Mahoney's purchase of Shorthorns, Scotch and Hereford thoroughbred bulls in 1873, and the Swan interest was so well pleased with the result of these operations that in the fall of 1882 they went to England and the result was the purchase of nearly four hundred cows to establish a herd of thoroughbreds at or near Cheyenne, and again in 1883 they brought over another large shipment, reaching to nearly, or quite, three hundred head. They are probably using in their different herds five hundred thoroughbred Hereford bulls, and a thousand or more high grade Herefords.

Messrs. Culver & Mahoney ought to be very proud of the great results that have come from their purchase in 1873, and we hope that the Hereford breeders of America will appreciate these efforts of Messrs. C. & M. and that they will not be unmindful of their obligations to Mr. Geo. W. Rust for opening up such a mine. Mr. Rust says, near the close of his article, "I should be pleased to gather from the 'Gazette' the real experience of the other firms which are quoted as not buying Herefords for the same financial reasons, and still more gladly would I give if the true inwardness of the thing could be arrived at, the private opinions and the bottom experiences of those who did buy Herefords and are suspected of being constrained to stay with them for the same financial reasons." Perhaps the "Gazette" and its Shorthorn patrons were satisfied with what Mr. Rust had already done, and asked him to leave the matter for Mr. Miller or someone else to develop.

Near by Messrs. Culver & Mahoney's home Mr. Geo. Zweck has a herd of cattle, and the first bulls that we sent to the plains in 1873 were sold to Mr. Geo. Zweck, a man who is fully posted in the cattle business. These bulls were Plato 590, of American Hereford Record, Duke of Beaufort 744, A. H. R. All of these bulls were in Mr. Zweck's possession doing good service up to the fall of 1882.

Plato when sold was three years old, and in the summer or fall of 1882 for some cause that we do not now recollect, died, but not because he was not vigorous, at the age of 12 years. Both of the other bulls were still in service, the Duke of Beaufort, 10 years old, and Hervey, 11 years old.

In 1876 or 1877 we sold to Mr. Zweck another bull, and Mr. Zweck won as a premium for the best show of grade Herefords another Hereford bull that we offered at the Colorado State Fair, taking five bulls in all. One of these last bulls Mr. Zweck lost by eating poison weed, and the other wore out to a good old age in his service; the other, Major, dropped in 1877, by Old Success, worked over ten years and was a bull in which Mr. Zweck took a great deal of pride. From these bulls we understand that Mr. Zweck has made very satisfactory sales of bulls, from that time to the present, and at last report we had of him he had something over 1,000 good grade Hereford heifers, the get of these bulls, and he was negotiating in the fall of 1882 for ten thoroughbred Hereford bulls to put into his herd. We urged upon Mr. Zweck from time to time that he get new bulls for this herd, and have said to him that it would inure very largely to his income if he would do so. Mr. Rust refers to Mr. Geo. Zweck in the following language:

"A neighbor of Mr. Culver's, Mr. Geo. Zweck, says the way to breed good Herefords is to use Shorthorn cows, but perhaps Zweck ought not to be quoted, although he has handled and bred the Herefords and their grades quite extensively for many years, for George is now getting on the fence with some leaning towards Shorthorns, and has lost caste with T. L. Miller, and perhaps other authorities, and Geo. F. Morgan declares that Mr. Miller asserts that Zweck never had any pure Herefords from him. However this may be, Mr. Zweck has had three or four bulls from Mr. Miller. One of them, Plato, had little short horns, standing forward and inclined inward, and everybody always said he had Shorthorn blood in him, although Zweck always stoutly maintained that he was straight Hereford and obtained at headquarters of the breed. Another Hereford bull owned by Mr. Zweck was a prize bull offered by Mr. Miller for the best display of Hereford calves at the Colorado State Fair. Zweck made a large display and carried off the prize in high glee. He is especially displeased at being told now that his calf was not pure."

To this our answer at the time in the "Breeders' Journal" was:

"Now, all that Mr. Rust says about Mr. Zweck's bulls not being pure is made up from Shorthorn sources. Mr. Zweck is not on the fence. Mr. Morgan never said that Mr. Miller asserted that he never sold Mr. Zweck any thoroughbred bulls; Mr. Miller never asserted any such thing, and Mr. George Rust never believed when he gave these statements currency that they were true; and we here give the pedigrees of each of these four bulls, and if Mr. Zweck will give us the name of the other bull we will publish that.

"Plato (4843) 590, A.H.R. Bred by G. W. Byers, Nevada, Ohio; property of Geo. Zweck, Longmont, Colo.; dropped Oct. 27, 1870. Sire, Guelph 461 (2023); dam, Beauty 2d 3; g.d. Beauty 1st 7; g.g.d. Duchess 15; g.g.g.d. Countess 31; g.g.g.g.d. Venus 4th 61; g.g.g.g.g.d. Venus 3d 109; g.g.g.g.g.g.d. Venus 2d 175; g.g.g.g.g.g.g.d. Venus 251.

"Duke of Beaufort (4527) 744. Bred by J. Humphries, Elyria, Ohio; property of Geo. Zweck, Longmont, Colo.; dropped March 15, 1872. Sire Marquis of Bath 745 (4764); dam Louisa 598; g.d. Princess Louisa 595; g.g.d. Princess Royal 488; g.g.g.d. Victoria 478; g.g.g.g.d. ——— 491; g.g.g.g.g.d. ——— 503; g.g.g.g.g.g.d. ——— 518.

"Hervey (4644) 815. Bred by T. L. Miller, Beecher, Ill.; property of Geo. Zweck, Longmont, Colo.; dropped March 6, 1873. Sire Sir Charles 543 (3434); dam Sophia 754; g.d. Blossom 532; g.g.d. Duchess 15; g.g.g.d. Countess 31; g.g.g.g.d. Venus 4th 61; g.g.g.g.g.d. Venus 3d 109; g.g.g.g.g.g.d. Venus 2d 175; g.g.g.g.g.g.g.d. Venus 251.

"Major 1027. Bred by T. L. Miller, Beecher, Ill.; property of Geo. Zweck, Longmont, Colo.; dropped Feb. 2, 1877. Sire Success 2 (5031); dam Fanny 541; g.d. Princess Mary 786; g.g.d. Princess Royal 488; g.g.g.d. Victoria 478; g.g.g.g.d. ——— 491; g.g.g.g.g.d. ——— 503; g.g.g.g.g.g.d. ——— 518.

"Every Hereford breeder, or any man that is

at all acquainted with Hereford breeding, will see by these pedigrees that they were among the very best breeding in England or America, and with these remarks we leave Mr. Rust to make what he can out of it.

"Near by Messrs. Culver & Mahoney's and Mr. Zweck's, say twenty-five miles distant, on the road to Denver, lived a Mr. Church, who bred Herefords before Messrs. Culver & Mahoney or Mr. Zweck, and we presume their preference for Herefords was largely from what they saw of the results of Church's work. Of Mr. Church Mr. Rust says, in speaking of a cow that had come into the possession of Mr. Culver from the Church herd:

"'Speaking of the Church Hereford cow brings to mind the bad luck of the Church herd. It was a very fine one, but the grass on the home range did not appear to maintain it well and so Mr. Church moved it over onto the rich grass on the slope west of the range, and the next winter most of them died. I don't know that Shorthorns would have done any better under the same circumstances, but no breed could have done much worse.'

"We would say in reference to Mr. Church that he bought from Mr. Stone of Canada, a Shorthorn and a Hereford bull, as early as 1870, possibly earlier, and he went over into York State and bought two Hereford cows, one of which had a Hereford bull calf by her side. This comprised Mr. Church's investment in Herefords, and his experience as between the Hereford and Shorthorn led him to discard the Shorthorn and use Herefords solely. His ranch was twelve miles north of Denver on the road to Longmont. He bred them on that range up to the fall of 1874, we think, and the range becoming short he moved them east of Denver to Hugo; but before moving, he took out two or three car loads of three-year-old grade Hereford steers that he sold. They went to Buffalo and sold on the market with no other feed than they had taken from the short range, at seven cents a pound live weight, and brought an average of some $90 at three years old.

"When Mr. Church moved his cattle to Hugo other ranchmen made great objections to their coming onto that range; but they went, and in the spring following at the round-ups the Church cattle were in so much better condition than any others that nearly all of the ranchmen on that range and in those round-ups became converted to the Hereford breed of cattle. Mr. Church stayed on the range near Hugo until he moved his cattle to the west side of the mountains. We have always understood that the object in moving was to reach a point where he could keep his own herd by themselves and make better improvement, and get the benefit of his own bulls more fully than he could on the range near Hugo.

"We have met a gentleman from the range since this letter of Mr. Rust's was published, and have learned that Mr. Church, in moving his cattle, reached a point of the range which the Indians had burnt over, and relying upon the hardihood of the cattle, he started to drive them through this burnt range and get beyond it to where he could find feed; that he drove for some six days without feed or water, and then, not knowing how much further he would have to go, he turned and drove them back. This resulted in a great loss, and the next winter was a severe one, with heavy snows, which prevented the cattle getting any feed, and a large portion of the remaining cattle were lost, owing largely to their sufferings on the burnt range. This we suppose to be a true version of the Church loss, but, of course, Mr. Rust did not understand anything about these facts, or he would have stated them and not left his Shorthorn friends to build an argument on the loss being occasioned by ordinary circumstances. So much for that part of Mr. Rust's statement.

"There is in Estes Park, perhaps thirty or forty miles from Messrs. Culver & Mahoney's, another herd of Hereford cattle, known as the 'James herd of Herefords.' There have been connected in this herd with Mr. James Messrs. Ewart & Hart, and the last season Messrs. Ewart & Hart bought out the James interest in the Herefords, as we understand. Mr. Hart was at Beecher last fall with a view to buying Hereford bulls to give fresh blood to this herd. He did not buy at Beecher, but we understand that he bought elsewhere, and he proposed to sell the bulls that they had been using.

"In the May number of the 'Journal,' on page 305, Messrs. Ewart and Hart offer a choice lot of grade Hereford bulls well bred, well marked, and in fine condition, and four choice thoroughbred Hereford bulls for sale; and in sending their advertisements they said nothing of any losses that might have been had."

In the winter Mr. James was at Beecher looking for Hereford bulls, and (¶ 368) was buying a large number of cows and heifers in Iowa, largely of the Shorthorn blood, to take to Colorado to establish a new herd. He did not buy his Hereford bulls at Beecher, but he bought them of Mr. T. J. Lewis, of Odebolt, Iowa.

Mr. Rust says: "The James herd of Herefords and high grades in Estes Park is said to have lost 100 head out of 140 during the past winter," and he says he "does not know that Shorthorns would have done any better under the same circumstances, but no breed could have done much worse."

We wrote a letter of inquiry to Mr. James which, with his reply, we give as follows:

BEECHER, ILL., May 31, 1884.

Mr. W. E. James, Hotel-keeper, Estes Park, Colo.:

Dear Sir:—I have a report written by Geo. W. Rust, from Boulder, that the James herd of Herefords and high grades have lost 100 out of 140 head of cattle. Please advise me what foundation there is for such report, and the circumstances and oblige. I understand you bought a lot of thoroughbred bulls from Mr. T. J. Lewis, of Odebolt, Iowa. Any information in reference to the cattle interest will be appreciated. My bulls have all been sold that are old enough for service.

Very truly yours, T. L. MILLER.

ESTES PARK, COLO., June 7, 1884.

Mr. T. L. Miller, Beecher, Ill.:

My Dear Sir:—Yours of May 31st just at hand and contents noted. I know no reason for such a report, as there has been no such loss in Estes Park. In fact, 100 head will more than cover the whole loss of the Park. My loss was twenty-seven head out of 144 head of Iowa heifers, and most of those were heifers that got injured in shipping. Yet, we have had a very hard winter for this part of the country. I believe the loss of cattle to be very large in North Park, as the snowfall was very heavy in that Park and laid on the ground for about three months. Our Hereford cattle have done very well indeed the past winter, and there has been a lively inquiry for young animals, both thoroughbred and grades, but they could not be found sufficient for the demand, which ought, and shortly will, stimulate the breeding of more Hereford cattle.

Yes, I with my friend, Mr. P. J. Pauly, Jr., bought fifteen head of thoroughbred Herefords from T. J. Lewis, of Odebolt, Iowa, and I wrote you from Marshalltown, Iowa, of my purchase, stating that I should not be back to Beecher on that trip. I think I wrote you on the 29th day of December. Glad to see that the demand has been good for thoroughbred stock. I remain yours sincerely in anything that can be truthfully said or done for the white-face cattle.

Yours respectfully, W. E. JAMES.

In Vol. V, page 593, of the "Breeders' Journal," we republished the following letter from Mr. W. E. Campbell to the "Breeders' Gazette":

"Much has been said of late through your columns in reference to Shorthorns and Herefords as range cattle by Geo. W. Rust, and others who have had little practical experience with rival breeds upon the plains, where cattle are compelled to rustle for a living, not wholly when the air is soft and balmy and the grass green and abundant, but when it is dry and hard and even covered with snow for days and perhaps weeks at a time, the thermometer indicating 20 degrees or more below zero, driven before pitiless winds or forced to seek shelter beneath the rugged banks of hollows and ravines, with only snow for a bed and an angry sky for a cover. These are the conditions through which all range cattle are compelled to pass or perish; and this is why our ranchmen have been testing the merits of rival breeds, and have generally adopted the Herefords on account of their hardihood, activity and self-reliance in time of need. This, coupled with their aptitude to fatten on grass without other feed, and their wonderful impressiveness as sires, will always make the white-faces more popular than any other breed with our ranchmen.

"I was once as staunch an advocate of Shorthorns as my friend Mr. Rust. Being familiar with them from early childhood I almost lived and swore by them, and I was one of the first men to attempt the improvement of Texas and range cattle in the Indian Territory by the use of thoroughbred Shorthorn bulls. My attention was first called to the Herefords by the early articles of T. L. Miller in a paper published in your city, and though I thought his claims very extravagant and unreasonable, I decided to test their merits at my own expense for my own satisfaction and, notwithstanding the abuse and criticism I then received for using my own sweet counsel, I have never had cause to regret my experiments, one of which I will relate to your readers just as the facts occurred, and they can draw their own conclusions.

"After making a number of small tests that were decidedly favorable to the Herefords, I determined to test the breeds thoroughly on a rigid and extended plan. I therefore went East and bought a carload of twenty-six Shorthorns and a carload of twenty-five Herefords and shipped them on the same train to the terminus of the railroad, and then drove them some ninety miles to my ranch, where they were all turned loose with my ranch cattle. When the heat of summer came the Shorthorns could be seen standing along the streams or in the shade, while the Herefords were busy grazing or rustling around, endeavoring to obey the first and most important commands of scripture—in reference to multiplying and replenishing the earth. Both breeds were allowed to remain on the open range the entire winter without any artificial food or shelter of any kind, and were compelled to rustle for a living or die. The winter proved to be one of unusual severity, and before spring came almost fifty per cent of my beloved Shorthorns had died, and the remainder were but reeling skeletons. With the Herefords the test was perfectly satisfactory, and every one of the twenty-five showed up in good shape, a number of them being fit for the butcher's block by the 10th of May. This and previous tests satisfied me that Mr. Miller's extravagant claims for the white-faces were nothing short of facts when applied to them as range cattle; and I have been a friend to them ever since.

"Notwithstanding the foregoing I am still a friend of the Shorthorns, and think them a grand and useful race of cattle in their proper place; but experience compels me to differ with my friend Rust, and to say most emphatically that their place is not on the plains. The Hereford bull is king of the range, and Mr. Rust will live to hear him bellow triumphantly over every grazing region from the snow-capped peaks of Mexico on the south to the British possessions on the north.

"(Signed) W. E. Campbell (¶ 369)."

Few men have had the long and varied experience of Mr. Campbell, and at this writing (1898) he is still ranching and breeding Herefords in the vicinity of his early range experience. It is a pleasure to us, giving the utmost satisfaction, to recall the countless friends of the Hereford, who have become their staunch advocates after ample experience with other breeds, that cannot be marred by even a very few contrary incidents. We have yet to learn of a single Hereford man who has abandoned his breed for another.

We commend this correspondence of Mr. Campbell's, selected for republication here from countless articles of less merit, because he was and is a master of the range business and he deals directly with facts. We ever had a contempt for flowery or caustic rhetoric such as is here used by Mr. Rust and has been used by Shorthorn advocates constantly and by some ephemeral champions of the Herefords.

This correspondence came out in the "Breeders' Gazette" and was republished in the "Breeders' Journal" in 1884:

To the "Gazette":

Theories invented by professional "quill drivers" look very well on paper; but as I am not skilled in this profession I am compelled to rely wholly upon facts and actual experience, to judge of the superiority of any breed as range cattle, and therefore will not attempt to theorize upon Mr. G. W. Rust's theories in your issue of October 2, but will cheerfully give further details as per his request, though I fear his controversies with Mr. T. L. Miller have so embittered him against the Herefords that it is impossible for him to look upon them or their friends with any degree of fairness. So far as I could learn, the twenty-five Herefords and the twenty-six Shorthorns referred to in my former article were fed and handled in substantially the same manner up to the time of their purchase, and were allowed to run in open pasture for sixty or seventy days prior to their shipment, after which they were turned in with my range cattle as heretofore stated. Now it may be that the twenty-five Herefords rustled round, and selected all the choicest morsels of grass and therewith put on a thick armor of tallow which withstood the fiercest storms of winter, while the unsuspecting Shorthorns were quietly snoozing.

But be this as it may, twelve of the Shorthorns died and the remaining fourteen were as poor as crows, and had not shed their old coat when the spring round-ups came.

On the other hand, every one of the twenty-five Herefords lived and were in good flesh, and as sleek as moles at the roundup. If one breed is as hardy as the other, as Mr. Rust claims, why was the mortality so great among the Shorthorns when there was none at all among the Herefords, on the same range and under precisely the same circumstances?

I am willing to admit my experience is somewhat limited, as I have only handled cattle on the plains for sixteen short years, and that tests made by such an amateur as myself are very tame affairs, when compared with the theories of one who has had long experience in cattle matters in the way of quill-driving. (¶ 370.)

In conclusion I will say, for five or six years past I have been steadily supplanting my Shorthorn bulls with Whitefaces, and last year alone I disposed of over 200 head of the aristocratic "red, white and roans," and hope to never own another. The Herefords are good enough for me, either on the broad prairie ranges, in our tame pastures, or in the feed yard.

By the way, I will state that I fed a car load of Shorthorn bulls last winter, including one disabled Hereford, all of which had been in service on my ranch up to November 1st. The Hereford received an injury about October 15th which resulted in a stricture, and in consequence had to be thrown, cut open and operated upon by a veterinarian at three different times. Notwithstanding these disadvantages and the tortures connected therewith, he laid on flesh much more rapidly than the Shorthorns, and when the lot were sold in Kansas City, the Hereford brought one and three-quarter cents more per pound than the Shorthorns, and all were sold to the same party.

Mr. Rust will in all probability theorize again, and theoretically prove to your readers that these surgical operations were an advantage to the Hereford, that it relieved him of a vast amount of bad blood which enabled him to fatten more rapidly and evenly, and to lay on a superabundance of flesh in the most valuable parts, and had it not been for this favorable chain of circumstances the Shorthorns might have out-sold him in the market.

W. E. CAMPBELL.

THAT EXPERIENCE OF MR. CAMPBELL'S.

To the "Gazette":

I have not, I believe, manifested any disposition to underrate the experience or question the accuracy of your correspondent, Mr. W. E. Campbell. And having mentioned my name, in relating his experience with Hereford and Shorthorn cattle, in such a way as to invite my examination of his statements, I think he should have done me the courtesy to have treated what I said in reply with the same candor with which I treated what he had said.

The only object I have in the discussion of these questions, and the only object anyone should have, is to get at the facts. If there is one breed of cattle superior to all others, it would be interesting to ascertain which breed it is; but we will never arrive at that knowledge without a most careful scrutiny of all the facts and experiences which lead up to it. If there are localities where one breed is more successful than other breeds, or purposes for which one breed is better adapted than other breeds, these facts should be shown, and they can only be arrived at by a patient and candid examination of the experiences of individuals. What I have said about the comparative claims of the several breeds has been dictated by a desire to bring out the real bottom facts, in each case, and not from any desire to advance personal pecuniary interests, for I have none whatever in the issue. (¶ 371.)

Mr. Campbell states that he turned out a car load each of Hereford and Shorthorn bulls; the Herefords grazed well during the summer, while the Shorthorns did not. The next winter was one of unusual severity, and about half the Shorthorns died, while all the Herefords came through in excellent shape. This I believe was about the substance of the statement of Mr. Campbell to which my attention was directed, and I have no doubt every word of it is true, and had it stood alone by itself I might have been disposed to accept it as conclusive on the point of indicating a superiority of one breed over another in hardiness, but it did not stand alone, and I was bound to consider it in connection with other facts and experiences. In the first place, it appeared from Mr. Campbell's own statement that he had for many years previously used Shorthorn bulls, and while he claimed to have found the Herefords more satisfactory, he did not state that in his long experience of sixteen years, he had ever before experienced losses anywhere approximating the case under review, so that even in his own experience this case seemed to be exceptional. Then I knew scores and scores of men who had used Shorthorn bulls with satisfaction upon the plains, and without any special loss, and some who preferred them to Herefords after having given both a trial. And under all the circumstances I could not but regard the case detailed by Mr. Campbell as exceptional and its results as due to something else besides a difference in the hardiness of the breeds. And I suggest that perhaps it was caused by some difference in the manner in which the two lots of bulls had been raised and treated; that animals reared mainly upon corn, are not well adapted for immediate range use, as compared with animals raised upon grass. But it was mere theory, it is true, in my supposition, that this may have been the trouble here, for it may have been something else, but my theory appears, for all I can see, to be as good as Mr. Campbell's, for he does not say he had any positive knowledge on this subject, or that having it in mind he made special inquiry on that point in purchasing. He simply does not know or has not been informed that there was any difference in the previous handling, and on the strength of that theorizes there might have been. But Mr. Campbell, if he desires to impress his conclusions upon the public as good, safe rules for other men to follow in their business, will surely not attempt to deny that the manner in which bulls are raised is a matter of some importance, and the attempt to sneer at it in mere theory will, I am satisfied, prejudice his position more than mine. People in this country who have examined and tested the matter to their satisfaction, are firmly convinced that the manner in which bulls are raised does have very much to do with their adaptability for plains use, and if the matter is new to Mr. Campbell I trust he will not dismiss it as a mere theory but, like an impartial investigator, look into it with the hope that after sixteen years' experience he may yet find "some strange things under the sun."

There was another thing referred to in my previous letter, to which Mr. Campbell makes no reply. I referred to the necessity in all cases of comparisons, where accurate results were desired, that there should be an equality of blood. Whether it was secured in this case or not, the public does not know, for Mr. Campbell gives no information as to where or of whom he purchased either class of bulls or whether they were both the best and most robust of their kind. I know Mr. Campbell has an ambition to own as good Herefords as anybody, and I understand his herd will compare favorably with the best in the land, and if I am not mistaken in the matter he has had an ambition for place as a successful exhibitor of Herefords where the competition was the strongest. This is, of course, all right, and I am glad it is so. But when he was using Shorthorns, did he have an equal ambition to own the best of that breed, and was his experience based upon careful tests as to what the best could do?

Mr. Campbell expresses a fear that my "controversies with T. L. Miller have so embittered me against the Herefords that it is impossible for me to look upon them or their friends with any degree of fairness," and yet, in the very article to which he is replying, I had referred to having seen a few days before "a grand herd of Hereford cows and calves at grass" at George Morgan's, and in the issue before that, had reviewed that Hereford herd at length, and in such terms as to elicit from Mr. Morgan personally the expression that it was the fairest and best article on a Hereford herd of cattle, which had appeared in the American press for many a day (¶ 372). However that may be, Mr. Campbell may rest assured that no man can so embitter me against any breed of cattle, that I cannot and will not recognize their merit when I see them. My experience is not so great, perhaps, as that of Mr. Campbell, and I may not perceive some things so quickly, but while I may not agree with him in many things, I trust he will not regard me as seeing but one side of any question. I should have had no controversies with T. L. Miller, nor had occasion to say many things which have been said, if he had not sought by persistent misquotation to make me appear as denying all merit in the Shorthorn, just as lately he would place me in the attitude of denying all merit in the Hereford. I am certainly entitled to standing room somewhere, and as I have never been the enemy of either I have the right to occupy my own ground as the friend of both, criticizing according to what I believe fair and just, and according to merit in all cases, whether as to the cattle themselves or the methods of those who are pushing their claims before the public, and entirely in this spirit of independence and impartiality I must again repeat that so far as I have been able to make up my mind, from the experience of a large number of people whose opinions I have sought on this question (and who have not been altogether harmonious in their views, some holding to one breed and some to another), I cannot see that there is any perceptible grounds for honest preference between Herefords and Shorthorns for plains use on the score of hardiness; that is, where animals themselves are in all respects equal and equally conditioned for the work. But it is undeniable that there are some Herefords better for this purpose than some Shorthorns; and it is equally true that there are some Shorthorns better for this purpose than some Herefords. But while I believe both to be sufficiently hardy, there may be some difference as to their feeding qualities, to which I have sought to direct the attention of Hereford breeders, as the question of importance. And this has been the main purpose I have had in view, and have continually brought to the front, and I do not think because I have urged the consideration of this question that my friendship for the Herefords should be challenged. How will they feed? Not wholly upon the plains, but as compared with each other, how will they feed when taken to the States to be finished? For their behavior there will be a very important element in determining values.

GEO. W. RUST.

THE PLAINS CATTLE PROBLEM.

To the "Gazette":

It is not creditable to Mr. W. E. Campbell's candor that he still declines to consider on its merits and continues to denounce as mere theory what I had said about the manner of rearing and previous treatment affecting the hardihood and usefulness of bulls brought upon the plains. It is not a mere theory, and the fact can be abundantly verified from the experience of any ranchman who has cared to investigate the matter. I regard the instance cited by Mr.

Campbell some time since, from his own experience, where two lots of bulls, Herefords and Shorthorns, were turned upon the range together—the Herefords all thriving and proving useful and valuable, while the Shorthorns did poorly and many of them dying the first winter—I regard this experience of Mr. Campbell as abundantly establishing the fact that the manner in which bulls have been raised and treated has very much to do with the value of bulls when brought upon the range. I say his experience proves this position, because there appears to be no other reason why the results attending the introduction of these two lots of bulls should have been essentially different, the experience of others not justifying the presumption that there is such a material difference in the breeds as would account for the results. And if necessary I could name many cases in which the facts support my position in this matter. Not to go outside of my own county I can say that Mr. Carey Culver purchased of the Illinois breeder Mr. Gillette, last December, some Shorthorn bulls which had been raised upon grass; turned them into enclosed fields where they roughed it all winter with the other cattle, and came through to the spring in fine shape and increased in weight. His brother, Robert Culver, brought in at the same time Galloway bulls purchased at public sale at Chicago, and evidently highly fed, and he had to feed them all winter; one of them could scarcely be brought through at all, and none of them were of any account until midsummer. At the same time Mr. Hart brought in some Hereford bulls, purchased of an Iowa breeder of high standing, who had probably pushed them along on corn, and they had to be kept up and fed all winter, and they came through in such wretched shape that Geo. Zweck, who kept one of them for its use until it was taken away July 1, was unable to secure calves from him, and had to send his cows to another bull. These three lots of bulls, representing these three breeds, came into this county within about ten days of each other last December, and were all kept within a radius of ten miles, and the difference I have related of their hardihood and vigor shows that the previous treatment to which they were subjected, was the important matter in determining their immediate and permanent usefulness, and that so far as mere hardiness is concerned the breed counts for nothing (¶ 373).

I am afraid Mr. Campbell does not care to profit either by his own experience or the experience of others else he would hasten to examine so important a matter as this as soon as suggested, instead of attempting to "sneer it down the wind" as a mere theory. And while unwilling to open his own eyes to all the facts, he does not appear to be willing that the public shall have the benefit of them. Believing that he had the very best Herefords which he could buy, I asked whether he was equally liberal and careful in his selections of Shorthorns, and whether the unfavorable experiences he relates of Shorthorns were based upon trials with the best Shorthorns he could buy. But in his reply he evades this point and fails to give the public the means of estimating at its worth the personal experience which he had advanced as showing the relative merits of these breeds of cattle. In a discusssion, where the demonstration of truthful facts should be the only object, this sort of evasion or concealment of important matters is not to be commended. No one should endeavor to win his case on technicalities. And besides I am told that Mr. Campbell, in giving his experience, has only given a part of it, and that a long time ago he confessed to Mr. Pliny Nichols, in Iowa, after having purchased two Hereford bulls, that one of them died before he got him home, or very shortly after, and that he had previously owned still another which had done him no good. As a candid gentleman Mr. Campbell should have stated such matters as these. To withhold them is to treat the public unfairly. In getting no benefit from two out of three, it would be interesting to know whether it was considered due to thê breed or to the manner in which these individuals had been handled before he got them.

In his last letter Mr. Campbell remarks:

"In the language of Napoleon I will say, 'I have but one lamp by which my feet are guided, and that is the lamp of experience.'"

I do not believe Napoleon ever said it. Patrick Henry had been in his grave many years before the great Corsican donned the purple, and I do not believe he ever had occasion to rob the dead American of his laurels. Mr. Campbell is unfortunate. His literary experience and observation do not appear to have been more careful and accurate than his observations and experiences with cattle, and both are equally in need of correction. I am not at all surprised at the confession of Mr. Campbell that he has been studying the cattle question by lamplight. It was quite evident that his range of vision was very limited, and the light he was enjoying very feeble. Perhaps he cannot stand a stronger light, but if he can, I would advise him to throw away his lamp, and come out into the daylight where he can see everything, and some things at least, in their true relations to each other. GEO. W. RUST.

MORE OF MR. CAMPBELL'S EXPERIENCE.

To the "Gazette":

In reply to Mr. G. W. Rust's letter which appeared in your issue of December 4th, I will say my former letter was written away from home while awaiting the arrival of a train, hence Mr. Rust's previous letter was not before me. I have since referred to it and find every essential point fully answered by well-authenticated facts. Had the gentleman but known my views in reference to the proper manner of breeding and rearing bulls for range use, he would have saved himself the trouble of writing a whole column to convince the public of my stupidity in not rushing headlong, at breakneck speed, to catch on to his wise suggestion. I have never deemed it necessary to dispute self-evident facts, which seems to be Mr. Rust's strong forte, from which he discharges his heavy theoretical guns. Now, in order to disabuse the gentleman's perverted mind and to set him right on the point in question, I will quote the following from one of my letters that went the rounds of the public press years ago, at which time (according to Mr. Rust's views) I was a competent judge:

"The weather by this time has become exceedingly hot, and the aristocratic bovine dukes and princes that have so suddenly been deprived of elegant quarters and epicurean diets begin to languish and dwindle away until there is nothing left but an unsightly rack of hair and bones. The result is nothing more than could have been expected under the circumstances. These animals have been bred and fed to sell, without any regard to their future usefulness. Many of them have been in the stalls all their lives, up to the very day of their shipment, and, as a natural result, have accumulated soft, spongy flesh and but little sinew, bone or muscle, so essential to all range cattle. Such experience has led many ranchmen to believe that thoroughbreds are not hardy or profitable for the plains, when in fact the fault should have been placed upon the breeder's shoulders. The truth is, no cattle are more hardy than the thoroughbreds, when properly bred and reared, and none so profitable to the ranchman. The stall-fed, pampered bulls that have never felt the summer's heat or winter's cold, never have and never can give satisfaction when taken to the plains. Our ranchmen cannot use too much care in the selection of their bulls. In reality the bulls are half their herds, for upon them depends the number and quality of their increase. They should select good, hardy young bulls, showing plenty of masculine vigor and constitution, and such only as carry plenty of flesh and show an aptitude to fatten without being tied up in the stalls and stuffed. Bulls of this character that have been raised in open air always have been profitable and given the best of satisfaction to western ranchmen."

So much for my stupidity and Mr. Rust's recent discovery. He contends that one breed is as hardy as another, which he knows is not true. Would it not be ridiculous to contend that the beautiful little Jerseys were as hardy as the shaggy-coated West Highlanders or Galloways? I assure you the difference between the Shorthorns and Herefords is as well established, and this marked distinction has resulted from the different manner in which they were bred and reared. Furthermore, Mr. Rust's own writings justify this conclusion. I am glad he has named Mr. Gillette in this connection, and he might have added the names of William Warfield, the Messrs. Potts, Col. Harris of this state, and a few other intelligent gentlemen who have labored hard and unceasingly to build up and improve the Shorthorns by breeding for individual worth, constitution and the butcher's block, which is the ultimatum of all beef animals. Unfortunately, while these gentlemen were trying to improve the Shorthorn race, thousands of their fellow breeders were as energetically breeding down, breeding pedigrees in-and-in, and individual merit, constitution and vitality out, until the entire race was almost engulfed and the country flooded with wheezing, coughing, consumptive weeds, unfit for use in any herd. This is very near the language once applied to Shorthorns by Mr. Rust, and fully explains why Shorthorns have proven so inferior to other breeds when put to the test in the open plains.

Mr. Rust is very anxious to know how it happened that I lost one of my first Hereford bulls, and why it was that a previous one for a time did no good. He says, "As a candid gentleman, Mr. Campbell should have stated such matters as these." Now that I have been placed on the witness stand, I shall give my evidence, from which I trust Mr. Rust will extract solid comfort. The first animal referred to was killed in my absence by an irritable, half-witted wretch by the name of Magee, who lived on an adjoining farm. The bull was a good one and I wished I had more of the same sort. The latter, that he alleges did me no good, was temporarily disabled by the formation of a fibrous tumor, which I removed by the use of the knife (I have done the same with a number of Shorthorns), after which he proved a vigorous server and an excellent getter. After leaving my hands he was used for several years by Geo. Hendricks, who only parted with him to prevent in-breeding. He was next purchased by Col. W. R. Colcord, and now stands at the head of his herd of Shorthorns, and though well advanced in years he is still active as and vigorous as a calf, and is giving perfect satisfaction. The Colonel is as clever a gentleman as ever left the blue grass regions of Kentucky, and if desired I have no doubt he would be pleased to give further particulars concerning this aged bovine monarch of the plains.

Mr. Rust, like Geo. Morgan's friend, "Red, White and Roan," evades the principal questions at issue, and now that the hunting and trout-fishing has been monopolized by Mr. "Red, White and Roan," he pitches into me personally, and virtually says my literary attainments are inferior to his, and that my hat band is smaller than the one worn by him, which is certainly very strong argument in favor of Shorthorns as range cattle, all of which I am willing to concede rather than neglect the cattle or quarrel over so trifling a matter. He is not even satisfied with this, but accuses me of studying the cattle business by lamplight. A grave charge indeed, but nevertheless true. I confess even more. I have studied it by the glimmering light of tallow candles, in lonely and isolated dugouts, far beyond the reach of civilization; by silvery starlight, while making my tedious nightly rounds guarding slumbering herds, when the country was infested by hostile savages; by brilliant sunlight, when my herds were slowly wending their way northward through the burning sands of a southern clime.

Yes, I have studied the cattle business by the light of as fierce and vivid death-dealing lightning as ever flashed from an angry sky, and a time, too, when comrades were laid low in death by the fury of the storm. When the artillery of heaven made the very earth tremble by the force of her cannonading and peals of thunder, that scattered my herds in the wildest and most terrific stampedes. Yes, my lessons in the cattle business were all learned in the stern school of experience, and of course cannot be compared to Mr. Rust's theories or "book larnin'."

W. E. CAMPBELL.

MATTERS AFFECTING USEFULNESS OF PLAINS CATTLE.

To the "Gazette":

There is, I presume, no occasion for further discussion between W. E. Campbell and myself, he having, in the matter at issue between us, acknowledged in his last letter that the position I had been endeavoring to maintain was correct. I had stated that the manner in which bulls were raised had more to do than anything else, with their practical usefulness and vigor upon the plains, and that to this, more than the matter of breed, was due the conflicting experience of individuals with the different kinds of bulls. Mr. Campbell would not listen to any such talk, and denounced it as mere theory. After considerable discussion back and forth, which I trust has not been wearisome to your readers, Mr. Campbell concedes that the position I had assumed was correct; that the manner in which bulls are reared—as to whether pampered, housed and forced to early and excessive growth, on corn and other concentrated foods—does affect their usefulness and capacities for plains life. And he even goes further and submits an extract written by himself and published years ago, in which the same idea was advanced. This is all right; I claim no originality in the idea. The "theory" is a sound one, and so long as it is admitted, I am quite willing that Mr. Campbell shall himself have the credit of having first conceived it. It is certainly worthy of the "experience" of which we have heard so much, and of the study of the cattle problem under the discouraging and appalling conditions he describes. Having brought him to the point of publicly confessing what he knew all the time to be facts, I congratulate him upon his candor and bid him good-by.

The difference in the breed of bulls for use on the plains, so far as the points of present discussion extend, appear to me to be much overestimated. The only point which has been seriously urged is one of hardiness. And considering the fact that the loss among plains cattle, with all the vicissitudes and inclemencies to which they are exposed, are practically as small, if not smaller, than the losses upon the farms and cultivated fields east, it cannot be contended that a matter of increased hardiness in plains cattle is one of very great importance, as compared with some other matters. A very trifling difference in quality or weight would cut much more figure in the profits of the ranchman than anything he could possibly secure through increased hardiness in cattle already hardy enough for the practical purposes of their surroundings. A very large proportion of these plains cattle, almost all of them in fact, save those brought from Texas and the regions bordering upon the Gulf, have a strong admixture of Shorthorn blood—not as much as they ought to have, but still more or less of it. And their hardiness, so far as the requirements of these plains are concerned, or capabilities of demonstrating the point, being sufficient, it is difficult to understand the necessity of looking further on that score. A good many Herefords have been introduced, and the influence of this blood can be seen here and there, but not so widely disseminated and tried as the Shorthorns; as a general thing the Hereford blood has had the advantage of being represented upon the range by animals more deeply bred than the animals representing the Shorthorn blood. That is to say, a great many of the animals brought upon the range and classed as of Shorthorn blood have had but a meager proportion of that blood, while of those brought upon the range and classed as Herefords, most of them have, until recently at any rate, been fairly crossed with that blood. There appears to be no lack of hardiness in the range animals showing Hereford blood, and in this respect they are fully equal to those owning Shorthorn blood, but, I do not believe, any better. If in the years to come experience shall finally indicate a permanent preference for one breed over another it will be, I am satisfied, not on the score of hardiness, but to difference in maturity, quality and kindly submission to feeding operations farther east.

But while the treatment of bulls before they are brought upon the plains has much to do with their usefulness, the manner in which the ranchman conducts his business will have much to do with the satisfaction derived from their use and is a factor which accounts for much of the diversity in the related experiences of individuals, with the various breeds. On one range, for instance, either belonging to the owner or his neighbors, an unusually large percentage of steers are maintained, which very materially reduces the calf crop, and this is sometimes wrongfully imputed to want of vigor on the part of the bulls. Another range is not favorably situated as to water, and animals have to travel far to get it, and become more widely scattered, and, in consequence, the bulls do not have as free and constant companionship with the cows as on ranges where the cattle find their food and water nearer together. Then some ranchmen turn their bulls upon the open range and expect them to stay there the whole year round, the same as other kinds of stock. The result is that the bulls are working more or less the whole year, which tells on them strongly, and then the calves come at all seasons, many are lost before the owner ever sees them, and the shortage in the crop is charged to the inefficiency of the bulls. If the owner happens to inspect the stock and compares experiences with some other ranchman who separates his bulls from the herd during a portion of the year, and feeds them during the most inclement season, he will be apt to acquire a very unfavorable impression concerning his own bulls and what they can accomplish. But about the last thing he will do, will be to ascribe the difference to the treatment which his bulls receive. It is very poor policy, especially with thoroughbred bulls of high type, to require them to make a living the whole year round on the range. If they rough it through, it detracts from their condition for service the next season, and with good bulls it does not pay. A great many ranchmen have discovered this and are careful to bring in their best bulls where they need not be exposed to the vicissitudes which may come to other cattle.

I have it from a personal acquaintance residing near Mr. Campbell that this is his practice, and that he is careful to gather up his thoroughbred Hereford bulls from the range at the close of each season and bring them into the ranch, where they can be cared for during the winter, and brought into proper condition for next season's work. The practice should be commended. No other course will secure the full advantages which should be secured from the use of thoroughbred bulls.

GEO. W. RUST.

OF THE SAME OPINION STILL.

To the "Gazette":

Mr. Geo. W. Rust, in his article in your issue of January 1, says: "There is, I presume, no occasion for further discussion between Mr. W. E. Campbell and myself." Thus far he is eminently correct. The Herefords have been shown to be superior to Shorthorns, not only as hardy, impressive range cattle, but as feeders also; and furthermore that they bring more per pound when sent to market. All these facts have been demonstrated by actual tests made by wealthy and reputable stockmen, the names of whom I have already given, and not by theories invented by correspondents who practically know nothing of the subject at issue.

To more fully explain my position I will state: Several years ago I bought a herd of Texas cows and calves for $18 for each cow and calf. Allowing the calves to be worth $6 per head, the cows cost $12 per head. Some of these cows escaped from the common herd and lodged with my fine herd some ten miles away, and thus were accidentally bred to Hereford bulls, and afterwards dropped calves that were as shapely and well marked as thoroughbreds and when yet yearlings weighed from 1,015 to

1,260 pounds each. The lightest, a heifer that had suffered severely from pinkeye and had been on grass all summer, was sold to a Chicago butcher for eight cents a pound, amounting to $81.20. I sold two others for $165 and the heaviest one for $100, which (to me at least) are satisfactory dividends on my $12 Texas cow investment, topped by a Hereford cross.

I have bred some good Shorthorn grades on the range, and have taken more prizes at fat stock shows with them than any other ranchman in the land. In fact, the only time my Shorthorn grades ever met defeat was when they came in competition with a carload of range cattle that had Hereford blood in their veins. The moment I saw them I knew defeat was inevitable, and I received a second premium very complacently. I have never seen any Shorthorn grades from the range that could compare with grade Herefords, either for weight or quality off grass, or that would bring as much per pound when sent to market, and these are the reasons why I prefer the Whitefaces. Furthermore, I am willing to show halfbreed Herefords, out of these little $12 Texas cows, against an equal number of half or even three-fourths bred Shorthorns, owned and bred on the range by exhibitor, at any railroad point within fifty or 100 miles from Caldwell, for a purse of $500, the money to be donated to any charitable enterprise the exhibitors or judges may designate.

Mr. Rust claims I have conceded his position as correct, though he knows such is not the case. In my last I showed that it was simply ridiculous to claim that one breed was as hardy as another, and gave good reasons for this conclusion, namely, the manner in which they (the breeds) have been bred and reared, not for the past twenty-four hours or twenty-four weeks, as Mr. Rust teaches, but for generations, or a term of fifty or one hundred years.

I further assigned reasons why the Shorthorns in particular were inferior to other breeds when put to the test on the open range. I stated that "the majority of Shorthorn breeders had followed the family or fashion craze for years, breeding pedigrees in-an-in, and individual merit, constitution and vitality out, until the country was flooded with wheezing, coughing weeds unfit for use in any herd." If this is conceding the correctness of Mr. Rust's opinion (and he has so stated) there is certainly no need of further argument. I therefore respond to his courteous "good-by" with an humble bow and a hearty farewell shake.

W. E. Campbell.

CHAPTER XLIII.

Conclusion—The Present and Future

OPEN BATTLE OF BREEDS BENEFICIAL TO ALL. BEWARE OF SUBTLE MANIPULATION.

It has come to be the settled principle of the advanced nations that they go to war only to obtain peace. That is to say, that in the case of the more intelligent and, therefore, leading nations, the sacrifices of war are followed by the establishment of a beneficent peace that is generally worth all the cost. Justifiable war is a duty.

The "Battle of the Breeds" has likewise been beneficial beyond measure to all sound cattle interests. Our position in the "Battle of the Breeds," and that of our predecessor, the Hereford champion, Mr. William H. Sotham, was, likewise, one of duty and not of pleasure. Each in turn maintained a sturdy fight, and put up with all the hardships and trials incident to battle. Yet, we dare say that the efforts of both were prompted by a well-founded knowledge that ultimately we must accomplish great good; not simply to the Hereford cause but to the whole cattle interest, and therefore to agriculture generally.

At the time we joined in the fight that Mr. Sotham had been maintaining singlehanded for over two score years, we found that the better class of Shorthorns were kept in the background, and the efforts of leading Shorthorn breeders were directed towards forcing upon the Shorthorn breeders an inferior type of Shorthorns and this at fabulous prices. Our efforts, directed as they were towards bringing a more compact type into public favor, by proving the Herefords and their type the more practical and profitable for beef production, were in no wise intended to menace the Shorthorn breed, but rather, it was a crusade directed particularly against the selfish wishes and ill-conceived plans of those Shorthorn breeders who were booming what Mr. Sotham aptly characterized as the "Bates mania."

Unfortunately for the Shorthorn interest, the breeding of the Bates family of Shorthorn cattle had gotten—as the saying was—into the hands of "men with more money than brains," or rather, as we would put it, into the possession of "men of means without practical cattle experience." Or, in other words, the Shorthorns were the pets and playthings of men who bred for pedigree rather than individuality. These men, and others less opulent, who aped them, were free handed in spending money with agricultural papers to puff and flatter them in their vain desires to appear as benefactors to American agriculture.

We have shown conclusively that beginning with the first agricultural paper, viz., "The Albany Cultivator," then the "Ohio Farmer," the "Michigan Farmer," the "National Live Stock Journal," the "Live Stock Record," the "Farmers' Magazine," etc., down to the "Breeders' Gazette," as they were each in turn established, all were controlled in the interest of the Bates Shorthorn, and the early files of the last named journal, though now unquestionably the leading agricultural journal of the world, prove that it was, in its incipiency, fostered by the Bates clique, though run ostensibly in the whole Shorthorn interest.

Most of these papers, still in existence, under different management, now emphatically condemn what they formerly advocated, and strenuously deny their partiality for any particular breed. Yet, regardless of denials, it must be noticed that the old "first love" prevails very marked in some instances. We can see some reason for this, in their self interest, because, although the Hereford is now acknowledged the beef breed par excellence, and his type has been as far as possible appropriated and adopted by the meritorious families of Shorthorns; yet, the Shorthorn advertisers predominate and there has been extreme effort on the part of certain of these editors to collect and publish favorable Shorthorn data that would be commendable, were they not by comparison extremely and unwarrantedly negligent in their efforts to collect and publish similar information in regard to the Herefords.

The Hereford men nowadays are very liberal advertisers, by reason of which they have been allowed to toot their own horns unmolested, but it is a most notable fact that even yet certain leading live stock papers of their own motion do comparatively nothing toward collecting and publishing Hereford data, while sparing no pains or expense to perfect and publish their knowledge of Shorthorns. Even the Angus have had a goodly share of editorial interest exhibited in them, and correspondents have been paid to dig up their family history in a connected and extended way that has not been accorded to the Herefords.

This state of affairs has made our Hereford History a necessity to the Hereford breed. As a matter of course, we have had to republish the history of many things that were unpleasant at the time of their occurrence, yet a history is a history, and we have always been in the habit of calling things by their right names and stating facts in their utmost baldness. If, in our work, we say or have said anything offensive of offending Shorthorn breeders or their offending friends, we entertain the sincere belief that it has been to the benefit of the Shorthorn breed, for in placing the Hereford prominently before the American beef raiser we have as is now everywhere conceded, compelled a modification and improvement of the Shorthorn breed of cattle.

To our certain knowledge, Scotch Shorthorns (for years the most meritorious family of the breed) in the days when we were making our hardest fight for the Herefords, so lacked appreciation that the breeders of these plebeian but splendid beef cattle, were compelled to make steers of their bulls, and place them in their feed lots. To meet the Herefords in the show ring, a demand sprung up for Scotch cattle when the victories of the Duke of Richmond and his get in the show ring were the only bright spots in the Shorthorn campaign. It was the Herefords that forced value into the Scotch blood for show ring purposes, yet for a while, though splendid Scotch animals brought hundreds each, at the same time, Bates' Dukes and Duchesses — manifestly inferior to the Scotch in every beefing particular—sold for as many thousands.

As the result of persistent Hereford aggressions, all things are changed to-day, and now we see in the Shorthorn breed the highest prices paid for individual excellence in animals that formerly were considered "unfashionable blood," their only danger now appearing to be in the direction of a "Scotch craze," based on the similar erratic lines of the "Bates mania," but the competition between breeds may, we think, be trusted to keep them within bounds.

The effective assistance rendered to the beef interests by the National Shows, and the impetus to State Shows, given by the intelligent use of the funds of the Hereford, Shorthorn, and Angus Associations, will, we trust, tend to keep down any craze in either of these breeds for special lines of blood not identified with the excellence of the individual. In this intelligent fostering of a great industry, we note with pride the Hereford breeders leading the way.

Were some of those who now most ardently support "individual merit" in Shorthorns, as against "pedigree fads," to rise up too quickly to criticize this, the closing work of our life, we should of necessity be obliged to quote somewhat further from their favoring connection with the grave errors of Shorthorn history, and if, in the quotations we have made, anything be said that is unpleasant to anyone that is or was connected with the Shorthorn interest, our plea in extenuation is that only by such plain statements of facts are the best interests of cattle breeding subserved.

The Hereford movement in America, in which we are accorded the leading position, not only brought the Hereford into notice but helped the Shorthorn breed as it never was able to help itself. This being conceded on all sides, we may be pardoned for making some suggestions.

Appreciation is all there is in life. The Hereford Society has in many instances been liberal in the matter of encouraging Hereford breeders by the disbursement of prize money, but when the matter is given thought, I am sure I will be endorsed in saying, that, though unintentional, perhaps, through their association they have been ungenerous in other directions.

In a previous chapter, the brief list of contributors of a testimonial purse to Wm. H. Sotham for his many years' championship of sound cattle interests is an instance. Though not at any time a man of wealth, there never lived a more independent man; and having some mind of our own, we can feel for the veteran when at a later meeting of the Hereford breeders, he appeared before it with the intention of returning the purse, because it had been said that it was a "charity." Nothing could have fired the old man's indignation further. Happily the leading Hereford breeders appreciated Mr. Sotham's work, and having his confidence, were able to convince him that paltry as the purse was, it was presented in the spirit of appreciation for his work, and a resolution was passed unanimously in accordance with these assurances.

That was in the days of individual subscription before the Association owned the Herd Book, or had any fixed source of revenue. To-day, it is different. With an overflowing treasury, the liberality of Hereford breeders should yet show itself in an appropriate monument, expressive of appreciation for the great work Mr. Sotham did. The Shorthorn Association should join in this movement, for Mr. Sotham was ever a friend of their best cattle and solicitous of the true interests of the Shorthorn breed. Likewise, the Angus had his good will, for they were favorites with him, while the little West Highlander was probably his ideal of a perfect beef animal. While Mr. Sotham advocated particularly the Hereford breed, it was because, as a breed, they had a higher standard of uniform excellence than any other breed; their standard if equalled at all by any other breed being approached only by its best specimens.

Our own experience at the recent meeting (1899) of the American Hereford Cattle Breeders' Association, leads us the more to believe that a man's work may not be appreciated at its full value during his lifetime, or while jealous contemporaries hold the machinery of organization. We fully appreciated Mr. Sotham's work during his lifetime, but, in the stress of the fight, we were unable to stop and bind up his wounds, although it was fully in our heart to do so; but time presses and waits for no man, and we are thus tardily doing our very best to have the works of Mr. Sotham appreciated. Feeling that Mr. T. F. B. Sotham would appreciate this recognition of his father's work for the improvement of American cattle we can state that our unpleasant experience at the recent meeting resulted unexpectedly in one source of gratification, for it led us to confer with Mr. Sotham, through which conference we agreed with him upon the revision and publication of our work, independent of any unfriendly censorship.

As was known to many members of the Hereford Cattle Breeders' Association, we agreed with the Executive Committee of the Association to write a History of Hereford Cattle for the Association. We entered into this work in no narrow mood, but with a love for the labor and a broad feeling of assurance that we were equipped better than anyone else to do this work, and in the firm belief that such a work was needed and would be appreciated. After spending a year upon the work, we found that at our time of life it was a much greater and more arduous task than we had estimated; several years having passed since we had retired from active business. While the money that we had agreed upon with the Executive Committee as compensation would have been a convenience, we, at no time, felt as though we were working for a money consideration. As we got into the work, reviewing our past struggles and triumphs, we renewed our youth in a revival of our interest in beef cattle improvement. The splendid enthusiasm of the Hereford men at their shows was encouraging.

When, therefore, we came to the meeting with several hundred pages of manuscript, ready to show what we had accomplished and to outline what we further wished to do in completion of the work, we suggested to the Executive Committee that we could use a part of the compensation, and were pained on meeting the committee to learn that our work was by them considered a mere dollars-and-cents matter, and that even when completed we could have no assurance that the work would be published. At any rate the Executive Committee crawled behind a technicality, and although they assumed unbridled authority in many directions in other matters they pleaded that they had no authority to turn over any part of the money until the entire work was completed.

To show how utterly devoid of sincerity the committee's plea was I would add that a little later, in honeyed phrases, they offered to take the data and uncompleted work as we then had it, and pay the full amount of money ($500) for it, uncompleted. The design of the Executive Committee ruler was so plainly evident that we would not, under any circumstances, allow our work to go into his hands to be revised and edited; and therefore we resolved to cancel our agreement with the Hereford Association. Our friends, however, were not content to let the matter remain in this condition, and it was brought before the meeting, but the Executive Committee, carrying in proxy votes the power of the Association, was enabled, by an empty subterfuge of alleged business principle, to defeat what was unquestionably the will of the Association.

It is beyond our comprehension how the American Hereford Cattle Breeders' Association continues to submit to the rule of a committee of three men who are known to be dominated by one of its members. The Executive Committee of the American Hereford Cattle Breeders' Association, as now (1899) constructed, amounts to one-man rule, and according to the Constitution and By-Laws that man, as the head of the Executive Committee, is the ruler of the Association. The forbearance and conciliatory spirit in the great majority of the Hereford Association are matters for wonder, as well as for great congratulation. The Constitution and By-Laws of the Association were unquestionably good and sufficient for the time when they were made. It was well to have then, and it is well to have now, and I hope we always will have the most stringent rules in regard to registration, and the firmest safeguards against those rules being thoughtlessly or hastily changed; but the other by-laws and all other business, outside of the registration rules, should be subject to the will of the majority at meetings, and not subject to the will of a single man and his clique, as it has been and will be under existing conditions, or until some leader rises up, and facing the unpleasant and demoralizing situation, declares the facts to each member of the Association. It has been demonstrated that the Executive Committee as at present constructed will not relinquish their power without a struggle, so that it is not difficult to prophesy that their narrow, selfish policy will, in time, make a revolution imperative.

When we think of what the Hereford breed of cattle is, in the importance of its relation to the beef interests of the world, we are amazed that a few Hereford breeders will come together once a year at the annual meeting to listen to the same old sort of a report, and to act upon cut-and-dried business prepared for them by the Executive Committee in the same old way, and to which cut-and-dried preparation they are limited in the action and business of the convention. Much against the will of an offensive oligarchy, the association elected Mr. T. F. B. Sotham (¶ 377) president at the recent meeting (1899). We mistake the character of Mr. Sotham, and he will not be true to his position as champion of the Herefords, or worthy of his father, if he does not throw light on this "clique" and the subtle errors of their way.

The Executive Committee passes upon all constitutional legislation to be submitted to the Association. If they decline to submit the proposition of a member that proposition cannot come before the meeting. If they condescend to present such proposition they present it in a form that must be adopted or rejected without the changing of a single word. This system except as concerns rules for registration is absurd, silly, and ought not to be tolerated. The idea of a body of men of the character, intelligence and wealth of the American Hereford Cattle Breeders' Association submitting to any such little narrow plan is preposterous, and I am convinced from much communication with the members of the Association that the system needs only to be aired properly by a well-meaning, loyal and influential breeder to insure a prompt and profitable change. We therefore, in closing, devote a little space to this matter.

Why should the Hereford interests submit to any such narrow control? We concede that the members of the present (1899) Executive Committee may be considered just as intelligent and worthy as any other members of the Association, but we insist they are not more so. There never was a body of men constituted that has more good material for effective work than the American Hereford Cattle Breeders' Association, and the Association falls very far short of its duty in not availing itself of this material. There is positively no good to be gained and an opportunity for much harm in giving any man a life tenure of office. The course for many years has been for the annual meeting to assemble in Chicago on one evening in November, ostensibly at 8 o'clock, although it is nearer 9 before the meeting is called to order; and when the clock nears 11 there has always been an urgent and very apparent desire on the part of the Executive Committeemen to have the business over and an adjournment agreed to. The business has consisted of the reports of the Executive Committee and Treasurer, the latter being a member of the Executive Committee. They report their work of the last year, and propose their work for the ensuing year, merely asking the sanction of a necessary appropriation. If they can possibly get these reports adopted and agreed to in full, at one stroke, the plain desire of the Executive Committee is then fully carried out. Committee matters of interest to breeders, discussions of what would be helpful to them in their work, matters that can only be elucidated by full and complete discussion, that would make the conventions of the Association interesting and profitable, are ignored, that a cut-and-dried program may be carried through in the most expeditious manner possible. The better acquaintance of Hereford breeders, the discussion of their experiences, the reading of appropriate papers, and the free and full discussion of pertinent Hereford matters, such as breeding problems, feeding experiments, judging systems, selection of judges, etc., would create and renew such *general interest* that the accruing benefits would make a general desire for the longest sessions convenient rather than the shortest farces possible.

We have no more pecuniary interest in the Herefords, and of more than eighty years of life we have spent thirty in the Hereford interest, and therefore know that we can say what we are saying without anyone rightly feeling that we have "an ax to grind." On this matter we write as we have in the rest of our work, *without any hidden meaning*. Straightforwardness has ever been our method. When we began preparing this work, knowing that it must be submitted to the Executive Committee, and knowing well each member of that committee, to meet what we conceived would be their requirements we were constrained to clothe our words with a little of that hateful subtlety which has ever characterized our opponents from without as well as within the Hereford camp. But when we felt obliged to cancel our agreement with the Executive Committee we were relieved of a distressing position, and we at once determined to renounce subtlety and all its works, and confine our efforts to those methods and to that system of language which had ever been inseparable from our work.

We know with what reluctance a Hereford breeder would take up the leadership of a crusade against existing, offensive and retarding conditions and dangerous tendencies in the American Hereford Cattle Breeders' Association. No one rushes, without grave provocation, into a family row. The Association, even under the prevailing system, has been, as it ought to be, in the advance of any similar organization. But the system makes for one man control and an autocrat generally resents the slightest criticism, and having control of the machinery of the Association uses it more or less effectively to the detriment of his critics. An octogenarian, however, out of business, speaking for the good of others in a cause he loves so well, need fear nothing, even if he speaks freely. Incidents illustrating our meaning may be enumerated, but one will suffice.

It was the habit of the Treasurer of the American Hereford Cattle Breeders' Association up to 1898 to give a personal bond, the strength of which up to that time, we are safe in saying, was never properly investigated. Now we do not question the validity or sufficiency of the bond, but call attention to the carelessness of the Association in regard to it.

Let us go further. The Treasurer's bond was passed upon by the Executive Committee, of which the Treasurer was the active, leading and ruling member. Again, a large surplus fund amounting to $20,000 had accumulated, which the Association ordered to be kept invested in U. S. bonds. With the consent of the Executive Committee, of which the Treasurer was THE PREVAILING MEMBER, this surplus was kept by the Treasurer for many years in unregistered United States Government bonds. These bonds needed no endorsement, being exactly the same as cash, thus placing the entire funds of our Association at the disposal of and for the use and benefit of the Treasurer without interest, upon a personal bond furnished to and passed upon by the Executive Committee, which was essentially himself. We consider it fortunate for the Hereford Association that our Treasurer was able to weather the panic of '93. We have no reason, however, to believe him any more honest than the treasurer of the Shorthorn Association, who, as is well known, became financially embarrassed, and, doing his best, turned over real estate of questionable value to the Shorthorn Association in lieu of a similar amount of cash. Again, in the office of the Treasurer there was no systematic accounting system. We are not specifically charging dishonesty on the part of the Treasurer. We merely call attention to the looseness of the Association's system and to the chances for loss occasioned by this delegation of all power to the Executive Committee trio, without a proper and annual accounting. Many another breeder tossed in the financial storm and paying high rates of interest would doubtless have been able to profit greatly by the free use of this great sum of ready money in the hands of the Treasurer.

Now, let us illustrate by a little unwritten history how badly this loose system acts for any individual who may have enough interest in the welfare of the Hereford breed and its Association to make suggestions for the improvement of the system of this all-powerful and criticism-hating committee.

The late Mr. G. W. Henry, of Chicago, and Mr. C. B. Smith (¶ 378), of Fayette, Mo., were particularly friendly in their relations. When Mr. Smith was elected President of the Association—like other presidents, he had to learn that the office of President of the American Hereford Cattle Breeders' Association was intended for a mere figurehead; that in fact the President had no duties, or at most very simple ones, for which ample preparations were made that they might be performed by the Executive Committee in case the figurehead neglected them. Mr. Smith, however, happily for Hereford interests, decided that in the estimation of the Association the President's office held some importance. He therefore took it upon himself to investigate somewhat of the condition and practice in the office of the Association. In the first place it was found that the Secretary and Treasurer were not the employes of, nor are they responsible to the Association. They belong, under the constitution, body and soul to the Executive Committee. They must make all reports to the Executive Committee, are responsible to that committee, and hold their offices by the power and command of that committee. In the second place he found the office was located at Independence, and that aside from its being the office of the Hereford Association it was the office of the Treasurer, and also of the Hereford cattle breeding firm of which the Treasurer was the active partner.

Further, it was discovered that the funds of the Association were kept in the private bank accounts of the Secretary and Treasurer, and that the machinery of the Hereford office was used to conduct the private commission business of the Secretary and the business of the Treasurer's firm. Further investigation showed a lack of system in the methods of the office, and particularly a very lax and unbusinesslike system of accounts. Still further investigation showed the unregistered bonds, equivalent to greenbacks, added to the cash in the hands of the Treasurer. Having the best of will, and we may say without fear of contradiction, a sincere friendship for both the Secretary and Treasurer, Mr. Smith was extremely reluctant to take action, but being himself an expert accountant and a business man, experienced with and high in the esteem of great corporations, he felt that action on his part was a duty, and thus he conferred with his friend, Mr. Henry, also a business man of high standing and familiar with corporations. In the kindliest and friendliest way possible they quietly but plainly laid some of the facts as they saw them before both Secretary and Treasurer, suggesting various changes in the system of accounting in both Secretary's and Treasurer's office, and insisting that the Treasurer should at once exchange the unregistered bonds, that were as negotiable as greenbacks, for bonds registered in the name of the American Hereford Breeders' Association; such bonds not being transferable without the signature of the President in addition to that of the Treasurer of the Association. Wilfully misconstruing the motives and distrusting the evident faith and good-will of Messrs. Smith and Henry, the Executive Committee appointed two breeders, neither of whom were accountants, as an auditing committee to pass upon the accounts of the Association, which committee made a report dated January 12, 1898. We do not call in question the honesty of that report, and it is not particularly our intention to call in question the honesty of the existing administration, but we do say that the appointment by the Executive Committee of an auditing committee to pass upon their own work is a farce too silly, after sober thought, to be countenanced by a body of men as intelligent as the members of the Hereford Cattle Breeders' Association.

We need not prolong the details of this matter, and would not have presented it at all in this, a history of Hereford cattle, were it not for our firm belief that the Hereford breed of cattle is to be the prevailing and leading breed in the improvement of the world's beef cattle; and that being the case, that the organization of the breeders of Hereford cattle must wield great influence upon the breed, and therefore any failures and mistakes would, in the end, have a bad effect upon the beef interest of the country, and therefore upon its agriculture.

We feel that most members of the American Hereford Cattle Breeders' Association take too lax a view of their duties and responsibilities, and of their individual importance in the great work of uplifting American agriculture. We say this with the welfare of the breeders of Hereford cattle sincerely at heart. We say this because the members of the American Hereford Cattle Breeders' Association annually receive from the Secretary a circular letter announcing the date of annual meeting and giving copies (sanctioned by the Executive Committee) of rules to be adopted or rejected at the coming meeting, and accompanying this they receive a blank proxy, and *too little importance has been attached by members to those proxies.* The great majority of the proxies fell into the waste basket, with the proposed rules, members promptly deciding there was nothing to be done at the meeting to interest them, or to be worth the expense incurred in attending. This would be well were it not that other members, noting the name of the Secretary printed in large letters at the head of the proxy, have felt that some attention should be paid to it, and though taking no personal interest in the meeting have endorsed their name on the proxy and returned it to the Secretary; and enough have usually done this to place the power in the hands of the Secretary to accomplish at the meeting whatever his will might be, regardless of the will of those Hereford breeders who think enough of the Hereford breed and its organization to pay their money to attend the meeting in person. These proxies the Secretary would, of course, divide among the Executive Committee trio, and their clique, in order that on the face of the returns the power of one man might not be too offensively visible.

We are glad that we can say that in most instances this power has been used in the interest of projects redounding ultimately to the good of the Hereford breed, for in the main no one can gainsay the fact that the Hereford Association has done a great deal of good, but *it might have done and can do more,* for this power unwittingly invested in the Secretary, who is the employee of the Executive Committee, has also been abused, to the detriment of the Association's best interests. Note the case which we have introduced, viz., that of ex-President Smith, for the death of Mr. Henry left Mr. Smith to work out these desirable results alone. Neither Mr. Smith nor Mr. Henry desired to make public the condition of the Association's office, preferring that the necessary changes be made without the knowledge of the general public, or even of any large number of members of the Association, and they were so made. Much that was thus secretly suggested by Messrs. Henry and Smith was as secretly adopted in the Hereford office. But nevertheless those who made the suggestions incurred the bitterest enmity of those who adopted those suggestions, and here lies the point that we wish to make, and we trust that it will not be lost in the coming and yet more prosperous years of the society. Let no one deceive himself; the open, boldly honest way may be the unpleasant way, but it is the only sensible, safe and enduring way.

Ex-President Smith and Mr. Henry made their suggestions, as we have said, quietly and in good faith as true friends of the Association's officers, and not in the slightest sense as their enemies. They made their suggestions in the truest interest of those officers of the Hereford breeders and the Hereford breed. They particularly wanted the officers, to whom the suggestions were made, continued in office; because they had always had faith in the honesty of those officers and because they wanted no hint of fraud or irregularity to get before the public or even before the Association itself, and thus important changes were made in the system and finances of the American Hereford Cattle Breeders' Association, diplomatically and without those who so thoroughly deserved it being publicly criticized. Mr. Henry, unfortunately for the Hereford breed, died at Kansas City, where he went to attend and put life (as none but he could) into a public sale of Herefords, and the resentment of the Treasurer and Secretary and their associates did not affect him. Ex-President Smith, therefore, came in for the brunt of their hatred. How they have been wreaking their spite upon Mr. Smith can best be judged by others. We have been too far from the scene of leading Hereford operations to discuss this, but we have been creditably informed that the spite has been exhibited in every possible way, and with most virulent venom.

The American Hereford Cattle Breeders' Association has never had a thorough or proper investigation and auditing of its accounts by an expert accountant since its organization, though we are advised that the Shorthorn Association has of late years had an expert accountant, appointed by an outside and unquestioned authority, namely, the First National Bank of Chicago, which duly appointed expert having at his first investigation gone to the very beginning of the Association's accounts has, to this sound foundation, annually thereafter added each succeeding year's accounts. Here is the one lone instance which we can copy with profit from our friends of the Shorthorn camp.

In discussing this matter we have referred to a point which we wish to illustrate, and this point is illustrated throughout the entire history of our work, as well as throughout the history of our predecessor, the Hereford champion, W. H. Sotham. This point is the fact that open war in the end produces greater, more lasting, and more beneficent results than any subtle scheme of diplomacy. Messrs. Smith and Henry accomplished great good quietly by diplomacy, but the result of that diplomacy, while beneficial to the Hereford Association, has been pointed to as the beneficent work of officers who would not inaugurate reforms of their own motion, but who were actually compelled by the force of circumstances to bring about this result; while at the same time, those officers have used their offices and their personal influence as far as possible to harm the diplomat. Verily, diplomacy will react, for there is no enemy like the one that poses as your friend; we ever preferred an open foe.

Here at last let me make the main point, which this entire matter is intended to bring out: Let the Association create wider and freer councils and broaden the scope of its work; let the annual meetings of the Hereford Association be of such interest as will bring out the largest attendance. Let the program be filled with discussions that will benefit every participant in the meeting. Let it be assumed that more than one, or three members, have wisdom and brains. Let these meetings discuss openly, frankly and with spirit every matter of interest appertaining to the good of the Hereford breed and the beef interest. Change the constitution and by-laws at once, that suggestions may be made and publicly discussed, that no individual need suffer for doing a good to the Association and breed. Let the majority rule. No loyal Hereford man need fear to present his case and abide by the decision of a true Hereford majority. No man is worthy of a Hereford office who is afraid of such majority and resorts to proxies.

We submit this matter reluctantly to a book that will go before the world as a History of Hereford Cattle, but we have endeavored to make this book of wider interest than an ordinary history; to make it valuable for the future as well as the past, and to give the light of the past for the better illumination of the future. There is no fear of the future if the menace of a "close corporation" be subdued. In this as in all other things we speak plainly, in words of no uncertain meaning. The Hereford breed of cattle have nothing to fear from any competitor on the score of absolute inherent merit. Intelligence and honesty in the future conduct of the Hereford breed will make their improvement steady, maintaining their present lead by keeping them apace with any improvement possible to other breeds, and consequently keeping them in the forefront of beef cattle for all time, in the future as in the past.

THE END.

TYPICAL MATURE BULL, "SPRING JACK" (14191) 108669.
(Photo from life.)

APPENDIX.

THE LATE MR. MILLER, OF ILLINOIS, U. S. A.

HOW HE BENEFITED HEREFORD BREEDERS.

It will doubtless be with a feeling of sincere regret that many breeders and others interested in Hereford cattle will receive the news of the death of that veteran champion of the breed in America, Mr. T. L. Miller, which took place at De Funiak Springs, Florida, U. S. A., on March 15, 1900, at the age of eighty-three.

Mr. Miller had, in spite of his great age, been in the enjoyment of excellent health, until a short time before his death, when he had a severe fall in the street, causing concussion of the brain. Although the serious nature of the accident was apparent from the first, he lingered on for some six weeks, until congestion of the lungs supervened, and was the immediate cause of death.

Mr. Miller was born in Connecticut, but early in life settled in Chicago, and engaged successfully in the insurance business. He then made one of those kaleidoscopic changes in business, so frequently made in America and the Colonies and so rarely with success in England, and exchanged city life for agricultural pursuits. For this purpose he removed to Beecher, where he had a large farm, about thirty miles from Chicago, built a large house and elaborate farm buildings. It was here, having discovered the merits of the breed, that he established and carried on for many years, his famous herd of Hereford cattle.

It is not too much to say that he was practically the pioneer of the breed in America, and it was mainly due to his persistent advocacy of their good qualities that American farmers realized how well suited to their purpose the Herefords were, both as a pure breed and also for crossing with native cattle. This result he achieved in the face of the strong opposition of the Shorthorn influence, which was at that time paramount in the United States.

Breeders in this country have cause to remember with gratitude his untiring and successful efforts to make the Herefords widely known on the other side of Atlantic, for his action was the means of bringing about the great boom, of some years back, which enabled those who were fortunate enough to possess pure pedigree herds to reap a golden harvest.

In 1883 he visited England, accompanied by Mrs. Miller and their niece, and bringing with him his rock-away carriage, pair of black horses, and American harness, with which to drive about the country and visit the principal breeders. This turn-out excited a great deal of interest and curiosity wherever it appeared. Among those who entertained Mr. and Mrs. Miller were Lord Bateman at Shobden Court, Mr. J. H. Arkwright, Hampton Court, Mr. T. Duckham, at Baysham Court, Messrs. Goode, at Ivingtonbury, Mr. J. Price, Pembridge, Mr. T. Powell, at the Bage, etc.

The home breeders availed themselves of Mr. Miller's presence in England, to show their appreciation of the valuable work he was doing, by presenting him with a public testimonial, in recognition of his efforts. On August 1st, he was entertained at a public dinner, at the Green Dragon Hotel, Hereford, at which a very large and representative gathering assembled, to do him honor, including Lord Glanusk—then Sir J. R. Bailey—who was in the chair, Lord Bateman, Lord Coventry, the county members, etc. The chairman then presented to him, on behalf of the subscribers, a beautifully illuminated address, signed by one hundred breeders, and a purse of £200. A silver cup was also presented to his able lieutenant, Mr. Geo. Morgan.

Characteristically, he declined to spend the money in silver plate, but bought with it a number of animals, which he took on with him and which were known as the "Testimonial Herd." Successful though he was, fortune was not always kind to him. Twice over, the costly and handsome buildings on his farm were burned to the ground. The first time they were struck

A MODEL BULL, "RED CROSS" (18040) 80076.
(Photo from life.)

by lightning and the second conflagration was believed to be the work of an incendiary. Ultimately he disposed of all his Herefords, and sold the Beecher property.

In order to escape the cold winters of the North, the last years of his life were spent in Florida. An indefatigable worker always, his energies were by no means confined to looking after his farm and the breeding of cattle. He became the mainstay of the church at Beecher and of its pastor at a time when the little community was too poor to provide for their own religious requirements. But his religious activities did not consist solely in giving liberal financial support, for both in Chicago and Beecher he was ever a regular and earnest Sunday school teacher. No stress of weather or badness of roads sufficed to keep him and his faithful wife—whom he always found an able seconder in all his plans—from his work in connection with the church. He contributed largely to the agricultural press, and eventually himself established a printing office and ran a newspaper mainly in the interest of Herefords for some years, and to which he subsequently added a monthly publication. It was also at Beecher, under his own editorship, much of the work being done by his own hands, that he started the "American Herd Book of Hereford Cattle," and continued until the work became too large to be carried on by private enterprise, and was therefore transferred to the management of the society in Chicago.

A FARM SALE.
Movable amphitheater. Weavergrace Farm of T. F. B. Sotham.

Altogether he was a fine specimen of the type of men who have made the United States the great and prosperous and progressive nation that it is to-day. A man of strong will and untiring energy, he put his hand to many things, and to none without some measure of success. At an age when most men are content to rest, he was still strenuously working, and the last work of his life, a task which he had only just completed when he met with his fatal accident, was to write a "History of Hereford Cattle," thus showing that his old interest in the "white faces" had not abated. Of him it may truly be said, as Chas. Dickens once said of himself, "that he never put one hand to anything, on which he could not throw his whole self."—*Hereford (Eng.) Times.*

RESOLUTIONS.

Unanimously adopted by a rising vote of the American Hereford Cattle Breeders' Association at Chicago, December 5, 1901:

WHEREAS, Through the mysterious working of a Divine Providence, one of the most prominent and enterprising of the pioneer members of this association, Mr. T. L. Miller, has been summoned to other fields; therefore, be it resolved by this association, in Annual Convention assembled:

First. That we freely and unhesitatingly accord to him the position of originator and leader in the propaganda of the Hereford in America.

Second. That to his enterprise and courage is largely due the position the Herefords have attained in this country.

Third. That as a breeder of Herefords and a citizen, we commend him and deplore his loss.

Fourth. That we condole with his family in their bereavement and sympathize with them in their affliction.

Fifth. That this association appropriate the sum of $500 to erect a monument as a tribute to his memory.

Sixth. That these resolutions be spread upon the minutes of this meeting.

CHARLES GUDGELL,
W. S. VAN NATTA,
T. F. B. SOTHAM,
Committee.

Chicago, Ill., December 5, 1901.

DEATH OF T. L. MILLER.

Mr. T. L. Miller died on Thursday, March 15, 1900, at De Funiak Springs, Fla., at the ripe age of 83 years. To the cattle breeding world he was universally known as T. L. Miller, of Beecher, Ill., and his Highland Stock Farm was the Mecca of all admirers of Hereford cattle for many years.

AN IDEAL 2-YEAR-OLD, "PRINCE BULBO" (17414).
(Photo from life.)

Supplementing, enlarging and succeeding the work of their first importer, the late Wm. H. Sotham, Mr. Miller may truthfully be said to have done more for the upbuilding and dissemination of the Hereford cattle than any other man who has connected himself with the breed. There were some differences between Mr. Sotham and Mr. Miller, mainly growing out of some methods of procedure during the thick of the "Battle of the Breeds," and because these two old generals were so alike and forceful in their temperament. Nevertheless, my father frequently told me before his death that Mr. Miller had done a noble and unparalleled service for the Hereford breed of cattle and for all sound cattle interests, while several years ago Mr. Miller wrote me that, "during the thick of the fight I could not stop to bind up the old man's wounds, but now that the battle is won, I will tell you what will no doubt interest you: All the information I had to make my fight for the Herefords I got from your father and his writings."

AT T. F. B. SOTHAM'S SALE, 1896.
Weavergrace Farm.

Mr. Miller was born in the East, in Massachusetts, I believe. After receiving a fair education, he was employed in a butcher shop, and later carried on the business for himself. He was always a lover of farm life, and after a very successful business career in Chicago, he retired to Highland Farm, at Beecher, Ill., a tract of 1,000 acres, which he improved by tiling and careful cultivation and embellished with the most extensive suite of farm buildings ever attempted up to that time. The famous Miller barn with its double-headed wind-mill was the talk of the live stock world until it was destroyed by fire, only to be succeeded by another more modern and convenient structure two hundred feet square. It is doubtful if this Miller barn has ever been equaled for capacity and convenience, nor can it ever be surpassed as an ideal cattle home.

No abler advocate ever championed the cause of improvement in cattle. Like my father, he was a fighter when it took a fighter to make an impression on existing erroneous prejudices and preferences. What the tenacious fighter in the courts of law of to-day adds to the value of an attorney, were the qualities essential to the successful consideration by the court of American cattle opinion, brought to the Hereford breed by T. L. Miller. There was jealousy of him among those who should have been his guileless friends, but pinned down to a fair estimate of

A SPLENDID SORT, "PROTECTOR" (19550).
(Photo from life.)

the work of T. L. Miller, no cattleman whose opinion is worth having, ever failed to give Mr. Miller that great measure of praise and appreciation he had fearlessly and honorably earned.

Breeders of Hereford cattle to-day, with their breed the acknowledged leading beef improver, with their organization leading all others in willing submission, with all the homage paid to breed and breeders that proves again the old adage that "nothing succeeds like success," cannot from the very nature of the present situation understand the bitterness of the fight that waged during "the battle of the breeds," when Herefords were denied a classification in the premium lists of the fairs, when space in the agricultural press was denied to Hereford writers, when the strong arm of a George Morgan was compelled to intervene to prevent bodily violence to such men as T. L. Miller. All well enough is it to-day to chime and re-echo in the effulgent light of a supreme peace secured by a universal acknowledgment of Hereford merit, but let us never forget that the Hereford breed owes the greater part of the security of its position to the efforts of T. L. Miller, in his incisive, insistent, never-faltering, effective methods, by which he compelled the cattle world to investigate and the powers that were to acknowledge the merits of Hereford cattle.

THE VETERAN FEEDER, JOHN LETHAM,
Goodenow, Ill., and his two champions. (Photo from life.)

The Hereford's merit is easy to see if only people have a chance to see and try for themselves. Chas. M. Culbertson, Adams Earl and W. S. Van Natta were all naturally Shorthorn men; the first Herefords they ever saw were T. L. Miller's. It came about in this way. The Tippecanoe County Fair was then, as now, held at Lafayette, Ind. Although there was no class for Herefords, Mr. Miller took his show herd there for exhibition. He was not allowed to compete with Shorthorns, but the Board of Directors met and voted him a special purse, and promised a class for Herefords next year. Messrs. Culbertson, Earl and Van Natta there, for the first time, saw the Hereford cattle. Mr. Culbertson and Mr. Van Natta bought their first Herefords soon after of Mr. Miller, and I think the same is true of Mr. Earl.

Under the able management of Geo. F. Morgan, Mr. Miller's herd won great distinction, beating herds of all breeds, including the imported Herefords of some of the contemporary breeders above mentioned. His great sire was "Success," whose lithograph in colors adorns thousands of farmers' homes throughout the country. I well remember with what pride and satisfaction Mr. Miller showed me the splendid matrons of his herd by "Old Success" on the occasion of my last visit to Highlands. Like my father, Mr. Miller loved Hereford cattle better than wealth, ease or life itself. No effort of labor, no sacrifice of time or means was too great for him to render in the interest of the Hereford breed. Not getting a fair hearing in the agricultural press he founded a paper of his own, "The Breeders' Journal," and maintained it at great sacrifice of time, labor and money—a paper that was widely read, and copies of which are kept by Hereford breeders to-day, as priceless souvenirs of a crisis long and happily past.

When the depression came, it caught Mr. Miller as it has caught many another honest,

A PROPER 2-YEAR OLD, "MAJESTIC" (20831).
(Photo from life.)

earnest worker, striving for principles and facts, rather than for greed and mammon. He had been trying for years to induce the Farwells to put Herefords on their great 3,000,000-acre range in Texas, had contracted with them (or one of them) to furnish them a thousand bulls at a fair price, when his difficulties overtook him, and Mr. John V. Farwell bought the entire herd at a song, singing it himself, and figuratively buried it on the great Capitol Ranch in Texas. No records were kept of the increase, and so when recently it was deemed advisable to restore the herd to the records, only a few of the remaining old cows could be identified, and thus it is that the blood of Mr. Miller's Highland Herefords was largely submerged in the flood derived from the latter importations.

Mr. Miller's last work was practically completed—a modern history of Hereford cattle. This is a work of vast importance to the breeders of Hereford cattle. It is the first authoritative history of the Herefords ever written by a loyal Hereford breeder. It is a labor of love. Hereford breeders have too long taken their literature from writers who are distinctively at heart advocates of other breeds. Mr. Miller's book will show that, beginning with Youatt, all the books ever written on the breeds of cattle have been biased toward and in favor of other breeds. In Mr. Miller's work, we will have the exploits of the breed portrayed by a Hereford advocate and breeder. I consider it a godsend that in the ripeness of his years he was spared with robust health, and keen, clear intellect, thus to complete and round out his work, and for the thousandth time again place the Hereford breed and fraternity under lasting obligations to him.

The immediate cause of Mr. Miller's death was an accidental fall. He caught his heel in the sidewalk and fell, never after gaining full consciousness. Four children survive him. Interment took place at Evanston, Ill., from the home of his daughter, on Sunday, March 18, 1900. I have, as President of the American Hereford Cattle Breeders' Association, requested of the family the privilege for our association of erecting the monument to mark his last resting place—a duty and privilege alike that is all that it is possible for us to do for him as a last mark of our respect and appreciation.—*T. F. B. Sotham, in Breeders' Gazette,* March 21, 1900.

ALBION (15027) 76960.
The great English sire and prize winner. (Photo from life.)

A GOOD YEARLING.
(Photo from life.)

A TRIBUTE TO T. L. MILLER.

For those of you who had the privilege of personal acquaintance with Mr. Miller, my heart is full of sympathy, realizing as I do that your loss must be greater than mine—and mine is great—for our acquaintance was one of correspondence. Nevertheless, the kindly nature, the broad, public spirit, the generous impulse which shone from those pages from the hand of T. L. Miller, with a reality and magnetism never to be forgotten, lent counsel and encouragement to one of the youngest breeders in the Hereford Association in his daily fight—not only in the "battle of the breeds," but in the "pioneer" work of developing and promoting the great live stock industry of the South. More than all others, Mr. Miller has been in sympathy and in touch with my work and I publish his first letter as an example of a noble, generous spirit, in the hope that its suggestions may help other breeders whose experience may not yet be ripened into years of silver-gray.

Mr. T. F. B. Sotham appreciates the necessity of steadfast devotion to one's aims and principles in the up-hill but winning fight for good cattle, and probably Mr. Sotham is best qualified of any breeder living to do justice to Mr. Miller—and Mr. Sotham has done so admirably.

After Mr. Miller's retirement from his active business to the milder climate of Florida, he saw the possibilities of the successful raising of good beef cattle in the South, and labored constantly for the advancement of this project. He believed, and he knew when he believed, that the breed par excellence best suited to the climatic and other conditions in the Southern States is the Hereford, that the South is eager for an improvement on their small stock and that the Hereford—"native" cross—is the one to produce the desired result. Mr. Miller had already secured agreements with the L. & N. railroad, the Plant System, and the Florida Central to give free transportation on pure-bred stock into that territory and expected other roads to do the same. He was also promoting Hereford interests in other equally broad-minded ways, and I am glad to have the opportunity of paying this tribute of admiration and respect to one about whom it may be said:

To those who know thee not, no words can paint;
And those who know thee, know all words are faint!

—*Murray Boocock, in Breeders' Gazette*, April 11, 1900.

Mr. Boocock (¶ 379) forwards the following letter for publication, written to him by Mr. Miller from De Funiak, Fla., under date of March 23, 1898:

"I notice your purchase of the $3,000-bull at Emporia. This brings you and your letter of July 1 before me again, and gives you prominence among Hereford breeders. How long have you been breeding Herefords? The late Hon. John Merryman, Cockeysville, Md., was one of the earlier breeders. I think he commenced in the fifties—his widow and son are both breeders now, as I presume you know.

"Permit me to call your attention to the fact that in Herefordshire, England, the breeders there carry a given number of cows, it may be twenty to sixty—as many as they care for—never more, never less—and after a cow has two to four calves she is turned off and a heifer has taken her place. Cows thus treated bring close to steer price for beef. Could Virginia or Pennsylvania adopt this plan they would find a profitable business. This refers to the average farmer, who would keep common cows and use a Hereford bull. If I were in Virginia or Pennsylvania, I would cultivate such trade.

"Say with twenty cows, there would be ten steers a year to go off. At two years old they would weigh 1,200 to 1,400 pounds, and when the heifers come of age he would turn off twenty beeves a year, or say close to $1,000; in this way he would improve his farm. He may let his calves run with the cows until six months old, or wean and use the milk, feeding skim milk to calves. There is no better butter cow than the Hereford.

"This is what Eastern farmers need. If you would cultivate such a trade and make Washington your market, I think you would find it a success."—*Breeders' Gazette,* April 11, 1900.

It was singularly appropriate that among the pall bearers at the funeral of the late T. L. Miller, held at Evanston, Ill., March 18, 1900, were Mr. T. F. B. Sotham, President of the Hereford Cattle Breeders' Association; Mr. George F. Morgan, the veteran Hereford breeder, who assisted Mr. Miller in his early days as a Hereford breeder, and Mr. Tom Smith, long manager of Mr. Miller's Highland Farm and herd at Beecher, Ill., and prominent now as a breeder of Hereford cattle.—*Breeders' Gazette,* March 28, 1900.

FAC-SIMILE OF LETTER FROM T. L. MILLER TO T. F. B. SOTHAM.

[illegible]
Apl. 14 - 99

Dear Son.

Rec'd yours of Mar 30.

I am reading up the record of my fight for the Herefords and have just met the following. Written in Feb '82 -

The History of Mr Sotham and his connection with the Herefords. although he did not receive an immediate and direct success his work had a great influence, and made the way easier for me. I have found all along those who were Hereford men from what they knew of his cattle.

I met a man of 70 years, who said of the Herefords. They are the best cattle I ever knew. When I asked what do you know of them his reply was. I bred and owned them 40 years ago in Vermont the cows were the best I ever owned, and the young things were always ready for the butcher.

I have used his cattle on the Rocky Mountains hauling goods where horses could hardly travel, I met a Mr Bowen at the Centennial from New York state. Who introduced himself and had a good word for them.

What do you know of them I asked. Says he I bred

three Years ago from the Sotham importation and When I asked him why he did not Continue them. he said the Short horn influence Was to Strong for him: and he Was a man of more than ordinary intelligence and Enterprise.

That Short horn influence has put its foot upon many a movement that. let alone. would have betterd the Country.

The Hon John Merryman of Maryland started his his herd from the Sotham importation and forty seven of this herd Went to Messrs A. A Crane Son of Osco in this State on Jany—

The Short horn men and Editors have called Mr Sotham a failure. (He Was not a failure; our Work has been Easier for What he did.)

He made a record that time cannot deface or dim.

He made a case and placed it on record that Was perfect in its detail & make up.

The Work that he did I have used and followed

The friends that he made have been my friends.

I have used the facts he gathered. When attacks have been made upon me.

I could not always, stop to say When the ammunition

came from. but I am now where I have reached a breathing place.
Gathering up this data of his own placing it on record. and passing it to his credit, a fight of 40 Years. from the prime of life to old age

Starting with the best Race of cattle. and the best Specimens of that race. for it will be noticed that in the record we are bringing up. None of his opponents question the merit of his cattle — they try to rob him by claiming the merit as belonging to a Shorthorn cross. Starting as I may say with the best breed and the best specimens of the breed, with a friend in the person of the Hon Erastus Corning of ample means, with those prospects and the skill necessary to carry out the Enterprise to a large success. Those Short horn men were enabled by force of numbers and the control of the press and the New York Agl Society. to destroy those prospects and retard ~~the~~ this Hereford Movement for 40 years.

But While they were ~~[illegible]~~ enabled to do this. they have never been able to break his spirit He has stood for 40 Years as defiant and aggressive as though he had Erastus Corning and a herd of Herefords behind him. he has stood in their path as faithfully and fearlessly as the angel stood in the path Of Balaam —

While I was in the thick of the fight, I could not stop to bind up his wounds; I could not stop as perhaps I ought, to notice his aid.

He has perhaps had occasion to feel that I neglected him, and that same jealousy that has kept his face towards his old enemies led him to strike at me occasionally. I should have been glad that it was not so, but I have no enmity.—

I at my first breathing place gather up his record and give it to this generation; and I will place the Hereford flag in his hands and the Hereford crown on his head.

Mr. Sam.—That is what I wrote & published in Feb '82—

It is [illegible] [illegible] that I from 1889—and during this Albany controversy took the Albany Cultivator and kept the files.

Truly Yours

T. L. Miller

A BIG ONE WITH SMOOTHNESS AND QUALITY, "HAPPY HAMPTON" (16097) 90058.
(Photo from life.)

COPY OF FOREGOING FAC-SIMILE LETTER.

De Funiak Springs, Fla., April 14, 1899.
T. F. B. Sotham, Esq., Chillicothe, Mo.
Dear Tom:

I rec. yours of Mar. 30.

I am reading up the record of my fight for the Herefords, and have just met the following, written in Feb., '82.

The History of Mr. Sotham and his connection with the Herefords. Although he did not receive an immediate and direct success, his work had a great influence, and made the way easier for us. I have found all along those who were Hereford men from what they knew of his cattle.

I met a man of seventy years, who said of the Herefords, they are the best cattle I ever knew. When I asked, what do you know of them, his reply was, I bred and owned them forty years ago in Vermont. The cows were the best I ever owned, and the young things were always ready for the butcher.

I have met his cattle on the Rocky Mountains, carrying goods where horses could hardly travel. I met a Mr. Bowen at the Centennial from New York state, who introduced himself and had a good word for them. What do you know of them, I asked. Says he, I bred them years ago from the Sotham importation, and when I asked him why he did not continue them, he said the Shorthorn influence was too strong for him; and he was a man of more than ordinary intelligence and enterprise.

That Shorthorn influence has put its foot upon many a movement that let alone would have bettered the country.

The Hon. John Merryman of Maryland started his herd from the Sotham importation, and forty-seven of this herd went to Messrs. A. A. Crane & Son of Osco in this state, in Jan.

The Shorthorn men and editors have called Mr. Sotham a failure. He was not a failure; our work has been easier for what he did. He made a record that time cannot deface or dim.

He made a case and placed it on record that was perfect in its detail and make-up. The work that he did I have used and felt. The friends that he made have been my friends. I have used the facts he gathered. When attacks have been made upon me, I could not always stop to say where my ammunition came from, but I am now where I have reached a breathing place.

Gathering up this data of his, I am placing it on record, and passing it to his credit, a fight of forty years from the prime of life to old age.

Starting with the best race of cattle and the best specimens of that race, for it will be noticed that in the record we are bringing up none of his opponent's question the merit of his cattle, they try to rob him by claiming the merits as belonging to a Shorthorn cross. Starting as I may say with the best breed and the best specimens of the breed, with a friend in the person of the Hon. Erastus Corning of ample means; with those prospects and the skill necessary to carry out the enterprise to a large success, those Shorthorn men were enabled, by force of numbers and the control of the press and the New York Agricultural Society, to destroy those prospects and retard this Hereford movement for 40 years.

But while they were able to do this, they have never been able to break his spirit. He has stood for forty years as defiant and aggressive as though he had Erastus Corning and a herd of Herefords behind him. He has stood in their path as faithfully and fearlessly as the angel stood in the path of Balaam.

While I was in the thick of the fight, I could not stop to bind up his wounds; I could not stop, as perhaps I ought, to notice his aid.

He has perhaps had occasion to feel that I neglected him, and that same fearlessness that has kept his face towards his old enemies, led him to strike at me occasionally. I should have been glad that it were not so, but I have no enmity.

I at my first breathing place, gather up his record and give it to his generation, and I will place the Hereford flag in his hands and the Hereford crown on his head.

Well, Tom, that is what I wrote and published in Feb., '82. It is a singular fact that I from 1839—and during this Albany controversy—took the Albany Cultivator, and kept the files. Truly yours. T. L. Miller.

THE ILLUSTRATIONS OF THE APPENDIX.

In accordance with the announcement at the introduction, in addition to the copious illustrations pertinent to the body of this work a number of full-page reproductions of photographs from life of choice specimen Herefords of different ages and both sexes are herewith included. It has been thought best to confine these specimen illustrations to photographs from life rather than from drawings, as it is believed that they will give more accurate representations of Hereford anatomy, and thus train the inexperienced eye to a knowledge of

UNIFORMITY OF DESIRABLE TYPE AND CHARACTER MARKS THE WORK OF A GREAT BREEDER.
(Photo from life.)

the formation of a good Hereford, so that all readers of Mr. Miller's History may reap full benefit of its teachings, and be enabled to take fullest advantage of the merits of Hereford blood.

It is believed the illustrations comparing prime beef with that found in the ordinary markets will serve a particularly useful purpose.

Few people realize that it is impossible to have a real prime beefsteak on their tables unless the animal from which it was taken possessed some of the blood of our improved beef breeds.

CHANGE IN ADMINISTRATION OF THE AMERICAN HEREFORD CATTLE BREEDERS' ASSOCIATION.

In Chapter XLIII of this history, on "The Present and Future of the Hereford," Mr. Miller criticizes the old administration of the Hereford Association affairs by an Executive Committee of three, that was provided for in the Association's by-laws. Members of this committee were elected for terms of three years, one member being elected each year, their terms thus interlapping and forming an endless chain.

Since Mr. Miller's death, a private investigation by prominent members of the Association showed that this committee and the by-laws creating it were unauthorized by law, and a subsequent appeal to the courts of Illinois (under whose laws the Association was chartered), brought about an official investigation which declared the committee illegal, resulting in the resignation of the entire Executive Committee and its officers in January, 1902, and the assumption of the administrative duties of the Association by its legally authorized Board of Directors.

Adhering to the charter of the Association, the law vests all administrative authority in a Board of five directors, who shall be elected annually for terms of one year, and the laws of Illinois compel this Board to assume and discharge these duties, making them responsible for the conduct of the Association.

Coming into this responsibility unexpectedly and without desire or effort on their part, the Board of Directors reluctantly assumed control, and at this time (July, 1902) have corrected much of the error Mr. Miller essayed to point out.

The unfortunate dissensions which brought about the aforesaid investigation, and the resultant shifting of authority, naturally engendered considerable feeling in the Association, so that the Board of Directors was obliged to assume its rightful, but unsought and unexpected power, under very unpleasant circumstances. It is due them to say that they took up their duties, inspired by that loyalty to Hereford interests that is common to the Hereford fraternity in general, and which has ever been the great source of the Association's strength. There is a reasonable hope entertained generally by the members of the Association, that a wise administration by the directors will result in a fair and lasting reunion into one intelligent and aggressive association of the factions caused by the temporary breach.

Mr. Miller pointed out the evils of the proxy system, which unfortunately without new laws cannot legally be entirely abolished. The proxy system has been the root of most Association evils, not only in the Hereford society, but in countless others, and the tendency on the part of many members of the Hereford Association is to refuse to give proxies. The trouble about proxies is that they are too often given for one purpose and voted for another. Proxies are largely used by the holder to promote his own selfish interests. It is safe to say that hundreds of proxies have been voted directly opposite from the way the members giving them would have voted, had they been personally present. As a general proposition, those members who do not care to go to the expense or trouble of attending the annual meeting, ought to be satisfied with the will of the majority of those who do attend, and not put a weapon in the hands of the minority to defeat the will of those members who do pay their money and expend their time to attend. If every member of the Association who cannot attend the annual meetings in person will refuse to be represented by proxy, Hereford dissensions will be annihilated for all time, for, as Mr. Miller truly said: "No loyal Hereford man need fear to present his case and abide by the decision of a true Hereford majority." And, "No man is worthy of a Hereford office who is afraid of such majority and resorts to proxies."

A GRAND OLD BREEDING MATRON, WORTH HER WEIGHT IN GOLD, "PEARL 5TH" (V. 15, P. 8) 80073.
(Photo from life.)

A PRIME AGED COW, "RANEE" (V. 25, P. 242).

IN HER PRIME. "TRUTHFUL" (V. 28, P. 152), AT EIGHT YEARS.
(Photo from life.)

A MODEL YEARLING. "SISTER PERILLA" (V. 25, P. 230).
(Photo from life.)

A TRAINLOAD OF GRADE HEREFORD STEER CALVES, SELECTED BY T. F. B. SOTHAM.

... in the middle and eastern states.

CARLOAD OF TEXAS-BRED GRADE HEREFORD STEERS, SELECTED BY T. F. B. SOTHAM, FED BY D. W. BLACK, OF LYNDON, OHIO.
Winners of the Grand Champion Prize for best carload of fifteen cattle of any breed, at the International Live Stock Exposition, Chicago, 1901. Average age, 2 years and 5 months; average weight, 1600 pounds.

GRADE HEREFORD BEEF—CARCASSES OF THE GRAND CHAMPION CARLOAD.

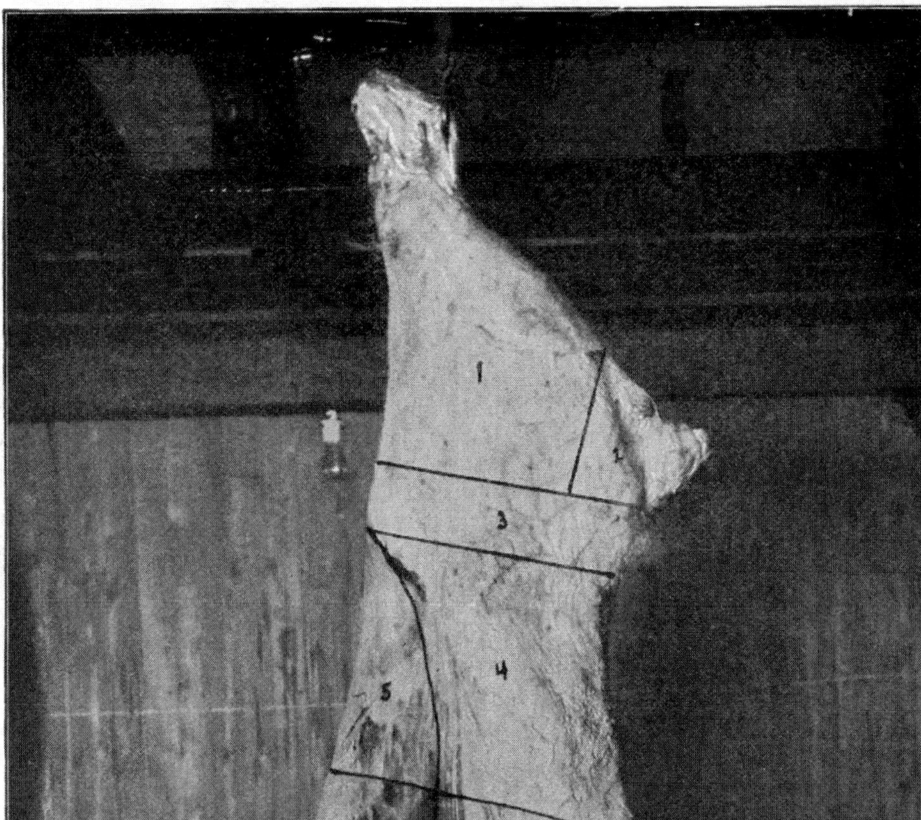

BUTCHER'S DIAGRAM. A SIDE OF GOOD GRADE HEREFORD BEEF.

Showing method of cutting up high-class carcasses. 1. The Round. 2. The Rump. 3. The Sirloin or Broad Loin. 4. The Short Loin. 5. The Flank. 6. The Navel Piece. 7. The Ribs. 8. The Chuck, or Shoulder Roast. 9. The Brisket. 10. The Neck. 11. The Shank.

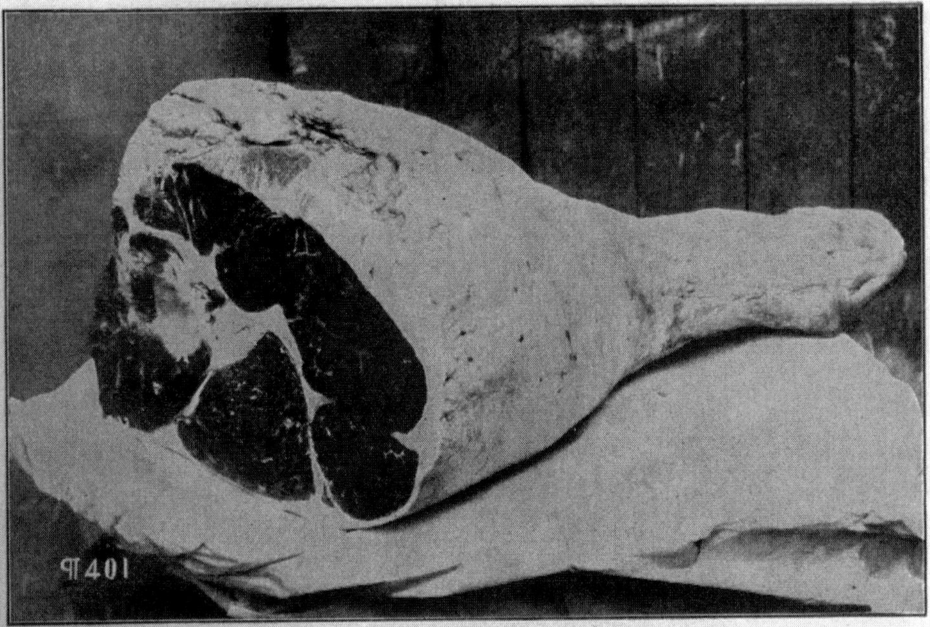

BUTCHER'S TERMS—"THE ROUND."
Taken from a good grade Hereford steer. (See Fig. 1 of butcher's diagram.)

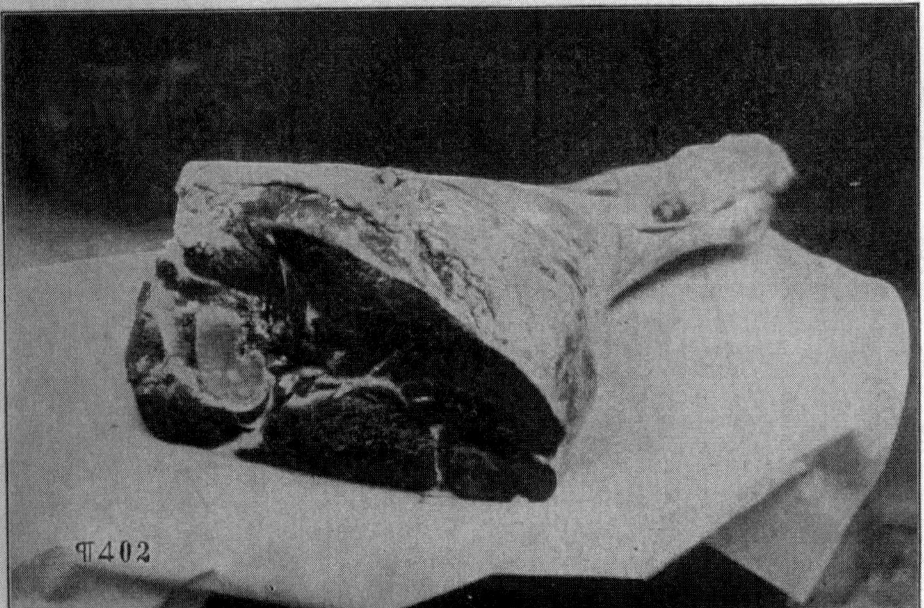

BUTCHER'S TERMS—"THE ROUND."
Taken from a steer classed as common butcher's stock. (See Fig. 1 of butcher's diagram.)

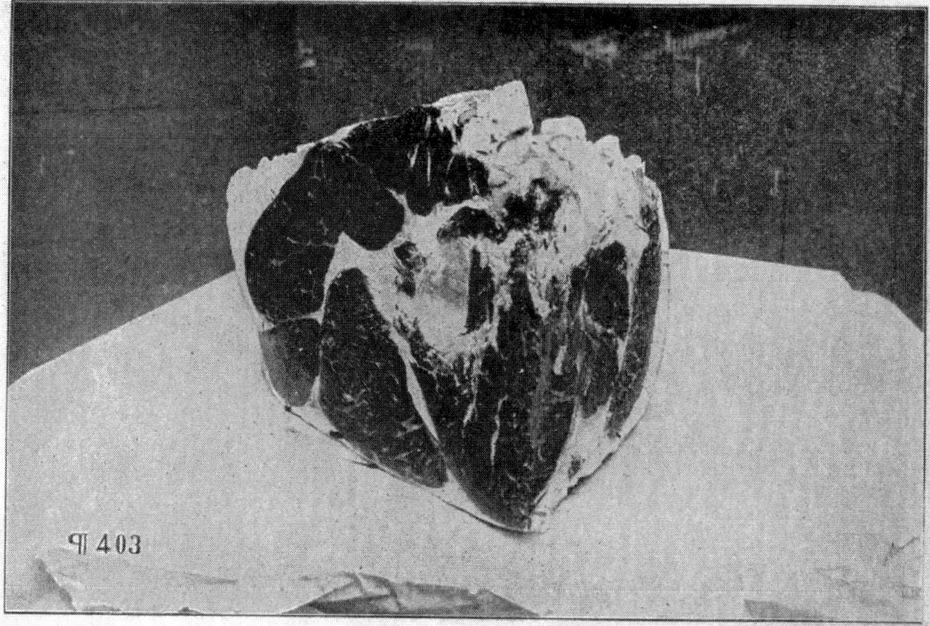

BUTCHER'S TERMS—"THE RUMP."
From a good grade Hereford steer. (See Fig. 2 of butcher's diagram.)

BUTCHER'S TERMS—"THE RUMP."
From a steer classed as common butcher's stock. (See Fig. 2 of butcher's diagram.)

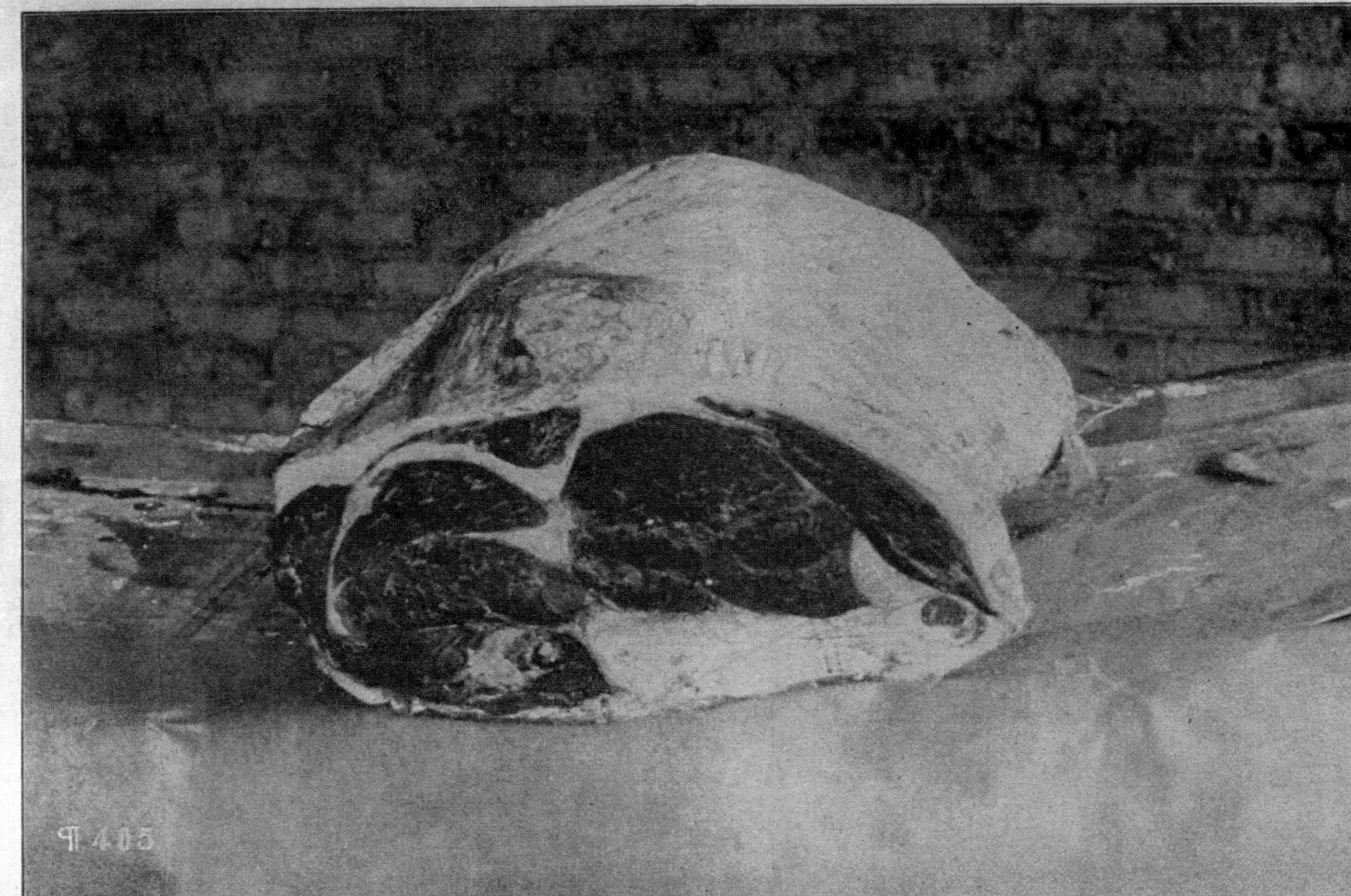

BUTCHER'S TERMS—"THE WHOLE LOIN." SHOWING SIRLOIN OR BROAD LOIN END.
From a good grade Hereford steer. (See Figs. 3 and 4 of butcher's diagram.)

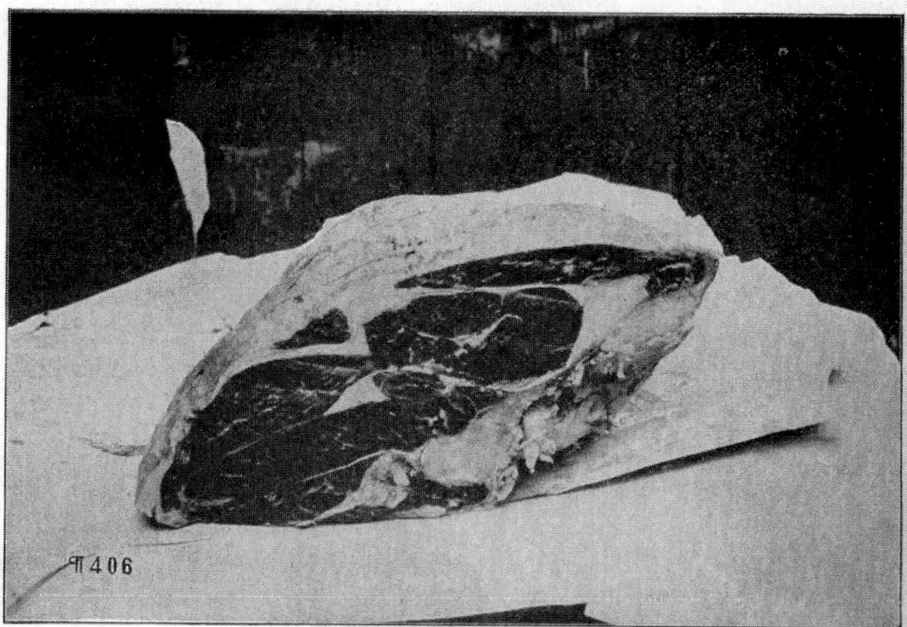

BUTCHER'S TERMS—"THE LOIN."
From a good grade Hereford steer. (See Fig. 3 of butcher's diagram.)

BUTCHER'S TERMS—"THE LOIN."
From a steer classed as common butcher's stock. (See Fig. 3 of butcher's diagram.)

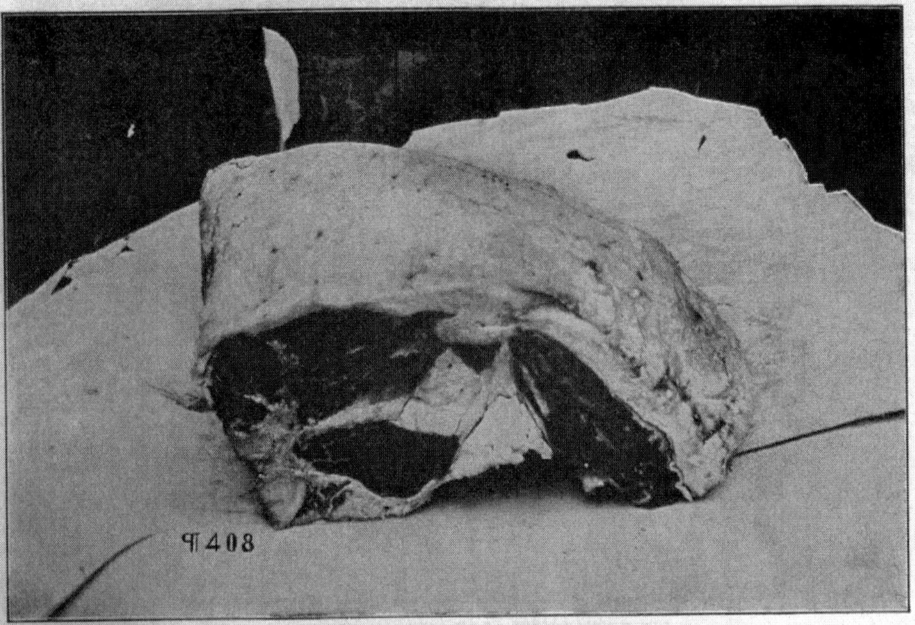

BUTCHER'S TERMS—"THE PORTERHOUSE."
From a good grade Hereford steer. (See Fig. 4 of butcher's diagram.)

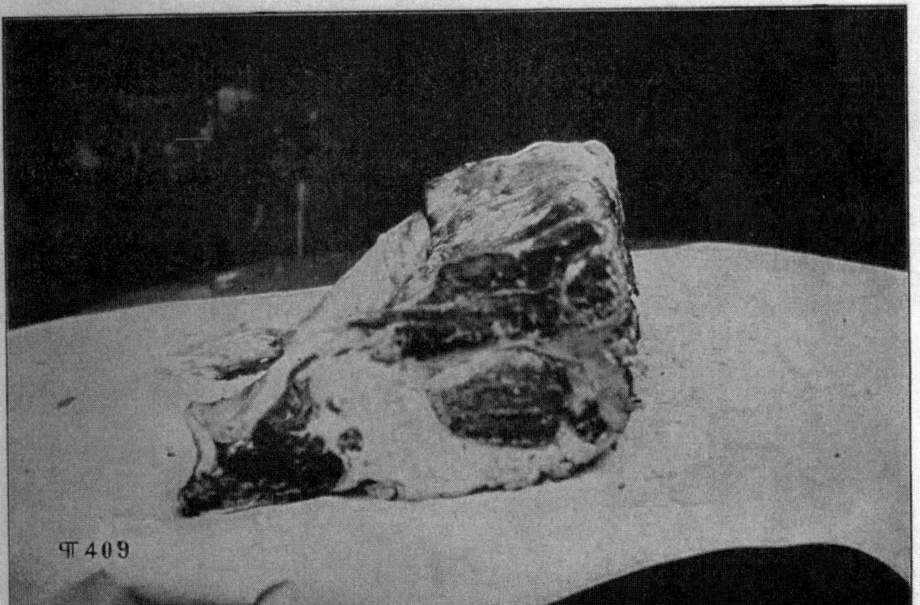

BUTCHER'S TERMS—"THE PORTERHOUSE."
From a steer classed as common butcher's stock. (See Fig. 4 of butcher's diagram.)

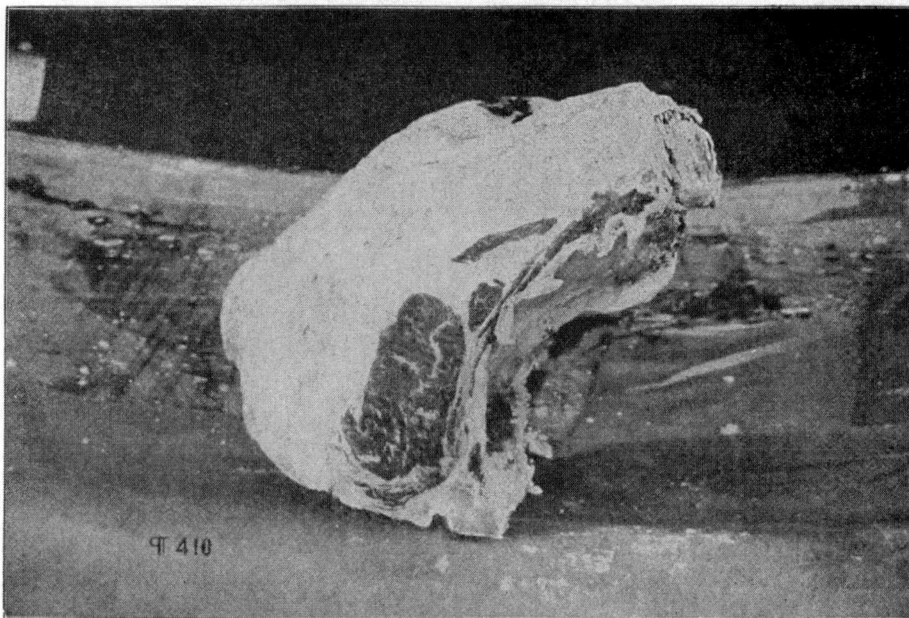

BUTCHER'S TERMS—"THE PORTERHOUSE END."
(The cut that divides the fore quarter from the hind. See Fig. 4 of butcher's diagram.)
From a good grade Hereford steer.

BUTCHER'S TERMS—"THE PORTERHOUSE END."
(The cut that divides the fore quarter from the hind. See Fig. 4 of butcher's diagram.)
From a steer classed as common butcher's stock.

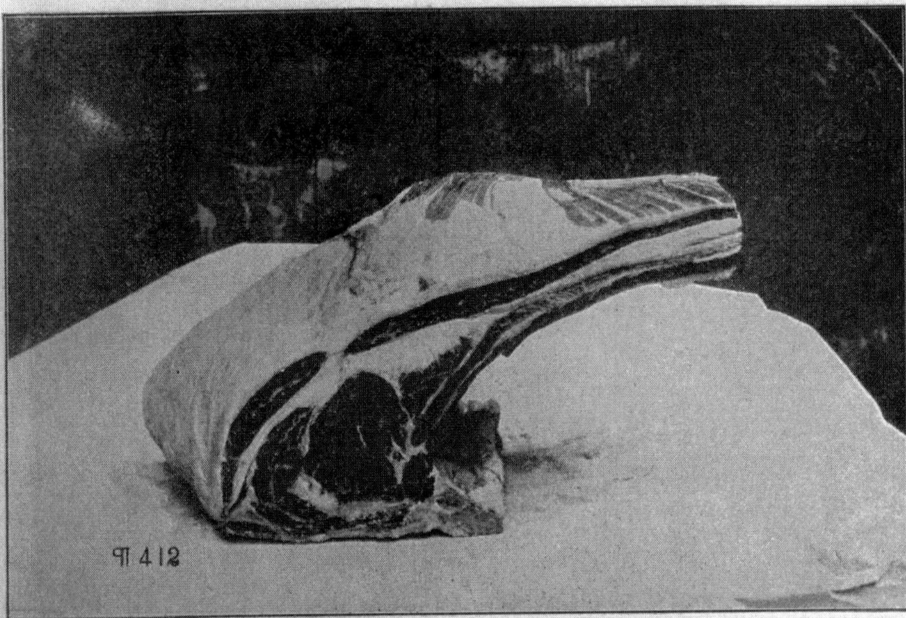

BUTCHER'S TERMS—"THE RIBS."
From a good grade Hereford steer. (See Fig. 7 of butcher's diagram.)

BUTCHER'S TERMS—"THE RIBS."
From a steer classed as common butcher's stock. (See Fig. 7 of butcher's diagram.)

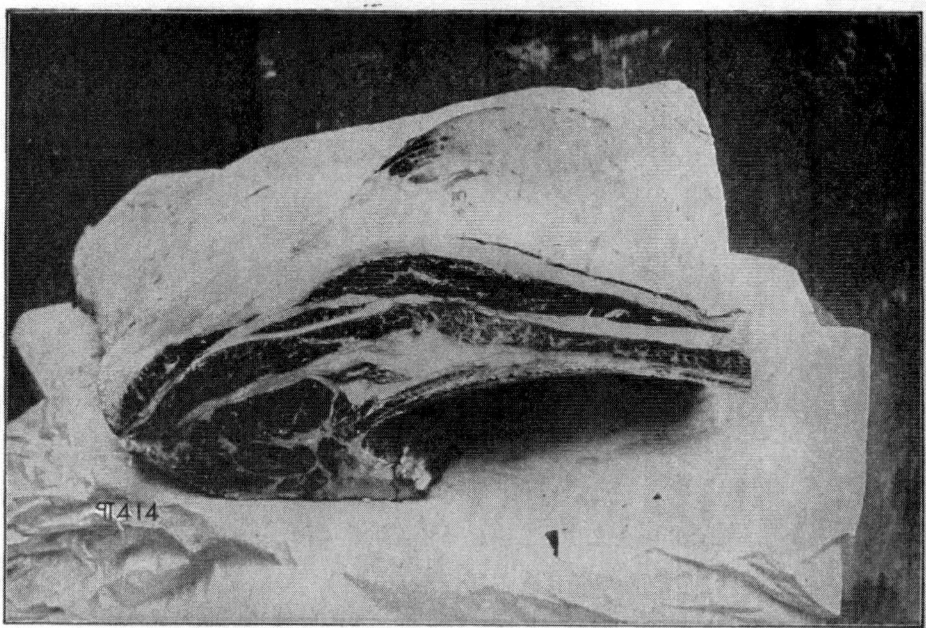

BUTCHER'S TERMS—"THE CHUCK, OR SHOULDER ROAST."
From a good grade Hereford steer. (See Fig. 8 of butcher's diagram.)

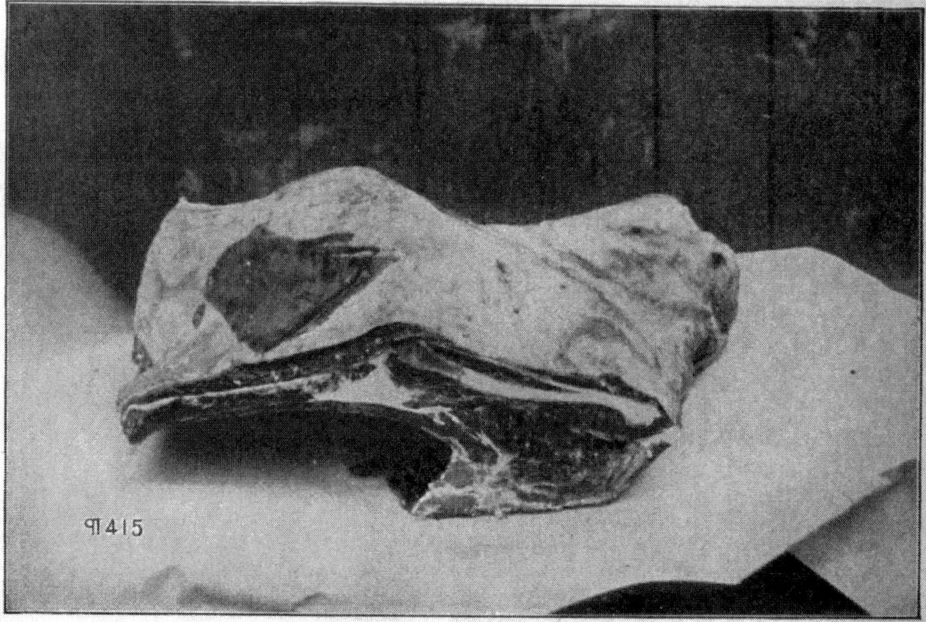

BUTCHER'S TERMS—"THE CHUCK, OR SHOULDER ROAST."
From a steer classed as common butcher's stock. (See Fig. 8. of butcher's diagram.)

BUTCHER'S TERMS—"THE BRISKET."
From a good grade Hereford steer. (See Fig. 9 of butcher's diagram.)

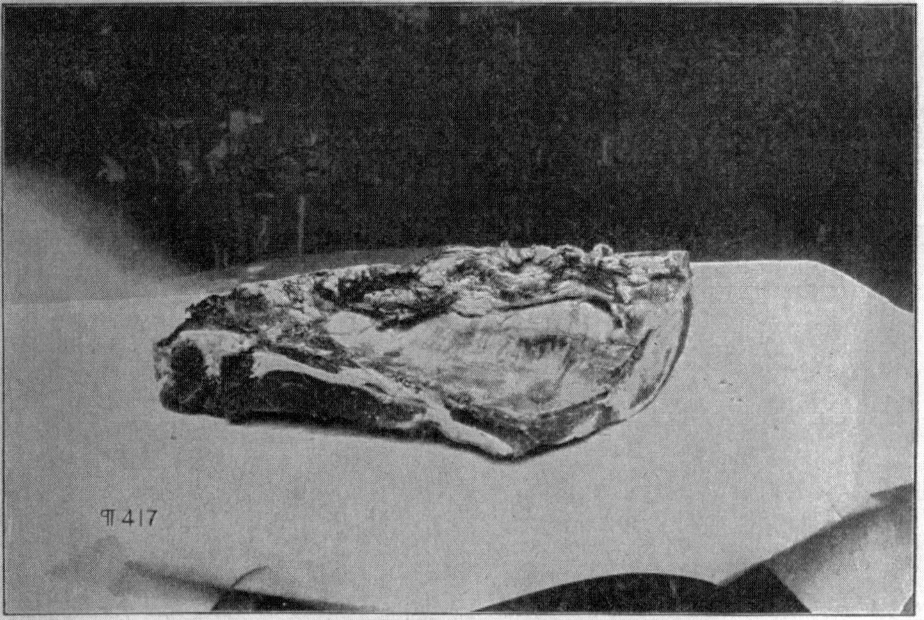

BUTCHER'S TERMS—"THE BRISKET."
From a steer classed as common butcher's stock. (See Fig. 9 of butcher's diagram.)

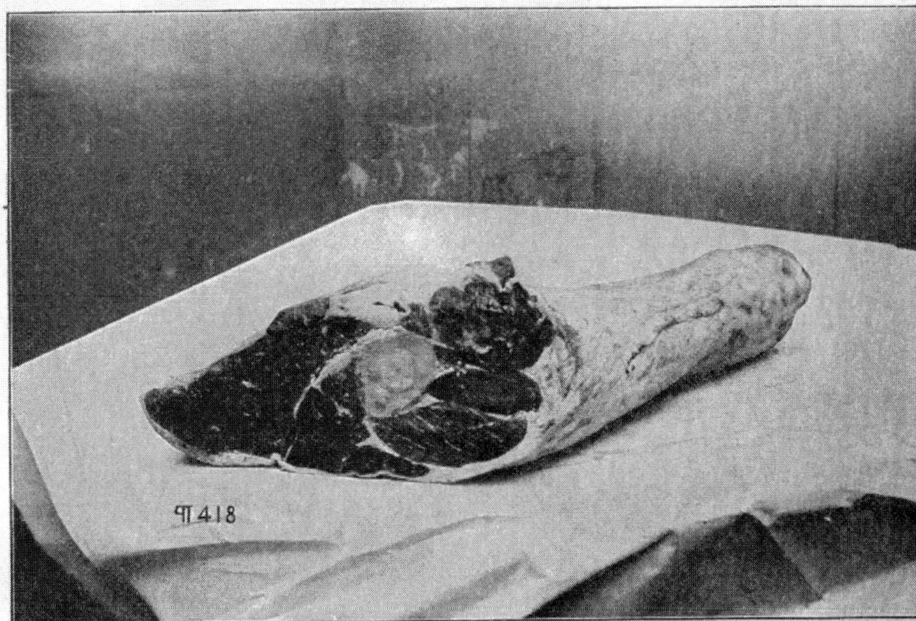

BUTCHER'S TERMS—"THE SHANK."
From a good grade Hereford steer. (See Fig. 11 of butcher's diagram.)

BUTCHER'S TERMS—"THE SHANK."
From a steer classed as common butcher's stock. (See Fig. 11 of butcher's diagram.)

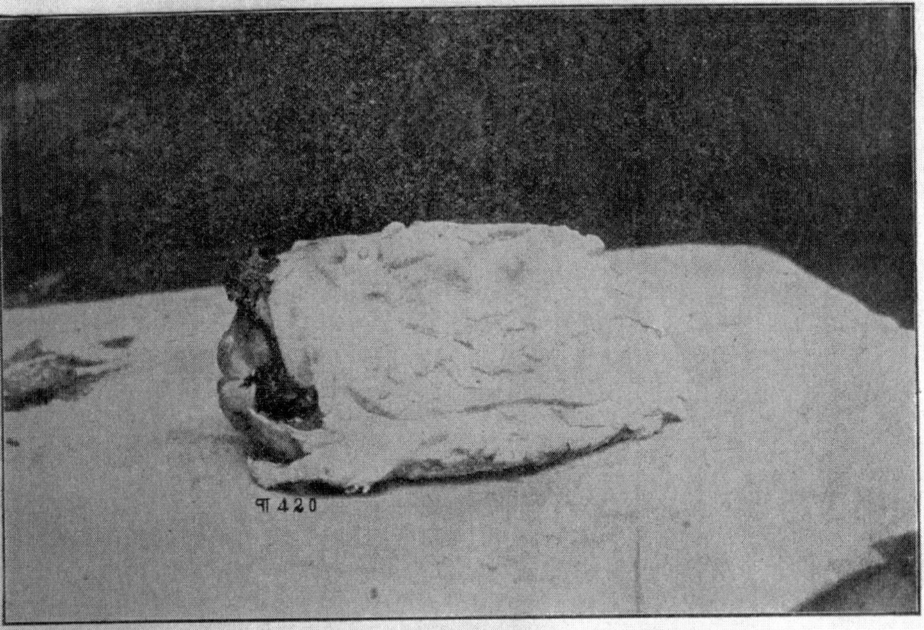

BUTCHER'S TERMS—"THE KIDNEY FAT."
From a good grade Hereford steer.

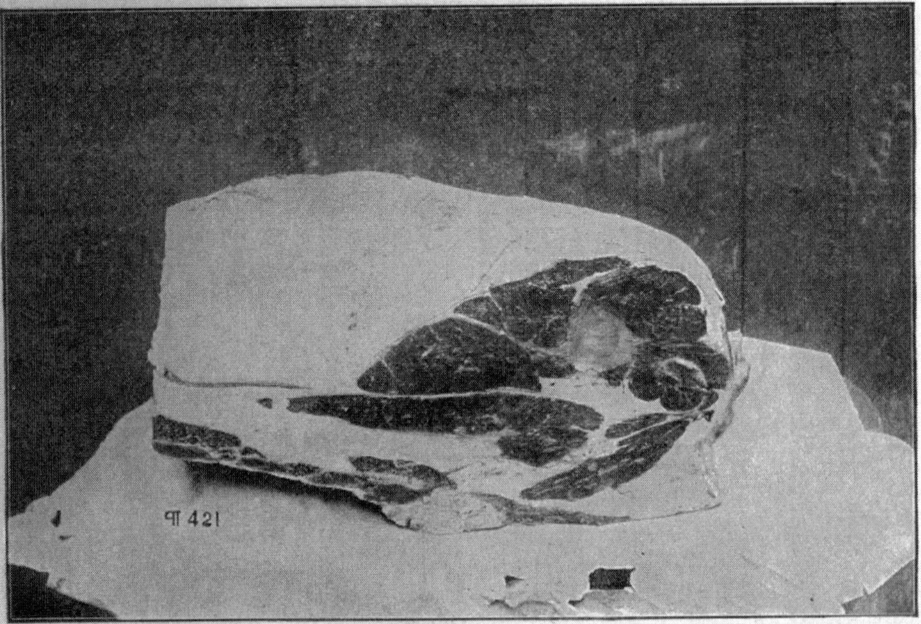

BUTCHER'S TERMS—"THE SHOULDER CLOD."
From a good grade Hereford steer. The shoulder is ordinarily classed as cheap meat, but in a good Hereford carcass this shoulder is better meat than the loin of an ordinary carcass.

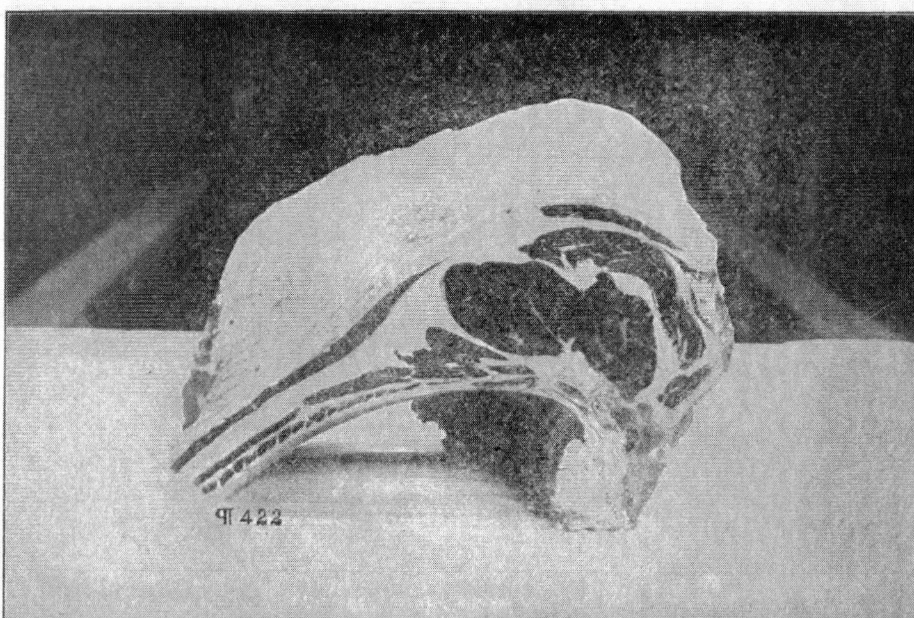

CHOICE HEREFORD ROAST.

Selected by Louis Pfaelzer, the celebrated Chicago butcher, as a model of that class of beef demanded by the very best American and English trade. The supply of this class of beef never equals the demand and it always commands high prices.

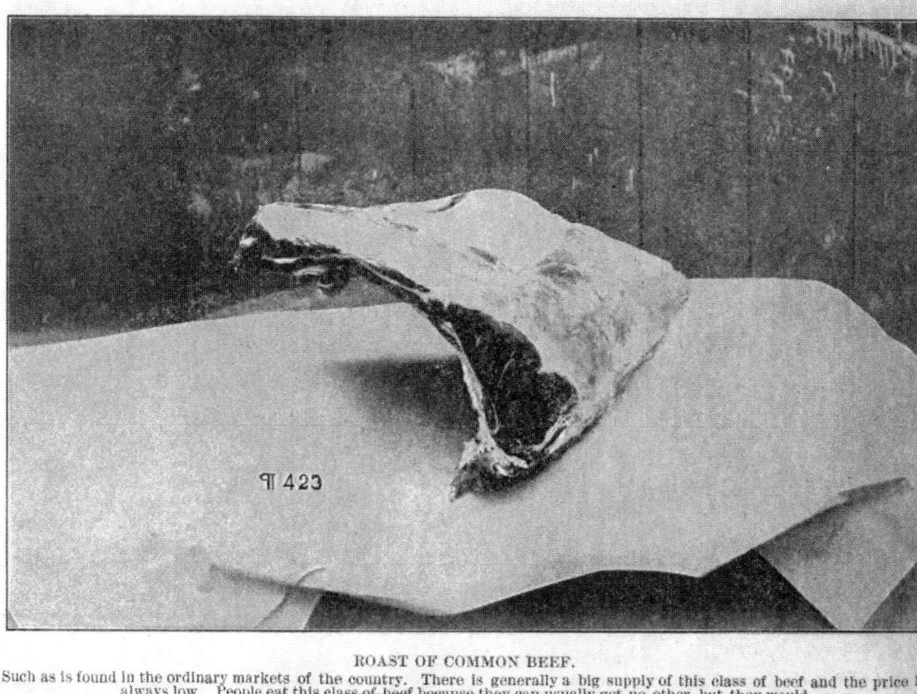

ROAST OF COMMON BEEF.

Such as is found in the ordinary markets of the country. There is generally a big supply of this class of beef and the price is always low. People eat this class of beef because they can usually get no other, but they would not eat it if they knew and could get the other.

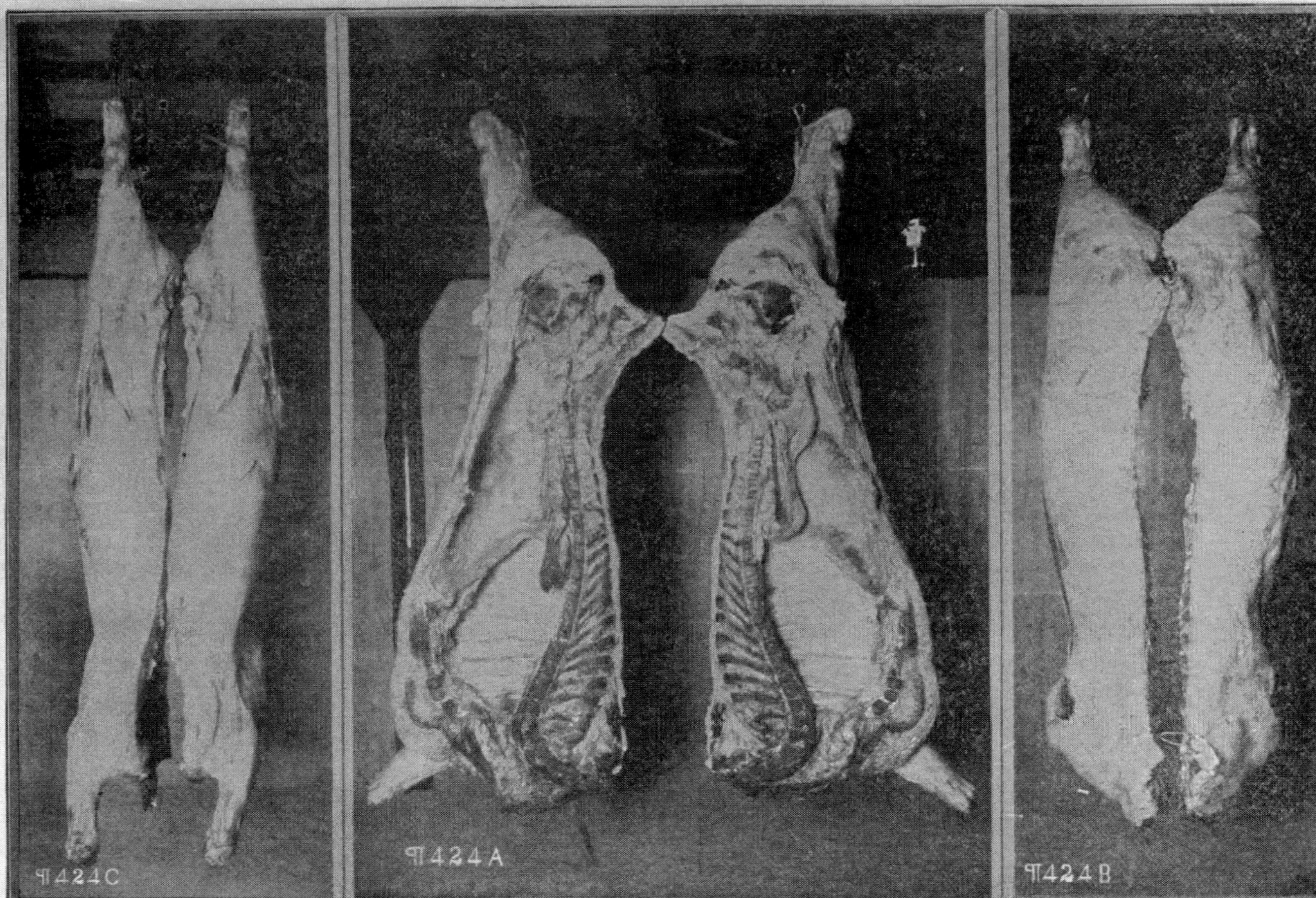

A MODEL CARCASS OF PRIME BEEF FROM A GOOD GRADE HEREFORD STEER.
Back, belly and interior views of the same carcass that will help the reader to know good beef when it is seen hanging in the shambles.

THE CELEBRATED HEREFORD SIRE, "CORRECTOR" 48976, BRED AND OWNED BY T. F. B. SOTHAM, CHILLICOTHE, MO.

INDEX.

NOTE.—The inverted P, or paragraph mark (¶), is used in this work to indicate a portrait or picture. Wherever this sign (¶) is used, in brackets, followed by a number, it indicates the number of an engraving that appears in this work. These pictures or engravings are arranged consecutively, with few exceptions, throughout the work.

CHAPTER IV.

CHAPTER V.

CHAPTER VI.

CHAPTER VII.

CHAPTER VIII.

CHAPTER IX.

CHAPTER X.

CHAPTER XI.

CHAPTER XII.

CHAPTER XIII.

CHAPTER XIV.

CHAPTER XV.

CHAPTER XVI.

CHAPTER XVII.

CHAPTER XVIII.

CHAPTER XIX.

CHAPTER XX.

CHAPTER XXI.

CHAPTER XXII.

CHAPTER XXIII.

CHAPTER XXIV.

CHAPTER XXV.

CHAPTER XXVI.

CHAPTER XXVII.

CHAPTER XXVIII.

CHAPTER XXIX.

CHAPTER XXX.

CHAPTER XXXI.

CHAPTER XXXII.

CHAPTER XXXIII.

CHAPTER XXXIV.

CHAPTER XXXV.

CHAPTER XXXVI.

CHAPTER XXXVII.

CHAPTER XXXVIII.

CHAPTER XXXIX.

CHAPTER XL.

CHAPTER XLI.

CHAPTER XLII.

CHAPTER XLIII.

APPENDIX.

P. F. PETTIBONE & CO. PRINTERS CHICAGO.